P9-BYR-792

The Materials Science *of* Thin Films

The Materials Science *of* Thin Films

Milton Ohring

Stevens Institute of Technology
Department of Materials Science and Engineering
Hoboken, New Jersey

Academic Press
San Diego New York Boston
London Sydney Tokyo Toronto

This book is printed on acid-free paper. ∞

Designed by Elizabeth E. Tustian

ACADEMIC PRESS
A Division of Harcourt Brace & Company
525 B Street, Suite 1900, San Diego, California 92101-4495

United Kingdom Edition published by
ACADEMIC PRESS LIMITED
24–28 Oval Road, London NW1 7DX

Library of Congress Cataloging-in-Publication Data

Ohring, Milton, date.
The materials science of thin films / Milton Ohring.
p. cm.
Includes bibliographical references and index.
ISBN 0-12-524990-X (Alk. paper)
1. Thin films. I. Title.
TA418.9.T45047 1991
620′.44—dc20 91-9664
CIP

Printed in the United States of America
98 MV 7 6

Contents

Chapter 3

Physical Vapor Deposition . . . 79

Chapter 4

Chemical Vapor Deposition . . . 147

Chapter 5

Film Formation and Structure . . . 195

Foreword

It is a distinct pleasure for me to write a foreword to this new textbook by my long-time friend, Professor Milt Ohring.

There have been at least 200 books written on various aspects of thin film science and technology, but this is the first true textbook, specifically intended for classroom use in universities. In my opinion there has been a crying need for a real textbook for a long time. Most thin film courses in universities have had to use many books written for relatively experienced thin film scientists and engineers, often supplemented by notes prepared by the course instructor. *The Materials Science of Thin Films*, a true textbook, complete with problems after each chapter, is available to serve as a nucleus for first courses in thin film science and technology.

In addition to his many years of experience teaching and advising graduate students at Stevens Institute of Technology, Professor Ohring has been the coordinator of an on-premises, M.S. degree program offered by Stevens at the AT&T Bell Laboratories in Murray Hill and Whippany, New Jersey. This ongoing cooperative program has produced over sixty M.S. graduates to date. Several of these graduates have gone on to acquire Ph.D. degrees. The combination of teaching, research, and industrial involvement has provided Professor Ohring with a broad perspective of thin film science and technology and tremendous insight into the needs of students entering this exciting field. His insight and experience are quite evident in this textbook.

John L. Vossen

Preface

Thin-film science and technology play a crucial role in the high-tech industries that will bear the main burden of future American competitiveness. While the major exploitation of thin films has been in microelectronics, there are numerous and growing applications in communications, optical electronics, coatings of all kinds, and in energy generation and conservation strategies. A great many sophisticated analytical instruments and techniques, largely developed to characterize thin films and surfaces, have already become indispensable in virtually every scientific endeavor irrespective of discipline. When I was called upon to offer a course on thin films, it became a genuine source of concern to me that there were no suitable textbooks available on this unquestionably important topic. This book, written with a materials science flavor, is a response to this need. It is intended for

1. Science and engineering students in advanced undergraduate or first-year graduate level courses on thin films
2. Participants in industrial in-house courses or short courses offered by professional societies
3. Mature scientists and engineers switching career directions who require an overview of the field.

Readers should be reasonably conversant with introductory college chemistry and physics and possess a passive cultural familiarity with topics commonly treated in undergraduate physical chemistry and modern physics courses.

It is worthwhile to briefly elaborate on this book's title and the connection between thin films and the broader discipline of materials science and engineering. A dramatic increase in our understanding of the fundamental nature of materials throughout much of the twentieth century has led to the development of materials science and engineering. This period witnessed the emergence of polymeric, nuclear, and electronic materials, new roles for metals and ceramics, and the development of reliable methods to process these materials in bulk and thin-film form. Traditional educational approaches to the study of materials have stressed *structure–property* relationships in bulk solids, typically utilizing metals, semiconductors, ceramics; and polymers, taken singly or collectively as illustrative vehicles to convey principles. The same spirit is adopted in this book except that *thin solid films* are the vehicle. In addition, the basic theme has been expanded to include the multifaceted *processing–structure–properties–performance* interactions. Thus the original science core is preserved but enveloped by the engineering concerns of *processing* and *performance*. Within this context, I have attempted to weave threads of commonality among seemingly different materials and properties, as well as to draw distinctions between materials that exhibit outwardly similiar behavior. In particular, parallels and contrasts between films and bulk materials are recurring themes.

An optional introductory review chapter on standard topics in materials establishes a foundation for subsequent chapters. Following a second chapter on vacuum science and technology, the remaining text is broadly organized into three categories. Chapters 3 and 4 deal with the principles and practices of film deposition from the vapor phase. Chapters 5–9 deal with the processes and phenomena that influence the structural, chemical, and physical attributes of films, and how to characterize them. Topics discussed include nucleation, growth, crystal perfection, epitaxy, mass transport effects, and the role of stress. These are the common thin-film concerns irrespective of application. The final portion of the book (Chapters 10–14) is largely devoted to specific film properties (electrical, magnetic, optical, mechanical) and applications, as well as to emerging materials and processes. Although the first nine chapters may be viewed as core subject matter, the last five chapters offer elective topics intended to address individual interests. It is my hope that instructors using this book will find this division of topics a useful one.

Much of the book reflects what is of current interest to the thin-film research and development communities. Examples include chapters on chemical vapor deposition, epitaxy, interdiffusion and reactions, metallurgical and protective coatings, and surface modification. The field is evolving so rapidly that even the classics of yesteryear, e.g., Maissel and Glang, *Handbook of Thin Film*

Technology (1970) and Chopra, *Thin Film Phenomena* (1969), as well as more recent books on thin films, e.g., Pulker, *Coatings on Glass* (1984), and Eckertova, *Physics of Thin Films* (1986), make little or no mention of these now important subjects.

As every book must necessarily establish its boundaries, I would like to point out the following: (1) Except for coatings (Chapter 12) where thicknesses range from several to as much as hundreds of microns (1 micron or 1 μm = 10^{-6} meter), the book is primarily concerned with films that are less than 1 μm thick. (2) Only films and coatings formed from the gas phase by physical (PVD) or chemical vapor deposition (CVD) processes are considered. Therefore spin and dip coating, flame and plasma spraying of powders, electrolytic deposition, etc., will not be treated. (3) The topic of polymer films could easily justify a monograph of its own, and hence will not be discussed here. (4) Time and space simply do not allow for development of all topics from first principles. (Nevertheless, I have avoided using the unwelcome phrase "It can be shown that . . . ," and have refrained from using other textbooks or the research literature to fill in missing steps of derivations.) (5) A single set of units (e.g., CGS, MKS, SI, etc.) has been purposely avoided to better address the needs of a multifaceted and interdisciplinary audience. Common usage, commercial terminology, the research literature and simple bias and convenience have all played a role in the ecumenical display of units. Where necessary, conversions between different systems of units are provided.

At the end of each chapter are problems of varying difficulty, and I believe a deeper sense of the subject matter will be gained by considering them. Three very elegant problems (i.e. 9-6, -7, -8) were developed by Professor W. D. Nix, and I thank him for their use.

By emphasizing immutable concepts, I hope this book will be spared the specter of rapid obsolescence. However, if this book will in some small measure help spawn new technology rendering it obsolete, it will have served a useful function.

Milton Ohring

Acknowledgments

At the top of my list of acknowledgments I would like to thank John Vossen for his advice and steadfast encouragement over a number of years. This book would not have been possible without the wonderfully extensive intellectual and physical resources of AT & T Bell Laboratories, Murray Hill, NJ, and the careful execution of the text and figures at Stevens Institute. In particular Bell Labs library was indispensable and I am indebted to AT & T for allowing me to use it. My long association with Bell Labs is largely due to my dear friend L. C. Kimerling (Kim), and I thank him and AT & T for supporting my efforts there. I am grateful to the many Bell Labs colleagues and students in the Stevens Institute of Technology/Bell Labs "On Premises Approved Program (OPAP)," who planted the seed for a textbook on thin films. In this regard D. C. Jacobson should be singled out for his continuous help with many aspects of this work. The following Bell Labs staff members contributed to this book through helpful comments and discussions, and by contributing figures, problems, and research papers: J. C. Bean, J. L. Benton, W. L. Brown, F. Capasso, G. K. Celler, A. Y. Cho, J. M. Gibson, H. J. Gossmann, R. Hull, R. W. Knoell, R. F. Kopf, Y. Kuk, H. S. Luftman, S. Nakahara, M. B. Panish, J. M. Poate, S. M. Sze, K. L. Tai, W. W. Tai, H. Temkin, L. F. Thompson, L. E. Trimble, M. J. Vasile, and R. Wolfe. I appreciate their time and effort spent on my behalf.

Some very special people at Stevens enabled the book to reach fruition. They include Pat Downes for expertly typing a few versions of the complete text during evenings that she could have spent more pleasantly; Eleanor Gehler, for kindly undertaking much additional typing; Kamlesh Patel for his professional computerized drafting of the bulk of the figures; Chris Rywalt, Manoj Thomas, and Tao Jen for carefully rendering the remainder of the figures; Mehboob Alam and Warren Moberly for their computer help in compiling the index, and drafting the cover, respectively; Dick Widdicombe, Bob Ehrlich, Dan Schwarcz, Lauren Snyder, and Noemia Carvalho for many favors; Professor R. Weil for helpful comments; Profs. W. Carr, H. Salwen, and T. Hart for their expert and generous assistance on several occasions; Professor B. Gallois and G. M. Rothberg for support; and those at Stevens responsible for granting my sabbatical leave in 1988. My sincere thanks to all of you.

Lastly, I am grateful to several anonymous reviewers for many pertinent comments and for uncovering textual errors. They are absolved of all responsibility for any shortcomings that remain.

This book is lovingly dedicated to Ahrona, Avi, Noam, and Feigel, who in varying degrees had to contend with a less that a full-time husband and father for too many years.

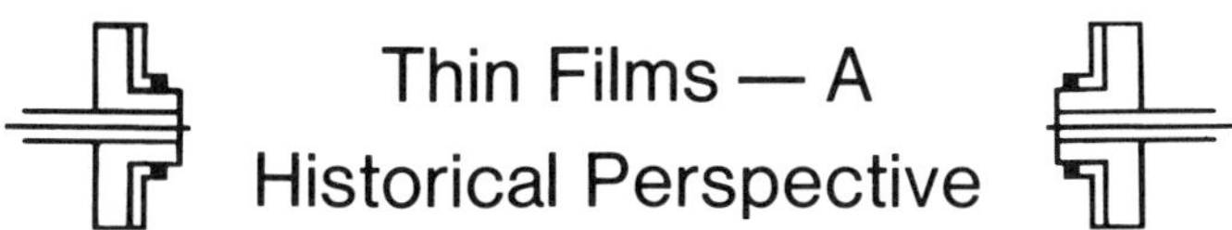

Thin Films — A Historical Perspective

Thin-film technology is simultaneously one of the oldest arts and one of the newest sciences. Involvement with thin films dates to the metal ages of antiquity. Consider the ancient craft of gold beating, which has been practiced continuously for at least four millenia. Gold's great malleability enables it to be hammered into leaf of extraordinary thinness while its beauty and resistance to chemical degradation have earmarked its use for durable ornamentation and protection purposes. The Egyptians appear to have been the earliest practitioners of the art of gold beating and gilding. Many magnificent examples of statuary, royal crowns, and coffin cases which have survived intact attest to the level of skill achieved. The process involves initial mechanical rolling followed by many stages of beating and sectioning composite structures consisting of gold sandwiched between layers of vellum, parchment, and assorted animal skins. Leaf samples from Luxor dating to the Eighteenth Dynasty (1567–1320 B.C.) measured 0.3 microns in thickness. As a frame of reference for the reader, the human hair is about 75 microns in diameter. Such leaf was carefully applied and bonded to smoothed wax or resin-coated wood surfaces in a mechanical (cold) gilding process. From Egypt the art spread as indicated by numerous accounts of the use of gold leaf in antiquity.

Today, gold leaf can be machine-beaten to 0.1 micron and to 0.05 micron when beaten by a skilled craftsman. In this form it is invisible sideways and quite readily absorbed by the skin. It is no wonder then that British gold beaters were called upon to provide the first metal specimens to be observed in the transmission electron microscope. Presently, gold leaf is used to decorate such diverse structures and objects as statues, churches, public buildings, tombstones, furniture, hand-tooled leather, picture frames and, of course, illuminated manuscripts.

Thin-film technologies related to gold beating, but probably not as old, are mercury and fire gilding. Used to decorate copper or bronze statuary, the cold mercury process involved carefully smoothing and polishing the metal surface, after which mercury was rubbed into it. Some copper dissolved in the mercury, forming a very thin amalgam film that left the surface shiny and smooth as a mirror. Gold leaf was then pressed onto the surface cold and bonded to the mercury-rich adhesive. Alternately, gold was directly amalgamated with mercury, applied, and the excess mercury was then driven off by heating, leaving a film of gold behind. Fire gilding was practiced well into the nineteenth century despite the grave health risk due to mercury vapor. The hazard to workers finally became intolerable and provided the incentive to develop alternative processes, such as electroplating.

The history of gold beating and gilding is replete with experimentation and process development in diverse parts of the ancient world. Practitioners were concerned with the purity and cost of the gold, surface preparation, the uniformity of the applied films, adhesion to the substrate, reactions between and among the gold, mercury, copper, bronze (copper-tin), etc., process safety, color, optical appearance, durability of the final coating, and competitive coating technologies. As we shall see in the ensuing pages, modern thin-film technology addresses these same generic issues, albeit with a great compression of time. And although science is now in the ascendancy, there is still much room for art.

REFERENCES

1. L. B. Hunt, *Gold Bull.* **9,** 24 (1976).
2. O. Vittori, *Gold Bull.* **12,** 35 (1979).
3. E. D. Nicholson, *Gold Bull.* **12,** 161 (1979).

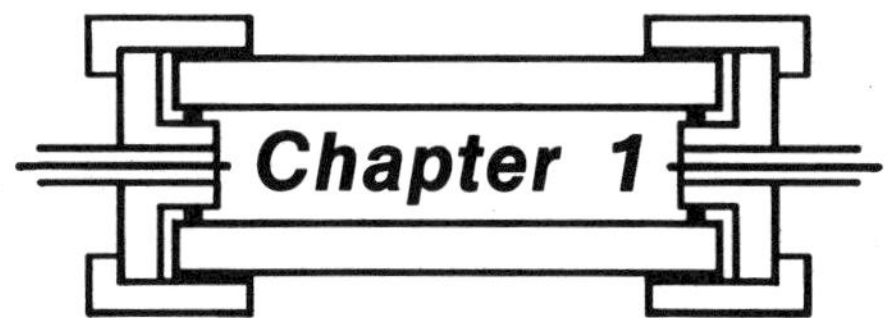

A Review of Materials Science

1.1. Introduction

A cursory consideration of the vast body of solid substances reveals what outwardly appears to be an endless multitude of external forms and structures possessing a bewildering variety of properties. The branch of study known as materials science and engineering evolved in part to classify those features that are common among the structure and properties of different materials in a manner somewhat reminiscent of chemical or biological classification schemes. This dramatically reduces the apparent variety. From this perspective, it turns out that solids can be classified as typically belonging to one of only four categories (metallic, ionic, covalent, or van der Waals), depending on the nature of the electronic structure and resulting interatomic bonding forces.

Similar divisions occur with respect to the structure of solids. Solids are either internally crystalline or noncrystalline. Those that are crystalline can be further subdivided according to one of 14 different geometric arrays or lattices, depending on the placement of the atoms. When properties are considered, there are similar simplifying categorizations. Thus, materials are either good, intermediate, or poor conductors of electricity, and they are either mechanically brittle or can easily be stretched without fracture, and they are either optically reflective or transparent, etc. It is, of course, easier to recognize that property differences exist than to understand why they exist. Nevertheless, much progress has been made in this subject as a result of the research of the past 50 years. Basically, the richness in the diversity of materials properties occurs because countless combinations of the admixture of chemical compositions, bonding types, crystal structures, and morphologies are available naturally or can be synthesized.

In this chapter various aspects of structure and bonding in solids are reviewed for the purpose of providing the background to better understand the remainder of the book. In addition, several topics dealing with thermodynamics and kinetics of atomic motion in materials are also included. These will later have relevance to aspects of the stability, formation, and solid-state reactions in thin films. Much of this chapter is a condensed adaptation of standard treatments of bulk materials, but it is equally applicable to thin films. Nevertheless, many distinctions between bulk materials and films exist, and they will be stressed where possible. Readers already familiar with concepts of materials science may wish to skip this chapter; those who seek deeper and broader coverage should consult the bibliography for recommended texts on this subject.

1.2. Structure

1.2.1. Crystalline Solids

Many solid materials possess an ordered internal crystal structure despite external appearances that are not what we associate with the term *crystalline*—i.e., clear, transparent, faceted, etc. Actual crystal structures can be imagined to arise from a three-dimensional array of points geometrically and repetitively distributed in space such that each point has identical surroundings. There are only 14 ways to arrange points in space having this property, and the resulting point arrays are known as Bravais lattices. They are shown in Fig. 1-1 with lines intentionally drawn in to emphasize the symmetry of the lattice. Only a single cell for each lattice is reproduced here, and the point array actually stretches in an endlessly repetitive fashion in all directions. If an atom or group of two or more atoms is now placed at each Bravais lattice point, a physically real crystal structure emerges. Thus, if individual copper atoms populated every point of a face-centered cubic (FCC) lattice whose cube edge dimension, or so-called lattice parameter, were 3.615 Å, the material known as

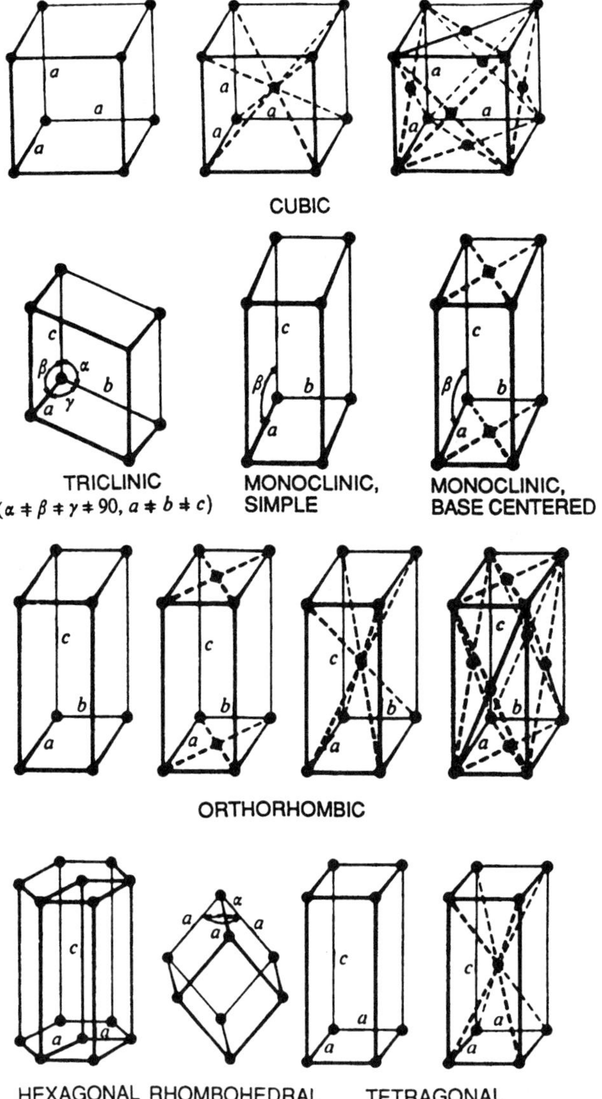

Figure 1-1. The 14 Bravais space lattices.

metallic copper would be generated; and similarly for other types of lattices and atoms.

The reader should realize that just as there are no lines in actual crystals, there are no spheres. Each sphere in the Cu crystal structure represents the atomic nucleus surrounded by a complement of 28 core electrons [i.e., $(1s)^2$

$(2s)^2$ $(2p)^6$ $(3s)^2$ $(3p)^6$ $(3d)^{10}$] and a portion of the free-electron gas contributed by 4s electrons. Furthermore, these spheres must be imagined to touch in certain crystallographic directions, and their packing is rather dense. In FCC structures the atom spheres touch along the direction of the face diagonals, i.e., [110], but not along the face edge directions, i.e., [100]. This means that the planes containing the three face diagonals shown in Fig. 1-2a, i.e., the (111) plane, are close-packed. On this plane the atoms touch each other in much the same way as a racked set of billiard balls on a pool table. All other planes in the FCC structure are less densely packed and thus contain fewer atoms per unit area.

Placement of two identical silicon atoms at each FCC point would result in the formation of the diamond cubic silicon structure (Fig. 1-2c), whereas the rock-salt structure (Fig. 1-2b) is generated if sodium–chlorine groups were substituted for each lattice point. In both cases the positions and orientation of the two atoms in question must be preserved from point to point.

In order to quantitatively identify atomic positions as well as planes and directions in crystals, simple concepts of coordinate geometry are utilized. First, orthogonal axes are arbitrarily positioned with respect to a cubic lattice (e.g., FCC) such that each point can now be identified by three coordinates

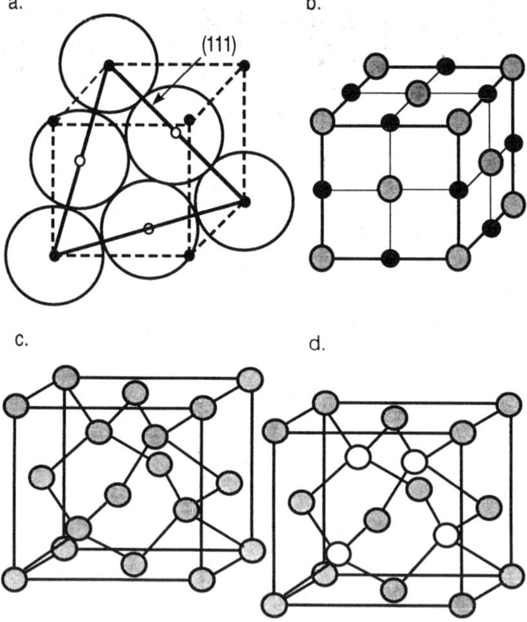

Figure 1-2. (a) (111) plane in FCC lattice; (b) rock-salt structure, e.g., NaCl; Na ●, Cl ◉; (c) diamond cubic structure, e.g., Si, Ge; (d) zinc blende structure, e.g., GaAs.

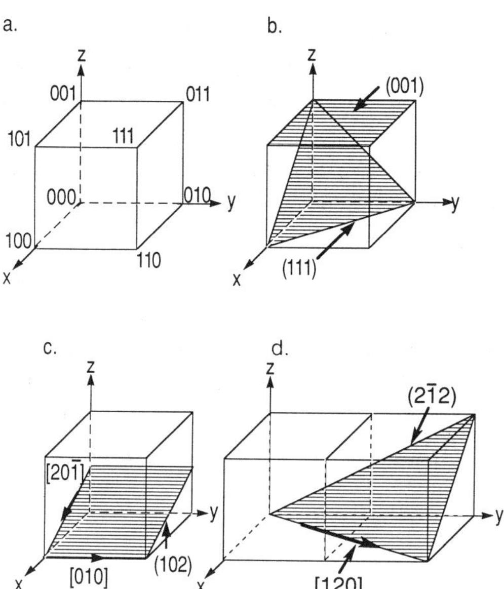

Figure 1-3. (a) coordinates of lattice sites; (b) Miller indices of planes; (c, d) Miller indices of planes and directions.

(Fig. 1-3a). If the center of the coordinate axes is taken as $x = 0$, $y = 0$, $z = 0$, or (0, 0, 0), then the coordinates of other nearest equivalent cube corner points are (1, 0, 0) (0, 1, 0) (1, 0, 0), etc. In this framework the two Si atoms referred to earlier, situated at the center of the coordinate axes, would occupy the (0, 0, 0) and (1/4, 1/4, 1/4) positions. Subsequent repetitions of this oriented pair of atoms at each FCC lattice point generate the diamond cubic structure in which each Si atom has four nearest neighbors arranged in a tetrahedral configuration. Similarly, substitution of the motif (0, 0, 0) Ga and (1/4, 1/4, 1/4) As for each point of the FCC lattice would result in the zinc blende GaAs crystal structure (Fig. 1-2d).

Specific crystal planes and directions are frequently noteworthy because phenomena such as crystal growth, chemical reactivity, defect incorporation, deformation, and assorted properties are not isotropic or the same on all planes and in all directions. Therefore, it is important to be able to identify accurately and distinguish crystallographic planes and directions. A simple recipe for identifying a given plane in the cubic system is the following:

1. Determine the intercepts of the plane on the three crystal axes in number of unit cell dimensions.
2. Take reciprocals of those numbers.
3. Reduce these reciprocals to smallest integers by clearing fractions.

The result is a triad of numbers known as the Miller indices for the plane in question, i.e., (h, k, l). Several planes with identifying Miller indices are indicated in Fig. 1-3. Note that a negative index is indicated above the integer with a minus sign.

Crystallographic directions shown in Fig. 1-3 are determined by the components of the vector connecting any two lattice points lying along the direction. If the coordinates of these points are u_1, v_1, w_1 and u_2, v_2, w_2, then the components of the direction vector are $u_1 - u_2, v_1 - v_2, w_1 - w_2$. When reduced to smallest integer numbers, they are placed within brackets and are known as the Miller indices for the direction, i.e., $[hkl]$. In this notation the direction cosines for the given directions are $h/\sqrt{h^2+k^2+l^2}$, $k/\sqrt{h^2+k^2+l^2}$, $l/\sqrt{h^2+k^2+l^2}$. Thus, the angle α between any two directions $[h_1, k_1, l_1]$ and $[h_2, k_2, l_2]$ is given by the vector dot product

$$\cos\alpha = \frac{h_1h_2 + k_1k_2 + l_1l_2}{\sqrt{h_1^2+k_1^2+l_1^2}\,\sqrt{h_2^2+k_2^2+l_2^2}}. \tag{1-1}$$

Two other useful relationships in the crystallography of cubic systems are given without proof.

1. The Miller indices of the direction normal to the (hkl) plane are $[hkl]$.
2. The spacing between individual (hkl) planes is $a = a_0/\sqrt{h^2+k^2+l^2}$, where a_0 is the lattice parameter.

As an illustrative example, we shall calculate the angle between any two neighboring tetrahedral bonds in the diamond cubic lattice. The bonds lie along [111]-type directions that are specifically taken here to be $[\bar{1}\bar{1}1]$ and [111]. Therefore, by Eq. 1-1,

$$\cos\alpha = \frac{(1)(-1) + 1(-1) + (1)(1)}{\sqrt{1^2+1^2+1^2}\,\sqrt{(-1)^2+(-1)^2+1^2}} = -\frac{1}{3} \quad \text{and}$$

$$\alpha = 109.5°.$$

These two bond directions lie in a common (110)-type crystal plane. The precise indices of this plane must be $(\bar{1}10)$ or $(1\bar{1}0)$. This can be seen by noting that the dot product between each bond vector and the vector normal to the plane in which they lie must vanish.

We close this brief discussion with some experimental evidence in support of the internal crystalline structure of solids. X-ray diffraction methods have very convincingly supplied this evidence by exploiting the fact that the spacing between atoms is comparable to the wavelength (λ) of X-rays. This results in easily detected emitted beams of high intensity along certain directions when incident X-rays impinge at critical diffraction angles (θ). Under these conditions the well-known Bragg relation

$$n\lambda = 2a\sin\theta \tag{1-2}$$

holds, where n is an integer.

In bulk solids large diffraction effects occur at many values of θ. In thin films, however, very few atoms are present to scatter X-rays into the diffracted beam when θ is large. For this reason the intensities of the diffraction lines or spots will be unacceptably small unless the incident beam strikes the film surface at a near-glancing angle. This, in effect, makes the film look thicker. Such X-ray techniques for examination of thin films have been developed and will be discussed in Chapter 6. A drawback of thin films relative to bulk solids is the long counting times required to generate enough signal for suitable diffraction patterns. This thickness limitation in thin films is turned into great advantage, however, in the transmission electron microscope. Here electrons must penetrate through the material under observation, and this can occur only in thin films or specially thinned specimens. The short wavelength of the electrons employed enables high-resolution imaging of the lattice structure as well as diffraction effects to be observed. As an example, consider the remarkable electron micrograph of Fig. 1-4, showing atom positions in a thin

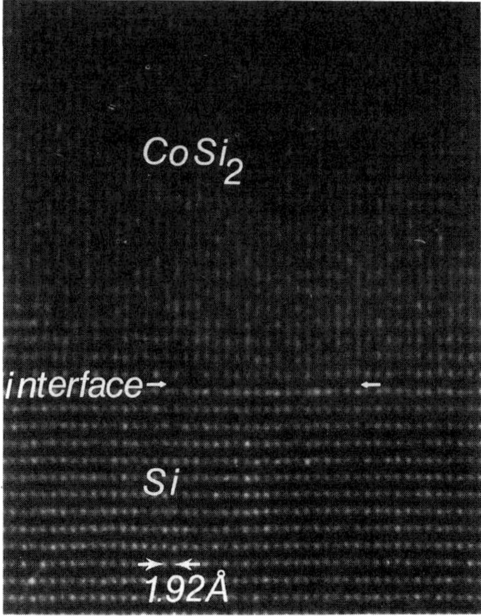

Figure 1-4. High resolution lattice image of epitaxial $CoSi_2$ film on (111) Si ($\langle 11\bar{2}\rangle$ projection). (Courtesy J. M. Gibson, AT & T Bell Laboratories).

film of cobalt silicide grown with perfect crystalline registry (epitaxially) on a silicon wafer. The silicide film–substrate was mechanically and chemically thinned *normal* to the original film plane to make the cross section visible. Such evidence should leave no doubt as to the internal crystalline nature of solids.

1.2.2. Amorphous Solids

In some materials the predictable long-range geometric order characteristic of crystalline solids breaks down. Such materials are the noncrystalline amorphous or glassy solids exemplified by silica glass, inorganic oxide mixtures, and polymers. When such bulk materials are cooled from the melt even at low rates, the more random atomic positions that we associate with a liquid are frozen in place within the solid. On the other hand, while most metals cannot be amorphized, certain alloys composed of transition metal and metalloid combinations (e.g., Fe–B) can be made in glassy form but only through extremely rapid quenching of melts. The required cooling rates are about 10^6 °C/sec, and therefore heat transfer considerations limit bulk glassy metals to foil, ribbon, or powder shapes typically ~ 0.05 mm in thickness or dimension. In general, amorphous solids can retain their structureless character practically indefinitely at low temperatures even though thermodynamics suggests greater stability for crystalline forms. Crystallization will, however, proceed with release of energy when these materials are heated to appropriate elevated temperatures. The atoms then have the required mobility to seek out equilibrium lattice sites.

Thin films of amorphous metal alloys, semiconductors, oxides, and chalcogenide glasses have been readily prepared by common physical vapor deposition (evaporation and sputtering) as well as chemical vapor deposition (CVD) methods. Vapor quenching onto cryogenically cooled glassy substrates has made it possible to make alloys and even pure metals—the most difficult of all materials to amorphize—glassy. In such cases, the surface mobility of depositing atoms is severely restricted, and a disordered atomic configuration has a greater probability of being frozen in.

Our present notions of the structure of amorphous inorganic solids are extensions of models first established for silica glass. These depict amorphous SiO_2 to be a random three-dimensional network consisting of tetrahedra joined at the corners but sharing no edges or faces. Each tetrahedron contains a central Si atom bonded to four vertex oxygen atoms, i.e., $(SiO_4)^{-4}$. The oxygens are, in turn, shared by two Si atoms and are thus positioned as the pivotal links between neighboring tetrahedra. In crystalline quartz the tetrahe-

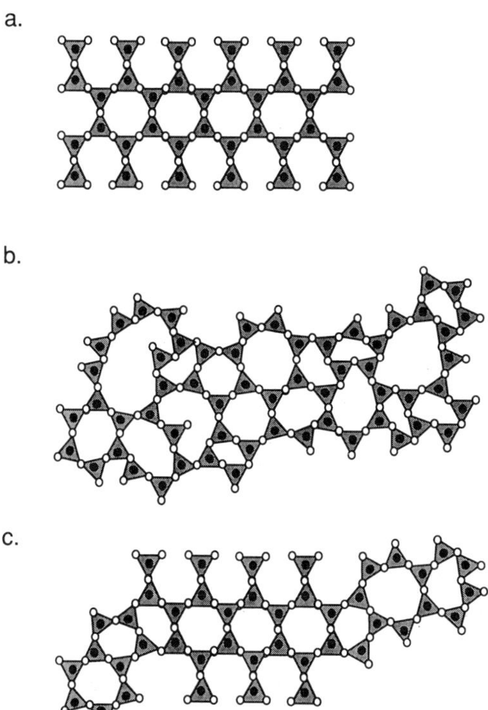

Figure 1-5. Schematic representation of (a) crystalline quartz; (b) random network (amorphous); (c) mixture of crystalline and amorphous regions. (Reprinted with permission from John Wiley and Sons, E. H. Nicollian and J. R. Brews, *MOS Physics and Technology*, Copyright © 1983, John Wiley and Sons).

dra cluster in an ordered six-sided ring pattern, shown schematically in Fig. 1-5a, should be contrasted with the completely random network depicted in Fig. 1-5b. In actuality, the glassy solid structure is most probably a compromise between the two extremes consisting of a considerable amount of short-range order and microscopic regions (i.e., less than 100 Å in size) of crystallinity (Fig. 1-5c). The loose disordered network structure allows for a considerable amount of "holes" or "vacancies" to exist, and it, therefore, comes as no surprise that the density of glasses will be less than that of their crystalline counterparts. In quartz, for example, the density is 2.65 g/cm^3, whereas in silica glass it is 2.2 g/cm^3. Amorphous silicon, which has found commercial use in thin-film solar cells, is, like silica, tetrahedrally bonded and believed to possess a similar structure. We return later to discuss further structural aspects and properties of amorphous films in various contexts throughout the book.

1.3. Defects in Solids

The picture of a perfect crystal structure repeating a particular geometric pattern of atoms without interruption or mistake is somewhat exaggerated. Although there are materials—carefully grown silicon single crystals, for example—that have virtually perfect crystallographic structures extending over macroscopic dimensions, this is not generally true in bulk materials. In thin crystalline films the presence of defects not only serves to disrupt the geometric regularity of the lattice on a microscopic level, it also significantly influences many film properties, such as chemical reactivity, electrical conduction, and mechanical behavior. The structural defects briefly considered in this section are grain boundaries, dislocations, and vacancies.

1.3.1. Grain Boundaries

Grain boundaries are surface or area defects that constitute the interface between two single-crystal grains of different crystallographic orientation. The atomic bonding, in particular grains, terminates at the grain boundary where more loosely bound atoms prevail. Like atoms on surfaces, they are necessarily more energetic than those within the grain interior. This causes the grain boundary to be a heterogeneous region where various atomic reactions and processes, such as solid-state diffusion and phase transformation, precipitation, corrosion, impurity segregation, and mechanical relaxation, are favored or accelerated. In addition, electronic transport in metals is impeded through increased scattering at grain boundaries, which also serve as charge recombination centers in semiconductors. Grain sizes in films are typically from 0.01 to 1.0 μm in dimension and are smaller, by a factor of more than 100, than common grain sizes in bulk materials. For this reason, thin films tend to be more reactive than their bulk counterparts. The fraction of atoms associated with grain boundaries is approximately $3a/l$, where a is the atomic dimension and l is the grain size. For $l = 1000$ Å, this corresponds to about 5 in 1000.

Grain morphology and orientation in addition to size control are not only important objectives in bulk materials but are quite important in thin-film technology. Indeed a major goal in microelectronic applications is to eliminate grain boundaries altogether through epitaxial growth of single-crystal semiconductor films onto oriented single-crystal substrates. Many special techniques involving physical and chemical vapor deposition methods are employed in this effort, which continues to be a major focus of activity in semiconductor technology.

1.3.2. Dislocations

Dislocations are line defects that bear a definite crystallographic relationship to the lattice. The two fundamental types of dislocations—the edge and the screw—are shown in Fig. 1-6 and are represented by the symbol ⊥ . The edge dislocation results from wedging in an extra row of atoms; the screw dislocation requires cutting followed by shearing of the perfect crystal lattice. The geometry of a crystal containing a dislocation is such that when a simple closed traverse is attempted about the crystal axis in the surrounding lattice, there is a closure failure; i.e., one finally arrives at a lattice site displaced from the starting position by a lattice vector, the so-called Burgers vector **b**. The individual cubic cells representing the original undeformed crystal lattice are now distorted somewhat in the presence of dislocations. Therefore, even without application of external forces on the crystal, a state of internal stress exists around each dislocation. Furthermore, the stresses differ around edge and screw dislocations because the lattice distortions differ. Close to the dislocation axis the stresses are high, but they fall off with distance (r) according to a $1/r$ dependence.

Dislocations are important because they have provided a model to help explain a great variety of mechanical phenomena and properties in all classes of crystalline solids. An early application was the important process of plastic deformation, which occurs after a material is loaded beyond its limit of elastic response. In the plastic range, specific planes shear in specific directions relative to each other much as a deck of cards shear from a rectangular prism

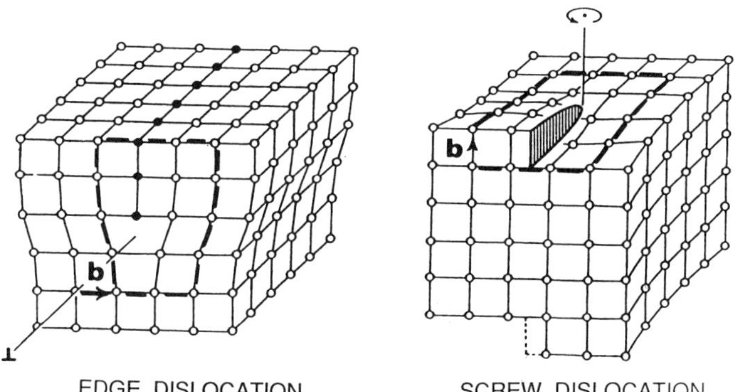

Figure 1-6. (left) Edge dislocation; (right) screw dislocation. (Reprinted with permission from John Wiley and Sons, H. W. Hayden, W. G. Moffatt, and J. Wulff, *The Structure and Properties of Materials*, Vol. III, Copyright © 1965, John Wiley and Sons).

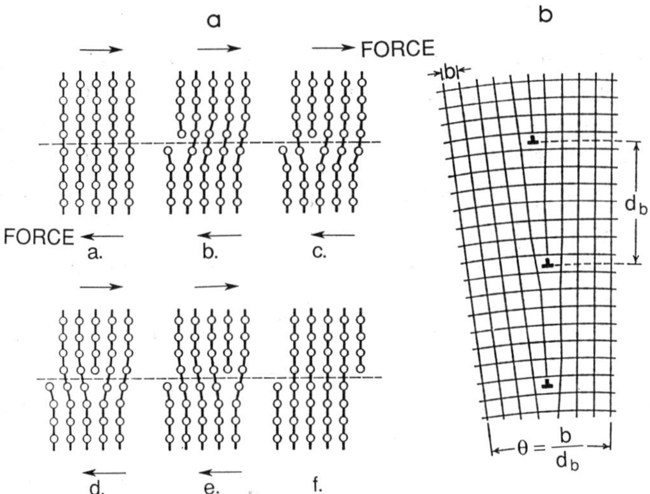

Figure 1-7. (a) Edge dislocation motion through lattice under applied shear stress. (Reprinted with permission from J. R. Shackelford, Introduction to Materials Science for Engineers, Macmillan, 1985). (b) Dislocation model of a grain boundary. The crystallographic misorientation angle θ between grains is b/d_b.

to a parallelepiped. Rather than have rows of atoms undergo a rigid group displacement to produce the slip offset step at the surface, the same amount of plastic deformation can be achieved with less energy expenditure. This alternative mechanism requires that dislocations undulate through the crystal, making and breaking bonds on the slip plane until a slip step is produced, as shown in Fig. 1-7a. Dislocations thus help explain why metals are weak and can be deformed at low stress levels. Paradoxically, dislocations can also explain why metals work-harden or get stronger when they are deformed. These explanations require the presence of dislocations in great profusion. In fact, a density of as many 10^{12} dislocation lines threading 1 cm^2 of surface area has been observed in highly deformed metals. Many deposited polycrystalline metal thin films also have high dislocation densities. Some dislocations are stacked vertically, giving rise to so-called small-angle grain boundaries (Fig. 1-7b). The superposition of externally applied forces and internal stress fields of individual or groups of dislocations, arrayed in a complex three-dimensional network, sometimes makes it more difficult for them to move and for the lattice to deform easily.

The role dislocations play in thin films is varied. As an example, consider the deposition of atoms onto a single-crystal substrate in order to grow an epitaxial single-crystal film. If the lattice parameter in the film and substrate differ, then some geometric accommodation in bonding may be required at the interface, resulting in the formation of interfacial dislocations. The latter are unwelcome defects particularly if films of high crystalline perfection are required. For this reason, a good match of lattice parameters is sought for epitaxial growth. Substrate steps and dislocations should also be eliminated where possible prior to growth. If the substrate has screw dislocations emerging normal to the surface, depositing atoms may perpetuate the extension of the dislocation spiral into the growing film. Like grain boundaries in semiconductors, dislocations can be sites of charge recombination or generation as a result of uncompensated "dangling bonds." Film stress, thermally induced mechanical relaxation processes, and diffusion in films are all influenced by dislocations.

1.3.3. Vacancies

The last type of defect considered is the vacancy. Vacancies are point defects that simply arise when lattice sites are unoccupied by atoms. Vacancies form because the energy ε_f required to remove atoms from interior sites and place them on the surface is not particularly high. This low energy, coupled with the increase in the statistical entropy of mixing vacancies among lattice sites, gives rise to a thermodynamic probability that an appreciable number of vacancies will exist, at least at elevated temperature. The fraction f of total sites that will be unoccupied as a function of temperature T is predicted to be approximately

$$f = e^{-\varepsilon_f/kT}, \tag{1-3}$$

reflecting the statistical thermodynamic nature of vacancy formation. Noting that k is the gas constant and ε_f is typically 1 eV/atom gives $f = 10^{-5}$ at 1000 K.

Vacancies are to be contrasted with dislocations, which are not thermodynamic defects. Because dislocation lines are oriented along specific crystallographic directions, their statistical entropy is low. Coupled with a high formation energy due to the many atoms involved, thermodynamics would predict a dislocation content of less than one per crystal. Thus, although it is possible to create a solid devoid of dislocations, it is impossible to eliminate vacancies.

Vacancies play an important role in all processes related to solid-state diffusion, including recrystallization, grain growth, sintering, and phase transformations. In semiconductors, vacancies are electrically neutral as well as charged and can be associated with dopant atoms. This leads to a variety of normal and anomalous diffusional doping effects.

1.4. Bonding of Materials

Widely spaced isolated atoms condense to form solids due to the energy reduction accompanying bond formation. Thus, if N atoms of type A in the gas phase (g) combine to form a solid (s), the binding energy E_b is released according to the equation

$$NA_g \rightarrow NA_s + E_b. \quad (1\text{-}4)$$

Energy E_b must be supplied to reverse the equation and decompose the solid. The more stable the solid, the higher is its binding energy. It has become the custom to picture the process of bonding by considering the energetics within and between atoms as the interatomic distance progressively shrinks. In each isolated atom, the electron energy levels are discrete, as shown on the right-hand side of Fig. 1-8a. As the atoms approach one another, the individual levels split, as a consequence of an extension of the Pauli exclusion principle, to a collective solid; namely, no two electrons can exist in the same quantum state. Level splitting and broadening occur first for the valence or outer electrons, since their electron clouds are the first to overlap. During atomic attraction, electrons populate these lower energy levels, reducing the overall energy of the solid. With further dimensional shrinkage, the overlap increases and the inner charge clouds begin to interact. Ion-core overlap now results in strong repulsive forces between atoms, raising the system energy. A compromise is reached at the equilibrium interatomic distance in the solid where the system energy is minimized. At equilibrium, some of the levels have broadened into bands of energy levels. The bands span different ranges of energy, depending on the atoms and specific electron levels involved. Sometimes as in metals, bands of high energy overlap. Insulators and semiconductors have energy gaps of varying width between bands where electron states are not allowed. The whys and hows of energy-level splitting, band structure evolution, and implications with regard to property behavior are perhaps the most fundamental and difficult questions in solid-state physics. We briefly return to the subject of electron-band structure after introducing the classes of solids.

An extension of the ideas expressed in Fig. 1-8a is commonly made by simplifying the behavior to atoms as a whole, in which case the potential energy of interaction $V(r)$ is plotted as a function of interatomic distance r in Fig. 1-8b. The generalized behavior shown is common for all classes of solid materials, regardless of the type of bonding or crystal structure. Although the mathematical forms of the attractive or repulsive portions are complex, a number of qualitative features of these curves are not difficult to understand.

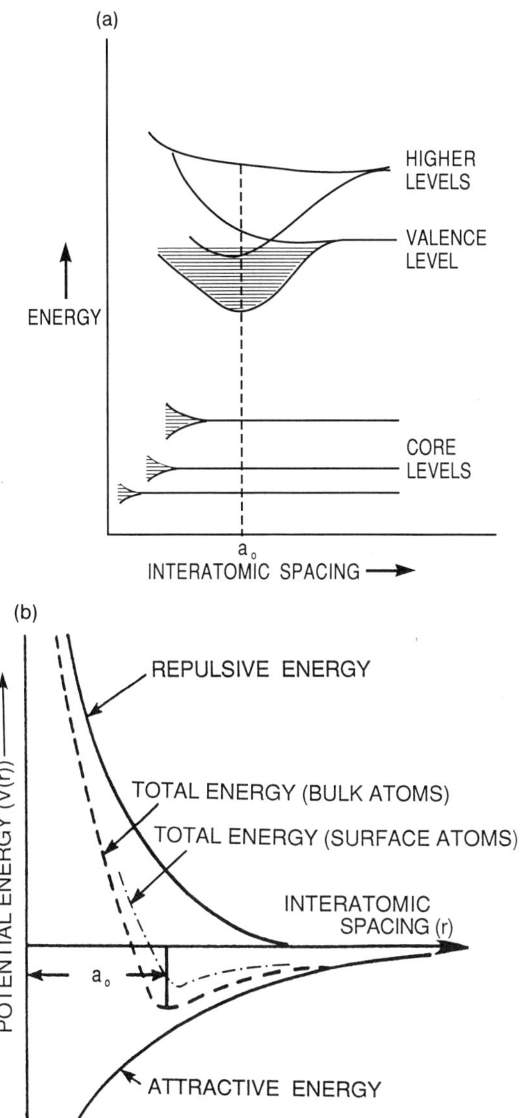

Figure 1-8. Splitting of electron levels (a) and energy of interaction between atoms (b) as a function of interatomic spacing. $V(r)$ vs. r shown schematically for bulk and surface atoms.

For example, the energy at the equilibrium spacing $r = a_0$ is the binding energy. Solids with high melting points tend to have high values of E_b. The curvature of the potential energy is a measure of the elastic stiffness of the solid. To see this, we note that around a_0 the potential energy is approximately harmonic or parabolic. Therefore, $V(r) = (1/2)K_s r^2$, where K_s is related to the spring constant (or elastic modulus). Narrow wells of high curvature are associated with large values of K_s, broad wells of low curvature with small values of K_s. Since the force F between atoms is given by $F = -dV/dr$, $F = -K_s r$, which has its counterpart in Hooke's law—i.e., that stress is linearly proportional to strain. Thus, in solids with high K_s values, correspondingly larger stresses develop under loading. Interestingly, a purely parabolic behavior for V implies a material with a coefficient of thermal expansion equal to zero. In real materials, therefore, some asymmetry or anharmonicity in $V(r)$ exists.

For the most part, atomic behavior within a thin solid film can also be described by a $V(r)$–r curve similar to that for the bulk solid. The surface atoms are less tightly bound, however, which is reflected by the dotted line behavior in Fig. 1-8b. The difference between the energy minima for surface and bulk atoms is a measure of the surface energy of the solid. From the previous discussion, surface layers would tend to be less stiff and melt at lower temperatures than the bulk. Slight changes in equilibrium atomic spacing or lattice parameter at surfaces may also be expected.

Despite apparent similarities, there are many distinctions between the four important types of solid-state bonding and the properties they induce. A discussion of these individual bonding categories follows.

1.4.1. Metallic

The so-called metallic bond occurs in metals and alloys. In metals the outer valence electrons of each atom form part of a collective free-electron cloud or gas that permeates the entire lattice. Even though individual electron–electron interactions are repulsive, there is sufficient electrostatic attraction between the free-electron gas and the positive ion cores to cause bonding.

What distinguishes metals from all other solids is the ability of the electrons to respond readily to applied electric fields, thermal gradients, and incident light. This gives rise to high electrical and thermal conductivities as well as high reflectivities. Interestingly, comparable properties are observed in liquid metals, indicating that aspects of metallic bonding and the free-electron model are largely preserved even in the absence of a crystal structure. Metallic electrical resistivities typically ranging from 10^{-5} to 10^{-6} ohm-cm should be contrasted with the much, much larger values possessed by other classes of solids.

Furthermore, the temperature coefficient of resistivity is positive. Metals thus become poorer electrical conductors as the temperature is raised. The reverse is true for all other classes of solids. The conductivity of pure metals is always reduced with low levels of impurity alloying, which is also contrary to the usual behavior in other solids. The effect of both temperature and alloying element additions on metallic conductivity is to increase electron scattering, which in effect reduces the net component of electron motion in the direction of the applied electric field. On the other hand, in ionic and semiconductor solids production of more charge carriers is the result of higher temperatures and solute additions.

The bonding electrons are not localized between atoms; thus, metals are said to have nondirectional bonds. This causes atoms to slide by each other and plastically deform more readily than is the case, for example, in covalent solids, which have directed atomic bonds.

Examples of thin-metal-film applications include Al contacts and interconnections in integrated circuits, and ferromagnetic alloys for data storage applications. Metal films are also used in mirrors, in optical systems, and as decorative coatings of various components and packaging materials.

1.4.2. Ionic

Ionic bonding occurs in compounds composed of strongly electropositive elements (metals) and strongly electronegative elements (nonmetals). The alkali halides (NaCl, LiF, etc.) are the most unambiguous examples of ionically bonded solids. In other compounds, such as oxides, sulfides, and many of the more complex salts of inorganic chemistry (e.g., nitrates, sulfates, etc.), the predominant, but not necessarily exclusive, mode of bonding is ionic in character. In the rock-salt structure of NaCl, for example, there is an alternating three-dimensional checkerboard array of positively charged cations and negatively charged anions. Charge transfer from the 3s electron level of Na to the 3p level of Cl creates a single isolated NaCl molecule. In the solid, however, the transferred charge is distributed uniformly among nearest neighbors. Thus, there is no preferred directional character in the ionic bond since the electrostatic forces between spherically symmetric inert gaslike ions is independent of orientation.

Much success has been attained in determining the bond energies in alkali halides without resorting to quantum mechanical calculation. The alternating positive and negative ionic charge array suggests that Coulombic pair interactions are the cause of the attractive part of the interatomic potential, which varies simply as $-1/r$. Ionic solids are characterized by strong electrostatic bonding forces and, thus, relatively high binding energies and melting points. They are poor conductors of electricity because the energy required to transfer electrons from anions to cations is prohibitively large. At high temperatures, however, the charged ions themselves can migrate in an electric field, resulting in limited electrical conduction. Typical resistivities for such materials can range from 10^6 to 10^{15} ohm-cm.

Among the ionic compounds employed in thin-film technology are MgF_2, ZnS, and CeF_3, which are used in antireflection coatings on optical components. Assorted thin-film oxides and oxide mixtures such as $Y_3Fe_5O_{12}$, $Y_3Al_5O_{12}$, and $LiNbO_3$ are employed in components for integrated optics. Transparent electrical conductors such as In_2O_3–SnO_2 glasses, which serve as heating elements in window defrosters on cars as well as electrical contacts over the light exposed surfaces of solar cells, have partial ionic character.

1.4.3. Covalent

Covalent bonding occurs in elemental as well as compound solids. The outstanding examples are the elemental semiconductors Si, Ge, and diamond, and the III-V compound semiconductors such as GaAs and InP. Whereas elements at the extreme ends of the periodic table are involved in ionic bonding, covalent bonds are frequently formed between elements in neighboring columns. The strong directional bonds characteristic of the group IV elements are due to the hybridization or mixing of the s and p electron wave functions into a set of orbitals which have high electron densities emanating from the atom in a tetrahedral fashion. A pair of electrons contributed by neighboring atoms makes a covalent bond, and four such shared electron pairs complete the bonding requirements.

Covalent solids are strongly bonded hard materials with relatively high melting points. Despite the great structural stability of semiconductors, relatively modest thermal stimulation is sufficient to release electrons from filled valence bonding states into unfilled electron states. We speak of electrons being promoted from the valence band to the conduction band, a process that increases the conductivity of the solid. Small dopant additions of group III elements like B and In as well as group V elements like P and As take up regular or substitutional lattice positions within Si and Ge. The bonding requirements are then not quite met for group III elements, which are one electron short of a complete octet. An electron deficiency or hole is thus created in the valence band.

For each group V dopant an excess of one electron beyond the bonding octet can be promoted into the conduction band. As the name implies, semiconductors lie between metals and insulators insofar as their ability to conduct electricity is concerned. Typical semiconductor resistivities range from 10^{-3} to 10^{5} ohm-cm. Both temperature and level of doping are very influential in altering the conductivity of semiconductors. Ionic solids are similar in this regard.

The controllable spatial doping of semiconductors over very small lateral and transverse dimensions is a critical requirement in processing integrated circuits. Thin-film technology is thus simultaneously practiced in three dimensions in these materials. Similarly, there is a great necessity to deposit compound semiconductor thin films in a variety of optical device applications. Other largely covalent materials such as SiC, TiC, and BN have found coating applications where hard, wear-resistant surfaces are required. They are usually deposited by chemical vapor deposition methods and will be discussed at length in Chapter 12.

1.4.4. van der Waals Forces

A large group of solid materials are held together by weak molecular forces. This so-called van der Waals bonding is due to dipole–dipole charge interactions between molecules that, though electrically neutral, have regions possessing a net positive or negative charge distribution. Organic molecules such as methane and inert gas atoms are weakly bound together in the solid by these charges. Such solids have low melting points and are mechanically weak. Thin polymer films used as photoresists or for sealing and encapsulation purposes contain molecules that are typically bonded by van der Waals' forces.

1.4.5. Energy-Band Diagrams

A common graphic means of distinguishing between different classes of solids involves the use of energy-band diagrams. Reference to Fig. 1-8a shows how individual energy levels broaden into bands when atoms are brought together to form solids. What is of interest here are the energies of electrons at the equilibrium atomic spacing in the crystal. For metals, insulators, and semiconductors the energy-band structures at the equilibrium spacing are schematically indicated in Fig. 1-9a, b, c. In each case the horizontal axis can be loosely interpreted as some macroscopic distance within the solid with much larger than atomic dimensions. This distance spans a region within the homogeneous bulk interior where the band energies are uniform from point to point. The

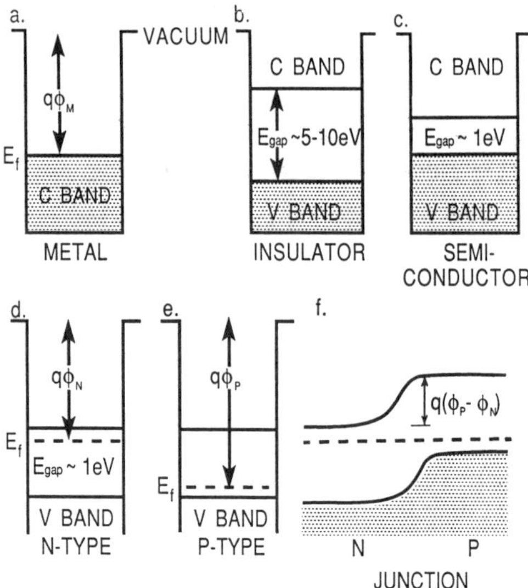

Figure 1-9. Schematic band structure for (a) metal; (b) insulator, (c) semiconductor; (d) *N*-type semiconductor; (e) *P*-type semiconductor; (f) *P–N* semiconductor junction.

uppermost band shown is called the conduction band because once electrons access its levels, they are essentially free to conduct electricity.

Metals have high conductivity because the conduction band contains electrons from the outset. One has to imagine that there are a mind-boggling 10^{22} electrons per cubic centimeter ($\sim$ one per atom) in the conduction band, all of which occupy different quantum states. Furthermore, there are enormous numbers of states all at the same energy level, a phenomenon known as degeneracy. Lastly, the energy levels are extremely closely spaced and compressed within a typical 5-eV conduction-band energy width. The available electrons occupy states within the band up to a certain level known as the Fermi energy E_f. Above E_f are densely spaced excited levels, but they are all vacant. If electrons are excited sufficiently (e.g., by photons or through heating), they can gain enough energy to populate these states or even leave the metal altogether (i.e., by photo- and thermionic emission) and enter the vacuum. As indicated in Fig. 1-9a, the energy difference between the vacuum level and E_f is equal to $q\phi_M$, where ϕ_M is the work function in volts and q is the electronic charge. Even under very tiny electric fields, the electrons in states at E_f can easily move into the unoccupied levels above it, resulting in a net current flow. For this reason, metals have high conductivities.

At the other extreme are insulators, in which the conduction band normally has no electrons. The valence electrons used in bonding completely fill the valence band. A large energy gap E_g ranging from 5 to 10 eV separates the filled valence band from the empty conduction band. There are normally no states and therefore no electrons within the energy gap. In order to conduct electricity, electrons must acquire sufficient energy to span the energy gap, but for all practical cases this energy barrier is all but insurmountable.

Pure (intrinsic) semiconductors at very low temperatures have a band structure like that of insulators, but E_g is smaller; e.g., $E_g = 1.1$ eV in Si and 0.68 eV in Ge. When a semiconductor is doped, new states are created within the energy gap. The electron (or hole) states associated with donors (or acceptors) are usually only a small fraction of an electron volt from the bottom of the conduction band (or top of the valence band). It now takes very little stimulation to excite electrons or holes to conduct electricity. The actual location of E_f with respect to the band diagram depends on the type and amount of doping atoms present. In an intrinsic semiconductor, E_f lies in the middle of the energy gap, because E_f is strictly defined as that energy level for which the probability of occupation is 1/2. If the semiconductor is doped with donor atoms to make it *N*-type, E_f lies above the midgap energy, as shown in Fig. 1-9d. If acceptor atoms are the predominant dopants, E_f lies below the midgap energy and a *P*-type semiconductor results (Fig. 1-9e).

Band diagrams have important implications in thin-film systems where composite layers of different materials are involved. A simple example is the *P–N* junction, which is shown in Fig. 1-9f without any applied electric fields. A condition ensuring thermodynamic equilibrium for the electrons is that E_f must be constant throughout the system. This is accomplished through electron transfer from the *N* side with high E_f (low ϕ_N) to the *P* side with low E_f (high ϕ_P). An internal built-in electric field is established due to this charge transfer resulting in both valence- and conduction-band bending in the junction region. In the bulk of each semiconductor, the bands are unaffected as previously noted. Similar band bending occurs in thin-film metal–semiconductor contacts, semiconductor superlattices, and in metal–oxide semiconductor (MOS) structures over dimensions comparable to the film thicknesses involved. Reference to some of these thin-film structures will be made in later chapters.

1.5. Thermodynamics of Materials

Thermodynamics is definite about events that are impossible. It will say, for example, that reactions or processes are thermodynamically impossible. Thus, gold films do not oxidize and atoms do not normally diffuse up a concentration gradient. On the other hand, thermodynamics is noncommittal about permissible reactions and processes. Thus, even though reactions are thermodynamically favored, they may not occur in practice. Films of silica glass should revert to crystalline form at room temperature according to thermodynamics, but the solid-state kinetics are so sluggish that for all practical purposes amorphous SiO_2 is stable. A convenient measure of the extent of reaction feasibility is the free-energy function G defined as

$$G = H - TS, \tag{1-5}$$

where H is the enthalpy, S the entropy, and T the absolute temperature. Thus, if a system changes from some initial (i) to final (f) state at constant temperature due to a chemical reaction or physical process, a free-energy change $\Delta G = G_f - G_i$ occurs given by

$$\Delta G = \Delta H - T\Delta S, \tag{1-6}$$

where ΔH and ΔS are the corresponding enthalpy and entropy changes. A consequence of the second law of thermodynamics is that spontaneous reactions occur at constant temperature and pressure when $\Delta G < 0$. This condition implies that systems will naturally tend to minimize their free energy and successively proceed from a value G_i to a still lower, more negative value G_f until it is no longer possible to reduce G further. When this happens, $\Delta G = 0$. The system has achieved equilibrium, and there is no longer a driving force for change.

On the other hand, for a process that cannot occur, $\Delta G > 0$. Note that neither the sign of ΔH nor of ΔS taken individually determines reaction direction; rather it is the sign of the combined function ΔG that is crucial. Thus, during condensation of a vapor to form a solid film, $\Delta S < 0$ because fewer atomic configurations exist in the solid. The decrease in enthalpy, however, more than offsets that in entropy, and the net change in ΔG is negative.

The concept of minimization of free energy as a criterion for both stability in a system and forward change in a reaction or process is a central theme in materials science. The following discussion will develop concepts of thermodynamics used in the analysis of chemical reactions and phase diagrams. Subsequent applications will be made to such topics as chemical vapor deposition, interdiffusion, and reactions in thin films.

1.5.1. Chemical Reactions

The general chemical reaction involving substances A, B, and C in equilibrium is

$$a\mathrm{A} + b\mathrm{B} \rightleftarrows c\mathrm{C}. \tag{1-7}$$

The free-energy change of the reaction is given by

$$\Delta G = cG_{\mathrm{C}} - aG_{\mathrm{A}} - bG_{\mathrm{B}}, \tag{1-8}$$

where a, b, and c are the stoichiometric coefficients. It is customary to denote the free energy of individual reactant or product atomic or molecular species by

$$G_i = G_i^{\circ} + RT \ln a_{\mathrm{i}}. \tag{1-9}$$

G_i° is the free energy of the species in its reference or standard state. For solids this is usually the stable pure material at 1 atm at a given temperature. The activity a_{i} may be viewed as an effective thermodynamic concentration and reflects the change in free energy of the species when it is not in its standard state. Substitution of Eq. 1-9 into Eq. 1-8 yields

$$\Delta G = \Delta G^{\circ} + RT \ln \frac{a_{\mathrm{C}}^{c}}{a_{\mathrm{A}}^{a} a_{\mathrm{B}}^{b}}, \tag{1-10}$$

where $\Delta G^{\circ} = cG_{\mathrm{C}}^{\circ} - aG_{\mathrm{A}}^{\circ} - bG_{\mathrm{B}}^{\circ}$. If the system is now in equilibrium, $\Delta G = 0$ and a_{i} is the equilibrium value $a_{\mathrm{i(eq)}}$. Thus,

$$0 = \Delta G^{\circ} + RT \ln \left\{ \frac{a_{\mathrm{C(eq)}}^{c}}{a_{\mathrm{A(eq)}}^{a} a_{\mathrm{B(eq)}}^{b}} \right\} \tag{1-11}$$

or

$$-\Delta G^{\circ} = RT \ln K, \tag{1-12}$$

where the equilibrium constant K is defined by the quantity in braces. Equation 1-12 is one of the most frequently used equations in chemical thermodynamics and will be helpful in analyzing CVD reactions.

Combining Eqs. 1-10 and 1-11 gives

$$\Delta G = RT \ln \left\{ \frac{\left(\dfrac{a_{\mathrm{C}}}{a_{\mathrm{C(eq)}}} \right)^{c}}{\left(\dfrac{a_{\mathrm{A}}}{a_{\mathrm{A(eq)}}} \right)^{a} \left(\dfrac{a_{\mathrm{B}}}{a_{\mathrm{B(eq)}}} \right)^{b}} \right\}. \tag{1-13}$$

Each term $a_{\mathrm{i}} / a_{\mathrm{i(eq)}}$ represents a supersaturation of the species if it exceeds 1, and a subsaturation if it is less than 1. Thus, if there is a supersaturation of reactants and a subsaturation of products, $\Delta G < 0$. The reaction proceeds spontaneously as written with a driving force proportional to the magnitude of ΔG. For many practical cases the a_{i} differ little from the standard-state activities, which are taken to be unity. Therefore, in such a case Eq. 1-10 yields

$$\Delta G = \Delta G^{\circ}. \tag{1-14}$$

Quantitative information on the feasibility of chemical reactions is thus provided by values of $\Delta G°$, and these are tabulated in standard references on thermodynamic data. The reader should be aware that although much of the data are the result of measurement, some values are inferred from various connecting thermodynamic laws and relationships. Thus, even though the vapor pressure of tungsten at room temperature cannot be directly measured, its value is nevertheless "known." In addition, the data deal with equilibrium conditions only, and many reactions are subject to overriding kinetic limitations despite otherwise favorable free-energy considerations.

A particularly useful representation of $\Delta G°$ data for formation of metal oxides as a function of temperature is shown in Fig. 1-10 and is known as an Ellingham diagram. As an example of its use, consider two oxides of importance in thin-film technology, SiO_2 and Al_2O_3, with corresponding oxidation reactions

$$Si + O_2 \rightarrow SiO_2; \qquad \Delta G°_{SiO_2}, \tag{1-15a}$$

$$(4/3)Al + O_2 \rightarrow (2/3)Al_2O_3; \qquad \Delta G°_{Al_2O_3}. \tag{1-15b}$$

Through elimination of O_2 the reaction

$$(4/3)Al + SiO_2 \rightarrow (2/3)Al_2O_3 + Si \tag{1-16}$$

results, where $\Delta G° = \Delta G°_{Al_2O_3} - \Delta G°_{SiO_2}$. Since the $\Delta G°-T$ curve for Al_2O_3 is more negative or lower than that for SiO_2, the reaction is thermodynamically favored as written. At 400 °C, for example, $\Delta G°$ for Eq. 1-16 is $-233 - (-180) = -53$ kcal/mole. Therefore, Al films tend to reduce SiO_2 films, leaving free Si behind, a fact observed in early field effect transistor structures. This was one reason for the replacement of Al film gate electrodes by polycrystalline Si films. As a generalization then, the metal of an oxide that has a more negative $\Delta G°$ than a second oxide will reduce the latter and be oxidized in the process. Further consideration of Eqs. 1-12 and 1-15b indicates that

$$K = \frac{(a_{Al_2O_3})^{2/3}}{(a_{Al})^{4/3} P_{O_2}} = \exp - \frac{\Delta G°}{RT}. \tag{1-17}$$

The Al_2O_3 and Al may be considered to exist in pure standard states with unity activities while the activity of O_2 is taken to be its partial pressure P_{O_2}. Therefore, $\Delta G° = RT \ln P_{O_2}$. If Al were evaporated from a crucible to produce a film, then the value of P_{O_2} in equilibrium with both Al and Al_2O_3 can be calculated at any temperature when $\Delta G°$ is known. If the actual oxygen partial pressure exceeds the equilibrium pressure, then Al ought to oxidize. If the reverse is true, Al_2O_3 would be reduced to Al. At 1000 °C, $\Delta G° = -202$

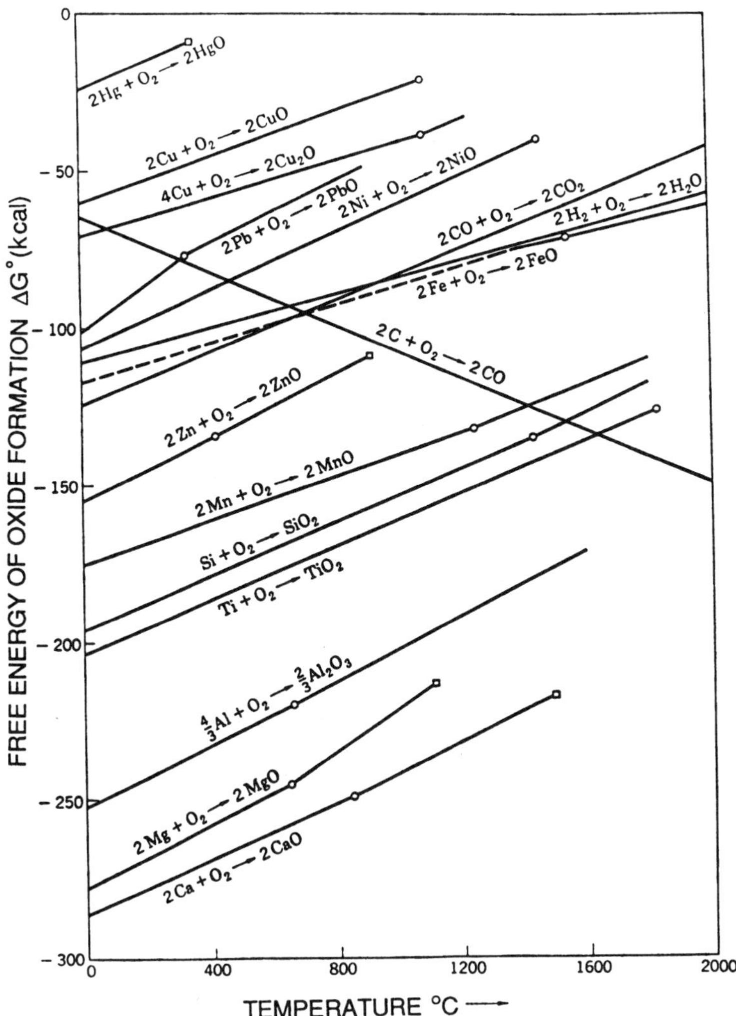

Figure 1-10. Standard free energy of oxide formation vs. temperature: ○ Melting point of metal; □ boiling point of metal (1 atm). (Reprinted with permission from A. G. Guy, *Introduction to Materials Science*, McGraw-Hill, Inc., 1972).

kcal and $P_{O_2} = 2 \times 10^{-35}$ atm. Since this value is many orders of magnitude below actual oxygen partial pressures in vacuum systems, Al would be expected to oxidize. It does to some extent, and a thin film of alumina probably forms on the surface of the molten aluminum source. Nevertheless, oxide-free films can be deposited in practice.

Similar Ellingham plots of free energy of formation versus temperature exist for sulfides, carbides, nitrides, and chlorides. In Chapter 4 we consider such a diagram for Si–H–Cl compounds, which is useful for the thermodynamic analysis of Si CVD.

1.5.2. Phase Diagrams

The most widespread method for representing the conditions of chemical equilibrium for inorganic systems as a function of initial composition, temperature, and pressure is through the use of phase diagrams. By phases we not only mean the solid, liquid, and gaseous states of pure elements and compounds but a material of variable yet homogeneous composition, such as an alloy, is also a phase. Although phase diagrams generally contain a wealth of thermodynamic information on systems in equilibrium, they can readily be interpreted without resorting to complex thermodynamic laws, functions, or equations. They have been experimentally determined for many systems by numerous investigators over the years and provide an invaluable guide when synthesizing materials.

There are a few simple rules for analyzing phase diagrams. The most celebrated of these is the Gibbs phase rule, which, though deceptively simple, is arguably the most important linear algebraic equation in physical science. It can be written as

$$f = n + 2 - \psi, \tag{1-18}$$

where n is the number of components (i.e., different atomic species), ψ is the number of phases, and f is the number of degrees of freedom or variance in the system. The number of intensive variables that can be independently varied without changing the phase equilibrium is equal to f.

1.5.2.1. One-Component System.

As an application to a one-component system, consider the P–T diagram given for carbon in Fig. 1-11. Shown are the regions of stability for different phases of carbon as a function of pressure and temperature. Within the broad areas, the single phases diamond and graphite are stable. Both P and T variables can be *independently* varied to a greater or lesser extent without leaving the single-phase field. This is due to the phase rule, which gives $f = 1 + 2 - 1 = 2$. Those states that lie on any of the lines of the diagram represent two-phase equilibria. Now $f = 1 + 2 - 2 = 1$. This means, for example, that in order to change but maintain the equilibrium along the diamond–graphite line, only one variable, either T or P, can be independently varied; the corresponding variables P or T must change in a dependent fashion. At a point where three phases coexist (not shown), $f = 0$. Any change of T or P will destroy the three-phase equilib-

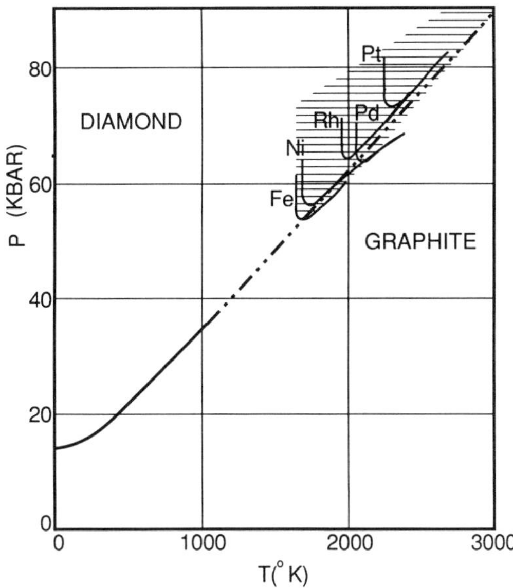

Figure 1-11. Portion of the pressure–temperature diagram for carbon showing stability regions of diamond and graphite. Shaded areas represent regions of diamond formation in the indicated metal solvents. (Reprinted with permission from R. C. DeVries, *Ann. Rev. Mater. Sci.* **17**, 161, 1987).

rium, leaving instead either one or two phases. The diagram suggests that pressures between 10^4 to 10^5 bars (~ 10,000–100,000 atm) are required to transform graphite into diamond. In addition, excessively high temperatures (~ 2000 K) are required to make the reaction proceed at appreciable rates. It is exciting, therefore, that diamond thin films have been deposited by decomposing methane in a microwave plasma at low pressure and temperature, thus avoiding the almost prohibitive pressure–temperature regime required for bulk diamond synthesis.

1.5.2.2. Two-Component Systems. When two elements or compounds are made to combine, many very important materials, less-well-known than the compounds of inorganic chemistry, can be produced. Binary metal alloys and compound semiconductors such as Ni–Cr (nichrome) and GaAs are examples that have important bulk as well as thin-film uses. The resultant phases that form as a function of initial reactant proportions and temperature are depicted on binary equilibrium phase diagrams. Collections of these have been published for metal, semiconductor, and ceramic systems and are among the most frequently consulted references in the field of materials. Unless noted other-

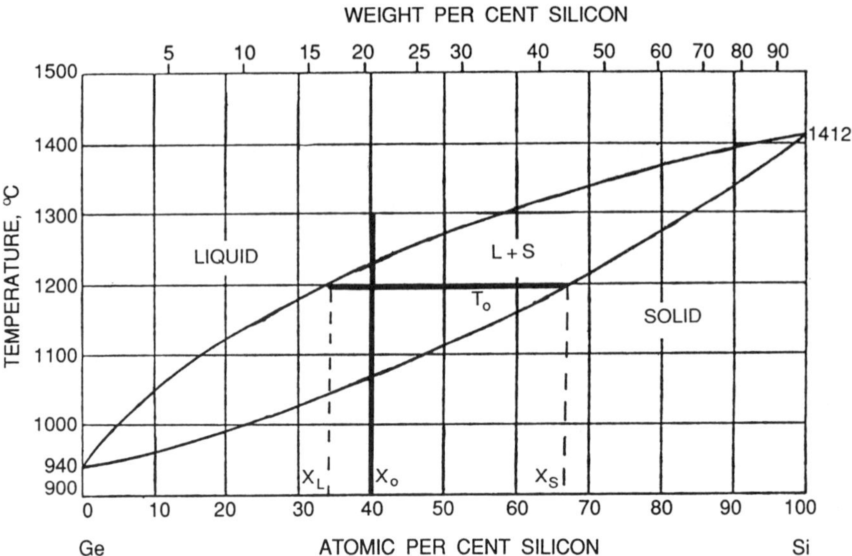

Figure 1-12. Ge–Si equilibrium phase diagram. (Reprinted with permission from M. Hansen, *Constitution of Binary Alloys*, McGraw–Hill, Inc. 1958).

wise, these diagrams hold at atmospheric pressure, in which case the variance is reduced by 1. The Gibbs phase rule now states $f = n + 1 - \psi$ or $f = 3 - \psi$. Thus, at most three phases can coexist in equilibrium.

To learn how to interpret binary phase diagrams, let us first consider the Ge–Si system shown in Fig. 1-12. Such a system is interesting because of the possibility of creating semiconductors having properties intermediate to those of Ge and Si. On the horizontal axis, the overall composition is indicated. Pure Ge and Si components are represented at the extreme left and right axes, respectively, and compositions of initial mixtures of Ge and Si are located in between. Such compositions are given in either weight or atomic percent. The following set of rules will enable a complete equilibrium phase analysis for an initial alloy composition X_0 heated to temperature T_0.

1. Draw a vertical line at composition X_0. Any point on this line represents a state of this system at the temperature indicated on the left-hand scale.
2. The chemical compositions of the resulting phases depend on whether the point lies (a) in a one-phase field, (b) in a two-phase field, or (c) on a sloping or horizontal (isothermal) boundary between phase fields.

   a. For states within a single-phase field., i.e., L (liquid), S (solid), or a compound, the phase composition or chemical analysis is always the same as the initial composition.

b. In a two-phase region, i.e., L + S, $\alpha + \beta$, etc., a horizontal tie line is first drawn through the state point extending from one end of the two-phase field to the other as shown in Fig. 1-12. On either side of the two-phase field are the indicated single-phase fields (L and S). The compositions of the two phases in question are given by projecting the ends of the tie line vertically down and reading off the values. For example, if $X_0 = 40$ at% Si and $T_0 = 1200$ °C, $X_L = 34$ at% Si and $X_S = 67$ at% Si.

c. State points located on either a sloping or a horizontal boundary cannot be analyzed; phase analyses can only be made above or below the boundary lines according to rules a and b. Sloping boundaries are known as liquidus or solidus lines when L/L + S or L + S/S phase field combinations are respectively involved. Such lines also represent solubility limits and are, therefore, associated with the processes of solution or rejection of phases (precipitation) from solution. The horizontal isothermal boundaries indicate the existence of phase transformations involving three phases. The following common reactions occur at these critical isotherms, where α, β and γ are solid phases:

1. Eutectic: $L \rightarrow \alpha + \beta$
2. Eutectoid: $\gamma \rightarrow \alpha + \beta$
3. Peritectic: $L + \alpha \rightarrow \gamma$

3. The relative amount of phases present depends on whether the state point lies in (a) a one-phase field or (b) a two-phase field.

a. Here the one phase in question is homogeneous and present exclusively. Therefore, the relative amount of this phase is 100%.

b. In the two-phase field the lever rule must be applied to obtain the relative phase amounts. From Fig. 1-12, state X_0, T_0, and the corresponding tie line, the relative amounts of L and S phases are given by

$$\%L = \frac{X_S - X_0}{X_S - X_L} \times 100; \qquad \%S = \frac{X_0 - X_L}{X_S - X_L} \times 100, \tag{1-19}$$

where %L plus %S = 100. (Substitution gives %L = (67 − 40)/(67 − 34) × 100 = 81.8, and %S = (40 − 34)/(67 − 34) × 100 = 18.2.) Equation 1-19 represents a definition of the lever rule that essentially ensures conservation of mass in the system. The tie line and lever rule can be applied *only* in a two-phase region; they make no sense in a one-phase region. Such analyses do reveal information on phase compositions and amounts, yet they say nothing about the physical appearance or shape that phases actually take. Phase morphology is dependent on issues related to nucleation and growth.

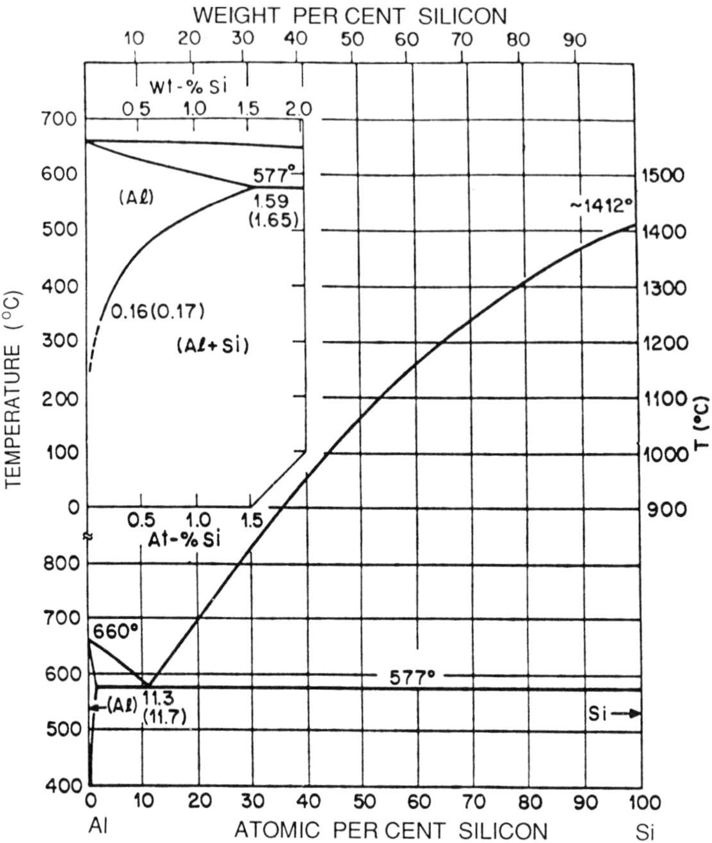

Figure 1-13. Al–Si equilibrium phase diagram. (Reprinted with permission from M. Hansen, *Constitution of Binary Alloys*, McGraw-Hill, Inc. 1958).

Before leaving the Ge–Si system, note that L represents a broad liquid solution field where Ge and Si atoms mix in all proportions. Similarly, at lower temperatures, Ge and Si atoms mix randomly but on the atomic sites of a diamond cubic lattice to form a substitutional solid solution. The lens-shaped L + S region separating the single-phase liquid and solid fields occurs in many binary systems, including Cu–Ni, Ag–Au, Pt–Rh, Ti–W, and Al_2O_3–Cr_2O_3.

A very common feature on binary phase diagrams is the eutectic isotherm. The Al–Si system shown in Fig. 1-13 is an example of a system undergoing a eutectic transformation at 577 °C. Alloy films containing about 1 at% Si are used to make contacts to silicon in integrated circuits. The insert in Fig. 1-13 indicates the solid-state reactions for this alloy involve either the formation of an Al-rich solid solution above 520 °C or the rejection of Si below this temperature in order to satisfy solubility requirements. Although this particular alloy cannot undergo a eutectic transformation, all alloys containing more than 1.59 at% Si can. When crossing the critical isotherm from high temperature, the reaction

$$\mathrm{L}(11.3\ \mathrm{at\%\ Si}) \xrightarrow{577\ ^\circ\mathrm{C}} \mathrm{Al}(1.59\ \mathrm{at\%\ Si}) + \mathrm{Si} \tag{1-20}$$

occurs. Three phases coexist at the eutectic temperature, and therefore $f = 0$. Any change in temperature and/or phase composition will drive this very special three-phase equilibrium into single- (i.e., L) or two-phase fields (i.e., L + Al, L + Si, Al + Si), depending on composition and temperature.

The important GaAs system shown in Fig. 1-14 contains two independent side-by-side eutectic reactions at 29.5 and 810 °C. For the purpose of analysis one can consider that there are two separate eutectic subsystems, Ga–GaAs and GaAs–As. In this way complex diagrams can be decomposed into simpler units. The critical eutectic compositions occur so close to either pure component that they cannot be resolved on the scale of this figure. The prominent central vertical line represents the stoichiometric GaAs compound, which melts at 1238 °C. Phase diagrams for several other important 3–5 semiconductors,

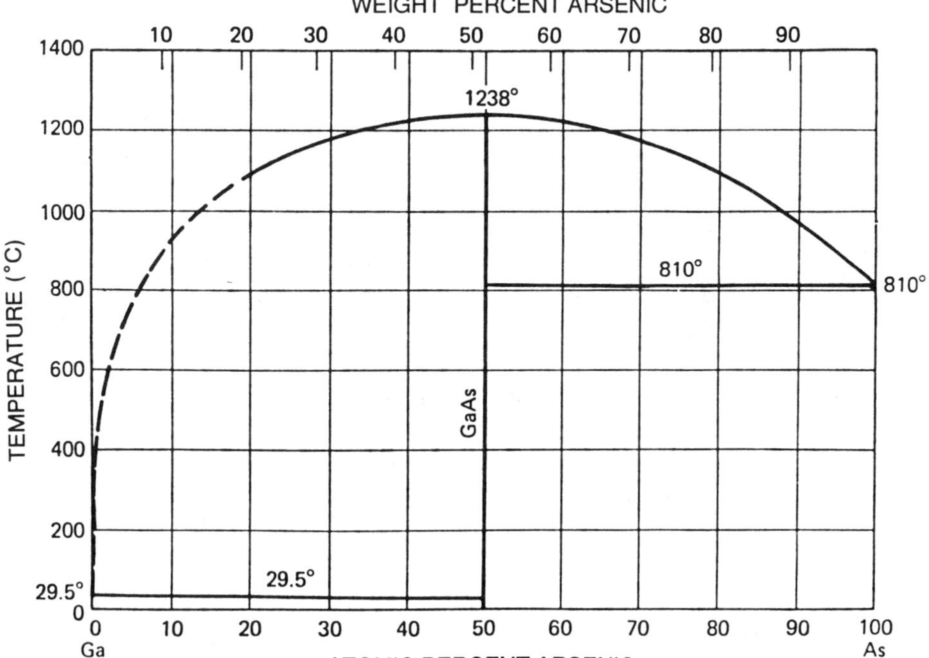

Figure 1-14. Ga–As equilibrium phase diagram. (Reprinted with permission from M. Hansen, *Constitution of Binary Alloys*, McGraw–Hill, Inc. 1958).

(e.g., InP, GaP, and InAs) have very similar appearances. These compound semiconductors are common in other ways. For example, one of the components (e.g., Ga, In) has a low melting point coupled with a rather low vapor pressure, whereas the other component (e.g., As, P) has a higher melting point and a high vapor pressure. These properties complicate both bulk and thin-film single-crystal growth processes.

We end this section on phase diagrams by reflecting on some distinctions in their applicability to bulk and thin-film materials. High-temperature phase diagrams were first determined in a systematic way for binary metal alloys. The traditional processing route for bulk metals generally involves melting at a high temperature followed by solidification and subsequent cooling to the ambient. It is a reasonable assumption that thermodynamic equilibrium is attained in these systems, especially at elevated temperatures. Atoms in metals have sufficient mobility to enable stable phases to nucleate and grow within reasonably short reaction times. This is not generally the case in metal oxide systems, however, because of the tendency of melts to form metastable glasses due to sluggish atomic motion.

In contrast, thin films do not generally pass from a liquid phase through a vertical succession of phase fields. For the most part, thin-film science and technology is characterized by low-temperature processing where equilibrium is difficult to achieve. Depending on what is being deposited and the conditions of deposition, thin films possessing varying degrees of thermodynamic stability can be readily produced. For example, single-crystal silicon is the most stable form of this element below the melting point. Nevertheless, chemical vapor deposition of Si from chlorinated silanes at 1200 °C will yield single-crystal films, and amorphous films can be produced below 600 °C. In between, polycrystalline Si films of varying grain size can be deposited. Since films are laid down an atomic layer at a time, the thermal energy of individual atoms impinging on a massive cool substrate heat sink can be transferred to the latter at an extremely rapid rate. Deprived of energy, the atoms are relatively immobile. It is not surprising, therefore, that metastable and even amorphous semiconductor and alloy films can be evaporated or sputtered onto cool substrates. When such films are heated, they crystallize and revert to the more stable phases indicated by the phase diagram.

Interesting issues related to binary phase diagrams arise with multicomponent thin films that are deposited in layered structures through sequential deposition from multiple sources. For example, "strained layer superlattices" of Ge–Si have been grown by molecular beam epitaxy (MBE) techniques (see Chapter 7). Films of Si and Si + Ge solid–solution alloy, typically tens of angstroms thick, have been sequentially deposited such that the resultant composite film is a single crystal with strained lattice bonds. The resolution of distinct layers as revealed by the transmission electron micrograph of Fig. 14-17 is suggestive of a two-phase mixture. On the other hand, a single crystal implies a single phase even if it possesses a modulated chemical composition. Either way, the superlattice is not in thermodynamic equilibrium because the Ge–Si phase diagram unambiguously predicts a stable solid solution at low temperature. Equilibrium can be accelerated by heating, which results in film homogenization by interatomic diffusion. In thin films, phases such as solid solutions and compounds are frequently accessed *horizontally* across the phase diagram during an isothermal anneal. This should be contrasted with bulk materials, where equilibrium phase changes commonly proceed *vertically* downward from elevated temperatures.

1.6. KINETICS

1.6.1. Macroscopic Transport

Whenever a material system is not in thermodynamic equilibrium, driving forces arise naturally to push it toward equilibrium. Such a situation can occur, for example, when the free energy of a microscopic system varies from point to point because of compositional inhomogeneities. The resulting atomic concentration gradients generate time-dependent, mass-transport effects that reduce free-energy variations in the system. Manifestations of such processes include phase transformations, recrystallization, compound growth, and degradation phenomena in both bulk and thin-film systems. In solids, mass transport is accomplished by diffusion, which may be defined as the migration of an atomic or molecular species within a given matrix under the influence of a concentration gradient. Fick established the phenomenological connection between concentration gradients and the resultant diffusional transport through the equation

$$J = -D\frac{dC}{dx}. \tag{1-21}$$

The minus sign occurs because the vectors representing the concentration gradient dC/dx and atomic flux J are oppositely directed. Thus an increasing concentration in the positive x direction induces mass flow in the negative x direction, and vice versa. The units of C are typically atoms/cm^3. The diffusion coefficient D, which has units of cm^2/sec, is characteristic of both the diffusing species and the matrix in which transport occurs. The extent of observable diffusion effects depends on the magnitude of D. As we shall later note, D increases in exponential fashion with temperature according to a Maxwell–Boltzmann relation; i.e.,

$$D = D_0 \exp - E_D/RT, \tag{1-22}$$

where D_0 is a constant and RT has the usual meaning. The activation energy for diffusion is E_D (cal/mole).

Solid-state diffusion is generally a slow process, and concentration changes occur over long periods of time; the steady-state condition in which concentrations are time-independent rarely occurs in bulk solids. Therefore, during one-dimensional diffusion, the mass flux across plane x of area A exceeds that which flows across plane $x + dx$. Atoms will accumulate with time in the volume $A\,dx$, and this is expressed by

$$JA - \left(J + \frac{dJ}{dx}dx\right)A = -\frac{dJ}{dx}A\,dx = \frac{dc}{dt}A\,dx. \tag{1-23}$$

Substituting Eq. 1-21 and assuming that D is a constant independent of C or x gives

$$\frac{\partial C(x,t)}{\partial t} = D\frac{\partial^2 C(x,t)}{\partial x^2}. \tag{1-24}$$

The non-steady-state heat conduction equation is identical if temperature is substituted for C and the thermal diffusivity for D. Many solutions for both diffusion and heat conduction problems exist for media of varying geometries, constrained by assorted initial and boundary conditions. They can be found in the books by Carslaw and Jaeger, and by Crank, listed in the bibliography. Since complex solutions to Eq. 1-24 will be discussed on several occasions (e.g., in Chapters 8, 9, and 13), we introduce simpler applications here.

Consider an initially pure thick film into which some solute diffuses from the surface. If the film dimensions are very large compared with the extent of diffusion, the situation can be physically modeled by the following conditions:

$$C(x,0) = 0 \quad \text{at } t = 0 \qquad \text{for } \infty > x > 0, \tag{1-25a}$$

$$C(\infty,t) = 0 \quad \text{at } x = \infty \qquad \text{for } t > 0. \tag{1-25b}$$

The second boundary condition that must be specified has to do with the nature of the diffusant distribution maintained at the film surface $x = 0$. Two simple cases can be distinguished. In the first, a thick layer of diffusant provides an essentially limitless supply of atoms maintaining a *constant* surface concentration C_0 for all time. In the second case, a very thin layer of diffusant provides an *instantaneous* source S_0 of surface atoms per unit area. Here the surface concentration diminishes with time as atoms diffuse into the underlying film. These two cases are respectively described by

$$C(0, t) = C_0 \tag{1-26a}$$

$$\int_0^\infty C(x, t)\, dx = S_0 \tag{1-26b}$$

Expressions for $C(x, t)$ satisfying these conditions are respectively

$$C(x, t) = C_0 \operatorname{erfc} \frac{x}{\sqrt{4\,Dt}} = C_0 \left(1 - \operatorname{erf} \frac{x}{\sqrt{4\,Dt}}\right), \tag{1-27a}$$

$$C(x, t) = \frac{S_0}{\sqrt{\pi\,Dt}} \exp - \frac{x^2}{4\,Dt}, \tag{1-27b}$$

and these represent the simplest mathematical solutions to the diffusion equation. They have been employed to determine doping profiles and junction

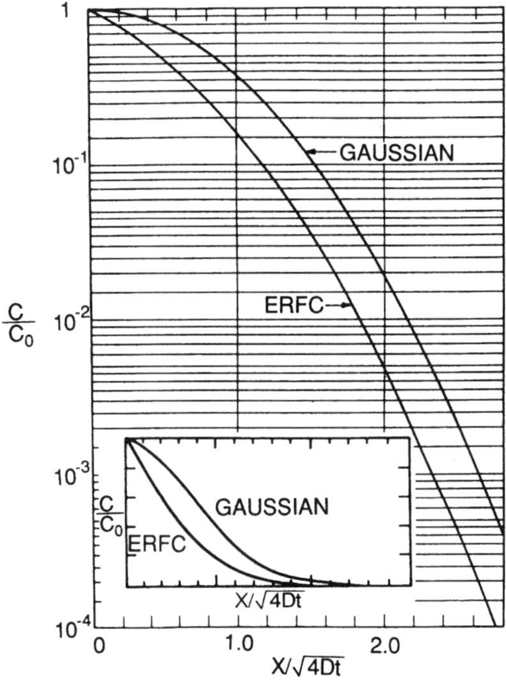

Figure 1-15. Normalized Gaussian and Erfc curves of C/C_0 vs. $x/\sqrt{4\,Dt}$. Both logarithmic and linear scales are shown. (Reprinted with permission from John Wiley and Sons, from W. E. Beadle, J. C. C. Tsai, and R. D. Plummer, *Quick Reference Manual for Silicon Integrated Circuit Technology*, Copyright © 1985, Bell Telephone Laboratories Inc. Published by J. Wiley and Sons).

depths in semiconductors. The error function erf $x/2\sqrt{Dt}$, defined by

$$\operatorname{erf}\frac{x}{2\sqrt{Dt}} = \frac{2}{\sqrt{\pi}}\int_0^{x/2\sqrt{Dt}} e^{-Z^2}\,dZ, \tag{1-28}$$

is a tabulated function of only the upper limit or argument $x/2\sqrt{Dt}$. Normalized concentration profiles for the Gaussian and Erfc solutions are shown in Fig. 1-15. It is of interest to calculate how these distributions spread with time. For the erfc solution, the diffusion front at the arbitrary concentration of $C(x,t)/C_0 = 1/2$ moves parabolically with time as $x = 2\sqrt{Dt}\,\operatorname{erfc}^{-1}(1/2)$ or $x = 0.96\sqrt{Dt}$. When $\sqrt{Dt}$ becomes large compared with the film dimensions, the assumption of an infinite matrix is not valid and the solutions do not strictly hold. The film properties may also change appreciably due to interdiffusion. To limit the latter and ensure the integrity of films, D should be kept small, which in effect means the maintenance of low temperatures. This subject will be discussed further in Chapter 8.

1.6.2. Atomistic Considerations

Macroscopic changes in composition during diffusion are the result of the random motion of countless individual atoms unaware of the concentration gradient they have helped establish. On a microscopic level, it is sufficient to explain how atoms execute individual jumps from one lattice site to another, for through countless repetitions of unit jumps macroscopic changes occur. Consider Fig. 1-16a, showing neighboring lattice planes spaced a distance a_0 apart within a region where an atomic concentration gradient exists. If there are n_1 atoms per unit area of plane 1, then at plane 2, $n_2 = n_1 + (dn/dx)a_0$,

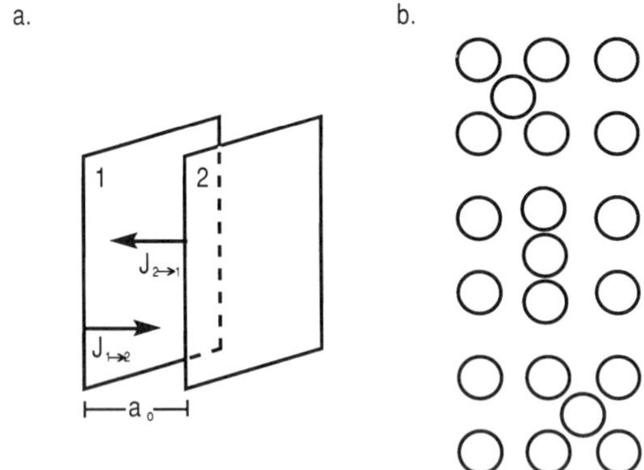

Figure 1-16. (a) Atomic diffusion fluxes between neighboring crystal planes. (b) Atomistic view of atom jumping into a neighboring vacancy.

where we have taken the liberty of assigning a continuum behavior at discrete planes. Each atom vibrates about its equilibrium position with a characteristic lattice frequency ν, typically 10^{13} sec $^{-1}$. Very few vibrational cycles have sufficient amplitude to cause the atom to actually jump into an adjoining lattice position, thus executing a direct atomic interchange. This process would be greatly encouraged, however, if there were neighboring vacant sites. The fraction of vacant lattice sites was previously given by $e^{-\varepsilon_f/kT}$ (see Eq. 1-3). In addition, the diffusing atom must acquire sufficient energy to push the surrounding atoms apart so that it can squeeze past and land in the so-called activated state shown in Fig. 1-16b. This step is the precursor to the downhill jump of the atom into the vacancy. The number of times per second that an atom successfully reaches the activated state is $\nu e^{-\varepsilon_j/kT}$, where ε_j is the vacancy jump or migration energy per atom. Here the Boltzmann factor may be interpreted as the fraction of all sites in the crystal that have an activated state configuration. The atom fluxes from plane 1 to 2 and from plane 2 to 1 are then, respectively, given as

$$J_{1\to 2} = \frac{1}{6}\nu \exp - \frac{\varepsilon_f}{kT} \exp - \frac{\varepsilon_j}{kT}(Ca_0), \tag{1-29a}$$

$$J_{2\to 1} = \frac{1}{6}\nu \exp - \frac{\varepsilon_f}{kT} \exp - \frac{\varepsilon_j}{kT}\left(C + \frac{dC}{dx}a_0\right)a_0, \tag{1-29b}$$

where we have substituted Ca_0 for n and used the factor of $1/6$ to account for bidirectional jumping in each of the three coordinate directions. The net flux J_N is the difference or

$$J_N = -\frac{1}{6}a_0^2\nu \exp - \frac{\varepsilon_f}{kT} \exp - \frac{\varepsilon_j}{kT}\left(\frac{dC}{dx}\right). \tag{1-30}$$

By association with Fick's law, D can be expressed as

$$D = D_0 \exp - E_D/RT \tag{1-31}$$

with $D_0 = (1/6)a_0^2\nu$ and $E_D = (\varepsilon_f + \varepsilon_j)N_A$, where N_A is Avogadro's number.

Although the above model is intended for atomistic diffusion in the bulk lattice, a similar expression for D would hold for transport through grain boundaries or along surfaces and interfaces of films. At such nonlattice sites, energies for defect formation and motion are expected to be less, leading to higher diffusivities. Dominating microscopic mass transport is the Boltzmann factor $\exp - E_D/RT$, which is ubiquitous when describing the temperature dependence of the rate of many processes in thin films. In such cases the kinetics can be described graphically by an Arrhenius plot in which the

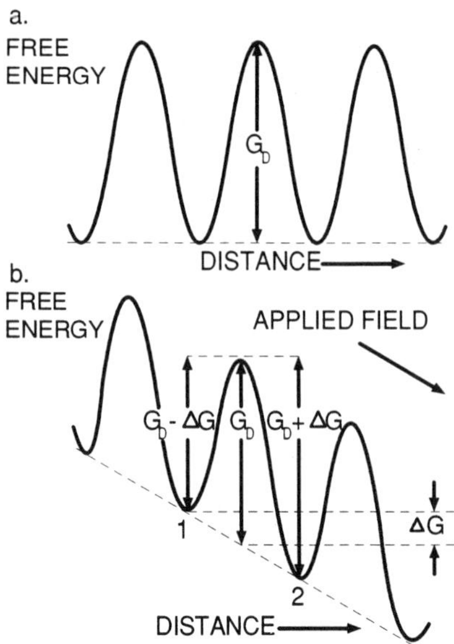

Figure 1-17. (a) Free-energy variation with atomic distance in the absence of an applied field. (b) Free-energy variation with atomic distance in the presence of an applied field.

logarithm of the rate is plotted on the ordinate and the reciprocal of the absolute temperature is plotted along the abscissa. The slope of the resulting line is then equal to $-E_D/R$, from which the characteristic activation energy can be extracted.

The discussion to this point is applicable to motion of both impurity and matrix atoms. In the latter case we speak of self-diffusion. For matrix atoms there are driving forces other than concentration gradients that often result in transport of matter. Examples are forces due to stress fields, electric fields, and interfacial energy gradients. To visualize their effect, consider neighboring atomic positions in a crystalline solid where no fields are applied. The free energy of the system has the periodicity of the lattice and varies schematically, as shown in Fig. 1-17a. Imposition of an external field now biases the system such that the free energy is lower in site 2 relative to 1 by an amount $2\,\Delta G$. A free-energy gradient exists in the system that lowers the energy barrier to motion from $1 \rightarrow 2$ and raises it from $2 \rightarrow 1$. The rate at which atoms move from 1 to 2 is given by

$$\mathbf{r}_{12} = \nu \exp\left(-\frac{G_D - \Delta G}{RT}\right) \text{sec}^{-1}. \tag{1-32a}$$

Similarly,

$$\mathbf{r}_{21} = \nu \exp\left(-\frac{G_D + \Delta G}{RT}\right) \sec^{-1}, \tag{1-32b}$$

and the *net* rate r_N is given by the difference or

$$r_N = \nu \exp - \frac{G_D}{RT}\left[\exp\frac{\Delta G}{RT} - \exp - \frac{\Delta G}{RT}\right] = 2\nu \exp - \frac{G_D}{RT}\sinh\frac{\Delta G}{RT}. \tag{1-33}$$

When $\Delta G = 0$, the system is in thermodynamic equilibrium and $r_N = 0$, so no net atomic motion occurs. Although G_D is typically a few electron volts or so per atom (1 eV = 23,060 cal/mole), ΔG is much smaller in magnitude since it is virtually impossible to impose external forces on solids comparable to the interatomic forces. In fact, $\Delta G/RT$ is usually much less than unity, so $\sinh \Delta G/RT \approx \Delta G/RT$. This leads to commonly observed linear diffusion effects. But when $\Delta G/RT \approx 1$, nonlinear diffusion effects are possible. By multiplying both sides of Eq. 1-33 by a_0, we obtain the atomic velocity v:

$$v = a_0 r_N = \left[a_0^2 \nu e^{-G_D/RT}\right]\frac{2\,\Delta G}{a_0 RT}. \tag{1-34}$$

The term in brackets is essentially the diffusivity D with G_D a diffusional activation energy. (The distinction between G_D and E_D need not concern us here.) The term $2\,\Delta G/a_0$ is a measure of the molar free-energy gradient or applied force F. Therefore, the celebrated Nernst–Einstein equation results:

$$v = DF/RT. \tag{1-35}$$

Application of this equation will be made subsequently to various thin-film mass transport phenomena, e.g., electric-field-induced atomic migration (electromigration), stress relaxation, and grain growth. The drift of charge carriers in semiconductors under an applied field can also be modeled by Eq. 1-35. In some instances, larger generalized forces can be applied to thin films relative to bulk materials because of the small dimensions involved.

Chemical reaction rate theory provides a common application of the preceding ideas. In Fig. 1-18 the reactants at the left are envisioned to proceed toward the right following the reaction coordinate path. Along the way, intermediate activated states are accessed by surmounting the free-energy barrier. Through decomposition of the activated species, products form. If C_R is the concentration of reactants at coordinate position 1 and C_P the concentration of products at 2, then the net rate of reaction is proportional to

$$r_N \approx C_R \exp\left(-\frac{G^*}{RT}\right) - C_P \exp\left(-\frac{G^* + \Delta G}{RT}\right), \tag{1-36}$$

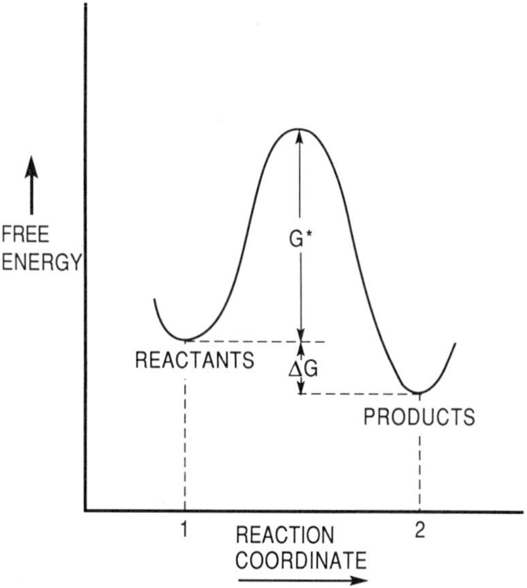

Figure 1-18. Free-energy path for thermodynamically favored reaction 1 → 2.

where G^* is the free energy of activation. As before, the Boltzmann factors represent the probabilities of surmounting the respective energy barriers faced by reactants proceeding in the forward direction, or products in the reverse direction. When chemical equilibrium prevails, the competing rates are equal and $r_N = 0$. Therefore,

$$\frac{C_P}{C_R} = \exp\frac{\Delta G}{RT} = \frac{\exp - G_P/RT}{\exp - G_R/RT}. \tag{1-37}$$

For the reaction to proceed to the right $\Delta G = G_R - G_P$ must be positive. By comparison with Eq. 1-12, it is apparent that the left-hand side is the equilibrium constant and ΔG may be associated with $-\Delta G°$. This expression is perfectly general, however, and applies, for example, to electron energy-level populations in semiconductors and lasers, as well as magnetic moment distributions in solids. In fact, whenever thermal energy is a source of activation energy, Eq. 1-37 is valid.

1.7. Nucleation

When the critical lines separating stable phase fields on equilibrium phase diagrams are crossed, new phases appear. Most frequently, a decrease in temperature is involved, and this may, for example, trigger solidification or solid-state phase transformations from now unstable melts or solid matrices. When such a transformation occurs, a new phase of generally different structure and composition emerges from the prior parent phase or phases. The process known as nucleation occurs during the very early stages of phase change. It is important in thin films because the grain structure that ultimately develops in a given deposition process is usually strongly influenced by what happens during film nucleation and subsequent growth.

Simple models of nucleation are first of all concerned with thermodynamic questions of the energetics of the process of forming a single stable nucleus. Once nucleation is possible, it is usual to try to specify how many such stable nuclei will form within the system per unit volume and per unit time—i.e., nucleation rate. As an example, consider the *homogeneous* nucleation of a spherical solid phase of radius r from a prior supersaturated vapor. Pure homogeneous nucleation is rare but easy to model since it occurs without benefit of complex heterogeneous sites such as exist on an accommodating substrate surface. In such a process the gas-to-solid transformation results in a reduction of the chemical free energy of the system given by $(4/3)\pi r^3 \Delta G_V$, where ΔG_V corresponds to the change in chemical free energy per unit volume. For the condensation reaction vapor (v) → solid (s), Eq. 1-13 indicates that

$$\Delta G_V = \frac{kT}{\Omega}\ln\frac{P_s}{P_v} = -\frac{kT}{\Omega}\ln\frac{P_v}{P_s}, \tag{1-38}$$

where P_s is the vapor pressure above the solid, P_v is the pressure of the supersaturated vapor, and Ω is the atomic volume. A more instructive way to write Eq. 1-38 is

$$\Delta G_V = -(kT/\Omega)\ln(1+S), \tag{1-39}$$

where S is the vapor supersaturation defined by $(P_v - P_s)/P_s$. Without supersaturation, ΔG_V is zero and nucleation is impossible. In our example, however, $P_v > P_s$ and ΔG_V is negative, which is consistent with the notion of energy reduction. Simultaneously, new surfaces and interfaces form. This results in an increase in the surface free energy of the system given by $4\pi r^2 \gamma$, where γ is the surface energy per unit area. The total free-energy change in forming the nucleus is thus given by

$$\Delta G = (4/3)\pi r^3 \Delta G_V + 4\pi r^2 \gamma, \tag{1-40}$$

and minimization of ΔG with respect to r yields the equilibrium size of $r = r^*$. Thus, $d\Delta G/dr = 0$, and $r^* = -2\gamma/\Delta G_V$. Substitution in Eq. 1-40

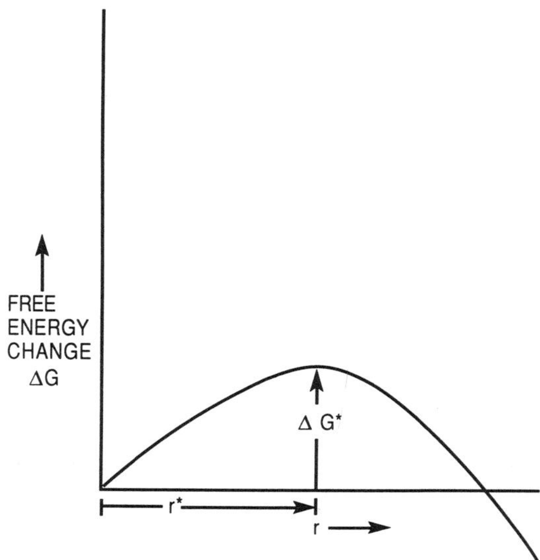

Figure 1-19. Free-energy change (ΔG) as a function of cluster ($r^* > r$) or stable nucleus ($r > r^*$) size. r^* is critical nucleus size, and ΔG^* is critical free-energy barrier for nucleation.

gives $\Delta G^* = 16\pi\gamma^3/3(\Delta G_V)^2$. The quantities r^* and ΔG^* are shown in Fig. 1-19, where it is evident that ΔG^* represents an energy barrier to the nucleation process. If a solid-like spherical cluster of atoms momentarily forms by some thermodynamic fluctuation, but with radius less than r^*, the cluster is unstable and will shrink by losing atoms. Clusters larger than r^* have surmounted the nucleation energy barrier and are stable. They tend to grow larger while lowering the energy of the system.

The nucleation rate $\dot{N}$ is essentially proportional to the product of three terms, namely,

$$\dot{N} = N^* A^* \omega \; (\text{nuclei}/\text{cm}^2\text{-sec}). \tag{1-41}$$

N^* is the equilibrium concentration (per cm^2) of stable nuclei, and ω is the rate at which atoms impinge (per cm^2-sec) onto the nuclei of critical area A^*. Based on previous experience of associating the probable concentration of an entity with its characteristic energy through a Boltzmann factor, it is appropriate to take $N^* = n_s e^{-\Delta G^*/kT}$, where n_s is the density of all possible nucleation sites. The atom impingement flux is equal to the product of the concentration of vapor atoms and the velocity with which they strike the nucleus. In the next chapter we show that this flux is given by $\alpha(P_v - P_s)N_A/\sqrt{2\pi MRT}$, where M is the atomic weight and α is the sticking coefficient. The nucleus area is simply $4\pi r^2$, since gas atoms impinge over the entire spherical surface.

Upon combining terms, we obtain

$$\dot{N} = n_s \exp - \frac{\Delta G^*}{kT} 4\pi r^2 \frac{\alpha(P_v - P_s)N_A}{\sqrt{2\pi MRT}}. \qquad (1\text{-}42)$$

The most influential term in this expression is the exponential factor. It contains ΔG^*, which is, in turn, ultimately a function of S. When the vapor supersaturation is sufficiently large, homogeneous nucleation in the gas is possible. This phenomenon causes one of the more troublesome problems associated with chemical vapor deposition processes since the solid particles that nucleate settle on and are incorporated into growing films destroying their integrity.

Heterogeneous nucleation of films is a more complicated subject in view of the added interactions between deposit and substrate. The nucleation sites in this case are kinks, ledges, dislocations, etc., which serve to stabilize nuclei of differing size. The preceding capillarity theory will be used again in Chapter 5 to model heterogeneous nucleation processes. Suffice it to say that when $\dot{N}$ is high during deposition, many crystallites will nucleate and a fine-grained film results. On the other hand, if nucleation is suppressed, conditions favorable to single-crystal growth are fostered.

1.8. Conclusion

At this point we conclude this introductory sweep through several relevant topics in materials science. If the treatment of structure, bonding, thermodynamics, and kinetics has introduced the reader to or elevated his or her prior awareness of these topics, it has served the intended purpose. Threads of this chapter will be woven into the subsequent fabric of the discussion on the preparation and properties of thin films.

Exercises

1. An FCC film is deposited on the (100) plane of a single-crystal FCC substrate. It is determined that the angle between the [100] directions in the film and substrate is 63.4°. What are the Miller indices of the plane lying in the film surface?

2. Both Au, which is FCC, and W, which is body-centered cubic (BCC) have a density of 19.3 g/cm^3. Their respective atomic weights are 197.0 and 183.9.

a. What is the lattice parameter of each metal?
b. Assuming both contain hard sphere atoms, what is the ratio of their diameters?

3. a. Comment on the thermodynamic stability of a thin-film superlattice composite consisting of alternating Si and $Ge_{0.4}Si_{0.6}$ film layers shown in Fig. 14-17 given the Ge–Si phase diagram (Fig. 1-12).
b. Speculate on whether the composite is a single phase (because it is a single crystal) or consists of two phases (because there are visible film interfaces).

4. Diffraction of 1.5406-Å X-rays from a crystallographically oriented (epitaxial) relaxed bilayer consisting of AlAs and GaAs yields two closely spaced overlapping peaks. The peaks are due to the (111) reflections from both films. The lattice parameters are a_0(AlAs) = 5.6611 Å and a_0(GaAs) = 5.6537 Å. What is the peak separation in degrees?

5. The potential energy of interaction between atoms in an ionic solid as a function of separation distance is given by $V(r) = -A/r + Br^{-n}$, where A, B, and n are constants.

a. Derive a relation between the equilibrium lattice distance a_0 and A, B, and n.
b. The force constant between atoms is given by $K_s = d^2V/dr^2 |_{r=a_0}$. If Young's elastic modulus (in units of force/area) is essentially given by K_s/a_0, show that it varies as a_0^{-4} in ionic solids.

6. What is the connection between the representations of electron energy in Figs. 1-8a and 1-9? Illustrate for the case of an insulator. If the material in Fig. 1-8a were compressed, how would E_g change? Would the electrical conductivity change? How?

7. A 75 at% Ga–25 at% As melt is cooled from 1200 °C to 0 °C in a crucible.

a. Perform a complete phase analysis of the crucible contents at 1200 °C, 1000 °C, 600 °C, 200 °C, 30 °C, and 29 °C. What phases are present? What are their chemical compositions, and what are the relative amounts of these phases? Assume equilibrium cooling.

b. A thermocouple immersed in the melt records the temperature as the crucible cools. Sketch the expected temperature–time cooling response.

c. Do a complete phase analysis for a 75 at% As–25 at% Ga melt at 1000 °C, 800 °C, and 600 °C.

8. A quartz (SiO_2) crucible is used to contain Mg during thermal evaporation in an effort to deposit Mg thin films. Is this a wise choice of crucible material? Why?

9. A solar cell is fabricated by diffusing phosphorous (N dopant) from a constant surface source of 10^{20} atoms/cm^3 into a P-type Si wafer containing 10^{16} B atoms/cm^3. The diffusivity of phosphorous is 10^{-12} cm^2/sec, and the diffusion time is 1 hour. How far from the surface is the junction depth—i.e., where $C_N = C_P$?

10. A brass thin film of thickness d contains 30 wt% Zn in solid solution within Cu. Since Zn is a volatile species, it readily evaporates from the free surface ($x = d$) at elevated temperature but is blocked at the substrate interface, $x = 0$.

a. Write boundary conditions for the Zn concentration at both film surfaces.

b. Sketch a time sequence of the expected Zn concentration profiles across the film during dezincification. (Do not solve mathematically.)

11. Measurements on the electrical resistivity of Au films reveal a three-order-of-magnitude reduction in the equilibrium vacancy concentration as the temperature drops from 600 to 300 °C.

a. What is the vacancy formation energy?

b. What fraction of sites will be vacant at 1080 °C?

12. During the formation of SiO_2 for optical fiber fabrication, soot particles 500 Å in size nucleate homogeneously in the vapor phase at 1200 °C. If the surface energy of SiO_2 is 1000 ergs/cm^2, estimate the value of the supersaturation present.

13. An ancient recipe for gilding bronze statuary alloyed with small amounts of gold calls for the following surface modification steps.

(1) Dissolve surface layers of the statue by applying weak acids (e.g., vinegar).

(2) After washing and drying, heat the surface to as high a temperature as possible but not to the point where the statue deforms or is damaged.
(3) Repeat step 1.
(4) Repeat step 2.
(5) Repeat this cycle until the surface attains the desired golden appearance.

Explain the chemical and physical basis underlying this method of gilding.

REFERENCES

A. General Overview

1. M. F. Ashby and D. R. H. Jones, *Engineering Materials*, Vols. 1 and 2, Pergamon Press, Oxford (1980 and 1986).
2. C. R. Barrett, W. D. Nix, and A. S. Tetelman, *The Principles of Engineering Materials*, Prentice Hall, Englewood Cliffs, NJ (1973).
3. O. H. Wyatt and D. Dew Hughes, *Metals, Ceramics and Polymers*, Cambridge University Press, London (1974).
4. J. Wulff, et al., *The Structure and Properties of Materials*, Vols. 1–4, Wiley, New York (1964).
5. M. Ohring, *Engineering Materials Science*, Academic Press, San Diego (1995).
6. L. H. Van Vlack, *Elements of Materials Science and Engineering*, Addison-Wesley, Reading, MA (1989).

B. Structure

1. B. D. Cullity, *Elements of X-ray Diffraction*, Addison-Wesley, Reading, MA (1978).
2. C. S. Barrett and T. B. Massalski, *The Structure of Metals*, McGraw-Hill, New York (1966).
3. G. Thomas and M. J. Goringe, *Transmission Electron Microscopy of Materials*, Wiley, New York (1979).

C. Defects

1. J. Friedel, *Dislocations*, Pergamon Press, New York (1964).

2. A. H. Cottrell, *Mechanical Properties of Matter*, Wiley, New York (1964).
3. D. Hull, *Introduction to Dislocations*, Pergamon Press, New York (1965).

D. Classes of Solids

a. Metals

1. A. H. Cottrell, *Theoretical Structural Metallurgy*, St. Martin's Press, New York (1957).
2. A. H. Cottrell, *An Introduction to Metallurgy*, St. Martin's Press, New York (1967).

b. Ceramics

1. W. D. Kingery, H. K. Bowen, and D. R. Uhlmann, *Introduction to Ceramics*, Wiley, New York (1976).

c. Glass

1. R. H. Doremus, *Glass Science*, Wiley, New York (1973).

d. Semiconductors

1. S. M. Sze, *Semiconductor Devices—Physics and Technology*, Wiley, New York (1985).
2. A. S. Grove, *Physics and Technology of Semiconductor Devices*, Wiley, New York (1967).
3. J. M. Mayer and S. S. Lau, *Electronic Materials Science: For Integrated Circuits in Si and GaAs*, Macmillan, New York (1990).

E. Thermodynamics of Materials

1. R. A. Swalin, *Thermodynamics of Solids*, Wiley, New York (1962).
2. C. H. Lupis, *Chemical Thermodynamics of Materials*, North-Holland, New York (1983).

F. Diffusion, Nucleation, Phase Transformations

1. P. G. Shewmon, *Diffusion in Solids*, McGraw-Hill, New York (1963).
2. J. Verhoeven, *Fundamentals of Physical Metallurgy*, Wiley, New York (1975).
3. D. A. Porter and K. E. Easterling, *Phase Transformations in Metals and Alloys*, Van Nostrand Reinhold, Berkshire, England (1981).

G. Mathematics of Diffusion

1. H. S. Carslaw and J. C. Jaeger, *Conduction of Heat in Solids*, Oxford University Press, London (1959).
2. J. Crank, *The Mathematics of Diffusion*, Oxford University Press, London (1964).

Vacuum Science and Technology

Virtually all thin-film deposition and processing methods as well as techniques employed to characterize and measure the properties of films require a vacuum or some sort of reduced-pressure environment. For this reason the relevant aspects of vacuum science and technology are discussed at this point. It is also appropriate in a broader sense because this subject matter is among the most undeservedly neglected in the training of scientists and engineers. This is surprising in view of the broad interdisciplinary implications of the subject and the ubiquitous use of vacuum in all areas of scientific research and technological endeavor. The topics treated in this chapter will, therefore, deal with:

2.1. Kinetic Theory of Gases
2.2. Gas Transport and Pumping
2.3. Vacuum Pumps and Systems

2.1. Kinetic Theory of Gases

2.1.1. Molecular Velocities

The well-known kinetic theory of gases provides us with an atomistic picture of the state of affairs in a confined gas (Refs. 1, 2). A fundamental assumption is that the large number of atoms or molecules of the gas are in a continuous state of random motion, which is intimately dependent on the temperature of the gas. During their motion the gas particles collide with each other as well as with the walls of the confining vessel. Just how many molecule–molecule or molecule–wall impacts occur depends on the concentration or pressure of the gas. In the perfect or ideal gas approximation, there are no attractive or repulsive forces between molecules. Rather, they may be considered to behave like independent elastic spheres separated from each other by distances that are large compared with their size. The net result of the continual elastic collisions and exchange of kinetic energy is that a steady-state distribution of molecular velocities emerges given by the celebrated Maxwell–Boltzmann formula

$$f(v) = \frac{1}{n}\frac{dn}{dv} = \frac{4}{\sqrt{\pi}}\left|\frac{M}{2RT}\right|^{3/2} v^2 \exp - \frac{Mv^2}{2RT}. \qquad (2\text{-}1)$$

This centerpiece of the kinetic theory of gases states that the fractional number of molecules $f(v)$, where n is the number per unit volume in the velocity range v to $v + dv$, is related to their molecular weight (M) and absolute temperature (T). In this formula the units of the gas constant R are on a per-mole basis.

Among the important implications of Eq. 2-1, which is shown plotted in Fig. 2-1, is that molecules can have neither zero nor infinite velocity. Rather, the most probable molecular velocity in the distribution is realized at the maximum

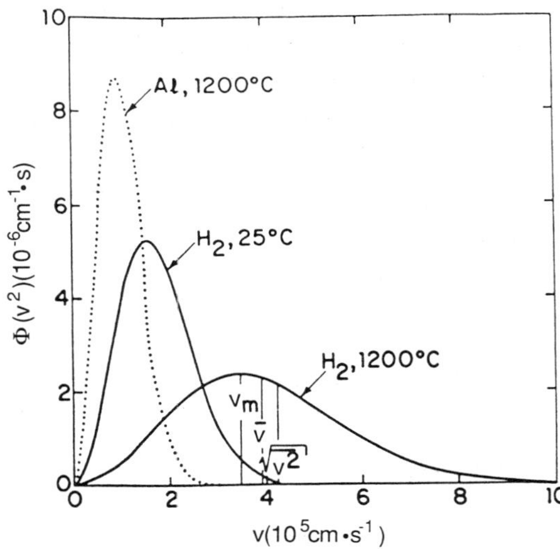

Figure 2-1. Velocity distributions for Al vapor and H_2 gas. (Reprinted with permission from Ref. 1).

value of $f(v)$ and can be calculated from the condition that $df(v)/dv = 0$. Since the net velocity is always the resultant of three rectilinear components v_x, v_y, and v_z, one or even two, but of course not all three, of these may be zero simultaneously. Therefore, a similar distribution function of molecular velocities in each of the component directions can be defined; i.e.,

$$f(v_x) = \frac{1}{n}\frac{dn_x}{dv_x} = \left|\frac{M}{2\pi RT}\right|^{1/2} \exp - \frac{Mv_x^2}{2RT}, \tag{2-2}$$

and similarly for the y and z components.

A number of important results emerge as a consequence of the foregoing equations. For example, the most probable (v_m), average ($\bar{v}$), and mean square ($\overline{v^2}$) velocities are given, respectively, by

$$v_m = \sqrt{\frac{2RT}{M}}, \tag{2-3a}$$

$$\bar{v} = \frac{\int_0^\infty v f(v)\, dv}{\int_0^\infty f(v)\, dv} = \sqrt{\frac{8RT}{\pi M}}, \tag{2-3b}$$

$$\overline{v^2} = \frac{\int_0^\infty v^2 f(v)\, dv}{\int_0^\infty f(v)\, dv} = \frac{3RT}{M}; \qquad \left(\overline{v^2}\right)^{1/2} = \sqrt{\frac{3RT}{M}}. \tag{2-3c}$$

These velocities, which are noted in Fig. 2-1, simply depend on the molecular weight of the gas and the temperature. In air at 300 K, for example, the average molecular velocity is 4.6×10^4 cm/sec, which is almost 1030 miles per hour. However, the kinetic energy of any collection of gas molecules is solely dependent on temperature. For a mole quantity it is given by $(1/2)M\overline{v^2} = (3/2)RT$ with $(1/2)RT$ partitioned in each of the coordinate directions.

2.1.2. Pressure

Momentum transfer from the gas molecules to the container walls gives rise to the forces that sustain the pressure in the system. Kinetic theory shows that the gas pressure P is related to the mean-square velocity of the molecules and, thus, alternatively to their kinetic energy or temperature. Thus,

$$P = \frac{1}{3}\frac{nM}{N_A}\overline{v^2} = \frac{nRT}{N_A}, \tag{2-4}$$

where N_A is Avogadro's number. From the definition of n it is apparent that Eq. 2-4 is also an expression for the perfect gas law. Pressure is the most widely quoted system variable in vacuum technology, and this fact has generated a large number of units that have been used to define it under various circumstances. Basically, two broad types of pressure units have arisen in practice. In what we shall call the scientific system (or coherent unit system (Ref. 2)), pressure is defined as the rate of change of the normal component of momentum of impinging molecules per unit area of surface. Thus, the pressure is normally defined as a force per unit area, and examples of these units are dynes/cm^2 (CGS) or newtons/meter^2 (N/m^2) (MKS). Vacuum levels are now commonly reported in SI units or pascals; 1 pascal (Pa) = 1 N/m^2. Historically, however, pressure was, and still is, measured by the height of a column of liquid, e.g., Hg or H_2O. This has led to a set of what we shall call practical or noncoherent units such as millimeters and microns of Hg, torr, atmospheres, etc., which are still widely employed by practitioners as well as by equipment manufacturers. Definitions of some units together with important conversions include

$$1 \text{ atm} = 1.013 \times 10^6 \text{ dynes/cm}^2 = 1.013 \times 10^5 \text{ N/m}^2 = 1.013 \times 10^5 \text{ Pa}$$

$$1 \text{ torr} = 1 \text{ mm Hg} = 1.333 \times 10^3 \text{ dynes/cm}^2 = 133.3 \text{ N/m}^2 = 133.3 \text{ Pa}$$

$$1 \text{ bar} = 0.987 \text{ atm} = 750 \text{ torr}.$$

The mean distance traveled by molecules between successive collisions, called the mean-free path λ_{mfp}, is an important property of the gas that depends on the pressure. To calculate λ_{mfp}, we note that each molecule presents a target area πd_c^2 to others, where d_c is its *collision* diameter. A binary collision occurs each time the center of one molecule approaches within a distance d_c of the other. If we imagine the diameter of one molecule increased to $2d_c$ while the other molecules are reduced to points, then in traveling a distance λ_{mfp} the former sweeps out a cylindrical volume $\pi d_c^2 \lambda_{mfp}$. One collision will occur under the conditions $\pi d_c^2 \lambda_{mfp} n = 1$. For air at room temperature and atmospheric pressure, $\lambda_{mfp} \approx 500$ Å, assuming $d_c \approx 5$ Å.

A molecule collides in a time given by λ_{mfp}/v and under the previous conditions, air molecules make about 10^{10} collisions per second. This is why gases mix together rather slowly even though the individual molecules are moving at great speeds. The gas particles do not travel in uninterrupted linear trajectories. As a result of collisions, they are continually knocked to and fro, executing a zigzag motion and accomplishing little net movement. Since n is directly proportional to P, a simple relation for ambient air is

$$\lambda_{\mathrm{mfp}} = 5 \times 10^{-3}/P, \tag{2-5}$$

with λ_{mfp} given in cm and P in torr. At pressures below 10^{-3} torr, λ_{mfp} is so large that molecules effectively collide only with the walls of the vessel.

2.1.3. Gas Impingement on Surfaces

A most important quantity that plays a role in both vacuum science and vapor deposition is the gas impingement flux Φ. It is a measure of the frequency with which molecules impinge on, or collide with, a surface, and should be distinguished from the previously discussed molecular collisions in the gas phase. The number of molecules that strike an element of surface (perpendicular to a coordinate direction) per unit time and area is given by

$$\Phi = \int_0^\infty v_x \, dn_x . \tag{2-6}$$

Upon substitution of Eq. 2-2, we get

$$\Phi = \frac{n}{(2\pi)^{1/2}} \left| \frac{M}{RT} \right|^{1/2} \int_0^\infty v_x \exp - \frac{M v_x^2}{2RT} \, dv_x = n \sqrt{\frac{RT}{2\pi M}} . \tag{2-7}$$

The use of Eq. 2-3b yields $\Phi = (1/4) n \bar{v}$, and substitution of the perfect gas law (Eq. 2-4) converts Φ into the more recognizable form

$$\Phi / N_{\mathrm{A}} = P / \sqrt{2\pi MRT} \text{ moles/cm}^2\text{-sec}. \tag{2-8}$$

A useful expansion of this formula is

$$\Phi = 3.513 \times 10^{22} \frac{P}{\sqrt{MT}} \text{ molecules/cm}^2\text{-sec}, \tag{2-9}$$

when P is expressed in torr.

As an application of the foregoing development, consider the problem of gas escaping the vessel through a hole of area A into a region where the gas concentration is zero. The rate at which molecules leave is given by ΦA, and this corresponds to a volume flow per second ($\dot{V}$) given by $\Phi A / n$ cm^3/sec.

Upon substitution of Eqs. 2-4 and 2-9, we have

$$\dot{V} = 3.64 \times 10^3 (T/M)^{1/2} A \text{ cm}^3/\text{sec}. \tag{2-10}$$

For air at 298 K this corresponds to $11.7A$ L/sec, where A has units of cm^2. In essence, we have just calculated what is known as the conductance of a circular aperture, a quantity that will be utilized later because of its significance in pumping gases.

As a second application, consider the question of how long it takes for a surface to be coated by and contaminated with a monolayer of gas molecules. This issue is highly important when one attempts to deposit or grow films under extremely clean conditions. The same concern arises during surface analysis of films performed at very low pressures to minimize surface contamination arising from the vacuum chamber environment. Here one must make certain that the analysis time is shorter than that required for impurities to accumulate. This characteristic contamination time τ_c is essentially the inverse of the impingement flux. Thus, for complete monolayer coverage of a surface containing some 10^{15} atoms/cm^2, the use of Eq. 2-9 yields

$$\tau_c = \frac{10^{15}}{3.513 \times 10^{22}} \frac{(MT)^{1/2}}{P} = 2.85 \times 10^{-8} \frac{(MT)^{1/2}}{P} \text{ sec}, \tag{2-11}$$

with P measured in torr. In air at atmospheric pressure and ambient temperature a surface will acquire a monolayer of gas in 3.49×10^{-9} sec, assuming all impinging atoms stick. On the other hand, at 10^{-10} torr a surface will stay clean for about 7.3 h.

A condensed summary of the way system pressure affects the gas density, mean-free path, incidence rates, and monolayer formation times is conveniently displayed in Fig. 2-2. The pressure range spanned in all thin-film research, development, and technological activities discussed in this book is over 13 orders of magnitude. Since all of the quantities depicted vary directly (on the left-hand axis) or inversely (on the right-hand axis) with pressure, a log-log plot connecting the variables is linear with a slope of 1. The pressure scale is arbitrarily subdivided into corresponding low-, medium-, high-, and ultrahigh-vacuum domains, each characterized by different requirements with respect to vacuum hardware (e.g., pumps, gauges, valves, gaskets, feedthroughs, etc.). Of the film deposition processes, evaporation requires a vacuum between the high and ultrahigh regimes, whereas sputtering and low-pressure chemical vapor deposition are accomplished at the border between the medium- and high-vacuum ranges. Of the analytical instruments, electron microscopes operate in high vacuum, and surface analytical equipment

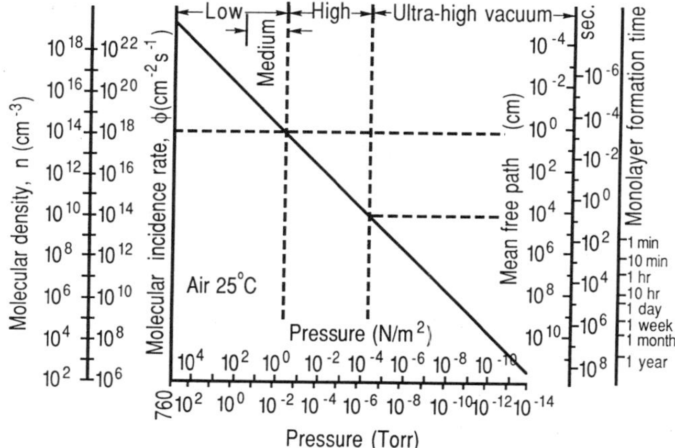

Figure 2-2. Molecular density, incidence rate, mean-free path, and monolayer formation time as a function of pressure. (Reprinted with permission from Ref. 2).

have the most stringent cleanliness requirements and are operative only under ultrahigh-vacuum conditions.

2.2. Gas Transport and Pumping

2.2.1. Gas Flow Regimes

In order to better design, modify, or appreciate reduced-pressure systems, one must understand concepts of gas flow (Refs. 2–4). An incomplete understanding of gas flow limitations frequently results in less efficient system performance as well as increased expense. For example, a cheap piece of tubing having the same length and diameter dimensions as the diffusion pump to which it is attached will cut the pumping speed of the latter to approximately half of its rated value. In addition to the effectively higher pump cost, a continuing legacy of such a combination will be the longer required pumping time to reach a given level of vacuum each time the system is operated.

Whenever there is a net directed movement of gas in a system under the influence of attached pumps, gas flow is said to occur. Under such conditions the gas experiences a pressure drop. The previous discussion on kinetic theory of gases essentially assumed an isolated sealed system. Although gas molecules certainly move, and with high velocity at that, there is no *net* gas flow and no establishment of pressure gradients in such a system. When gas does flow, however, it is appropriate to distinguish between different regimes of flow. These regimes depend on the geometry of the system involved as well as the pressure, temperature, and type of gas under consideration. At one extreme we have free molecular flow, which occurs at low gas densities. The chambers of high-vacuum evaporators and analytical equipment, such as Auger electron spectrometers and electron microscopes, operate within the molecular flow regime. Here the mean-free path between intermolecular collisions is large compared with the dimensions of the system. Kinetic theory provides an accurate picture of molecular motion under such conditions. At higher pressures the mean-free path is reduced, and successive intermolecular collisions predominate relative to collisions with the walls of the chamber. At this extreme the so-called viscous flow regime is operative. An important example of such flow occurs in atmospheric chemical vapor deposition reactors. Compared with molecular flow, viscous flow is quite complex. At low gas velocities the flow is *laminar* where layered, parallel flow lines may be imagined. At the walls of a tube, for example, the laminar flow velocity is zero, but it increases to a maximum at the axis. At higher velocity the flowing gas layers are no longer parallel but swirl and are influenced by any obstacles in the way. In this *turbulent* flow range, cavities of lower pressure develop between layers. More will be said about viscous flow in Chapter 4.

Criteria for distinguishing between the flow regimes are based on the magnitude of the Knudsen number Kn, which is defined by the ratio of the

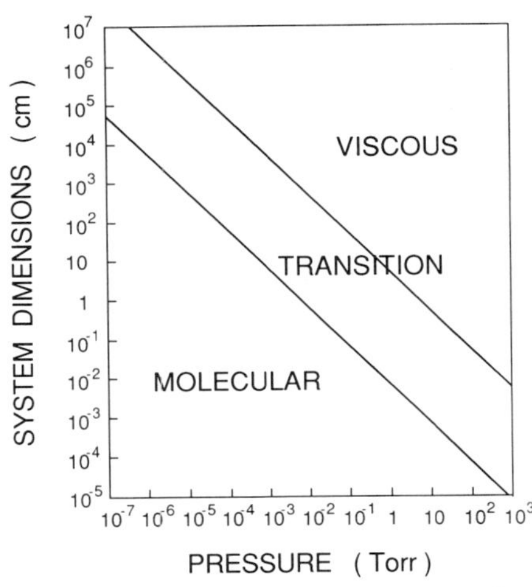

Figure 2-3. Dominant gas flow regimes as a function of system dimensions and pressure.

chamber (pipe) diameter D_p to gas mean-free path, i.e., $Kn = D_p/\lambda_{\mathrm{mfp}}$*. Thus for

$$\text{molecular flow} \qquad Kn < 1, \tag{2-12a}$$

$$\text{intermediate flow} \qquad 1 < Kn < 110, \tag{2-12b}$$

$$\text{viscous flow} \qquad Kn > 110. \tag{2-12c}$$

In air these limits can be alternatively expressed by $D_p P < 5 \times 10^{-3}$ cm-torr for molecular flow, and $D_p P > 5 \times 10^{-1}$ cm-torr in the case of viscous flow through the use of Eq. 2-5. Figure 2-3 serves to map the dominant flow regimes on this basis. Note that flow mechanisms may differ in various parts of the same system. Thus, although molecular flow will occur in the high-vacuum chamber, the gas may flow viscously in the piping near the exhaust pumps.

2.2.2. Conductance

Let us reconsider the molecular flow of gas through an orifice of area A that now separates two large chambers maintained at low pressures, P_1 and P_2. From a phenomenological standpoint, a flow driven by the pressure difference is expected; i.e.,

$$Q = C(P_1 - P_2). \tag{2-13}$$

Here Q is defined as the gas throughput with units of pressure $\times$ volume per second (e.g., torr-L/sec). The constant of proportionality C is known as the conductance and has units of L/sec. Alternately, viewing flow through the orifice in terms of kinetic theory, we note that the molecular impingements in each of the two opposing directions do not interfere with each other. Therefore, the net gas flow at the orifice plane is given by the difference, or $(\Phi_1 - \Phi_2)A$.

Through the use of Eq. 2-10 it is easily shown that the conductance of the orifice is

$$C = 3.64\sqrt{T/M}\,A \qquad \text{or} \qquad 11.7A \text{ L/sec}$$

for air at 298 K. Note in the choice of terms the analogy to electrical circuits. If $P_1 - P_2$ is associated with the voltage difference, Q may be viewed as a current. Conductances of other components where the gas flow is in the molecular regime can be similarly calculated or measured. Results for a number of important geometric shapes are given in Fig. 2-4 (Ref. 5). Note that C is simply a function of the geometry for a specific gas at a given temperature. This is not true of viscous flow, where C also depends on

*Note that Kn is also defined as $\lambda_{\mathrm{mfp}}/D_p$ in the literature.

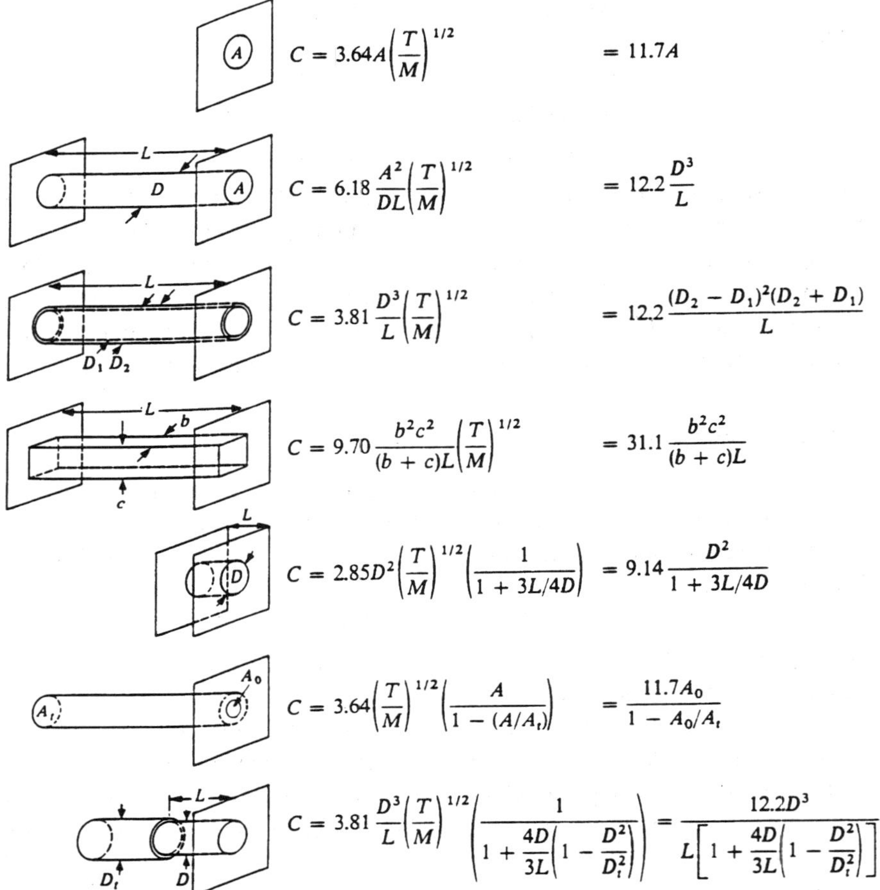

Figure 2-4. Conductances of various geometric shapes for molecular flow of air at 25°C. Units of C are L/sec. (Reprinted with permission from Ref. 5).

pressure. When conductances are joined in series, the system conductance C_{sys} is given by

$$\frac{1}{C_{sys}} = \frac{1}{C_1} + \frac{1}{C_2} + \frac{1}{C_3} + \cdots. \tag{2-14}$$

Clearly C_{sys} is lower than that of any individual conductance. When connected in parallel

$$C_{sys} = C_1 + C_2 + C_3 + \cdots. \tag{2-15}$$

As an example, (Ref. 6), consider the conductance of the cold trap assembly of Fig. 2-5 that isolates a vacuum system above from the pump below.

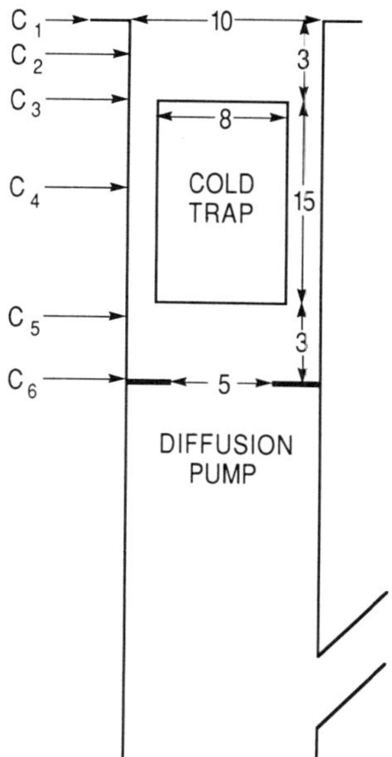

Figure 2-5. Cold trap assembly. (Adapted from R. W. Roberts, *An Outline of Vacuum Technology*, G.E. Report No. 64-RL-3394C, 1964, with permission from General Electric Company).

Contributions to the total conductance come from

C_1 = conductance of aperture of 10 cm diameter

$$= 11.7A = 11.7\pi(5)^2 = 919 \text{ L/sec}$$

C_2 = conductance of pipe 3 cm long

$$= 12.2D^3/L = 12.2(10)^3/3 = 4065 \text{ L/sec}$$

C_3 = conductance of annular aperture

$$= 11.7A_{\text{ann}} = 11.7(0.25)\pi(10^2 - 8^2) = 331 \text{ L/sec}$$

C_4 = conductance of annular pipe

$$= 12.2\frac{(D_2 - D_1)^2(D_1 + D_2)}{L} = \frac{12.2(10 - 8)^2(10 + 8)}{15} = 58.6 \text{ L/sec}$$

$C_5 = C_2 = 4065$ L/sec

C_6 = conductance of aperture in end of pipe/diffusion pump

$$= 11.7\frac{AA_0}{A - A_0} = \frac{11.7\pi(2.5)^2\pi 5^2}{\pi(5^2 - (2.5)^2)} = 303 \text{ L/sec}$$

Therefore,

$$\frac{1}{C_{\text{Total}}} = \sum_{i=1}^{6} \frac{1}{C_i} \qquad \text{and} \qquad C_{\text{Total}} = 40 \text{ L/sec}$$

upon evaluation. Strictly speaking, C_3 and C_4 should be multiplied by a correction factor of 1.27, which would have the effect of increasing C_{Total} to 51.1 L/sec. As we shall soon see, it is always desirable to have as large a conductance as possible. Clearly, the overall conductance is severely limited in this case by the annular region between the concentric pipes.

2.2.3. Pumping Speed

Pumping is the process of removing gas molecules from the system through the action of pumps. The pumping speed S is defined as the volume of gas passing the plane of the inlet port per unit time when the pressure at the pump inlet is P. Thus,

$$S = Q/P. \tag{2-16}$$

Although the throughput Q can be measured at any plane in the system, P and S refer to quantities measured at the pump inlet.

Although conductance and pumping speed have the same units and may even be equivalent numerically, they have different physical meanings. Conductance implies a component of a given geometry across which a pressure differential exists. Pumping speed refers to any plane that may be considered to be a pump for preceding portions of the system. To apply these ideas, consider a pipe of conductance C connecting a chamber at pressure P to a pump at pressure P_p as shown in Fig. 2-6a. Therefore, $Q = C(P - P_p)$. Elimination of Q through

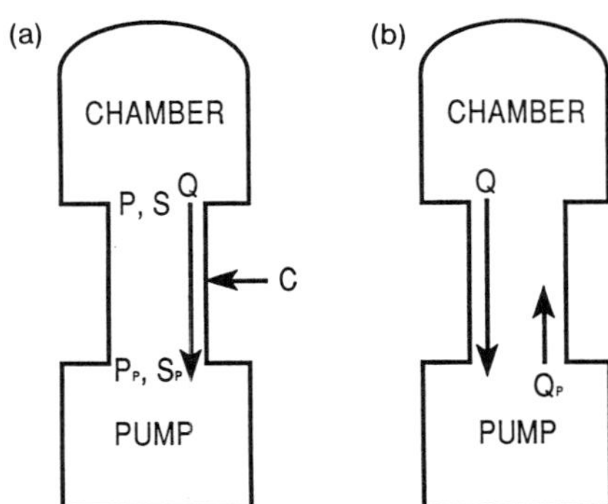

Figure 2-6. Chamber–pipe–pump assembly: (a) no outgassing; (b) with outgassing.

the use of Eq. 2-16 yields

$$S = \frac{S_p}{1 + S_p/C}, \tag{2-17}$$

where S_p is the intrinsic speed at the pump inlet ($S_p = Q/P_p$) and S is the effective pumping speed at the base of the chamber. The latter never exceeds S_p or C and is, in fact, limited by the smaller of these quantities. If, for example, $C = S_p$ in magnitude, then $S = S_p/2$ and the effective pumping speed is half the rated value for the pump. The lesson, therefore, is to keep conductances large by making ducts between the pump and chamber as short and wide as possible.

Real pumps outgas or release gas into the system as shown in Fig. 2-6b. Account may be taken of this by including an oppositely directed extra throughput term Q_p such that

$$Q = S_p P - Q_p = S_p P \left(1 - \frac{Q_p}{S_p P}\right). \tag{2-18}$$

When $Q = 0$, the ultimate pressure of the pump, P_0, is reached and $Q_p = S_p P_0$. The effective pumping speed is then

$$S = Q/P = S_p(1 - P_0/P), \tag{2-19}$$

and falls to zero as the ultimate pressure of the pump is reached.

An important issue in vacuum systems is the time required to achieve a given pressure. The pump-down time can be calculated by noting that the throughput may be defined as the time (t) derivative of the product of volume and pressure; i.e., $Q = -d(VP)/dt = -V(dP/dt)$. Employing Eq. 2-18, we write

$$-V\frac{dP(t)}{dt} = S_p P - Q_p,$$

where Q_p includes pump as well as chamber outgassing. Upon integration

$$\frac{P(t) - Q_p/S_p}{P_i - Q_p/S_p} = \frac{P(t) - P_0}{P_i - P_0} = \exp -\frac{S_p t}{V}, \tag{2-20}$$

where it is assumed that initially $P = P_i$. During pump-down the pressure thus exponentially decays to P_0 with time constant given by V/S_p. At high pressures where viscous flow is involved, S_p is a function of P, and, therefore, Eq. 2-10 is not strictly applicable in such cases.

2.3. Vacuum Pumps and Systems

2.3.1. Pumps

The vacuum systems employed to deposit and characterize thin films contain an assortment of pumps, tubing, valves, and gauges to establish and measure the required reduced pressures (Ref. 7). Of these components pumps are generally the most important, and only they will be discussed at any length. Vacuum pumps may be divided into two broad categories: *gas transfer* pumps and *entrapment* pumps. Gas transfer pumps remove gas molecules from the pumped volume and convey them to the ambient in one or more stages of compression. Entrapment pumps condense or chemically bind molecules at walls situated within the chamber being pumped. In contrast to gas transfer pumps, which remove gas permanently, some entrapment pumps are reversible and release trapped (condensed) gas back into the system upon warm-up.

Gas transfer pumps may be further subdivided into positive-displacement and kinetic vacuum pumps. Rotary mechanical and Roots pumps are important examples of the positive-displacement variety. Diffusion and turbomolecular pumps are the outstanding examples of kinetic vacuum pumps. Among the entrapment pumps commonly employed are the adsorption, sputter-ion, and cryogenic pumps. Each pump is used singly or in combination in a variety of pumping system configurations. Pumps do not remove the gas molecules by exerting an attractive pull on them. The molecules are unaware that pumps exist. Rather, the action of pumps is to limit, interfere with, or alter natural molecular motion. We start this brief survey of some of the more important pumps with the positive-displacement types.

2.3.1.1. Rotary Mechanical Pump. The rotary piston and rotary vane pumps are the two most common devices used to attain reduced pressure. In the rotary piston pump shown in Fig. 2-7a, gas is drawn into space A as the keyed shaft rotates the eccentric and piston. There the gas is isolated from the inlet after one revolution, then compressed and exhausted during the next cycle. Piston pumps are often employed to evacuate large systems and to back Roots blower pumps.

The rotary vane pump contains an eccentrically mounted rotor with spring-loaded vanes. During rotation the vanes slide in and out within the cylindrical interior of the pump, enabling a quantity of gas to be confined, compressed, and discharged through an exhaust valve into the atmosphere. Compression ratios of up to 10^6 can be achieved in this way. Oil is employed as a sealant as

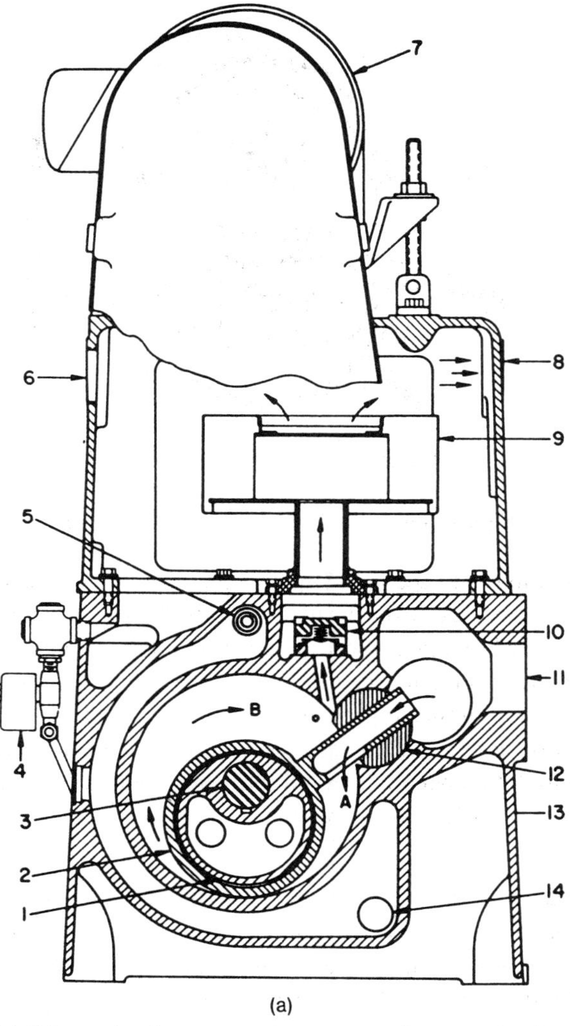

(a)

Figure 2-7. (a) Schematic of a rotary piston pump: 1. eccentric; 2. piston; 3. shaft; 4. gas ballast; 5. cooling water inlet; 6. optional exhaust; 7. motor; 8. exhaust; 9. oil mist separator; 10. poppet valve; 11. inlet; 12. hinge bar; 13. casing; 14. cooling water outlet. (Courtesy of Stokes Vacuum Inc.)

well as a lubricant between components moving within tight clearances of both types of rotary pumps. To calculate the pumping speed, let us assume that a volume of gas, V_0 (liters), is enclosed between the rotor and pump stator housing and swept into the atmosphere for each revolution of the rotor. The intrinsic speed of the pump will then be $S_p = V_0 f_s$, where f_s is the rotor

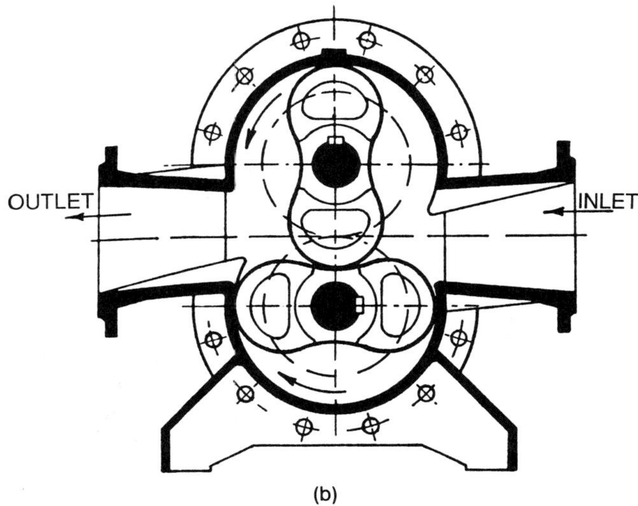

(b)

Figure 2-7. (b) Schematic of the Roots pump.

speed in revolutions per second. Typical values of S_p for vane pumps range from 1 to 300 L/sec, and from 10 to 500 L/sec for piston pumps. At elevated pressures the actual pumping speed S is constant but eventually becomes zero at the ultimate pump pressure in a manner suggested by Eq. 2-19. Single-stage vane pumps have an ultimate pressure of 10^{-2} torr, and two-stage pumps can reach 10^{-4} torr. Rotary pumps are frequently used to produce the minimal vacuum required to operate both oil diffusion and turbomolecular pumps, which can then attain far lower pressures.

2.3.1.2. Roots Pump. An important variant of the positive-displacement pump is the Roots pump, shown in Fig. 2-7b, where two figure-eight-shaped lobes rotate in opposite directions against each other. The extremely close tolerances eliminate the need to seal with oil. These pumps have very high pumping speeds, and even though they can attain ultimate pressures below 10^{-5} torr, a forepump (e.g., rotary mechanical) is required. Maximum pumping is achieved in the pressure range of 10^{-3} to 20 torr, where speeds of up to several thousand liters per second can be attained. This combination of characteristics has made Roots pumps popular in low-pressure chemical vapor deposition (LPCVD) systems where large volumes of gas continuously pass through reactors maintained at $\sim$ 1 torr.

2.3.1.3. Diffusion Pump. In contrast to mechanical pumps, the diffusion pump shown in Fig. 2-8 has no moving parts. Diffusion pumps are designed to

(a)

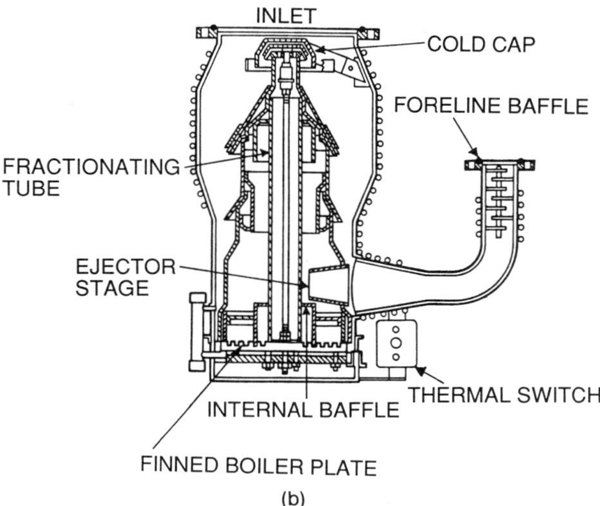

(b)

Figure 2-8. (a) Diffusion pump; (b) schematic of pump interior (Courtesy of Varian Associates, Vacuum Products Division)

operate in the molecular flow regime and can function over pressures ranging from well below 10^{-10} torr to about 10^{-2} torr. Because they cannot discharge directly into the atmosphere, a mechanical forepump is required to maintain an outlet pressure of about 0.1 torr. Since the pump inlet is essentially like the orifice of Fig. 2-4, a pumping speed of $11.7A$ L/sec would be theoretically expected for air at room temperature. Actual pumping speeds are typically only 0.4 of this value.

Diffusion pumps have been constructed with pumping speeds ranging from a few liters per second to over 20,000 L/sec. Pumping is achieved through the action of a fluid medium (typically silicone oil) that is boiled and vaporized in a multistage jet assembly. As the oil vapor stream emerges from the top nozzles, it collides with and imparts momentum to residual gas molecules, which happen to bound into the pump throat. These molecules are thus driven toward the bottom of the pump and compressed at the exit side where they are exhausted. A region of reduced gas pressure in the vicinity of the jet is produced, and more molecules from the high-vacuum side move into this zone, where the process is repeated. Several jets working in series serve to enhance the pumping action.

A serious problem associated with diffusion pumps is the backstreaming of oil into the chamber. Such condensed oil can contaminate both substrate and deposit surfaces, leading to poor adhesion and degraded film properties. Oil vapor dissociated on contact with hot filaments or by electrical discharges also leaves carbonaceous or siliceous deposits that can cause electrical leakage or even high-voltage breakdown. For these reasons, diffusion pumps are not used in surface analytical equipment such as Auger electron and secondary ion mass spectrometers or in ultrahigh-vacuum deposition systems. Nevertheless, diffusion-pumped systems are widely used in nonelectronic (e.g., decorative, optical, tool) coating applications. To minimize backstreaming, attempts are made to condense the oil before it enters the high-vacuum chamber. Cold caps on top of the uppermost jet together with refrigerated traps and optically dense baffles are used for this purpose, but at the expense of somewhat reduced conductance and pumping speed.

2.3.1.4. Turbomolecular Pump. The drive to achieve the benefits of oil-less pumping has spurred the development and use of turbomolecular pumps. Like the diffusion pump, the turbomolecular pump imparts a preferred direction to molecular motion, but in this case the impulse is caused by impact with a rapidly whirling turbine rotor spinning at rates of 20,000 to 30,000 revolutions per minute. The turbomolecular pump of Fig. 2-9 is a vertical, axial flow

(a)

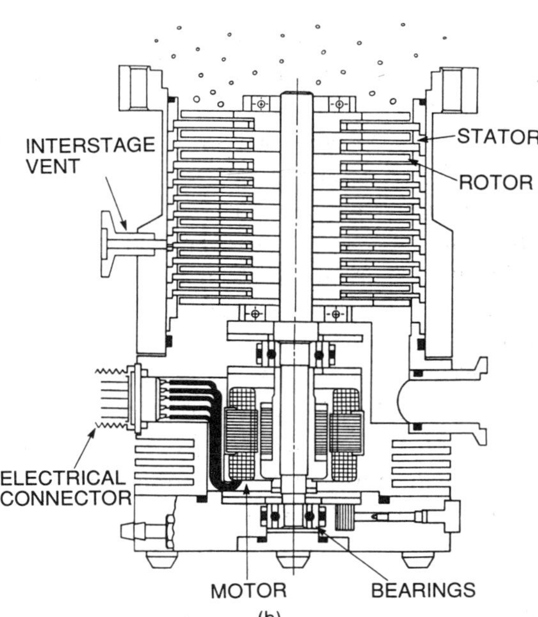

(b)

Figure 2-9. (a) Turbomolecular pump; (b) schematic of pump interior (Courtesy of Varian Associates, Vacuum Products Division)

compressor consisting of many rotor–stator pairs or stages mounted in series. Gas captured by the upper stages is transferred to the lower stages, where it is successively compressed to the level of the fore-vacuum pressure. The maximum compression varies linearly with the circumferential rotor speed, but exponentially with the square root of the molecular weight of the gas. Typical compression ratios for hydrocarbons, N_2 and H_2, are 10^{10}, 10^9, and 10^3, respectively. Since the partial pressure of a given gas specie on the low-pressure (i.e., vacuum chamber) side of the pump is equal to that on the high-pressure (exhaust) side divided by the compression ratio, only H_2 will fail to be pumped effectively. An important consequence of the very high compression is that oil backstreaming is basically reduced to negligible levels. In fact, no traps or baffles are required, and the turbomolecular pump can be backed by a rotary pump and effectively achieve oil-less pumping. Turbomolecular pumps are expensive, but are increasingly employed in all sorts of thin-film deposition and characterization equipment. Typical characteristics include pumping speeds of 10^3 L/sec and ultimate pressures below 10^{-10} torr.

2.3.1.5. Cryopumps. Cryopumps are capable of generating a very clean vacuum in the pressure range of 10^{-3} to 10^{-10} torr. These are gas entrapment pumps, which rely on the condensation of vapor molecules on surfaces cooled below 120 K. Temperature-dependent van der Waals forces are responsible for physically binding or sorbing gas molecules. Several kinds of surfaces are employed to condense gas. These include (1) untreated bare metal surfaces, (2) a surface cooled to 20 K containing a layer of precondensed gas of higher boiling point (e.g., Ar or CO_2 for H_2 or He sorption), and (3) a microporous surface of very large area within molecular sieve materials, such as activated charcoal or zeolite. The latter are the working media of the common sorption pumps, which achieve forepressures of about 10^{-3} torr by surrounding a steel canister containing sorbent with a Dewar of liquid nitrogen. Cryopumps designed to achieve ultrahigh vacuum (Fig. 2-10) have panels that are cooled to 20 K by closed-cycle refrigerators. These cryosurfaces cannot be directly exposed to the room-temperature surfaces of the chamber because of the radiant heat load, so they are surrounded by liquid-nitrogen-cooled shrouds.

The starting or forepressure, ultimate pressure, and pumping speed of cryopumps are important characteristics. Cryopumps require an initial forepressure of about 10^{-3} torr in order to prevent a prohibitively large thermal load on the refrigerant and the accumulation of a thick condensate on the cryopanels. The ultimate pressure (P_{ult}) attained for a given gas is reached when the impingement rate on the cryosurface, maintained at temperature T,

(a)

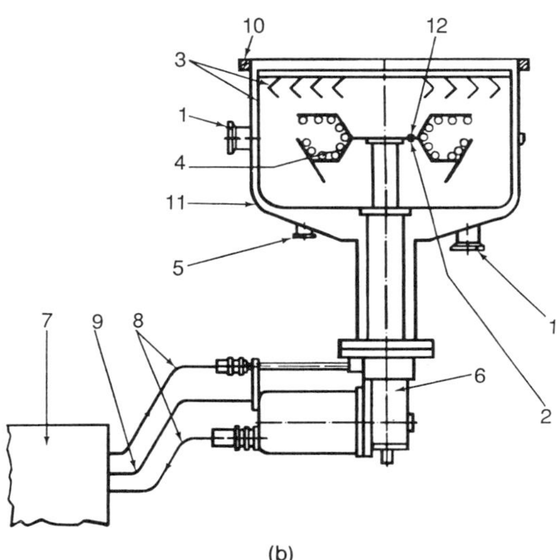

(b)

Figure 2-10. (a) Cryopump; (b) schematic of pump interior: 1. fore-vacuum port; 2. temperature sensor; 3. 77 K shield; 4. 20 K condenser with activated charcoal; 5. port for gauge head and pressure relief valve; 6. cold head; 7. compressor unit; 8. helium supply and return lines; 9. electrical supply cable; 10. high-vacuum flange; 11. pump housing; 12. temperature measuring instrument. (Courtesy of Balzers, High Vacuum Products)

equals that on the vacuum chamber walls held at 300 K. Therefore from Eq. 2-8,

$$P_{ult} = P_s(T)\sqrt{300/T}, \tag{2-21}$$

where $P_s(T)$ is the saturation pressure of the pumped gas. As an example, for N_2 at 20 K, the P_s value is about 10^{-11} torr, so $P_{ult}(N_2) = 3.9 \times 10^{-11}$ torr.

Because of high vapor pressures at 20 K, H_2, as well as He and Ne, cannot be effectively cryopumped. Of all high-vacuum pumps, cryopumps have the highest pumping speeds since they are limited only by the rate of gas impingement. Therefore, the pumping speed is given by Eq. 2-10, which for air at 20 K is equal to 3 L/sec for each square centimeter of cooled surface. Although they are expensive, cryopumps offer the versatility of serving as the main pump or, more frequently, acting in concert with other conventional pumps (e.g., turbopumps). They are, therefore, becoming increasingly popular in thin-film research and processing equipment.

2.3.1.6. *Sputter-Ion Pumps.* The last pump we consider is the sputter-ion pump shown in Fig. 2-11, which relies on sorption processes initiated by ionized gas to achieve pumping. The gas ions are generated in a cold cathode electrical discharge between stainless steel anode and titanium cathode arrays maintained at a potential difference of a few kilovolts. Electrons emitted from the cathode are trapped in the applied transverse magnetic field of a few thousand gauss, resulting in a cloud of high electron density (e.g., 10^{13} electrons/cm^3). After impact ionization of residual gas molecules, the ions travel to the cathode and knock out or sputter atoms of Ti. The latter deposit elsewhere within the pump, where they form films that getter or combine with reactive gases such as nitrogen, oxygen, and hydrogen. These gases and corresponding Ti compounds are then buried by fresh layers of sputtered metal.

Similar pumping action occurs in the Ti sublimation pump, where Ti metal is *thermally* evaporated (sublimed) onto cryogenically cooled surfaces. A combination of physical cryopumping and chemical sorption processes then ensues. Sputter-ion pumps display a wide variation in pumping speeds for different gases. For example, hydrogen is pumped several times more effectively than oxygen, water, or nitrogen and several hundred times faster than argon. Unlike cryopumping action, the gases are permanently removed. These pumps are quite expensive and have a finite lifetime that varies inversely with the operating pressure. They have been commonly employed in oil-less ultra-

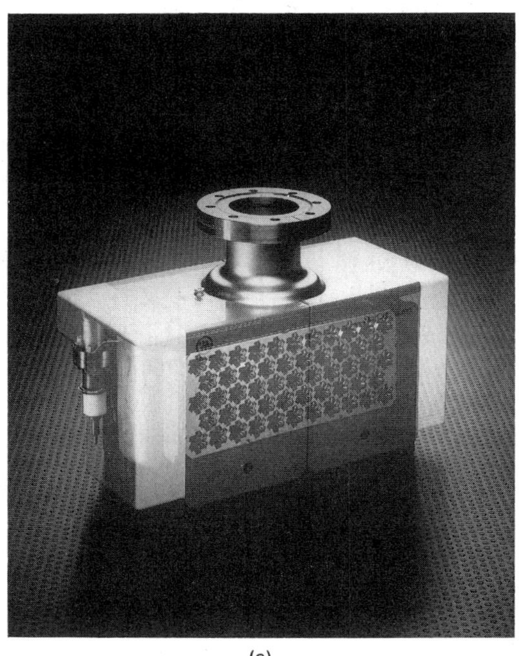

(a)

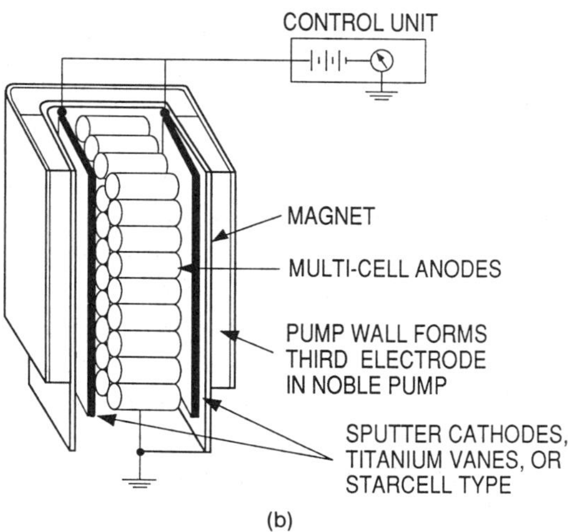

(b)

Figure 2-11. (a) Sputter-ion pump; (b) schematic of pump interior. (Courtesy of Varian Associates, Vacuum Products Division)

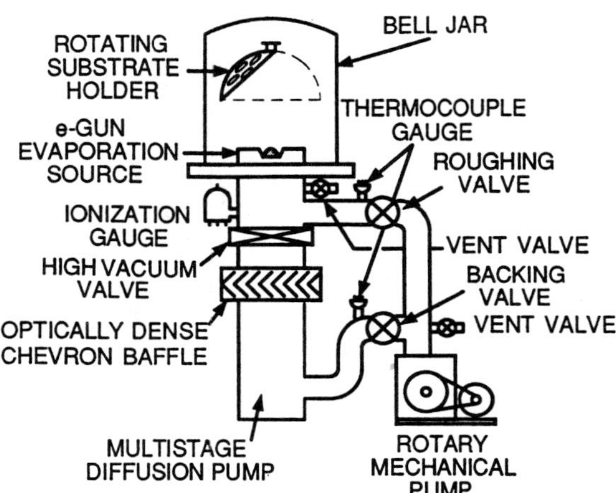

Figure 2-12. Schematic of vacuum deposition system.

high (10^{-10} torr) vacuum deposition and surface analytical equipment, but are being supplanted by turbomolecular and cryopumps.

2.3.2. Systems

The broad variety of applications requiring a low-pressure environment is reflected in a corresponding diversity of vacuum system design. One such system shown in Fig. 2-12 is employed for vacuum evaporation of metals. The basic pumping system consists of a nominal 15-cm diameter, multistage oil-diffusion pump backed by a 17-cfm (8.0 L/sec) rotary mechanical pump. In order to sequentially coat batch lots, the upper chamber must be vented to air in order to load substrates. To minimize the pumping cycle time, however, we desire to operate the diffusion pump continuously, thus avoiding the wait involved in heating or cooling pump oil. This means that the pump must always view a vacuum of better than $\sim 10^{-1}$ torr above and be backed by a similar pressure at the exhaust. A dual vacuum-pumping circuit consisting of three valves, in addition to vent valves, is required to accomplish these ends.

When starting cold, the high-vacuum and roughing valves are closed and the backing valve is open. Soon after the oil heats up, a high vacuum is reached above the diffusion pump. The backing valve is then closed, thus isolating the diffusion pump, and the roughing valve is opened, enabling the rotary pump to evacuate the chamber to a tolerable vacuum of about 10^{-1} torr. Finally, the roughing valve is shut, and both the backing and high-vacuum valves are opened, allowing the diffusion pump to bear the main pumping burden. By reversal of the valving, the system can be alternately vented or pumped rapidly and efficiently. This same operational procedure is followed in other diffusion-pumped systems, such as electron microscopes, where ease of specimen exchange is a requirement. To eliminate human error, pump-down cycles are now automated or computerized through the use of pressure sensors and electrically actuated valves. In other oil-less vacuum systems a similar valving arrangement exists between the involved fore- and main pumps.

Components worthy of note in the aforementioned evaporator are the high-vacuum valve and the optically dense baffle, both of which are designed to have a high conductance. Cryogenic cooling of the baffle helps prevent oil from backstreaming or creeping into the vacuum chamber. To ensure proper pressure levels for the functioning of the diffusion pump, thermocouple gauges are located in both the roughing and backing forelines. They operate from 10^{-3} torr to 1 atm. Ionization gauges, on the other hand, are sensitive to vacuum levels spanning the range 10^{-3} to lower than 10^{-10} torr and are, therefore, located to measure chamber pressure. Virtually all quoted vacuum pressures in thin-film deposition, processing, and characterization activities are derived from ionization gauges.

An actual vacuum deposition system is shown in Fig. 2-13.

Figure 2-13. Vacuum deposition system. (Courtesy of Cooke Vacuum Products)

2.3.3. System Pumping Considerations

During the pump-down of a system, gas is removed from the chamber by (1) volume pumping and (2) pumping of species outgassed from internal surfaces. For volume pumping it is a relatively simple matter to calculate the time required to reach a given pressure. As an example, let us estimate the time required to evacuate a cylindrical bell jar, 46 cm in diameter and 76 cm high, from atmospheric pressure to a forepressure of 10^{-1} torr. If an 8-L/sec mechanical pump is used, then substitution of $S_p = 8$ L/sec, $V = (\pi/4)(46)^2(76)/1000 = 126.3$ L, $P(t) = 10^{-1}$ torr, $P_i = 760$ torr, and $P_0 \approx 10^{-4}$ torr in Eq. 2-20 yields a pump-down time of 2.35 min. This value is comparable to typical forepumping times in clean, tight systems.

It is considerably more difficult to calculate pumping times in the high-vacuum regime where the system pressure depends on outgassing rates. There are two sources of this gas: (1) permeation and diffusion through the system walls and (2) desorption from the chamber surfaces and vacuum hardware. Specific vacuum materials, surface condition (smooth, porous, degree of cleanliness, etc.), and bakeout procedures critically affect the gas evolution rate. If the latter is known, however, it is possible to determine the necessary pumping speed at the required operating pressure through the use of Eq. 2-16. For example, suppose the vacuum system described has a total surface area of 1.5 m^2, including all accessories. If the gas evolution rate (throughput) q_0 is assumed to be 1.5×10^{-4} (torr-L/sec)/m^2, then maintenance of a pressure $P = 7.5 \times 10^{-7}$ torr requires an effective pumping speed of $S = 1.5q_0/P = 300$ L/sec. This value of S is necessary only to pump the quantity of gas arising through gas evolution from the walls, and is clearly a lower bound for the effective pumping speed of the system.

Lastly, it is appropriate to comment on vacuum system leaks. There is scarcely a thin-film technologist who has not struggled with them. No vacuum apparatus is absolutely vacuum-tight and, in principle, does not have to be. What is important, however, is that the leak rate be small and not influence the required working pressure, gas content, or ultimate system pressure. Leak rates are given in throughput units, e.g. torr-L/sec, and measured by noting the pressure rise in a system after isolating the pumps. The leak tightness of high-vacuum systems can be generally characterized by the following leakage rates (Ref. 7):

Very leak tight—$< 10^{-6}$ torr-L/sec

Adequately leak tight—$\sim 10^{-5}$ torr-L/sec

Not leak tight—$> 10^{-4}$ torr-L/sec

One way to distinguish between gas leakage and outgassing from the vessel walls and hardware is to note the pressure rise with time. Gas leakage causes a linear rise in pressure, whereas outgassing results in a pressure rise that becomes gradually smaller and tends to a limiting value. The effect of leakage throughput on pumping time can be accounted for by inclusion in Eq. 2-20.

EXERCISES

1. Consider a mole of gas in a chamber that is *not* being pumped. What is the probability of a self-pumping action such that all of the gas molecules will congregate in one-half of the chamber and leave a perfect vacuum in the other half?

2. A 1-m^3 cubical-shaped vacuum chamber contains O_2 molecules at a pressure of 10^{-4} atm at 300 K.

a. How many molecules are there in the chamber?
b. What is the ratio of maximum potential energy to average kinetic energy of these molecules?
c. What fraction of gas molecules has a kinetic energy in the x direction exceeding RT? What fraction exceeds $2RT$?

3. In many vacuum systems there is a gate valve consisting of a gasketed metal plate that acts to isolate the chamber above from the pumps below.

a. A sample is introduced into the chamber at 760 torr while the isolated pumps are maintained at 10^{-6} torr. For a 15-cm-diameter opening, what force acts on the valve plate to seal it?
b. The chamber is forepumped to a pressure of 10^{-2} torr. What force now acts on the valve plate?

4. Supersonic molecular beams have a velocity distribution given by

$$f(v) = Av^3 \exp - \frac{M(v - v_0)^2}{2RT},$$

where v_0, the stream velocity, is related to the Mach number.

a. What does a plot of $f(v)$ vs. v look like?
b. What is the average gas speed in terms of v_0, M, and T? *Note:*

$$\int_0^\infty x^{2n} e^{-ax^2}\,dx = \frac{\Gamma(n + 1/2)}{2a^{n+1/2}}; \qquad \int_0^\infty x^{2n+1} e^{-ax^2}\,dx = \frac{n!}{2a^{n+1}}.$$

Assume $v_0 = 0$.

5. The trap in Fig. 2-5 is filled with liquid N_2 so that the entire trap surface is maintained at 80 K. What effect does this have on conductance?

6. Two identical lengths of piping are to be joined by a curved 90° elbow section or a sharp right-angle elbow section. Which overall assembly is expected to have a higher conductance? Why?

7. Show that the conductance of a pipe joining two large volumes through apertures of area A and A_0 is given by $C = 11.7AA_0/(A - A_0)$. [*Hint:* Calculate the conductance of the assembly in both directions.]

8. A chamber is evacuated by two sorption pumps of identical pumping speed. In one configuration the pumps are attached in parallel so that both pump simultaneously. In the second configuration they pump in serial or sequential order (one on and one off). Comment on the system pumping characteristics (pressure vs. time) for both configurations.

9. It is common to anneal thin films under vacuum in a closed-end quartz tube surrounded by a furnace. Consider pumping on such a cylindrical tube of length L, diameter D, and conductance C that outgasses uniformly at a rate q_0 (torr-L/cm^2-sec). Derive an expression for the steady-state pressure distribution along the tube axis. [*Hint:* Equate the gas load within length dx to throughput through the same length.]

10. After evacuation of a chamber whose volume is 30 L to a pressure of 1×10^{-6} torr, the pumps are isolated. The pressure rises to 1×10^{-5} torr in 3 min.

a. What is the leakage rate?
b. If a diffusion pump with an effective speed of 40 L/sec is attached to the chamber, what ultimate pressure can be expected?

11. Select any instrument or piece of equipment requiring high vacuum during operation (e.g., electron microscope, evaporator, Auger spectrometer, etc.). Sketch the layout of the vacuum-pumping components within the system. Explain how the gauges that measure the system pressure work.

12. A system of volume equal to 1 m^3 is evacuated to an ultimate pressure of 10^{-7} torr employing a 200 L/sec pump. For a reactive evaporation process, 100 cm^3 of gas (STP) must be continuously delivered through the system per minute.

a. What is the ultimate system pressure under these conditions?
b. What conditions are necessary to maintain this process at 10^{-2} torr?

13. In a tubular low-pressure chemical vapor deposition (LPCVD) reactor, gas is introduced at one end at a rate of 75 torr-L/min. At the other end is a vacuum pump of speed S_p. If the reactor must operate at 1 torr, what value of S_p is required?

REFERENCES

1.* R. Glang, in *Handbook of Thin Film Technology*, eds. L. I. Maissel and R. Glang, McGraw-Hill, New York (1970).
2.* A. Roth, *Vacuum Technology*, North-Holland, Amsterdam (1976).
3.* S. Dushman, *Scientific Foundations of Vacuum Techniques*, Wiley, New York (1962).
4.* R. Glang, R. A. Holmwood, and J. A. Kurtz, in *Handbook of Thin Film Technology*, eds. L. I. Maissel and R. Glang, McGraw-Hill, New York (1970).
5. J. M. Lafferty, *Techniques of High Vacuum*, GE Report No. 64-RL-3791G (1964).
6. R. W. Roberts, *An Outline of Vacuum Technology*, GE Report No. 64-RL 3394C (1964).
7.* "Vacuum Technology: Its Foundations, Formulae and Tables," in *Product and Vacuum Technology Reference Book*, Leybold-Heraeus, San Jose, CA (1986).

*Recommended texts or reviews.

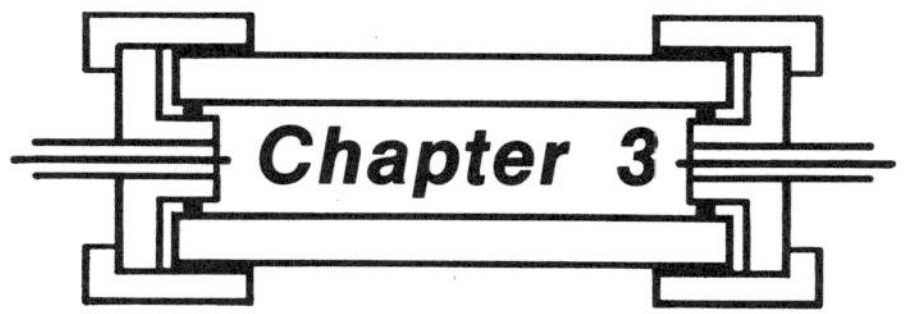

Physical Vapor Deposition

3.1. Introduction

In this chapter we focus on evaporation and sputtering, two of the most important methods for depositing thin films. The objective of these deposition processes is to controllably transfer atoms from a source to a substrate where film formation and growth proceed atomistically. In evaporation, atoms are removed from the source by thermal means, whereas in sputtering they are dislodged from solid target (source) surfaces through impact of gaseous ions. The earliest experimentation in both of these deposition techniques can apparently be traced to the same decade of the nineteenth century. In 1852, Grove (Ref. 1) observed metal deposits sputtered from the cathode of a glow discharge. Five years later Faraday (Ref. 2), experimenting with exploding fuselike metal wires in an inert atmosphere, produced evaporated thin films.

Advances in the development of vacuum-pumping equipment and the fabrication of suitable Joule heating sources, first made from platinum and then tungsten wire, spurred the progress of evaporation technology. Scientific interest in the phenomenon of evaporation and the properties of thin metal films was soon followed by industrial production of optical components such as mirrors, beam splitters, and, later, antireflection coatings. Simultaneously, sputtering was used as early as 1877 to coat mirrors. Later applications included the coating of flimsy fabrics with Au and the deposition of metal films on wax masters of phonograph records prior to thickening. Up until the late 1960s, evaporation clearly surpassed sputtering as the preferred film deposition technique. Higher deposition rates, better vacuum, and, thus, cleaner environments for film formation and growth, and general applicability to all classes of materials were some of the reasons for the ascendancy of evaporation methods. However, films used for magnetic and microelectronic applications necessitated the use of alloys, with stringent stoichiometry limits, which had to conformally cover and adhere well to substrate surfaces. These demands plus the introduction of radio frequency (RF), bias, and magnetron variants, which extended the capabilities of sputtering, and the availability of high-purity targets and working gases, helped to promote the popularity of sputter deposition. Today the decision of whether to evaporate or sputter films in particular applications is not always obvious and has fostered a lively competition between these methods. In other cases, features of both have been forged into hybrid processes.

Physical vapor deposition (PVD), the term that includes both evaporation and sputtering, and chemical vapor deposition (CVD), together with all of their variant and hybrid processes, are the basic film deposition methods treated in this book. Some factors that distinguish PVD from CVD are:

1. Reliance on solid or molten sources
2. Physical mechanisms (evaporation or collisional impact) by which source atoms enter the gas phase
3. Reduced pressure environment through which the gaseous species are transported
4. General absence of chemical reactions in the gas phase and at the substrate surface (reactive PVD processes are exceptions)

The remainder of the chapter is divided into the following sections:

3.2. The Physics and Chemistry of Evaporation
3.3. Film Thickness Uniformity and Purity
3.4. Evaporation Hardware and Techniques
3.5. Glow Discharges and Plasmas
3.6. Sputtering
3.7. Sputtering Processes
3.8. Hybrid and Modified PVD Processes

Additional excellent reading material on the subject can be found in Refs. 3–6. The book by Chapman is particularly recommended for its entertaining and very readable presentation of the many aspects relating to phenomena in rarefied gases, glow discharges, and sputtering.

3.2. The Physics and Chemistry of Evaporation

3.2.1. Evaporation Rate

Early attempts to quantitatively interpret evaporation phenomena are connected with the names of Hertz, Knudsen, and, later, Langmuir (Ref. 3). Based on experimentation on the evaporation of mercury, Hertz, in 1882, observed that evaporation rates were:

1. Not limited by insufficient heat supplied to the surface of the molten evaporant
2. Proportional to the difference between the equilibrium pressure P_e of Hg at the given temperature and the hydrostatic pressure P_h acting on the evaporant.

Hertz concluded that a liquid has a specific ability to evaporate at a given temperature. Furthermore, the maximum evaporation rate is attained when the number of vapor molecules emitted corresponds to that required to exert the equilibrium vapor pressure while none return. These ideas led to the basic equation for the rate of evaporation from both liquid and solid surfaces, namely,

$$\Phi_e = \frac{\alpha_e N_A (P_e - P_h)}{\sqrt{2\pi MRT}}, \tag{3-1}$$

where Φ_e is the evaporation flux in number of atoms (or molecules) per unit area per unit time, and α_e is the coefficient of evaporation, which has a value between 0 and 1. When $\alpha_e = 1$ and P_h is zero, the maximum evaporation rate is realized. By analogy with Eq. 2-9, an expression for the maximum value of Φ_e is

$$\Phi_e = 3.513 \times 10^{22} \frac{P_e}{\sqrt{MT}} \text{ molecules/cm}^2\text{-sec}. \tag{3-2}$$

when P_e is expressed in torr. A useful variant of this formula is

$$\Gamma_e = 5.834 \times 10^{-2} \sqrt{M/T}\, P_e \text{ g/cm}^2\text{-sec}, \tag{3-3}$$

where Γ_e is the mass evaporation rate. At a pressure of 10^{-2} torr, a typical value of Γ_e for many elements is approximately 10^{-4} g/cm^2-sec of evaporant. The key variable influencing evaporation rates is the temperature, which has a profound effect on the equilibrium vapor pressure.

3.2.2. Vapor Pressure of the Elements

A convenient starting point for expressing the connection between temperature and vapor pressure is the Clausius–Clapeyron equation, which for both solid–vapor and liquid–vapor equilibria can be written as

$$\frac{dP}{dT} = \frac{\Delta H(T)}{T \Delta V}. \tag{3-4}$$

The changes in enthalpy, $\Delta H(T)$, and volume, ΔV, refer to differences between the vapor (v) and the particular condensed phase (c) from which it originates, and T is the transformation temperature in question. Since $\Delta V = V_v - V_c$, and the volume of vapor normally considerably exceeds that of the condensed solid or liquid phase, $\Delta V = V_v$. If the gas is assumed to be perfect, $V_v \approx RT/P$, and Eq. 3-4 may be rewritten as

$$\frac{dP}{dT} = \frac{P \Delta H(T)}{RT^2}. \tag{3-5}$$

As a first approximation, $\Delta H(T) = \Delta \mathrm{He}$, the molar heat of evaporation (a constant), in which case simple integration yields

$$\ln P \approx -\frac{\Delta \mathrm{He}}{RT} + I, \tag{3-6}$$

where I is a constant of integration. Through substitution of the latent heat of vaporization for $\Delta \mathrm{He}$, the boiling point for T, and 1 atm for P, I can be evaluated for the liquid–vapor transformation. For practical purposes, Eq. 3-6 adequately describes the temperature dependence of the vapor pressure in many materials. It is rigorously applicable only over a small temperature range, however. To extend the range of validity, we must account for the temperature dependence of $\Delta H(T)$. For example, careful evaluation of thermodynamic data reveals that the vapor pressure of liquid Al is given by (Ref. 3)

$$\log P_{(\mathrm{torr})} = 15{,}993/T + 12.409 - 0.999 \log T - 3.52 \times 10^{-6} T. \tag{3-7}$$

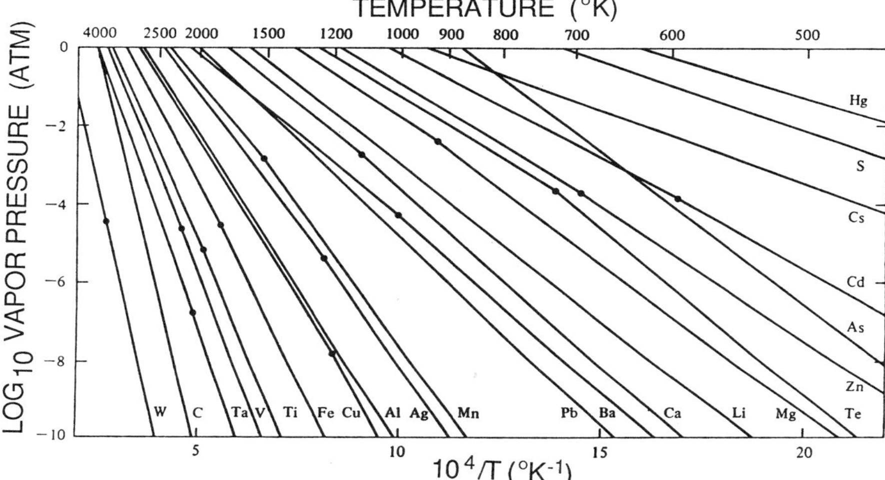

Figure 3-1. Vapor pressures of selected elements. Dots correspond to melting points. (From Ref. 7).

The Arrhenius character of log P vs. $1/T$ is essentially preserved, since the last two terms on the right-hand side are small corrections.

Vapor-pressure data for many other metals have been similarly obtained and conveniently represented as a function of temperature in Fig. 3-1. Similarly, vapor-pressure data for elements important in the deposition of semiconductor films are presented in Fig. 3-2. Much of the data represent direct measurements of the vapor pressures. Other values are inferred indirectly from thermodynamic relationships and identities using a limited amount of experimental data. Thus the vapor pressures of refractory metals such as W and Mo can be unerringly extrapolated to lower temperatures, even though it may be impossible to measure them directly.

Two modes of evaporation can be distinguished in practice, depending on whether the vapor effectively emanates from a liquid or solid source. As a rule of thumb, a melt will be required if the element in question does not achieve a vapor pressure greater than 10^{-3} torr at its melting point. Most metals fall into this category, and effective film deposition is attained only when the source is heated into the liquid phase. On the other hand, elements such as Cr, Ti, Mo, Fe, and Si reach sufficiently high vapor pressures below the melting point and, therefore, sublime. For example, Cr can be effectively deposited at high rates from a solid metal source because it attains vapor pressures of 10^{-2} torr some 500 °C below the melting point. The operation of the Ti sublimation pump mentioned in Chapter 2 is, in fact, based on the sublimation from heated Ti

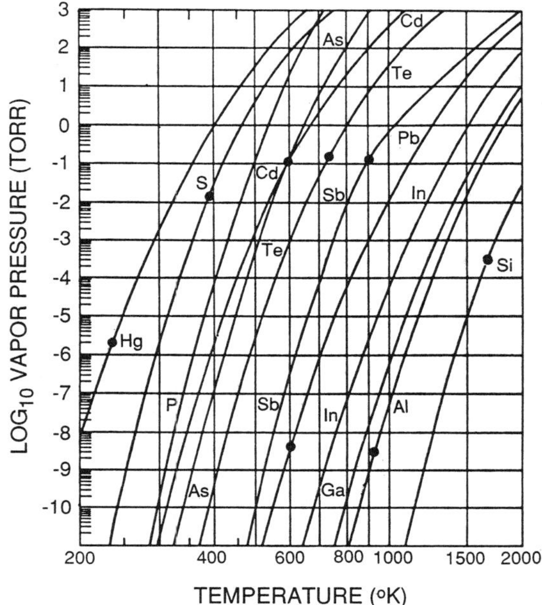

Figure 3-2. Vapor pressures of elements employed in semiconductor materials. Dots correspond to melting points. (Adapted from Ref. 8).

filaments. A third example is carbon, which is used to prepare replicas of the surface topography of materials for subsequent examination in the electron microscope. The carbon is sublimed from an arc struck between graphite electrodes.

3.2.3. Evaporation of Compounds

While metals essentially evaporate as atoms and occasionally as clusters of atoms, the same is not true of compounds. Very few inorganic compounds evaporate without molecular change, and, therefore, the vapor composition is usually different from that of the original solid or liquid source. A consequence of this is that the stoichiometry of the film deposit will generally differ from that of the source. Mass spectroscopic studies of the vapor phase have shown that the processes of molecular association as well as dissociation frequently occur. A broad range of evaporation phenomena in compounds occurs, and these are categorized briefly in Table 3-1.

Table 3-1. Evaporation of Compounds

Reaction Type	Chemical Reaction	Examples	Comments
Evaporation without dissociation	$MX(s \text{ or } l) \rightarrow MX(g)$	SiO, B_2O_3 GeO, SnO, AlN CaF_2, MgF_2	Compound stoichiometry maintained in deposit
Decomposition	$MX(s) \rightarrow M(s) + (1/2)X_2(g)$	Ag_2S, Ag_2Se	Separate sources are required to deposit these compounds
	$MX(s) \rightarrow M(l) + (1/n)X_n(g)$	III-V semiconductors	
Evaporation with dissociation			Deposits are metal-rich; separate sources are required to deposit these compounds
a. Chalcogenides X = S, Se, Te	$MX(s) \rightarrow M(g) + (1/2)X_2(g)$	CdS, CdSe CdTe	
b. Oxides	$MO_2(s) \rightarrow MO(g) + (1/2)O_2(g)$	SiO_2, GeO_2 TiO_2, SnO_2 ZrO_2	Metal-rich discolored deposits; dioxides are best deposited in O_2 partial pressure (reactive evaporation)

Note M = metal, X = nonmetal.
Adapted from Ref. 3.

3.2.4. Evaporation of Alloys

Evaporated metal alloy films are widely utilized for a variety of electronic, magnetic, and optical applications as well as for decorative coating purposes. Important examples of such alloys that have been directly evaporated include Al–Cu, Permalloy (Fe–Ni), nichrome (Ni–Cr), and Co–Cr. Atoms in metals of such alloys are generally less tightly bound than atoms in the inorganic compounds discussed previously. The constituents of the alloys, therefore, evaporate nearly independently of each other and enter the vapor phase as single atoms in a manner paralleling the behavior of pure metals. Metallic melts are solutions and as such are governed by well-known thermodynamic laws. When the interaction energy between A and B atoms of a binary AB alloy melt are the same as between A–A and B–B atom pairs, then no preference is shown for atomic partners. Such is the environment in an ideal solution. Raoult's law, which holds under these conditions, states that the vapor pressure of component B in solution is reduced relative to the vapor pressure of pure B ($P_B(0)$) in proportion to its mole fraction X_B. Therefore,

$$P_B = X_B P_B(0). \tag{3-8}$$

Metallic solutions usually are not ideal, however. This means that either more or less B will evaporate relative to the ideal solution case, depending on whether the deviation from ideality is positive or negative, respectively. A positive deviation occurs because B atoms are physically bound less tightly to the solution, facilitating their tendency to escape or evaporate. In real solutions

$$P_B = a_B P_B(0), \tag{3-9}$$

where a_B is the effective thermodynamic concentration of B known as the activity. The activity is, in turn, related to X_B through an activity coefficient γ_B; i.e.,

$$a_B = \gamma_B X_B. \tag{3-10}$$

By combination of Eqs. 3-2, 3-9, and 3-10, the ratio of the fluxes of A and B atoms in the vapor stream above the melt is given by

$$\frac{\Phi_A}{\Phi_B} = \frac{\gamma_A X_A P_A(0)}{\gamma_B X_B P_B(0)} \sqrt{\frac{M_B}{M_A}}. \tag{3-11}$$

Practical application of this equation is difficult because the melt composition changes as evaporation proceeds. Therefore, the activity coefficients, which can sometimes be located in the metallurgical literature, but just as frequently not, also change with time. As an example of the use of Eq. 3-11, consider the problem of estimating the approximate Al–Cu melt composition required to evaporate films containing 2 wt% Cu from a single crucible heated to 1350 K. Substituting gives

$$\frac{\Phi_{Al}}{\Phi_{Cu}} = \frac{98/M_{Al}}{2/M_{Cu}}, \quad \frac{P_{Al}(0)}{P_{Cu}(0)} = \frac{10^{-3}}{2 \times 10^{-4}} \quad \text{and assuming } \gamma_{Cu} = \gamma_{Al},$$

$$\frac{X_{Al}}{X_{Cu}} = \frac{98}{2} \frac{2 \times 10^{-4}}{10^{-3}} \sqrt{\frac{63.7}{27.0}} = 15.$$

This suggests that the original melt composition should be enriched to 13.6 wt% Cu in order to compensate for the preferential vaporization of Al. It is, therefore, feasible to evaporate such alloys from one heated source. If the alloy melt is of large volume, fractionation-induced melt composition changes are minimal. A practical way to cope with severe fractionation is to evaporate from dual sources maintained at different temperatures.

3.3. Film Thickness Uniformity and Purity

3.3.1. Deposition Geometry

In this section aspects of the deposition geometry, including the characteristics of evaporation sources, and the orientation and placement of substrates are discussed. The source–substrate geometry, in turn, influences the ultimate film uniformity, a concern of paramount importance, which will be treated subsequently. Evaporation from a point source is the simplest of situations to model. Evaporant particles are imagined to emerge from an infinitesimally small region (dA_e) of a sphere of surface area A_e with a uniform mass evaporation rate as shown in Fig. 3-3a. The total evaporated mass $\overline{M}_e$ is then given by the double integral

$$\overline{M}_e = \int_0^t \int_{A_e} \Gamma_e \, dA_e \, dt. \tag{3-12}$$

Of this amount, mass $d\overline{M}_s$ falls on the substrate of area dA_s. Since the projected area dA_s on the surface of the sphere is dA_c, with $dA_c = dA_s \cos\theta$, the proportionality $d\overline{M}_s : \overline{M}_e = dA_c : 4\pi r^2$ holds. Finally,

$$\frac{d\overline{M}_s}{dA_s} = \frac{\overline{M}_e \cos\theta}{4\pi r^2} \tag{3-13}$$

is obtained. On a per unit time basis we speak of deposition rate $\dot{R}$ (atoms/cm^2-sec), a related quantity referred to later in the book. The deposi-

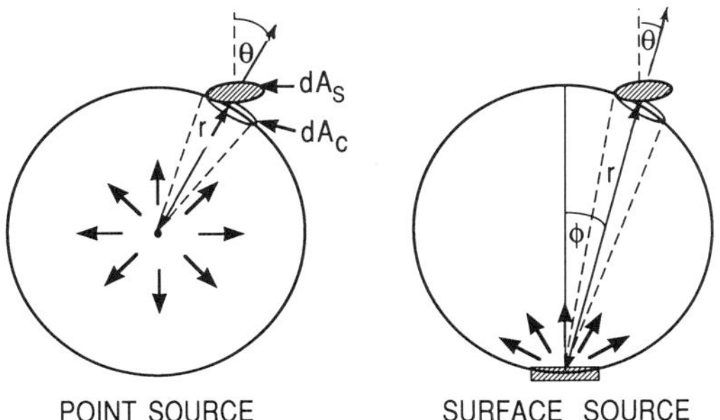

Figure 3-3. Evaporation from (a) point source, (b) surface source.

tion varies with the geometric orientation of the substrate and with the inverse square of the source–substrate distance. Substrates placed tangent to the surface of the receiving sphere would be coated uniformly, since $\cos\theta = 1$.

An evaporation source employed in the pioneering research by Knudsen made use of an isothermal enclosure with a very small opening through which the evaporant atoms or molecules effused. These effusion or Knudsen cells are frequently employed in molecular-beam epitaxy deposition systems, where precise control of evaporation variables is required. Kinetic theory predicts that the molecular flow of the vapor through the opening is directed according to a cosine distribution law, and this has been verified experimentally. The mass deposited per unit area is given by

$$\frac{d\overline{M}_s}{dA_s} = \frac{\overline{M}_e \cos\phi \cos\theta}{\pi r^2} \tag{3-14}$$

and now depends on two angles (emission and incidence) that are defined in Fig. 3-3b. Evaporation from a small area or surface source is also modeled by Eq. 3-14. Boat filaments and wide crucibles containing a pool of molten material to be evaporated approximate surface sources in practice.

From careful measurements of the angular distribution of film thickness, it has been found that, rather than a $\cos\phi$ dependence, a $\cos^n\phi$ evaporation law is more realistic. As shown in Fig. 3-4, n is a number that determines the geometry of the lobe-shaped vapor cloud and the angular distribution of evaporant flux from a source. When n is large, the vapor flux is highly directed. Physically n is related to the evaporation crucible geometry and scales directly with the ratio of the melt depth (below top of crucible) to the

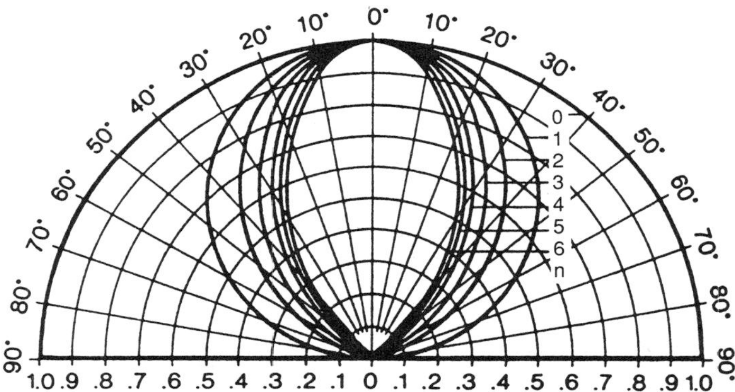

Figure 3-4. Calculated lobe-shaped vapor clouds with various cosine exponents. (From Ref. 9).

melt surface area. Deep narrow crucibles with large n have been employed to confine evaporated radioactive materials to a narrow angular spread in order to minimize chamber contamination. The corresponding deposition equation is (Ref. 9)

$$\frac{d\overline{M}_s}{dA_s} = \frac{\overline{M}_e(n+1)\cos^n\phi \cos\theta}{2\pi r^2} \qquad (n \geq 0). \tag{3-15}$$

As the source becomes increasingly directional, the surface area effectively exposed to evaporant shrinks (i.e., $2\pi r^2$, πr^2, and $2\pi r^2/(n+1)$ for point, $\cos\phi$, and $\cos^n\phi$ sources, respectively).

3.3.2. Film Thickness Uniformity

While maintaining thin-film thickness uniformity is always desirable, but not necessarily required, it is absolutely essential for microelectronic and many optical coating applications. For example, thin-film, narrow-band optical interference filters require a thickness uniformity of $\pm 1\%$. This poses a problem, particularly if there are many components to be coated or the surfaces involved are large or curved. Utilizing formulas developed in the previous section, we can calculate the thickness distribution for a variety of important source–substrate geometries. Consider evaporation from the point and small surface source onto a *parallel* plane-receiving substrate surface as indicated in the insert of Fig. 3-5. The film thickness d is given by $d\overline{M}_s/\rho\, dA_s$, where ρ is the density of the deposit. For the point source

$$d = \frac{\overline{M}_e \cos\theta}{4\pi\rho r^2} = \frac{\overline{M}_e h}{4\pi\rho r^3} = \frac{\overline{M}_e h}{4\pi\rho(h^2 + l^2)^{3/2}}. \tag{3-16}$$

The thickest deposit (d_0) occurs at $l = 0$, in which case $d_0 = \overline{M}_e/4\pi\rho h^2$, and, thus,

$$\frac{d}{d_0} = \frac{1}{\{1 + (l/h)^2\}^{3/2}}. \tag{3-17}$$

Similarly, for the surface source

$$d = \frac{\overline{M}_e \cos\theta \cos\phi}{\pi\rho r^2} = \frac{\overline{M}_e}{\pi\rho r^2}\frac{h}{r}\frac{h}{r} = \frac{\overline{M}_e h^2}{\pi\rho(h^2 + l^2)^2}, \tag{3-18}$$

since $\cos\theta = \cos\phi = h/r$. When normalized to the thickest dimensions, or $d_0 = \overline{M}_e/\pi\rho h^2$,

$$\frac{d}{d_0} = \frac{1}{\{1 + (l/h)^2\}^2}. \tag{3-19}$$

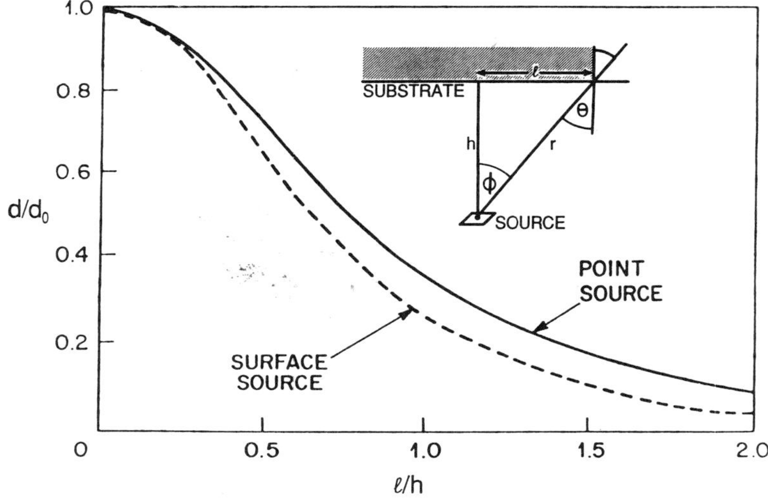

Figure 3-5. Film thickness uniformity for point and surface sources. (Insert) Geometry of evaporation onto parallel plane substrate.

A comparison of Eqs. 3-17 and 3-19 is made in Fig. 3-5, where it is apparent that less thickness uniformity can be expected with the surface source.

A couple of practical examples (Ref. 10) will demonstrate how these film thickness distributions are used in designing source–substrate geometries for coating applications. In the first example suppose it is desired to coat a 150-cm-wide strip utilizing two evaporation sources oriented as shown in the insert of Fig. 3-6. If a thickness tolerance of $\pm 10\%$ is required, what should the distance between sources be and how far should they be located from the substrate? A superposition of solutions for two individual surface sources (Eq. 3-19) gives the thickness variation shown graphically in Fig. 3-6 as a function of the relative distance r from the center line for various values of the source spacing D. All pertinent variables are in terms of dimensionless ratios r/h_v and D/h_v. The desired tolerance requires that d/d_0 stay between 0.9 and 1.1, and this can be achieved with $D/h_v = 0.6$ yielding a maximum value of $r/h_v = 0.87$. Since $r = 150/2 = 75$ cm, $h_v = 75/0.87 = 86.2$ cm. The required distance between sources is therefore $2D = 2 \times 0.6 \times 86.2 = 103.4$ cm. There are other solutions, of course, but we are seeking the *minimum* value of h_v. It is obvious that the uniformity tolerance can always be realized by extending the source–substrate distance, but this wastes evaporant.

As a second example, consider a composite optical coating where a $\pm 1\%$ film thickness variation is required in each layer. The substrate is rotated to even out source distribution anomalies and minimize preferential film growth that can adversely affect coating durability and optical properties. Since multiple films of different composition will be sequentially deposited, the necessary fixturing requires that the sources be offset from the axis of rotation by a distance $R = 20$ cm. How high above the source should a 25-cm-diameter substrate be rotated to maintain the desired film tolerance? The film thickness distribution in this case is a complex function of the three-dimensional geometry, which, fortunately, has been graphed in Fig. 3-7. Reference to this figure indicates that the curve $h_v/R = 1.33$ in conjunction with $r/R = 0.6$ will generate a thickness deviation ranging from about -0.6 to $+0.5\%$. On this basis, the required distance is $h_v = 1.33 \times 20 = 26.6$ cm.

A clever way to achieve thickness uniformity, however, is to locate both the surface evaporant source and the substrates on the surface of a sphere as shown in Fig. 3-8. In this case, $\cos\theta = \cos\phi = r/2r_0$, and Eq. 3-14 becomes

$$\frac{d\overline{M}_s}{dA_s} = \frac{\overline{M}_e}{\pi r^2}\frac{r}{2r_0}\frac{r}{2r_0} = \frac{\overline{M}_e}{4\pi r_0^2}. \tag{3-20}$$

The resultant deposit thickness is a constant clearly independent of angle. Use is made of this principle in the planetary substrate fixtures that hold silicon wafers to be coated with metal (metallized) by evaporation. To further promote uniform coverage, the planetary fixture is rotated during deposition. Physi-

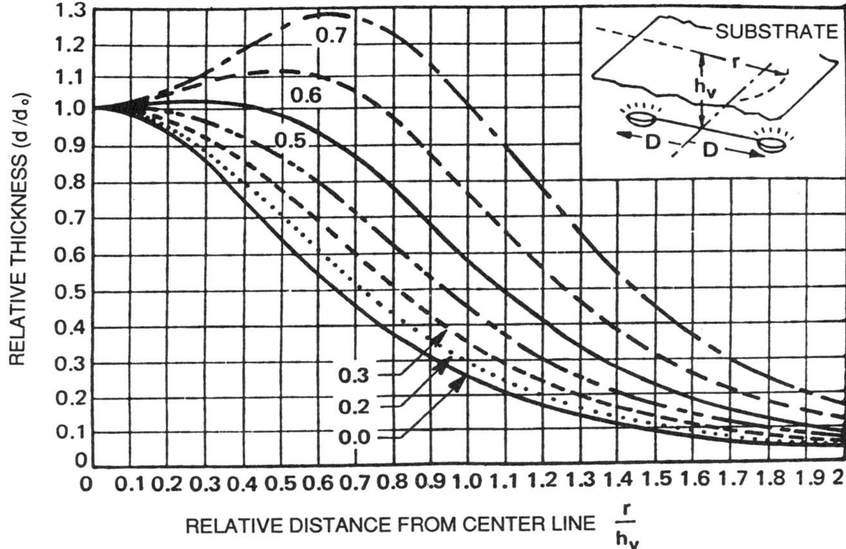

Figure 3-6. Film thickness uniformity across a strip employing two evaporation sources for various values of D/h_v. (From Ref. 10).

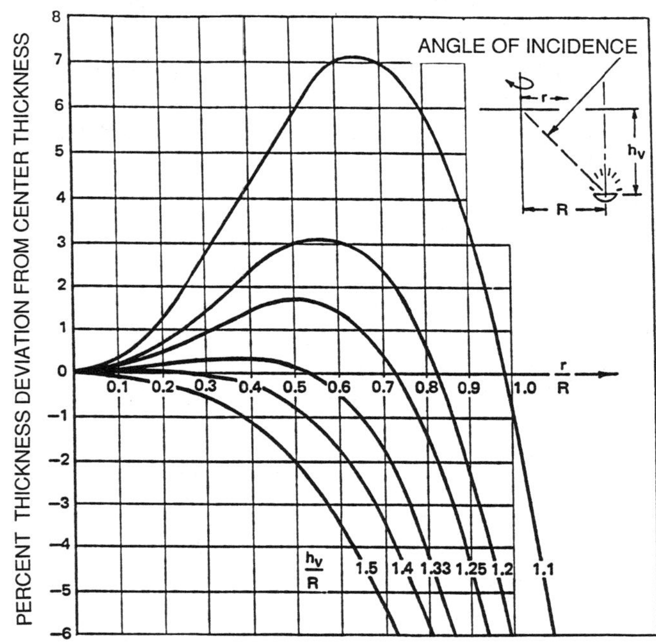

Figure 3-7. Calculated film thickness variation across the radius of a rotating disk. (From Ref. 10).

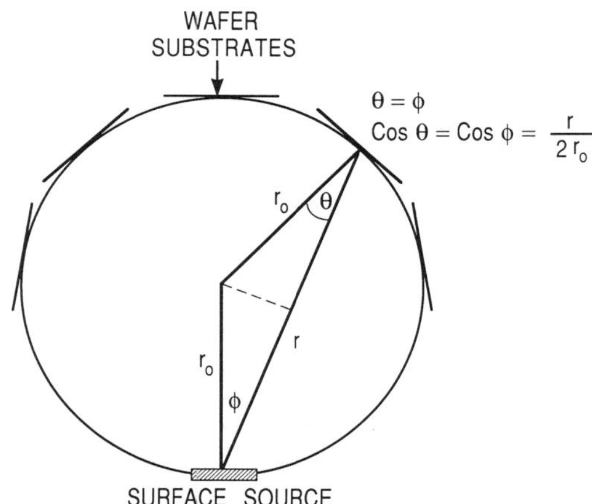

Figure 3-8. Evaporation scheme to achieve uniform deposition. Source and substrates lie on sphere of radius r_0.

cally, deposition uniformity is achieved because short source–substrate distances are offset by unfavorably large vapor emission and deposition angles. Alternately, long source–substrate distances are compensated by correspondingly small emission and reception angles. For sources with a higher degree of directionality (i.e., where $\cos^n\phi$ rather than $\cos\phi$ is involved), the reader can easily show that thickness uniformity is no longer maintained.

Two principal methods for optimizing film uniformity over large areas involve varying the geometric location of the source and interposing static as well as rotating shutters between evaporation sources and substrates. Computer calculations have proven useful in locating sources and designing shutter contours to meet the stringent demands of optical coatings. Film thickness uniformity cannot, however, be maintained beyond $\pm 1\%$ because of insufficient mechanical stability of both the stationary and rotating hardware.

In addition to the parallel source–substrate configuration, calculations of thickness distributions have also been made for spherical as well as conical, parabolic, and hyperbolic substrate surfaces (Ref. 9). Similarly, cylindrical, wire, and ring evaporation source geometries have been treated (Ref. 11). For the results, interested readers should consult the appropriate references.

3.3.3. Conformal Coverage

An issue related to film uniformity is step or, more generally, conformal coverage, and it arises primarily in the fabrication of integrated circuits. The required semiconductor contact and device interconnection metallization depositions frequently occur over a terrain of intricate topography where microsteps, grooves, and raised stripes abound. When the horizontal as well as

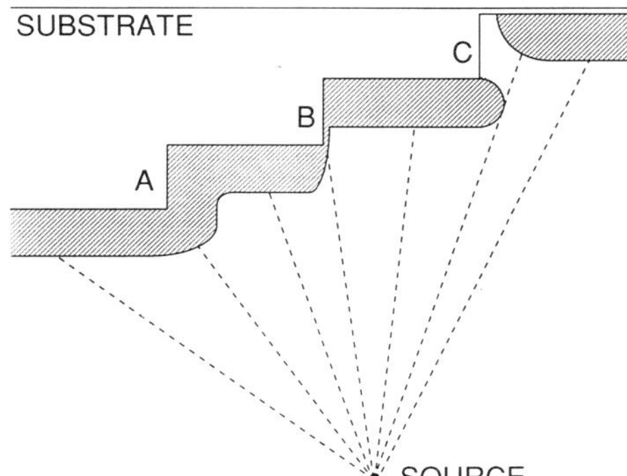

Figure 3-9. Schematic illustration of film coverage of stepped substrate: (A) uniform coverage; (B) poor sidewall coverage; (C) lack of coverage—discontinuous film.

vertical surfaces of substrates are coated to the same thickness, we speak of conformal coverage. On the other hand, coverage will not be uniform when physical shadowing effects cause unequal deposition on the top and sidewalls of steps. Inadequate step coverage can lead to minute cracks in the metalization, which have been shown to be a major source of failure in device reliability testing. Thinned regions on conducting stripes exhibit greater Joule heating, which sometimes fosters early burnout. Step coverage problems have been shown to be related to the profile of the substrate step as well as to the evaporation source–substrate geometry. The simplest model of evaporation from a point source onto a stepped substrate results in either conformal coverage or a lack of deposition in the step shadow, as shown schematically in Fig. 3-9. Line-of-sight motion of evaporant atoms and sticking coefficients of unity can be assumed in estimating the extent of coverage.

More realistic computer modeling of step coverage has been performed for the case in which the substrate is located on a rotating planetary holder (Ref. 12). In Fig. 3-10 coverage of a 1-μm-wide, 1-μm-high test pattern with 5000 Å

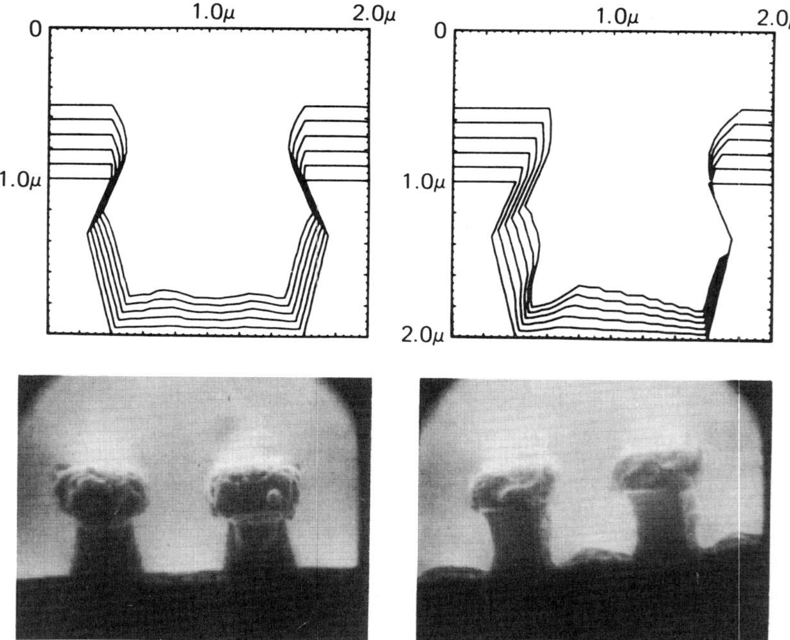

Figure 3-10. Comparison of simulated and experimental Al film coverage of 1-μm line step and trench features. (Left) Orientation of most symmetric deposition. (Right) Orientation of most asymmetric deposition. (Reprinted with permission from Cowan Publishing Co., from C. H. Ting and A. R. Neureuther, *Solid State Technology* **25**, 115, 1982).

of evaporated Al is simulated and compared with experiment. In the symmetric orientation the region between the pattern stripes always manages to "see" the source, and this results in a small plateau of full film thickness. In the asymmetric orientation, however, the substrate stripes cast a shadow with respect to the source biasing the deposition in favor of unequal sidewall coverage. In generating the simulated film profiles, the surface migration of atoms was neglected, a valid assumption at low substrate temperatures. Heating the substrate increases surface diffusion of depositing atoms, thus promoting the filling of potential voids as they form. Interestingly, similar step coverage problems exist in chemical-vapor-deposited SiO_2 and silicon nitride films.

3.3.4. Film Purity

The chemical purity of evaporated films depends on the nature and level of impurities that (1) are initially present in the source, (2) contaminate the source from the heater, crucible, or support materials, and (3) originate from the residual gases present in the vacuum system. In this section only the effect of residual gases on film purity will be addressed. During deposition the atoms and molecules of both the evaporant and residual gases impinge on the substrate in parallel, independent events. The evaporant vapor impingement rate is $\rho N_A \dot{d}/M_a$ atoms/cm^2-sec, where ρ is the density and $\dot{d}$ is the deposition rate (cm/sec). Simultaneously, gas molecules impinge at a rate given by Eq. 2-9. The ratio of the latter to the former is the impurity concentration C_i:

$$C_i = 5.82 \times 10^{-2} \frac{P}{\sqrt{M_g T}} \frac{M_a}{\rho \dot{d}}. \tag{3-21}$$

Terms M_a and M_g refer to evaporant and gas molecular weights, respectively, and P is the residual gas vapor pressure in torr.

Table 3-2 illustrates the combined role that deposition rate and residual pressure play in determining the oxygen level that can be incorporated into thin tin films (Ref. 13). Although the concentrations are probably overestimated because the sticking probabiltiy of O_2 is about 0.1 or less, the results have several important implications. To produce very pure films, it is important to deposit at very high rates while maintaining very low background pressures of residual gases such as H_2O, CO_2, CO, O_2, and N_2. Neither of these requirements is too formidable for vacuum evaporation, where deposition rates from electron-beam sources can reach 1000 Å/sec at chamber pressures of $\sim 10^{-8}$ torr.

On the other hand, in sputtering processes, discussed later in the chapter,

Table 3-2. Maximum Oxygen Concentration in Tin Films Deposited at Room Temperature

P_{O_2} (torr)	Deposition Rate (A/sec)			
	1	10	100	1000
10^{-9}	10^{-3}	10^{-4}	10^{-5}	10^{-6}
10^{-7}	10^{-1}	10^{-2}	10^{-3}	10^{-4}
10^{-5}	10	1	10^{-1}	10^{-2}
10^{-3}	10^{3}	10^{2}	10	1

From Ref. 13.

deposition rates are typically more than an order of magnitude less, and chamber pressures five orders of magnitude higher than for evaporation. Therefore, the potential exists for producing films containing high gas concentrations. For this reason sputtering was traditionally not considered to be as "clean" a process as evaporation. Considerable progress has been made in the last two decades, however, with the commercial development of high-deposition-rate magnetron sputtering systems, operating at somewhat lower gas pressures in cleaner vacuum systems. In the case of aluminum films, comparable purities appear to be attained in both processes. Lastly, Table 3-2 suggests that very high oxygen incorporation occurs at residual gas pressures of 10^{-3} torr. Advantage of this fact is taken in reactive evaporation processes where intentionally introduced gases serve to promote reactions with the evaporant metal and control the deposit stoichiometry.

The presence of gaseous impurities within metal films sometimes has a pronounced effect in degrading many of its properties. Oxygen and nitrogen incorporation has been observed to reduce both the electrical conductivity and optical reflectivity as well as increase the hardness of Al films (Ref. 14).

3.4. Evaporation Hardware and Techniques

3.4.1. Resistance-Heated Evaporation Sources

This section will primarily be devoted to a brief description of the most widely used methods for heating evaporants. Clearly, heaters must reach the temperature of the evaporant in question while having a negligible vapor pressure in comparison. Ideally, they should not contaminate, react, or alloy with the evaporant or release gases such as oxygen, nitrogen, or hydrogen at the evaporation temperature. These requirements have led to the development and use of suitable resistance and electron-beam-heated sources.

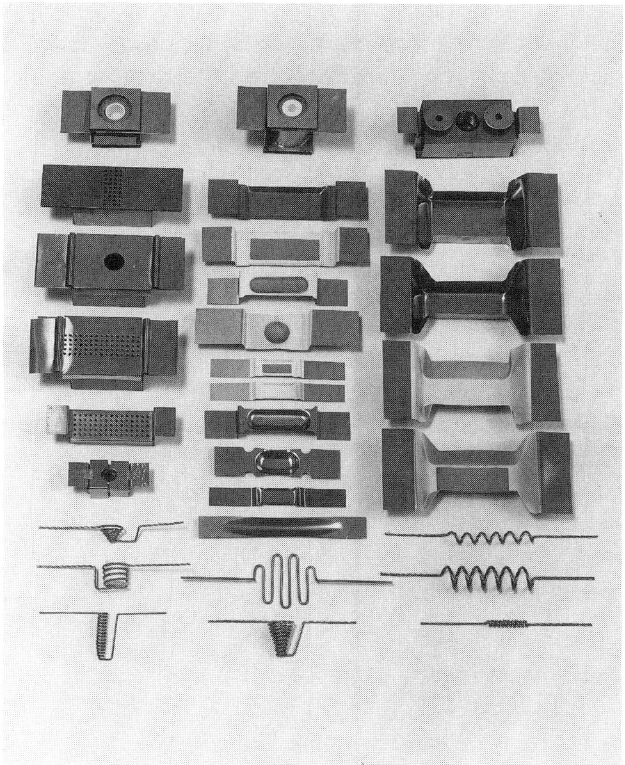

Figure 3-11. Assorted resistance heated evaporation sources. (Courtesy of R. D. Mathis Company).

Resistively heated evaporation sources are available in a wide variety of forms utilizing refractory metals singly or in combination with inert oxide or ceramic compound crucibles. Some of these are shown in Fig. 3-11. They can be divided into the following important categories.

3.4.1.1. Tungsten Wire Sources. Tungsten wire sources are in the form of individual or multiply stranded wires twisted into helical or conical shapes. Helical coils are used for metals that wet tungsten readily, whereas the conical baskets are better adapted to contain poorly wetting materials. In the former case, metal evaporant wire is wrapped around or hung from the tungsten strands, and the molten beads of metal are retained by surface tension forces.

3.4.1.2. Refractory Metal Sheet Sources. Tungsten, tantalum, and molybdenum sheet metal sources, like the wire filaments, are self-resistance heaters that require low-voltage, high-current power supplies. These sources have been fabricated into a variety of shapes, including dimpled strip, boat, canoe, and deep-folded configurations. Folded boat sources have been used to evaporate MgF_2 by containing the bulk salt and melting it prior to vaporization. Powder mixtures of metals and metal oxides used for coating ophthalmic lenses have been similarly evaporated from deep-folded boats in batch-type evaporators.

3.4.1.3. Sublimation Furnaces. Efficient evaporation of sulfides, selenides, and some oxides is carried out in sublimation furnaces. The evaporant materials in powder form are pressed and sintered into pellets and heated by surrounding radiant heating sources. Spitting and ejection of particles caused by evolution of gases occluded within the source compacts are avoided through the use of baffled heating assemblies. These avoid direct line-of-sight access to substrates, and evaporation rates from such sources tend to be constant over extended periods of time. The furnaces are typically constructed of sheet tantalum, which is readily cut, bent, and spot-welded to form heaters, radiation shields, supports, and current bus strips.

3.4.1.4. Crucible Sources. The most common sources are cylindrical cups composed of oxides, pyrolytic BN, graphite, and refractory metals, which are fabricated by hot-pressing powders or machining bar stock. These crucibles are normally heated by external tungsten wire heating elements wound to fit snugly around them.

Other crucible sources rely on high-frequency induction rather than resistance heating. In a configuration resembling a transformer, high-frequency currents are induced in either a conducting crucible or evaporant charge serving as the secondary, resulting in heating. The powered primary is a coil of water-cooled copper tubing that surrounds the crucible. Aluminum evaporated from BN or BN/TiB_2 composite crucibles, in order to metallize integrated circuits, is an important example of the use of induction heating.

Another category of crucible source consists of a tungsten wire resistance heater in the form of a conical basket encased in Al_2O_3 or refractory oxide to form an integral crucible–heater assembly. Such crucibles frequently serve as evaporant sources in laboratory scale film deposition systems.

3.4.2. Electron-Beam Evaporation

Disadvantages of resistively heated evaporation sources include possible contamination by crucibles, heaters, and support materials and the limitation of relatively low input power levels. This makes it difficult to deposit pure films or evaporate high-melting-point materials at appreciable rates. Electron-beam heating eliminates these disadvantages and has, therefore, become the most widely used vacuum evaporation technique for preparing highly pure films. In principle, this type of source enables evaporation of virtually all materials at almost any rate. As shown in Fig. 3-12, the evaporant charge is placed in either a water-cooled crucible or in the depression of a water-cooled copper hearth. The purity of the evaporant is ensured because only a small amount of charge melts or sublimes so that the effective crucible is the unmelted skull material next to the cooled hearth. For this reason there is no contamination of the evaporant by Cu. Multiple-source units are available for the sequential or parallel deposition of more than one material.

In the most common configuration of the gun source, electrons are thermionically emitted from heated filaments, which are shielded from direct line of sight of the evaporant charge and substrate. Film contamination from the heated electron source is eliminated in this way. The filament cathode assembly potential is biased negatively with respect to a nearby grounded anode by anywhere from 4 to 20 kV, and this serves to accelerate the electrons. In addition, a transverse magnetic field is applied, which serves to deflect the electron beam in a 270° circular arc and focus it on the hearth and evaporant charge at ground potential. The reader can verify the electron trajectory through the use of the right-hand rule. This states that if the thumb is in the direction of the initial electron emission, and the middle finger lies in the direction of the magnetic field (north to south), then the forefinger indicates the direction of the force on the electron and its resultant path at any instant.

It is instructive to estimate the total power that must be delivered by the electron beam to the charge in order to compensate for the following heat losses incurred during evaporation of 10^{18} atoms/cm^2-sec (Ref. 5).

1. The power density P_s (watts/cm^2) that must be supplied to account for the heat of sublimation ΔH_s (eV) is

$$P_s = 10^{18}(1.6 \times 10^{-19})\,\Delta H_s = 0.16\,\Delta H_s. \tag{3-22a}$$

2. The kinetic energy of evaporant is $(3/2)kT_s$ per atom so that the required power density P_k is

$$P_k = 10^{18}\,(3/2)(1.38 \times 10^{-23})T_s = 2.07 \times 10^{-5}T_s, \tag{3-22b}$$

where T_s is the source temperature.

3. The radiation heat loss density is

$$P_r = 5.67 \times 10^{-12}\varepsilon(T_s^4 - T_0^4), \tag{3-22c}$$

where ε is the source emissivity at T_s, and $T_0 \approx 293$ K.

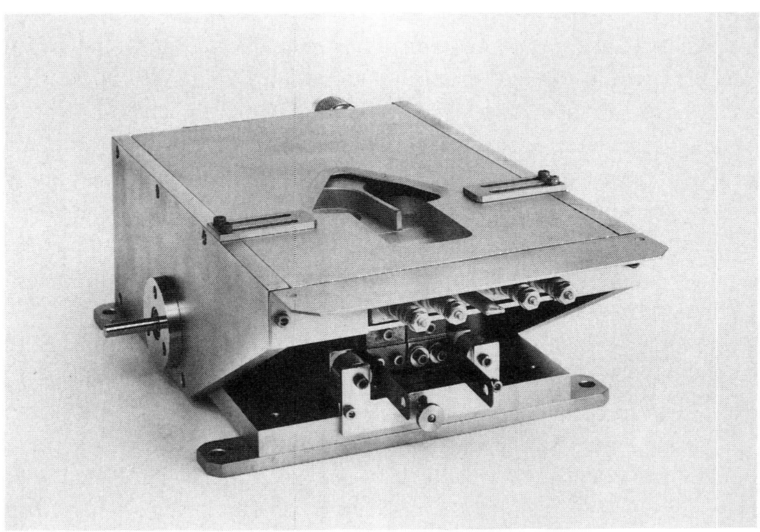

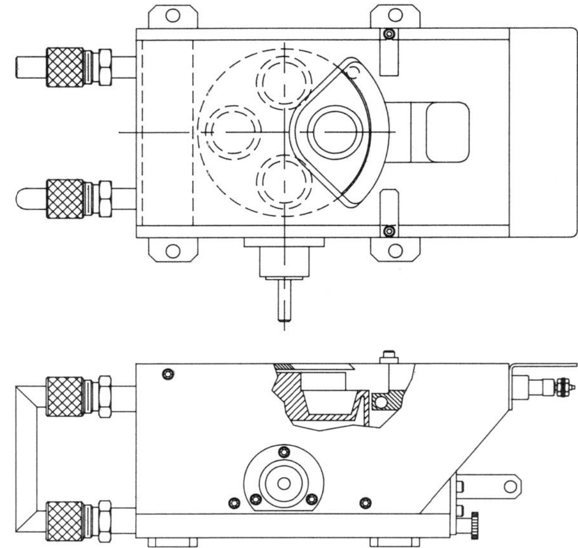

Figure 3-12. Multihearth electron beam evaporation unit with accompanying schematics. (Courtesy of Temescal unit of Edwards High Vacuum International, a division of the BOC Group, Inc.).

4. Heat conduction through a charge of thickness l into the hearth dissipates a power density P_c equal to

$$P_c = \kappa\left(\frac{T_s - T_0}{l}\right), \tag{3-22d}$$

where κ is the thermal conductivity of the charge. For the case of Au at T_s = 1670 K, where $\Delta H_s \approx 3.5$ eV, $\varepsilon \sim 0.4$, $l = 1$ cm, and $\kappa = 3.1$ W/cm-K, the corresponding values are $P_s = 0.56$ W/cm^2, $P_k = 0.034$ W/cm^2, P_r = 17.6 W/cm^2, and $P_c = 4.3$ kW/cm^2. Clearly the overwhelming proportion of the power delivered by the electron beam is conducted through the charge to the hearth. In actuality, power densities of $\sim$ 10 kW/cm^2 are utilized in melting metals, but such levels would damage dielectrics, which require perhaps only 1–2 kW/cm^2. To optimize evaporation conditions, provision is made for altering the size of the focal spot and for electromagnetically scanning the beam.

3.4.3. Deposition Techniques

By now, films of virtually all important materials have been prepared by physical vapor deposition techniques. A practical summary (Refs. 15–17) of vacuum evaporation methods is given in Table 3-3, where recommended heating sources and crucible materials are listed for a number of metals, alloys, oxides, and compounds. Prior to settling on a particular vapor phase deposition process, both PVD and CVD options should be investigated together with the numerous hybrid variants of these methods (see Section 3.8). Paramount attention should be paid to film quality and properties, and the requirements and costs necessary to achieve them. If, after all, vacuum evaporation is selected, modestly equipped laboratories may wish to consider the resistively heated sources before the more costly electron-beam or induction heating alternatives.

3.5. Glow Discharges and Plasmas

3.5.1. Introduction

A perspective of much of the contents of the remainder of the chapter can be had by considering the simplified sputtering system shown in Fig. 3-13a. The target is a plate of the materials to be deposited or the material from which a film is synthesized. Because it is connected to the negative terminal of a dc or RF power supply, the target is also known as the cathode. Typically, several

Table 3-3. Evaporation Characteristics of Materials

Material	Minimum[a] Evap. Temp	State of Evaporation	Recommended Crucible Material	e-beam[b] Deposition Rate (Å/s)	e-beam[b] Power (kW)
Aluminum	1010	Melts	BN	20	5
Aluminum oxide	1325	Semimelts		10	0.5
Antimony	425	Melts	BN, Al_2O_3	50	0.5
Arsenic	210	Sublimes	Al_2O_3	100	0.1
Beryllium	1000	Melts	Graphite, BeO	100	1.5
Beryllium oxide		Melts		40	1.0
Boron	1800	Melts	Graphite, WC	10	1.5
Boron carbide		Semimelts		35	1.0
Cadmium	180	Melts	Al_2O_3, quartz	30	0.3
Cadmium sulfide	250	Sublimes	Graphite	10	0.25
Calcium fluoride		Semimelts		30	0.05
Carbon	2140	Sublimes		30	1.0
Chromium	1157	Sublimes	W	15	0.3
Cobalt	1200	Melts	Al_2O_3, BeO	20	2.0
Copper	1017	Melts	Graphite, Al_2O_3	50	0.2
Gallium	907	Melts	Al_2O_3, graphite		
Germanium	1167	Melts	Graphite	25	3.0
Gold	1132	Melts	Al_2O_3, BN	30	6.0
Indium	742	Melts	Al_2O_3	100	0.1
Iron	1180	Melts	Al_2O_3, BeO	50	2.5
Lead	497	Melts	Al_2O_3	30	0.1
Lithium fluoride	1180	melts (viscous)	Mo, W	10	0.15
Magnesium	327	sublimes	graphite	100	0.04
Magnesium fluoride	1540	semimelts	Al_2O_3	30	0.01
Molybdenum	2117	melts		40	4.0
Nickel	1262	melts	Al_2O_3	25	2.0
Permalloy	1300	melts	Al_2O_3	30	2.0
Platinum	1747	melts	graphite	20	4.0
Silicon	1337	melts	BeO	15	0.15
Silicon dioxide	850	semimelts	Ta	20	0.7

[a]Temperature (°C) at which vapor pressure is 10^{-4} torr.
[b]For 10 kV, copper hearth, source–substrate distance of 40 cm.
Adapted from Refs. 16 and 17.

Table 3-3. *Continued.*

Material	Minimum[a] Evap. Temp	State of Evaporation	Recommended Crucible Material	e-beam[b] Deposition Rate (Å/s)	e-beam[b] Power (kW)
Silicon monoxide	600	sublimes	Ta	20	0.1
Tantalum	2590	semimelts		100	5.0
Tin	997	melts	Al_2O_3, graphite	10	2.0
Titanium	1453	melts		20	1.5
Titanium dioxide	1300	melts	W	10	1.0
Tungsten	2757	melts		20	5.5
Zinc	250	sublimes	Al_2O_3	50	0.25
Zinc selenide	660	sublimes	quartz		
Zinc sulfide	300	sublimes	Mo		
Zirconium	1987	melts	W	20	5.0

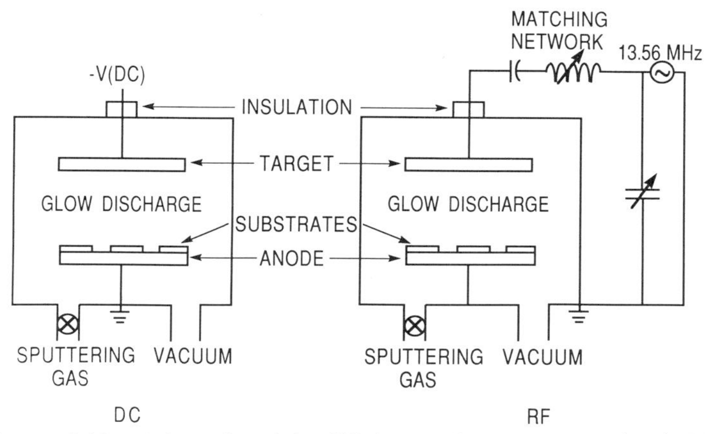

Figure 3-13. Schematics of simplified sputtering systems: (a) dc, (b) RF.

kilovolts are applied to it. The substrate that faces the cathode may be grounded, electrically floating, biased positively or negatively, heated, cooled, or some combination of these. After evacuation of the chamber, a gas, typically argon, is introduced and serves as the medium in which a discharge is initiated and sustained. Gas pressures usually range from a few to 100 mtorr. After a visible glow discharge is maintained between the electrodes, it is observed that a current flows and that a film condenses on the substrate (anode). In vacuum, of course, there is no current flow and no film deposition. Microscopically, positive ions in the discharge strike the cathode plate and eject neutral target atoms through momentum transfer. These atoms enter and pass through the discharge region to eventually deposit on the growing film. In addition, other particles (secondary electrons, desorbed gases, and negative ions) as well as radiation (X-rays and photons) are emitted from the target. In the electric field the negatively charged ions are accelerated toward the substrate to bombard the growing film.

From this simple description, it is quite apparent that compared to the predictable rarefied gas behavior in an evaporation system, the glow discharge is a very busy and not easily modeled environment. Regardless of the type of sputtering, however, roughly similar discharges, electrode configurations, and gas–solid interactions are involved. Therefore, issues common to all glow discharges will be discussed prior to the detailed treatment required of specific sputtering processes and applications.

3.5.2. DC Glow Discharges

The manner in which a glow discharge progresses in a low-pressure gas using a high-impedance dc power supply is as follows (Refs. 4–6). A very small current flows at first due to the small number of initial charge carriers in the system. As the voltage is increased, sufficient energy is imparted to the charged particles to create more carriers. This occurs through ion collisions with the cathode, which release secondary electrons, and by impact ionization of neutral gas atoms. With charge multiplication, the current increases rapidly, but the voltage, limited by the output impedance of the power supply, remains constant. This regime is known as the Townsend discharge. Large numbers of electrons and ions are created through avalanches. Eventually, when enough of the electrons generated produce sufficient ions to regenerate the same number of initial electrons, the discharge becomes self-sustaining. The gas begins to glow now, and the voltage drops, accompanied by a sharp rise in current. At this state "normal glow" occurs. Initially, ion bombardment of the cathode is not uniform but is concentrated near the cathode edges or at other surface irregularities. As more power is applied, the bombardment increasingly spreads over the entire surface until a nearly uniform current density is achieved. A further increase in power results in higher voltage and current density levels. The "abnormal discharge" regime has now been entered, and this is the operative domain for sputtering and other discharge processes (e.g., plasma etching of thin films). At still higher currents, low-voltage arcs propagate.

Adjacent to the cathode there is a highly luminous layer known as the

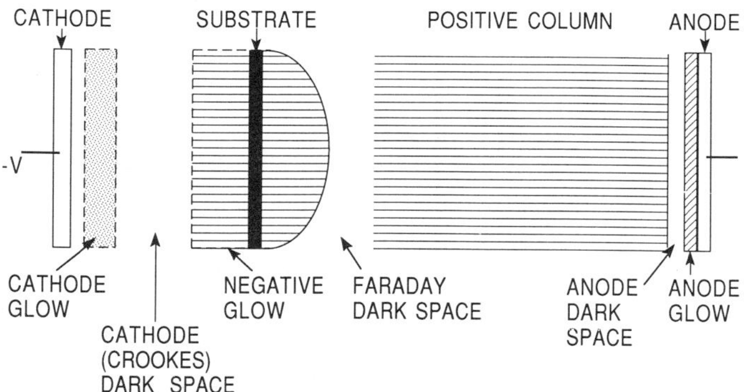

Figure 3-14. Luminous regions of the dc glow discharge.

cathode glow. The light emitted depends on the incident ions and the cathode material. In the cathode glow region, neutralization of the incoming discharge ions and positive cathode ions occurs. Secondary electrons start to accelerate away from the cathode in this area and collide with neutral gas atoms located some distance away from the cathode. In between is the Crookes dark space, a region where nearly all of the applied voltage is dropped. Within the dark space the positive gas ions are accelerated toward the cathode.

The next distinctive region is the "negative glow," where the accelerated electrons acquire enough energy to impact-ionize the neutral gas molecules. Beyond this is the Faraday dark space and finally the positive column. The sequence of these discharge regimes is schematically depicted in Fig. 3-14. The substrate (anode) is placed inside the negative glow, well before the Faraday dark space so that the latter as well as the positive column do not normally appear during sputtering.

3.5.3. Discharge Species (Ref. 6)

A discharge is essentially a plasma—i.e., a partially ionized gas composed of ions, electrons, and neutral species that is electrically neutral when averaged over all the particles contained within. Moreover, the density of charged particles must be large enough compared with the dimensions of the plasma so that significant Coulombic interaction occurs. This interaction enables the charged species to behave in a fluidlike fashion and determines many of the plasma properties. The plasmas used in sputtering are called glow discharges. In them the particle density is low enough, and the fields are sufficiently strong so that neutrals are not in equilibrium with electrons. Typically, the degree of ionization or ratio between numbers of ions and neutrals is about 10^{-4}.

Therefore, at pressures of 10 mtorr, the perfect gas law indicates that about 3×10^9 ions as well as electrons/cm^3 will be present at 25°C. Measurements on glow discharges yield average electron energies of about 2 eV. The effective temperature T associated with a given energy E is simply given by $T = E/k$, where k is the Boltzmann constant. Substituting, we find that electrons have an astoundingly high temperature of some 23,000 K. However, because there are so few of them, their heat content is small and the chamber walls do not heat appreciably. Neutrals and ions are not nearly as energetic; the former have energies of only 0.025 eV (or $T = 290$ K) and the latter, energies of ~ 0.04 eV (or $T = 460$ K). Ions have higher energies than neutrals because they acquire energy from the applied electric field.

Since surfaces (e.g., targets, substrates) are immersed in the plasma, they are bombarded by the species present. The neutral particle flux can be calculated from Eq. 2-8. Charged particle impingement results in an effective current density J_i given by the product of the particle flux and the charge q_i transported. Therefore,

$$J_i = n_i q_i \bar{v}_i / 4, \tag{3-23}$$

where n_i and $\bar{v}_i$ are the specie concentration and mean velocity, respectively. By Eq. 2-3b, $\bar{v}_i = (8kT/\pi m)^{1/2}$. For electrons $m = 9.1 \times 10^{-28}$ g, and if we assume $T = 23{,}000$ K and $n = 10^{10}/\text{cm}^3$, $J_{\text{electron}} \sim 38$ mA/cm^2. The ions, present in the same amounts as electrons, are much heavier and have a lower effective temperature than the electrons. This accounts for their very low velocity compared with that of electrons. For example, $\bar{v}_{\text{ion}} = 5.2 \times 10^4$ cm/sec for Ar ions as well as neutral atoms, whereas for electrons $\bar{v}_{\text{electron}} = 9.5 \times 10^7$ cm/sec. The ion current is correspondingly reduced relative to the electron current by the ratio of these velocities, so $J_{\text{ion}} = 21$ μA/cm^2.

The implication of this simple calculation is that an isolated surface within the plasma charges negatively initially. Subsequently, additional electrons are repelled and positive ions are attracted. Therefore, the surface continues to charge negatively at a decreasing rate until the electron flux equals the ion flux and there is no net steady-state current. We can then expect that both the anode and cathode in the glow discharge will be at a negative potential with respect to the plasma. Of course, the application of the large external negative potential alters the situation, but the voltage distribution in a dc glow discharge under these conditions is shown schematically in Fig. 3-15. A sheath develops around each electrode with a net positive space charge. The lower electron density in the sheath means less ionization and excitation of neutrals. Hence, there is less luminosity there than in the glow itself. Electric fields (derivative of the potential) are restricted to the sheath regions. The plasma itself is not at a

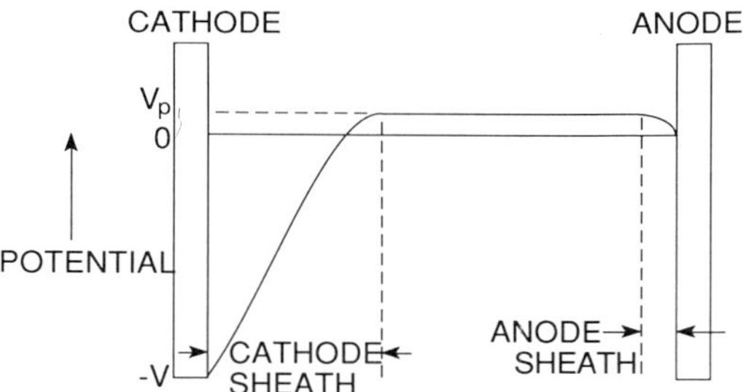

Figure 3-15. Voltage distribution across dc glow discharge.

potential intermediate between that of the electrodes but is typically some 10 V positive with respect to the anode at zero potential. Sheath width dimensions depend on the electron density and temperature. Using the values given earlier for the electrically isolated surface, we find that the sheath width is about 100 μm. It is at the sheath–plasma interface that ions begin to accelerate on their way to the target during sputtering; electrons, however, are repelled from both sheath regions. All of these unusual charge effects stem from the fact that the fundamental plasma particles (electrons and ions) have such different masses and, hence, velocities and energies.

3.5.4. Collision Processes

Collisions between electrons and all the other species (charged or neutral) within the plasma dominate the properties of the glow discharge. Collisions are elastic or inelastic, depending on whether the internal energy of the colliding specie is preserved. In an elastic collision, exemplified by the billiard ball analogy of elementary physics, only kinetic energy is interchanged, and we speak of conservation of momentum and *kinetic* energy of translational motion. The *potential* energy basically resides within the electronic structure of the colliding ions, atoms, and molecules, etc., and increases in potential energy are manifested by ionization or other excitation processes. In an elastic collision, no atomic excitation occurs and potential energy is conserved. This is the reason why only kinetic energy is considered in the calculation. The well-known result for elastic binary collisions is

$$\frac{E_2}{E_1} = \frac{4M_1M_2}{(M_1 + M_2)^2}\cos^2\theta, \tag{3-24}$$

where 1 and 2 refer to the two particles of mass M_i and energy E_i. We assume M_2 is initially stationary and M_1 collides with it at an angle θ defined by the initial trajectory and the line joining their centers at contact. The quantity $4M_1M_2/(M_1+M_2)^2$ is known as the energy transfer function. When $M_1 = M_2$, it has a value of 1; i.e., after collision the initial moving projectile remains stationary, and all of its energy is efficiently transferred to the second particle, which speeds away. When, however, $M_1 \ll M_2$, reflecting, say, a collision between a moving electron and a stationary nitrogen molecule, then the energy transfer function is $\sim 4M_1/M_2$ and has a typical value of $\sim 10^{-4}$. Little kinetic energy is transferred in the collision of the light electron with the massive nitrogen atom.

Now consider inelastic collisions. The change in internal energy, ΔU, of the struck particle must now be accounted for in the condition requiring conservation of total energy. It is left as an exercise for the reader to demonstrate that the maximum fraction of kinetic energy transferred is given by

$$\frac{\Delta U}{(1/2)M_1 v_1^2} = \frac{M_2}{M_1+M_2}\cos^2\theta, \tag{3-25}$$

where v_1 is the initial velocity of particle 1. For the inelastic collision between an electron and nitrogen molecule, $\Delta U/(1/2)M_1 v_1^2 \approx 1$, when $\cos\theta = 1$. Therefore, contrary to an elastic collision, virtually all of an electron's kinetic energy can be transferred to the heavier species in the inelastic collision.

We now turn our attention to a summary of the rich diversity of inelastic collisions and chemical processes that occur in plasmas. It is well beyond the scope of this book to consider anything beyond a cataloging of reactions (Ref. 6). Suffice it to say that these reactions generally enhance film deposition and etching processes.

1. Ionization. The most important process in sustaining the discharge is electron impact ionization. A typical reaction is

$$e^- + Ar^\circ \rightarrow Ar^+ + 2e^-. \tag{3-26a}$$

The two electrons can now ionize more Ar°, etc. By this multiplication mechanism the glow discharge is sustained. The reverse reaction, in which an electron combines with the positive ion to form a neutral, also occurs and is known as recombination.

2. Excitation. In this case the energy of the electron excites quantitized transitions between vibrational, rotational, and electronic states, leaving the molecule in an excited state (denoted by an asterisk). An example is

$$e^- + O_2{}^\circ \rightarrow O_2^* + e^-. \tag{3-26b}$$

3. Dissociation. In dissociation the molecule is broken into smaller atomic or molecular fragments. The products (radicals) are generally much more chemically active than the parent gas molecule and serve to accelerate reactions. Dissociation of CF_4, for example, is relied on in plasma etching or film removal processes; i.e.,

$$e^- + CF_4 \rightarrow e^- + CF_3^* + F^*. \qquad (3\text{-}26c)$$

4. Dissociative Ionization. During dissociation one of the excited species may become ionized; e.g.,

$$e^- + CF_4 \rightarrow 2e^- + CF_3{}^+ + F^*. \qquad (3\text{-}26d)$$

5. Electron Attachment. Here neutral molecules become negative ions after capturing an electron. For example,

$$e^- + SF_6{}^\circ \rightarrow SF_6{}^-. \qquad (3\text{-}26e)$$

6. Dissociative Attachment.

$$e^- + N_2{}^\circ \rightarrow N^+ + N^- + e^-. \qquad (3\text{-}26f)$$

In addition to electron collisions, ion–neutral as well as excited or metastable–excited, and excited atom–neutral collisions occur. Some generic examples of these reactions are as follows:

7. Symmetrical Charge Transfer.

$$A + A^+ \rightarrow A^+ + A.$$

8. Asymmetric Charge Transfer.

$$A + B^+ \rightarrow A^+ + B.$$

9. Metastable–Neutral.

$$A^* + B \rightarrow B^+A + e^-.$$

10. Metastable–Metastable Ionization.

$$A^* + A^* \rightarrow A + A^+ + e^-.$$

Evidence for these uncommon gas-phase species and reactions has accumulated through real-time monitoring of discharges by mass as well as light emission spectroscopy. As a result, a remarkable picture of plasma chemistry has emerged. For example, a noble gas like Ar when ionized loses an electron and resembles Cl electronically as well as chemically. The fact that these species are not in equilibrium confounds the thermodynamic and kinetic descriptions of these reactions.

3.6. Sputtering

3.6.1. Ion – Surface Interactions

Critical to the analysis and design of sputtering processes is an understanding of what happens when ions collide with surfaces (Ref. 18). Some of the

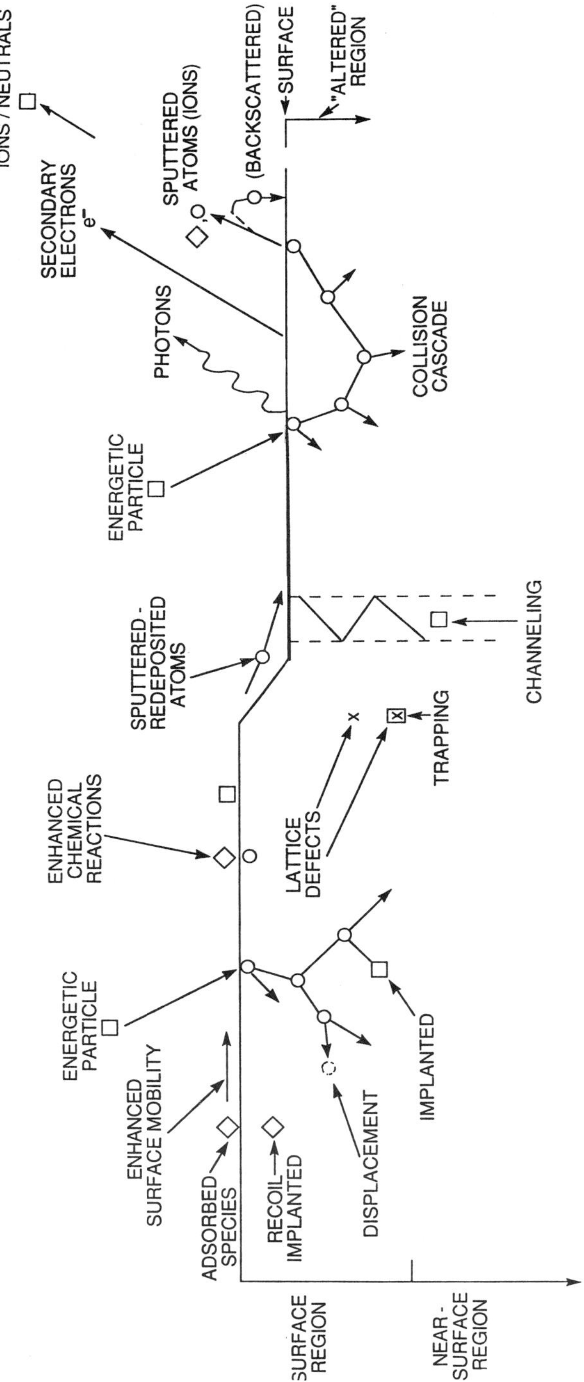

Figure 3-16. Depiction of energetic particle bombardment effects on surfaces and growing films. (From Ref. 18).

interactions that occur are shown schematically in Fig. 3-16. Each depends on the type of ion (mass, charge), the nature of surface atoms involved, and, importantly, on the ion energy. Several of these interactions have been capitalized upon in widely used thin-film processing, deposition, and characterization techniques. For example, ion implantation involves burial of ions under the target surface. Ion implantation of dopants such as B, P, As into Si wafers at ion energies ranging from tens to 100 keV is essential in the fabrication of devices in very large scale integrated (VLSI) circuits. Even higher energies are utilized to implant dopants into GaAs matrices. Ion fluxes, impingement times, and energies must be precisely controlled to yield desired doping levels and profiles. In contrast, ion-scattering spectroscopy techniques require that the incident ions be reemitted for measurement of energy loss. Rutherford backscattering (RBS) is the most important of these analytical methods and typically relies on 2-MeV He^+ ions. Through measurement of the intensity of the scattered ion signal, it is possible to infer the thickness and composition of films as well as subsurface compound layers. This subject is treated at length in Chapter 6. Secondary electrons as well as the products of core electron excitation—Auger electrons, X-rays, etc.—also form part of the complement of particles and radiation leaving the surface.

3.6.2. Sputter Yield

When the ion impact establishes a train of collision events in the target leading to the ejection of a matrix atom, we speak of sputtering. An impressive body of literature has been published indicating that sputtering is related to momentum transfer from energetic particles to the surface atoms of the target. Sputtering has, therefore, been aptly likened to ''atomic pool'' where the ion (cue ball) breaks up the close-packed rack of atoms (billiard balls), scattering some backward (toward the player). Even though atoms of a solid are bound to one another by a complex interatomic potential, whereas billiard balls do not interact, sputtering theory uses the idea of elastic binary collisions. Theoretical expressions for the sputter yield S, the most fundamental parameter characterizing sputtering, include the previously introduced energy transfer function. The sputter yield is defined as the number of atoms or molecules ejected from a target surface per incident ion and is a measure of the efficiency of sputtering.

Intuitively we expect S to be proportional to a product of the following factors (Ref. 19):

1. The number of atoms displaced toward the surface per primary collision. This term is given by $\overline{E}/2E_t$, where $\overline{E}$ is the mean energy of the struck

target atom and E_t is the threshold energy required to displace an atom. The factor of 2 is necessary because only half of the displaced atoms move toward the surface. The quantity $\bar{E}$ may be taken as an average of E_2, the kinetic energy transferred to the target atom, and E_t; i.e.,

$$\bar{E} = \frac{E_2 + E_t}{2}, \qquad \text{where } E_2 = \frac{4M_1M_2}{(M_1 + M_2)^2} E_1.$$

2. The number of atomic layers that contain these atoms and contribute to sputtering. Statistics show that the number of collisions required to slow an atom of energy $\bar{E}$ to E_b, the surface binding energy, is

$$N = \frac{\ln \bar{E}/E_b}{\ln 2}. \tag{3-27}$$

By a random walk model, the average number of contributing atomic layers is $1 + N^{1/2}$.
3. The number of target atoms per unit area n_A.
4. The cross section $\sigma_0 = \pi a^2$, where a is related to the Bohr radius of the atom a_b, and the atomic numbers Z_1, Z_2 of the incident ion and sputtered atom respectively; i.e.,

$$a = \frac{a_b}{\sqrt{Z_1^{2/3} + Z_2^{2/3}}}.$$

Combining terms gives

$$S = \frac{\bar{E}}{E_t}\left\{1 + \left(\frac{\ln \bar{E}/E_b}{\ln 2}\right)^{1/2}\right\}\sigma_0 n_A. \tag{3-28}$$

As an example consider the sputtering of Cu with 1-keV Ar ions. The calculated value of S will depend strongly on E_t, and for Ar incident on Cu experiment suggests that $E_t = 17$ eV. For Cu, $M_2 = 63.5$, $Z_2 = 29$, $a_b = 1.17$ Å, $n_A = 1.93 \times 10^{15}$ atoms/cm^2, and $E_b = 3.5$ eV. For Ar, $M_1 = 39.9$ and $Z_1 = 18$. Substitution shows that $\bar{E} = 483$ eV and $S = 2.6$. This calculated value compares with the measured sputter yield of 2.85, as indicated by the data of Table 3-4.

The currently accepted theory for the sputtering yield from collision cascades is due to Sigmund (Ref. 20) and predicts that

$$S = \frac{3\alpha}{4\pi^2}\frac{4M_1M_2}{(M_1 + M_2)^2}\frac{E_1}{E_b} \qquad (E_1 < 1\text{ keV}) \tag{3-29}$$

Table 3-4. Sputtering Yield Data for Metals (atoms/ion)

Sputtering Gas Energy (keV)	He 0.5	Ne 0.5	Ar 0.5	Kr 0.5	Xe 0.5	Ar 1.0	Ar Threshold Voltage (eV)
Ag	0.20	1.77	3.12	3.27	3.32	3.8	15
Al	0.16	0.73	1.05	0.96	0.82	1.0	13
Au	0.07	1.08	2.40	3.06	3.01	3.6	20
Be	0.24	0.42	0.51	0.48	0.35		15
C	0.07	—	0.12	0.13	0.17		
Co	0.13	0.90	1.22	1.08	1.08		25
Cu	0.24	1.80	2.35	2.35	2.05	2.85	17
Fe	0.15	0.88	1.10	1.07	1.00	1.3	20
Ge	0.08	0.68	1.1	1.12	1.04		25
Mo	0.03	0.48	0.80	0.87	0.87	1.13	24
Ni	0.16	1.10	1.45	1.30	1.22	2.2	21
Pt	0.03	0.63	1.40	1.82	1.93		25
Si	0.13	0.48	0.50	0.50	0.42	0.6	
Ta	0.01	0.28	0.57	0.87	0.88		26
Ti	0.07	0.43	0.51	0.48	0.43		20
W	0.01	0.28	0.57	0.91	1.01		33

From Refs. 4 and 6.

and

$$S = 3.56\alpha \frac{Z_1 Z_2}{Z_1^{2/3} + Z_2^{2/3}} \left(\frac{M_1}{M_1 + M_2} \right) \frac{S_n(E)}{E_b} \qquad (E_1 > 1 \text{ keV}). \quad (3\text{-}30)$$

These equations depend on two complex quantities, α and $S_n(E)$. The parameter α, a measure of the efficiency of momentum transfer in collisions, increases monotonically from 0.17 to 1.4 as M_1/M_2 ranges from 0.1 to 10. The reduced stopping power, $S_n(E)$, is a measure of the energy loss per unit length due to nuclear collisions. It is a function of the energy as well as masses and atomic numbers of the atoms involved. At high energy, S is relatively constant because $S_n(E)$ tends to be independent of energy.

The sputter yields for a number of metals are entered in Table 3-4. Values for two different energies (0.5 keV and 1.0 keV) as well as five different inert gases (He, Ne, Ar, Kr, and Xe) are listed. It is apparent that S values typically span a range from 0.01 to 4 and increase with the mass and energy of the sputtering gas.

3.6.3. Sputtering of Alloys

In contrast to the fractionation of alloy melts during evaporation, with subsequent loss of deposit stoichiometry, sputtering allows for the deposition of films having the same composition as the target source. This is the primary reason for the widespread use of sputtering to deposit metal alloy films. We note, however, that each alloy component evaporates with a different vapor pressure and sputters with a different yield. Why then is film stoichiometry maintained during sputtering and not during evaporation? One reason is the generally much greater disparity in vapor pressures compared with the difference in sputter yields under comparable deposition conditions. Second, and perhaps more significant, melts homogenize readily due to rapid atomic diffusion and convection effects in the liquid phase; during sputtering, however, minimal solid-state diffusion enables the maintenance of the required altered target surface composition.

Consider now sputtering effects (Ref. 5) on a binary alloy target surface containing a number of A atoms (n_A) and B atoms (n_B), such that the total number is $n = n_A + n_B$. The target concentrations are $C_A = n_A/n$ and $C_B = n_B/n$, with sputter yields S_A and S_B. Initially, the ratio of the sputtered atom fluxes (ψ) is given by

$$\frac{\psi_A}{\psi_B} = \frac{S_A C_A}{S_B C_B}. \tag{3-31}$$

If n_g sputtering gas atoms impinge on the target, the total number of A and B atoms ejected are $n_g C_A S_A$ and $n_g C_B S_B$, respectively. Therefore, the target surface concentration ratio is modified to

$$\frac{C'_A}{C'_B} = \frac{C_A}{C_B}\frac{(1 - n_g S_A/n)}{(1 - n_g S_B/n)} \tag{3-32}$$

instead of C_A/C_B. If $S_A > S_B$, the surface is enriched in B atoms, which now begin to sputter in greater profusion; i.e.,

$$\frac{\psi'_A}{\psi'_B} = \frac{S_A C'_A}{S_B C'_B} = \frac{S_A C_A}{S_B C_B}\frac{(1 - n_g S_A/n)}{(1 - n_g S_B/n)}. \tag{3-33}$$

Progressive change in the target surface composition alters the sputtered flux ratio to the point where it is equal to C_A/C_B, which is the same as the original target composition. Simultaneously, the target surface reaches the value $C'_A/C'_B = C_A S_B / C_B S_A$, which is maintained thereafter. A steady-state transfer of atoms from the bulk target to the plasma ensues, resulting in stoichiometric film deposition. This state of affairs persists until the target is consumed. Conditioning of the target by sputtering a few hundred atom layers is required to reach steady-state conditions. As an explicit example, consider the deposition of Permalloy films having atomic ratio 80 Ni–20 Fe from a target of this same composition. For 1-keV Ar, the sputter yields are $S_{Ni} = 2.2$ and

$S_{Fe} = 1.3$. The target surface composition is altered in the steady state to $C'_{Ni}/C'_{Fe} = 80(1.3)/20(2.2) = 2.36$, which is equivalent to 70.2 Ni and 29.8 Fe.

3.6.4. Thermal History of the Substrate (Ref. 21)

One of the important issues related to sputtering is the temperature rise in the substrate during film deposition. Sputtered atoms that impinge on the substrate are far more energetic than similar atoms emanating from an evaporation source. During condensation, this energy must be dissipated by the substrate, or else it may heat excessively, to the detriment of the quality of the deposited film. To address the question of substrate heating, we start with an equation describing the heat power balance, namely,

$$\rho c d(dT/dt) = P - L. \tag{3-34}$$

The term on the left is the net thermal energy per unit area per unit time (in typical units of watts/cm^2) retained by a substrate whose density, heat capacity, effective thickness, and rate of temperature rise are given by ρ, c, d, and dT/dt, respectively.

The incident power flux P has three important components:

1. Heat of condensation of atoms, ΔH_c (eV/atom).
2. Average kinetic energy of incident adatoms, $\bar{E}_k$ (eV/atom).
3. Plasma heating from bombarding neutrals and electrons. The plasma energy is assumed to be E_p (eV/atom).

Table 3-5 contains values for these three energies during magnetron sputtering at 1 keV (Ref. 22). For a deposition rate $\dot{d}$ (Å/min),

$$P = \frac{2.67 \times 10^{-29}\dot{d}(\Delta H_c + \bar{E}_k + E_p)}{\Omega} \text{ watts/cm}^2, \tag{3-35}$$

where Ω is the condensate atomic volume in cm^3/atom. The L term represents the heat loss to the substrate holder by conduction or to cooler surfaces in the chamber by radiation. For the moment, let us neglect L and calculate the temperature rise of a thermally isolated substrate. Substituting Eq. 3-35 into Eq. 3-34 and integrating, we obtain

$$T(t) = \frac{2.67 \times 10^{-29}\dot{d}(\Delta H_c + \bar{E}_k + E_p)t}{\rho c d\Omega}. \tag{3-36}$$

Consider Al deposited at a rate of 10,000 Å/min on a Si wafer 0.050 cm thick. For Al, $\Delta H_c + \bar{E}_k + E_p = 13$ eV/atom and $\Omega = 16 \times 10^{-24}$

Table 3-5. Energies Associated with Magnetron Sputttering

Metal	Heat of Condensation (eV/atom)	Kinetic Energy of Sputtered Atoms (eV/atom)	Plasma (eV/atom)	Estimated Flux (eV/atom)	Measured Flux (eV/atom)
Al	3.33	6	4	13	13
Ti	4.86	8	9	22	20
V	5.29	7	8	20	19
Cr	4.11	8	4	16	20
Fe	2.26	9	4	15	21
Ni	4.45	11	4	19	15
Cu	3.50	6	2	12	17
Zr	6.34	13	7	26	41
Nb	6.50	13	8	28	28
Mo	6.88	13	6	26	47
Rh	5.60	13	4	23	43
Cd	1.16	4	1	6	8
In	2.52	4	2	9	20
Hf	6.33	20	7	33	63
Ta	8.10	21	9	38	68
W	8.80	22	9	40	73
Au	3.92	13	2	19	23

From Ref. 22.

cm^3/atom, and for Si, $\rho = 2.3$ g/cm^3 and $c = 0.7$ J/g-°C. In depositing a film 1 μm thick, $t = 60$ sec, and the temperature rise of the substrate is calculated to be 162 °C. Higher deposition rates and substrates of smaller thermal mass will result in proportionately higher temperatures.

The temperature will not reach values predicted by Eq. 3-36 because of L. For simplicity we only consider heat loss by radiation. If the front and rear substrate surfaces radiate to identical temperature sinks at T_0 with equal emissivity ε, then $L = 2\sigma\varepsilon(\mathrm{T}^4 - \mathrm{T}_0^4)$, where σ, the Stefan–Boltzmann constant, equals 5.67×10^{-12} W/cm^2-K^4. Substitution in Eq. 3-34 and direct integration, after separation of variables, yields

$$t = \frac{1}{2\alpha^{3/2}\beta^{1/2}}\left[\tan^{-1}\sqrt{\frac{\beta}{\alpha}}\,T - \tan^{-1}\sqrt{\frac{\beta}{\alpha}}\,T_0 + \frac{1}{2}\ln\left(\frac{\sqrt{\alpha} + \sqrt{\beta}\,T}{\sqrt{\alpha} + \sqrt{\beta}\,T_0}\cdot\frac{\sqrt{\alpha} - \sqrt{\beta}\,T_0}{\sqrt{\alpha} - \sqrt{\beta}\,T}\right)\right], \qquad (3\text{-}37)$$

where $\alpha = \sqrt{(2\sigma\varepsilon T_0^4 + P)/\rho c d}$ and $\beta = \sqrt{2\sigma\varepsilon/\rho c d}$.

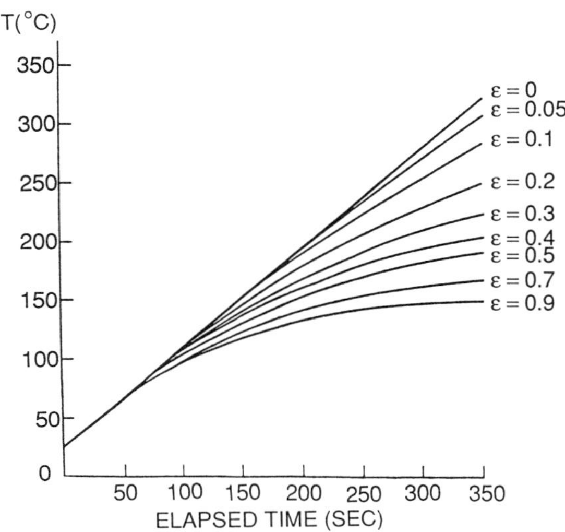

Figure 3-17. Temperature–time response for film–substrate combination under the influence of a power flux of 250 mW/cm^2. Deposition rate ~ 1 μm/min. (Reprinted with permission from Cowan Publishing Co., from L. T. Lamont, *Solid State Technology* **22(9)**, 107, 1979).

Equation 3-37 expresses the time it takes for a substrate to reach temperature T starting from T_0, assuming radiation cooling. For short times Eq. 3-36 holds, whereas for longer times the temperature equilibrates to a radiation-limited value dependent on the incident power flux and substrate emissivity. Sputter deposition for most materials at the relatively high rate of 1 μm/min generates a typical substrate power flux of ~ 250 mW/cm^2. The predicted rate of film heating is shown in Fig. 3-17. If substrate bias (Section 3.7.5.) is also applied, temperature increases can be quite substantial. In Al films, temperatures in excess of 200 °C have been measured. This partially accounts for enhanced atom mobility and step coverage during application of substrate bias.

Finally, we briefly consider sputter etching, a process that occurs at the target during sputtering. Films utilized in microelectronic applications must be etched in order to remove material and expose patterned regions for subsequent film deposition or doping processes. In the VLSI regime etching is carried out in plasmas and reactive gas environments, where the films involved essentially behave like sputtering targets. At the same power level, sputter etching rates tend to be lower, by more than an order of magnitude, than film deposition rates. This means that etching requires high power levels that frequently range from 1 to 2 W/cm^2. The combination of high power levels and long etching times cause substrates to reach high radiation-limited temperatures. In Al, for example, temperature increases well in excess of 300 °C have been measured during etching.

3.7. Sputtering Processes

For convenience we divide sputtering processes into four cateogories: (1) dc, (2) RF, (3) magnetron, (4) reactive. We recognize, however, that there are important variants within each category (e.g., dc bias) and even hybrids between categories (e.g., reactive RF). Targets of virtually all important materials are commercially available for use in these sputtering processes. A selected number of target compositions representing the important classes of solids, together with typical sputtering applications for each are listed in Table 3-6.

In general, the metal and alloy targets are fabricated by melting either in vacuum or under protective atmospheres, followed by thermomechanical processing. Refractory alloy targets (e.g., Ti–W) are hot-pressed via the powder metallurgy route. Similarly, nonmetallic targets are generally prepared by hot-pressing of powders. The elemental and metal targets tend to have purities of 99.99% or better, whereas those of the nonmetals are generally less pure, with a typical upper purity limit of 99.9%. In addition, less than theoretical densities are achieved during powder processing. These metallurgical realities are sometimes reflected in emission of particulates, release of trapped gases, nonuniform target erosion, and deposited films of inferior quality. Targets are available in a variety of shapes (e.g., disks, toroids, plates, etc.) and sizes. Prior to use, they must be bonded to a cooled backing plate to avoid thermal cracking. Metal-filled epoxy cements of high thermal conductivity are employed for this purpose.

3.7.1. DC Sputtering

Virtually everything mentioned in the chapter so far has dealt with dc sputtering, also known as diode or cathodic sputtering. There is no need to further discuss the system configuration (Fig. 3-13), the discharge environment (Section 3.5), the ion–surface interactions (Section 3.6.1), or intrinsic sputter yields (Section 3.6.2). It is worthwhile, however, to note how the relative film deposition rate depends on the sputtering pressure and current variables. At low pressures, the cathode sheath is wide and ions are produced far from the target; their chances of being lost to the walls are great. The mean-free

Table 3-6. Sputtering Targets

Material	Application
1. *Metals*	
Aluminum	Metallization for integrated circuits, front surface mirrors
Chromium	Adhesion layers, resistor films (with SiO lithography master blanks
Germanium	Infrared filters
Gold	Contacts, reflecting films
Iron, nickel	Ferromagnetic films
Palladium, platinum	Contacts
Silver	Reflective films, contacts
Tantalum	Thin-film capacitors
Tungsten	Contacts
2. *Alloys*	
Al–Cu, Al–Si, Al–Cu–Si	Metallization for integrated circuits
Co–Fe, Co–Ni, Fe–Tb, Fe–Ni, Co–Ni–Cr	Ferromagnetic films
Ni–Cr	Resistors
Ti–W	Diffusion barriers in integrated circuits
Gd–Co	Magnetic bubble memory devices
3. *Oxides*	
Al_2O_3	Insulation, protective films for mirrors
$BaTiO_3$, $PbTiO_3$	Thin-film capacitors
CeO_2	Antireflection coatings
In_2O_3–SnO_2	Transparent conductors
$LiNbO_3$	Piezoelectric films
SiO_2	Insulation
SiO	Protective films for mirrors, infrared filters
Ta_2O_5, TiO_2, ZrO_2, HfO_2, MgO	Dielectric films for multilayer optical coatings
Yttrium aluminum oxide (YAG), yttrium iron oxide (YIG), $Gd_3Ga_5O_{12}$	Magnetic bubble memory devices
YVO_3–Eu_2O_3	Phosphorescent coating on special currency papers
$Cu_3Ba_2YO_7$	High temperature superconductors
4. *Fluorides*	
CaF_2, CeF_3, MgF_2, ThF_4, Na_3AlF_6 (cryolite)	Dielectric films for multilayer optical coatings (antireflection coatings, filters, etc.)
5. *Borides*	
TiB_2, ZrB_2	Hard, wear-resistant coatings
LaB_6	Thermionic emitters
6. *Carbides*	
SiC	High-temperature semiconduction
TiC, TaC, WC	Hard, wear-resistant coatings

Table 3-6. *Continued.*

Material	Application
7. *Nitrides*	
Si_3N_4	Insulation, diffusion barriers
TaN	Thin-film resistors
TiN	Hard coatings
8. *Silicides*	
$MoSi_2$, $TaSi_2$, $TiSi_2$, WSi_2	Contacts, diffusion barriers in integrated circuits
9. *Sulfides*	
CdS	Photoconductive films
MoS_2, TaS_2	Lubricant films for bearings and moving parts
ZnS	Multilayer optical coatings
10. *Selenides, tellurides*	
CdSe, PbSe, CdTe	Photoconductive films
ZnSe, PbTe	Optical coatings
MoTe, MoSe	Lubricants

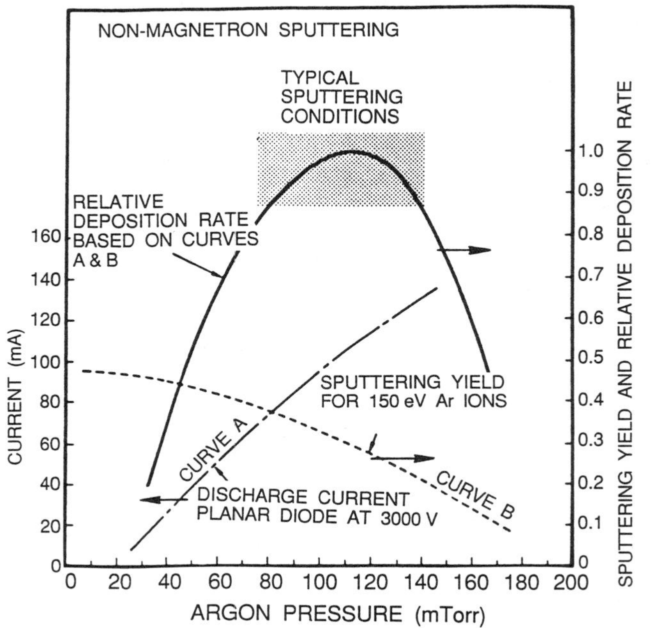

Figure 3-18. Influence of working pressure and current on deposition rate for nonmagnetron sputtering. (From Ref. 23).

electron path between collisions is large, and electrons collected by the anode are not replenished by ion-impact-induced cathode secondary emission. Therefore, ionization efficiencies are low, and self-sustained discharges cannot be maintained below about 10 mtorr. As the pressure is increased at a fixed voltage, the electron mean-free path is decreased, more ions are generated, and larger currents flow. But if the pressure is too high, the sputtered atoms undergo increased collisional scattering and are not efficiently deposited. The trade-offs in these opposing trends are shown in Fig. 3-18, and optimum operating conditions are shaded in. In general, the deposition rate is proportional to the power consumed, or to the square of the current density, and inversely dependent on the electrode spacing.

3.7.2. RF Sputtering

RF sputtering was invented as a means of depositing insulating thin films. Suppose we wish to produce thin SiO_2 films and attempt to use a quartz disk 0.1 cm thick as the target in a conventional dc sputtering system. For quartz $\rho \approx 10^{16}$ Ω-cm. To draw a current density J of 1 mA/cm^2, the cathode needs a voltage $V = 0.1\rho J$. Substitution gives an impossibly high value of 10^{12} V, which indicates why dc sputtering will not work. If we set a convenient level of $V = 100$ V, it means that a target with a resistivity exceeding 10^6 Ω-cm could not be dc-sputtered.

Now consider what happens when an ac signal is applied to the electrodes. Below about 50 kHz, ions are sufficiently mobile to establish a complete discharge at each electrode on each half-cycle. Direct current sputtering conditions essentially prevail at both electrodes, which alternately behave as cathodes and anodes. Above 50 kHz two important effects occur. Electrons oscillating in the glow region acquire enough energy to cause ionizing collisions, reducing the need for secondary electrons to sustain the discharge. Secondly, RF voltages can be coupled through any kind of impedance so that the electrodes need not be conductors. This makes it possible to sputter any material irrespective of its resistivity. Typical RF frequencies employed range from 5 to 30 MHz. However, 13.56 MHz has been reserved for plasma processing by the Federal Communications Commission and is widely used.

RF sputtering essentially works because the target *self*-biases to a negative potential. Once this happens, it behaves like a dc target where positive ion bombardment sputters away atoms for subsequent deposition. Negative target bias is a consequence of the fact that electrons are considerably more mobile than ions and have little difficulty in following the periodic change in the electric field. In Fig. 3-13b we depict an RF sputtering system schematically,

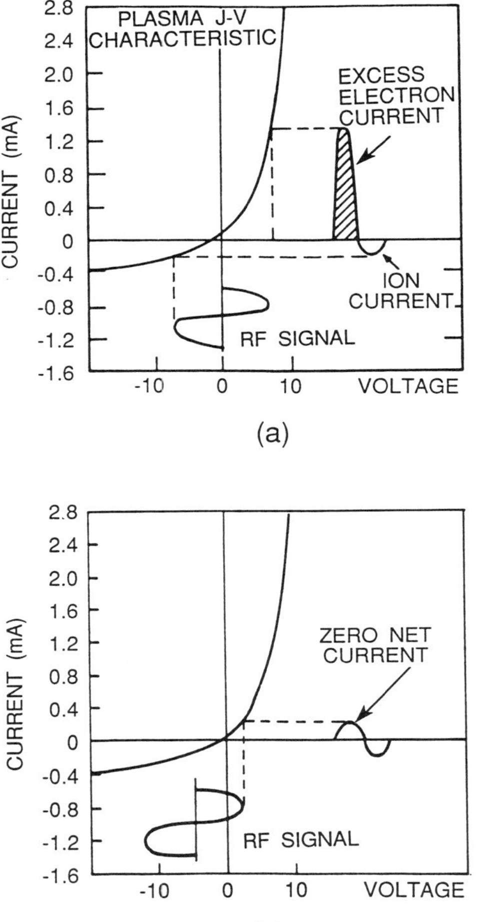

Figure 3-19. Formation of pulsating negative sheath on capacitively coupled cathode of RF discharge (a) Net current/zero self-bias voltage. (b) Zero current/nonzero self-bias voltage. (From Ref. 4).

where the target is capacitively coupled to the RF generator. The disparity in electron and ion mobilities means that isolated positively charged electrodes draw more electron current than comparably isolated negatively charged electrodes draw positive ion current. For this reason the discharge current–voltage characteristics are asymmetric and resemble those of a leaky rectifier or diode. This is indicated in Fig. 3-19, and even though it applies to a dc discharge, it helps to explain the concept of self-bias at RF electrodes.

As the pulsating RF signal is applied to the target, a large initial electron current is drawn during the positive half of the cycle. However, only a small ion current flows during the second half of the cycle. This would enable a *net* current averaged over a complete cycle to be different from zero; but this cannot happen because no charge can be transferred through the capacitor. Therefore, the operating point on the characteristic shifts to a negative voltage —the target bias—and no *net* current flows.

The astute reader will realize that since ac electricity is involved, both electrodes should sputter. This presents a potential problem because the resultant film may be contaminated as a consequence. For sputtering from only one electrode, the sputter target must be an insulator and be capacitively coupled to the RF generator. The equivalent circuit of the sputtering system can be thought of as two series capacitors—one at the target sheath region, the other at the substrate—with the applied voltage divided between them. Since capacitive reactance is inversely proportional to the capacitance or area, more voltage will be dropped across the capacitor of a smaller surface area. Therefore, for efficient sputtering the area of the target electrode should be small compared with the total area of the other, or *directly* coupled, electrode. In practice, this electrode consists of the substrate stage and system ground, but it also includes baseplates, chamber walls, etc. It has been shown that the ratio of the voltage across the sheath at the (small) capacitively coupled electrode (V_c) to that across the (large) directly coupled electrode V_d is given by (Ref. 24)

$$V_c / V_d = (A_d / A_c)^4, \tag{3-38}$$

where A_c and A_d are the respective electrode areas. In essence, a steady-state voltage distribution prevails across the system similar to that shown in Fig. 3-15. The fourth-power dependence means that a large value of A_d is very effective in raising the target sheath potential while minimizing ion bombardment of grounded fixtures.

3.7.3. Magnetron Sputtering

3.7.3.1. Electron Motion in Parallel Electric and Magnetic Fields. Let us now examine what happens when a magnetic field of strength **B** is superimposed on the electric field $\mathscr{E}$ between the target and substrate. Such a situation arises in magnetron sputtering as well as in certain plasma etching configurations. Electrons within the dual field environment experience the well-known Lorentz force in addition to electric field force, i.e.,

$$\mathbf{F} = \frac{m\,dv}{dt} = -q(\mathscr{E} + \mathbf{v} \times \mathbf{B}), \tag{3-39}$$

where q, m and v are the electron charge, mass, and velocity, respectively.

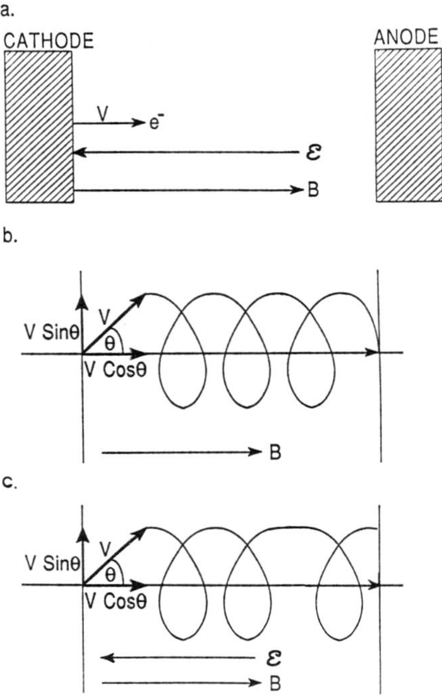

Figure 3-20. Effect of $\mathscr{E}$ and **B** on electron motion. (a) Linear electron trajectory when $\mathscr{E} \parallel \mathbf{B}$ ($\theta = 0$); (b) helical orbit of constant pitch when $\mathbf{B} \neq 0$, $\mathscr{E} = 0$, ($\theta \neq 0$); (c) helical orbit of variable pitch when $\mathscr{E} \parallel \mathbf{B}$ ($\theta \neq 0$).

First consider the case where **B** and $\mathscr{E}$ are parallel as shown in Fig. 3-20a. When electrons are emitted exactly normal to the target surface and parallel to both fields, then $\mathbf{v} \times \mathbf{B}$ vanishes; electrons are only influenced by the $\mathscr{E}$ field, which accelerates them toward the anode. Next consider the case where the $\mathscr{E}$ field is neglected but **B** is still applied as shown in Fig. 3-20b. If an electron is launched from the cathode with velocity v at angle θ with respect to **B**, it experiences a force $qvB \sin\theta$ in a direction perpendicular to **B**. The electron now orbits in a circular motion with a radius r that is determined by a balance of the centrifugal $(m(v \sin\theta)^2/r)$ and Lorentz forces involved, i.e., $r = mv \sin\theta \,/\, qB$. The electron motion is helical; in corkscrew fashion it spirals down the axis of the discharge with constant velocity $v \cos\theta$. If the magnetic field were not present, such off-axis electrons would tend to migrate out of the discharge and be lost at the walls.

The case where electrons are launched at an angle to parallel, uniform $\mathscr{E}$ and **B** fields is somewhat more complex. Corkscrew motion with constant radius occurs, but because of electron acceleration in the $\mathscr{E}$ field, the pitch of the helix lengthens with time (Fig. 3-20c). Time varying $\mathscr{E}$ fields complicate matters further and electron spirals of variable radius can occur. Clearly, magnetic fields prolong the electron residence time in the plasma and thus enhance the probability of ion collisions. This leads to larger discharge currents and increased sputter deposition rates. Comparable discharges in a simple diode-sputtering configuration operate at higher currents and pressures. Therefore, applied magnetic fields have the desirable effect of reducing electron bombardment of substrates and extending the operating vacuum range.

3.7.3.2. Perpendicular Electric and Magnetic Fields. In magnetrons, electrons ideally do not even reach the anode but are trapped near the target, enhancing the ionizing efficiency there. This is accomplished by employing a magnetic field oriented parallel to the target and perpendicular to the electric field, as shown schematically in Fig. 3-21. Practically, this is achieved by placing bar or horseshoe magnets behind the target. Therefore, the magnetic field lines first emanate normal to the target, then bend with a component *parallel* to the target surface (this is the magnetron component) and finally return, completing the magnetic circuit. Electrons emitted from the cathode are initially accelerated toward the anode, executing a helical motion in the process; but when they encounter the region of the parallel magnetic field, they are bent in an orbit back to the target in very much the same way that electrons are deflected toward the hearth in an e-gun evaporator. By solving the coupled differential equations resulting from the three components of Eq. 3-39, we readily see that the parameric equations of motion are

$$y = \frac{q\mathscr{E}}{m\omega_c^2}(1 - \cos \omega_c t), \tag{3-40a}$$

$$x = \frac{\mathscr{E} t}{B}\left(1 - \frac{\sin \omega_c t}{\omega_c t}\right), \tag{3-40b}$$

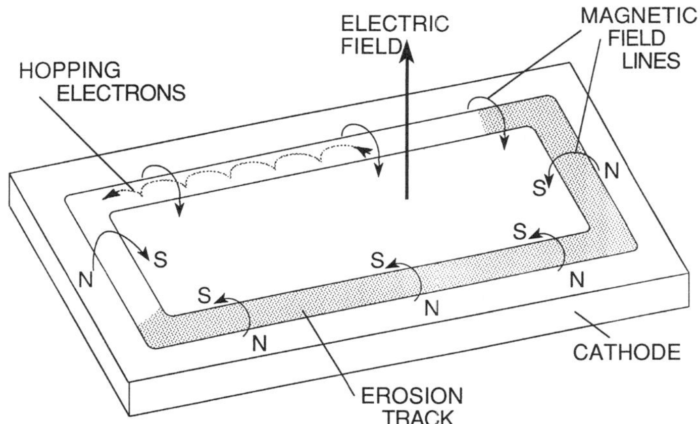

Figure 3-21. Applied fields and electron motion in the planar magnetron.

where y and x are the distances above and along the target, and $\omega_c = qB/m$. These equations describe a cycloidal motion that the electrons execute within the cathode dark space where both fields are present. If, however, electrons stray into the negative glow region where the $\mathscr{E}$ field is small, the electrons describe a circular motion before collisions may drive them back into the dark space or forward toward the anode. By suitable orientation of target magnets, a "race track" can be defined where the electrons hop around at high speed. Target erosion by sputtering occurs within this track because ionization of the working gas is most intense above it.

Magnetron sputtering is presently the most widely commercially practiced sputtering method. The chief reason for its success is the high deposition rates achieved (e.g., up to 1 μm/min for Al). These are typically an order of magnitude higher than rates attained by conventional sputtering techniques. Popular sputtering configurations utilize planar, toroidal (rectangular cross section), and toroidal-conical (trapezoidal cross section) targets (i.e., the S-gun). In commercial planar magnetron sputtering systems, the substrate plane translates past the parallel facing target through interlocked vacuum chambers to allow for semicontinuous coating operations. The circular (toridal-conical) target, on the other hand, is positioned centrally within the chamber, creating a deposition geometry approximating that of the analogous planar (ring) evaporation source. In this manner wafers on a planetary substrate holder can be coated as uniformly as with e-gun sources.

3.7.4. Reactive Sputtering

In reactive sputtering, thin films of compounds are deposited on substrates by sputtering from metallic targets in the presence of a reactive gas, usually mixed with the inert working gas (invariably Ar). The most common compounds reactively sputtered (and the reactive gases employed) are briefly listed:

1. Oxides (oxygen)—Al_2O_3, In_2O_3, SnO_2, SiO_2, Ta_2O_5
2. Nitrides (nitrogen, ammonia)—TaN, TiN, AlN, Si_3N_4
3. Carbides (methane, acetylene, propane)—TiC, WC, SiC
4. Sulfides (H_2S)—CdS, CuS, ZnS
5. Oxycarbides and oxynitrides of Ti, Ta, Al, and Si

Irrespective of which of these materials is considered, during reactive sputtering the resulting film is either a solid solution alloy of the target metal doped with the reactive element (e.g., $TaN_{0.01}$), a compound (e.g., TiN), or some mixture of the two. Westwood (Ref. 25) has provided a useful way to visualize the conditions required to yield alloys or compounds. These two

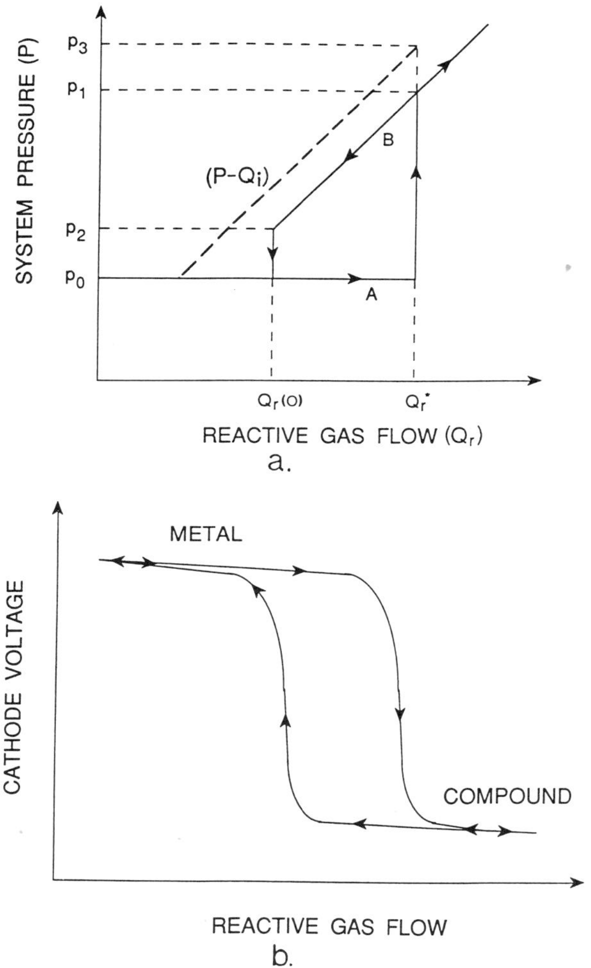

Figure 3-22. (a) Generic hysteresis curve for system pressure vs. reactive gas flow rate during reactive sputtering. Dotted line represents behavior with inert gas. (From Ref. 25). (b) Hysteresis curve of cathode voltage vs. reactive gas flow rate at constant discharge current.

regimes are distinguished in Fig. 3-22a, illustrating the generic hysteresis curve for the total system pressure (P) as a function of the flow rate of *reactive* gas (Q_r) into the system. First, however, consider the dotted line representing the variation of P with flow rate of an *inert* sputtering gas (Q_i). Clearly, as Q_i increases, P increases because of the constant pumping speed (see Eq. 2-16). An example of this characteristic occurs during Ar gas sputtering of Ta. Now consider what happens when reactive N_2 gas is introduced into the system. As Q_r increases from $Q_r(0)$, the system pressure essentially remains at the initial value P_0 because N_2 reacts with Ta and is removed from the gas phase. But beyond a critical flow rate Q_r^*, the system pressure jumps to the new value P_1. If no reactive sputtering took place, P would be somewhat higher (i.e., P_3). Once the equilibrium value of P is established, subsequent changes in Q_r cause P to increase or decrease linearly as shown. As Q_r decreases sufficiently, P again reaches the initial pressure.

The hysteresis behavior represents two stable states of the system with a rapid transition between them. In state A there is little change in pressure, while for state B the pressure varies linearly with Q_r. Clearly, all of the reactive gas is incorporated into the deposited film in state A—the doped metal and the atomic ratio of reactive gas dopant to sputtered metal increases with Q_r. The transition from state A to state B is triggered by compound formation on the metal target. Since ion-induced secondary electron emission is usually much higher for compounds than for metals, Ohm's law suggests that the *plasma* impedance is effectively lower in state B than in state A. This effect is reflected in the hysteresis of the target voltage with reactive gas flow rate, as schematically depicted in Fig. 3-22b.

The choice of whether to employ compound targets and sputter directly or sputter reactively is not always clear. If reactive sputtering is selected, then there is the option of using simple dc diode, RF, or magnetron configurations. Many considerations go into making these choices, and we will address some of them in turn.

3.7.4.1. Target Purity. It is easier to manufacture high-purity metal targets than to make high-purity compound targets. Since hot pressed and sintered compound powders cannot be consolidated to theoretical bulk densities, incorporation of gases, porosity, and impurities is unavoidable. Film purity using elemental targets is high, particularly since high-purity reactive gases are commercially available.

3.7.4.2. Deposition Rates. Sputter rates of metals drop dramatically when compounds form on the targets. Decreases in deposition rate well in excess of 50% occur because of the lower sputter yield of compounds relative to metals. The effect is very much dependent on reactive gas pressure. In dc discharges, sputtering is effectively halted at very high gas pressures, but the limits are also influenced by the applied power. Conditioning of the target in pure Ar is required to restore the pure metal surface and desired deposition rates. Where high deposition rates are a necessity, the reactive sputtering mode of choice is either dc or RF magnetron.

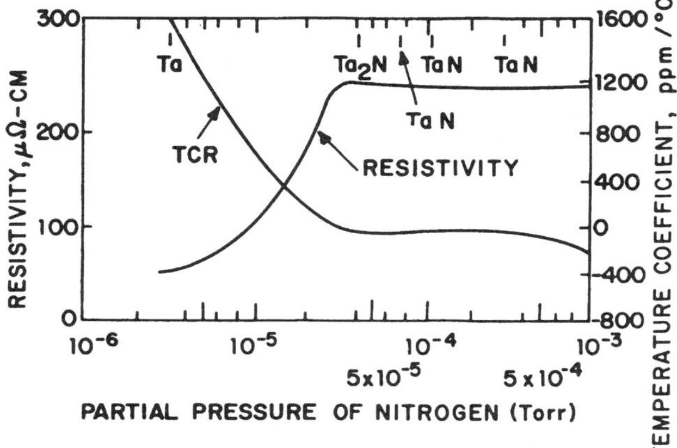

Figure 3-23. Influence of nitrogen on composition, resistivity, and temperature coefficient of resistivity of Ta films. (From Ref. 26).

3.7.4.3. Stoichiometry and Properties. Considerable variation in the composition and properties of reactively sputtered films is possible, depending on operating conditions. The case of tantalum nitride is worth considering in this regard. One of the first electronic applications of reactive sputtering involved deposition of TaN resistors employing dc diode sputtering at voltages of 3–5 kV, and pressures of about 30×10^{-3} torr. The dependence of the resistivity of "tantalum nitride" films is shown in Fig. 3-23, where either Ta, Ta_2N, TaN, or combinations of these form as a function of N_2 partial pressure. Color changes accompany the varied film stoichiometries. For example, in the case of titanium nitride films, the metallic color of Ti gives way to a light gold, then a rose, and finally a brown color with increasing nitrogen partial pressure.

3.7.5. Bias Sputtering

In bias sputtering, electric fields near the substrate are modified in order to vary the flux and energy of incident charged species. This is achieved by applying either a negative dc or RF bias to the substrate. With target voltages of −1000 to −3000 V, bias voltages of −50 to −300 V are typically used. Due to charge exchange processes in the anode dark space, very few discharge ions strike the substrate with full bias voltage. Rather a broad low energy distribution of ions and neutrals bombard the growing film. The technique has been utilized in all sputtering configurations (dc, RF, magnetron, and reactive).

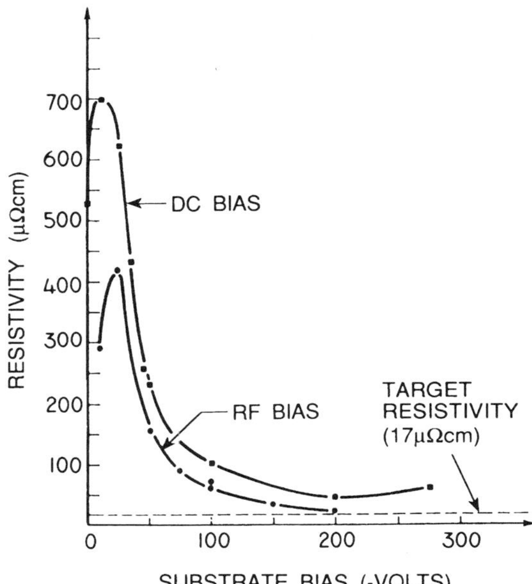

Figure 3-24. Resistivity of Ta films vs. substrate bias voltage; dc bias (3000 Å thick). (From Ref. 27). RF bias (1600 Å thick). (From Ref. 28).

Bias sputtering has been effective in altering a broad range of properties in deposited films. As specific examples we cite (Refs. 4–6).

a. *Resistivity*—A significant reduction in resistivity has been observed in metal films such as Ta, W, Ni, Au, and Cr. The similar variation in Ta film resistivity with dc or RF bias shown in Fig. 3-24 suggests that a common mechanism, independent of sputtering mode, is operative.
b. *Hardness and Residual Stress*—The hardness of sputtered Cr has been shown to increase (or decrease) with magnitude of negative bias voltage applied. Residual stress is similarly affected by bias sputtering.
c. *Dielectric Properties*—Increasing RF bias during RF sputtering of SiO_2 films has resulted in decreases in relative dielectric constant, but increases in resistivity.
d. *Etch Rate*—The wet chemical etch rate of reactively sputtered silicon nitride films is reduced with increasing negative bias.
e. *Optical Reflectivity*—Unbiased films of W, Ni, and Fe appear dark gray or black, whereas bias-sputtered films display metallic luster.
f. *Step Coverage*—Substantial improvement in step coverage of Al accompanies application of dc substrate bias.

g. *Film morphology*—The columnar microstructure of RF-sputtered Cr is totally disrupted by ion bombardment and replaced instead by a compacted, fine-grained structure (Ref. 18).
h. *Density*—Increased film density has been observed in bias-sputtered Cr (Ref. 18). Lower pinhole porosity and corrosion resistance are manifestations of the enhanced density.
i. *Adhesion*—Film adhesion is normally improved with ion bombardment of substrates during initial stages of film formation.

Although the details are not always clearly understood, there is little doubt that bias controls the film gas content. For example, chamber gases (e.g., Ar, O_2, N_2, etc.) sorbed on the growing film surface may be resputtered during low-energy ion bombardment. In such cases both weakly bound physisorbed gases (e.g., Ar) or strongly attached chemisorbed species (e.g., O or N on Ta) apparently have large sputtering yields and low sputter threshold voltages. In other cases, sorbed gases may have anomalously low sputter yields and will be incorporated within the growing film. In addition, energetic particle bombardment prior to and during film formation and growth promotes numerous changes and processes at a microscopic level, including removal of contaminants, alteration of surface chemistry, enhancement of nucleation and renucleation (due to generation of nucleation sites via defects, implanted, and recoil-implanted species), higher surface mobility of adatoms, and elevated film temperatures with attendant acceleration of atomic reaction and interdiffusion rates. Film properties are then modified through roughening of the surface, elimination of interfacial voids and subsurface porosity, creation of a finer, more isotropic grain morphology, and elimination of columnar grains—in a way that strongly dramatizes structure–property relationships in practice.

There are few ways to broadly influence such a wide variety of thin-film properties, in so simple and cheap a manner, than by application of substrate bias.

3.7.6. Evaporation versus Sputtering

Now that the details of evaporation and sputtering have been presented, we compare their characteristics with respect to process variables and resulting film properties. Distinctions in the stages of vapor species production, transport through the gas phase, and condensation on substrate surfaces for the two PVD processes are reviewed in tabular form in Table 3-7.

Table 3-7. Evaporation versus Sputtering

Evaporation	Sputtering
A. Production of Vapor Species	
1. Thermal evaporation mechanism	1. Ion bombardment and collisional momentum transfer
2. Low kinetic energy of evaporant atoms (at 1200 K, $E = 0.1$ eV)	2. High kinetic energy of sputtered atoms ($E = 2$–30 eV)
3. Evaporation rate (Eq. 3-2) (for $M = 50$, $T = 1500$ K, and $P_e = 10^{-3}$) $\approx 1.3 \times 10^{17}$ atoms/cm^2-sec.	3. Sputter rate (at 1 mA/cm^2 and $S = 2$) $\approx 3 \times 10^{16}$ atoms/cm^2-sec
4. Directional evaporation according to cosine law	4. Directional sputtering according to cosine law at high sputter rates
5. Fractionation of multicomponent alloys, decomposition, and dissociation of compounds	5. Generally good maintenance of target stoichiometry, but some dissociation of compounds.
6. Availability of high evaporation source purities	6. Sputter targets of all materials are available; purity varies with material
B. The Gas Phase	
1. Evaporant atoms travel in high or ultrahigh vacuum ($\sim 10^{-6}$–10^{-10} torr) ambient	1. Sputtered atoms encounter high-pressure discharge region ($\sim$ 100 mtorr)
2. Thermal velocity of evaporant 10^5 cm/sec	2. Neutral atom velocity $\sim 5 \times 10^4$ cm/sec
3. Mean-free path is larger than evaporant–substrate spacing. Evaporant atoms undergo no collisions in vacuum	3. Mean-free path is less than target–substrate spacing. Sputtered atoms undergo many collisions in the discharge
C. The Condensed Film	
1. Condensing atoms have relatively low energy	1. Condensing atoms have high energy
2. Low gas incorporation	2. Some gas incorporation
3. Grain size generally larger than for sputtered film	3. Good adhesion to substrate
4. Few grain orientations (textured films)	4. Many grain orientations

3.8. Hybrid and Modified PVD Processes

This chapter concludes with a discussion of several PVD processes that are more complex than the conventional ones considered up to this point. They demonstrate the diversity of process hybridization and modification possible in producing films with unusual properties. Ion plating, reactive evaporation, and ion-beam-assisted deposition will be the processes considered first. In the first two, the material deposited usually originates from a heated evaporation source. In the third, well-characterized ion beams bombard films deposited by evaporation or sputtering. The chapter closes with a discussion of ionized cluster-beam deposition. This process is different from others considered in this chapter in that film formation occurs through impingement of collective groups of atoms from the gas phase rather than individual atoms.

3.8.1. Ion Plating

Ion plating, developed by Mattox (Ref. 29), refers to evaporated film deposition processes in which the substrate is exposed to a flux of high-energy ions capable of causing appreciable sputtering before and during film formation. A schematic representation of a diode-type batch, ion-plating system is shown in Fig. 3-25a. Since it is a hybrid system, provision must be made to sustain the plasma, cause sputtering, and heat the vapor source. Prior to deposition, the substrate, negatively biased from 2 to 5 kV, is subjected to inert-gas ion bombardment at a pressure in the millitorr range for a time sufficient to sputter-clean the surface and remove contaminants. Source evaporation is then begun without interrupting the sputtering, whose rate must obviously be less than that of the deposition rate. Once the interface between film and substrate has formed, ion bombardment may or may not be continued. To circumvent the relatively high system pressures associated with glow discharges, high-vacuum ion-plating systems have also been constructed. They rely on directed ion beams targeted at the substrate. Such systems, which have been limited thus far to research applications, are discussed in Section 3.8.3.

Perhaps the chief advantage of ion plating is the ability to promote extremely good adhesion between the film and substrate by the ion and particle bombardment mechanisms discussed in Section 3.7.5. A second important advantage is the high ''throwing power'' when compared with vacuum evaporation. This results from gas scattering, entrainment, and sputtering of the film, and enables deposition in recesses and on areas remote from the source–substrate line of sight. Relatively uniform coating of substrates with complex shapes is thus achieved. Lastly, the quality of deposited films is frequently enhanced. The continual bombardment of the growing film by high-energy ions or neutral atoms and molecules serves to peen and compact it to near bulk densities. Sputtering of loosely adhering film material, increased surface diffusion, and reduced shadowing effects serve to suppress undesirable columnar growth.

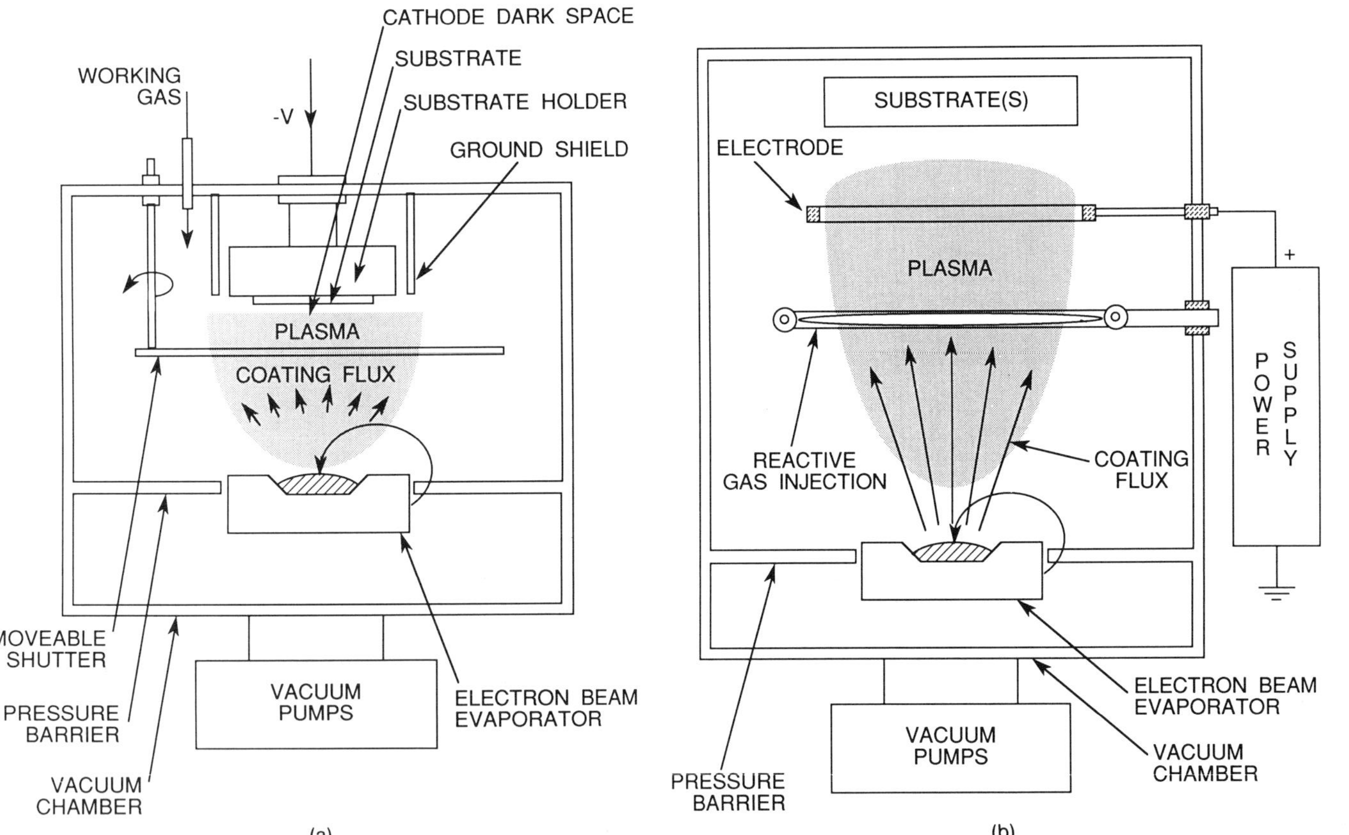

Figure 3-25. Hybrid PVD process: (a) Ion plating. (From Ref. 29). (b) Activated reactive evaporation. (From Ref. 30). (c) Ion-beam-assisted deposition. (From Ref. 31).

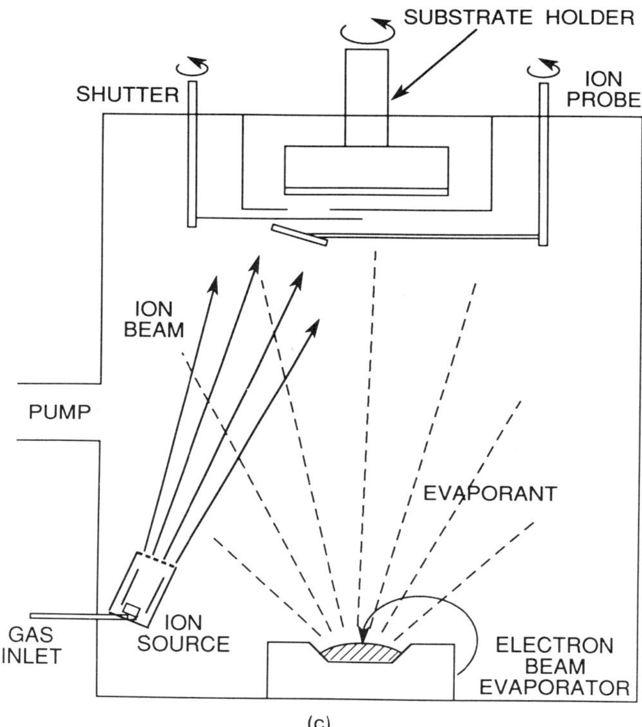

(c)

Figure 3-25. *Continued.*

A major use of ion plating has been to coat steel and other metals with very hard films for use in tools and wear-resistant applications. For this purpose, metals like Ti, Zr, Cr, and Si are electron-beam-evaporated through an Ar plasma in the presence of reactive gases such as N_2, O_2, and CH_4, which are simultaneously introduced into the system. This variant of the process is known as reactive ion plating (RIP), and coatings of nitrides, oxides, and carbides have been deposited in this manner.

3.8.2. Reactive Evaporation Processes

In reactive evaporation the evaporant metal vapor flux passes through and reacts with a gas (at $1-30 \times 10^{-3}$ torr) introduced into the system to produce compound deposits. The process has a history of evolution in which evaporation was first carried out without ionization of the reactive gas. In the more recent activated reactive evaporation (ARE) processes developed by Bunshah and co-workers (Ref. 30), a plasma discharge is maintained directly within the reaction zone between the metal source and substrate. Both the metal vapor and reactive gases, such as O_2, N_2, CH_4, C_2H_2, etc., are, therefore, ionized increasing their reactivity on the surface of the growing film or coating, promoting stoichiometric compound formation. One of the process configurations is illustrated in Fig. 3-25b, where the metal is melted by an electron beam. A thin plasma sheath develops on top of the molten pool. Low-energy secondary electrons from this source are drawn upward into the reaction zone by a circular wire electrode placed above the melt biased to a positive dc potential (20–100 V), creating a plasma-filled region extending from the electron-beam gun to near the substrate. The ARE process is endowed with considerable flexibility, since the substrates can be grounded, allowed to float electrically, or biased positively or negatively. In the latter variant ARE is quite similar to RIP. Other modifications of ARE include resistance-heated evaporant sources coupled with a low-voltage cathode (electron) emitter–anode assembly. Activation by dc and RF excitation has also been employed to sustain the plasma, and transverse magnetic fields have been applied to effectively extend plasma electron lifetimes.

Before considering the variety of compounds produced by ARE, we recall that thermodynamic and kinetic factors are involved in their formation. The high negative enthalpies of compound formation of oxides, nitrides, carbides, and borides indicate no thermodynamic obstacles to chemical reaction. The rate-controlling step in simple reactive evaporation is frequently the speed of the chemical reaction at the reaction interface. The actual physical location of the latter may be the substrate surface, the gas phase, the surface of the metal evaporant pool, or a combination of these. Plasma activation generally lowers the energy barrier for reaction by creating many excited chemical species. By eliminating the major impediment to reaction, ARE processes are thus capable of deposition rates of a few thousand angstroms per minute.

A partial list of compounds synthesized by ARE methods includes the oxides αAl_2O_3, V_2O_3, TiO_2, indium–tin oxide; the carbides TiC, ZrC, NbC, Ta_2C, W_2C, VC, HfC; and the nitrides TiN, MoN, HfN, and cubic boron nitride. The extremely hard TiN, TiC, Al_2O_3, and HfN compounds have found extensive use as coatings for sintered carbide cutting tools, high-speed drills, and gear cutters. As a result, they considerably increase wear resistance and extend tool life. In these applications ARE processing competes with the CVD methods discussed in Chapters 4 and 12. The fact that no volatile metal-bearing compound is required as in CVD is an attractive advantage of ARE. Most significantly, these complex compound films are synthesized at relatively low temperatures; this is a unique feature of plasma-assisted deposition processes.

3.8.3. Ion-Beam-Assisted Deposition Processes (Ref. 31)

We noted in Section 3.7.5 that ion bombardment of biased substrates during sputtering is a particularly effective way to modify film properties. Process control in plasmas is somewhat haphazard, however, because the direction, energy, and flux of the ions incident on the growing film cannot be regulated. Ion-beam-assisted processes were invented to provide independent control of the deposition parameters and, particularly, the characteristics of the ions bombarding the substrate. Two main ion source configurations are employed. In the dual-ion-beam system, one source provides the inert or reactive ion beam to *sputter* a target in order to yield a flux of atoms for deposition onto the substrate. Simultaneously, the second ion source, aimed at the substrate, supplies the inert or reactive ion beam that bombards the depositing film. Separate film-thickness-rate and ion-current monitors, fixed to the substrate holder, enable the two incident beam fluxes to be independently controlled.

In the second configuration (Fig. 3-25c), an ion source is used in conjunction with an *evaporation* source. The process, known as ion-assisted deposition (IAD), combines the benefits of high film deposition rate and ion bombardment. The energy flux and direction of the ion beam can be regulated independently of the evaporation flux. In both configurations the ion-beam angle of incidence is not normal to the substrate and can lead to anisotropic film properties. Substrate rotation is, therefore, recommended if isotropy is desired.

Broad-beam (Kaufman) ion sources, the heart of ion-beam-assisted deposition systems, were first used as ion thrusters for space propulsion (Ref. 32). Their efficiency has been optimized to yield high-ion-beam fluxes for given power inputs and gas flows. They contain a discharge chamber that is raised to a potential corresponding to the desired ion energy. Gases fed into the chamber become ionized in the plasma, and a beam of ions is extracted and accelerated through matching apertures in a pair of grids. Current densities of several mA/cm^2 are achieved. (Note that 1 mA/cm^2 is equivalent to 6.25×10^{15} ions/cm^2-sec or several monolayers per second.) The resulting beams have a low-energy spread (typically 10 eV) and are well collimated, with divergence angles of only a few degrees. Furthermore, the background pressure is quite low ($\sim 10^{-4}$ torr) compared with typical sputtering or etching plasmas.

Examples of thin-film property modification as a result of IAD are given in Table 3-8. The reader should appreciate the applicability to all classes of solids and to a broad spectrum of properties. For the most part, ion energies are lower than those typically involved in sputtering. Bombarding ion fluxes are generally smaller than depositing atom fluxes. Perhaps the most promising

Table 3-8. Property Modification by Ion Bombardment during Film Deposition

Film material	Ion species	Property modified	Ion energy (eV)	Ion/Atom Arrival Rate Ratio
Ge	Ar^+	Stress, adhesion	65–3000	2×10^{-4} to 10^{-1}
Nb	Ar^+	Stress	100–400	3×10^{-2}
Cr	Ar^+, Xe^+	Stress	3,400–11,500	8×10^{-3} to 4×10^{-2}
Cr	Ar^+	Stress	200–800	$\sim 7 \times 10^{-3}$ to 2×10^{-2}
SiO_2	Ar^+	Step coverage	500	0.3
SiO_2	Ar^+	Step coverge	~ 1–80	~ 4.0
AlN	N_2^+	Preferred orientation	300–500	0.96 to 1.5
Au	Ar^+	Coverage at 50 Å thickness	400	0.1
GdCoMo	Ar^+	Magnetic anisotropy	~ 1–150	~ 0.1
Cu	Cu^+	Improved epitaxy	50–400	10^{-2}
BN	$(B-N-H)^+$	Cubic structure	200–1000	~ 1.0
ZrO_2, SiO_2, TiO_2	Ar^+, O_2^+	Refractive index, amor → crys	600	2.5×10^{-2} to 10^{-1}
SiO_2, TiO_2	O_2^+	Refractive index	300	0.12
SiO_2, TiO_2	O_2^+	Optical transmission	30–500	0.05 to 0.25
Cu	N^+, Ar^+	Adhesion	50,000	10^{-2}
Ni on Fe	Ar^+	Hardness	10,000–20,000	~ 0.25

From Ref. 32.

application of ion bombardment is the enhancement of the density and index of refraction of optical coatings. This subject is treated again in Chapter 11.

3.8.4. Ionized Cluster Beam (ICB) Deposition (Ref. 33)

The idea of employing energetic ionized clusters of atoms to deposit thin films is due to T. Takagi. In this novel technique, vapor-phase aggregates or clusters, thought to contain a few hundred to a few thousand atoms, are

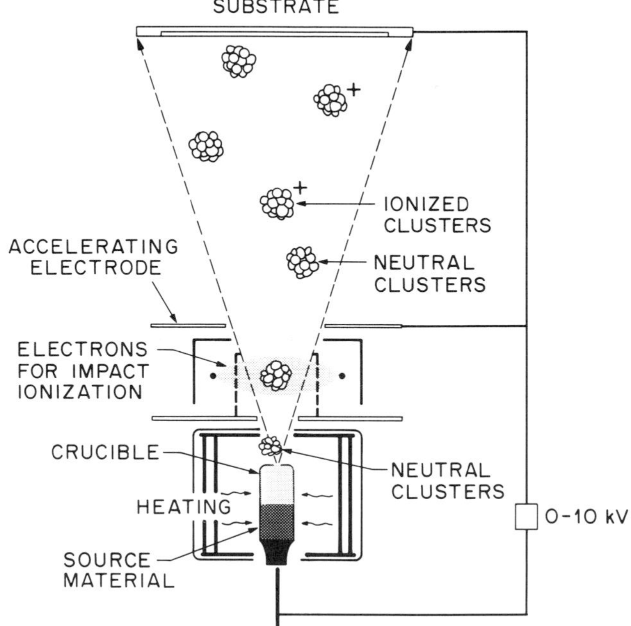

Figure 3-26. Schematic diagram of ICB system. (Courtesy of W. L. Brown, AT&T Bell Laboratories. Reprinted with permission of the publisher from Ref. 34).

created, ionized, and accelerated toward the substrate as depicted schematically in Fig. 3-26. As a result of impact with the substrate, the cluster breaks apart, releasing atoms to spread across the surface. Cluster production is, of course, the critical step and begins with evaporation from a crucible containing a small aperture or nozzle. The evaporant vapor pressure is much higher (10^{-2}–10 torr) than in conventional vacuum evaporation. For cluster formation the nozzle diameter must exceed the mean-free path of vapor atoms in the crucible. Viscous flow of atoms escaping the nozzle then results in an adiabatic supersonic expansion and the formation of stable cluster nuclei. Optimum expansion further requires that the ratio of the vapor pressure in the crucible to that in the vacuum chamber exceed 10^4 to 10^5.

The arrival of ionized clusters with the kinetic energy of the acceleration voltage (0–10 kV), and neutral clusters with the kinetic energy of the nozzle ejection velocity, affects film nucleation and growth processes in the following ways:

1. The local temperature at the point of impact increases.
2. Surface diffusion of atoms is enhanced.

3. Activated centers for nucleation are created.
4. Coalescence of nuclei is fostered.
5. At high enough energies, the surface is sputter-cleaned, and shallow implantation of ions may occur.
6. Chemical reactions between condensing atoms and the substrate or gas-phase atoms are favored.

Moreover, the magnitude of these effects can be modified by altering the extent of electron impact ionization and the accelerating voltage.

Virtually all classes of film materials have been deposited by ICB (and variant reactive process versions), including pure metals, alloys, intermetallic compounds, semiconductors, oxides, nitrides, carbides, halides, and organic compounds. Special attributes of ICB-prepared films worth noting are strong adhesion to the substrate, smooth surfaces, elimination of columnar growth morphology, low-temperature growth, controllable crystal structures, and, importantly, very high quality single-crystal growth (epitaxial films). Large Au film mirrors for CO_2 lasers, ohmic metal contacts to Si and GaP, electromigration- (Section 8.4) resistant Al films, and epitaxial Si, GaAs, GaP, and InSb films deposited at low temperatures are some examples indicative of the excellent properties of ICB films. Among the advantages of ICB deposition are vacuum cleanliness ($\sim 10^{-7}$ torr in the chamber) of evaporation and energetic ion bombardment of the substrate, two normally mutually exclusive features. In addition, the interaction of slowly moving clusters with the substrate is confined, limiting the amount of damage to both the growing film and substrate. Despite the attractive features of ICB, the formation of clusters and their role in film formation are not well understood. Recent research (Ref. 34), however, clearly indicates that the total number of atoms agglomerated in large metal clusters is actually very small (only 1 in 10^4) and that only a fraction of large clusters is ionized. The *total* energy brought to the film surface by ionized clusters is, therefore, quite small. Rather, it appears that individual atomic ions, which are present in much greater profusion than are ionized clusters, are the dominant vehicle for transporting energy and momentum to the growing film. In this respect, ICB deposition belongs to the class of processes deriving benefits from the ion-beam-assisted film growth mechanisms previously discussed.

Exercises

1. Employing Figs. 3-1 and 3-2, calculate values for the molar heat of vaporization of Si and Ga.

2. Design a laboratory experiment to determine a working value of the heat of vaporization of a metal employing common thin-film deposition and characterization equipment.

3. Suppose Fe satisfactorily evaporates from a surface source, 1 cm^2 in area, which is maintained at 1550 °C. Higher desired evaporation rates are achieved by raising the temperature 100 °C. But doing this will burn out the source. Instead, the melt area is increased without raising its temperature. By what factor should the source area be enlarged?

4. A molecular-beam epitaxy system contains separate Al and As effusion evaporation sources of 4 cm^2 area, located 10 cm from a (100) GaAs substrate. The Al source is heated to 1000 °C, and the As source is heated to 300 °C. What is the growth rate of the AlAs film in Å/sec? [*Note:* AlAs basically has the same crystal structure and lattice parameter (5.661 Å) as GaAs.]

5. How far from the substrate, in illustrative problem on p. 90, would a *single* surface source have to be located to maintain the same deposited film thickness tolerance?

6. An Al film was deposited at a rate of 1 $\mu m/min$ in vacuum at 25 °C, and it was estimated that the oxygen content of the film was 10^{-3}. What was the partial pressure of oxygen in the system?

7. Alloy films of Ti–W, used as diffusion barriers in integrated circuits, are usually sputtered. The Ti–W, phase diagram resembles that of Ge–Si (Fig. 1-13) at elevated temperatures.

   a. Comment on the ease or feasibility of evaporating a 15 wt% Ti–W alloy.
   b. During sputtering with 0.5-keV Ar, what composition will the target surface assume in the steady state?

8. In order to deposit films of the alloy YBa_2Cu_3, the metals Y, Ba, and Cu are evaporated from three *point* sources. The latter are situated at the corners of an equilateral triangle whose side is 20 cm. Directly above the centroid of the source array, and parallel to it, lies a small substrate; the deposition system geometry is thus a tetrahedron, each side being 20 cm long.

   a. If the Y source is heated to 1740 K to produce a vapor pressure of 10^{-3} torr, to what temperature must the Cu source be heated to maintain film stoichiometry?

b. Rather than a point source, a surface source is used to evaporate Cu. How must the Cu source temperature be changed to ensure deposit stoichiometry?
c. If the source configuration in part (a) is employed, what *minimum* O_2 partial pressure is required to deposit stoichiometric $YBa_2Cu_3O_7$ superconducting films by a reactive evaporation process? The atomic weights are Y = 89, Cu = 63.5, Ba = 137, and O = 16.

9. One way to deposit a thin metal film of known thickness is to heat an evaporation source to dryness (i.e., until no metal remains in the crucible). Suppose it is desired to deposit 5000 Å of Au on the internal spherical surface of a hemispherical shell measuring 30 cm in diameter.

a. Suggest two different evaporation source configurations (source type and placement) that would yield uniform coatings.
b. What weight of Au would be required for each configuration, assuming evaporation to dryness?

10. Suppose the processes of electron impact ionization and secondary emission of electrons by ions control the current J in a sputtering system according to the Townsend equation (Ref. 19)

$$J = \frac{J_0 \exp \alpha d}{1 - \gamma[\exp(\alpha d) - 1]},$$

where J_0 = primary electron current density from external source
α = number of ions per unit length produced by electrons
γ = number of secondary electrons emitted per incident ion
d = interelectrode spacing.

a. If the film deposition rate during sputtering is proportional to the product of J and S, calculate the proportionality constant for Cu in this system if the deposition rate is 200 Å/min for 0.5-keV Ar ions. Assume $\alpha = 0.1$ ion/cm, $\gamma = 0.08$ electron/ion, $d = 10$ cm, and $J_0 = 100$ mA/cm^2.
b. What deposition rate can be expected for 1-keV Ar if $\alpha = 0.15$ ion/cm and $\gamma = 0.1$ electron/ion.

11. In a dc planar magnetron system operating at 1000 V, the anode–cathode spacing is 10 cm. What magnetic field should be applied to trap electrons within 1 cm of the target?

12. At what sputter deposition rate of In on a Si substrate will the film melt within 1 min? The melting point of In is 155 °C.

13. a. During magnetron sputtering of Au at 1 keV, suppose there are two collisions with Ar atoms prior to deposition. What is the energy of the depositing Au atoms? (Assume Ar is stationary in a collision.)

b. The probability that gas-phase atoms will travel a distance x *without* collision is $\exp - x/\lambda$, where λ is the mean-free path between collisions. Assume λ for Au in Ar is 5 cm at a pressure of 1 mtorr. If the target–anode spacing is 12 cm, at what operating pressure will 99% of the sputtered Au atoms undergo gas-phase collisions prior to deposition?

14. For a new application it is desired to continuously coat a 1-m-wide steel strip with a 2-μm-thick coating of Al. The x-y dimensions of the steel are such that an array of electron-beam gun evaporators lies along the y direction and maintains a uniform coating thickness across the strip width. How fast should the steel be fed in the x direction past the surface sources, which can evaporate 20 g of Al per second? Assume that Eq. 3-18 holds for the coating thickness along the x direction, that the source–strip distance is 30 cm, and that the steel sheet is essentially a horizontal substrate 40 cm long on either side of the source before it is coiled.

15. Select the appropriate film deposition process (evaporation, sputtering, etc., sources, targets, etc.) for the following applications:

a. Coating a large telescope mirror with Rh
b. Web coating of potato chip bags with Al films
c. Deposition of Al–Cu–Si thin-film interconnections for integrated circuits
d. Deposition of TiO_2–SiO_2 multilayers on artificial gems to enhance color and reflectivity

16. Theory indicates that the kinetic energy (E) and angular spread of neutral atoms sputtered from a surface are given by the distribution function

$$F(E,\theta) = CS\frac{E}{(E+U)^3}\cos\theta,$$

where U = binding energy of surface atoms
C = constant
θ = angle between sputtered atoms and the surface normal.

a. Sketch the dependence of $f(E,\theta)$ vs. E for two values of U.
b. Show that the maximum in the energy distribution occurs at $E = U/2$.

17. a. To better visualize the nucleation of clusters in the ICB process, schematically indicate the free energy of cluster formation vs. cluster size as a function of vapor supersaturation (see Section 1.7).

b. What vapor supersaturation is required to create a 1000-atom cluster of Au if the surface tension is 1000 ergs/cm^2?

c. If such a cluster is ionized and accelerated to an energy of 10 keV, how much energy is imparted to the substrate by each cluster atom?

References

1. W. R. Grove, *Phil. Trans. Roy. Soc., London A* **142**, 87 (1852).
2. M. Faraday, *Phil. Trans.* **147**, 145 (1857).
3.* R. Glang, in *Handbook of Thin Film Technology*, eds. L. I. Maissel and R. Glang, McGraw-Hill, New York (1970).
4.* J. L. Vossen and J. J. Cuomo, in *Thin Film Processes*, eds. J. L. Vossen and W. Kern, Academic Press, New York (1978).
5.* W. D. Westwood, in *Microelectronic Materials and Processes*, ed. R. A. Levy, Kluwer Academic, Dordrecht (1989).
6.* B. N. Chapman, *Glow Discharge Processes*, Wiley, New York (1980).
7. C. H. P. Lupis, *Chemical Thermodynamics of Materials*, North-Holland, Amsterdam (1983).
8. R. E. Honig, *RCA Rev.* **23**, 567 (1962).
9.* H. K. Pulker, *Coatings on Glass*, Elsevier, New York, (1984).
10. Examples taken from *Physical Vapor Deposition*, Airco-Temescal (1976).
11. L. Holland, *Vacuum Deposition of Thin Films*, Wiley, New York (1956).
12. C. H. Ting and A. R. Neureuther, *Solid State Technol.* **25(2)**, 115 (1982).
13. H. L. Caswell, in *Physics of Thin Films*, Vol. 1, ed. G. Hass, Academic Press, New York (1963).
14. L. D. Hartsough and D. R. Denison, *Solid State Technology* **22(12)**, 66 (1979).
15. *Handbook—The Optical Industry and Systems Directory*, H-11 (1979).
16. E. B. Grapper, *J. Vac. Sci. Technol.* **5A(4)**, 2718 (1987); **8**, 333 (1971).

*Recommended texts or reviews.

17. P. Archibald and E. Parent, *Solid State Technol.* **19(7)**, 32 (1976).
18. D. M. Mattox, *J. Vac. Sci. Technol.* **A7(3)**, 1105 (1989).
19.* A. B. Glaser and G. E. Subak-Sharpe, *Integrated Circuit Engineering*, Addison-Wesley, Reading, MA (1979).
20. P. Sigmund, *Phys. Rev.* **184**, 383 (1969).
21. L. T. Lamont, *Solid State Technol.* **22(9)**, 107 (1979).
22. J. A. Thornton, *Thin Solid Films* **54**, 23 (1978).
23. J. A. Thornton, in *Thin Film Processes*, eds. J. L. Vossen and W. Kern, Academic Press, New York (1978).
24. H. R. Koenig and L. I. Maissel, *IBM J. Res. Dev.* **14**, 168 (1970).
25.* W. D. Westwood, in *Physics of Thin Films*, Vol. 14, eds. M. H. Francombe and J. L. Vossen, Academic Press, New York (1989).
26.* L. I. Maissel and M. H. Francombe, *An Introduction to Thin Films*, Gordon and Breach, New York, (1973).
27. L. I. Maissel and P. M. Schaible, *J. Appl. Phys.* **36**, 237 (1965).
28. J. L. Vossen and J. J. O'Neill, *RCA Rev.* **29**, 566 (1968).
29. D. M. Mattox, *J. Vac. Sci. Technol.* **10**, 47 (1973).
30.* R. F. Bunshah and C. Deshpandey, in *Physics of Thin Films*, Vol. 13, eds. M. H. Francombe and J. L. Vossen, Academic Press, New York (1987).
31. J. M. E. Harper and J. J. Cuomo, *J. Vac. Sci. Technol.* **21(3)**, 737 (1982).
32. J. M. E. Harper, J. J. Cuomo, R. J. Gambino, and H. R. Kaufman, in *Ion Beam Modification of Surfaces*, eds. O. Auciello and R. Kelly, Elsevier, Amsterdam (1984).
33.* T. Takagi, in *Physics of Thin Films*, Vol. 13, eds. M. H. Francombe and J. L. Vossen, Academic Press, New York (1987).
34. W. L. Brown, M. F. Jarrold, R. L. McEachern, M. Sosnowski, G. Takaoka, H. Usui and I. Yamada, *Nuclear Instruments and Methods in Physics Research*, to be published (1991).

Chemical Vapor Deposition

4.1. Introduction

Chemical vapor deposition (CVD) is the process of chemically reacting a volatile compound of a material to be deposited, with other gases, to produce a nonvolatile solid that deposits atomistically on a suitably placed substrate. High-temperature CVD processes for producing thin films and coatings have found increasing applications in such diverse technologies as the fabrication of solid-state electronic devices, the manufacture of ball bearings and cutting tools, and the production of rocket engine and nuclear reactor components. In particular, the need for high-quality epitaxial semiconductor films for both Si bipolar and MOS transistors, coupled with the necessity to deposit various insulating and passivating films at low temperatures, has served as a powerful impetus to spur development and implementation of CVD processing methods. A schematic view of the MOS field effect transistor structure in Fig. 4-1 indicates the extent to which the technology is employed. Above the plane of the base P–Si wafer, all of the films with the exception of the gate oxide and Al metallization are deposited by some variant of CVD processing. The films include polysilicon, dielectric SiO_2, and SiN.

Among the reasons for the growing adoption of CVD methods is the ability to produce a large variety of films and coatings of metals, semiconductors, and

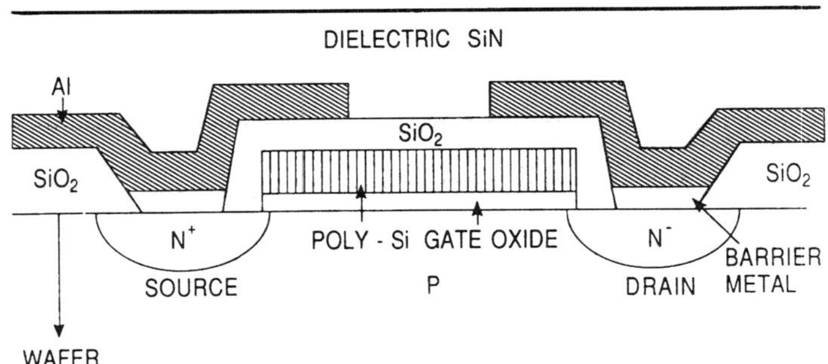

Figure 4-1. Schematic view of MOS field effect transistor cross section.

compounds in either a crystalline or vitreous form, possessing high purity and desirable properties. Furthermore, the capability of controllably creating films of widely varying stoichiometry makes CVD unique among deposition techniques. Other advantages include relatively low cost of the equipment and operating expenses, suitability for both batch and semicontinuous operation, and compatibility with other processing steps. Hence, many variants of CVD processing have been researched and developed in recent years, including low-pressure (LPCVD), plasma-enhanced (PECVD), and laser-enhanced (LECVD) chemical vapor deposition. Hybrid processes combining features of both physical and chemical vapor deposition have also emerged.

In this chapter, a number of topics related to the basic chemistry, physics, engineering, and materials science involved in CVD are explored. Practical concerns of chemical vapor transport, deposition processes, and equipment involved are discussed. The chapter is divided into the following sections:

4.2. Reaction Types
4.3. Thermodynamics of CVD
4.4. Gas Transport
4.5. Growth Kinetics
4.6. CVD Processes and Systems

Recommended review articles and books dealing with these aspects of CVD can be found in Refs. 1–7.

To gain an appreciation of the scope of the subject, we first briefly categorize the various types of chemical reactions that have been employed to deposit films and coatings (Refs. 1–3). Corresponding examples are given for each by indicating the essential overall chemical equation and approximate reaction temperature.

4.2. Reaction Types

4.2.1. Pyrolysis

Pyrolysis involves the thermal decomposition of such gaseous species as hydrides, carbonyls, and organometallic compounds on hot substrates. Commercially important examples include the high-temperature pyrolysis of silane to produce polycrystalline or amorphous silicon films, and the low-temperature decomposition of nickel carbonyl to deposit nickel films.

$$SiH_{4(g)} \rightarrow Si_{(s)} + 2H_{2(g)} \qquad (650\ ^\circ C), \tag{4-1}$$

$$Ni(CO)_{4(g)} \rightarrow Ni_{(s)} + 4CO_{(g)} \qquad (180\ ^\circ C). \tag{4-2}$$

Interestingly, the latter reaction is the basis of the Mond process, which has been employed for over a century in the metallurgical refining of Ni.

4.2.2. Reduction

These reactions commonly employ hydrogen gas as the reducing agent to effect the reduction of such gaseous species as halides, carbonyl halides, oxyhalides, or other oxygen-containing compounds. An important example is the reduction of $SiCl_4$ on single-crystal Si wafers to produce epitaxial Si films according to the reaction

$$SiCl_{4(g)} + 2H_{2(g)} \rightarrow Si_{(s)} + 4HCl_{(g)} \qquad (1200\ ^\circ C). \tag{4-3}$$

Refractory metal films such as W and Mo have been deposited by reducing the corresponding hexafluorides, e.g.,

$$WF_{6(g)} + 3H_{2(g)} \rightarrow W_{(s)} + 6HF_{(g)} \qquad (300\ ^\circ C), \tag{4-4}$$

$$MoF_{6(g)} + 3H_{2(g)} \rightarrow Mo_{(s)} + 6HF_{(g)} \qquad (300\ ^\circ C). \tag{4-5}$$

Tungsten films deposited at low temperatures have been actively investigated as a potential replacement for aluminum contacts and interconnections in integrated circuits. Interestingly, WF_6 gas reacts directly with exposed silicon surfaces, depositing thin W films while releasing the volatile SiF_4 by-product. In this way silicon contact holes can be selectively filled with tungsten while leaving neighboring insulator surfaces uncoated.

4.2.3. Oxidation

Two examples of important oxidation reactions are

$$SiH_{4(g)} + O_{2(g)} \rightarrow SiO_{2(s)} + 2H_{2(g)} \qquad (450\ ^\circ C), \qquad (4\text{-}6)$$

$$4PH_{3(g)} + 5O_{2(g)} \rightarrow 2P_2O_{5(s)} + 6H_{2(g)} \qquad (450\ ^\circ C). \qquad (4\text{-}7)$$

The deposition of SiO_2 by Eq. 4-6 is often carried out at a stage in the processing of integrated circuits where higher substrate temperatures cannot be tolerated. Frequently, about 7% phosphorous is simultaneously incorporated in the SiO_2 film by the reaction of Eq. 4-7 in order to produce a glass film that flows readily to produce a planar insulating surface, i.e., "planarization."

In another process of technological significance, SiO_2 is also produced by the oxidation reaction

$$SiCl_{4(g)} + 2H_{2(g)} + O_{2(g)} \rightarrow SiO_{2(g)} + 4HCl_{(g)} \qquad (1500\ ^\circ C). \quad (4\text{-}8)$$

The eventual application here is the production of optical fiber for communications purposes. Rather than a thin film, the SiO_2 forms a cotton-candy-like deposit consisting of soot particles less than 1000 Å in size. These are then consolidated by elevated temperature sintering to produce a fully dense silica rod for subsequent drawing into fiber. Whether silica film deposition or soot formation occurs is governed by process variables favorable to heterogeneous or homogeneous nucleation, respectively. Homogeneous soot formation is essentially the result of a high $SiCl_4$ concentration in the gas phase.

4.2.4. Compound Formation

A variety of carbide, nitride, boride, etc., films and coatings can be readily produced by CVD techniques. What is required is that the compound elements exist in a volatile form and be sufficiently reactive in the gas phase. Examples of commercially important reactions include

$$SiCl_{4(g)} + CH_{4(g)} \rightarrow SiC_{(s)} + 4HCl_{(g)} \qquad (1400\ ^\circ C), \qquad (4\text{-}9)$$

$$TiCl_{4(g)} + CH_{4(g)} \rightarrow TiC_{(s)} + 4HCl_{(g)} \qquad (1000\ ^\circ C), \qquad (4\text{-}10)$$

$$BF_{3(g)} + NH_{3(g)} \rightarrow BN_{(s)} + 3HF_{(g)} \qquad (1100\ ^\circ C) \qquad (4\text{-}11)$$

for the deposition of hard, wear-resistant surface coatings. Films and coatings of compounds can generally be produced through a variety of precursor gases and reactions. For example, in the much studied SiC system, layers were first produced in 1909 through reaction of $SiCl_4 + C_6H_6$ (Ref. 8). Subsequent reactant combinations over the years have included $SiCl_4 + C_3H_8$, $SiBr_4 + C_2H_4$, $SiCl_4 + C_6H_{14}$, $SiHCl_3 + CCl_4$, and $SiCl_4 + C_6H_5CH_3$, to name a few, as well as volatile organic compounds containing both silicon and carbon in the same molecule (e.g., CH_3SiCl_3, CH_3SiH_3, $(CH_3)_2SiCl_2$, etc.). Although the deposit is nominally SiC in all cases, resultant properties generally differ because of structural, compositional, and processing differences.

Impermeable insulating and passivating films of Si_3N_4 that are used in integrated circuits can be deposited at 750 °C by the reaction

$$3SiCl_2H_{2(g)} + 4NH_{3(g)} \rightarrow Si_3N_{4(s)} + 6H_{2(g)} + 6HCl_{(g)}. \qquad (4\text{-}12)$$

The necessity to deposit silicon nitride films at lower temperatures has led to alternative processing involving the use of plasmas. Films can be deposited below 300 °C with SiH_4 and NH_3 reactants, but considerable amounts of hydrogen are incorporated into the deposits.

4.2.5. Disproportionation

Disproportionation reactions are possible when a nonvolatile metal can form volatile compounds having different degrees of stability, depending on the temperature. This manifests itself in compounds, typically halides, where the metal exists in two valence states (e.g., GeI_4 and GeI_2) such that the lower-valent state is more stable at higher temperatures. As a result, the metal can be transported into the vapor phase by reacting it with its volatile, higher-valent halide to produce the more stable lower-valent halide. The latter disproportionates at lower temperatures to produce a deposit of metal while regenerating the higher-valent halide. This complex sequence can be simply described by the reversible reaction

$$2GeI_{2(g)} \underset{600\ ^\circ C}{\overset{300\ ^\circ C}{\rightleftharpoons}} Ge_{(s)} + GeI_{4(g)} \qquad (4\text{-}13)$$

and realized in systems where provision is made for mass transport between hot and cold ends. Elements that have lent themselves to this type of transport reaction include aluminum, boron, gallium, indium, silicon, titanium, zirconium, beryllium, and chromium. Single-crystal films of Si and Ge were grown by disproportionation reactions in the early days of CVD experimentation on semiconductors employing reactors such as that shown in Fig. 4-2. The enormous progress made in this area is revealed here.

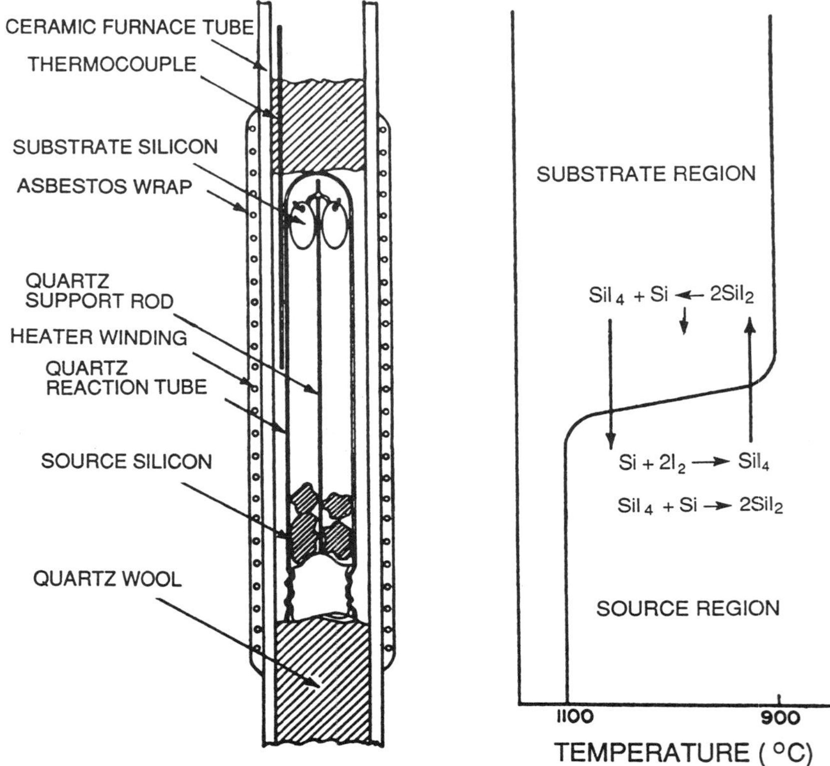

Figure 4-2. Experimental reactor for epitaxial growth of Si films. (E. S. Wajda, B. W. Kippenhan, W. H. White, *IBM J. Res. Dev.* **7**, 288, © 1960 by International Business Machines Corporation, reprinted with permission).

4.2.6. Reversible Transfer

Chemical transfer or transport processes are characterized by a reversal in the reaction equilibrium at source and deposition regions maintained at different temperatures within a single reactor. An important example is the deposition of single-crystal (epitaxial) GaAs films by the chloride process according to the reaction

$$As_{4(g)} + As_{2(g)} + 6GaCl_{(g)} + 3H_{2(g)} \underset{850\ ^\circ C}{\overset{750\ ^\circ C}{\rightleftarrows}} 6GaAs_{(s)} + 6HCl_{(g)}. \quad (4\text{-}14)$$

Here $AsCl_3$ gas from a bubbler transports Ga toward the substrates in the form of GaCl vapor. Subsequent reaction with As_4 causes deposition of GaAs.

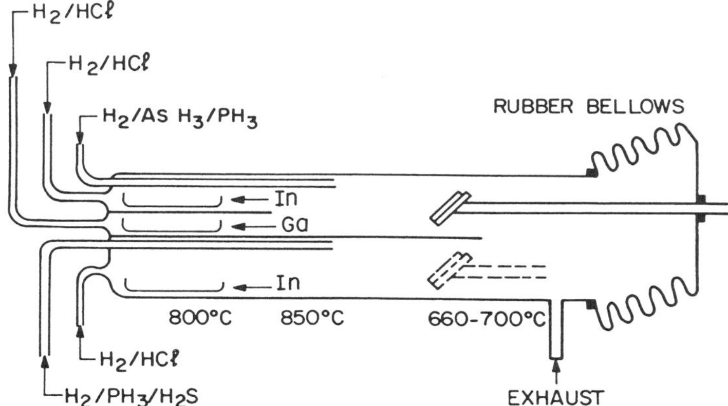

Figure 4-3. Schematic of atmospheric CVD reactor used to grow GaAs and other compound semiconductor films by the hydride process. (Reprinted with permission from Ref. 10).

Alternatively, in the hydride process, As is introduced in the form of AsH_3 (arsine), and HCl serves to transport Ga. Both processes essentially involve the same gas-phase reactions and are carried out in similar reactors, shown schematically in Fig. 4-3. What is significant is that single-crystal, *binary* (primarily GaAs and InP but also GaP and InAs) as well as *ternary* (e.g., (Ga, In)As and Ga(As, P)) compound films have been grown by these vapor phase epitaxy (VPE) processes. Similarly, in addition to binary and ternary semiconductor films, *quaternary* epitaxial films containing controlled amounts of Ga, In, As, and P have been deposited by the hydride VPE process. Combinations of gas mixtures and more complex reactors are required in this case to achieve the desired stoichiometries. The resulting films are the object of intense current research and development activity in a variety of optoelectronic devices (e.g., lasers and detectors). For quaternary alloy deposition by the hydride process, single-crystal InP substrates are employed. Gas-phase source reactions include

$$2AsH_3 \rightleftarrows As_2 + 3H_2,$$

$$2PH_3 \rightleftarrows P_2 + 3H_2,$$

$$2HCl + 2In \rightleftarrows 2InCl + H_2,$$

$$2HCl + 2Ga \rightleftarrows 2GaCl + H_2. \tag{4-15}$$

Table 4-1. CVD Films and Coatings

Deposited Material	Substrate	Input Reactants	Deposition Temperature (°C)	Crystallinity
Si	Single-crystal Si	Either $SiCl_2H_2$, $SiCl_3H$, or $SiCl_4 + H_2$	1050–1200	E
Si		$SiH_4 + H_2$	600–700	P
Ge	Single-crystal Ge	$GeCl_4$ or $GeH_4 + H_2$	600–900	E
SiC	Single-crystal Si	$SiCl_4$, toluene, H_2	1100	P
AlN	Sapphire	$AlCl_3$, NH_3, H_2	1000	E
$In_2O_3 : Sn$	Glass	In-chelate, $(C_4H_9)_2Sn(OOCH_3)_2$, H_2O, O_2, N_2	500	A
ZnS	GaAs, GaP	Zn, H_2S, H_2	825	E
CdS	GaAs, sapphire	Cd, H_2S, H_2	690	E
Al_2O_3	Si,	$Al(CH_3)_3 + O_2$	275–475	A
	cemented carbide	$AlCl_3$, CO_2, H_2	850–1100	A
SiO_2	Si	$SiH_4 + O_2$	450	A
		$SiCl_2H_2 + 2N_2O$	900	
Si_3N_4	SiO_2	$SiCl_2H_2 + NH_3$	~ 750	A
SiNH	SiO_2	$SiH_4 + NH_3$ (plasma)	300	A
TiO_2	Quartz	$Ti(OC_2H_5)_4 + O_2$	450	A
TiC	Steel	$TiCl_4$, CH_4, H_2	1000	P
TiN	Steel	$TiCl_4$, N_2, H_2	1000	P
BN	Steel	BCl_3, NH_3, H_2	1000	P
TiB_2	Steel	$TiCl_4$, BCl_3, H_2	> 800	P

Note: E = epitaxial; P = polycrystalline; A = amorphous.
Adapted from Refs. 1, 2, 3.

Deposition reactions at substrates include

$$2GaCl + As_2 + H_2 \rightleftarrows 2GaAs + 2HCl,$$
$$2GaCl + P_2 + H_2 \rightleftarrows 2GaP + 2HCl,$$
$$2InCl + P_2 + H_2 \rightleftarrows 2InP + 2HCl,$$
$$2InCl + As_2 + H_2 \rightleftarrows 2InAs + 2HCl. \tag{4-16}$$

Aspects of the properties of these deposited films will be discussed in Chapter 7.

The previous examples are but a small sample of the total number of film and coating deposition reactions that have been researched in the laboratory as well as developed for commercial applications. Table 4-1 contains a brief list of CVD processes for depositing elemental and compound semiconductors and assorted compounds. The entries are culled from various review articles (Refs. 1–6) on the subject, and the interested reader is encouraged to consult these articles to obtain the primary references for specific details on process variables.

In carefully examining the aforementioned categories of CVD reactions, the discerning reader will note two common features:

1. All of the chemical reactions can be written in the simplified generalized form

$$aA_{(g)} + bB_{(g)} \rightarrow cC_{(s)} + dD_{(g)}, \tag{4-17}$$

   where A, B, . . . refer to the chemical species and a, b, . . . to the corresponding stoichiometric coefficients. A single solid and mixture of gaseous species categorizes each heterogeneous reaction.
2. Many of the reactions are reversible, and this suggests that standard concepts of chemical thermodynamics may prove fruitful in analyzing them.

There is a distinction between chemical vapor *deposition* and chemical vapor *transport* reactions that should be noted. In the former, one or more gaseous species enter the reactor from gas tanks or liquid bubbler sources maintained *outside* the system. The reactants then combine at the hot substrate to produce the solid film. In chemical vapor transport reactions, solid or liquid sources are contained *within* closed or open reactors. In the latter case, externally introduced carrier or reactant gases flow over the sources and cause them to enter the vapor stream where they are transported along the reactor. Subsequently, deposition of solid from the gas phase occurs at the substrates. Both chemical vapor deposition and transport reactions are, however, described by the same type of chemical reaction. As far as thermodynamic analyses are concerned, no further distinction will be made between them, and the generic term CVD will be used for both.

4.3. Thermodynamics of CVD

4.3.1. Reaction Feasibility

Thermodynamics addresses several important issues with respect to CVD. Whether a given chemical reaction is feasible is perhaps the most important of these. Once it is decided that a reaction is possible, thermodynamic calculation can frequently provide information on the partial pressures of the involved gaseous species and the direction of transport in the case of reversible reactions. Importantly, it provides an upper limit of what to expect under specified conditions. Thermodynamics does not, however, consider questions related to the speed of the reaction and resulting film growth rates. Indeed, processes that are thermodynamically possible frequently proceed at such low rates, due to vapor transport kinetics and vapor-solid reaction limitations, as to be unfeasible in practice. Furthermore, the use of thermodynamics implies that chemical equilibrium has been attained. Although this may occur in a closed system, it is generally not the case in an open or flow reactor where gaseous reactants and products are continuously introduced and removed. Thus, CVD may be presently viewed as an empirical science with thermodynamic guidelines.

Provided that the free-energy change ΔG can be approximated by the *standard* free-energy change $\Delta G°$, many simple consequences of thermodynamics with respect to CVD can be understood. For example, consider the requirements for suitable chemical reactions in order to grow single-crystal films. In this case, it is essential that a single nucleus form as an oriented seed for subsequent growth. According to elementary nucleation theory, a small negative value of ΔG_V, the chemical free energy per unit volume, is required to foster a low nucleation rate of large critical-sized nuclei (see Section 1.7). This, in turn, would require a $\Delta G°$ value close to zero. When this happens, large amounts of reactants and products are simultaneously present. If ΔG_V were large, the likelihood of a high rate of heterogeneous nucleation, or even homogeneous nucleation of solid particles within the gas phase, would be enhanced. The large driving force for chemical reaction tends to promote polycrystal formation in this case.

As an example, we follow the thought process involved in the design of a CVD reaction to grow crystalline Y_2O_3 films. Following the treatment by Laudise (Ref. 11), consider the reaction

$$2YCl_{3(g)} + (3/2)O_{2(g)} \rightleftarrows Y_2O_{3(s)} + 3Cl_{2(g)} . \qquad (4\text{-}18)$$

At 1000 K, $\Delta G° = -59.4$ kcal/mole, corresponding to log $K = +13$. The reaction is thus too far to the right for practical film growth. If the chloride is replaced by a bromide or an iodide, the situation is worse. YBr_3 and YI_3 are expected to be less stable than YCl_3, making $\Delta G°$ even more negative. The situation is improved by adding a gas-phase reaction with a positive value of

$\Delta G°$: e.g.,

$$CO_{2(g)} \rightleftarrows CO_{(g)} + (1/2)O_{2(g)}; \qquad \Delta G° = +46.7 \text{ kcal/mole}. \quad (4\text{-}19)$$

Thus, the possible overall reaction is now

$$2YCl_{3(g)} + 3CO_{2(g)} \rightleftarrows Y_2O_{3(s)} + 3CO_{(g)} + 3Cl_{2(g)}, \quad (4\text{-}20)$$

and $\Delta G° = -59.4 + 3(46.7) = +80.7$ kcal/mole. The equilibrium now falls too far to the left, but by substituting YBr_3 and Br_2 for YCl_3 and Cl_2, we change the sign of $\Delta G°$ once again. Thus, for

$$2YBr_{3(g)} + 3CO_{2(g)} \rightleftarrows Y_2O_{3(s)} + 3CO_{(g)} + 3Br_{2(g)}, \quad (4\text{-}21)$$

$\Delta G° = -27$ kcal/mole.

Although a value of $\Delta G°$ closer to zero would be more desirable, this reaction yields partial pressures of YBr_3 equal to 10^{-2} atm when the total pressure is 2 atm. Growth in other systems has occurred at such pressures. Pending availability of YBr_3 in readily volatile form and questions related to the operating temperature and safety of handling reactants and products, Eq. 4-21 appears to be a potential candidate for successful film growth. For analysis of chemical reactions, good values of thermodynamic data are essential. Several sources of this information are listed among the references (Refs. 12, 13).

4.3.2. Conditions of Equilibrium

Thermodynamics can provide us with much more than a prediciton of whether a reaction will proceed. Under certain circumstances, it can yield quantitative information on the operating intensive variables that characterize the equilibrium. The problem is to evaluate the partial pressures or concentrations of the involved species within the reactor, given the reactant compositions and operating temperature. In practice, the calculation is frequently more complicated than initially envisioned, because in situ mass spectroscopic analysis of operating reactors has, surprisingly, revealed the presence of unexpected species that must be accounted for. For example, in the technologically important deposition of Si films, at least eight gaseous compounds have been identified during the reduction of chlorosilanes. The Si–Cl–H system has been much studied, and the following example illustrates the method of calculation (Refs. 14, 15). The most abundant chemical species in this system are $SiCl_4$, $SiCl_3H$, $SiCl_2H_2$, $SiClH_3$, SiH_4, $SiCl_2$, HCl, and H_2. These eight gaseous species are connected by the following six equations of chemical equilibrium:

$$SiCl_{4(g)} + 2H_{2(g)} \rightleftarrows Si_{(s)} + 4HCl_{(g)}; \qquad K_1 = \frac{(a_{Si})P_{HCl}^4}{P_{SiCl_4}P_{H_2}^2}, \tag{4-22a}$$

$$SiCl_3H_{(g)} + H_{2(g)} \rightleftarrows Si_{(s)} + 3HCl_{(g)}; \qquad K_2 = \frac{(a_{Si})P_{HCl}^3}{P_{SiCl_3H}P_{H_2}}, \tag{4-22b}$$

$$SiCl_2H_{2(g)} \rightleftarrows Si_{(s)} + 2HCl_{(g)}; \qquad K_3 = \frac{(a_{Si})P_{HCl}^2}{P_{SiCl_2H_2}}, \tag{4-22c}$$

$$SiClH_{3(g)} \rightleftarrows Si_{(s)} + HCl_{(g)} + H_{2(g)}; \qquad K_4 = \frac{(a_{Si})P_{HCl}P_{H_2}}{P_{SiClH_3}}, \tag{4-22d}$$

$$SiCl_{2(g)} + H_{2(g)} \rightleftarrows Si_{(s)} + 2HCl_{(g)}; \qquad K_5 = \frac{(a_{Si})P_{HCl}^2}{P_{SiCl_2}P_{H_2}}, \tag{4-22e}$$

$$SiH_{4(g)} \rightleftarrows Si_{(s)} + 2H_{2(g)}; \qquad K_6 = \frac{(a_{Si})P_{H_2}^2}{P_{SiH_4}}. \tag{4-22f}$$

The activity of solid Si, a_{Si}, is taken to be unity.

To solve for the eight unknown partial pressures, we need two more equations relating these quantities. The first specifies that the total pressure in the reactor, which is equal to the sum of the individual partial pressures, is fixed, say at 1 atm. Therefore,

$$P_{SiCl_4} + P_{SiCl_3H} + P_{SiCl_2H_2} + P_{SiClH_3} + P_{SiH_4} + P_{SiCl_2} + P_{HCl} + P_{H_2} = 1. \tag{4-23}$$

The final equation involves the Cl/H molar ratio, which may be taken to be fixed if neither Cl or H atoms are effectively added or removed from the system. Therefore,

$$\left(\frac{Cl}{H}\right) = \frac{4P_{SiCl_4} + 3P_{SiCl_3H} + 2P_{SiCl_2H_2} + 2P_{SiCl_2} + P_{SiClH_3} + P_{HCl}}{2P_{H_2} + P_{SiCl_3H} + 2P_{SiCl_2H_2} + 3P_{SiClH_3} + P_{HCl} + 4P_{SiH_4}}. \tag{4-24}$$

The numerator represents the total amount of Cl in the system and is equal to the sum of the Cl contributed by each specie. For example, the mass of Cl in $SiCl_4$ is given by $m_{Cl} = 4M_{Cl}(m_{SiCl_4}/M_{SiCl_4})$, where m and M refer to the mass and molecular weight, respectively. But by the perfect gas law,

$$\frac{m_{\mathrm{SiCl_4}}}{M_{\mathrm{SiCl_4}}} = \frac{P_{\mathrm{SiCl_4}}V}{RT},$$

and, therefore,

$$\text{number of moles of Cl} = \left(\frac{m_{\mathrm{Cl}}}{M_{\mathrm{Cl}}}\right) = \frac{4P_{\mathrm{SiCl_4}}V}{RT}.$$

Similarly, for all other terms in the numerator and denominator.

The common factor V/RT, involving the volume V and the temperature T of the reactor, cancels, and all that is left is given by Eq. 4-24. There are now eight independent equations relating the eight unknown partial pressures, which can be determined, at least in principle. First, however, the individual equilibrium constants K_i must be specified, and this requires a slight excursion requiring additional thermodynamic calculation. The value of K_i is fixed by specifying T and $\Delta G°$. A convenient summary of thermodynamic data in the Si–Cl–H system is given in Fig. 4-4, where the free energies of compound formation are plotted versus temperature in an Ellingham-type diagram. Each line represents the equation $\Delta G° = \Delta H° - T\Delta S°$, from which $\Delta H°$, $\Delta S°$, and $\Delta G°$ can be calculated for the compound in question at any temperature. For example, consider the formation reactions for $SiCl_4$ and HCl at 1500 K.

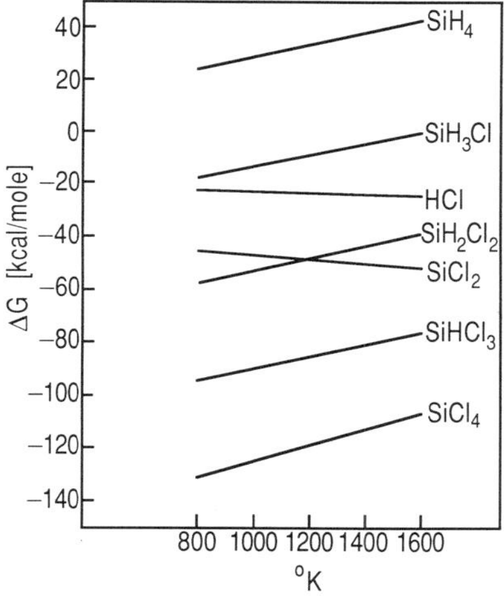

Figure 4-4. Free energies of formation of important gaseous species in the Si–Cl–H system in the temperature range 800–1600 K. (Reprinted by permission of the publisher, The Electrochemical Society, Inc. from Ref. 14).

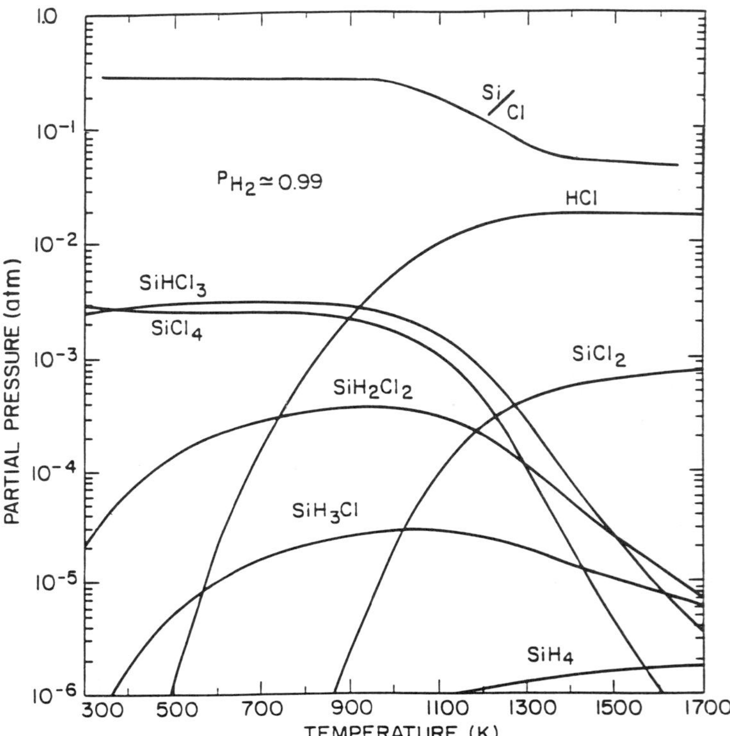

Figure 4-5. Composition of gas phase versus reactor temperature. Total pressure = 1 atm, Cl/H = 0.01. (E. Sirtl, L. P. Hunt and D. H. Sawyer, *J. Electrochem. Soc.* **121**, 919(1974).

From Fig. 4-4,

$$Si + 2Cl_2 \rightleftarrows SiCl_4, \qquad \Delta G^\circ = -106 \text{ kcal/mole},$$

$$(1/2)H_2 + (1/2)Cl_2 \rightleftarrows HCl, \qquad \Delta G^\circ = -25 \text{ kcal/mole},$$

$$SiCl_4 + 2H_2 \rightleftarrows Si + 4HCl, \qquad \Delta G^\circ = +106 + 4(-25) = +6 \text{ kcal/mole}.$$

Therefore $K_1 = \exp - 6000/(1.99)1500 = 0.13$, and similarly for other values of K.

The results of the calculation are shown in Fig. 4-5 for a molar ratio of [Cl/H] = 0.01, which is typical of conditions used for epitaxial deposition of Si. Through application of an equation similar to 4-24, the molar ratio of [Si/Cl] was obtained and is schematically plotted in the same figure. A reactor operating temperature in the vicinity of 1400 K is suggested because the Si content in the gas phase is then minimized. Such temperatures are employed in practice. Analogous calculations have also been made for the case where [Cl/H] = 0.1, which is typical of conditions favoring deposition of polycrystalline Si. At equivalent temperatures, the [Si/Cl] ratios are somewhat higher than obtained for epitaxial deposition, reflecting the greater Si gas concentration operative during polycrystal growth. In both cases, hydrogen is by far the most abundant species in the gas phase.

Similar multispecies thermodynamic analyses involving simultaneous nonlinear equations have been made in a variety of semiconductor, oxide, nitride, and carbide systems. As a second example, consider the deposition of silicon carbide utilizing two independent source gases CH_4 and $SiCl_4$ (Ref. 16). These react according to Eq. 4-9. There are now four atomic species (C, Cl, H, and Si) rather than three for the deposition of Si. The Gibbs phase rule (Eq. 1-18) can be useful in analyzing this situation. Since $n = 4$ and $\psi = 2$ (solid SiC and gas), $f = 4$. This suggests that, in addition to fixing the temperature and pressure of the reactor, it is now necessary to specify two input molar ratios (i.e., [H/Cl], [Si/C]) to uniquely determine the system. The results of a thermodynamic analysis shown in Fig. 4-6 predicts that stoichiometric SiC will form for only those combinations of H/Cl and Si/C that fall on the darkened line. In actuality, SiC can be synthesized over a fairly broad range of input gas ratios at 1400 °C, as seen in the experimentally determined "phase diagram."

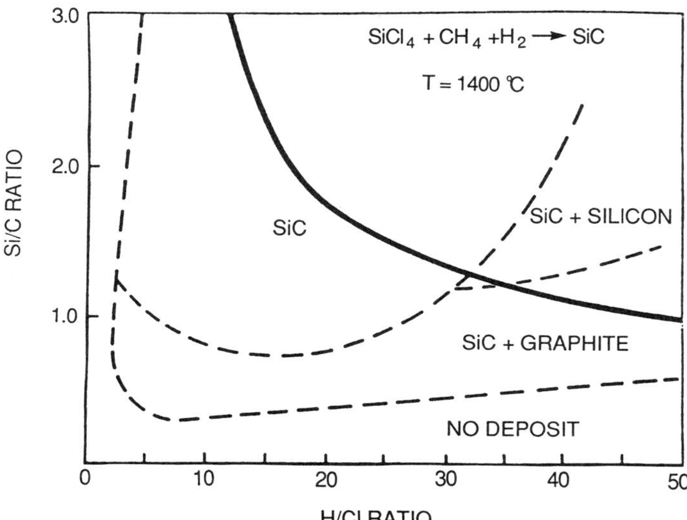

Figure 4-6. Map of decomposition products for various [Si/C] and [H/Cl] ratios at 1400 °C. (Reprinted with permission from JOM, (formerly *J. of Metals*) **28**(6), 6 (1976), a publication of the Minerals, Metals & Materials Society).

This behavior is indicative of complex thermodynamic and reaction kinetics factors that are not easily accounted for.

4.4. Gas Transport

Gas transport is the process by which volatile species flow from one part of a reactor to another. It is important to understand the nature of gas transport phenomena in CVD systems for the following reasons:

1. The deposited film or coating thickness uniformity depends on the delivery of equal amounts of reactants to all substrate surfaces.
2. High deposition growth rates depend on optimizing the flow of reactants through the system and to substrates.
3. More efficient utilization of process gases can be achieved as a result.
4. The computer modeling of CVD processes can be facilitated, enabling improved reactor design and better predictive capability with regard to performance.

At the outset, it is important to distinguish between diffusion and bulk flow processes in gases. Diffusion involves the motion of individual atomic or molecular species, whereas in bulk transport processes, such as viscous flow or convection, parts of the gas move as a whole. Different driving forces and resulting transport equations define and characterize these two broad types of gas flow. Each of these will now be discussed briefly as a prelude to considering the combinations of flow that take place in actual CVD reactors.

4.4.1. Viscous Flow

The viscous flow regime is operative when gas transport occurs at pressures of roughly 0.01 atm and above in reactors of typical size. This is also the pressure range characteristic of CVD systems. At typical flow velocities of tens of cm/sec, the reactant gases exhibit what is known as laminar or streamline flow. The theory of fluid mechanics provides a picture of what occurs under such circumstances. We shall consider the simplest of flow problems—that parallel to a flat plate.

As shown in Fig. 4-7, the flow velocity has a uniform value v_0, but only prior to impinging on the leading edge of the plate. However, as flow progresses, velocity gradients must form because the gas clings to the plate. Far away, the velocity is still uniform, but drops rapidly to zero at the plate

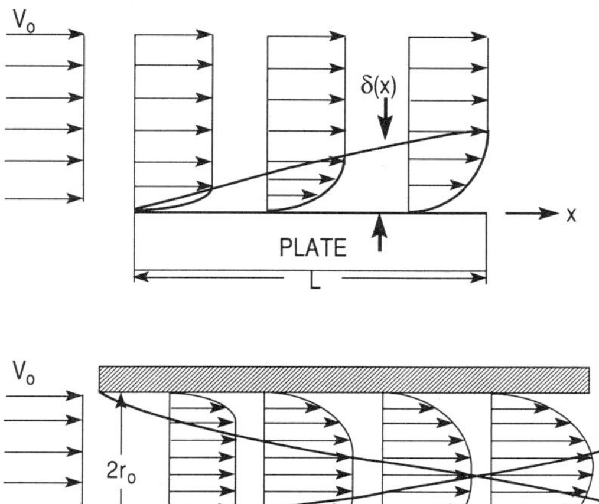

Figure 4-7. Laminar gas flow patterns: (top) flow across flat plate; (bottom) flow through circular pipe.

surface, creating a boundary layer. The latter grows with distance along the plate and has a thickness $\delta(x)$ given by $\delta(x) = 5x/\sqrt{\mathrm{Re}_x}$, where Re_x is the Reynolds number, defined as $\mathrm{Re}_x = v_0 \rho x/\eta$. The quantities η and ρ are the gas viscosity and density, respectively. More will be said about η, but note that the viscosity essentially establishes the frictional viscous forces that decelerate the gas at the plate surface.

The average boundary-layer thickness over the whole plate is

$$\bar{\delta} = \frac{1}{L}\int_0^L \delta(x)\, dx = \frac{10}{3}\frac{L\sqrt{\eta}}{\sqrt{v_o \rho L}} = \frac{10}{3}\frac{L}{\sqrt{\mathrm{Re}_L}}, \tag{4-25}$$

where Re_L is defined as $\mathrm{Re}_L = \rho v_0 L/\eta$. Because both gaseous reactants and products must pass through the boundary layer separating the laminar stream and film deposit, low values of $\bar{\delta}$ are desirable in enhancing mass-transport rates. This can be practically achieved by increasing the gas flow rate (v_0), which raises the value of Re. Typical values of Re in CVD reactors range up to a few hundred. If, however, Re exceeds approximately 2100, a transition from laminar to turbulent flow occurs. The resulting erratic gas eddies and swirls are not conducive to uniform defect-free film growth, and are to be avoided.

It is instructive to now consider gas flow through a tube of circular cross section. The initial uniform axial flow velocity is altered after the gas enters the tube. Boundary layers develop at the walls and grow with distance along the tube, as shown in Fig. 4-7. The Reynolds number is now given by $\sim 2\rho v_0 r_0/\eta$, where r_0 is the tube radius. Beyond a certain critical entry length $L_e \approx 0.07 r_0 \mathrm{Re}$, the flow is fully developed and the velocity profile no longer changes. At this point, the boundary layers around the tube circumference have merged, and the whole cross section consists of "boundary layer." The axial flow is now described by the Hagen–Poiseuille relation

$$\dot{V} = \frac{\pi r_0^4}{8\eta} \frac{\Delta P}{\Delta x} \ \mathrm{cm^3/sec}, \tag{4-26}$$

where $\dot{V}$ is the volumetric flow rate and $\Delta P/\Delta x$ is the pressure gradient driving force for viscous flow. The volumetric flow rate is defined as the volume of gas that moves per unit time through the cross section and is related to the average gas velocity $\bar{v}$ by $\dot{V} = \pi r_0^2 \bar{v}$. Within the tube, the gas velocity $v(r)$ assumes a parabolic profile as a function of the radial distance r from the center, given by $\bar{v}(r) = v_{\max}(1 - r^2/r_0^2)$, where $v_{\max}$ is the maximum gas velocity. The gas flux J is given by the product of the concentration of the species in question and the velocity with which it moves:

$$J_i = C_i \bar{v}_i. \tag{4-27}$$

Upon substitution of $C_i = P_i/RT$ from the perfect gas law, and $\bar{v}_i = \dot{V}/\pi r_0^2$, we have

$$J_i = \frac{P_i}{RT} \frac{r_0^2}{8\eta} \frac{\Delta P_i}{\Delta x}. \tag{4-28}$$

Provided the molar flux of any gaseous species in a chemical reaction is known, the fluxes of other species can be determined from the stoichiometric coefficients if equilibrium conditions prevail.

Viscous flow is characterized by the coefficient of viscosity η. The kinetic theory of gases predicts that η varies with temperature as $T^{1/2}$ but is independent of pressure. Experimental data bear out the lack of a pressure dependance at least to several atmospheres, but indicate that η varies as T^n, with n having values between 0.6 and 1.0. Gas viscosities typically range between 0.01 centipoise (cP) at 0 °C and 0.1 cP at 1000 °C (1 poise = 1 dyne-sec cm^{-2}.)

4.4.2. Diffusion

The phenomenon of diffusion occurs in gases and liquids as well as in solids. If two different gases are initially separated and then allowed to mix, each will move from regions of higher to lower concentration, thus increasing the entropy of the system. The process by which this occurs is known as diffusion and is characterized by Fick's law (Section 1.6). Elementary kinetic theory of gases predicts that the diffusivity D depends on pressure and temperature as $D \sim T^{3/2}/P$. It is therefore usual to represent D in gases by

$$D = D_0 \frac{P_0}{P} \left(\frac{T}{T_0} \right)^n, \tag{4-29}$$

where n is experimentally found to be approximately 1.8. The quantity D_0, the value of D measured at standard temperature T_0 (273 K) and pressure P_0 (1 atm), depends on the gas combination in question. Typical D_0 values at temperatures of interest span the range 0.1–10 cm^2/sec and are many orders of magnitude higher than even the largest values for diffusivity in solids. If the gas composition is reasonably dilute so that the perfect gas law applies, $C = P/RT$ and Eq. 1-21 can be equivalently expressed by

$$J_i = -\frac{D}{RT} \frac{dP_i}{dx}. \tag{4-30}$$

This formula can be applied to the diffusion of gas through the stagnant boundary layer of thickness δ adjacent to the substrate. The flux is then given by

$$J_i = -\frac{D(P_i - P_{i0})}{\delta RT}. \tag{4-31}$$

Here P_i is the vapor pressure in the bulk gas and P_{i0} is the vapor pressure at the surface.

Since D varies inversely with pressure, gas mass-transfer rates can be enhanced by reducing the pressure in the reactor. Advantage of this fact is taken in low-pressure CVD (LPCVD) systems, which are now extensively employed in semiconductor processing. Their operation will be discussed in Section 4.6.3.

As an example that integrates both thermodynamics and diffusion in a CVD process, consider the deposition of CdTe films by close-spaced vapor transport (CSVT) (Ref. 17). In this process, mass is transferred from a solid CdTe source at temperature T_1 located a very short distance l (typically 1 mm) from the substrate maintained at T_2 ($T_1 > T_2$). Our objective is to establish conditions necessary to derive an expression for the film growth rate. We assume that chemical equilibrium prevails at the respective temperatures. The basic reaction is

$$CdTe_{(s)} \rightleftarrows Cd_{(g)} + 1/2Te_{2(g)}, \tag{4-32}$$

for which $\Delta G = 68.64 - 44.94 \times 10^{-3}T$ kcal/mole. Therefore, the equations

$$P_{Cd}(T_1)P_{Te_2}^{1/2}(T_1) = \exp - \frac{\Delta G(T_1)}{RT_1} = K(T_1), \qquad (4\text{-}33a)$$

$$P_{Cd}(T_2)P_{Te_2}^{1/2}(T_2) = \exp - \frac{\Delta G(T_2)}{RT_2} = K(T_2) \qquad (4\text{-}33b)$$

express the equilibria at source and substrate. If the concentrations of the gas-phase species vary linearly with distance, the individual mass fluxes (in units of moles/cm^2-sec) are expressed by

$$J_{Cd} = \frac{D_{Cd}}{l}\left(\frac{P_{Cd}(T_1)}{RT_1} - \frac{P_{Cd}(T_2)}{RT_2}\right), \qquad (4\text{-}34a)$$

$$J_{Te_2} = \frac{D_{Te_2}}{l}\left(\frac{P_{Te_2}(T_1)}{RT_1} - \frac{P_{Te_2}(T_2)}{RT_2}\right). \qquad (4\text{-}34b)$$

Note the use of the perfect gas law and the neglect of the temperature dependence of D. Maintenance of stoichiometry requires that

$$J_{Cd} = 2J_{Te_2}, \qquad (4\text{-}35)$$

the factor of 2 arising because Te_2 is a dimer. The film growth rate is obtained from the relation

$$\dot{G}(\mu m/min) = \frac{J_{Cd}M_{CdTe}(60 \times 10^4)}{\rho}, \qquad (4\text{-}36)$$

where M and ρ are the molar mass and density of CdTe, respectively. If T_1 exceeds T_2 by approximately 100 °C, then $P_i(T_1) \gg P_i(T_2)$, where i refers to both Cd and Te_2. By neglecting the T_2 terms in Eqs. 4-34a and b, we write

$$\frac{P_{Cd}(T_1)}{P_{Te_2}(T_1)} = \frac{2D_{Te_2}}{D_{Cd}} = 1.1. \qquad (4\text{-}37)$$

The value of 1.1 is derived from kinetic theory of gases, which suggests that $D_{Cd} = 1.85D_{Te_2}$ in H_2, He, or Ar ambients. Equations 4-33, 4-34, 4-35, and 4-37 enable all the partial pressures to be determined. By knowing the value of D_{Cd} or D_{Te_2}, $\dot{G}$ can be evaluated.

Since its inception in 1963, CSVT has been used to grow a wide variety of semiconductor films for experimental purposes, including CdS, CdSe, GaP, GaAs, $GaAs_xP_{1-x}$, $Hg_{1-x}Cd_xTe$, InP, ZnS, ZnSe, and ZnTe.

4.4.3. Convection

Convection is a bulk gas flow process that can be distinguished from both diffusion and viscous flow. Whereas gas diffusion involves the statistical motion of atoms and molecules driven by concentration gradients, convection arises from the response to gravitational, centrifugal, electric, and magnetic forces. It is manifested in CVD reactors when there are vertical gas density or temperature gradients. An important example occurs in cold-wall reactors, such as depicted in Figure 4-13, where heated susceptors are surrounded above as well as on the sides by the cooler walls. Cooler, denser gases then lie above hotter, less dense gases. The resultant convective instability causes an overturning of the gas by bouyancy effects. Subsequently, a complex coupling of mass and heat transfer serves to reduce both density and temperature gradients in the system. Another example of convective flow occurs in two-temperature-zone, vertical reactors. In the disproportionation process considered previously, it is immaterial whether the hotter zone is physically located above or below the cooler zone insofar as thermodynamics is concerned. But efficient gas flow considerations mandate the placement of the cooler region on top to enhance gas circulation as shown in Fig. 4-2.

Note that film growth is limited by viscous, diffusive, and convective mass transport fluxes, which, in turn, are driven by gas pressure gradients. In open reactors the metered gas (volumetric) flow rates establish these pressure gradients. In closed reactors the latter arise because of imposed temperature differences that locally alter the equilibrium partial pressures.

4.5. Growth Kinetics

The growth kinetics of CVD films depends on several factors associated with the gas–substrate interface, including

a. Transport of reactants through the boundary layer to the substrate
b. Adsorption of reactants at the substrate
c. Atomic and molecular surface diffusion, chemical reactions, and incorporation into the lattice
d. Transport of products away from the substrate through the boundary layer

The intimate microscopic details of these steps are usually unknown, and, therefore, the growth kinetics are frequently modeled in macroscopic terms. This is a simpler approach since it makes no atomistic assumptions but yet is capable of predicting deposit growth rates and uniformity within reactors.

4.5.1. Growth Rate Uniformity

In what follows, a relatively simple approach to the analysis of epitaxial growth of Si in a horizontal reactor is considered. In particular, we are interested in knowing how uniform the deposit will be as a function of distance along the reactor. Although the specific reaction considered is the hydrogen reduction of chlorosilane, the results can be broadly applied to other CVD processes as well.

In this treatment (Ref. 18), the reactor configuration is shown in Fig. 4-8a and the following is assumed:

1. The gas has a constant velocity component along the axis tube.
2. The whole system is at constant temperature.
3. The reactor extends a large distance in the z direction so that the problem reduces to one of two dimensions.

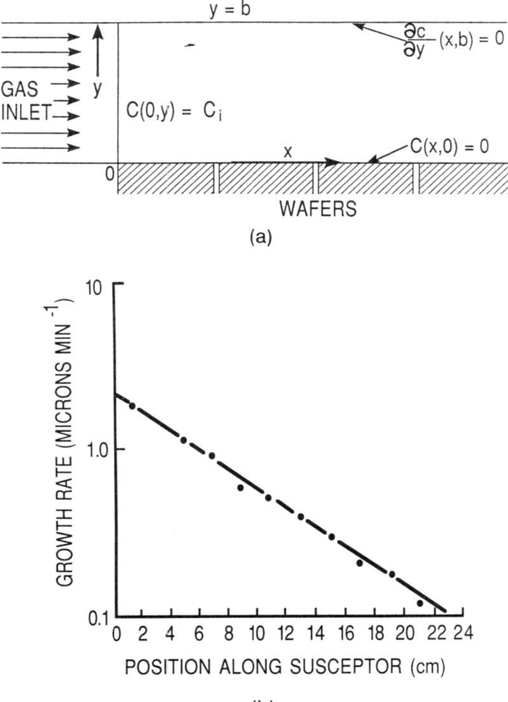

Figure 4-8. (a) Horizontal reactor geometry. (b) Variation of growth rate with position along susceptor. Reactor conditions: $\bar{v} = 7.5$ cm/sec, $b = 1.4$ cm, $T = 1200$ °C, and $C_i = 3.1 \times 10^{-5}$ g/cm^3. (Reprinted from P. C. Rundle, *Int'l J. Electronics* **24**, 405, © 1968 Taylor and Francis, Ltd.).

The flow is simply treated by assuming the mass flux **J** vector at any point to be composed of two terms:

$$\mathbf{J} = C(x, y)\bar{v} - D\nabla C(x, y). \tag{4-38}$$

The first term represents a bulk viscous (plug) flow where the source of concentration $C(x, y)$ moves as a whole with drift velocity $\bar{v}$. The second term is due to diffusion of individual gas molecules, with diffusivity D, along concentration gradients. Since two-dimensional diffusion is involved, both x and y components of ∇C must be considered.

By taking the difference of the mass flux into and out of an elemental volume (as was done in Chapter 1) and equating it to the mass accumulation, we obtain

$$\frac{\partial C(x, t)}{\partial t} = D\left(\frac{\partial^2 C(x, y)}{\partial x^2} + \frac{\partial^2 C(x, y)}{\partial y^2}\right) - \bar{v}\frac{\partial C(x, y)}{\partial x}. \tag{4-39}$$

Only the steady-state solutions are of concern, so $\partial C(x, t)/\partial t = 0$. The resulting equation is subject to three conditions:

$$C = 0 \qquad \text{when } y = 0, \quad x > 0, \tag{4-40a}$$

$$\frac{\partial C}{\partial y} = 0 \qquad \text{when } y = b, \quad x \geq 0, \tag{4-40b}$$

$$C = C_i \qquad \text{at } x = 0, \quad b \geq y \geq 0. \tag{4-40c}$$

The first condition assumes that the chemical reaction is complete at the substrate surface $y = 0$ and, therefore, the concentration of the Si containing source gas is zero there. The second condition implies that there is no net diffusive mass flux at the top of the reactor. Gas molecules impinging at surface $y = b$ are merely reflected back into the system. The final boundary condition states that the input source gas concentration is C_i, a constant.

Equations 4-39 and 4-40 specify a boundary value problem, and the well-known techniques of partial differential equations involving separation of variables or Laplace transform methods gives as the solution

$$C(x, y) = \frac{4C_i}{\pi}\sum_{n=0}^{\infty}\frac{1}{2n+1}\, sin\left[(2n+1)\frac{\pi y}{2b}\right] \times \exp\left\{\frac{\bar{v}}{D} - \sqrt{\frac{\bar{v}^2}{D^2} + (2n+1)^2\frac{\pi^2}{b^2}}\right\}x. \tag{4-41}$$

To obtain a more readily usable form of the solution, we assume as a first approximation that $\bar{v}b \geq D\pi$. Except for short distances into the reactor or small values of x, only the first term in the series need be retained. These simplications give

$$C(x, y) = \frac{4C_i}{\pi}\sin\left(\frac{\pi y}{2b}\right)\exp - \frac{\pi^2 Dx}{4\bar{v}b^2}. \tag{4-42}$$

The flux of source gas to the substrate surface is given by

$$J(x) = -D\frac{\partial C(x, y)}{\partial y}\bigg|_{y=0} \text{ g/cm}^2\text{sec}. \tag{4-43}$$

The resultant deposit growth rate $\dot{G}(x)$ is related to $J(x)$ through simple material constants by

$$\dot{G}(x) = \frac{M_{\text{Si}}}{\rho M_s}J(x) \text{ cm/sec} \tag{4-44}$$

where M_{Si} and M_s are the molecular weights of the Si and source gas, respectively, and ρ is the density of Si. Combining Eqs. (4-42)–(4-44) yields

$$\dot{G}(x) = \frac{2C_i M_{\text{Si}}}{b\rho M_s}D\exp - \frac{\pi^2 Dx}{4\bar{v}b^2}. \tag{4-45}$$

The values of D, $\bar{v}$, and C_i are strictly those pertaining to the mean temperature of the reactor, T. An exponential decay in the Si growth rate with distance along the reactor is predicted. This is not too surprising, since the input gases are progressively depleted of reactants. The implicit boundary condition requiring that $C = 0$ at $x \approx \infty$ accounts for this loss. Despite the extreme simplicity of the assumptions, the model provides rather good agreement with experimental data on the variation of Si growth rate with distance, as indicated in Fig. 4-8b.

From the standpoint of reactor performance, high growth rates and uniform deposition are the two most important concerns, assuming film quality is not compromised. The equation for $\dot{G}(x)$ provides design guidelines, but they are not always simple to implement. Tilting the susceptor as shown in Fig. 4-13 is an effective way to improve growth uniformity. The increase in gas flow velocity in the tapered space above susceptors within horizontal or barrel reactors serves to decrease δ. Enhanced transport across the stagnant layer then compensates for reactant depletion. Another simple remedy is to continuously increase the temperature downstream within the reactor.

4.5.2. Temperature Dependence

It is instructive to reproduce the treatment of kinetics of film growth by Grove (Ref. 19), to understand the effect of temperature. The essentials of this simple model are shown in Fig. 4-9, where the environment in the vicinity of the gas-growing film interface is shown. A drop in concentration of the reactant from C_g in the bulk of the gas to C_s at the interface occurs. The corresponding mass flux is given by

$$J_{gs} = h_g(C_g - C_s), \tag{4-46}$$

where h_g is the gas-phase mass-transfer coefficient, to be defined later. The flux consumed by the reaction taking place at the surface of the growing film is approximated by

$$J_s = k_s C_s, \tag{4-47}$$

where first-order kinetics are assumed, and k_s is the rate constant for surface reaction. In the steady state, $J_{gs} = J_s$, so

$$C_s = \frac{C_g}{1 + k_s/h_g}. \tag{4-48}$$

This formula predicts that the surface concentration drops to zero if $k_s \gg h_g$, a condition referred to as mass-transfer control. In this case, low gas transport through the boundary layer limits the otherwise rapid surface reaction.

Conversely, surface reaction control dominates where $h_g \gg k_s$, in which case C_s approaches C_g. Here the surface reaction is sluggish even though sufficient reactant gas is available. The film growth rate $\dot{G}$ is given by $\dot{G} = J_s/N_0$, where N_0 is the atomic density or number of atoms incorporated

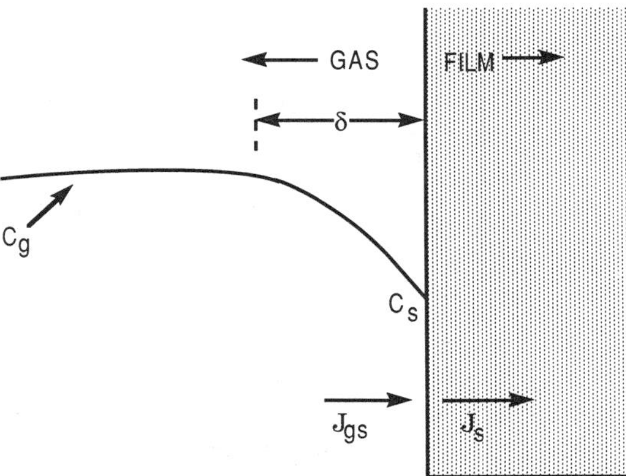

Figure 4-9. Model of growth process. Gas flows normal to plane of paper. (Reprinted with permission from John Wiley and Sons, from A. S. Grove, *Physics and Technology of Semiconductor Devices*, Copyright © 1967, John Wiley and Sons).

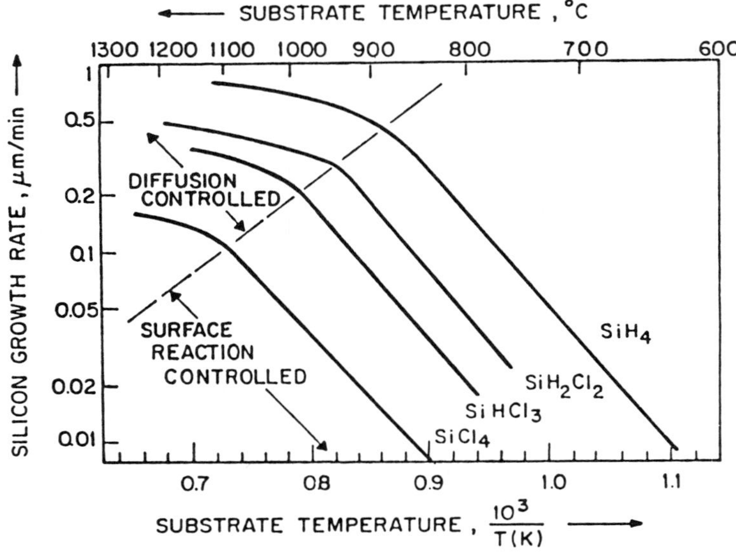

Figure 4-10. Deposition rate of Si from four different precursor gases as a function of temperature. (From W. Kern in *Microelectronic Materials & Processes*, ed. by R. A. Levy, reprinted by permission of Kluwer Academic Publishers, 1989).

into the film per unit volume. Therefore,

$$\dot{G} = \frac{k_s h_g}{k_s + h_g} \frac{C_g}{N_0}. \tag{4-49}$$

The temperature dependence of $\dot{G}$ hinges on the properties of k_s and h_g. A Boltzmann factor behavior dominates the temperature dependence of k_s; i.e., $k_s \sim \exp - E/RT$, where E is the characteristic activation energy involved. Comparison of Eqs. 4-31 and 4-46 reveals that h_g is related to D/δ. Since D_g varies as T^2 at most and δ is weakly dependent on T, h_g is relatively insensitive to variations in temperature. At low temperatures, film growth is surface-reaction-controlled; i.e., $\dot{G} = k_s C_g / N_0$. At high temperatures, however, the mass transfer or diffusion-controlled regime is accessed where $\dot{G} = h_g C_g / N_0$. The predicted behavior is borne out by growth rate data for epitaxial Si, as shown in the Arrhenius plots of Fig. 4-10. Actual film growth processes are carried out in the gas diffusion-controlled region, where the temperature response is relatively flat. At lower temperatures the same activation energy of about 1.5 eV is obtained irrespective of the chlorosilane used. Migration of Si adatoms is interpreted to be the rate-limiting step in this temperature regime.

Table 4-2. Influence of Process Conditions on Kinetics of Si Deposition From $SiHCl_3-H_2$ and $SiCl_4-H_2$ Mixtures

Variable	Diffusion Controlled		Reaction Controlled
1. Linear flow rate	Low (1a)	Medium	High (1b)
2. Mole fraction chlorosilane	Low (2a)	Medium	High (2b)
3. Substrate temperature	High (3a)	Medium	Low (3b)
4. Temperature gradient (near surface)	Low	Medium	High
5. Surface site density	High	Medium	Low
6. Silicon surface per reaction volume	High	Medium	Low

1a < 0.3 cm/sec 1b. > 3 cm/sec
2a < 0.01 2b. > 0.1
3a > 1550 K 3b < 1300 K
From Ref. 15.

Regardless of the different types of equipment and deposition conditions employed, general outlines can be given as to what extent either diffusion-controlled or surface-controlled processes dominate. In Table 4-2 the influence of the most important variables is shown. Depending on whether the balance of conditions lies to the left or to the right of center, a reasonable prediction of operative kinetic mechanism can be made.

4.5.3. Thermodynamic Considerations

The previous discussion implies that all reaction rates increase with temperature. Though generally true, it is sometimes observed that higher reactor temperatures lead to lower film growth rates in certain systems. This apparent paradox can be explained by considering the reversibility of chemical reactions. In Chapter 1 the net rate for a forward exothermic reaction (and reverse endothermic reaction) was given by Eq. 1-36 and modeled in Fig. 1-18. Recall that exothermic reactions mean that the sign of $\Delta H°$ is negative; the reactants have more energy than the products. For endothermic reactions, $\Delta H°$ is positive. The individual forward and reverse reaction components are now shown in Fig. 4-11a on a common plot. Clearly, the activation energy barrier (or slope) for the reverse reaction exceeds that for the forward reaction. The *net* reaction rate or difference between the individual rates is also indicated. Interestingly, it reaches a maximum and then drops with temperature. A practical manifestation of this is etching—the reverse of deposition—in the high-temperature range.

In Fig. 4-11b the alternate case is considered—a forward endothermic deposition reaction and a reverse exothermic reaction. Here, the net reaction

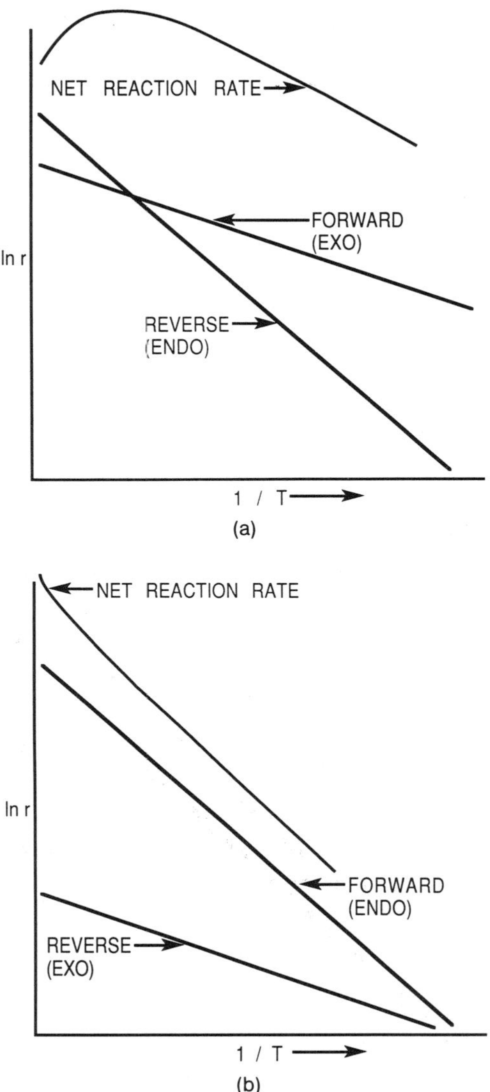

Figure 4-11. Chemical reaction energetics. (From Ref. 20.) (a) Activation energy for forward exothermic reaction is less than for reverse endothermic reaction. (b) Activation energy for forward endothermic reaction is greater than for reverse exothermic reaction.

rate increases monotonically with temperature, and film growth rates will always increase with temperature. It is left for the reader to show that $\Delta H°$ for the reduction of $SiCl_4$ by H_2 (Eq. 4-3) is endothermic as written; e.g., $\Delta H° = +60$ kcal/mole. There is actually a thermodynamic driving force that tends to transport Si from the cooler regions (i.e., walls) to the hottest part of the reactor. This is where inductively heated substrates are placed and where film growth rates are highest. Epitaxial Si is most efficiently deposited in *cold-wall* reactors for this reason.

The opposite is true when $\Delta H° < 0$. Reversible reactions (Eq. 4-14) for the deposition of GaAs are exothermic, and *hot-wall* reactors are employed in this case to prevent deposition on the walls.

These results are a direct consequence of the well-known van't Hoff equation

$$\frac{d \ln K_{eq}}{dT} = \frac{\Delta H°}{RT^2}, \tag{4-50}$$

where K_{eq} is the reaction equilibrium constant.

4.5.4. Structure

The actual film and coating structural morphologies that develop during CVD are the result of a complex sequence of atomic migration events on substrates leading to observable nucleation and growth processes (Ref. 21). Since the kinetic details are similar in all film formation and growth processes irrespective of whether deposition is chemical or physical in nature, they will be treated within a common framework in Chapter 5. Perhaps the two most important variables affecting growth morphologies are vapor supersaturation and substrate temperature. The former influences the film nucleation rate, whereas the latter affects the growth rate. In concert they influence whether epitaxial films, platelets, whiskers, dendrites, coarse-grained polycrystals, fine-grained polycrystals, amorphous deposits, gas-phase powder, or some combination of these form. Thus, single-crystal growth is favored by low gas supersaturation and high substrate temperatures, whereas amorphous film formation is promoted at the opposite extremes. An example of the effect of substrate temperature on the structure of deposited Si films is shown in Fig. 4-12 (Ref. 22). Decomposition of silane at temperatures of about 600 °C and below yields amorphous films with no detectable structure. Polysilicon deposited from 600 to 650 °C has a columnar structure with grain sizes ranging from 0.03 to 0.3 μm and possesses a {110} preferred orientation. Larger Si crystallites form at higher temperatures, and eventually single-crystal film growth can be achieved at 1200 °C.

Figure 4-12. Morphology of poly-Si deposited from SiH_4 on an SiO_2 substrate: (top) columnar grains deposited above 650 °C, (middle) fine-grained poly Si; (bottom) partly amorphous structure deposited at 625 °C. Deposition temperatures below 600 °C produce an amorphous film. Courtesy of R. B. Marcus Bellcore Corp. (Reprinted with permission from John Wiley and Sons, from R. B. Marcus and T. T. Sheng, *Transmission Electron Microscopy of Silicon VLSI Circuits and Structures*, Copyright © 1983, John Wiley and Sons).

4.6. CVD Processes and Systems

The great variety of materials deposited by CVD methods has inspired the design and construction of an equally large number of processes and systems. These have been broadly categorized and described by such terms as low and high temperature, atmospheric and low pressure, cold and hot wall, closed and open in order to differentiate them. Incorporation of physical deposition features such as plasmas and evaporation sources has further enriched and expanded the number of potential CVD processes. Within a specific category, the variations in design and operating variables frequently make it difficult to compare performance of individual systems or reactors, even when depositing the same material. Regardless of process type, however, the associated equipment must have the following capabilities:

1. Deliver and meter the reactant and diluent gases into the reactor.
2. Supply heat to the substrates so that reaction and deposition can proceed efficiently.
3. Remove the by-product and depleted gases.

We begin with a discussion of atmospheric pressure processes and distinguish between the low- and high-temperature reactors employed. More information exists on CVD applications and equipment for microelectronic device fabrication and the subsequent discussion will reflect this bias.

4.6.1. Low-Temperature Systems

In the fabrication of both Si bipolar and MOS integrated circuits there is an important need to deposit thin films of SiO_2, phosphosilicate (PSG) and borophosphosilicate (BPSG) glasses, and silicon nitride films in order to insulate, passivate, or hermetically seal various parts of the underlying circuitry. At the present time, an upper temperature limit that can be tolerated is ~ 450 °C because the Al metallization used for device contacts and interconnections begins to react with Si beyond this point. Several types of atmospheric pressure, low-temperature reactors have been devised for the purpose of depositing these insulator films. They include resistance-heated rotary reactors of radial configuration and reactors featuring a close-spaced nozzle geometry. In the latter, gases impinge on wafers translated past the nozzles by a metal conveyer belt. Films of SiO_2 are deposited at 325–450 °C from $SiH_4 + O_2$ mixtures diluted with N_2. Co-oxidation with PH_3 yields PSG, and PH_3 +

B_2H_6 mixtures generate BPSG. As noted in Section 4.6.3, LPCVD processes have largely surpassed atmospheric CVD methods for depositing such films.

4.6.2. High-Temperature Systems

There is need to reduce semiconductor processing temperatures, but the growth of high-quality epitaxial thin films can only be achieved by high-temperature CVD methods. This is true of Si as well as compound semiconductors. High-temperature atmospheric systems are also extensively employed in metallurgical coating operations. The reactors can be broadly divided into hot-wall and cold-wall types. Hot-wall reactors are usually tubular in form, and heating is accomplished by surrounding the reactor with resistance elements. An example of such a reactor for the growth of single-crystal compound semiconductor films by the hydride process was given in Fig. 4-3. Higher temperatures are maintained in the source and reaction zones ($\sim$ 800–850 °C) relative to the deposition zone (700 °C). Prior to deposition, the substrate is sometimes

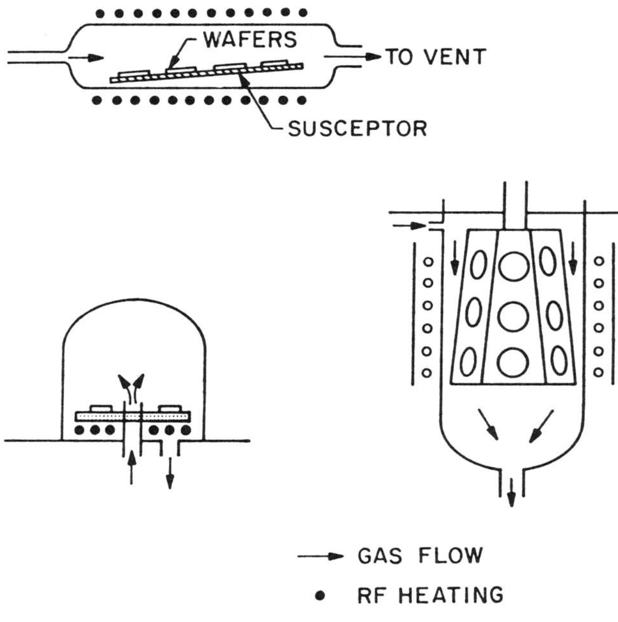

Figure 4-13. Schematic diagrams of reactors employed in epitaxial Si deposition: (top) horizontal; (lower left) pancake; (lower right) barrel. (Reprinted with permission from John Wiley and Sons, from S. M. Sze, *Semiconductor Devices: Physics and Technology*, Copyright © 1985, John Wiley and Sons).

etched by raising its temperatures to 900 °C. Provision for multiple temperature zones is essential for efficient transport of matrix as well as dopant atoms. By programming flow rates and temperatures, the composition, doping level and layer thickness can be controlled, making it possible to grow complex multilayer structures for device applications.

Cold-wall reactors are utilized extensively for the deposition of epitaxial Si films. Substrates are placed in good thermal contact with SiC-coated graphite susceptors, which can be inductively heated while the nonconductive chamber walls are air- or water-cooled. Three popular cold-wall reactor configurations are depicted in Fig. 4-13 (Ref. 23). Of note in both the horizontal and barrel reactors are the tilted susceptors. This feature compensates for reactant depletion, which results in progressively thinner deposits downstream as previously discussed. In contrast to the other types, the wafer substrates lie horizontal in the pancake reactor. Incoming reactant gases flow radially over the substrates where they partially mix with the product gases. Cold-wall reactors typically operate with H_2 flow rates of 100–200 (standard liters per minute) and 1 vol% of $SiCl_4$. Silicon crystal growth rates of 0.2 to 3 μm/min are attained under these conditions. Substantial radiant heat loss from the susceptor surface and consumption of large quantities of gas, 60% of which is exhausted without reacting at the substrate, limit the efficiency of these reactors.

4.6.3. Low-Pressure CVD

One of the more recent significant developments in CVD processing has been the introduction of low-pressure reactor systems for use in the semiconductor industry. Historically, LPCVD methods were first employed to deposit polysilicon films with greater control over stoichiometry and contamination problems. In practice, large batches of wafers, say 100 or more, can be processed at a time. This coupled with generally high deposition rates, improved film thickness uniformity, better step coverage, lower particle density, and fewer pinhole defects has given LPCVD important economic advantages relative to atmospheric CVD processing in the deposition of dielectric films.

The gas pressure of $\sim$ 0.5 to 1 torr employed in LPCVD reactors distinguishes it from conventional CVD systems operating at 760 torr. To compensate for the low pressures, the input reactant gas concentration is correspondingly increased relative to the atmospheric reactor case. Low gas pressures primarily enhance the mass flux of gaseous reactants and products through the boundary layer between the laminar gas stream and substrates. According to Eq. 4-31, the mass flux of the gaseous specie is directly proportional to D/δ.

Since the diffusivity varies inversely with pressure, D is roughly 1000 times higher in the case of LPCVD. This more than offsets the increase in δ, which is inversely proportional to the square root of the Reynolds number. In an LPCVD reactor, the gas flow velocity is generally a factor of 10–100 times higher, the gas density a factor of 1000 lower, and the viscosity unchanged relative to the atmospheric CVD case. Therefore, Re is a factor of 10 to 100 times lower, and δ is about 3 to 10 times larger. Because the change in D dominates that of δ, a mass-transport enhancement of over an order of magnitude can be expected for LPCVD. The increased mean-free path of the gas molecules means that substrate wafers can be stacked closer together, resulting in higher throughputs. When normalized to the same reactant partial pressure, LPCVD film growth rates exceed those for conventional atmospheric CVD.

The commercial LPCVD systems commonly employ horizontal hot-wall reactors like that shown in Fig. 4-14. These consist of cylindrical quartz tubes heated by wire-wound elements. Large mechanical pumps as well as blower booster pumps are required to accommodate the gas flow rates employed—e.g., 50–500 standard cm^3/min at 0.5 torr—and maintain the required operating pressure. One significant difference between atmospheric and LPCVD systems concerns the nature of deposition on reactor walls. Dense adherent deposits accumulate on the hot walls of LPCVD reactors, whereas thinner, less adherent films form on the cooler walls of the atmospheric reactors. In the latter case, particulate detachment and incorporation in films is a problem, especially on horizontally placed wafers. It is less of a problem for LPCVD reactors where vertical stacking is employed. Typically, 100 wafers, 15 cm in

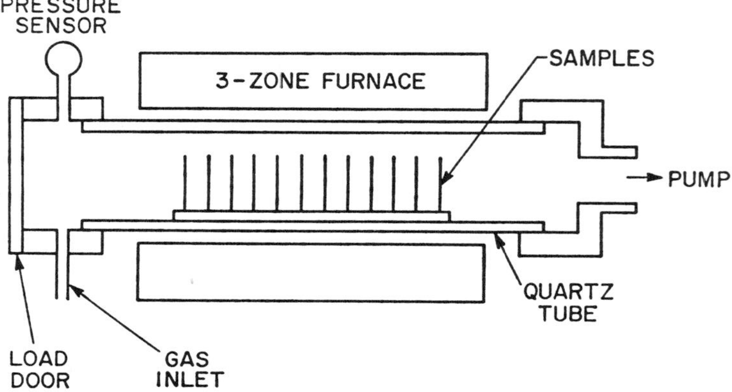

Figure 4-14. Schematic diagram of hot-wall reduced pressure reactor (From Ref. 24).

diameter, can be processed per hour in this reactor. In addition to polysilicon and dielectric films, silicides and refractory metals have been deposited by LPCVD methods.

4.6.4. Plasma-Enhanced CVD

In PECVD processing, glow discharge plasmas are sustained within chambers where simultaneous CVD reactions occur. The reduced-pressure environment utilized is somewhat reminiscent of LPCVD systems. Generally, the radio frequencies employed range from about 100 kHz to 40 MHz at gas pressures between 50 mtorr to 5 torr. Under these conditions, electron and positive-ion densities number between 10^9 and $10^{12}/cm^3$, and average electron energies range from 1 to 10 eV. This energetic discharge environment is sufficient to decompose gas molecules into a variety of component species, such as electrons, ions, atoms, and molecules in ground and excited states, free radicals, etc. The net effect of the interactions among these reactive molecular fragments is to cause chemical reactions to occur at much lower temperatures than in conventional CVD reactors without benefit of plasmas. Therefore, previously unfeasible high-temperature reactions can be made to occur on temperature-sensitive substrates.

In the overwhelming majority of the research and development activity in PECVD processing, the discharge is excited by an rf field. This is due to the

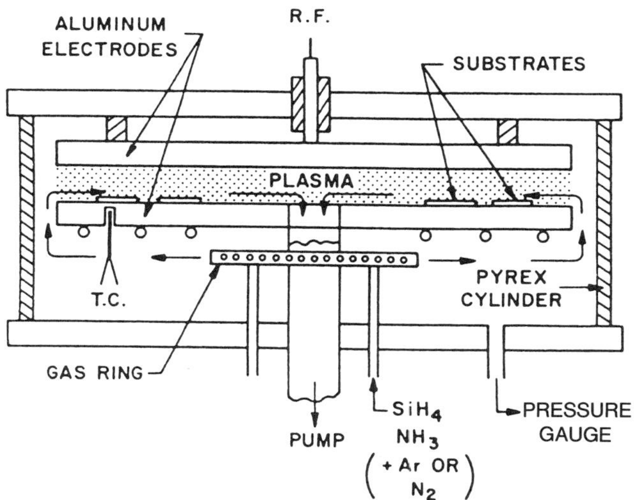

Figure 4-15. Typical cylindrical, radial flow, silicon nitride deposition reactor (From Ref. 26).

fact that most of the films deposited by this method are dielectrics, and dc discharges are not feasible. The tube or tunnel reactors employed can be coupled inductively with a coil or capacitively with electrode plates. In both cases, a symmetric potential develops on the walls of the reactor. High wall potentials are avoided to minimize sputtering of wall atoms and their incorporation into growing films.

A major commercial application of PECVD processing has been to deposit silicon nitride films in order to passivate and encapsulate completely fabricated microelectronic devices. At this stage the latter cannot tolerate temperatures much above 300 °C. A parallel-plate, plasma deposition reactor of the type shown in Fig. 4-15 is commonly used for this purpose. The reactant gases first flow through the axis of the chamber and then radially outward across rotating substrates that rest on one plate of an rf-coupled capacitor. This diode configuration enables a reasonably uniform and controllable film deposition to occur. The process is carried out at low pressures to take advantage of enhanced mass transport, and typical deposition rates of about 300 Å/min are attained at power levels of 500 W. Silicon nitride is normally prepared by reacting silane with ammonia in an argon plasma, but a nitrogen discharge with

Table 4-3. Physical and Chemical Properties of Silicon Nitride Films from $SiH_4 + NH_3$

Property	Si_3N_4 1 atm CVD 900 °C	$Si_3N_4(H)$ LPCVD 750 °C	$Si_xN_yH_z$ PECVD 300 °C
Density (g/cm^3)	2.8–3.1	2.9–3.1	2.5–2.8
Refractive index	2.0–2.1	2.01	2.0–2.1
Dielectric constant	6–7	6–7	6–9
Dielectric breakdown field (V/cm)	10^7	10^7	6×10^6
Bulk resistivity (Ω cm)	10^{15}–10^{17}	10^{16}	10^{15}
Stress at 23 °C on Si (dynes/cm^2)	1.5×10^{10} (T)	10^{10} (T)	$1 - 8 \times 10^9$ (C)
Color transmitted	None		Yellow
H_2O permeability	Zero		Low–none
Thermal stability	Excellent		Variable > 400 °C
Si/N ratio	0.75	0.75	0.8–1.0
Etch rate, 49% HF (23 °C)	80 Å/min		1500–3000 Å/min
Na^+ penetration	< 100 Å		< 100 Å
Step coverage	Fair		Conformal

Note: T = tensile; C = compressive.
Adapted from Refs. 24, 25.

Table 4-4. PECVD Reactants and Products, Deposition Temperatures, and Rates

Deposit	T (K)	Rate (cm/sec)	Reactants
a-Si	573	10^{-8}–10^{-7}	SiH_4; SiF_4–H_2; Si(s)–H_2
c-Si	673	10^{-8}–10^{-7}	SiH_4–H_2; SiF_4–H_2; Si(s)–H_2
a-Ge	673	10^{-8}–10^{-7}	GeH_4
c-Ge	673	10^{-8}–10^{-7}	GeH_4–H_2; Ge(s)–H_2
a-B	673	10^{-8}–10^{-7}	B_2H_6; BCl_3–H_2; BBr_3
a-P, c-P	293–473	10^{-5}	P(s)–H_2
As	< 373	10^{-6}	AsH_3; As(s)–H_2
Se, Te, Sb, Bi	373	10^{-7}–10^{-6}	Me–H_2
Mo			$Mo(CO)_4$
Ni			$Ni(CO)_4$
C (graphite)	1073–1273	10^{-5}	C(s)–H_2; C(s)–N_2
CdS	373–573	10^{-6}	Cd–H_2S
Oxides			
SiO_2	523	10^{-8}–10^{-6}	$Si(OC_2H_5)_4$; SiH_4–O_2, N_2O
GeO_2	523	10^{-8}–10^{-6}	$Ge(OC_2H_5)_4$; GeH_4–O_2, N_2O
SiO_2/GeO_2	1273	3×10^{-4}	$SiCl_4$–$GeCl_4$ + O_2
Al_2O_3	523–773	10^{-8}–10^{-7}	$AlCl_3$–O_2
TiO_2	473–673	10^{-8}	$TiCl_4$–O_2; metallorganics
B_2O_3			$B(OC_2H_5)_3$–O_2
Nitrides			
$Si_3N_4(H)$	573–773	10^{-8}–10^{-7}	SiH_4–N_2, NH_3
AlN	1273	10^{-6}	$AlCl_3$–N_2
GaN	873	10^{-8}–10^{-7}	$GaCl_3$–N_2
TiN	523–1273	10^{-8}–5×10^{-6}	$TiCl_4$–H_2 + N_2
BN	673–973		B_2H_6–NH_3
P_3N_5	633–673	5×10^{-6}	P(s)–N_2; PH_3–N_2
Carbides			
SiC	473–773	10^{-8}	SiH_4–C_nH_m
TiC	673–873	5×10^{-8}–10^{-6}	$TiCl_4$–CH_4 + H_2
B_xC	673	10^{-8}–10^{-7}	B_2H_6–CH_4

From Ref. 27.

silane can also be used. As much as 25 at% hydrogen can be incorporated in plasma silicon nitride, which may, therefore, be viewed as a ternary solid solution. This should be contrasted with the stoichiometric compound Si_3N_4, formed by reacting silane and ammonia at 900 °C in an atmospheric CVD reactor. It is instructive to further compare the physical and chemical property differences in three types of silicon nitride, and this is done in Table 4-3. Although Si_3N_4 is denser, more resistant to chemical attack, and has higher resistivity and dielectric breakdown strength, SiNH tends to provide better step coverage.

Some elements, such as carbon and boron, in addition to metals, oxides, nitrides, and silicides, have been deposited by PECVD methods. Operating temperatures and nominal deposition rates are included in Table 4-4. An important recent advance in PECVD relies on the use of microwave—also called electron cyclotron resonance (ECR)—plasmas. As the name implies, microwave energy is coupled to the natural resonant frequency of the plasma electrons in the presence of a static magnetic field. The condition for energy absorption is that the microwave frequency ω_m (commonly 2.45 GHz) be equal to qB/m, where all terms were previously defined in connection with magnetron sputtering (Section 3.7.3). Physically, plasma electrons then undergo one circular orbit during a single period of the incident microwave. Whereas rf plasmas contain a charge density of $\sim 10^{10}$ cm^{-3} in a 10^{-2}-to-1-torr environment, the ECR discharge is easily generated at pressures of 10^{-5} to 10^{-3} torr. Therefore, the degree of ionization is about 1000 times higher than in the rf plasma. This coupled with low-pressure operation, controllability of ion energy, low-plasma sheath potentials, high deposition rates, absence of source contamination (no electrodes!), etc., has made ECR plasmas attractive for both film deposition as well as etching processes. A reactor that has been employed for the deposition of SiO_2, Al_2O_3, SiN, and Ta_2O_5 films is shown in Fig. 4-16. A significant benefit of microwave plasma processing is the ability to produce high-quality films at low substrate temperatures.

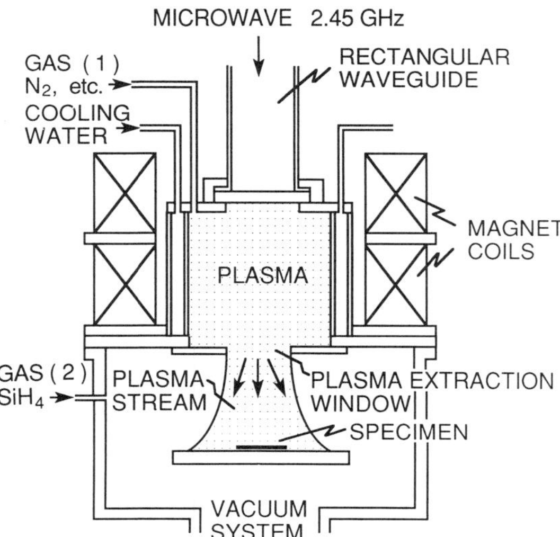

Figure 4-16. ECR plasma deposition reactor. (From Ref. 28, with permission from Noyes Publications).

4.6.5. Laser-Enhanced CVD

Laser or, more generally, optical chemical processing involves the use of monochromatic photons to enhance and control reactions at substrates. Two mechanisms are involved during laser-assisted deposition, and these are illustrated in Fig. 4-17. In the pyrolytic mechanism the laser heats the substrate to decompose gases above it and enhance rates of chemical reactions there. Pyrolytic deposition requires substrates that melt above the temperatures necessary for gas decomposition. Photolytic processes, on the other hand, involve direct dissociation of molecules by energetic photons. Ultraviolet light sources are required because many useful parent molecules (e.g., SiH_4, Si_2H_6, Si_3H_8, N_2O) require absorption of photons with wavelengths of less than 220 nm to initiate dissociation reactions. The only practical continuous-wave laser is the frequency-doubled Ar^+ at 257 nm with a typical power of 20 mW. Such power levels are too low to enable high deposition rates over large areas but are sufficient to "write" or initiate deposits where the scanned light beam hits the substrate. Similar direct writing of materials has been accomplished by pyrolytic processes. Both methods have the potential for local deposition of metal to repair integrated circuit chips.

A number of metals such as Al, Au, Cr, Cu, Ni, Ta, Pt, and W have been

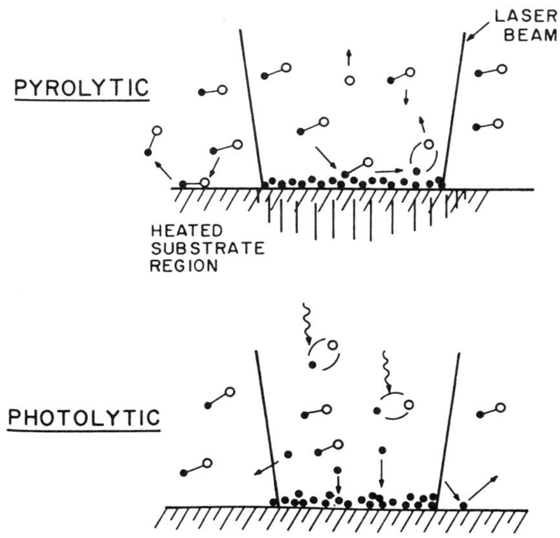

Figure 4-17. Mechanisms of laser-assisted deposition. (Reproduced with permission from Ref. 29, © 1985 by Annual Reviews Inc.).

deposited through the use of laser processing. For photolytic deposition, organic metal dialkyl and trialkyls have yielded electrically conducting deposits. Carbonyls and hydrides have been largely employed for pyrolytic depositions. There is frequently an admixture of pyrolytic and photolytic deposition processes occurring simultaneously with deep UV sources. Alternatively, pyrolytic deposition is accompanied by some photodissociation of loosely bound complexes if the light source is near the UV.

Dielectric films have also been deposited in low-pressure photosensitized CVD processes (Ref. 30). The photosensitized reaction of silane and hydrazine yields silicon nitride films, and SiO_2 films have been produced from a gas mixture of SiH_4, N_2O, and N_2. In SiO_2, deposition rates of 150 Å/min at temperatures as low as 50 °C have been reported (Ref. 23), indicating the exciting possibilities inherent in such processing.

4.6.6. Metalorganic CVD (MOCVD) (Ref. 31)

Also known as OMVPE (organometallic vapor phase epitaxy), MOCVD has presently assumed considerable importance in the deposition of epitaxial compound semiconductor films. Unlike the previous CVD variants, which differ on a physical basis, MOCVD is distinguished by the chemical nature of the precursor gases. As the name implies, metalorganic compounds like trimethyl-gallium (TMGa), trimethyl-indium (TMIn), etc, are employed. They are reacted with group V hydrides, and during pyrolysis the semiconductor compound forms; e.g.,

$$(CH_3)_3Ga_{(g)} + AsH_{3(g)} \rightarrow GaAs_{(s)} + 3CH_{4(g)} \tag{4-51}$$

Group V organic compounds TMAs, TEAs (triethyl-arsenic), TMP, TESb, etc., also exist, so that all-organic pyrolysis reactions have been carried out. The great advantage of using metalorganics is that they are volatile at moderately low temperatures; there are no troublesome liquid Ga or In sources in the reactor to control for transport to the substrate. Carbon contamination of films is a disadvantage, however. Since all constituents are in the vapor phase, precise electronic control of gas flow rates and partial pressures is possible. This, combined with pyrolysis reactions that are relatively insensitive to temperature, allows for efficient and reproducible deposition. Utilizing computer-controlled gas exchange and delivery systems, epitaxial multilayer semiconductor structures with sharp interfaces have been grown in reactors such as shown in Fig. 4-18. In addition to GaAs, other III-V as well as II-VI and IV-VI compound semiconductor films have been synthesized. Table 4-5 lists

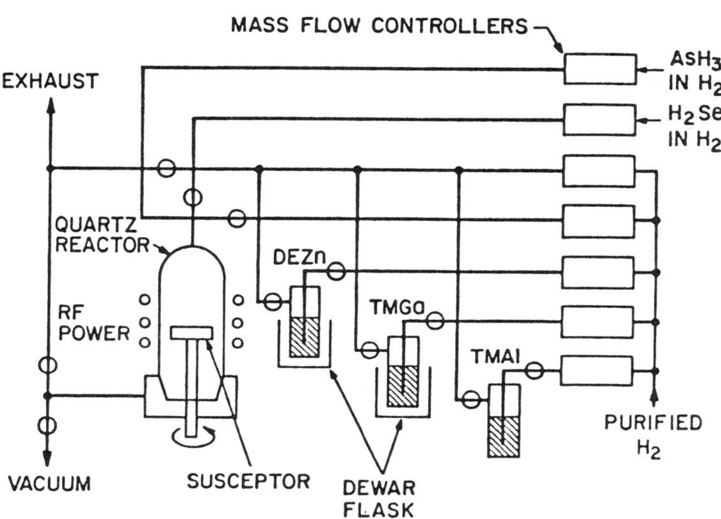

Figure 4-18. Schematic diagram of a vertical atmospheric-pressure MOCVD reactor. (Reprinted with permission. From R. D. Dupuis, *Science* **226**, 623, 1984).

Table 4-5. Organo Metallic Precursors and Semiconductor Films Grown by MOCVD

Compound	Reactants	Vapor Pressure* of OM precursor a	b	Growth Temperature (°C)
AlAs	TMAl + AsH_3	8.224	2135	700
AlN	TMAl + NH_3			1250
GaAs	TMGa + AsH_3	8.50	1824	650–750
GaN	TMGa + NH_3			800
GaP	TMGa + PH_3			750
GaSb	TEGa + TMSb	9.17	2532	500–550
		7.73	1709	
InAs	TEIn + AsH_3			650–700
InP	TEIn + PH_3			725
ZnS	DEZn + H_2S	8.28	2190	
ZnSe	DEZn + H_2Se			
CdS	DMCd + H_2S	7.76	1850	
HgCdTe	Hg + DMCd + DMTe	7.97	1865	
CdTe	DMCd + DMTe			

*$\log P(\text{torr}) = a - b/T$ K
Adapted from Ref. 31.

some films formed on insulating and semiconducting substrates together with corresponding reactants and film growth temperatures.

Film growth rates ($\dot{G}$) and compositions directly depend on gas partial pressures and flow rates ($\dot{V}$). For $Al_xGa_{1-x}As$ films,

$$\dot{G} = K(T)\left[P_{Ga}\dot{V}_{Ga} + 2P_{Al}\dot{V}_{Al}\right], \tag{4-51}$$

$$x = \frac{2P_{Al}\dot{V}_{Al}}{2P_{Al}\dot{V}_{Al} + P_{Ga}\dot{V}_{Ga}}. \tag{4-52}$$

In these equations $K(T)$ is a temperature-dependent constant, and the factor of 2 enters because trimethyl-aluminum is a dimer. MOCVD has been particularly effective in depositing films for a variety of visible and long-wavelength lasers as well as quantum well structures. The use of these precursor gases is not only limited to semiconductor technology; volatile organo-Y, Ba, and Cu compounds have been explored in connection with the deposition of high-temperature superconducting films having the nominal composition $YBa_2Cu_3O_7$.

4.6.7. Safety

The safe handling of gases employed in CVD systems is a concern of paramount importance. Because the reactant or product gases are typically toxic, flammable, pyrophoric, or corrosive, and frequently possess a combination of these attributes, they present particular hazards to humans. Exposure of reactor hardware and associated gas-handling equipment to corrosive environments also causes significant maintenance problems and losses due to downtime. Table 4-6 lists gases commonly employed in CVD processes together with some of their characteristics. A simple entry in the table does not accurately reflect the nature of the gas in practice. Silane, for example, more so than other gases employed in the semiconductor industry, has an ominous and unpredictable nature. It is stable but pyrophoric, so it ignites on contact with air. If it accumulates in a stagnant airspace, however, the resulting mixture may explode upon ignition. In simulation tests of leaks, high flow rates of silane have resulted in violent explosions. For this reason, silane cylinders are stored outside buildings in concrete bunkers. The safety problems are magnified in low-pressure processing where concentrated gases are used. For example, in the deposition of polysilicon, pure silane is used during LPCVD, whereas only 3% silane is employed in atmospheric CVD processing.

Corrosive attack of gas-handling equipment (e.g., valves, regulators, piping) occurs in virtually all CVD systems. The problems are particularly acute in LPCVD processing because of the damage to mechanical pumping systems.

Table 4-6. Hazardous Gases Employed in CVD

Gas	Corrosive	Flammable	Pyrophoric	Toxic	Bodily Hazard
Ammonia (NH_3)	x			x	eye and respiratory irritation
Arsine (AsH_3)		x		x	Anemia, kidney damage death
Boron Trichloride (BCl_3)	x				
Boron Trifluoride (BF_3)	x				
Chlorine (Cl_2)	x			x	Eye and respiratory irritation
Diborane (B_2H_6)		x	x	x	Respiratory irritation
Dichlorosilane (SiH_2Cl_2)	x	x			
Germane (GeH_4)		x		x	
Hydrogen chloride (HCl)	x				
Hydrogen fluoride (HF)	x				Severe burns
Hydrogen (H_2)		x			
Phosphine (PH_3)		x	x	x	Respiratory irritation, death
Phosphorous pentachloride (PCl_5)	x				
Silane (SiH_4)		x	x	x	
Silicon tetrachloride ($SiCl_4$)	x				
Stibine (SbH_3)		x		x	

Since many reactors operate at high temperatures, the effluent gases are very hot and capable of further downstream reactions in the pumping hardware. Furthermore, the exhaust stream generally contains corrosive species such as acids, water, oxidizers, unreacted halogenated gases, etc., in addition to large quantities of abrasive particulates. In semiconductor processing, for example, SiO_2 and Si_3N_4 particles are most common. All of these products are ingested by the mechanical pumps, and the chamber walls become coated with precipitates or particulate crusts. The oils used are degraded through polymerization and incorporation of solids. The lubrication of moving parts and hardware is thus hampered, and they tend to corrode and wear out more readily. All of this is a small price to pay for the wonderful array of film materials that CVD has made possible.

EXERCISES

1. a. Write a balanced chemical equation for the CVD reaction that produces Al_2O_3 films from the gas mixture consisting of $AlCl_3 + CO_2 + H_2$.

b. If a 2-μm-thick coating is to be deposited on a 2-cm-diameter substrate placed within a tubular reactor 50 cm long and 5 cm in diameter, calculate the minimum weight of $AlCl_3$ precursor required.

c. Repeat parts (a) and (b) for VC films from a $VCl_4 + C_6H_5CH_3 + H_2$ gas mixture.

2. Consider the generic reversible CVD reaction

$$A_g \underset{T_2}{\overset{T_1}{\rightleftarrows}} B_s + C_g (T_2 > T_1)$$

at 1 atm pressure ($P_A + P_C = 1$), where the free energy of the reaction is $\Delta G° = \Delta H° - T\,\Delta S°$. Through consideration of the equilibria at T_1 and T_2,

a. derive an expression for $\Delta P_A = P_A(T_1) - P_A(T_2)$ as a function of T, ΔH, and ΔS.

b. plot ΔP_A as a function of ΔH.

c. comment on the gas transport direction and magnitude as a function of the sign and value of ΔH.

3. In growing epitaxial Ge films by the disproportionation reaction of Eq.

4-13, the following thermodynamic data apply:

$$I_{2(g)} = 2I_{(g)} \qquad \Delta G° = -38.4T \text{ cal/mole}$$

$$Ge_{(s)} + I_{2(g)} = GeI_{2(g)} \qquad \Delta G° = -1990 - 11.2T \text{ cal/mole}$$

$$Ge_{(s)} + GeI_{4(g)} = 2GeI_{2(g)} \qquad \Delta G° = 36300 - 57.5T \text{ cal/mole}$$

a. What is $\Delta G°$ for the reaction $Ge_{(s)} + 2I_{2(g)} = GeI_{4(g)}$?
b. Suggest a reactor design. Which region is hotter; which is cooler?
c. Roughly estimate the operating temperature of the reactor.
d. Suggest how you would change the reactor conditions to deposit polycrystalline films.

4. a. At 1200 °C the following growth rates of Si films were observed using the indicated Si–Cl–H precursor gas. The same CVD reactor was employed for all gases.

Precursor	Growth Rate (μm/min)
SiH_4	1
SiH_2Cl_2	0.5
$SiHCl_3$	0.3
$SiCl_4$	0.15

b. The density of poly-Si nuclei on an SiO_2 substrate at 1000 °C was observed to be 10^{10} cm^{-2} for SiH_4, 5×10^7 cm^{-2} for SiH_2Cl_2, and 3×10^6 cm^{-2} for $SiHCl_3$.

Are the observations made in (a) and (b) consistent? From what you know about these gases explain the two findings.

5. Plot $\ln P^4_{HCl}/P_{SiCl_4}P^2_{H_2}$ vs. $1/T$ K for the temperature range 800 to 1500 K, using the results of Fig. 4-5.

a. What is the physical significance of the slope of this Arrhenius plot?
b. Calculate ΔH for the reaction given by Eq. 4-22a, using data in Fig. 4-4.

6. Assume you are involved in a project to deposit ZnS and CdS films for infrared optical coatings. Thermodynamic data reveal

1. $H_2S_{(g)} + Zn_{(g)} \rightarrow ZnS_{(s)} + H_{2(g)}$
   $\Delta G = -76{,}400 + 82.1T - 5.9T \ln T$ (cal/mole)
2. $H_2S_{(g)} + Cd_{(g)} \rightarrow CdS_{(s)} + H_{2(g)}$
   $\Delta G = -50{,}000 + 85.2T - 6.64T \ln T$ (cal/mole)

a. Are these reactions endothermic or exothermic?

b. In practice, reactions 1 and 2 are carried out at 680 °C and 600 °C, respectively. From the vapor pressures of Zn and Cd at these temperatures, estimate the P_{H_2}/P_{H_2S} ratio for each reaction, assuming equilibrium conditions.

c. Recommend a reactor design to grow either ZnS or CdS, including a method for introducing reactants and heating substrates.

7. It is observed that when WF_6 gas passes over a substrate containing exposed areas of Si and SiO_2:

1. W selectively deposits over Si and not over SiO_2.
2. Once a continuous film of W deposits (i.e., ~ 100–150 Å), the reaction is self-limiting and no more W deposits.

Suggest a possible way to subsequently produce a thicker W deposit.

8. The disproportionation reaction Si + $SiCl_4$ = $2SiCl_2$ ($\Delta G^\circ = 83{,}000 + 3.64T \log T - 89.4T$ (cal/mole)) is carried out in a closed tubular atmospheric pressure reactor whose diameter is 15 cm. Deposition of Si occurs on a substrate maintained at 750 °C and located 25 cm away from the source, which is heated to 900 °C. Assuming thermodynamic equilibrium prevails at source and substrate, calculate the flux of $SiCl_2$ transported to the substrate if the gas viscosity is 0.08 cP. [Hint: See problem 2.]

9. Find the stoichiometric formula for the following films:

a. PECVD silicon nitride containing 20 at% H with a Si/N ratio of 1.2.
b. LPCVD silicon nitride containing 6 at% H with a Si/N ratio of 0.8.
c. LPCVD SiO_2 with a density of 2.2 g/cm^3, containing 3×10^{21} H atoms/cm^3.

10. Tetrachlorosilane diluted to 0.5% mole in H_2 gas flows through a 12-cm-diameter, tubular, atmospheric CVD reactor at a velocity of 20 cm/sec. Within the reactor is a flat pallet bearing Si wafers resting horizontally. If the viscosity of the gas is 0.03 cP at 1200 °C,

a. what is the Reynolds number for the flow?
b. estimate the boundary layer thickness at a point 5 cm down the pallet.
c. If epitaxial Si films deposit at a rate of 1 μm/min, estimate the diffusivity of Si through the boundary layer.

11. Polysilicon deposits at a rate of 30 Å/min at 540 °C. What deposition rate can be expected at 625 °C if the activation energy for film deposition is 1.65 eV?

12. Consider a long tubular CVD reactor in which one-dimensional steady-state diffusion and convection processes occur together with a homogeneous first-order chemical reaction. Assume the concentration $C(x)$ of a given species satisfies the ordinary differential equation

$$D\frac{d^2C}{dx^2} - v\frac{dC}{dx} - KC = 0,$$

where K is the chemical rate constant and x is the distance along the reactor.

a. If the boundary conditions are $C(x = 0) = 1$ and $C(x = 1\text{ m}) = 0$, derive an expression for $C(x)$.

b. If $C(x = 0) = 1$ and $dC/dx(x = 1\text{ m}) = 0$, derive an expression for $C(x)$.

c. Calculate expressions for the concentration profiles if $D = 1000$ cm^2/sec, $v = 100$ cm/sec, and $K = 1\text{ sec}^{-1}$. [Hint: A solution to the differential equation is $\exp \alpha x$, where α is a constant.]

13. Select any film material (e.g., semiconductor, oxide, nitride, carbide metal alloy, etc.) that has been deposited or grown by both PVD and CVD methods. In a report, compare the resultant structures, stoichiometries, and properties. The *Journal of Vacuum Science and Technology* and *Thin Solid Films* are good references for such information.

References

1.* W. Kern and V. S. Ban, in *Thin Film Processes*, eds. J. L. Vossen and W. Kern, Academic Press, New York (1978).
2.* W. Kern, in *Microelectronic Materials and Processes*, ed. R. A. Levy, Kluwer Academic, Dordrecht (1989).
3.* K. K. Yee, *Int. Metals Rev.* **23**, 19 (1978).
4.* J. W. Hastie, *High Temperature Vapors – Science and Technology*, Academic Press, New York (1975).
5.* J. M. Blocher, in *Deposition Technologies for Films and Coatings*, ed. R. F. Bunshah, Noyes, Park Ridge, NJ (1982).
6.* W. A. Bryant, *J. Mat. Sci.* **12**, 1285 (1977).
7.* K. K. Schuegraf, *Handbook of Thin-Film Deposition Processes and Techniques*, Noyes, Park Ridge, NJ (1988).
8. J. Schlichting, *Powder Metal Int.* **12(3)**, 141 (1980).

*Recommended texts or reviews.

9. E. S. Wajda, B. W. Kippenhan, and W. H. White, *IBM J. Res. Dev.* **7**, 288 (1960).
10. T. Mizutani, M. Yoshida, A. Usui, H. Watanabe, T. Yuasa, and I. Hayashi, *Japan J. Appl. Phys.* **19**, L113 (1980).
11. R. A. Laudise, *The Growth of Single Crystals*, Prentice Hall, Englewood Cliffs, NJ (1970).
12. O. Kubaschewski and E. L. Evans, *Metallurgical Thermochemistry*, Pergamon Press, New York (1958).
13. D. R. Stull and H. Prophet, *JANAF Thermochemical Tables*, 2nd ed., U.S. GPO, Washington, DC (1971).
14. V. S. Ban and S. L. Gilbert, *J. Electrochem. Soc.* **122(10)**, 1382 (1975).
15. E. Sirtl, I. P. Hunt and D. H. Sawyer, *J. Electrochem. Soc.* **121**, 919 (1974).
16. J. E. Doherty, *J. Metals* **28(6)**, 6 (1976).
17. T. C. Anthony, A. L. Fahrenbruch and R. H. Bube, *J. Vac. Sci. Tech.* **A2(3)**, 1296 (1984).
18. P. C. Rundle, *Int. J. Electron.* **24**, 405 (1968).
19. A. S. Grove, *Physics and Technology of Semiconductor Devices*, Wiley, New York (1967).
20. W. S. Ruska, *Microelectronic Processing*, McGraw-Hill, New York (1987).
21. J. Bloem and W. A. P. Claassen, *Philips Tech. Rev.* **41**, 60 (1983, 1984).
22. R. B. Marcus and T. T. Sheng, *Transmission Electron Microscopy of Silicon VLSl Circuits and Structures*, Wiley, New York (1983).
23. S. M. Sze, *Semiconductor Devices – Physics and Technology*, Wiley, New York (1985).
24. A. C. Adams, in *VLSI Technology*, 2nd ed., ed. S. M. Sze, McGraw-Hill, New York (1988).
25. J. R. Hollahan and S. R. Rosler, in *Thin Film Processes*, ed. J. L. Vossen and W. Kern, Academic Press, New York (1978).
26. M. J. Rand, *J. Vac. Sci. Tech.* **16**(**2**), 420 (1979).
27. S. Veprek, *Thin Solid Films* **130**, 135 (1985).
28. S. Matuso in Ref. 7.
29. R. M. Osgood and H. H. Gilgen, *Ann. Rev. Mater. Sci.* **15**, 549 (1985).
30. R. L. Abber in Ref. 7.
31. G. B. Stringfellow, *Organo Vapor-Phase Epitaxy: Theory and Practice*, Academic Press, Boston (1989).
32. R. D. Dupuis, *Science* **226**, 623 (1984).

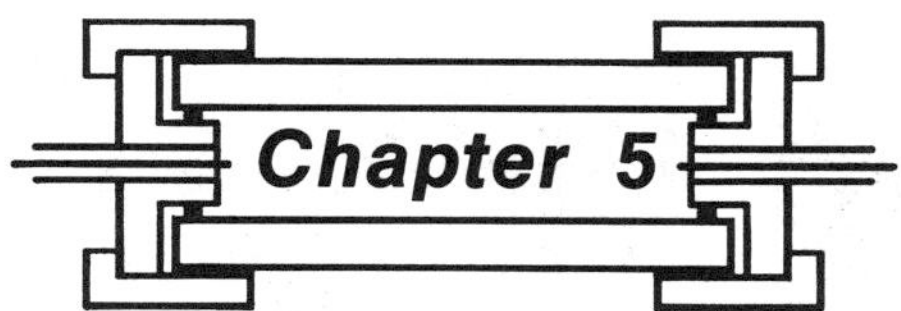

Film Formation and Structure

5.1. INTRODUCTION

Interest in thin-film formation processes dates at least to the 1920s. During research at the Cavendish Laboratories in England on evaporated thin films, the concept of formation of nuclei that grew and coalesced to form the film was advanced (Ref. 1). All phase transformations, including thin-film formation, involve the processes of nucleation and growth. During the earliest stages of film formation, a sufficient number of vapor atoms or molecules condense and establish a permanent residence on the substrate. Many such film birth events occur in this so-called nucleation stage. Although numerous high-resolution transmission electron microscopy investigations have focused on the early stages of film formation, it is doubtful that there is a clear demarcation between the end of nucleation and the onset of nucleus growth. The sequence of nucleation and growth events can be described with reference to the micrographs of Fig. 5-1. Soon after exposure of the substrate to the incident vapor, a uniform distribution of small but highly mobile clusters or islands is observed. In this stage the prior nuclei incorporate impinging atoms and subcritical clusters and grow in size while the island density rapidly saturates. The next stage involves merging of the islands by a coalescence phenomenon

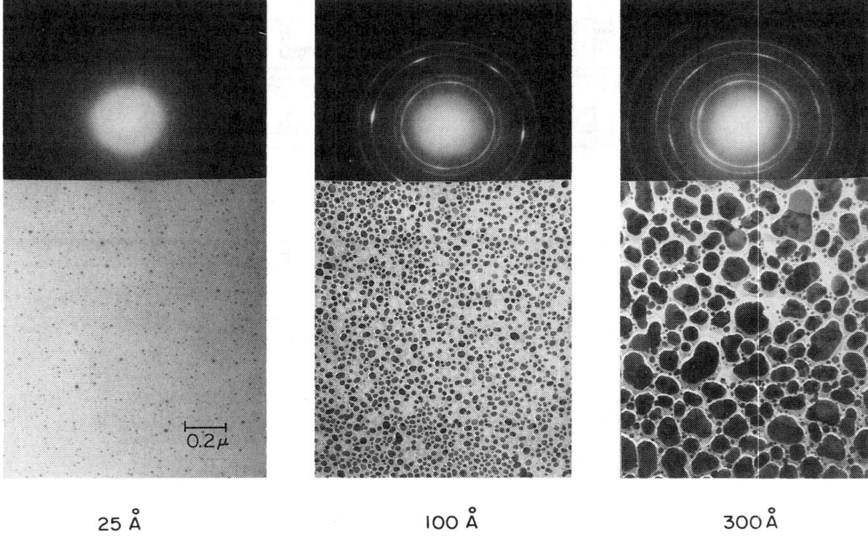

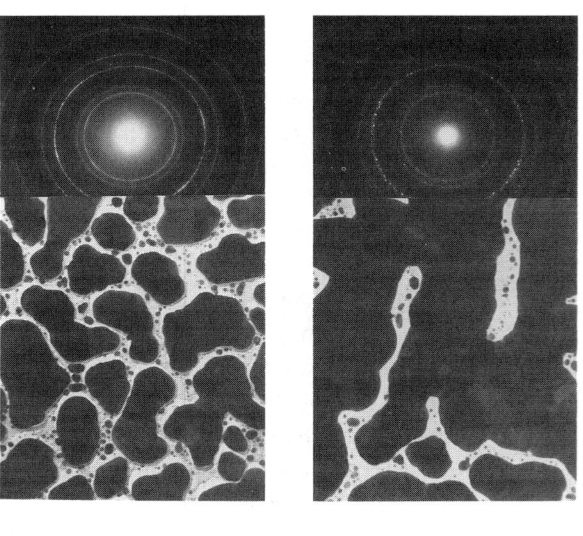

Figure 5-1. Nucleation, growth and coalescence of Ag films on (111) NaCl substrates. Corresponding diffraction patterns are shown. (Courtesy of R. W. Vook. Reprinted with permission from the Metals Society, from R. W. Vook, *Int. Metals Rev.* **27**, 209(1982).

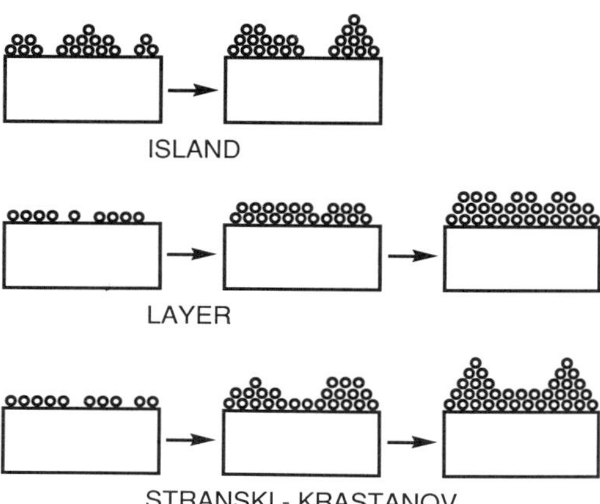

Figure 5-2. Basic modes of thin-film growth.

that is liquidlike in character especially at high substrate temperatures. Coalescence decreases the island density, resulting in local denuding of the substrate where further nucleation can then occur. Crystallographic facets and orientations are frequently preserved on islands and at interfaces between initially disoriented, coalesced particles. Coalescence continues until a connected network with unfilled channels in between develops. With further deposition, the channels fill in and shrink, leaving isolated voids behind. Finally, even the voids fill in completely, and the film is said to be continuous. This collective set of events occurs during the early stages of deposition, typically accounting for the first few hundred angstroms of film thickness.

The many observations of film formation have pointed to three basic growth modes: (1) island (or Volmer–Weber), (2) layer (or Frank–van der Merwe), and (3) Stranski–Krastanov, which are illustrated schematically in Fig. 5-2. Island growth occurs when the smallest stable clusters nucleate on the substrate and grow in three dimensions to form islands. This happens when atoms or molecules in the deposit are more strongly bound to each other than to the substrate. Many systems of metals on insulators, alkali halide crystals, graphite, and mica substrates display this mode of growth.

The opposite characteristics are displayed during layer growth. Here the extension of the smallest stable nucleus occurs overwhelmingly in two dimensions resulting in the formation of planar sheets. In this growth mode the atoms are more strongly bound to the substrate than to each other. The first complete monolayer is then covered with a somewhat less tightly bound second layer. Providing the decrease in bonding energy is continuous toward the bulk crystal value, the layer growth mode is sustained. The most important example of this growth mode involves single-crystal epitaxial growth of semiconductor films, a subject treated extensively in Chapter 7.

The layer plus island or Stranski–Krastanov (S.K.) growth mechanism is an intermediate combination of the aforementioned modes. In this case, after forming one or more monolayers, subsequent layer growth becomes unfavorable and islands form. The transition from two- to three-dimensional growth is not completely understood, but any factor that disturbs the monotonic decrease in binding energy characteristic of layer growth may be the cause. For example, due to film–substrate lattice mismatch, strain energy accumulates in the growing film. When released, the high energy at the deposit–intermediate-layer interface may trigger island formation. This growth mode is fairly common and has been observed in metal–metal and metal–semiconductor systems.

At an extreme far removed from early film formation phenomena is a regime of structural effects related to the actual grain morphology of polycrystalline films and coatings. This external grain structure together with the internal defect, void, or porosity distributions frequently determines many of the engineering properties of films. For example, columnar structures, which interestingly develop in amorphous as well as polycrystalline films, have a profound effect on magnetic, optical, electrical, and mechanical properties. In this chapter we discuss how different grain and deposit morphologies evolve as a function of deposition variables and how some measure of structural control can be exercised. Modification of the film structure through ion bombardment or laser processing both during and after deposition has been a subject of much research interest recently and is treated in Chapters 3 and 13. Subsequent topics in this chapter are:

5.2 Capillarity Theory
5.3 Atomistic Nucleation Processes
5.4 Cluster Coalescence and Depletion
5.5 Experimental Studies of Nucleation and Growth
5.6 Grain Structure of Films and Coatings
5.7 Amorphous Thin Films

References 1–5 are recommended sources for much of the subject matter in this chapter.

5.2. Capillarity Theory

5.2.1. Thermodynamics

Capillarity theory possesses the mixed virtue of yielding a conceptually simple qualitative model of film nucleation, which is, however, quantitatively inaccurate. The lack of detailed atomistic assumptions gives the theory an attractive broad generality with the power of creating useful connections between such variables as substrate temperature, deposition rate, and critical film nucleus size. An introduction to the thermodynamic aspects of *homogeneous* nucleation was given on p. 40 and is worth reviewing. In a similar spirit, we now consider the *heterogeneous* nucleation of a solid film on a planar substrate. Film-forming atoms or molecules in the vapor phase are assumed to impinge on the substrate, creating aggregates that either tend to grow in size or disintegrate into smaller entities through dissociation processes.

The free-energy change accompanying the formation of an aggregate of mean dimension r is given by

$$\Delta G = a_3 r^3 \Delta G_V + a_1 r^2 \gamma_{vf} + a_2 r^2 \gamma_{fs} - a_2 r^2 \gamma_{sv}. \tag{5-1}$$

The chemical free-energy change per unit volume, ΔG_V, drives the condensation reaction. According to Eq. 1-39, any level of gas-phase supersaturation generates a negative ΔG_V without which nucleation is impossible. There are several interfacial tensions, γ, to contend with now, and these are identified by the subscripts f, s, and v representing film, substrate, and vapor, respectively. For the cap-shaped nucleus in Fig. 5-3, the curved surface area $(a_1 r^2)$, the projected circular area on the substrate $(a_2 r^2)$, and the volume $(a_3 r^3)$ are involved, and the respective geometric constants are $a_1 = 2\pi(1 - \cos\theta)$, $a_2 = \pi \sin^2\theta$, $a_3 = \pi(2 - 3\cos\theta + \cos^3\theta)/3$. Consideration of the mechanical equilibrium among the interfacial tensions or forces yields Young's equation

$$\gamma_{sv} = \gamma_{fs} + \gamma_{vf}\cos\theta. \tag{5-2}$$

Therefore, the contact angle θ depends only on the surface properties of the involved materials. The three modes of film growth can be distinguished on the

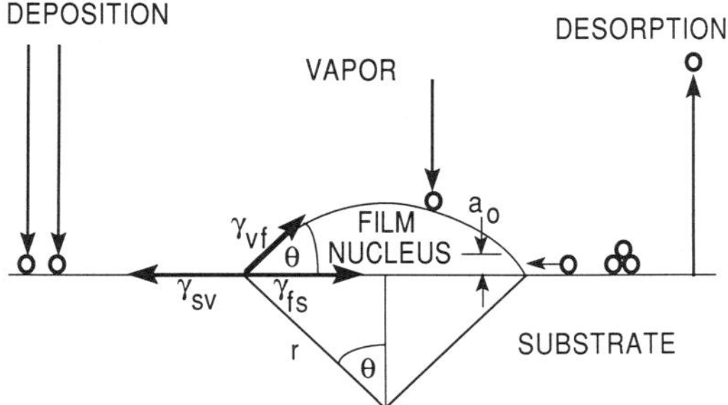

Figure 5-3. Schematic of basic atomistic processes on substrate surface during vapor deposition.

basis of Eq. 5-2. For island growth, $\theta > 0$, and therefore

$$\gamma_{sv} < \gamma_{fs} + \gamma_{vf}. \tag{5-3}$$

For layer growth the deposit "wets" the substrate and $\theta = 0$. Therefore,

$$\gamma_{sv} = \gamma_{fs} + \gamma_{vf}. \tag{5-4}$$

A special case of this condition is ideal homo- or *autoepitaxy*. Because the interface between film and substrate essentially vanishes, $\gamma_{fs} = 0$. Lastly, for S.K. growth,

$$\gamma_{sv} > \gamma_{fs} + \gamma_{vf}. \tag{5-5}$$

In this case, the strain energy per unit area of film overgrowth is large with respect to γ_{vf}, permitting nuclei to form above the layers. In contrast, a film strain energy that is small compared with γ_{vf} is characteristic of layer growth.

Returning now to Eq. 5-1, we note that any time a new interface appears there is an increase in surface free energy, hence the positive sign for the first two surface terms. Similarly, the loss of the circular substrate–vapor interface under the cap implies a reduction in system energy and a negative contribution to ΔG. The critical nucleus size r^* (i.e., the value of r when $d\,\Delta G/dr = 0$) is given by differentiation, namely,

$$r^* = \frac{-2(a_1\gamma_{vf} + a_2\gamma_{fs} - a_2\gamma_{sv})}{3a_3\,\Delta G_V}. \tag{5-6}$$

Correspondingly, ΔG evaluated at $r = r^*$ is

$$\Delta G^* = \frac{4(a_1\gamma_{vf} + a_2\gamma_{fs} - a_2\gamma_{sv})^3}{27a_3^2\,\Delta G_V^2}. \tag{5-7}$$

Both r^* and ΔG^* scale in the manner shown in Fig. 1-19. An aggregate smaller in size than r^* disappears by shrinking, lowering ΔG in the process. Critical nuclei grow to supercritical dimensions by further addition of atoms, a process that similarly lowers ΔG. In heterogeneous nucleation the accommodating substrate catalyzes vapor condensation and the energy barrier ΔG^* depends on the contact angle. After substitution of the geometric constants, it is easily shown that ΔG^* is essentially the product of two factors; i.e.,

$$\Delta G^* = \left(\frac{16\pi\gamma_{vf}^3}{3\,\Delta G_V^2}\right)\left(\frac{2 - 3\cos\theta + \cos^3\theta}{4}\right). \tag{5-8}$$

The first is the value for ΔG^* derived for homogeneous nucleation. It is modified by the second term, a wetting factor that has the value of zero for $\theta = 0$ and unity for $\theta = 180°$. When the film wets the substrate, there is no barrier to nucleation. At the other extreme of dewetting, ΔG^* is maximum and equal to that for homogeneous nucleation.

The preceding formalism provides a generalized framework for inclusion of other energy contributions. If, for example, the film nucleus is elastically strained throughout because of the bonding mismatch between film and substrate, then a term $a_3 r^3 \Delta G_s$, where ΔG_s is the strain energy per unit volume, would be appropriate. In the calculation for ΔG^*, the denominator of Eq. 5-7 would then be altered to $27a_3^2(\Delta G_V + \Delta G_s)^2$. Because the sign of ΔG_V is negative while ΔG_s is positive, the overall energy barrier to nucleation increases in such a case. If, however, deposition occurred on an initially strained substrate—i.e., one with emergent cleavage steps or screw dislocations—then stress relieval during nucleation would be manifested by a reduction of ΔG^*. Substrate charge and impurities similarly influence ΔG^* by affecting terms related to either surface and volume electrostatic, chemical, etc., energies.

5.2.2. Nucleation Rate

The nucleation rate is a convenient synthesis of terms that describes how many nuclei of critical size form on a substrate per unit time. Nuclei can grow through direct impingement of gas phase atoms, but this is unlikely in the earliest stages of film formation when nuclei are spaced far apart. Rather, the rate at which critical nuclei grow depends on the rate at which adsorbed monomers (adatoms) attach to it. In the model of Fig. 5-3, energetic vapor atoms that impinge on the substrate may immediately desorb, but usually they remain on the surface for a length of time τ_s given by

$$\tau_s = \frac{1}{\nu} \exp \frac{E_{\text{des}}}{kT}. \tag{5-9}$$

The vibrational frequency of the adatom on the surface is ν (typically 10^{12} sec^{-1}), and E_{des} is the energy required to desorb it back into the vapor. Adatoms, which have not yet thermally accommodated to the substrate, execute random diffusive jumps and, in the course of their migration, may form pairs with other adatoms, or attach to larger atomic clusters or nuclei. When this happens, it is unlikely that these atoms will return to the vapor phase. Changes in E_{des} are particularly expected at substrate heterogeneities, such as cleavage steps or ledges where the binding energy of adatoms is greater relative to a planar surface. The proportionately large number of atomic bonds available at these accommodating sites leads to higher E_{des} values. For this reason, a significantly higher density of nuclei is usually observed near cleavage steps and other substrate imperfections. The presence of impurities similarly alters E_{des} in a complex manner, depending on type and distribution of atoms or molecules involved.

We now exploit some of these microscopic notions in the capillarity theory of the nucleation rate $\dot{N}$. Reproducing Eq. 1-41, we obtain the expression for $\dot{N}$:

$$\dot{N} = N^* A^* \omega \text{ nuclei/cm}^2\text{-sec}. \tag{5-10}$$

(The Zeldovich factor, included in other treatments, is omitted here for simplicity.) Based on the thermodynamic probability of existence, the equilibrium number of nuclei of critical size per unit area of substrate is given by

$$N^* = n_s \exp - \Delta G^*/kT. \tag{5-11}$$

The quantity n_s represents the total nucleation site density. A certain number of these sites are occupied by adatoms whose surface density, n_a, is given by the product of the vapor impingement rate (Eq. 2-8) and the adatom lifetime, or

$$n_a = \tau_s P N_A / \sqrt{2\pi MRT}. \tag{5-12}$$

Surrounding the cap-shaped nucleus of Fig. 5-3 are adatoms poised to attach to the circumferential belt whose area is

$$A^* = 2\pi r^* a_0 \sin\theta. \tag{5-13}$$

Quantities r^* and θ were defined previously, and a_0 is an atomic dimension.

Lastly, the impingement rate onto area A^* requires adatom diffusive jumps on the substrate with a frequency given by $\nu \exp -E_S/kT$, where E_S is the activation energy for surface diffusion. The overall impingement flux is the product of the jump frequency and n_a, or

$$\omega = \frac{\tau_s P N_A \nu \exp - E_S/kT \ (\text{cm}^{-2}\text{sec}^{-1})}{\sqrt{2\pi MRT}}. \tag{5-14}$$

There is no dearth of adatoms that can diffuse to and be captured by the existing nuclei. During their residence time, adatoms are capable of diffusing a mean distance X from the site of incidence given by

$$X = \sqrt{2 D_S \tau_s}. \tag{5-15}$$

The surface diffusion coefficient D_S is essentially

$$D_S = (1/2) a_0^2 \nu \exp - E_S/kT$$

and therefore

$$X = a_0 \exp \frac{E_{\text{des}} - E_S}{2kT}. \tag{5-16}$$

Large values of E_{des} coupled with small values of E_S serve to extend the nucleus capture radius.

Upon substitution of Eqs. 5-11, 5-13, and 5-14 in Eq. 5-10, we obtain

$$\dot{N} = 2\pi r^* a_0 \sin\theta \frac{PN_A}{\sqrt{2\pi MRT}} n_s \exp\frac{E_{des} - E_S - \Delta G^*}{kT}. \quad (5\text{-}17)$$

The nucleation rate is a very strong function of the nucleation energetics, which are largely contained within the term ΔG^*. It is left to the reader to develop the steep dependence of $\dot{N}$ on the vapor supersaturation ratio. As noted previously, a high nucleation rate encourages a fine-grained, or even amorphous, structure, whereas a coarse-grained deposit develops from a low value of $\dot{N}$.

5.2.3. Nucleation Dependence on Substrate Temperature and Deposition Rate

Substrate temperature and deposition rate $\dot{R}$ (atoms/cm^2-sec) are among the chief variables affecting deposition processes. Calculating their effect on r^* and ΔG^* is both instructive and simple to do. First we define

$$\Delta G_V = -\frac{kT}{\Omega} \ln\frac{\dot{R}}{\dot{R}_e} \quad (5\text{-}18)$$

by analogy to Eq. 1-38, where $\dot{R}_e$ is the equilibrium evaporation rate from the film nucleus at the substrate temperature T, and Ω is the atomic volume. Assuming an inert substrate, i.e., $\gamma_{vf} = \gamma_{fs}$, direct differentiation of Eq. 5-6 yields (Ref. 3)

$$\left(\frac{dr^*}{dT}\right)_{\dot{R}} = \frac{2}{3}\left|\frac{\gamma_{vf}(\partial\,\Delta G_v/\partial T) - (a_1 + a_2)\,\Delta G_v(\partial\gamma_{vf}/\partial T)}{a_3\,\Delta G_V^2}\right|. \quad (5\text{-}19)$$

Assuming typical values, $\gamma_{vf} \approx 1000$ ergs/cm^2 and $\partial\gamma_{vf}/\partial T = -0.5$ erg/cm^2-K. An estimate for $\partial\,\Delta G_v/\partial T$ is the entropy change for vaporization, which is roughly 8×10^7 ergs/cm^3-K for many metals. If $|\Delta G_v| < 1.6 \times 10^{11}$ ergs/cm^3, then direct substitution gives

$$(\partial r^*/\partial T)_{\dot{R}} > 0. \quad (5\text{-}20)$$

Similarly, by the same assumptions and arguments,

$$(\partial\,\Delta G^*/\partial T)_{\dot{R}} > 0. \quad (5\text{-}21)$$

It is also simple to show that

$$(\partial r^*/\partial \dot{R})_T < 0. \tag{5-22}$$

Direct chain-rule differentiation using Eqs. 5-6 and 5-18 yields

$$\frac{\partial r^*}{\partial \dot{R}} = \left(\frac{\partial r^*}{\partial \Delta G_V}\right)\left(\frac{\partial \Delta G_V}{\partial \dot{R}}\right) = \left(-\frac{r^*}{\Delta G_V}\right)\left(-\frac{kT}{\Omega \dot{R}}\right).$$

Since ΔG_v is negative, the overall sign is negative. Similarly, it is easily shown that

$$(\partial \Delta G^*/\partial \dot{R})_T < 0. \tag{5-23}$$

The preceding four inequalities have interesting implications and summarize a number of common effects observed during film deposition. From Eq. 5-20 we note that a higher substrate temperature leads to an increase in the size of the critical nucleus. Also a discontinuous island structure is predicted to persist to a higher average coverage than at low substrate temperatures. The second inequality (Eq. 5-21) suggests that a nucleation barrier may exist at high substrate temperatures, whereas at lower temperatures it is reduced in magnitude. Also because of the exponential dependence of $\dot{N}$ on ΔG^*, the number of supercritical nuclei decreases rapidly with temperature. Thus, a continuous film will take longer to develop at higher substrate temperatures. From Eq. 5-22, it is clear that increasing the deposition rate results in smaller islands. Because ΔG^* is also reduced, nuclei form at a higher rate, suggesting that a continuous film is produced at lower average film thickness. If, for the moment, we associate a large r^* and ΔG^* with being conducive to large crystallites or even monocrystal formation, this is equivalent to high substrate temperatures and low deposition rates. Alternatively, low substrate temperatures and high deposition rates yield polycrystalline deposits. These simple guidelines summarize much practical deposition experience in a nutshell. A plot of these structural zones as a function of deposition rate and substrate temperature is shown in Fig. 5-4 for Cu films deposited on a (111) NaCl substrate. Maps of this sort have been graphed for other film–substrate combinations. In semiconductor systems a regime of amorphous film growth is additionally observed when $\dot{R}$ is large and T is low.

Despite the qualitative utility of capillarity theory, it is far from being quantitatively correct. For example, let us estimate the value of r^* for a typical metal film at 300 K. To simplify matters, the homogeneous nucleation formula, $r^* = -2\gamma/\Delta G_V$, is used, where $\Delta G_V = -kT/\Omega \ln P_v/P_s$ (Eq. 1-38). Assuming $\Omega = 20 \times 10^{-24}\text{cm}^3/\text{at}$, $\gamma = 1000$ ergs/cm^2, $P_v = 10^{-3}$ torr,

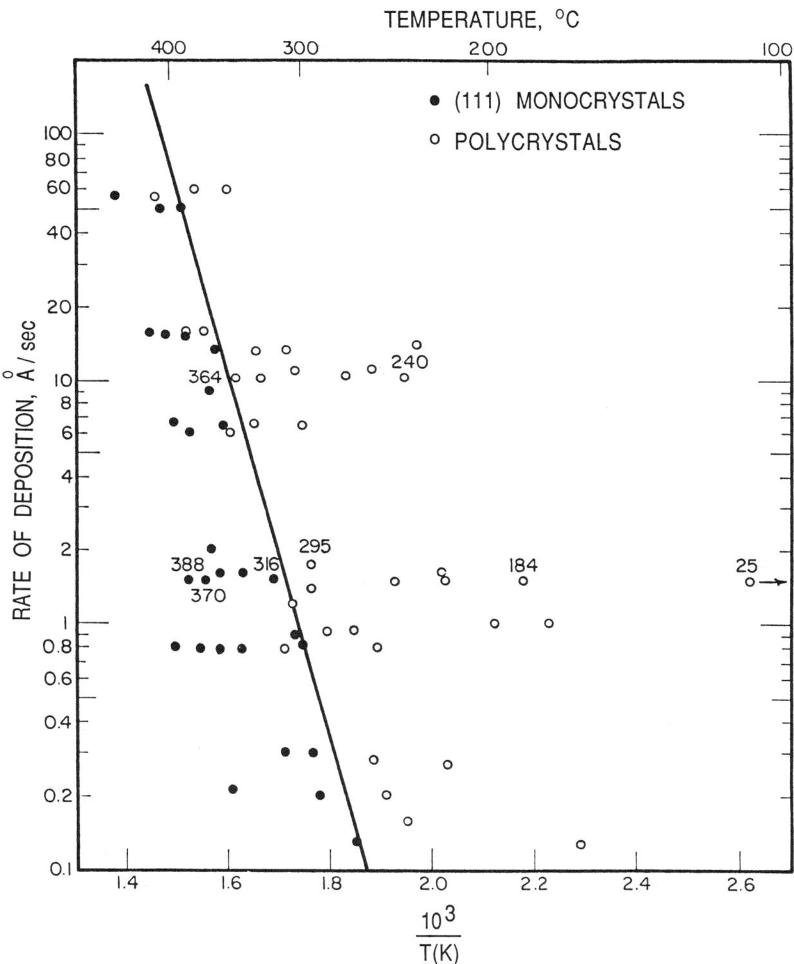

Figure 5-4. Dependence of microstructure on deposition rate and substrate temperature for Cu films on (111) NaCl. (Some substrate temperatures are noted.) (Reprinted with permission from the Metals Society, from R. W. Vook, *Int. Metals Rev.* **27**, 209, 1982).

and $P_s = 10^{-10}$ torr, then

$$r^* = \frac{-2(1000)}{-1.38 \times 10^{-16}(300/20 \times 10^{-24})\ln(10^{-3}/10^{-10})} = 6 \times 10^{-8} \text{ cm}.$$

A nucleus of this size contains just a few atoms, and it is therefore doubtful that continuum concepts like surface tension and nucleus radius have much significance in such a case. It is more realistic to interpret vapor condensation phenomena in terms of a theory of heterogeneous nucleation based on an atomistic model. This is considered in the next section.

5.3. Atomistic Nucleation Processes

5.3.1. The Walton – Rhodin Theory

Atomistic theories of nucleation describe the role of individual atoms and associations of small numbers of atoms during the earliest stages of film formation. An important milestone in the atomistic approach to nucleation was the theory advanced by Walton *et al.* (Ref. 6), which treated clusters as macromolecules and applied concepts of statistical mechanics in describing them. They introduced the critical dissociation energy E_{i*}, defined to be that required to disintegrate a *critical* cluster containing i atoms into i separate adatoms. The critical concentration of clusters per unit area of size i, N_{i*}, is then given by

$$\frac{N_{i*}}{n_0} = \left|\frac{N_1}{n_0}\right|^{i^*} \exp\frac{E_{i*}}{kT}, \tag{5-24}$$

which expresses the chemical equilibrium between the clusters and monomers. In Eq. 5-24, E_{i*} may be viewed as the negative of a cluster formation energy, n_0 is the total density of adsorption sites, and N_1 is the monomer density. The latter, by analogy to Eq. 5-12, is given by $N_1 = \dot{R}\tau_s$. Hence,

$$N_1 = \dot{R}\nu^{-1}\exp E_{\text{des}}/kT. \tag{5-25}$$

Lastly, the critical monomer supply rate is essentially given by the vapor impingement rate and the area over which adatoms are capable of diffusing before desorbing. Therefore, through Eq. 5-16,

$$\dot{R}X^2 = \dot{R}a_0^2\exp\frac{E_{\text{des}} - E_S}{kT}. \tag{5-26}$$

By combining Eqs. 5-24–5-26, we obtain the critical nucleation rate (cm^{-2} sec^{-1})

$$\dot{N}_{i*} = \dot{R}a_0^2 n_0 \left(\frac{\dot{R}}{\nu n_0}\right)^{i^*} \exp\left(\frac{(i^* + 1)E_{\text{des}} - E_S + E_{i*}}{kT}\right). \tag{5-27}$$

This expression for the nucleation rate of small clusters has been central to subsequent theoretical treatments and in variant forms has also been used to test much experimental data. It has the advantage of expressing the nucleation rate in terms of measurable parameters rather than macroscopic quantities such as ΔG^*, γ, or θ, which are not known with certainty, nor are they easily estimated in the capillarity theory. The uncertainties are now in i^* and E_{i^*}.

One of the important applications of this theory has been to the subject of epitaxy, where the crystallographic geometry of stable clusters has been related to different conditions of supersaturation and substrate temperature. In Fig. 5-5, experimental data for Ag on (100) NaCl are shown together with the atomistic model of evolution to stable clusters upon attachment of a single adatom. At high supersaturations and low temperatures, the nucleation rate is frequently observed to depend on the square of the deposition rate; i.e., $\dot{N}^* \propto \dot{R}^2$, indicating that $i^* = 1$. This means that a single adatom is, in effect, the critical nucleus. At higher temperatures, two or three atom nuclei are critical, and the stable clusters now assume the planar atomic pattern suggestive of (111) or (100) packing, respectively. Epitaxial films would then evolve over macroscopic dimensions provided the original nucleus orientation were preserved with subsequent deposition and cluster impingement.

A thermally activated nucleation rate whose activation energy depends on the size of the critical nucleus is predicted by Eq. 5-27. This suggests the existence of critical temperatures where the nucleus size and orientation undergo change. For example, the temperture $T_{1\rightarrow 2}$ at which there is a

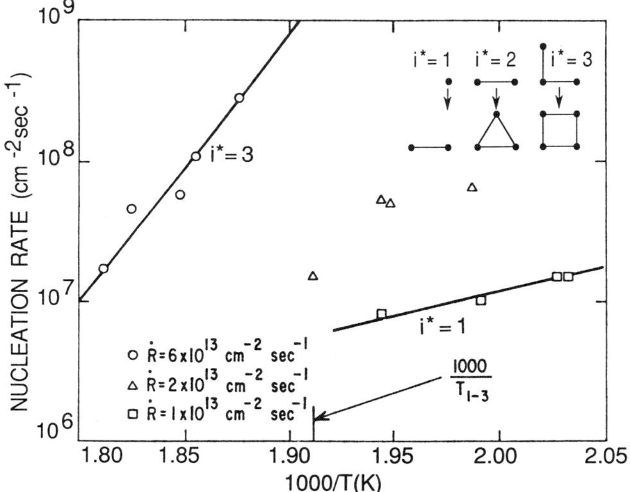

Figure 5-5. Nucleation rate of Ag on (100) NaCl as a function of temperature. Data for three different deposition rates are plotted. Also shown are smallest stable epitaxial clusters corresponding to critical nuclei. (From Ref. 6).

transition from a one- to two-atom nucleus is given by equating the respective nucleation rates:

$$\dot{N}_{i^*=1} = \dot{N}_{i^*=2}.$$

After substitution and some algebra, we have

$$T_{1\to 2} = -\frac{E_{\text{des}} + E_2}{k \ln(\dot{R}/\nu n_0)} \tag{5-28a}$$

or

$$\dot{R} = \nu n_0 \exp -\left|\frac{E_{\text{des}} + E_2}{kT_{1\to 2}}\right|. \tag{5-28b}$$

Equation 5-28b can be used to analyze the data in Fig. 5-4. The transition of polycrystalline Cu to single-crystal Cu films can be thought of as the onset of (111) epitaxy ($i^* = 2$) from atomic nuclei ($i^* = 1$). The Arrhenius plot of the line of demarcation between mono- and polycrystal deposits yields an activation energy of $E_{\text{des}} + E_2$ equal to 1.48 eV. At a deposition rate of 8.5×10^{14} atoms/cm^2-sec or 1 Å/sec, and $\nu n_0 = 6.9 \times 10^{27}$ atoms/cm^2-sec, the epitaxial transition temperature can be calculated to be 577 K. Equations similar to 5-28 can be derived for transitions between $i^* = 1$ and $i^* = 3$ or $i^* = 2$ and $i^* = 3$, etc.

It is instructive to compare these estimates of epitaxial growth temperatures with ones based on surface-diffusion considerations. For smooth layer-by-layer growth on high-symmetry crystal surfaces, adatoms must reach growth ledges

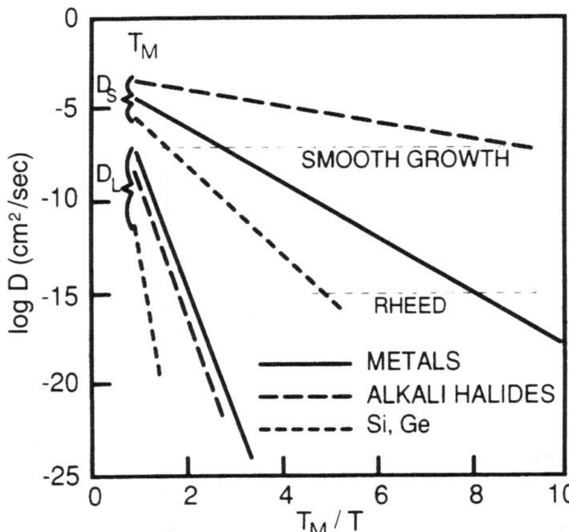

Figure 5-6. Lattice (D_L) and surface diffusivities (D_S) as a function of T_M/T for metals, semiconductors, and alkali halides. (From Ref. 7).

by diffusional hopping before other islands nucleate. Atoms must, therefore, migrate a distance of some 100–1000 atoms, the typical terrace width on well-oriented surfaces during the monolayer growth time of 0.1–1 sec. By Eq. 5-15 a surface diffusivity greater than $\sim 10^{-8}$ cm^2/sec is, therefore, required irrespective of the material deposited. This means a different critical epitaxial growth temperature (T_E), depending on whether semiconductor, metal, or alkali halide films are involved. In the Arrhenius plot of Fig. 5-6, D_S values are graphed versus T_M/T, where T_M is the melting point. Similar relationships for lattice, grain-boundary, and dislocation diffusivities will be introduced in Chapter 8. At the critical value of $D_S = 10^{-8}$, Fig. 5-6 indicates that $T_E \sim 0.5T_M$, $\sim 0.3T_M$, and $\sim 0.1T_M$ for layer growth on group IV semiconductors, metals, and alkali halides, respectively. These temperatures agree qualitatively with experimental observations.

5.3.2. Kinetic Models of Nucleation

More recent approaches to the modeling of nucleation processes have stressed the kinetic behavior of atoms and clusters containing a small number of atoms. Rate equations similar to those describing the kinetics of chemical reactions are used to express the time rate of change of cluster densities in terms of the processes that occur on the substrate surface. Because these theories and models are complex mathematically as well as physically, the discussion will stress the results without resorting to extensive development of derivations. It is appropriate to start with the fate of the mobile monomers. If coalescence is neglected, then

$$\frac{dN_1}{dt} = \dot{R} - \frac{N_1}{\tau_s} - K_1 N_1^2 - N_1 \sum_{i=2}^{\infty} K_i N_i. \tag{5-29}$$

The equation states that the time rate of change of the monomer density is given by the incidence rate, minus the desorption rate, minus the rate at which two monomers combine to form a dimer. This latter term follows second-order kinetics with a rate constant K_1. The last term represents the loss in monomer population due to their capture by larger clusters containing two or more atoms. There are similar equations describing the population of dimer, trimer, etc., clusters as they interact with monomers. The general form of the rate equation for clusters of size i is

$$dN_i/dt = K_{i-1}N_1N_{i-1} - K_iN_1N_i, \tag{5-30}$$

where the first term on the right expresses their increase by attachment of monomers to smaller $(i-1)$-sized clusters, and the second term their decrease when they react with monomers to produce larger $(i+1)$-sized clusters. There are i of these *coupled* rate equations to contend with, each one of which depends on direct impingement from the vapor as well as desorption through their link back to Eq. 5-25. Inclusion of diffusion terms, i.e., d^2N_i/dx^2, enables change in cluster shape to be accounted for. When cluster mobility and coalescence are also taken into account, a fairly complete chronology of nucleation events emerges but at the expense of greatly added complexity.

Transient as well as steady-state (i.e., where $dN_i/dt = 0$) solutions have been obtained for the preceding rate equations for a variety of physical situations and the results are summarized below starting with the case for which $i^* = 1$.

1. The nucleation and growth kinetics for the one-atom critical nucleus has been treated under simplifying assumptions by Robinson and Robins (Ref. 8). They divided the deposition conditions into two categories. At high temperatures and/or low deposition rates, the reevaporation rate from the surface will control the adatom density and exceed the rate of diffusive capture into growing nuclei. Here the adsorption–desorption equilibrium is rapidly established where $N_1 = \dot{R}\tau_s$, and *incomplete* condensation is said to occur. When reevaporation is not important, however, because the energy to desorb adatoms (E_{des}) is large, then we speak of *complete* condensation. This regime dominates at low tempertures and/or high deposition rates where the monomer capture rate by growing nuclei exceeds the rate at which they are lost due to reevaporation. These same condensation regimes also refer to situations where $i^* > 1$. The division between these high- and low-temperature regions is not abrupt like $T_{1\to 2}$ (Eq. 5-28a). It can be characterized by a dividing temperature T_D, calculated to be

$$T_D = \frac{2E_S - 3E_{\text{des}}}{k\ln\left[(C\alpha^2/\beta)(\dot{R}/\nu n_0)\right]}, \tag{5-31}$$

where C is the number of pair formation sites ($C = 4$ for a square lattice), and α and β are dimensionless constants with typical values of ~ 0.3 and 4, respectively. Relatively simple closed-form analytic expressions for the time (t) dependence of the transient density of stable nuclei $N(t)$ as well as the saturation value of $N(t = \infty)$ or N_S in terms of measurable quantities were obtained at temperatures above or below T_D.

For $T > T_D$,

$$N(t) = N_S \tanh\left(\frac{\dot{N}_o(0)\,t}{N_S}\right), \tag{5-32}$$

$$N_S = \left(\frac{Cn_0}{\beta\nu}\right)^{1/2} \dot{R}^{1/2} \exp\frac{E_{\text{des}}}{2kT}. \tag{5-33}$$

For $T < T_D$,

$$N(t) = N_S\left\{1 - \exp - 3\eta^2\frac{\dot{N}(0)\,t}{N_S^3}\right\}^{1/3}, \tag{5-34}$$

$$N_S = \left(\frac{Cn_0^2}{\alpha\beta\nu}\right)^{1/3}\dot{R}^{1/3}\exp\frac{E_S}{3kT}. \tag{5-35}$$

Clearly, as $t \to \infty$, $N(\infty) \to N_S$. For all temperatures,

$$\dot{N}(0) = \frac{C\dot{R}^2}{n_0\nu}\exp\frac{2E_{\text{des}} - E_S}{kT}. \tag{5-36}$$

Only $\dot{N}(0)$, the initial nucleation rate (i.e., $\dot{N}(0) = dN(t)/dt$ at $t = 0$) and η, which is equal to $(n_0/\alpha)\exp(E_S - E_{\text{des}}/kT)$, have not been defined before.

The equations predict that $N(t)$ increases with time, eventually saturating at the value N_S. In practice, the stable nucleus density is observed to increase approximately linearly with deposition time, then saturate at values ranging from $\sim 10^9$ to 10^{12}, depending on deposition rate and substrate temperature. With further deposition, N decreases due to nucleus coalescence. Furthermore, N_S is larger the lower the substrate temperature, as shown schematically in Fig. 5-7. This is an obvious consequence of Eqs. 5-33 and 5-35. When the temperature range spanned exceeds T_D, N_S data as a function of T directly yield $E_{\text{des}}/2$ on an Arrhenius plot. Similarly, $E_S/3$ can be extracted from such data when $T < T_D$. In this manner, values of E_{des} and E_S for noble metal–alkali halide systems have been obtained and the results summarized in Table 5-1. An experimental verification of Eq. 5-36 is shown in Fig. 5-8. That

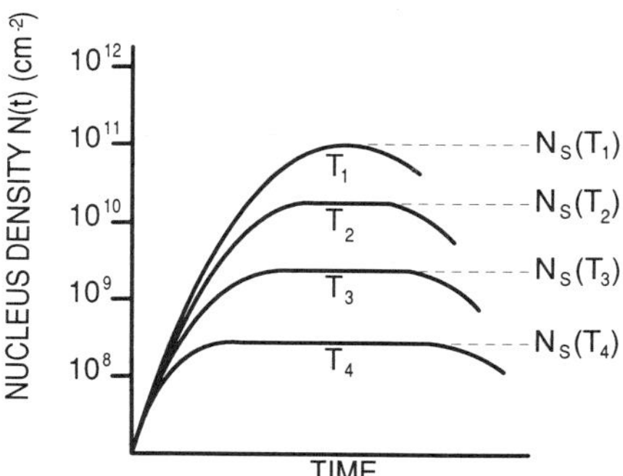

Figure 5-7. Schematic dependence of $N(t)$ with time and substrate temperature. ($T_4 > T_3 > T_2 > T_1$) From Ref. 4.

Table 5-1. Nucleation Parameter Values for Noble Metal–Alkali Halide Systems

System	E_{des} (eV)	E_S (eV)	ν (10^{12}/sec)
Au–NaCl	0.70	0.29	1.65
Au–KCl	0.68	0.25	1.26
Au–KBr	0.83	0.44	~ 24
Au–NaF	0.64	0.093	0.91
Ag–NaCl	0.63	0.093	~ 0.8
Ag–KCl	0.48	0.2	0.24
Ag–KBr	0.48	0.22	~ 0.1

From Ref. 5.

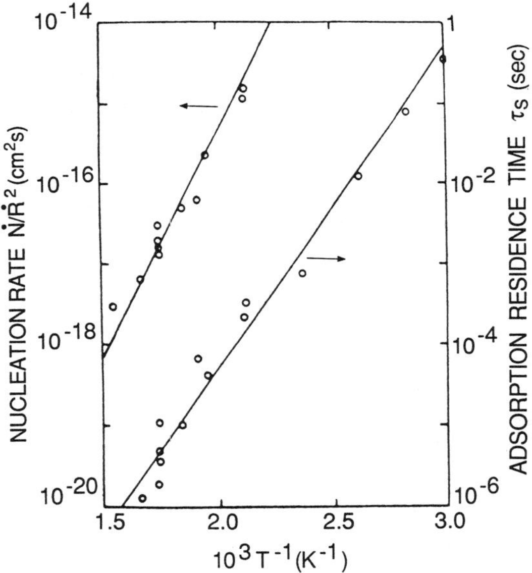

Figure 5-8. Plot of $\dot{N}/\dot{R}^2$ vs. $1/T$ (left) and τ_s vs. $1/T$ (right) for Au on KCl. (From Ref. 5, with permission from IOP Publishing Ltd.)

$\dot{N}(0)/\dot{R}^2$ is observed to vary linearly as a function of $1/T$ is confirmation that Au monomers are stable nuclei on KCl substrates.

2. Venables *et al.* (Ref. 5) have neatly summarized nucleation behavior for cases where i^* assumes any integer value. In general, the stable cluster density is given by

$$N_S = An_0 \left| \frac{\dot{R}}{n_o \nu} \right|^p \exp \frac{E}{kT}, \tag{5-37}$$

Table 5-2. Nucleation Parameters p and E in Eq. 5-37

Regime	3D Islands	2D Islands
Extreme incomplete	$p = (2/3)i^*$	i^*
	$E = (2/3)[E_{i^*} + (i^* + 1)E_{des} - E_S]$	$E_{i^*} + (i^* + 1)E_{des} - E_S$
Initially incomplete	$p = 2i^*/5$	$i^*/2$
	$E = (2/5)(E_{i^*} + i^*E_{des})$	$(1/2)(E_{i^*} + i^*E_{des})$
Complete	$p = i^*/(i^* + 2.5)$	$i^*/(i^* + 2)$
	$E = \dfrac{E_{i^*} + i^*E_S}{i^* + 2.5}$	$\dfrac{E_{i^*} + i^*E_S}{i^* + 2}$

From Ref. 5.

where A is a calculable dimensionless constant dependent on the substrate coverage. Parameters p and E depend on the condensation regime and are summarized in Table 5-2. Three regimes of condensation and two types of island nuclei are considered. The *complete* and *extreme incomplete* condensation categories parallel those considered previously, but intermediate incomplete condensation regimes may also be imagined, depending on deposition conditions. As a result of such generalized equations, experimental data for N_S have been tested as a function of the substrate temperature and deposition rate, and values for the energies of desorption, diffusion, and cluster binding have been extracted from them. The Walton *et al.* nucleation theory is seen to be a special case of the more general rate theory. For extremely incomplete condensation of 2-D islands, the p and E parameters are seen to vary in the same way as Eq. 5-27.

Although the major application of the kinetic model has been to island growth, the theory is also capable of describing S.K. growth.

5.4. Cluster Coalescence and Depletion

The results of the kinetic theories of nucleation indicate that in the initial stages of growth the density of stable nuclei increases with time up to some maximum level before decreasing. In this section, the coalescence processes that are operative beyond the cluster saturation regime are examined. Coalescence of nuclei is generally characterized by the following features:

1. A decrease in the total projected area of nuclei on the substrate occurs.
2. There is an increase in the height of the surviving clusters.
3. Nuclei with well-defined crystallographic facets sometimes become rounded.

4. The composite island generally reassumes a crystallographic shape with time.
5. When two islands of very different orientation coalesce, the final compound cluster assumes the crystallographic orientation of the larger island.
6. The coalescence process frequently appears to be liquidlike in nature with islands merging and undergoing shape changes after the fashion of liquid droplet motion.
7. Prior to impact and union, clusters have been observed to migrate over the substrate surface in a process described as cluster–mobility coalescence.

Several mass-transport mechanisms have been proposed to account for these coalescence phenomena, and these are discussed in turn.

5.4.1. Ostwald Ripening (Ref. 2)

Prior to coalescence there is a collection of islands of varied size, and with time the larger ones grow or "ripen" at the expense of the smaller ones. The time evolution of the distribution of island sizes has been considered both from a macroscopic surface diffusion–interface transfer viewpoint as well as from statistical models involving single atom processes. The former is simply driven by a desire to minimize the surface free energy of the island structure. To understand the process, consider two isolated islands of different size in close proximity, as shown in Fig. 5-9a. For simplicity they are assumed to be spherical with radii r_1 and r_2. The surface free energy per unit area of a given

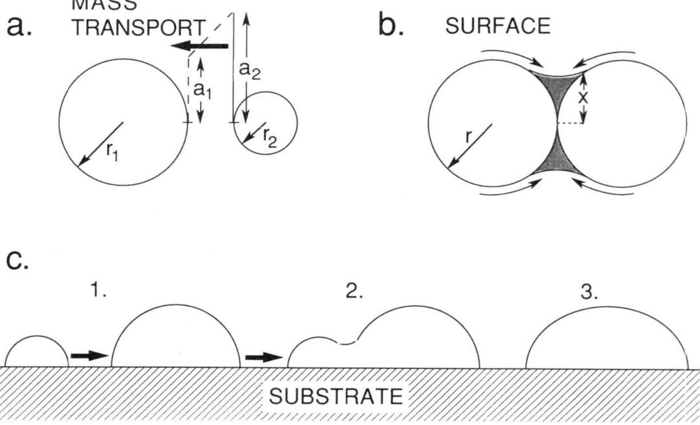

Figure 5-9. Coalescence of islands due to (a) Ostwald ripening, (b) sintering, (c) cluster migration.

island is γ, so the total energy $G_\gamma = 4\pi r_i^2 \gamma$. The island contains a number of atoms n_i given by $4\pi r_i^3/3\Omega$, where Ω is the atomic volume. Defining the free energy per atom μ_i as dG_γ/dn_i in this application, we have, after substitution,

$$\mu_i = \frac{8\pi r_i \gamma \, dr_i}{4\pi r_i^2 \, dr_i/\Omega} = \frac{2\Omega\gamma}{r_i}.$$

Making use of Eq. 1-9 ($\mu_i = \mu_0 + kT \ln a_i$), we see that the Gibbs–Thomson relation

$$kT \ln \frac{a_i}{a_\infty} = \frac{2\Omega\gamma}{r_i} \qquad \text{or} \qquad a_i = a_\infty \exp \frac{2\Omega\gamma}{r_i kT} \tag{5-38}$$

directly follows. This equation states that atoms in an island of radius r_i can be in equilibrium only with a substrate adatom activity or effective concentration a_i. The quantity a_∞ may be interpreted as the adatom concentration in equilibrium with a planar island ($r_i = \infty$) or, alternatively, with the vapor pressure of island atoms at temperature T. When the island surface is convex (r_i is positive), atoms have a greater tendency to escape, compared with atoms situated on a planar surface, because there are relatively fewer atomic bonds to attach to. Therefore, $a_i > a_\infty$. Conversely, at a concave island surface, r_i is negative and $a_\infty > a_i$. These simple ideas have significant implications not only with respect to Ostwald ripening but to the sintering mechanisms of coalescence that are treated in the next section.

The establishment of the concentration gradient of adatoms situated between the two particles of Fig. 5-9a can now be understood. Diffusion of individual adatoms will proceed from the smaller to larger island until the former disappears entirely. A mechanism has thus been established for coalescence without the islands having to be in direct contact. In a multi-island array the kinetic details are complicated, but ripening serves to establish a quasi-steady-state island size distribution that changes with time. Ostwald ripening processes never reach equilibrium during film growth since the theoretically predicted narrow distribution of crystallite sizes is generally not observed.

5.4.2. Sintering (Ref. 9)

Sintering is a coalescence mechanism involving islands in contact. It can be understood by referring to Fig. 5-10, depicting a time sequence of coalescence events between Au particles deposited on molybdenite (MoS_2) at 400 °C and photographed within the transmission electron microscope (TEM). Within tenths of a second a neck forms between islands and then successively thickens

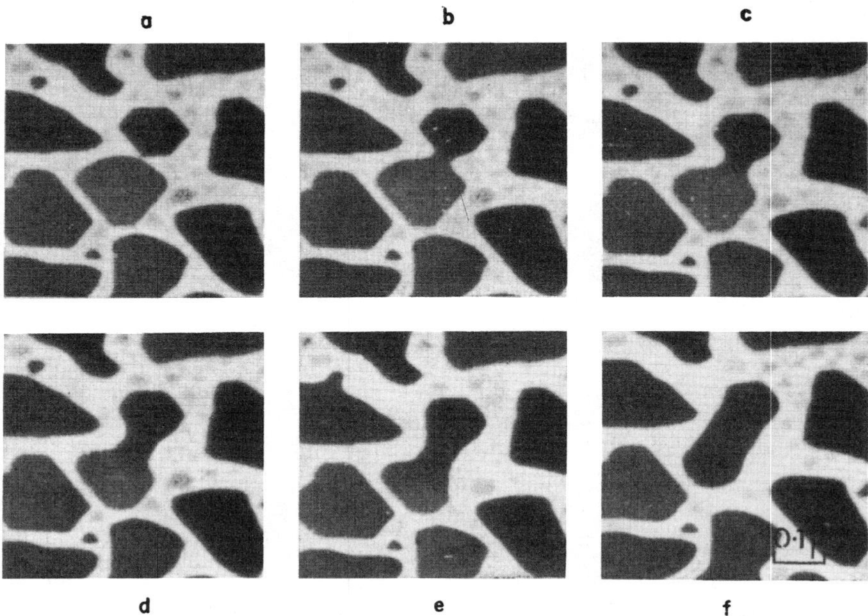

Figure 5-10. Successive electron micrographs of Au deposited on molybdenite at 400 °C illustrating island coalescence by sintering (a) arbitrary zero time, (b) .06 sec, (c) 0.18 sec, (d) 0.50 sec, (e) 1.06 sec, (f) 6.18 sec. (From Ref. 10).

as atoms are transported into the region. The driving force for neck growth is simply the natural tendency to reduce the total surface energy (or area) of the system. Since atoms on the convex island surfaces have a greater activity than atoms situated in the concave neck, an effective concentration gradient between these regions develops. This results in the observed mass transport into the neck. Variations in island surface curvature also give rise to local concentration differences that are alleviated by mass flow. Of the several mechanisms available for mass transport, the two most likely ones involve self-diffusion through the bulk or via the surface of the islands. In the case of sintering or coalescence of two equal spheres of radius r (Fig. 9b), theoretical calculations in the metallurgical literature have shown that the sintering kinetics is given by $x^n/r^m = A(T)t$, where x is the neck radius, $A(T)$ is a temperature-dependent constant, and n and m are constants. Explicit expressions for the bulk and surface diffusion mechanisms are

$$\frac{x^5}{r^2} = \frac{10\pi D_L \gamma \Omega t}{kT} \tag{5-39}$$

and

$$\frac{x^7}{r^3} = \frac{28 D_S \gamma (\Omega)^{4/3} t}{kT}, \tag{5-40}$$

respectively, and D_L and D_S are the lattice and surface diffusion coefficients, respectively. In principle, experimental determination of m and n and the activation energy for diffusion would serve to pinpoint the transport mechanism, but insufficient data have precluded such an analysis. A simple relative comparison between the two mechanisms can, however, be made to order to predict which dominates. The ratio of the times required to reach a neck radius $x = 0.1r$, for example, is given by

$$\frac{t_S}{t_L} = \frac{\pi r}{280} \frac{D_L}{D_S \Omega^{1/3}}. \tag{5-41}$$

In the case of Au at 400 °C, the ratio of D_L / D_S can be directly read off Fig. 5-6. At the value $T_M / T = 1336/673 = 1.99$, $D_L / D_S \sim 10^{-13}/10^{-6} = 10^{-7}$. Substituting $\Omega^{1/3} = 2.57 \times 10^{-8}$ cm and $r = 10^{-5}$ cm, $t_S / t_L = 4.4 \times 10^{-7}$. Therefore, surface diffusion is expected to control sintering coalescence. This is also true for any plausible combination of r and T values in films.

Surface energy and diffusion-controlled mass-transport mechanisms undoubtedly influence liquidlike coalescence phenomena involving islands in contact, yet other driving forces are probably also operative. For example, sintering mechanisms are unable to explain

1. Observed liquidlike coalescence of metals on substrates maintained at 77 K where atomic diffusion is expected to be negligible
2. Widely varying stabilities of irregularly shaped necks, channels, and islands possessing high curvatures at some points
3. A large range of times required to fill visually similar necks and channels
4. An observed enhanced coalescence in the presence of an applied electric field in the substrate plane

5.4.3. Cluster Migration (Ref. 2)

The last mechanism for coalescence considered deals with migration of clusters on the substrate surface (Fig. 9-c). Coalescence occurs as a result of collisions between separate islandlike crystallites (or droplets) as they execute random motion. Evidence provided by the field ion microscope, which has the capability of resolving individual atoms, has revealed the migration of dimer and trimer clusters. Electron microscopy has shown that crystallites with diameters of up to 50–100 Å can migrate as distinct entities, provided the substrate temperature is high enough. Interestingly, the mobility of metal particles can be significantly altered in different gas ambients. Not only do the clusters translate but they have been observed to rotate as well as even jump upon each other and sometimes reseparate thereafter! Cluster migration has been directly observed in many systems, e.g., Ag and Au on MoS_2, Au and Pd on MgO, Ag and Pt on graphite in so-called conservative systems, i.e., where the mass of the deposit remains constant because further deposition from the vapor has ceased. Observations of coalescence in a conservative system include a decreased density of particles, increased mean volume of particles, a particle size distribution that increases in breadth, and a decreased coverage of the substrate.

The surface migration of a cap-shaped cluster with projected radius r is characterized by an effective diffusion coefficient $D(r)$ with units of cm^2/sec. Presently there exist several formulas for the dependence of D on r based on models assumed for cluster migration. The movement of peripheral cluster atoms, the fluctuations of areas and surface energies on different faces of equilibrium-shaped crystallites, and the glide of crystallite clusters aided by dislocation motion are three such models. In each case, $D(r)$ is given by an expression of the form (Ref. 11)

$$D(r) = \frac{B(T)}{r^s} \exp - \frac{E_C}{kT}, \tag{5-42}$$

where $B(T)$ is a temperature-dependent constant and s is a number ranging from 1 to 3. It comes as no surprise that cluster migration is thermally activated with an energy E_C related to that for surface self-diffusion, and that it is more rapid the smaller the cluster. However, there is a lack of relevant experimental data that can distinguish among the mechanisms. In fact, it is difficult to distinguish cluster mobility coalescence from Ostwald ripening based on observed particle size distributions.

The interesting effect an applied electric field has in enhancing coalescence is worthy of brief comment. Chopra (Ref. 12) has explained the effect on the basis of the interaction of the field with electrically charged islands. The assumed island charge is derived from ionized vapor atoms and/or the potential at the substrate interface. For a spherical particle of radius r, which already possesses a surface free energy, the presence of a charge q contributes additional electrostatic energy (i.e., q^2/r). The increase in total energy is accommodated by an increase in surface area. Therefore, the sphere distorts into a flattened oblate spheroid, the exact shape being determined by the balance of various free energies. The net effect of charging is then to promote further coalescence by ripening, sintering, or cluster mobility processes. However, with greater charge or higher fields, the cluster may break up in much the same way that a charged droplet of mercury does.

5.5. Experimental Studies of Nucleation and Growth

A full complement of microscopic and surface analytical techniques has been employed to reveal the physical processes of nucleation and test the theories used to describe them. In this section we focus on just a few of the more important experimental techniques and the results of some studies.

5.5.1. Structural Characterization

The most widely used tool, particularly for island growth studies, is the conventional TEM. The technique consists of depositing metals such as Ag and Au onto cleaved alkali halide single crystals (e.g., LiF, NaCl, and KCl) in an ultrahigh-vacuum system for a given time at fixed $\dot{R}$ and T. Then the deposit is covered with carbon, and the substrate is dissolved away outside the system, leaving a rigid carbon replica that retains the metal clusters in their original crystal orientation. To enhance the very early stages of nucleation, the technique of decoration is practiced. Existing clusters are decorated with a lower-melting-point metal, such as Zn or Cd, that does not condense in the absence of prior noble metal deposition. This technique renders visible otherwise invisible clusters that may contain as few as two atoms, thus making possible more precise comparisons with theory. A disadvantage of these postmortem step-by-step observations is that much useful information on the structure and dynamical behavior of the intermediate stages of nucleation and growth is lost. For this reason, in situ techniques within both scanning (SEM) and transmission electron microscopes, modified to contain deposition sources and heated substrates, have been developed but at the expense of decreased resolution. The contamination of substrates from hydrocarbons present in the specimen chamber is a general problem in such studies necessitating the use of high-vacuum, oil-less pumping methods. In addition to direct imaging of film nucleation and growth, the electron-diffraction capability of the TEM provides additional information on the crystallography and orientation of deposits. Diffraction halos and continuous, as well as spotty, diffraction rings characterize stages of growth where clusters, which are initially randomly oriented, begin to acquire some preferred orientation as shown in Fig. 5-1. When a

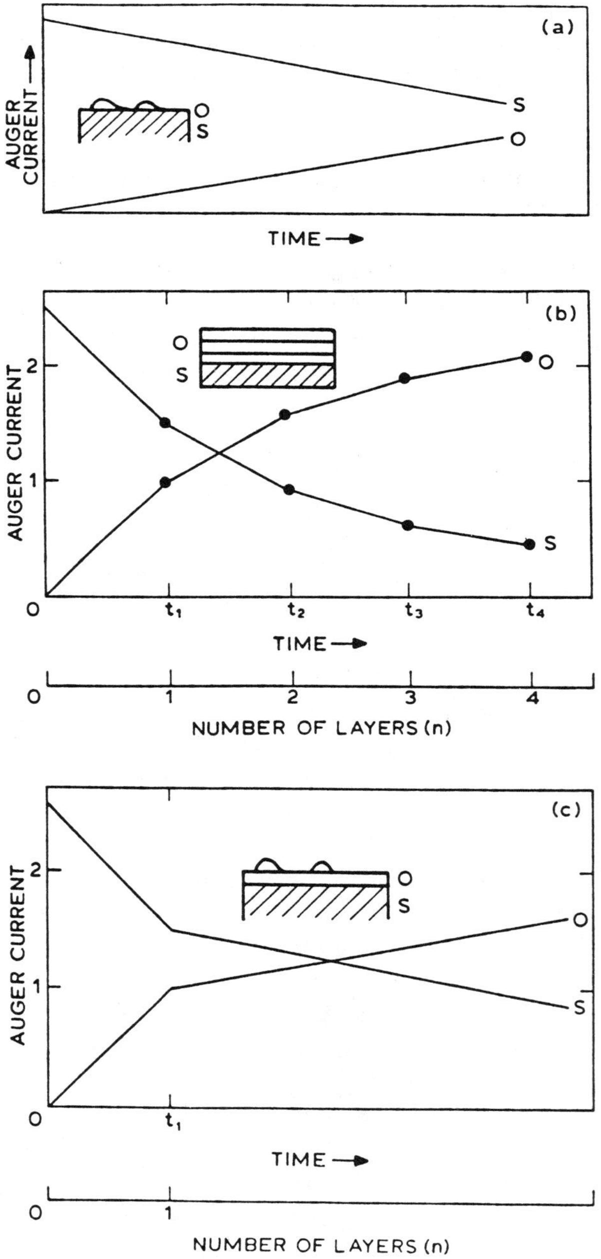

Figure 5-11. Schematic Auger signal currents as a function of time for the three growth modes: (a) island, (b) planar, (c) S.K. O = overlayer, S = substrate. (From Ref. 2).

continuous epitaxial single-crystal film eventually develops, then individual diffraction spots appear.

5.5.2. Auger Electron Spectroscopy (AES)

The AES technique is based on the measurement of the energy and intensity of the Auger electron signal emitted from atoms located within some 5–15 Å of a surface excited by a beam of incident electrons. The subject of AES will be treated in more detail in Chapter 6, but here it is sufficient to note that the Auger electron energies are specific to, or characteristic of, the atoms emitting them and thus serve to fingerprint them. The magnitude of the AES signal is related directly to the abundance of the atoms in question. Consider now the deposit substrate combinations corresponding to the three growth modes. If the AES signal from the film surface of each is continuously monitored during deposition at a constant rate, it will have the coverage or time dependence shown schematically in Fig. 5-11, assuming a sticking coefficient of unity (Ref. 13).

In the case of island growth, the signal from the deposit atoms builds slowly while that from the substrate atoms correspondingly falls. For S.K. growth the signal is ideally characterized by an initial linear increase up to one monolayer or sometimes a few monolayers. Then there is a sharp break, and the Auger amplitude rises slowly as islands, covering a relatively small part of the substrate, are formed. The interpretation of the AES signal in the case of layer growth is more complicated. During growth of the initial monolayer, the Auger signal is proportional to the deposition rate and sticking coefficient of adatoms as well as to the sensitivity in detecting specific elements. For the second and succeeding monolayers, the sticking coefficients change. This gives rise to slight deviations in slope of the AES signal each time a complete monolayer is deposited, and the overall response is therefore segmented as indicated in Fig. 5-11b. It would be misleading to suggest that all AES data fit one of the categories in Fig. 5-11. Complications in interpretation arise from atomic contamination, diffusion, and alloying between deposit and substrate, and transitions between two- and three-dimensional growth processes.

5.5.3. Some Results for Metal Films

Studies of the nucleation and growth of metals, especially the noble ones, on assorted substrates have long provided a base for understanding epitaxy and film formation processes. An appreciation of the scope of past as well as present research activity on metal–substrate systems can be gained by referring to Table 5-3. Only epitaxial Au film–substrate combinations are entered in this

Table 5-3. Substrates on Which Epitaxial Gold Deposits Have Been Observed

Substrate	Refs.	Substrate	Refs.	Substrate	Refs.
1. Au–Metal Halides					
CaF_2	(2)	KI	(10)	NaI	(1)
CdI_2	(1)	LiF	(7)	$PbCl_2$	(1)
KBr	(10)	NaBr	(2)	RbBr	(1)
KCl	(35)	NaCl	(131)	RbI	(1)
KF	(2)	NaF	(2)		
2. Au–Metals					
Ag	(42)	Fe	(3)	Pt	(5)
Al	(1)	Mo	(6)	W	(10)
Cr	(2)	Ni	(2)	Zn	(1)
Cu	(17)	Pb	(2)		
Cu_3Au	(1)	Pd	(17)		
3. Au–Selected Semiconductors and Chalcogenides					
C (graphite)	(5)	MoS_2	(20)	ZnS	(1)
Ge	(2)	PbS	(6)		
Si	(16)	PbSe	(3)		
GaSb	(3)	PbTe	(2)		
GaAs	(5)	SnTe	(1)		
4. Au–Carbonates, Oxides, Mica					
$CaCO_3$	(3)	MgO	(20)	Mica	(23)
Al_2O_3 (sapphire)	(2)	SiO_2 (quartz)	(4)		
$BaTiO_3$	(2)	ZnO	(2)		

Numbers of references dealing with particular Au–substrate system are in parentheses.
From Ref. 14.

recent compilation (Ref. 14), and yet they correspond to over 300 references to the research literature. Other epitaxial metal film–substrate systems have been comprehensively tabulated (Ref. 15) together with deposition methods and variables. The sheer numbers and varieties of metals and substrates involved point to the fact that epitaxy is a common phenomenon. In the overwhelming number of studies, island growth is involved. Perusal of Table 5-3 reveals that epitaxial Au films can be deposited on metallic, covalent, and ionic substrates. Although the majority of substrate materials listed have cubic crystal structures, this is not an essential requirement for epitaxy that occurs, for example, on hexagonal close-packed Zn as well as on monoclinic mica. That epitaxy is possible between materials of different chemical bonding and crystal structure means that its origins are not simple. The long-held belief that a small difference in lattice constant between film and substrate is essential for epitaxy is mistaken; small lattice mismatch is neither a necessary nor sufficient condition for epitaxy. The lattice parameter of the metal can either be larger or smaller than that of the substrate. Having said this, it is also true that the defect density in these metal film "island" epitaxial systems is very much larger than in the "planar" epitaxial semiconductor systems discussed in Chapter 7. Very close lattice matching is maintained in the planar epitaxial systems.

The following specific findings briefly characterize the numerous studies of epitaxy of metal films on ionic substrates (Ref. 12).

1. Substrate. The FCC metals generally grow with parallel orientations on (100), (110), and (111) surfaces of NaCl, but with the (111) plane parallel to the (100) mica cleavage plane. Complex relative positioning of atoms due to translational, and more frequently rotational movements, appears to be the significant variable in epitaxy rather than lattice parameter differences.

2. Temperature. High substrate temperatures facilitate epitaxy by (a) lowering supersaturation levels, (b) stimulating desorption of impurities, (c) enhancing surface diffusion of adatoms into equilibrium sites, and (d) promoting island coalescence. The concept of an epitaxial temperature T_E has been advanced for alkali halide substrates. Temperature T_E depends on the nature of the substrate as well as the deposition rate. For example, T_E for Ag on LiF, NaCl, KCl, and KI was determined to be 340 °C, 150 °C, 130 °C and 80 °C, respectively. The progressive decrease in T_E correlates with increases in lattice parameter and enhanced ionic (both positive and negative) polarizabilities. The latter facilitate attractive forces between metal and substrate atoms.

3. Deposition Rate. In general, low deposition rates, $\dot{R}$, foster epitaxy. It has been established that epitaxy occurs when $\dot{R} \leq \text{const } e^{-E/kT_E}$. This inequality is satisfied physically when the rate at which adatoms settle into equilibrium sites exceeds the rate at which adatoms collide with each other. Such an interpretation requires that E be a surface diffusion activation energy rather than $E_{des} + E_2$ in Eq. 5-28b. The reader should compare this criterion for T_E with those proposed earlier.

4. Contamination. The effect of contamination is a source of controversy. It has been reported that epitaxy of FCC metals is more difficult on ultrahigh-vacuum-cleaved alkali halide substrates than on air-cleaved crystals. Apparently air contamination increases the density of initial nuclei inducing earlier coalescence.

5.6. Grain Structure of Films and Coatings

5.6.1. Zone Models for Evaporated and Sputtered Coatings

Until now the chapter has largely focused on the early stages of the formation of both polycrystalline and single-crystal films. In this section the leap is made to the regime of the fully developed grain structure of thick polycrystalline films and coatings. As we have seen, condensation from the vapor involves incident atoms becoming bonded adatoms, which then diffuse over the film surface until they desorb or, more commonly, are trapped at low-energy lattice sites. Finally, incorporated atoms reach their equilibrium positions in the lattice by bulk diffusive motion. This atomic odyssey involves four basic processes: shadowing, surface diffusion, bulk diffusion, and desorption. The last three are quantified by the characteristic diffusion and sublimation activa-

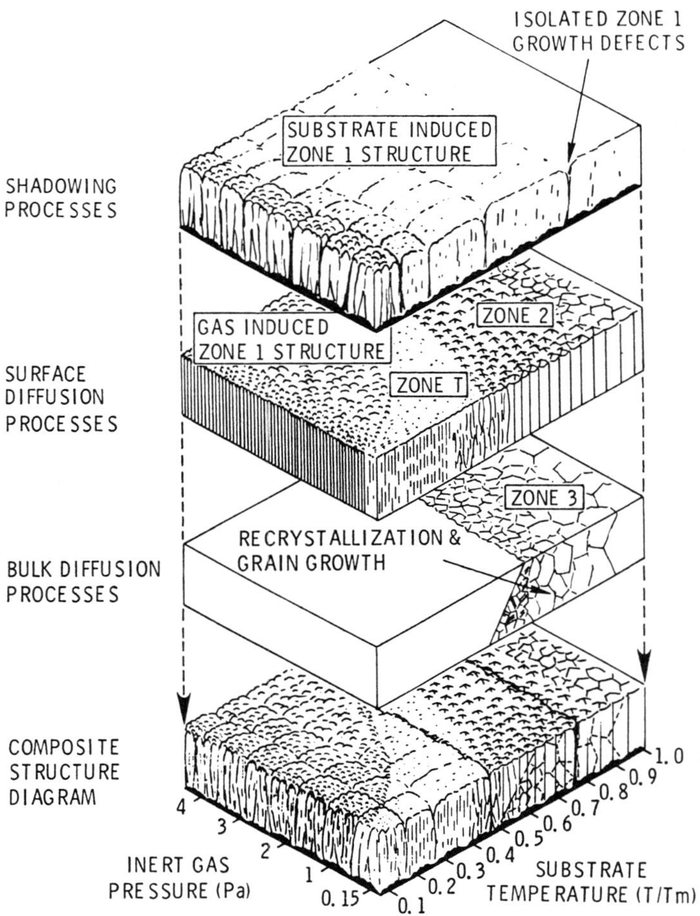

Figure 5-12. Schematic representation showing the superposition of physical processes which establish structural zones. (Reprinted with permission from Ref. 17, © 1977 Annual Reviews Inc.).

tion energies whose magnitudes scale directly with the melting point T_M of the condensate. Shadowing is a phenomenon arising from the geometric constraint imposed by the roughness of the growing film and the line-of-sight impingement of arriving atoms. The dominance of one or more of these four processes as a function of substrate temperature T_S is manifested by different structural morphologies. This is the basis of the zone structure models that have been developed to characterize film and coating grain structures.

The earliest of the zone models was proposed by Movchan and Demchishin (Ref. 16), based on observations of very thick evaporated coatings (0.3 to 2 mm) of metals (Ti, Ni, W, Fe) and oxides (ZrO_2 and Al_2O_3) at rates ranging from 12,000 to 18,000 Å/min. The structures were identified as belonging to one of three zones (1, 2, 3). A similar zone scheme was introduced by Thornton (Ref. 17) for sputtered metal deposits, but with four zones (1, T, 2, 3). His model is based on structures developed in 20- to 250-μm-thick magnetron sputtered coatings deposited at rates ranging from 50 to 20,000 Å/min. The exploded view of Fig. 5-12 illustrates the effect of the individual physical processes on structure and how they depend on substrate temperature

Table 5-4. Zone Structures in Thick Evaporated and Sputtered Coatings

Zone	T_S / T_M	Structural Characteristics	Film Properties
1 (E)	< 0.3	Tapered crystals, dome tops, voided boundaries.	High dislocation density, hard.
1 (S)	< 0.1 at 0.15 Pa to < 0.5 at 4 Pa	Voided boundaries, fibrous grains. Zone 1 is promoted by substrate roughness and oblique deposition.	Hard.
T (S)	0.1 to 0.4 at 0.15 Pa, $\sim$ 0.4 to 0.5 at 4 Pa	Fibrous grains, dense grain boundary arrays.	High dislocation density, hard, high strength, low ductility.
2 (E)	0.3 to 0.5	Columnar grains, dense grain boundaries.	Hard, low ductility.
2 (S)	0.4 to 0.7		
3 (E)	0.5–1.0	Large equiaxed grains, bright surface.	Low dislocation density, soft recrystallized grains.
3 (S)	0.6–1.0		

Note: (E) refers to evaporated. (S) refers to sputtered.

and inert sputtering gas (Ar) pressure. A comparison between zone structures and properties for evaporated and sputtered coatings is made in Table 5-4. In general, analogous structures evolve at somewhat lower temperatures in evaporated films than in sputtered films. Zone 1 structures, which appear in amorphous as well as crystalline deposits, are the result of shadowing effects that overcome limited adatom surface diffusion. In the zone 2 regime, structures are the result of surface diffusion-controlled growth. Lattice diffusion dominates at the highest substrate temperatures, giving rise to the equiaxed recrystallized grains of zone 3.

In contrast to metals, ceramic materials tend to have low hardness at low values of T_S / T_M, indicating that their strength is adversely affected by lattice and grain-boundary imperfections. Ceramics also become harder, not softer, in zones 2 and 3.

5.6.2. Zone Model for Evaporated Metal Films

In a recent study (Ref. 18), a zone model for thin evaporated metal films 1000 Å thick has been developed. The results for 10 elemental films are shown in Fig. 5-13a, where the maximum and minimum grain size variation with T_S is shown. For $T_S / T_M < 0.2$ ($T_M / T_S > 5$), the grains are equiaxed with a diameter of less than 200 Å. Within the range $0.2 < T_S / T_M < 0.3$, some grains larger than 500 Å appear surrounded by smaller grains. Columnar grains make their appearance at $T_S / T_M > 0.37$, and still higher temperatures promote lateral growth with grain sizes larger than the film thickness as shown schematically in Fig. 5-13b. Although the same zone classification scheme has been used for both sputtered and evaporated films, the grain morphology in zones 1 and T differ. Zones 1 and T (a transition zone) possess structures produced by continued renucleation of grains during deposition and subsequent grain growth. The result is the bimodal grain structure of zone T. Zone 2 structures are the result of granular epitaxy and grain growth. The variation in grain structure in zones 1, T, and 2 presumably arises because different grain boundaries become mobile at different temperatures. In zone 1, virtually all grain boundaries are immobile, whereas in zone 2 they are all mobile. Consequently, at higher temperatures the probability of any boundary sweeping across a grain and reacting to form another mobile boundary is increased. Coupled with enhanced surface diffusion, a decrease in porosity results in zone 3. Bulk grain growth and surface recrystallization occur at the highest temperatures with the largest activation energies. This is evident in Fig. 5-13a, which shows the steep dependence of grain size with T_S for $T_M / T_S < 3$.

(a)

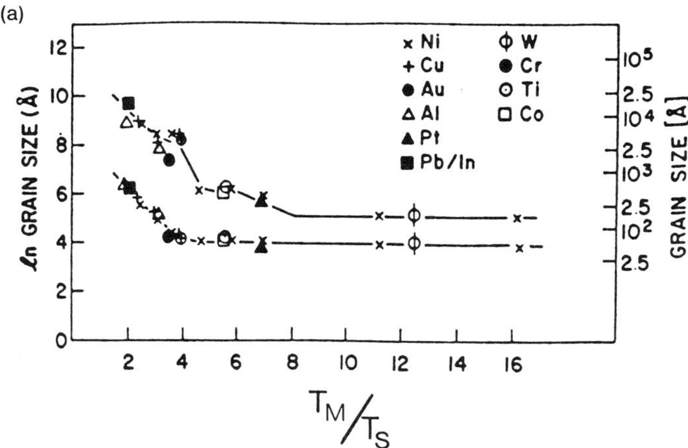

(b)

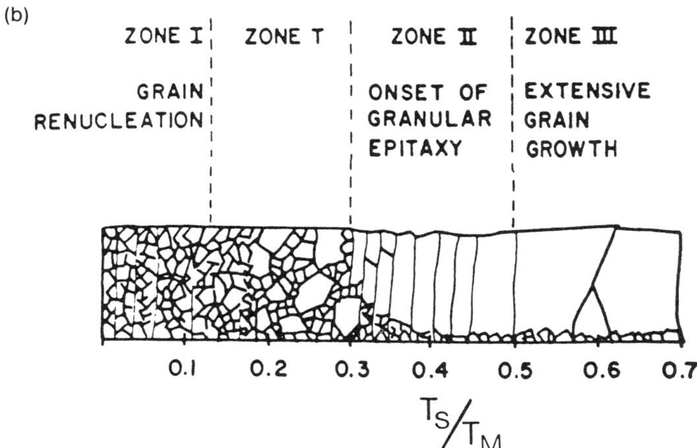

Figure 5-13. (a) Plot of maximum and minimum grain size variation with homologous substrate temperature for 10 different evaporated metals. (b) Zone model for evaporated metal films. (From Ref. 18).

5.6.3. Columnar Grain Structure

The columnar grain structure of thin films has been a subject of interest for several decades. This microstructure consisting of a network of low-density material surrounding an array of parallel rod-shaped columns of higher density has been much studied by transmission and scanning electron microscopy. As noted, columnar structures are observed when the mobility of deposited atoms is limited, and therefore their occurrence is ubiquitous. For example, columnar grains have been observed in high-melting-point materials (Cr, Be, Si, and Ge), in compounds of high binding energy (TiC, TiN, CaF_2, and PbS), and in non-noble metals evaporated in the presence of oxygen (Fe and Fe–Ni). Amorphous films of Si, Ge, SiO, and rare earth–transition metal alloys (e.g., Gd–Co), whose very existence depends on limited adatom mobility, are frequently columnar when deposited at sufficiently low temperature. Inasmuch as grain boundaries are axiomatically absent in amorphous films, it is more correct to speak of columnar *morphology* in this case. This columnar morphology is frequently made visible by transverse fracture of the film because of crack propagation along the weak, low-density intercolumnar regions. Magnetic, optical, electrical, mechanical, and surface properties of films are affected, sometimes strongly, by columnar structures. In particular, the magnetic anisotropy of seemingly isotropic amorphous Gd–Co films is apparently due to its columnar structure and interspersed voids. A collection of assorted electron micrographs of film and coating columnar structures is shown in Fig. 5-14. Particularly noteworthy are the structural similarities among varied materials deposited by different processes, suggesting common nucleation and growth mechanisms.

An interesting observation (Ref. 20) on the geometry of columnar grains has been formulated into the so-called tangent rule expressed by Eq. 5-43. Careful measurements on obliquely evaporated Al films reveal that the columns are oriented toward the vapor source, as shown in the microfractograph of Fig. 5-15. The angle β between the columns and substrate normal is universally observed to be somewhat less than the angle α, formed by the source direction and substrate normal. An experimental relation connecting values of α and β, obtained by varying the incident vapor angle over a broad range ($0 < \alpha < 90°$), was found to closely approximate

$$\tan \alpha = 2 \tan \beta . \tag{5-43}$$

The very general occurrence of the columnar morphology implies a simple nonspecific origin such as geometric shadowing, which affords an understanding of the main structural features.

Recently, a closer look has been taken of the detailed microstructure of columnar growth in sputtered amorphous Ge and Si, as well as TiB_2, WO_3, BN, and SiC thin films (Ref. 21). Interestingly, an evolutionary development of columnar grains ranging in size from ~ 20 to 4000 Å occurs. When prepared under low adatom mobility conditions ($T_S / T_M < 0.5$), three general structural units are recognized; nano-, micro-, and macrocolumns together with associated nano-, micro-, and macrovoid distributions. A schematic of

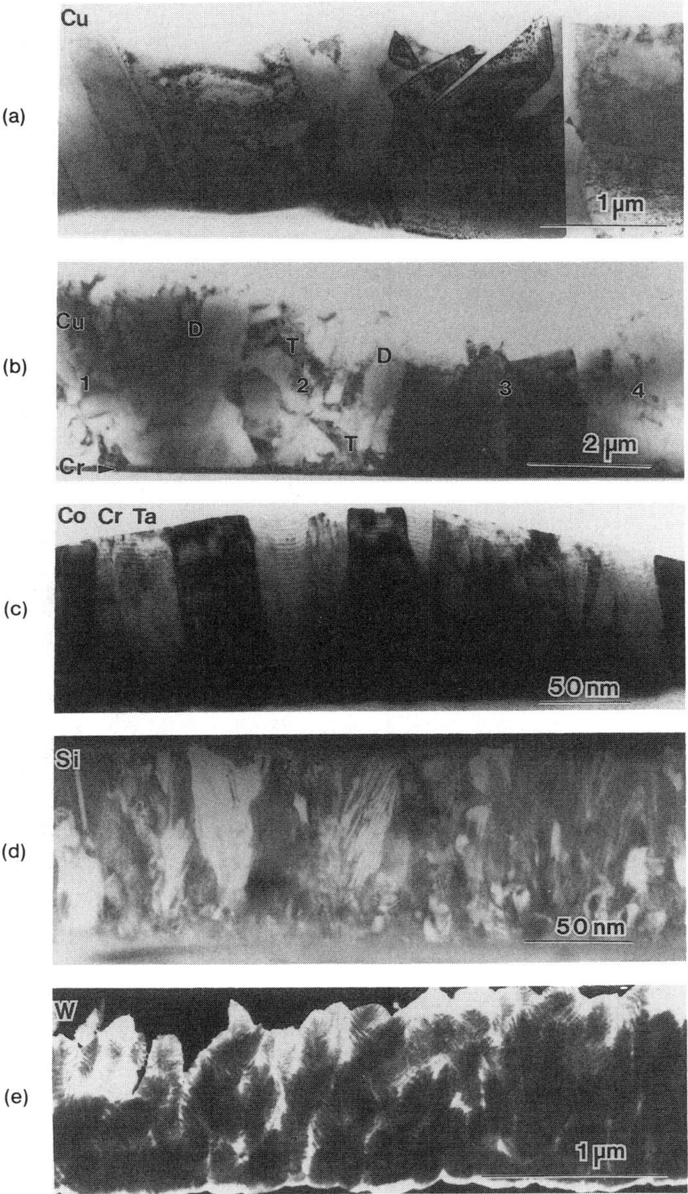

Figure 5-14. Representative set of cross-sectional transmission electron micrographs of thin films illustrating variants of columnar microstructures. (a) acid-plated Cu, (b) sputtered Cu, (c) sputtered Co–Cr–Ta alloy, (d) CVD silicon (also Fig. 4-12), (e) sputtered W, D = dislocation, T = twin. (Courtesy of D. A. Smith, IBM T. J. Watson Research Lab. Reprinted with permission from Trans–Tech Publication, from Ref. 19).

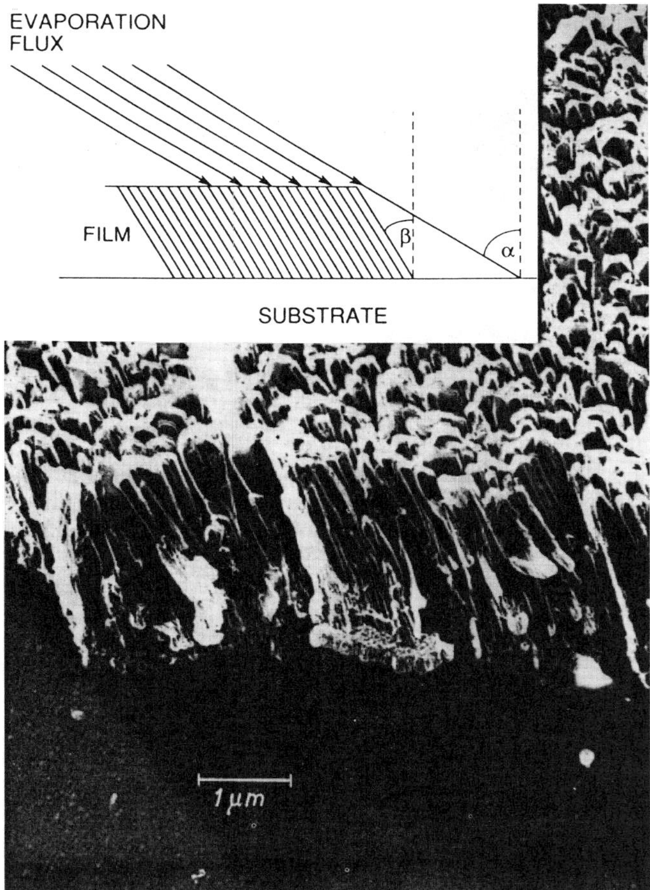

Figure 5-15. Electron micrograph of a replica of a ~ 2 μm-thick Al film. Inset shows deposition geometry. (From Ref. 20).

these interrelated, nested columns is shown in Fig. 5-16. It is very likely that the columnar grains of zones 1 and T in the Thornton scheme are composed of nano- and microcolumns.

Computer simulations (Ref. 22) have contributed greatly to our understanding of the origin of columnar grain formation and the role played by shadowing. By serially "evaporating" individual hard spheres (atoms) randomly onto a growing film at angle α, the structural simulations in Fig. 5-17 were obtained. The spheres were allowed to relax following impingement into the nearest triangular pocket formed by three previously deposited atoms, thus maximizing close atomic packing. The simulation shows that limited atomic mobility during low-temperature deposition reproduces features observed experimentally. As examples, film density decreases with increasing α, high-density columnlike regions appear at angles for which $\beta < \alpha$, and film densification is enhanced at elevated temperatures. Lastly, the column orientations agree well with the tangent rule. The evolution of voids occurs if those atoms exposed to the vapor beam shield or shadow unoccupied sites from direct impingement, and if post-impingement atom migration does not succeed in filling the voids. This self-shadowing effect is thus more pronounced the lower the atomic mobility and extent of lattice relaxation.

An important consequence of the columnar-void microstructure is the instability it engenders in optical coatings exposed to humid atmospheres. Under typical evaporation conditions ($\sim 10^{-6}$ torr, $T_S = 30$–300 °C and deposition rate of 300–3000 Å/s) dielectric films generally develop a zone 2 structure. Water from the ambient is then absorbed throughout the film by capillary action. The process is largely irreversible and alters optical properties such as

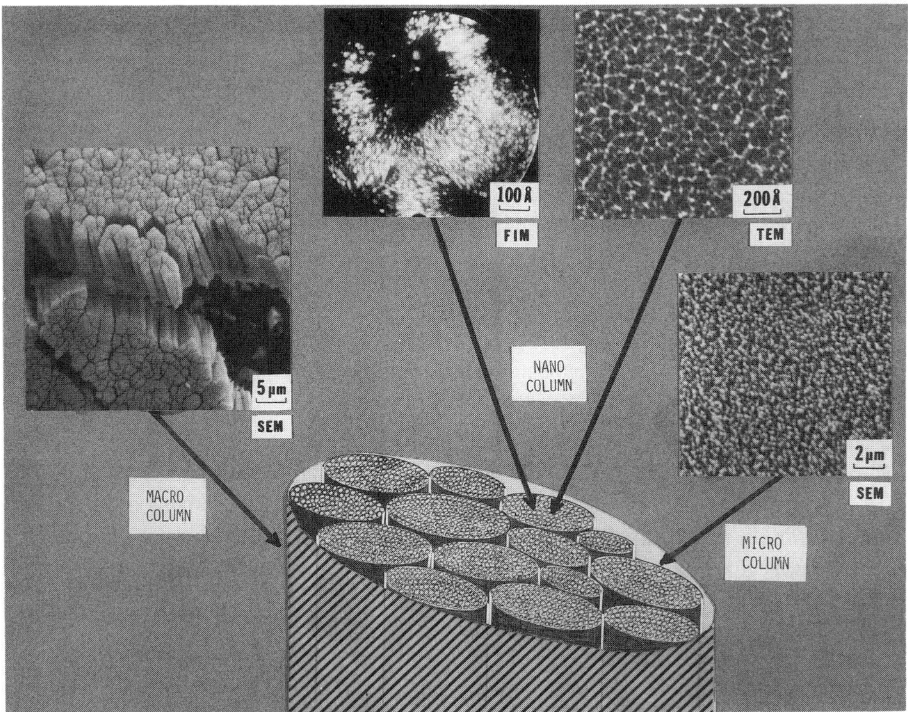

Figure 5-16. Schematic representation of macro, micro and nano columns for sputtered amorphous Ge films. (Courtesy of R. Messier, from Ref. 21).

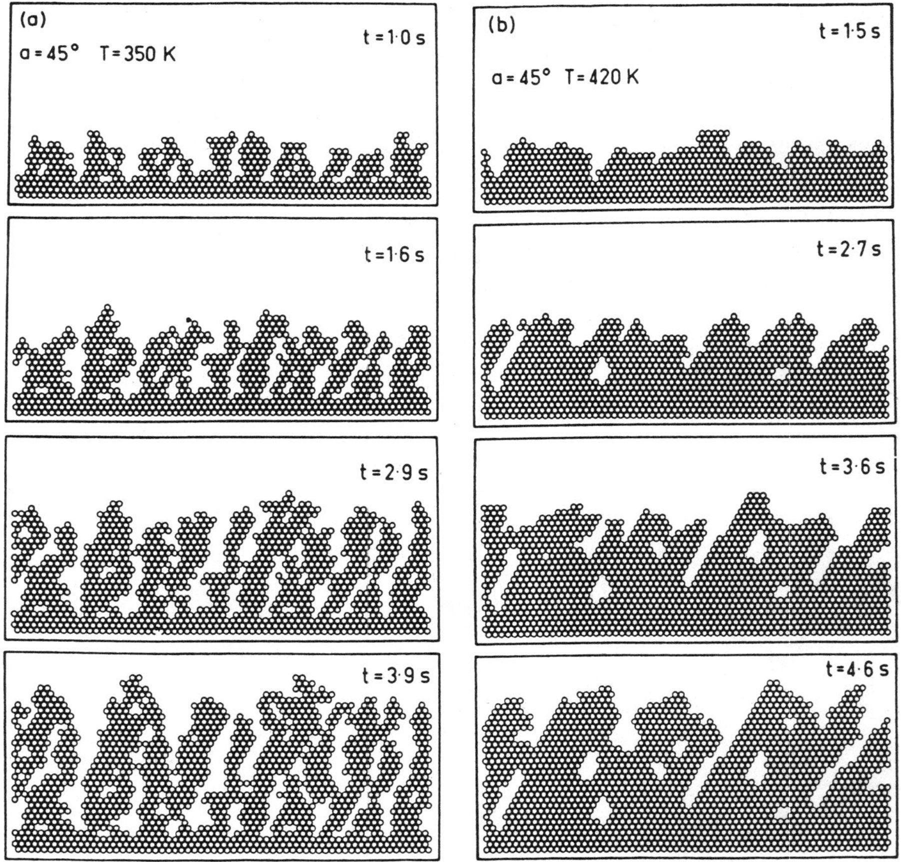

Figure 5-17. Computer-simulated microstructure of Ni film during deposition at different times for substrate temperatures of (a) 350 K and (b) 420 K. The angle of vapor deposition α is 45°. (From Ref. 22).

index of refraction and absorption coefficient. Moisture-induced degradation has plagued optical film development for many years. A promising remedy for this problem is ion bombardment, which serves to compact the film structure. This approach is discussed further in Chapters 3 and 11.

5.6.4. Film Density

A reduced film density relative to the bulk density is not an unexpected outcome of the zone structure of films and its associated porosity. Because of the causal structure-density and structure-property relationships, density is expected to strongly influence film properties. Indeed we have already alluded to the deleterious effect of lowered overall film densities on optical and mechanical properties. A similar degradation of film adhesion and chemical stability as well as electrical and magnetic properties can also be expected. Measurement of film density generally requires a simultaneous determination of film mass per unit area and thickness. Among the experimental findings related to film density are the following (Ref. 23):

1. The density of both metal and dielectric films increases with thickness and reaches a plateau value that asymptotically approaches that of the bulk density. The plateau occurs at different thicknesses, depending on material deposition method and conditions. In Al, for example, a density of 2.1 g/cm^3 at 250 Å rises to 2.58 g/cm^3 above 525 °C and then remains fairly constant thereafter. As a reference, bulk Al has a density of 2.70 g/cm^3. The gradient in film density is thought to be due to several causes, such as higher crystalline disorder, formation of oxides, greater trapping of vacancies and holes, pores produced by gas incorporation, and special growth modes that predominate in the early stages of film formation.

2. Metal films tend to be denser than dielectric films because of the larger void content in the latter. A quantitative measure of the effect of voids on density is the packing factor P, defined as

$$P = \frac{\text{volume of solid}}{\text{total volume of film (solid + voids)}}. \tag{5-44}$$

Typical values of P for metals are greater than 0.95, whereas for fluoride films (e.g., MgF_2, CaF_2) P values of approximately 0.7 are realized. However, by raising T_S for the latter, we can increase P to almost unity.

3. Thin-film condensation is apparently accompanied by the incorporation of large nonequilibrium concentrations of vacancies and micropores. Whereas bulk metals may perhaps contain a vacancy concentration of 10^{-3} at the melting point, freshly formed thin films can have excess concentrations of 10^{-2} at room temperature. In addition, microporosity on a scale much finer than imagined in zones 1 and T has been detected by TEM phase (defocus) contrast techniques (Ref. 24). Voids measuring 10 Å in size, present in densities of about 10^{17} cm^{-3} have been revealed in films prepared by evaporation as shown in Fig. 5-18. The small voids appear as white dots surrounded by black rings in the underfocused condition. Microporosity is evident both at grain boundaries and in the grain interior of metal films. In dielectrics a continuous network of microvoids appears to surround grain

Figure 5-18. Transmission electron micrograph showing microvoid distribution in evaporated Au films. (Courtesy of S. Nakahara, AT&T Bell Laboratories.)

boundaries. This crack network has also been observed in Si and Ge films, where closer examination has revealed that it is composed of interconnecting cylindrical voids. Limited surface diffusion, micro-self-shadowing effects, and stabilization by adsorbed impurities encourage the formation of microporosity. In addition to reducing film density, excess vacancies and microvoids may play a role in fostering interdiffusion in thin-film couples where the Kirkendall effect has been observed (see Chapter 8). The natural tendency to decrease the vacancy concentration through annihilation is manifested by such film changes as stress relaxation, surface faceting, adhesion failure, recrystallization and grain growth, formation of dislocation loops and stacking faults, and decrease in hardness.

5.7. Amorphous Thin Films

5.7.1. Systems, Structures, and Transformations

Amorphous or glassy materials have a structure that exhibits only short-range order or regions where a predictable placement of atoms occurs. However, within a very few atom spacings, this order breaks down, and no long-range correlation in the geometric positioning of atoms is preserved. Although bulk amorphous materials such as silica glasses, slags, and polymers are well known, amorphous metals were originally not thought to exist. An interesting aspect of thin-film deposition techniques is that they facilitate the formation of amorphous metal and semiconductor structures relative to bulk preparation methods.

As noted, production of amorphous films requires very high deposition rates and low substrate temperatures. The latter immobilizes or freezes adatoms on the substrate where they impinge and prevents them from diffusing and seeking out equilibrium lattice sites. By the mid-1950s Buckel (Ref. 25) produced amorphous films of pure metals such as Ga and Bi by thermal evaporation onto substrates maintained at liquid helium temperatures. Alloy metal films proved easier to deposit in amorphous form because each component effectively inhibits the atomic mobility of the other. This meant that higher substrate temperatures (~ 77 K) could be tolerated and that vapor quench rates did not have to be as high as those required to produce pure amorphous metal films. Although they are virtually impossible to measure, vapor quench rates in excess of 10^{10} °C/sec have been estimated. From laboratory curiosities, amorphous Si, Se, GdCo, and GeSe thin films have been exploited for such applications as solar cells, xerography, magnetic bubble memories, and high-resolution optical lithography, respectively.

Important fruits of the early thin-film work were realized in the later research and development activities surrounding the synthesis of *bulk* amorphous metals by quenching melts. Today continuously cast ribbon and strip of metallic glasses (Metglas) are commercially produced for such applications as soft magnetic transformer cores and brazing materials. Cooling rates of ~ 10^6 °C/sec are required to prevent appreciable rates of nucleation and growth of crystals. Heat transfer limitations restrict the thickness of these metal glasses to less than 0.1 mm. In addition to achieving the required quench rates, the alloy compositions are critical. Most of the presently known glass-forming binary alloys fall into one of four categories (Ref. 26):

1. Transition metals and 10–30 at% semimetals
2. Noble metals (Au, Pd, Cu) and semimetals
3. Early transition metals (Zr, Nb, Ta, Ti) and late transition metals (Fe, Ni, Co, Pd)
4. Alloys consisting of IIA metals (Mg, Ca, Be)

In common, many of the actual glass compositions correspond to where "deep" (low-temperature) eutectics are found on the phase diagram.

Amorphous thin films of some of these alloys as well as other metal alloys and virtually all elemental and compound semiconductors, semimetals, oxides, and chalcogenide (i.e., S-, Se-, Te-containing) glasses have been prepared by a variety of techniques. Amorphous Si films, for example, have been deposited by evaporation, sputtering, and chemical vapor deposition techniques. In addition, large doses of ion-implanted Ar or Si ions will amorphize surface layers of crystalline Si. Even during ion implantation of conventional dopants, local amorphous regions are created where the Si matrix is sufficiently damaged, much to the detriment of device behavior. Lastly, pulsed laser surface melting followed by rapid freezing has produced amorphous films in Si as well as other materials (see Chapter 13).

5.7.2. Au – Co and Ni – Zr Amorphous Films

It is instructive to consider amorphous Co–30Au films since they have been well characterized structurally and through resistivity measurements (Ref. 27). The films were prepared by evaporation from independently heated Co and Au sources onto substrates maintained at 80 K. Dark-field electron microscope images and corresponding diffraction patterns are shown side by side in Fig. 5-19. The as-deposited film is rather featureless with a smooth topography, and the broad halos in the diffraction pattern cannot be easily and uniquely assigned to the known lattice spacings of the crystalline alloy phases in this system. Both pieces of evidence point to the existence of an amorphous phase whose structural order does not extend beyond the next-nearest-neighbor distance. The question of whether so-called amorphous films are in reality microcrystalline is not always easy to resolve. In this case, however, the subsequent annealing behavior of these films was quite different from what is expected of fine-grained crystalline films. Heating to 470 K resulted in the face-centered cubic diffraction pattern of a single metastable phase, whereas at 650 K, lines corresponding to the equilibrium Co and Au phases appeared. Resistivity changes accompanying the heating of Co–38Au (an alloy similar to Co–30Au) revealed a two-step transformation as shown in Fig. 5-20. Beyond 420 K there is an irreversible change from the amorphous structure to a metastable FCC crystalline phase, which subsequently decomposes into equilibrium phases above 550 K. The final two-phase structure is clearly seen in Fig. 5-19. The high resistivity of the amorphous films is due to the enhanced electron scattering by the disordered solid solution. Crystallization to the FCC structure reduces the resistivity, and phase separation, further still.

Both the amorphous and metastable phases are stable over a limited temperature range in which the resistivity of each can be cycled reversibly. Once the two-phase structure appears, it, of course, can never revert to less thermody-

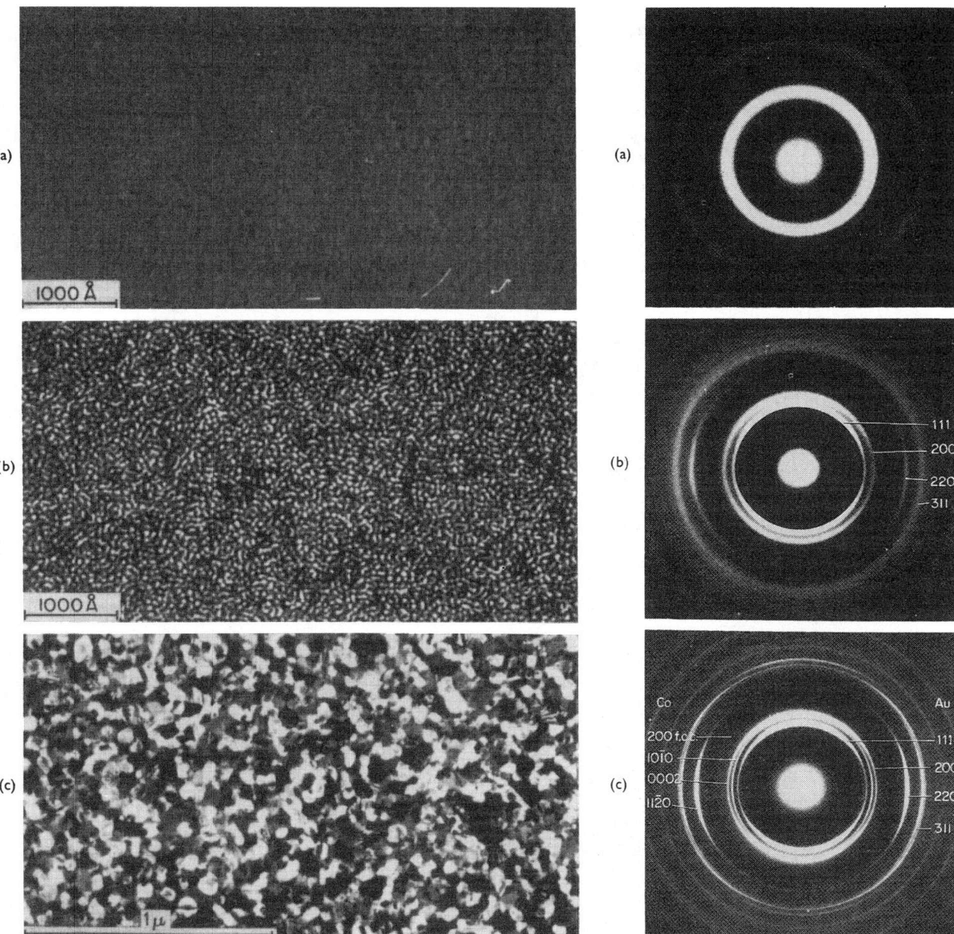

Figure 5-19. Electron micrographs and diffraction patterns of Co–30at%Au: (top) as deposited at 80 K, warmed to 300 K (amorphous); (middle) film warmed to 470 K (single-phase FCC structure); (bottom) film heated to 650 K (two-phase equilibrium). (From Ref. 27).

namically stable forms. This amorphous-crystalline transformation apparently proceeds in a manner first suggested by Ostwald in 1897. According to the so-called Ostwald rule, a system undergoing a reaction proceeds from a less stable to a final equilibrium state through a succession of intermediate metastable states of increasing stability. In this sense, the amorphous phase is akin to a quenched liquid phase. Quenched films exhibit other manifestations of thermodynamic instability. One is increased atomic solubility in amorphous

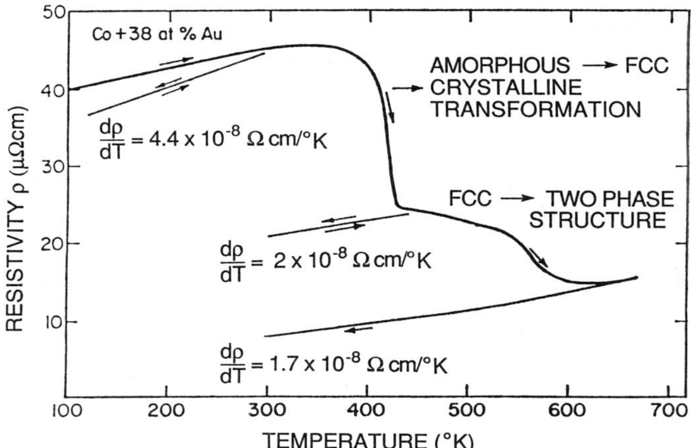

Figure 5-20. Resistivity of a Co–38at%Au film as a function of annealing temperature. Reversible values of $d\rho/dT$ in various structural states of the film are shown together with changes in ρ during phase transformation. (From Ref. 27).

or single-phase metastable matrices. For example, the equilibrium phase diagram for Ag–Cu is that of a simple eutectic with relatively pure terminal phases of Ag and Cu that dissolve less than 0.4 at% Cu and 0.1 at% Ag, respectively, at room temperature. These limits can be extended to 35 at% on both sides by vapor-quenching the alloy vapor. Similar solubility increases have been observed in the Cu–Mg, Au–Co, Cu–Fe, Co–Cu, and Au–Si alloy systems.

Confounding the notion that rapid quenching of liquids or vapors is required to produce amorphous alloy films is the startling finding that they can also be formed by solid-state reaction. Consider Fig. 5-21, which shows the result of annealing a bilayer couple consisting of pure polycrystalline Ni and Zr films at 300°C for 4 h. The phase diagram predicts negligible mutual solid solubility and extensive intermetallic compound formation; surprisingly, an amorphous NiZr alloy film is observed to form. Clearly, equilibrium compound phases have been bypassed in favor of amorphous phase nucleation and growth, as kinetic considerations dominate the transformation. The effect, also observed in Rh–Si, Si–Ti, Au–La, and Co–Zr systems, is not well understood. Apparently the initial bilayer film passes to the metastable amorphous state via a lower energy barrier than that required to nucleate stable crystalline compounds. However, the driving force for either transformation is similar. Unlike other amorphous films, extensive interdiffusion can be tolerated in NiZr without triggering crystallization.

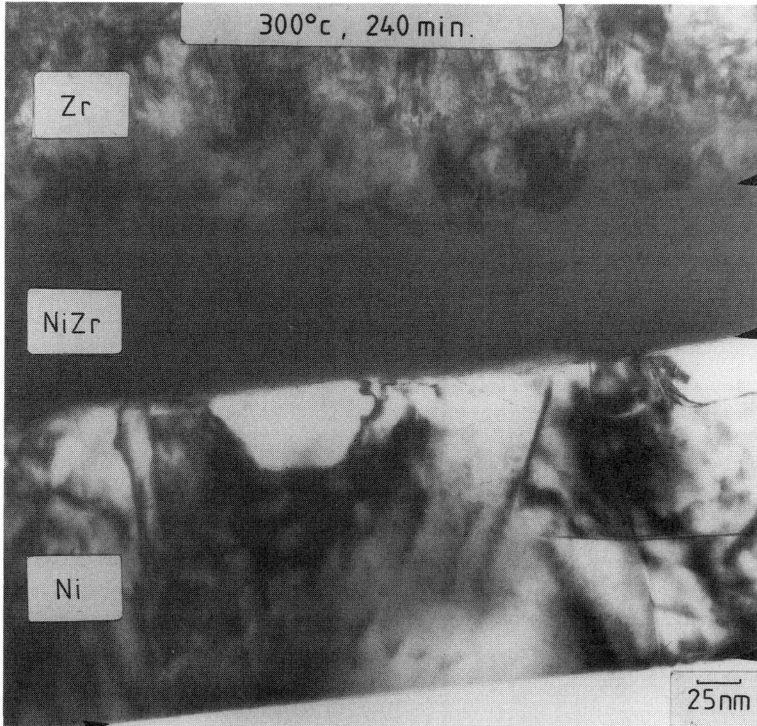

Figure 5-21. Cross-sectional electron micrograph of an amorphous Ni–Zr alloy film formed by annealing a crystalline bilayer film of Ni and Zr at 300 °C for 4 hours. (Courtesy of K. N. Tu, IBM Corp., T. J. Watson Research Lab., from Ref. 28).

5.7.3. A Model To Simulate Structural Effects in Thin Films

One of the outcomes of their research on quenched alloy films was an engaging mechanical model Mader and Nowick (Ref. 29) developed to better explain the experimental results. Many phenomena observed in pure and alloy thin-film structures are qualitatively simulated by this model. For this reason, it is valuable as a pedogogic tool and worth presenting here. The "atoms" composing the thin films were acrylic plastic balls of different sizes. They were rolled down a pinball-like runway tilted at 1.5° to the horizontal to simulate the random collision of evaporant atoms. A monolayer of these atoms was then "deposited" on either an "amorphous" or "crystalline" substrate. The former was a flat sheet of plastic, and the latter contained a perfect two-dimensional periodic array of interstices into which atoms could nest. Provision was made to alter the alloy composition by varying the ball feed. A magnetic vibrator simulated thermal annealing. To obtain diffraction patterns from the arrays, they prepared reduced negatives (with an array size of about 4 mm square). The balls appeared transparent on a dark background with a mean ball separation of ~ 0.13 mm. Fraunhofer optical diffraction patterns were generated by shining light from a He–Ne laser ($\lambda = 6328$ Å) on the negative mounted in contact with a 135-mm lens of a 35-mm camera. The resulting photographs are reproduced here.

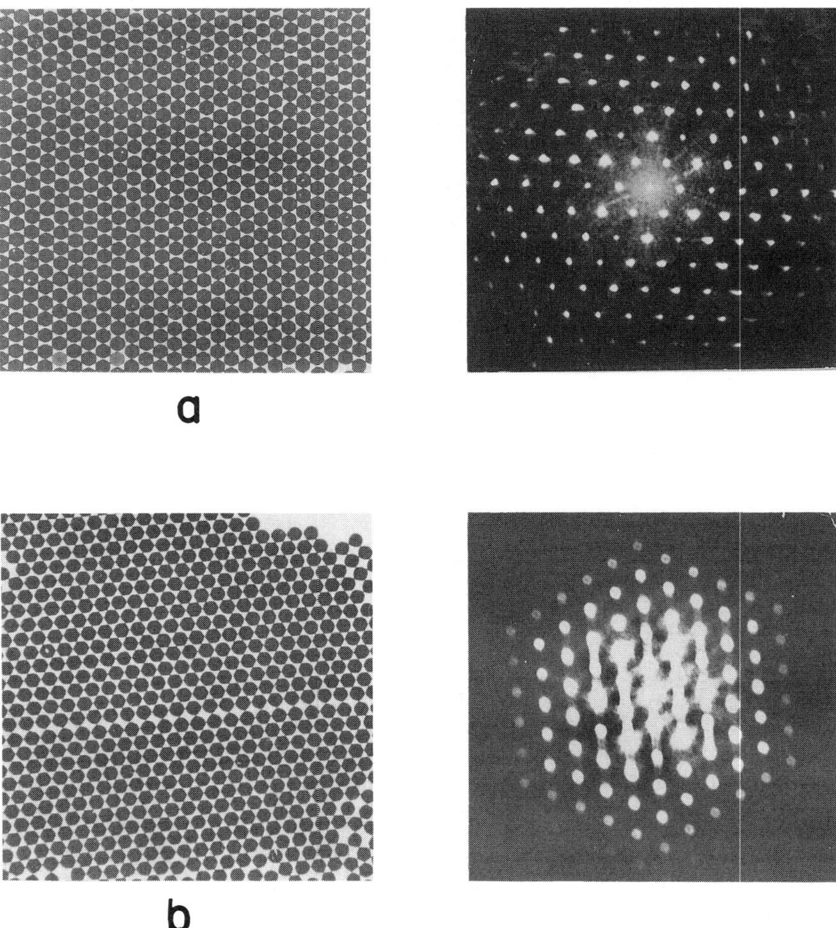

Figure 5-22. Atomic sphere film structures and corresponding Fraunhofer diffraction patterns for (a) perfect array, (b) stacking fault, (c) pure film; low deposition rate, (d) pure film; high deposition rate. (Reprinted with permission from the IBM Corp., from A. S. Nowick and S. R. Mader, *IBM J. Res. Dev.* **9**, 358, 1965).

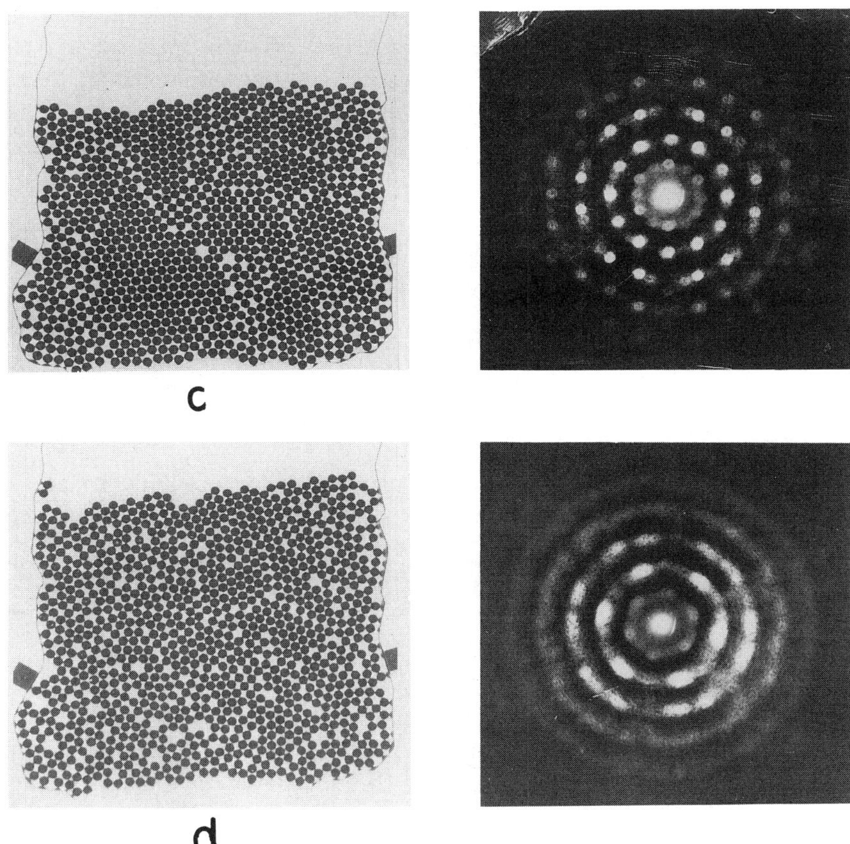

Figure 5-22. *Continued.*

The perfect array of spheres of one size is shown together with the corresponding diffraction pattern in Fig. 5-22a. A hexagonal pattern of sharp spots, very reminiscent of electron diffraction patterns of single-crystal films, is obtained, reflecting the symmetry of the close-packed array. After creation of a stacking fault in the structure, the diffraction pattern shows streaks (Fig. 5-22b). These run perpendicular to the direction of the fault in the structure.

The effect of deposition rate is shown in Figs. 5-22c and 5-22d. When the film is deposited "slowly," there are grains, vacancies, and stacking faults present in the array. Relative to Fig. 5-22a, the diffraction spots are broadened, a precurser to ring formation. In Fig. 5-22d, the film is deposited at a "high" rate and the grain structure is considerably finer and more disordered with numerous point defects, voids, and grain boundaries present. Now, semicontinuous diffraction rings appear, which are very much like the common X-ray Debye–Scherer rings characteristic of polycrystals. Interestingly, the intensity variation around the ring is indicative of preferred orientation. When the rapidly deposited films are annealed through vibration, the array densifies, vacancies are annihilated, faults are eliminated, and grains reorient, coalesce, and grow. The larger grains mean a return to the spotted diffraction pattern.

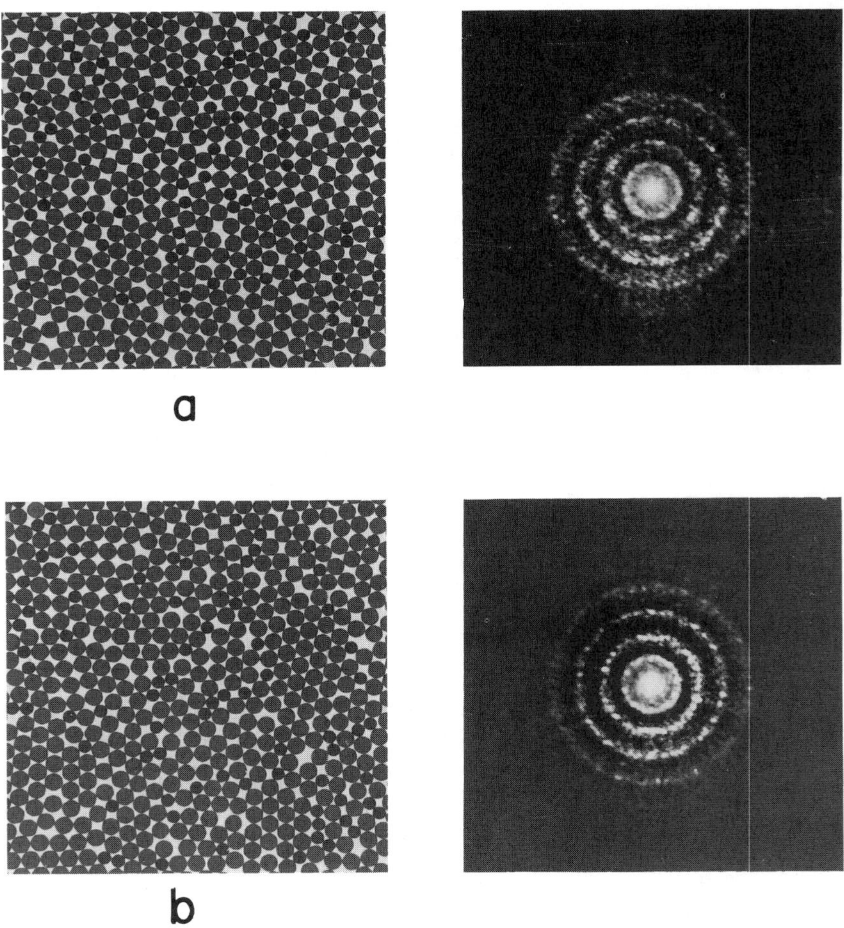

Figure 5-23. Atomic sphere film structure for concentrated alloy (50A–50B, 27% size difference: (a) as-deposited (amorphous); (b) vibration annealed. (Reprinted with permission from the IBM Corp., from A. S. Nowick and S. R. Mader, *IBM J. Res. Dev.* **9**, 358, 1965).

We now turn our attention to alloy films. For "concentrated" alloys containing equal numbers of large and small spheres with a size difference of 27%, the as-deposited structure is amorphous, as indicated in Fig. 5-23.. The diffraction pattern contains broad halos. Upon vibration annealing, the film densifies slightly, but the atomic logjam cannot be broken up. There is no appreciable change in its structure or diffraction pattern—it is still amorphous. For less concentrated alloys (~ 17%), however, the as-deposited structure is very fine grained but apparently crystalline.

All of the foregoing results were for films deposited on the smooth substrate. The "crystalline substrate" affords the opportunity to model epitaxy phenomena. Pure films deposit in almost perfect alignment with the substrate when deposited slowly. Imperfect regions are readily eliminated upon annealing and nearly perfect single crystals are obtained. Rapidly deposited films are less influenced by the underlying substrate and remain polycrystalline after annealing. Clearly epitaxial growth is favored by low deposition rates. The presence of alloying elements impeded epitaxy from occurring in accord with experience.

The foregoing represents a sampling of the simulations of the dependence of film structure on deposition variables. Readers interested in this as well as other mechanical models of planar arrays of atoms, such as the celebrated Bragg bubble raft model (Ref. 30), should consult the literature on the subject. Much insight can be gained from them.

Exercises

1. Under the same gas-phase supersaturation, cube-shaped nuclei are observed to form homogeneously in the gas and heterogeneously both on a flat substrate surface and at right-angle steps on this surface. For each of these three sites calculate the critical nucleus size and energy barrier for nucleation.

2. A cylindrical pill-like cluster of radius r nucleates on a dislocation that emerges from the substrate. The free-energy change per unit thickness is given by

$$\Delta G = \pi r^2 \Delta G_V + 2\pi\gamma r + A - B\ln r,$$

where $A - B\ln r$ represents the dislocation energy within the cluster.

a. Sketch ΔG vs. r (note at $r = 0$, $\Delta G = \infty$).

b. Determine the value of r^*.

c. Show that when $\Delta G_V B/\pi\gamma^2 > 1/2$, ΔG monotonically decreases with r, but when $\Delta G_V B/\pi\gamma^2 < 1/2$ there is a turnaround in the curve. (The latter case corresponds to a metastable state and associated energy barrier.)

3. Cap-shaped nuclei on substrates grow both by direct impingement of atoms from the vapor phase as well as by attachment of adatoms diffusing across the substrate surface.

a. In qualitative terms how will the ratio of the two mass fluxes depend on nucleus size, area density of nuclei, and deposition rate.
b. Write a quantitative expression for the flux ratio, making any reasonable assumptions you wish.

4. Two spherical nuclei with surface energy γ having radii r_1 and r_2 coalesce in the gas phase to form one spherical nucleus. If mass is conserved, calculate the energy reduction in the process. Suppose two spherical *caps* of different radii coalesce on a planar substrate to form one cap-shaped nucleus. Calculate the energy reduction.

5. Two spherical nuclei of radii r_1 and r_2 are separated by a distance l. If $r_1 \gg r_2$, derive an expression for the time it will take for the smaller nucleus to disappear by sequential atomic dissolution and diffusion to the larger nucleus by Ostwald ripening. Assume the diffusivity of atoms on the surface is D_S. Make simplifying assumptions as you see fit.

6. Assume that the two nuclei in Fig. 5-10 coalesce by a sintering mechanism.

a. By carefully measuring the neck width and plotting it as a function of time, determine the value of n in the general sintering kinetics formula.
b. From these data, estimate a value for the approximate diffusivity. Assume $\gamma = 1000$ ergs/cm$_2$, $T = 400$ °C, and $\Omega = 17 \times 10^{-24}$ cm^3/atom.

7. A film is deposited on a substrate by means of evaporation. In the expression for the rate of heterogeneous nucleation (Eq. 5-17), identify which terms are primarily affected by

a. raising the temperature of the evaporant source.
b. changing the substrate material.
c. doubling the source–substrate distance.

d. raising the substrate temperature.
e. improving the system vacuum.
In each case qualitatively describe the nature of the change.

8. From data shown in Fig. 5-5 calculate values for E_{des}, E_S, and E_{i*}. (For answers consult Ref. 3, page 8-23.)

9. Three different methods for estimating the temperature for epitaxial growth of films have been discussed in this chapter.

a. Comment on the similarities and differences in the respective approaches.
b. How well do they predict the experimental findings of Fig. 5-4?

10. Derive expressions for the epitaxial transition temperatures $T_{1\rightarrow 3}$ and $T_{2\rightarrow 3}$.

11. During examination of the grain structure of a film evaporated from a point source onto a large planar substrate, the following observations were made as a function of position:

1. There is a film thickness variation.
2. There is a grain size variation.
3. There is a variation in the angular tilt of columnar grains.

Explain the physical reasons for these observations.

12. The formation of three-dimensional crystallites from an amorphous matrix undergoing transformation by *nucleation* and *growth* processes follows the time (t) dependent kinetics given by

$$f(t) = 1 - \exp - \frac{\pi \dot{N} v^3 t^4}{3}.$$

$\dot{N}$ is the nucleation rate of crystallites (per unit volume), v is their growth velocity, and f is the fractional extent of transformation.

a. $\dot{N}$ is small near the critical transformation temperature and at low temperature, but larger in between. Why?
b. v is usually larger for higher temperatures. Why?
c. Schematically sketch $f(t)$ vs. t (or $\ln t$) at a series of temperatures. Note that an incubation time dependent on temperature is suggested.

13. a. Atoms on either side of a curved grain boundary (GB) reside on surfaces of different curvature, establishing a local chemical potential gradient that will drive GB migration. Use the Nernst–Einstein equation to show that the grain size will tend to grow with parabolic kinetics.

b. Part (a) is valid when the film grain size is smaller than the film thickness. Why? If the reverse is true, suggest why parabolic growth kinetics may not be observed.

References

1.* B. Lewis and J. C. Anderson, *Nucleation and Growth of Thin Films*, Academic Press, London (1978).
2.* R. W. Vook, *Int. Metals Rev.* **27**, 209 (1982).
3.* C. A. Neugebauer, in *Handbook of Thin-Film Technology*, eds. L. I. Maissel and R. Glang, McGraw Hill, New York (1970).
4.* K. Reichelt, *Vacuum* **38**, 1083 (1988).
5.* J. A. Venables, G. D. T. Spiller, and M. Hanbücken, *Rep. Prog. Phys.* **47**, 399 (1984).
6. D. Walton, T. N. Rhodin, and R. W. Rollins, *J. Chem. Phys.* **38**, 2698 (1963).
7. H. M. Yang and C. P. Flynn, *Phys. Rev. Lett.* **62**, 2476 (1989).
8. V. N. E. Robinson and J. L. Robins, *Thin Solid Films* **20**, 155 (1974).
9. R. M. German, *Powder Metallurgy Science*, Metal Powder Industries Federation, Princeton, NJ (1984).
10. D. W. Pashley and M. J. Stowell, *J. Vac. Sci. Tech.* **3**, 156 (1966).
11. D. Kashchiev, *Surface Science* **86**, 14 (1979).
12. K. L. Chopra, *Thin-Film Phenomena*, McGraw-Hill, New York (1969).
13. G. E. Rhead, *J. Vac. Sci. Tech.* **13**, 603 (1976).
14. R. W. Vook and B. Oral, *Gold Bull.* **20**, (1/2), 13 (1987).
15. E. Grunbaum, in *Epitaxial Growth B*, ed. J. W. Matthews, *Academic Press*, New York (1976).
16. B. A. Movchan and A. V. Demchishin, *Phys. Met. Metallogr.* **28**, 83 (1969).
17. J. A. Thornton, *Ann. Rev. Mater. Sci.* **7**, 239 (1977).
18. H. T. G. Hentzell, C. R. M. Grovenor, and D. A. Smith, *J. Vac. Sci. Tech.* **A2**, 218 (1984).
19. M. F. Chisholm and D. A. Smith, in *Advanced Techniques for Microstructural Characterization*, eds. R. Krishnan, T. R. Anantharaman, C. S. Pande, and O. P. Arora, Trans-Tech. Publ. Switzerland (1988).

*Recommended texts or reviews.

20. J. M. Nieuwenhuizen and H. B. Haanstra, *Philips Tech. Rev.* **27**, 87 (1966).
21. R. Messier, A. P. Giri, and R. Roy, *J. Vac. Sci. Tech.* **A2**, 500 (1984).
22. K. H. Müller, *J. Appl. Phys.* **58**, 2573 (1985).
23. H. Pulker, *Coatings on Glass*, Elsevier, Amsterdam (1984).
24. S. Nakahara, *Thin Sold Films* **64**, 149 (1979).
25. W. Buckel, *Z. Phys.* **138**, 136 (1954).
26. H. S. Chen, H. J. Leamy, and C. E. Miller, *Ann. Rev. Mater. Sci.* **10**, 363 (1980).
27. S. Mader, in *The Use of Thin Films in Physical Investigations*, ed. J. C. Anderson, Academic Press, New York (1966).
28. S. B. Newcomb and K. N. Tu, *Appl. Phys. Lett.* **48**, 1436 (1986).
29. A. S. Nowick and S. R. Mader, *IBM J. Res. Dev.* **9**, 358 (1965).
30. W. L. Bragg and J. F. Nye, *Proc. Roy. Soc.* **A190**, 474 (1947).

Characterization of Thin Films

6.1. Introduction

Scientific disciplines are identified and differentiated by the experimental equipment and measurement techniques they employ. The same is true of thin-film science and technology. For the first half of this century, interest in thin films centered around optical applications. The role played by films was largely a utilitarian one, necessitating measurement of film thickness and optical properties. However, with the explosive growth of thin-film utilization in microelectronics, there was an important need to understand the intrinsic nature of films. With the increasingly interdisciplinary nature of applications, new demands for film characterization and other property measurements arose. It was this necessity that drove the creativity and inventiveness that culminated in the development of an impressive array of commercial analytical instruments. These are now ubiquitous in the thin-film, coating, and broader scientific communities. In many instances, it was a question of borrowing and modifying existing techniques employed in the study of bulk materials (e.g., X-ray diffraction, microscopy, mechanical testing) to thin-film applications. In other cases well-known physical phenomena (e.g., electron spectroscopy, nuclear scattering, mass spectroscopy) were exploited. A partial list of the

Table 6-1. Analytical Techniques Employed in Thin-Film Science and Technology

Primary Beam	Energy Range	Secondary Signal	Acronym	Technique	Application
Electron	20–200 eV	Electron	LEED	Low-energy electron diffraction	Surface structure
	300–30,000 eV	Electron	SEM	Scanning electron microscopy	Surface morphology
	1 keV–30 keV	X-ray	EMP (EDX)	Electron microprobe	Surface region composition
	500 eV–10 keV	Electron	AES	Auger electron spectroscopy	Surface layer composition
	100–400 keV	Electron	TEM	Transmission electron microscopy	High-resolution structure
	100–400 keV	Electron, X-ray	STEM	Scanning TEM	Imaging, X-ray analysis
	100–400 keV	Electron	EELS	Electron energy loss spectroscopy	Local small area composition
Ion	0.5–2.0 keV	Ion	ISS	Ion-scattering spectroscopy	Surface composition
	1–15 keV	Ion	SIMS	Secondary ion mass spectroscopy	Trace composition vs. depth
	1–15 eV	Atoms	SNMS	Secondary neutral mass spectrometery	Trace composition vs. depth
	1 keV and up	X-ray	PIXE	Particle-induced X-ray emission	Trace composition
	5–20 keV	Electron	SIM	Scanning ion microscopy	Surface characterization
	> 1 MeV	Ion	RBS	Rutherford backscattering	Composition vs. depth
Photon	> 1 keV	X-ray	XRF	X-ray fluorescence	Composition (μm depth)
	> 1 keV	X-ray	XRD	X-ray diffraction	Crystal structure
	> 1 keV	Electron	ESCA, XPS	X-ray photoelectron spectroscopy	Surface composition
	Laser	Ions	—	Laser microprobe	Composition of irradiated area
	Laser	Light	LEM	Laser emission microprobe	Trace element analysis

From Ref. 1.

modern techniques employed in the characterization of electronic thin-film materials and devices is given in Table 6-1. Among their characteristics are the unprecedented structural resolution and chemical analysis capabilities over small lateral and depth dimensions. Some techniques only sense and provide information on the first few atom layers of the surface. Others probe more deeply, but in no case are depths much beyond a few microns accessible for analysis. Virtually all of these techniques require a high or ultrahigh vacuum ambient. Some are nondestructive, others are not. In common, they all utilize incident electron, ion, or photon beams. These interact with the surface and excite it in such a way that some combination of secondary beams of electrons, ions, or photons are emitted, carrying off valuable structural and chemical information in the process. A rich collection of acronyms has emerged to differentiate the various techniques. These abbreviations are now widely employed in the thin-film and surface science literature.

General testing and analysis of thin films is carried out with equipment and instruments which are wonderfully diverse in character. For example, consider the following extremes in their attributes:

1. *Size*—This varies from a portable desktop interferometer to the 50-ft long accelerator and beam line of a Rutherford backscattering (RBS) facility.
2. *Cost*—This ranges from the modest cost of test instruments required to measure electrical resistance of films to the approximate $1 million price tag of a commercial SIMS spectrometer.
3. *Operating Environment*—This varies from the ambient in the measurement of film thickness to the 10^{-10}-torr vacuum required for the measurement of film surface composition.
4. *Sophistication*—At one extreme is the manual scotch-tape film peel test for adhesion, and at the other is an assortment of electron microscopes and surface analytical equipment where operation and data gathering, analysis, and display are essentially computer-controlled.

What is remarkable is that films can be characterized structurally, chemically, and with respect to various properties with almost the same ease and precision that we associate with bulk measurement. This despite the fact that there are many orders of magnitude fewer atoms available in films. To appreciate this, consider AES analysis of a Si wafer surface layer containing 1 at% of an impurity. Only the top 10–15 Å is sampled, and since state-of-the-art systems have a lateral resolution of 500 Å, the total measurement volume corresponds to $(\pi/4)(500)^2(15) = 3 \times 10^6$ Å^3. In Si this corresponds to about 150,000 matrix atoms, and therefore only 1500 impurity atoms are detected in the analysis! Such measurements pose challenges in handling and experimental techniques, but the problems are usually not insurmountable.

This chapter will only address the experimental techniques and applications associated with determination of

1. Film thickness
2. Film morphology and structure
3. Film composition

These represent the common core of information required of all films and coatings irrespective of ultimate application. Within each of these three categories, only the most important techniques will be discussed. Beyond these broad characteristics there are a host of individual properties (e.g., hardness, adhesion, stress, electrical conductivity, reflectivity, etc.), that are specific to the particular application. The associated measurement techniques will therefore be addressed in the appropriate context throughout the book.

6.2. Film Thickness

6.2.1. Introduction

The thickness of a film is among the first quoted attributes of its nature. The reason is that thin-film properties and behavior depend on thickness. Historically, the use of films in optical applications spurred the development of techniques capable of measuring film thicknesses with high accuracy. In contrast, other important film attributes, such as structure and chemical composition, were only characterized in the most rudimentary way until relatively recently. In some applications, the actual film thickness, within broad limits, is not particularly crucial to function. Decorative, metallurgical, and protective films and coatings are examples where this is so. On the other hand, microelectronic applications generally require the maintenance of precise and reproducible film thicknesses as well as lateral dimensions. Even more stringent thickness requirements must be adhered to in optical applications, particularly in multilayer coatings.

The varied types of films and their uses have generated a multitude of ways to measure film thickness. A list of methods mentioned in this chapter is given in Table 6-2 together with typical measurement ranges and accuracies. Included are destructive and nondestructive methods. The overwhelming majority are applicable to films that have been prepared and removed from the deposition chamber. Only a few are suitable for real-time monitoring of film thickness during growth. We start with optical techniques, a subject that is covered extensively in virtually every book and reference on thin films (Refs. 2–4).

Table 6-2. A Summary of Selected Film Thickness Measurement Techniques

Method	Range	Accuracy or Precision	Comments
Multiple-beam FET	30–20,000 Å	10–30 Å	A step and reflective coating required
Multiple-Beam FECO	10–20,000 Å	2 Å	A step, reflective coating, and spectrometer required; accurate but time-consuming
VAMFO	800 Å–10 μm	0.02–0.05 %	For transparent films on reflective substrates; Nondestructive
CARIS	400 Å–20 μm	10 Å–0.1%	For transparent films; Nondestructive
Step gauge	500–15,000 Å	~ 200 Å	Values for SiO_2 on Si
Ellipsometry	A few Å to a few μm	1 Å	Transparent films; complicated mathematical analysis
Stylus	20 Å to no limit	A few Å to < 3%	Step required; simple and rapid
Weight measurement	< 1 Å to no limit		Accuracy depends on knowledge of film density
Crystal oscillator	< 1 Å to a few μm	< 1 Å to a few %	Nonlinear behavior at larger film thicknesses

6.2.2. Optical Methods for Measuring Film Thickness

Optical techniques for film thickness determinations are widely used for a number of reasons. They are applicable to both opaque and transparent films, yielding thickness values of generally high accuracy. In addition, measurements are quickly performed, frequently nondestructive, and utilize relatively inexpensive equipment. The single basic principle which most optical techniques rely on is the interference of two or more beams of light whose optical path difference is related to film thickness. The details of instrumentation differ, depending on whether opaque or transparent films are involved.

6.2.2.1. Interferometry of Opaque Films

Fringes of Equal Thickness (FET). For opaque films a sharp step down to a substrate plane must be first generated either during deposition through a mask or by subsequent etching. A neighboring pair of light rays reflected from the film–substrate will travel different lengths and interfere by an amount dependent on the step height. To capitalize on the effect, one uses, multiple-

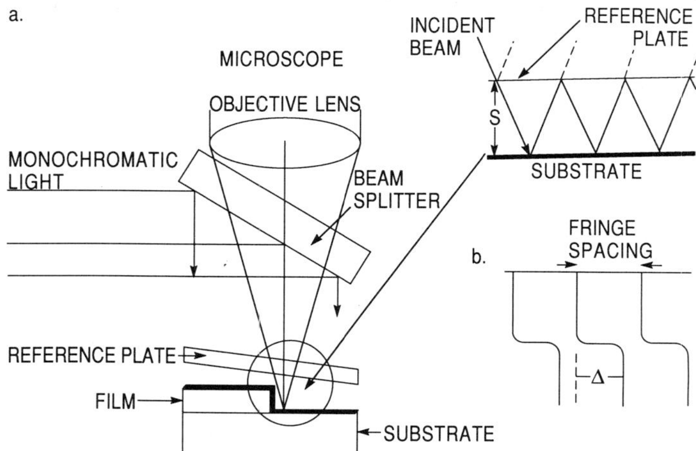

Figure 6-1. (a) Schematic view of experimental arrangement required to produce multiple-beam Fizeau fringes. (b) Fringe displacement at step.

beam interferometry, a technique developed by Tolansky (Ref. 5). This requires that the optical reflectance of both the film and substrate be very high as well as uniform. This is accomplished by evaporating a metal such as Al or, better yet, Ag over both film and substrate. Interference fringes are generated by placing a highly reflective, but semitransparent, optically flat reference plate very close to the film–step–substrate region as shown in Fig. 6-1a. The two highly reflective surfaces are tilted slightly off parallel, enabling the light beam to be reflected in a zig-zag fashion between them many times. A series of increasingly attentuated beams are now available for interference. This sharpens the resultant, so-called Fizeau fringes of equal thickness which can be viewed with a microscope.

The condition for constructive interference is that the optical path difference between successive beams be an integral number of wavelengths or

$$2S + \frac{2\,\delta\lambda}{2\pi} = n\lambda. \tag{6-1}$$

Here S is the distance between film and flat, λ is the wavelength of the monochromatic radiation employed, and n is an integer. The phase change accompanying reflection δ is assumed to be the same at both surfaces and is taken to be π because of the high film reflectances. Therefore,

$$S = \left(\frac{n-1}{2}\right)\lambda, \tag{6-2}$$

and the distance between maxima of successive fringes corresponds to $S = \lambda/2$. The existence of the step now displaces the fringe pattern abruptly by an amount Δ proportional to the film thickness d. As indicated in Fig. 6-1b, the film thickness is given by

$$d = \frac{\Delta}{\text{fringe spacing}} \frac{\lambda}{2}. \tag{6-3}$$

For highly reflective surfaces, the fringe width is about 1/40 of the fringe spacing. Displacements of about 1/5 of a fringe width can be detected. For the Hg green line ($\lambda = 5640$ Å) the resolution is therefore $(1/40)(1/5)(5640/2) \approx$ 14 Å. The resolution and ease of measurement are, respectively, influenced by the fraction of incident light reflected (R) and the fraction absorbed (A) by the film overlying the step. Raising R from 0.9 to 0.95 reduces the fringe width by half, whereas high A values reduce the fringe intensity.

Fringes of Equal Chromatic Order (FECO). Film preparation for measurement is identical to that for FET. Now, however, *white*, rather than monochromatic light, is employed, and the reflected light is spectrally analyzed by a spectrometer. Equation 6-1 still applies, but λ is no longer single-valued; i.e.,

$$2S = n\lambda_1 = (n+1)\lambda_2 = \cdots = (n+i)\lambda_{i+1}. \tag{6-4}$$

Adjacent lines correspond to different λ's and different orders, where $n + i$ is the chromatic order of a given line. When a step of height d is present, then

$$2(S+d) - 2S = 2d = n\,\Delta\lambda, \tag{6-5}$$

where $\Delta\lambda$ is the measurable wavelength shift of a fringe due to the resulting interference. To obtain the film thickness, we must know the order n. By Eq. 6-4, $n = \lambda_2/(\lambda_1 - \lambda_2)$, where λ_2 corresponds to the shorter wavelength. For fringes corresponding to λ_1 and λ_2,

$$d = \frac{\lambda_2}{2} \frac{\Delta\lambda}{\lambda_1 - \lambda_2}. \tag{6-6}$$

By analogy to Eq. 6-3 and Fig. 6-1b, it may be useful to think of $\lambda_1 - \lambda_2$ as the fringe spacing, and $\Delta\lambda$ as the fringe displacement, but with distances measured in an optical spectrometer rather than in a microscope. In general, the FECO technique is capable of higher accuracy than FET, especially for films that are very thin. The maximum resolution is about ± 5 Å but to attain this, precise positioning of the reference plate to align fringes is essential.

6.2.2.2. Interferometry of Transparent Films. A perfectly suitable method for measuring the thickness of transparent films is to first generate a step, metalize the film–substrate, and then proceed with either the FET or FECO techniques previously discussed. However, transparent films are ideally suited for interferometry because interference of light occurs naturally between beams reflected from the two film surfaces. This means that a step is no longer required. Since different interfaces (the air–film and film–substrate) are involved for the beams that interfere, precautions must be taken to account for phase changes on reflection. This subject is discussed at length in Chapter 11. The relevant Fig. 6-2 summarizes what happens when monochromatic light of wavelength λ is normally incident on a transparent film–substrate combination. If n_1 and n_2 are the respective film and substrate indices of refraction, the intensity of the reflected light undergoes oscillations as a function of the optical film thickness, or $n_1 d$. When $n_1 > n_2$, then maxima occur at film thicknesses equal to

$$d = \frac{\lambda}{4n_1}, \frac{3\lambda}{4n_1}, \frac{5\lambda}{4n_1}, \ldots . \tag{6-7}$$

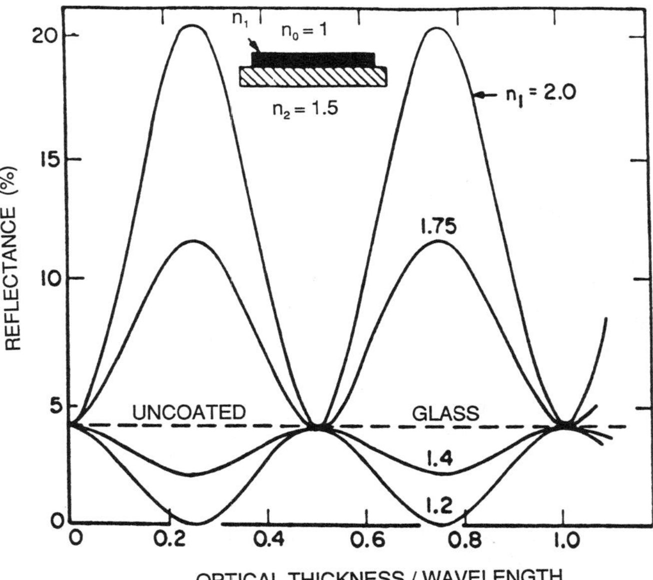

Figure 6-2. Calculated variation of reflectance (on air side) with normalized thickness ($n_1 d/\lambda$) for films of various refractive indices on a glass substrate of index 1.5. In Chapter 11 there is a fuller discussion of this figure. (From K. L. Chopra, *Thin Film Phenomena*, 1969, reproduced with permission from McGraw-Hill, Inc.).

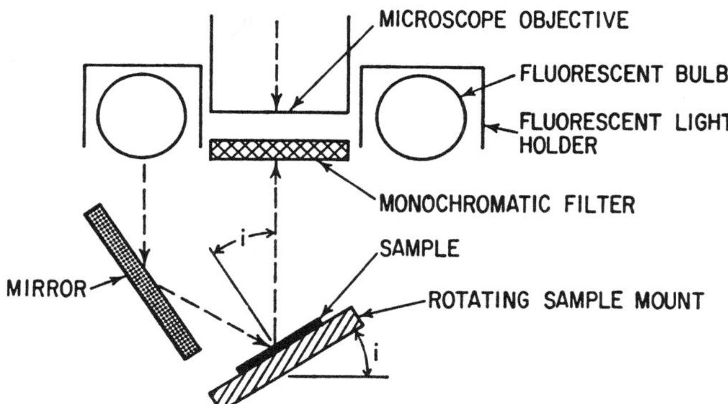

Figure 6-3. Schematic of experimental arrangement in the VAMFO technique. (From W. A. Pliskin & S. J. Zanin in *Handbook of Thin Film Technology*, edited by L. I. Maissel and R. Glang, © 1970, with permission from McGraw-Hill Inc.).

For values of d halfway between these, the reflected intensity is minimum. When $n_1 < n_2$, a reversal in intensity occurs at the same optical film thickness. In Fig. 6-2 these results are shown for the case of a glass substrate ($n_2 = 1.5$) and for films in which n_1 is either greater or less than 1.5. In order to exploit these concepts for the measurement of film thickness, we must devise experimental arrangements so that the intensity oscillations can be revealed.

VAMFO (Ref. 2). In the VAMFO (variable angle monochromatic fringe observation) method provision is made to vary the angle of incidence(i) of light on the film as shown in Fig. 6-3. In this arrangement a monochromatic light filter is employed to select a single wavelength for detection; monochromatic light can also be used. As the stage and sample are rotated, maxima (bright) and minima (dark) fringes are observed on the film surface. For a transparent film on an absorbing reflecting substrate, the film thickness is simply given by

$$d = (N\lambda \cos \theta)/2n_1. \tag{6-8}$$

Here, θ is the angle of *refraction* in the film, and N is the fringe order, which is measured by counting successive minima starting from perpendicular incidence. More accurate d values are obtained when the more easily detected intensity minima are measured. Then N assumes half-integer values 1/2, 3/2, 5/2, etc. The technique has the advantage of not requiring an optically flat substrate or a collimated light source. A disadvantage is that the refractive index of the film must be known at the wavelength of measurement. Otherwise, values must be assumed, and a series of successive approximations made until predicted and measured intensity mimina angles coincide. Corrections due to phase changes on reflection at the substrate must also be made. For further details, the reader is referred to the literature (Ref. 2) where applications to bilayer transparent films are also discussed.

CARIS (Ref. 2). In the technique known as CARIS (constant-angle reflection interference spectroscopy), the wavelength of the incident light rather than the angle of observation is systematically varied. The radiation is reflected from the film into a spectrometer with fringes being formed as a function of wavelength. For homogeneous films the thickness is determined by

$$d = \frac{\Delta N_f \lambda_1 \lambda_2}{2n_1(\lambda_1 - \lambda_2)\cos\theta}, \tag{6-9}$$

where ΔN_f is the number of fringes between wavelengths of λ_1 and λ_2. In applying Eq. 6-9, it is important to realize that n_1 varies with λ. The dispersion is greatest in the ultraviolet for most materials leading to large errors in this region. As with VAMFO, n_1 values, if unknown, can be initially assumed and then determined through successive approximations, taking into account phase changes accompanying reflection at interfaces. Electronic detection methods extend the capability of CARIS to measurement of thickness in semiconductor films that are transparent in the infrared. Bilayer film thicknesses can also be determined through complex analysis (Ref. 2).

Step Gauges. If there is a particular need to frequently measure the thickness of one kind of film, it may be advantageous to construct a step gauge. Films of different but independently known thickness are deposited on the substrate of interest and are arrayed sequentially. Interference colors observed when the specimen film is examined in reflected light are matched to the color of the step gauge standard. For example, a step gauge for SiO_2 films on a Si wafer has proven to be useful in estimating film thickness to approximately 200 Å. A simple way to prepare such a gauge is to etch a thick SiO_2 film in the shape of a wedge by slowly lifting the wafer from the etchant (dilute HF) in which it is immersed, at a constant velocity.

Ellipsometry (Refs. 2, 6). Also known as polarimetry and polarization spectroscopy, the technique of ellipsometry is a century old and has been used to obtain the thickness and optical constants of films. The method consists of measuring and interpreting the change of polarization state that occurs when a polarized light beam is reflected at non-normal incidence from a film surface. Shown in Figure 6-4 is the experimental arrangement for ellipsometer mea-

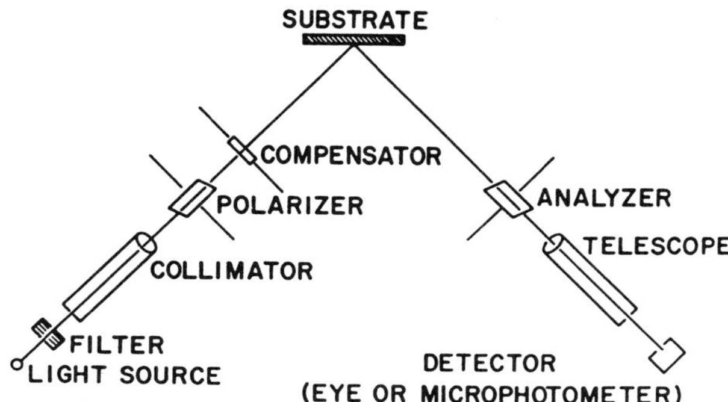

Figure 6-4. Experimental arrangement in ellipsometry. (From W. A. Pliskin & S. J. Zanin in *Handbook of Thin Film Technology*, edited by L. I. Maissel and R. Glang, © 1970, with permission from McGraw-Hill Inc.).

surements. The light source is first made monochromatic, collimated, and then linearly polarized. Upon passing through the compensator (usually a quarter-wave plate), the light is circularly polarized and then impinges on the specimen surface. After reflection, the light is transmitted through a second polarizer that serves as the analyzer. Finally, the light intensity is judged by eye or measured quantitatively by a photomultiplier detector. The polarizer and analyzer are rotated until light extinction occurs. The extinction readings enable the phase difference (Δ_e) and amplitude ratio ($\tan \psi$) of the two components of reflected light to be determined. From these, either the film thickness or the index of refraction can be obtained.

Space limitations do not allow for discussion of these relationships, but an appreciation of what is involved can be gained by referring to Chapter 11. There, only light normally incident on the film–substrate is considered, and, therefore, the distinction between parallel and perpendicular polarizations vanishes. But for light incident at an angle the two components of reflectivity, $r^{\parallel}$ and $r^{\perp}$, are distinct and the ratio of their amplitudes is given by

$$r^{\parallel}/r^{\perp} = \tan \psi e^{i\Delta_e}. \tag{6-10}$$

This fundamental equation of ellipsometry relates the film and substrate indices of refraction, film thickness, and phase changes during reflection at the film interfaces. Computer programs and graphical solutions exist to enable unknown n and d values to be extracted (Ref. 2). Provision can be made to accommodate partially absorbing films and substrates.

There has long been interest in ellipsometry as a tool in the semiconductor industry to monitor the in situ deposition and growth of thin films on well-polished, reflecting substrates, such as Si and GaAs. Because of its

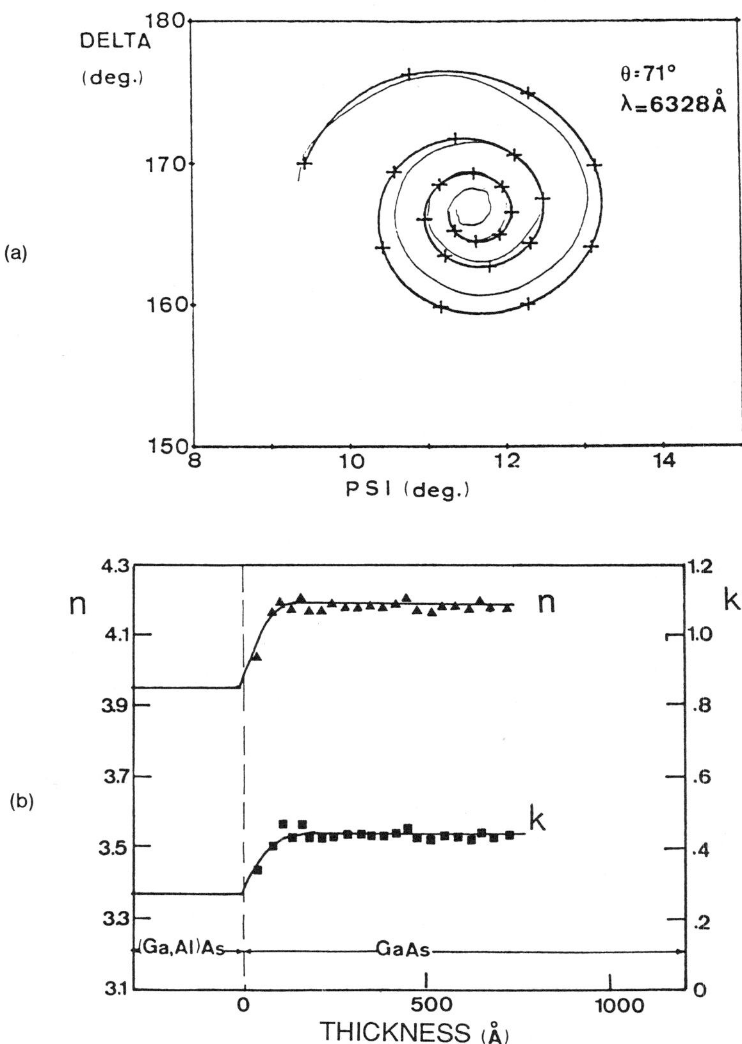

Figure 6-5. Ellipsometry measurement of (GaAl)As transition in the same organometallic CVD system. (a) Experimental trajectory of Δ_e and ψ, and comparison with theoretical growth model (solid line), with crosses corresponding to 100-Å increments in thickness. (b) Refractive index (n) and absorption coefficient (k) as a function of GaAs film thickness. (Reproduced with permission from Ref. 7, © 1987, Annual Reviews Inc.).

sensitivity in the submonolayer range and ability to function at high temperatures or pressures, through gases and liquids, etc., ellipsometry has been employed in plasma and CVD reactors, MBE equipment, and electrochemical cells. Advances in real-time data acquisition and reduction have enabled the measurement and control of thin-film growth rates.

An interesting example of ellipsometry techniques in monitoring real-time epitaxial growth in the GaAs–GaAlAs system is shown in Fig. 6-5. The experiment was carried out in a CVD reactor at 600 °C containing flows of AsH_3, $(CH_3)_3Ga$, and $(CH_3)_3Al$ gases past a GaAlAs film, illuminated by a He–Ne laser (6328 Å) at a 71° angle of incidence. After the Al species flow was stopped, growth of the GaAs film commenced. The resulting spiral trajectory of Δ_e and ψ with film thickness (or time) is shown together with changes in the optical constants of GaAs. Within 150 Å the GaAlAs → GaAs growth transition is apparently complete. The ability of ellipsometry to precisely monitor optical property changes in very thin films makes it attractive in multilayer film growth and etching studies.

6.2.3. Mechanical Techniques for Measuring Film Thickness

6.2.3.1. Stylus-Method Profilometry. The stylus method consists of measuring the mechanical movement of a stylus as it is made to trace the topography of a film–substrate step. A diamond needle stylus with a tip radius of ~ 10 μm serves as the electromagnetic pickup. The stylus force is adjustable from 1 to 30 mg, and vertical magnifications of a few thousand up to a million times are possible. Film thickness is directly read out as the height of the resulting step-contour trace. Several factors that limit the accuracy of stylus measurements are

1. *Stylus penetration and scratching of films.* This is sometimes a problem in very soft films (e.g., In, Sn).
2. *Substrate roughness.* This introduces excessive noise into the measurement, which creates uncertainty in the position of the step.
3. *Vibration of the equipment.* Proper shock mounting and rigid supports are essential to minimize background vibrations.

In modern instruments the leveling and measurement functions are computer-controlled. The vertical stylus movement is digitized, and the data can be processed to magnify areas of interest and yield best profile fits. Calibration profiles are available for standardization of measurements. The measurement range spans distances from 200 Å to 65 μm, and the vertical resolution is ~ 10 Å.

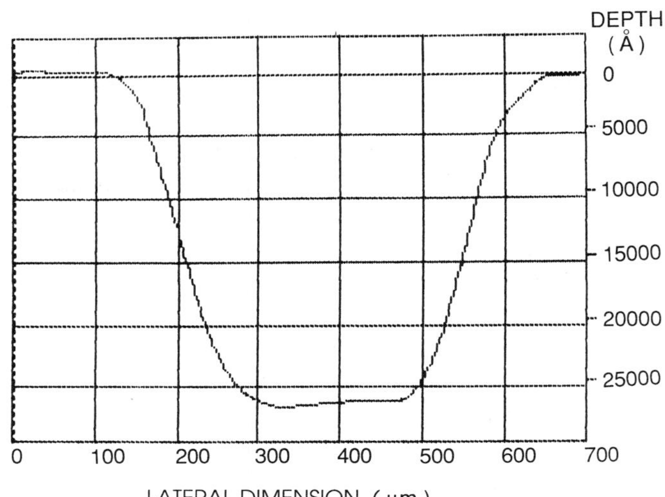

Figure 6-6. Profilometer trace of crater in InP produced during raster scanning of 12.5-keV Cs^+ ion beam across surface. (Courtesy of H. Luftman, AT&T Bell Laboratories.)

One of the important applications of stylus measurements is to determine the flatness and depth of a sputtered crater during depth-profiling analysis by AES or SIMS. In this technique a circular region of the film surface is sputtered away and an electron (AES) or ion (SIMS) probe beam ideally analyzes the flat bottom of the crater formed. Such a crater profile generated during SIMS analysis is shown in Fig. 6-6, where the total depth sputtered exceeds 2.5 μm. Since the sputtering times are known, this information can be converted to an equivalent depth scale for use in determination of precise concentration profiles. If the crater walls are slanted rather than vertical, the analyzing beam may not sample a well-defined flat surface but some portion of the sidewall. This leads to errors in the concentration depth profile that should be corrected for.

6.2.3.2. Weight Measurement. Measurement of the weight of the film deposit appears, at first glance, to be an easy direct way to determine film thickness d. Knowing the film mass m, the deposit area A, and film density ρ_f, we have

$$d = m/A\rho_f. \tag{6-11}$$

This simple method has been often used in ill-equipped laboratories where precision mass balances are more common than interferometers or stylus instruments. Values of d so obtained are imprecise because the film density is not known with certainty. The reason is that the film packing factor P, a measure of the void content, can be quite low; e.g., $P \approx 0.75$ for porous deposits. If handbook bulk values of ρ are used in Eq. 6-11, d would, of course, be underestimated. Furthermore, in cases where the substrate contains a great deal of relief in the form of roughness, cleavage steps, patterned topography, etc., the effective deposit area will be larger than the assumed projected area. In this case, the film thickness may be overestimated. For ultrathin films possessing an island structure, this method, as well as others noted previously, are problematical.

Even though gravimetric techniques have disadvantages, very delicate and novel microbalances have been constructed and widely employed to monitor film thickness during deposition. Microbalance designs have relied on such principles as the elongation of a thin quartz-fiber helix, the torsion of a wire, or the deflection of a pivot-mounted beam. Sensitive optical and, more commonly, electromechanical transducers and compensators for null measurements have been employed, enabling detection of $\sim 10^{-8}$ g. By utilizing very light, large-area substrates, we can measure deposits fractions of a monolayer thick. Typical equivalent film thicknesses of less than 10 Å for low-density materials (e.g., SiO_2) and 1 Å for high-density metals (e.g., Pt) are detectable. Microbalances made almost entirely of quartz can be degassed at elevated temperatures, making them suitable for ultrahigh-vacuum operation. The most important gravimetric technique involves the use of quartz crystal oscillators. It is to this almost universally employed technique for in situ monitoring of the thickness of physical vapor-deposited films that we now turn our attention.

6.2.3.3. Crystal Oscillators (Refs. 4, 8). Homogeneous elastic plates set into mechanical vibration have resonant frequencies that depend on their dimensions, elastic moduli, and, importantly, density. Additional mass in the form of a deposited thin film alters (lowers) the resonant frequency by effectively changing the properties of the composite vibrating plate. This is the principle that underlies the use of crystal oscillators to measure film thickness. In this method an AT quartz crystal, i.e., cut $\sim 35°$ with respect to the c axis, containing metal film electrodes on both wide faces is mounted within the deposition chamber close to the substrate. The fundamental frequency f of the shear mode is given by

$$f = v_q / 2 d_q, \tag{6-12}$$

where v_q is the elastic wave velocity and d_q is the plate thickness; d_q is also equal to half the wavelength of the transverse wave. If mass dm deposits on one of the crystal electrodes, the thickness increases by an amount given by Eq. 6-11. Combination of these two equations yields a frequency change given by

$$df = -\frac{f^2}{C\rho_f}\frac{dm}{A}, \tag{6-13}$$

where $C = v_q/2$ is defined as the frequency constant whose value is 1656 kHz-mm in AT cut quartz. In deriving Eq. 6-13, we have assumed that the addition of a small foreign film mass can be treated as an equivalent mass change of the quartz crystal. The formula is not rigorously correct because the elastic properties of the film are not the same as those of quartz, and A is generally not equal to the total crystal face area. These effects greatly complicate the frequency-response analysis. Nevertheless, as long as the accumulated mass deposited on the crystal does not shift the resonant frequency by a few percent of its original value, df varies linearly with dm.

To appreciate the kind of numbers involved, we note that a 6-MHz crystal is commonly employed. Since a frequency shift of 1 Hz is readily measurable, Eq. 6-13 reveals that this is equivalent to a mass of Al = 1.24×10^{-8} g, and if $A = 1$ cm^2, to a film thickness of 0.46 Å. This sensitivity is suitable for most applications. It can be enhanced, however, by more than an order of magnitude by employing thinner crystals with higher resonant frequencies and by detecting smaller frequency shifts.

The change in frequency is usually measured by beating the crystal signal against that from a reference (undeposited) crystal and counting the frequency difference. Quartz crystal oscillators are commonly employed for the measurement of deposition rate rather than film thickness. Therefore, commercial rate monitors contain circuitry to mathematically differentiate the frequency change with respect to time, display the rate, and provide feedback to control the power delivered to evaporation heaters. In these functions it is essential to eliminate uncertainty in the frequency shift measurement. A potentially important source of error arises from the temperature increase of the crystal due to radiant heat exposure from the evaporant source, and from the heat of condensation liberated by depositing atoms. Typically, temperature increases of a few degrees Celsius above that of the reference crystal result in frequency shifts of 10–100 Hz that are equivalent to a mass change of 10^{-7} to 10^{-6} g/cm^2 (Ref. 8). For this reason, crystals are enclosed in water-cooled shrouds having a small entrance aperture to sample the evaporant stream. Lastly, precise work requires a correction due to the geometry of the monitor relative to the substrate.

6.3. Structural Characterization

6.3.1. Introduction

Several levels of structural information are of interest to the thin-film scientist and technologist in research, process development, and reliability and failure analysis activities. The first broadly deals with the geometry of patterned films where issues of lateral or depth dimensions and tolerances, uniformity of thickness and coverage, completeness of etching, etc., are of concern. Beyond this, the film surface topography and morphology, including grain size and shape, existence of compounds, presence of hillocks or whiskers, evidence of film voids, microcracking or lack of adhesion, formation of textured surfaces, etc., are of concern. Somewhat more difficult to obtain, but crucial to microelectronic device fabrication and optical coating technology, are the cross-sectional views of multilayer structures where interfacial regions, substrate interactions, and geometry and perfection of electronic devices with associated conducting and insulating layers may be observed.

Lastly, and most complex of all, are diffraction patterns, the crystallographic information they convey, and the high-resolution lattice images of both plain-view and transverse film sections. Among the applications here are defect structures in films and devices, structure of grain boundaries, identification of phases, and a host of issues related to epitaxial structures—e.g., the crystallographic orientations involved, direct imaging of atoms at interfaces, interfacial quality and defects, perfection of quantum well and strained-layer superlattices. The transmission electron microscope (TEM) is required for these applications, whereas those of the previous paragraph are normally addressed by the scanning electron microscope (SEM) and, occasionally, by the reflection metallurgical microscope. There is an interesting distinction between the TEM and SEM. The former is a true microscope in that all image information is acquired simultaneously or in parallel. In the SEM, however, only a small portion of the total image is probed at any instant, and the image builds up serially by scanning the probe. Strictly speaking, the SEM is more like the scanning Auger electron and SIMS microprobes than a traditional microscope. In this section we treat only electron microscopy, a subject dealt with at length in the recommended references (Refs. 9, 10). We start with the SEM.

6.3.2. Scanning Electron Microscopy

Because seeing is believing and understanding, the SEM is perhaps the most widely employed thin-film and coating characterization instrument. A schematic of the typical SEM is shown in Fig. 6-7. Electrons thermionically emitted from a tungsten or LaB_6 cathode filament are drawn to an anode, focused by two successive condenser lenses into a beam with a very fine spot size (~ 50 Å). Pairs of scanning coils located at the objective lens deflect the beam either linearly or in raster fashion over a rectangular area of the specimen surface. Electron beams having energies ranging from a few thousand to 50 keV, with 30 keV a common value, are utilized. Upon impinging on the specimen, the primary electrons decelerate and in losing energy transfer it inelastically to

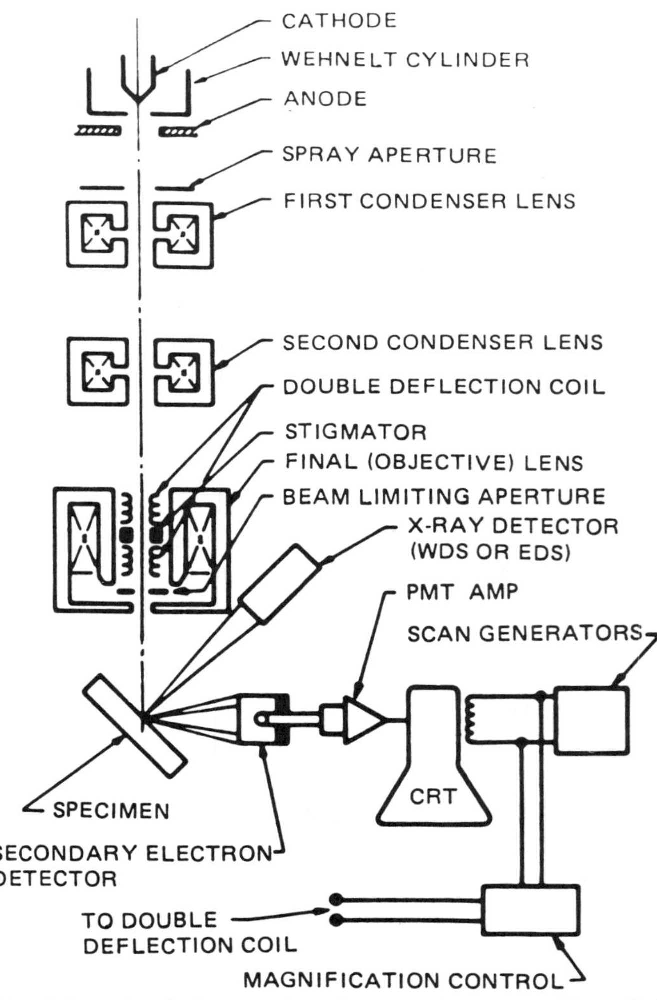

Figure 6-7. Schematic of the scanning electron microscope. (From Ref. 9, with permission from Plenum Publishing Corp.).

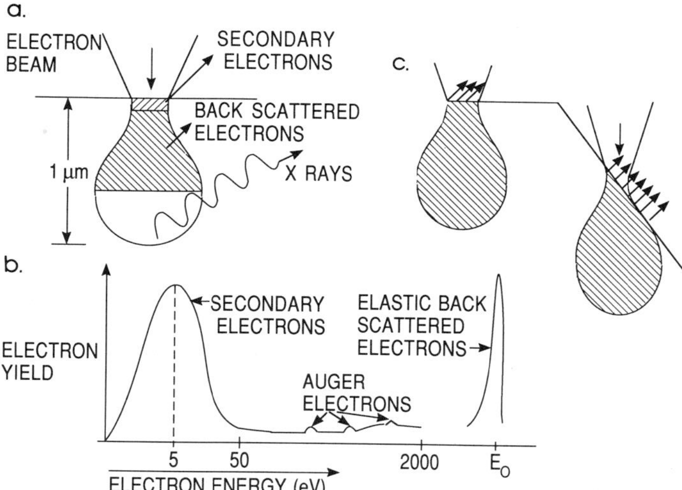

Figure 6-8. (a) Electron and photon signals emanating from tear-shaped interaction volume during electron-beam impingement on specimen surface. (b) Energy spectrum of electrons emitted from specimen surface. (c) Effect of surface topography on electron emission.

other atomic electrons and to the lattice. Through continuous random scattering events, the primary beam effectively spreads and fills a teardrop-shaped interaction volume (Fig. 6-8a) with a multitude of electronic excitations. The result is a distribution of electrons that manage to leave the specimen with an energy spectrum shown schematically in Fig. 6-8b. In addition, target X-rays are emitted, and other signals such as light, heat, and specimen current are produced, and the sources of their origin can be imaged with appropriate detectors.

The various SEM techniques are differentiated on the basis of what is subsequently detected and imaged.

6.3.2.1. *Secondary Electrons.* The most common imaging mode relies on detection of this very lowest portion of the emitted energy distribution. Their very low energy means they originate from a subsurface depth of no larger than several angstroms. The signal is captured by a detector consisting of a scintillator–photomultiplier combination, and the output serves to modulate the intensity of a CRT, which is rastered in synchronism with the raster-scanned primary beam. The image magnification is then simply the ratio of scan lengths on the CRT to that on the specimen. Resolution specifications quoted on research quality SEMs are ~ 50 Å. Great depth of focus enables images of beautiful three-dimensional quality to be obtained from nonplanar surfaces. The contrast variation observed can be understood with reference to Fig. 6-8c. Sloping surfaces produce a greater secondary electron yield because the portion of the interaction volume projected on the emission region is larger than on a flat surface. Similarly, edges will appear even brighter. Many examples of secondary electron SEM images have been reproduced in various places throughout the book.

6.3.2.2. Backscattered Electrons. Backscattered electrons are the high-energy electrons that are *elastically* scattered and essentially possess the same energy as the incident electrons. The probability of backscattering increases with the atomic number Z of the sample material. Since the backscattered fraction is not a very strong function of Z (varying very roughly as $\sim 0.05Z^{1/2}$ for primary electron beams employed in the SEM), elemental identification is

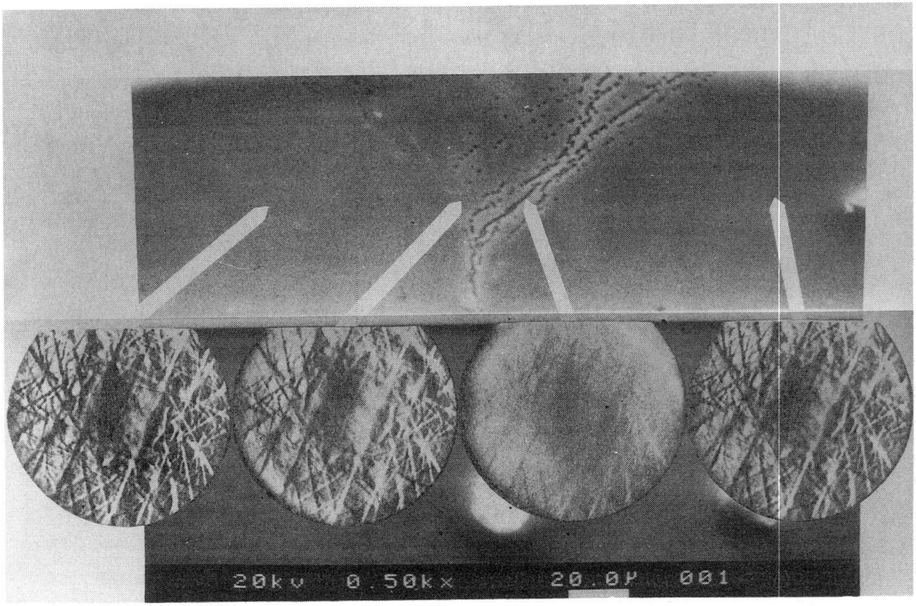

Figure 6-9. Secondary electron image and corresponding electron channeling patterns from (100) Si layers epitaxially regrown from the melt in the LEGO process (Section 7.5.2). The blurred channeling pattern coincides with zone of melt impingement where solidification defects (dislocations) form. (Unpublished research—D. Schwarcz and M. Ohring.)

not feasible from such information. Nevertheless, useful contrast can develop between regions of the specimen that differ widely in Z. Since the escape depth for high-energy backscattered electrons is much greater than for low-energy secondaries, there is much less topological contrast in the images.

An interesting phenomenon that makes use of backscattered electrons is electron channeling. In a single-crystal specimen (e.g., an epitaxial film), some of the crystal planes are oriented properly for electron diffraction to occur. To exploit the effect, the incident beam is made to rock about the normal to the surface over a range of Bragg angles. Electrons channel between lattice planes and are scattered in the forward and reverse directions. These latter electrons are intercepted by a broad-area detector, producing electron channeling patterns. An epitaxial Si layer regrown from melted polysilicon gave rise to the characteristic channeling patterns in Fig. 6-9. Where regrowth is less perfect due to the presence of defects, the pattern is somewhat blurred. The technique is difficult to quantify but is capable of nondestructively probing the crystallinity of regions that are several microns in extent.

6.3.2.3. Electron-Beam-Induced Current (EBIC). The EBIC mode is applicable to semiconductor devices. When the primary electron beam strikes the surface, electron–hole pairs are generated and the resulting current is collected to modulate the intensity of the CRT image. The technique is useful in spatially locating subsurface defects and failure sites within a junction region.

6.3.2.4. X-Rays. A SEM is like a large X-ray vacuum tube used in conventional X-ray diffraction systems. Electrons emitted from the filament (cathode) are accelerated to high energies where they strike the specimen target (anode). In the process, X-rays characteristic of atoms in the irradiated area are emitted. By an analysis of their energies, the atoms can be identified and by a count of the numbers of X-rays emitted the concentration of atoms in the specimen can be determined. This important technique, known as X-ray energy dispersive analysis (EDX), is discussed in more detail in Section 6.4.3.

6.3.3. Transmission Electron Microscopy

As the name implies, the transmission electron microscope is used to obtain structural information from specimens that are thin enough to transmit electrons. Thin films are, therefore, ideal for study, but they must be removed from electron-impenetrable substrates prior to insertion in the TEM. The two basic modes of TEM operation are differentiated by the schematic ray dia-

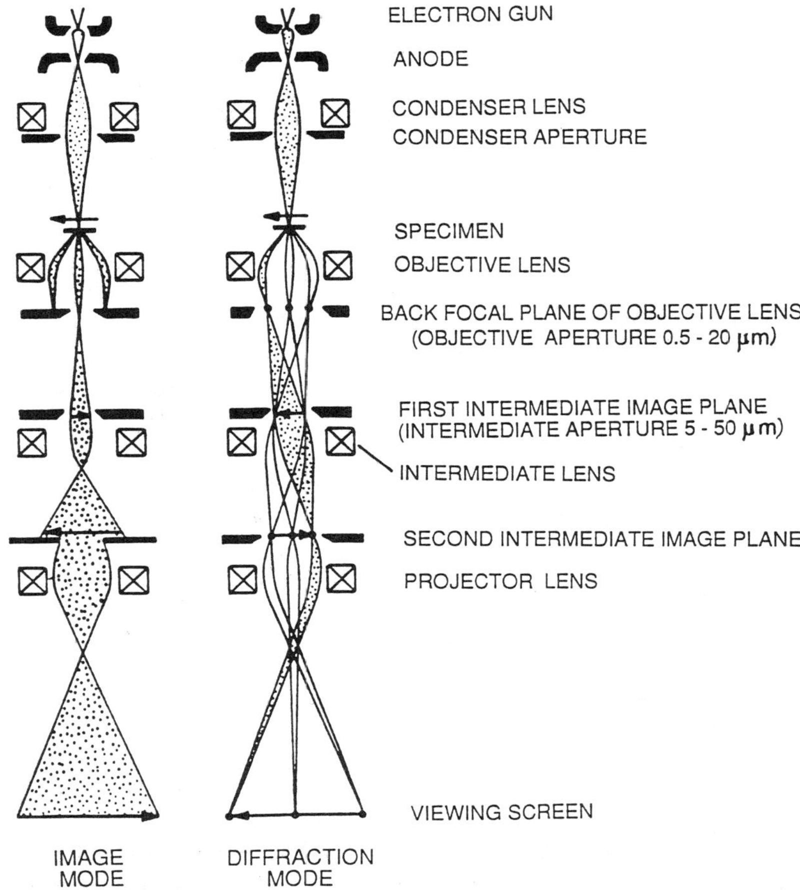

Figure 6-10. Ray paths in the TEM under imaging and diffraction conditions. (Reprinted with permission from John Wiley and Sons, from G. Thomas and M. J. Goringe, Transmission Electron Microscopy of Materials, Copyright © 1979, John Wiley and Sons).

grams of Fig. 6-10. Electrons thermionically emitted from the gun are accelerated to 100 keV or higher (1 MeV in some microscopes) and first projected onto the specimen by means of the condenser lens system. The scattering processes experienced by electrons during their passage through the specimen determine the kind of information obtained. Elastic scattering, involving no energy loss when electrons interact with the potential field of the ion cores, gives rise to diffraction patterns. Inelastic interactions between beam and matrix electrons at heterogeneities such as grain boundaries, dislocations, second-phase particles, defects, density variations, etc., cause complex absorption and scattering effects, leading to a spatial variation in the intensity of the transmitted beam. The generation of characteristic X-rays and Auger electrons also occurs, but these by-products are not usually collected.

The emergent primary and diffracted electron beams are now made to pass through a series of post-specimen lenses. The objective lens produces the first image of the object and is, therefore, required to be the most perfect of the lenses. Depending on how the beams reaching the back focal plane of the objective lens are subsequently processed distinguishes the operational modes. Basically, either magnified images are formed or diffraction patterns are obtained as shown in Fig. 6-10. A discussion of the analysis of diffraction effects is well beyond the scope of this book. Some notion of the correlation between structure and diffraction patterns can be gained from the model of thin films presented in Section 5.7.3.

Images can be formed in a number of ways. The bright-field image is obtained by intentionally excluding all diffracted beams and only allowing the central beam through. This is done by placing suitably sized apertures in the back focal plane of the objective lens. Intermediate and projection lenses then magnify this central beam. Dark-field images are also formed by magnifying a single beam; this time one of the diffracted beams is chosen by means of an aperture that blocks the central beam and the other diffracted beams. The micrograph of alternating 40 Å-wide films in the GaAs–$Al_{0.5}Ga_{0.5}As$ superlattice structure shown in Fig. 6-11 is a dark-field image employing the 200 diffracted beam. In both of these cases we speak of amplitude contrast because diffracted beams with their phase relationships are excluded from the imaging sequence. In a third method of imaging, the primary transmitted and one or more of the diffracted beams are made to recombine, thus preserving both their amplitudes and phases. This is the technique employed in high-resolution lattice imaging, enabling diffracting planes and arrays of individual atoms to be distinguished. The interface image of the epitaxial $CoSi_2$ film on Si (Fig. 1-4) is an example of this remarkable technique. Other examples of epitaxial film interfaces are shown in Figs. 7-16 and 14-17.

The high magnification of all TEM methods is a result of the small effective wavelengths (λ) employed. According to the de Broglie relationship,

$$\lambda = h / \sqrt{2mqV}, \tag{6-14}$$

where m and q are the electron mass and charge, h is Planck's constant, and V is the potential difference through which electrons are accelerated. Electrons of 100 keV energy have wavelengths of 0.037 Å and are capable of effectively transmitting through about 0.6 μm of Si.

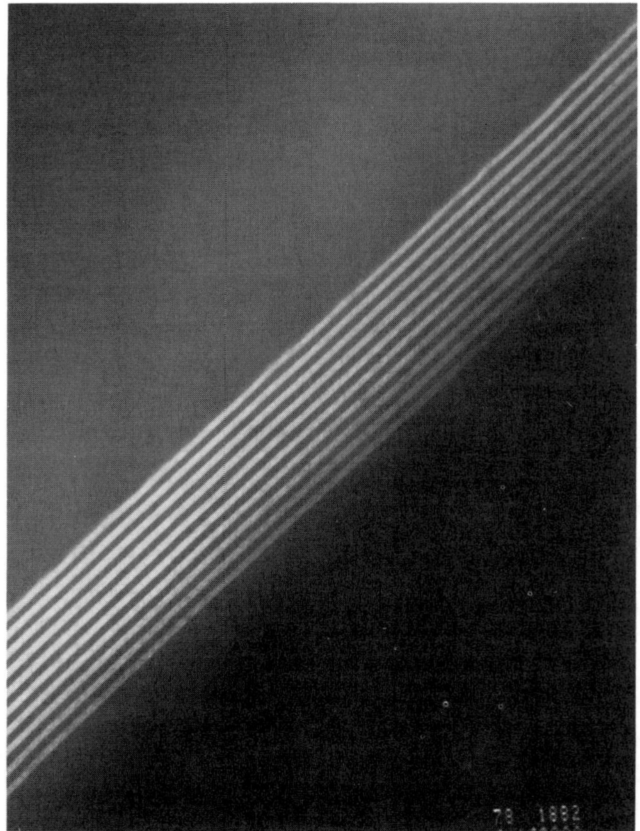

Figure 6-11. Dark-field TEM image of alternating 40-Å-wide GaAs–$Ga_{0.5}Al_{0.5}As$ superlattice films. Light bands contain Al. (Courtesy of S. Nakahara, AT&T Bell Laboratories.

The ability to prepare thin vertical sections of integrated circuits for TEM observation is one of the most important recent advances in technique. If one can imagine the plane of this page to be the surface thinned for conventional TEM work, transverse imaging requires head-on thinning and viewing of the ~ 75-μm-thick page edge. What is involved is the transverse cleavage of a number of wafer specimens, bonding these in an epoxy button, and thinning them by mechanical grinding and polishing. Finally, the resulting thin disk is mounted in an ion-milling machine where the specimen is further sputter thinned by ion bombardment until a hole appears. In VLSI applications, many specimens must be simultaneously mounted to enhance the probability of

Figure 6-12. A cross-section TEM bright-field image of field effect transistor structure after junction delineation. Junctions were formed by As implantation and diffusion. (Courtesy of R. B. Marcus, Bellcore Corp., Reprinted with permission from John Wiley and Sons, from R. B. Marcus and T. T. Sheng, *Transmission Electron Microscopy of Silicon VLSI Circuits and Structures*, Copyright © 1983, John Wiley and Sons).

capturing images from desired circuit features. An example of the vertical section of a field effect transistor is shown in Fig. 6-12 (Ref. 11).

6.3.4. X-Ray Diffraction

X-ray diffraction is a very important experimental technique that has long been used to address all issues related to the crystal structure of bulk solids, including lattice constants and geometry, identification of unknown materials, orientation of single crystals, and preferred orientation of polycrystals, defects, stresses, etc. Extension of X-ray diffraction methods to thin films has not been pursued with vigor for two main reasons: First the great penetrating power of X-rays means that with typical incident angles, their path length through films is too short to produce diffracted beams of sufficient intensity. Under such conditions the substrate, rather than the film, dominates the scattered X-ray signal; thus, diffraction peaks from films require long counting times. Second, the TEM provides similar diffraction information with the added capability of performing analysis over very small selected areas. Nevertheless, X-ray methods have advantages because they are nondestructive and do not require elaborate sample preparation or film removal from the substrate.

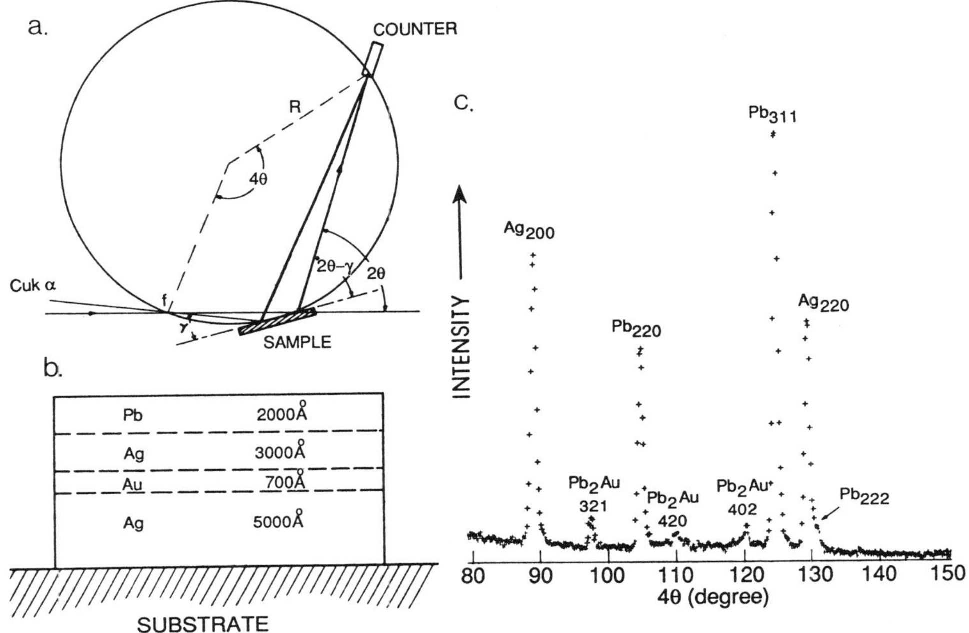

Figure 6-13. (a) Seeman–Bohlin diffraction geometry: f = effective location of X-ray source, γ = angle of incidence, θ = Bragg angle. (b) Schematic of layered film specimen. (c) A selected portion of the diffraction pattern with reflections from Ag, Pb, and Pb_2Au shown. (Courtesy of K. N. Tu, IBM T. J. Watson Research Laboratory, from Ref. 12).

What is required for a workable X-ray method is to make the film appear to be thicker to the beam than it actually is. This can be done by employing a grazing angle of incidence γ as shown in Fig. 6-13. Thus if $\gamma = 5°$, the film is effectively 12 times thicker. The Seeman–Bohlin diffraction geometry is employed with the focal point of the X-ray source, film specimen, and detector slit all located on the circumference of one great circle. Each of the diffracted peaks (at different angles) are sequentially swept through as the X-ray detector moves along the circumference. All the while, a servomotor rotates the detector to keep it aimed at the specimen and preserve the overall focusing geometry.

An example of the diffraction pattern obtained is shown in Fig. 6-13c. The specimen consists of consecutively evaporated polycrystalline films of Ag, Au, Ag, and Pb on fused quartz. After the composite structure was annealed at 200 °C for 24 h, Pb_2Au peaks emerged, indicating that Au atoms diffused through the Ag layer and then reacted with the Pb. Grain boundaries in Ag were the likely diffusion pathways because no penetration of single-crystal Ag films by Au was observed.

6.4. Chemical Characterization

6.4.1. Introduction

We now focus on chemical characterization of thin films. This includes identification of surface and interior atoms and compounds, as well as their lateral and depth spatial distributions. To meet these needs, we use an important subset of the analytical techniques listed in Table 6-1. Space limitation will restrict the discussion to include only the most popular methods (EDX, AES, XPS, RBS, and SIMS) and variants based on these. The justification for selecting these and not others is that they, together with the SEM and TEM, form the core of the diagnostic facilities associated with all phases of the research, development, processing, reliability, and failure analysis of thin-film electronic devices and integrated circuits. In VLSI technology some of these methods have gained wide acceptance as support tools for manufacturing lines. In addition, all of the associated equipment for these techniques is now commercially available, albeit at high cost. All excellent film characterization laboratories are outfitted with the total complement of this equipment.

Table 6-3 will assist the reader to distinguish among the various chemical analytical methods. The capabilities and limitations of each are indicated, and the comparative strengths and weaknesses for particular analytical applications can, therefore, be assessed. The following remarks summarize several of these distinctions:

1. AES, XPS, and SIMS are true surface analytical techniques, since the detected electrons and ions are emitted from surface layers less than ~ 15 Å deep. Provision is made to probe deeper, or depth profile, by sputter-etching the film and continuously analyzing the newly exposed surfaces.
2. EDX and RBS generally sample the total thickness of the thin film (~ 1 μm) and frequently some portion of the substrate as well. Unlike RBS with a depth resolution of ~ 200 Å, EDX has little depth resolution capability.
3. AES, XPS, and SIMS are broadly applicable to detecting, with few exceptions, all of the elements in the periodic table.
4. EDX can ordinarily only detect elements with $Z > 11$, and RBS is restricted to only selected combinations of elements whose spectra do not overlap.

Table 6-3. Summary of Major Chemical Characterization Techniques

Method	Elemental Sensitivity	Detection Limit (at%)	Lateral Resolution	Effective Probe Depth
Scanning Electron microscope–energy dispersive x-ray (SEM/EDX)	Na–U	~ 0.1	~ 1 μm	~ 1 μm
Auger Electron spectroscopy (AES)	Li–U	~ 0.1–1	500 Å	~ 15 Å
X-Ray Photoelectron spectroscopy (XPS)	Li–U	~ 0.1–1	~ 100 μm	~ 15 Å
Rutherford Backscattering (RBS)	He–U	~ 1	1 mm	~ 200 Å
Secondary ion mass spectrometry (SIMS)	H–U	~ 10^{-4}%	~ 1 μm	15 Å

5. The detection limits for AES, XPS, EDX, and RBS are similar, ranging from about ~ 0.1 to 1 at%. On the other hand, the sensitivity of SIMS is much higher and parts per million can be detected. Even lower concentration levels (~ 10^{-6} at%) are detectable in certain instances.
6. Quantitative chemical analysis with AES and XPS is problematical with composition error bounds of several atomic percent. EDX is better and SIMS significantly worse in this regard. Composition standards are essential for quantitative SIMS analysis.
7. Only RBS is quantitatively precise to within an atomic percent or so from first principles and without the use of composition standards. It is the only nondestructive technique that provides simultaneous depth and composition information.
8. The lateral spatial resolution of the region over which analyses can be performed is highest for AES (~ 500 Å) and poorest for RBS (~ 1 mm). In between are EDX (~ 1 μm), SIMS (several μm), and XPS (~ 0.1 mm). AES has the distinction of being able to sample the smallest volume for analysis.

9. Only XPS, and to a much lesser extent AES, are capable of readily providing information on the nature of chemical bonding and valence states.

The preceding characteristics earmark certain instruments for specific tasks. Suppose, for example, a film surface is locally discolored due to contamination, or contains a residue, and it is desired to identify the source of the unknown impurities. Assuming access to all instruments at equal cost, AES and EDX would be the techniques of choice. If only ultrathin surface layers are involved, EDX would probably be of little value. The presence of trace elements would necessitate the higher sensitivity of SIMS analysis. If preliminary examination pointed to the presence of Cl from an etching process, then evidence of the actual chemical compound formed would be obtained from XPS measurements. In a second example, a broad-area, thin-film metal bilayer structure is heated. Here we know which elements are initially present, but wish to determine the stoichiometry of intermetallic compounds formed as well as their thicknesses. This information is without question most unambiguously provided by RBS methods.

In what follows, the various techniques are considered individually where additional details of instrumentation, aspects of particular capabilities and limitations, and applications will be presented. First, however, it is essential to appreciate the scientific principles underlying each type of analysis. More detailed discussions of these characterization techniques are given in Refs. 13–19.

6.4.2. Electron Spectroscopy

We start with a discussion of atomic core electron spectroscopy since it is the basis for identification of the elements by EDX, AES, and XPS techniques. Consider the electronic structure of an unexcited atom schematically depicted in Fig. 6-14a. Both the K, L, M, etc., shell notation and the corresponding 1s, 2s, 2p, 3s, etc., electron states are indicated. Through excitation by an incident electron or photon, a hole or electron vacancy is created in the K shell (Fig. 6-14b).

In EDX an electron from an outer shell lowers its energy by filling the hole, and an X-ray is emitted in the process (Fig. 6-14c). If the electron transition occurs between L and K shells, $K\alpha$ X-rays are produced. Different X-rays are generated, e.g., K_β X-rays from $M \rightarrow K$, and $L\alpha$ X-rays from $M \rightarrow L$ transitions. There are two facts worth remembering about these X-rays.

1. The difference in energy between the levels involved in the electron transition is what determines the energy (or wavelength) of the emitted X-ray.

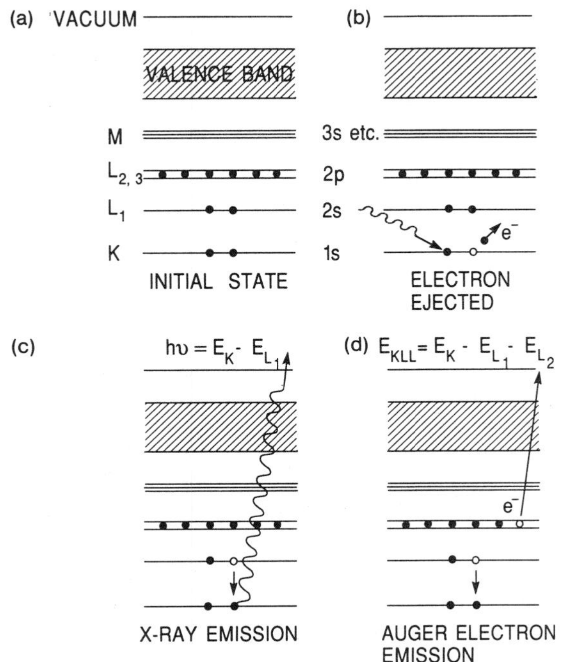

Figure 6-14. Schematic of electron energy transitions: (a) initial state; (b) incident photon (or electron) ejects K shell electron; (c) X-ray emission when 2s electron fills vacancy; (d) Auger electron emission. KLL transition shown.

For example,

$$E_{K\alpha_1} = \frac{hc}{\lambda_{K\alpha_1}} = E_K - E_{L_3}, \tag{6-15}$$

where h, c, and λ have their usual meaning.

2. The emitted X-rays are characteristic of the particular atom undergoing emission. Thus, each atom in the Periodic Table exhibits a unique set of K, L, M, etc., X-ray spectral lines that serve to unambiguously identify it. These characteristic X-rays are also known as fluorescent X-rays when excited by incident photons (e.g., X-rays and gamma rays).

There is, however, an alternative process by which the electron hole in Fig. 6-14b can be filled. This involves a complex transition in which three, rather than two, electron levels, as in EDX, participate. The Auger process, which is the basis of AES, first involves an electron transition from an outer level (e.g., L_1) to the K hole. The resulting excess energy is not channeled into the creation of a photon but is expended in ejecting an electron from yet a third level (e.g., L_2). As shown in Fig. 6-14d, the atom finally contains two electron holes after starting with a single hole. The electron that leaves the atom is known as an Auger electron, and it possesses an energy given by

$$E_{KLL} = E_K - E_{L_1} - E_{L_2} = E_K - E_{L_2} - E_{L_1}. \tag{6-16}$$

The last equality indicates KL_1L_2 and KL_2L_1 transitions are indistinguishable. Similarly, other common transitions observed are denoted by LMM and MNN. Since the K, L, and M energy levels in a given atom are unique, the Auger spectral lines are characteristic of the element in question. By measuring the energies of the Auger electrons emitted by a material, we can identify its chemical makeup.

To quantitatively illustrate these ideas, let us consider the X-ray and Auger excitation processes in titanium. The binding energies of each of the core electrons are indicated in Fig. 6-15, where electrons orbiting close to the nucleus are strongly bound with large binding energies. Electrons at the Fermi level are far from the pull of the nucleus and therefore taken to have zero binding energy, thus establishing a reference level. They would still have to acquire the work function energy to be totally free of the solid. Some notion of the rough magnitude of the core energy levels can be had from the well-known formula for hydrogen-like levels; i.e.,

$$E = 13.6Z^2/n^2 \text{ (eV)}, \tag{6-17}$$

where Z is the atomic number and n is the principal quantum number. For Ti ($Z = 22$), the calculated energy of the K shell ($n = 1$) is 6582 eV. Complex electron-electron interactions and shielding of the nucleus makes this formula far too simplistic for multielectron atoms. Both effects reduce electron binding energies relative to Eq. 6-17. Several of the prominent characteristic X-ray energies and wavelengths for Ti are $K\alpha_1$: $E_K - E_{L_3} = 4966.4 - 455.5 = 4511$ eV, $\lambda = 2.75$ Å; K_{β_1}: $E_K - E_{M_3} = 4966.4 - 34.6 = 4932$ eV, $\lambda = 2.51$ Å; $L\alpha$: $E_{L_3} - E_{M_{4,5}} = 455.5 - 3.7 = 452$ eV, $\lambda = 27.4$ Å. Similarly, a prominent Ti Auger spectral transition is LMM or

$$E_{LMM} = E_{L_3} - E_{M_3} - E_{M_4} = 455.5 - 34.6 - 3.7 = 417 \text{ eV}.$$

The question may arise: When do atoms with electron holes undergo X-ray transitions, and when do they execute Auger processes? The answer is that both processes go on simultaneously. In the low-Z elements, the probability is greater that an Auger transition will occur, whereas X-ray emission is favored for high Z elements. The fractional proportions of the characteristic X-ray

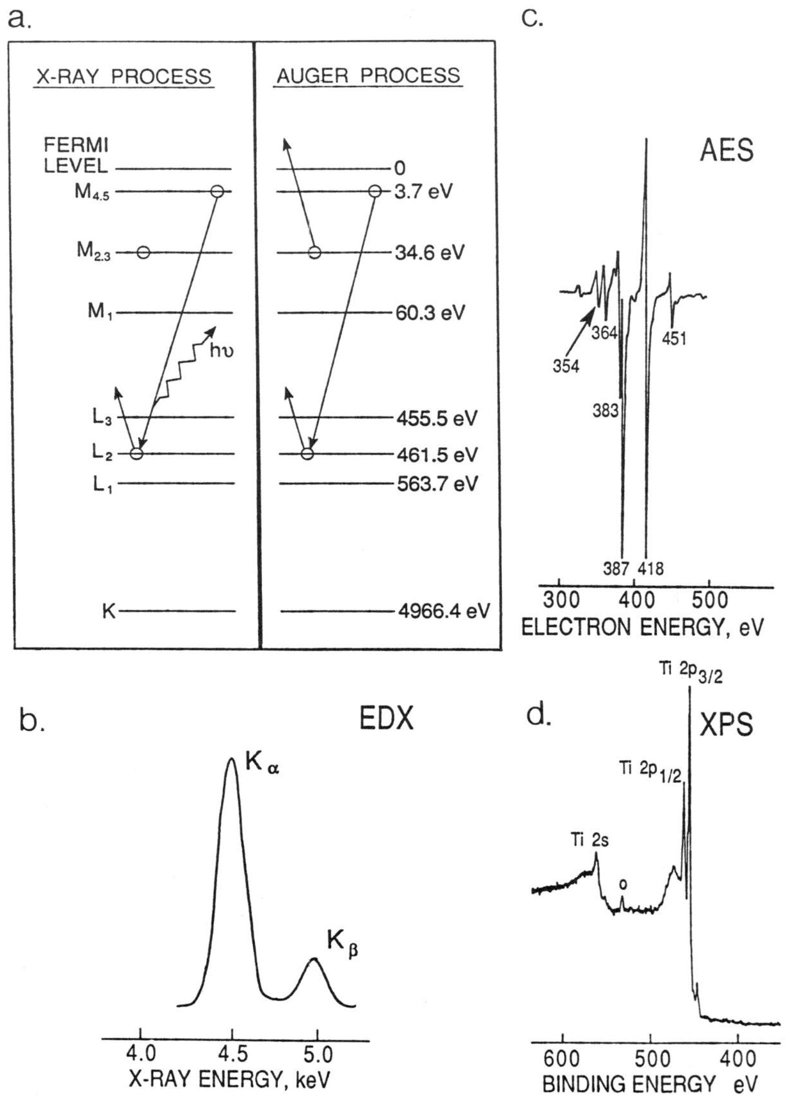

Figure 6-15. Electron excitation processes in Ti: (a) energy-level scheme; (b) EDX spectrum of Ti employing Si(Li) detector; (c) AES spectral lines for Ti ($dN(E)/dE$ vs. E); (d) a portion of the XPS spectrum for Ti (MgKα radiation).

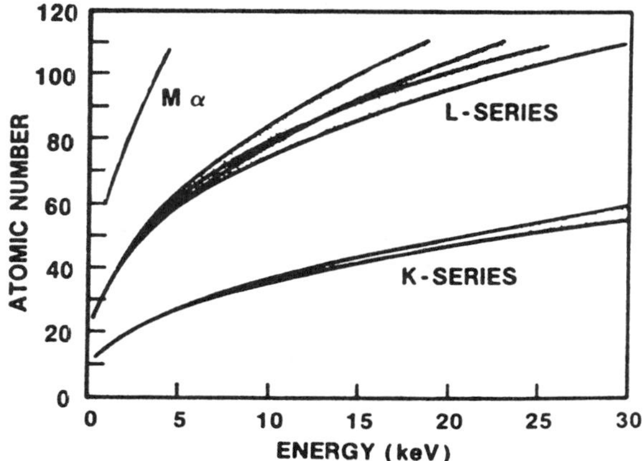

Figure 6-16. Characteristic X-ray emission energies of the elements. (Courtesy of Princeton Gamma Tech, Inc.)

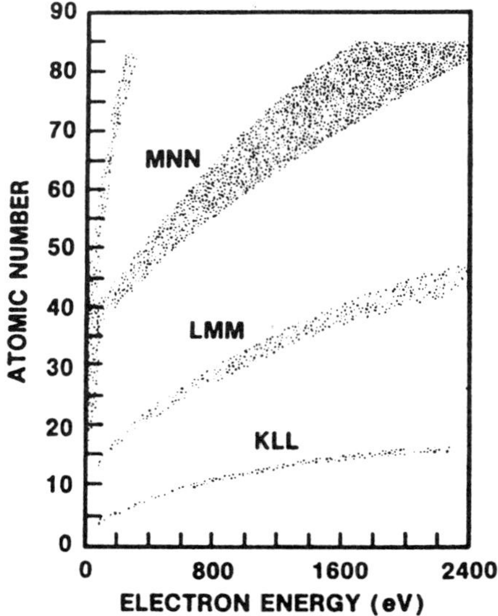

Figure 6-17. Principal Auger electron energies of elements. (Courtesy of Physical Electronics Industries, Inc.)

and Auger yields for K, L, M transitions can be found in standard references (Ref. 13).

The variation of the principal characteristic X-ray and Auger lines with atomic number is shown in Figs. 6-16 and 6-17, respectively. Commercial spectrometers typically operate within the energy range spanned in these figures. Therefore, in EDX, K X-ray transitions are conveniently measured in low-Z materials, and L series X-rays appear when high-Z elements are involved. Similarly, in AES, KLL and LMM transitions are involved for low-Z elements, and LMM and MNN lines appear for high-Z elements. Although keeping track of the particular shell involved is sometimes annoying, spectra from virtually all of the elements in the periodic table can be detected with a single excitation source and a single detector. Fortunately, the resolution of X-ray or electron detectors is such that prominent lines of neighboring elements do not seriously overlap. This facilitates spectral interpretation and atomic fingerprinting.

The basis for understanding XPS lies in the same atomic core electron scheme that we have been considering. Rather than incident electrons in the case of EDX and AES, relatively low-energy X-rays impinge on the specimen in this technique. The absorption of the photon results in the ejection of electrons via the photoelectric effect. Governing this process is the well-known equation expressed by

$$E_{\mathrm{KE}} = h\nu - E_B, \tag{6-18}$$

where E_{KE}, $h\nu$, and E_B are the energies of the ejected electron, incident photon, and the involved bound electron state. Since values of the binding energy are element-specific, atomic identification is possible through measurement of photoelectron energies.

6.4.3. X-Ray Energy-Dispersive Analysis (EDX)

6.4.3.1. Equipment. Most energy-dispersive X-ray analysis systems are interfaced to SEMs, where the electron beam serves to excite characteristic X-rays from the area of the specimen being probed. Attached to the SEM column is the liquid-nitrogen Dewar with its cooled Si(Li) detector aimed to efficiently intercept emitted X-rays. The Si(Li) detector is a reverse-biased Si diode doped with Li to create a wide depletion region. An incoming X-ray generates a photoelectron that eventually dissipates its energy by creating electron–hole pairs. The incident photon energy is linearly proportional to the number of pairs produced or equivalently proportional to the amplitude of the voltage pulse they generate when separated.

The pulses are amplified and then sorted according to voltage amplitude by a multichannel analyzer, which also counts and stores the number of pulses within given increments of the voltage (energy) range. The result is the characteristic X-ray spectrum shown for Ti in Fig. 6-15. Si(Li) detectors typically have a resolution of about 150 eV, so overlap of peaks occurs when they are not separated in energy by more than this amount. Overlap sometimes occurs in multicomponent samples or when neighboring elements in the periodic table are present.

Several variants of X-ray spectroscopy are worth mentioning. In X-ray *wavelength*-dispersive analysis (WDX), where wavelength rather than energy is dispersed, a factor of 20 or so improvement in X-ray linewidth resolution is possible. In this case, emitted X-rays, rather than entering a Si(Li) detector, are diffracted from single crystals with known interplanar spacings. From Bragg's law, each characteristic wavelength reflects constructively at different corresponding angles, which can be measured with very high precision. As the goniometer–detector assembly rotates, the peak is swept through as a function of angle. The electron microprobe (EMP) is an instrument specially designed to perform WDX analysis. This capability is also available on an SEM by attaching a diffractometer to the column. The high spectral resolution of WDX is offset by its relatively slow speed.

Characteristic X-rays can also be generated by using photons and energetic particles rather than electrons as the excitation source. For example, conventional X-ray tubes, and radioactive sources such as ^{241}Am (60-keV gamma ray, 26.4-keV X-ray) and ^{109}Cd (22.1-keV Ag–K X-ray) can excite *fluorescent* X-rays from both thin-film and thick specimens. Unlike electron-beam sources, they have virtually no lateral spatial resolution.

6.4.3.2. Quantification.

Quantitative analysis of an element in a multicomponent matrix is a complicated matter. The expected X-ray yield, $Y_x(d)$, originating from some depth d below the surface depends on a number of factors: $I_0(d)$, the intensity of the electron beam at d; C, the atomic concentration; σ, the ionization cross section; w_x, the X-ray yield; μ, the X-ray absorption coefficient; and ε, Ω, and θ, the detector efficiency, solid angle, and angle with respect to the beam, respectively. Therefore,

$$Y_x(d) \sim I_0(d) C w_x e^{-\mu d/\cos\theta} \varepsilon \, d\Omega/4\pi, \tag{6-19}$$

and the total signal detected is the sum contributed by all atomic species present integrated over the depth range. It is sometimes simpler to calibrate the yields against known composition standards. Excellent computer programs, both standardless and employing standards, are available for analysis, and compositions are typically computed to approximately 0.1 at%.

6.4.4. Auger Electron Spectroscopy (AES)

6.4.4.1. Equipment.

The typical AES spectrometer, shown schematically in Fig. 6-18, is housed within an ultrahigh vacuum chamber maintained at

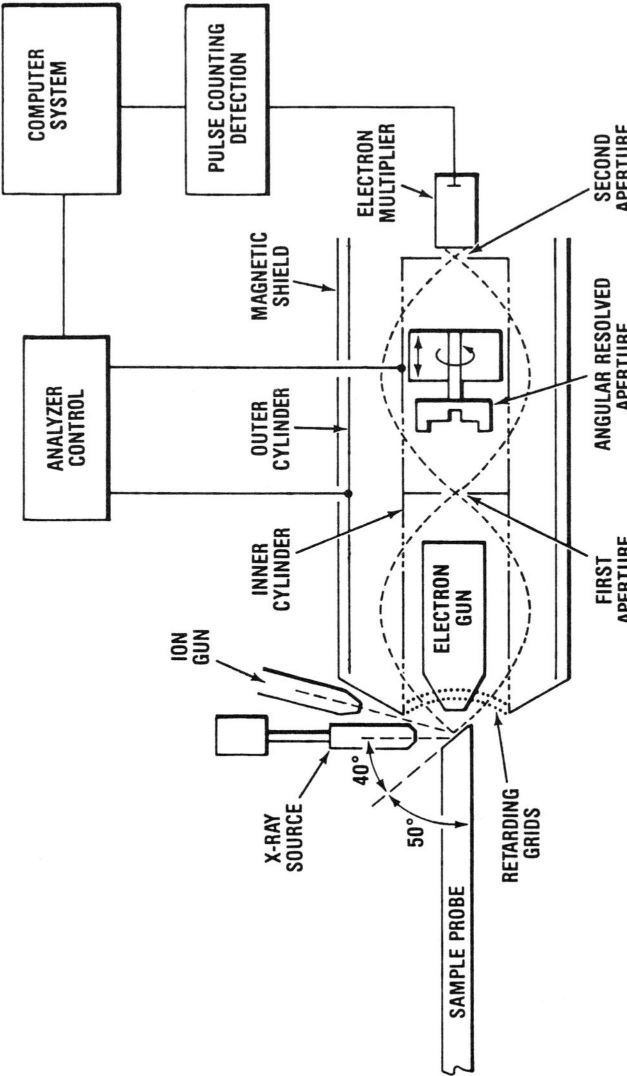

Figure 6-18. Schematic of spectrometer with combined AES and XPS capabilities. (Courtesy of Physical Electronics Industries, Inc.)

$\sim 10^{-10}$ torr. This level of cleanliness is required to prevent surface coverage by contaminants (e.g., C, O) in the system. The electron-gun source aims a finely focused beam of $\sim$ 2-keV electrons at the specimen surface, where it is scanned over the region of interest. Emitted Auger electrons are then energy-analyzed by a cylindrical (or hemispherical in some systems) analyzer. The latter consists of coaxial metal cylinders (or hemispheres) raised to different potentials. The electron pass energy E is proportional to the voltage on the outer cylinder, and the incremental energy range ΔE of transmitted electrons determines the resolution ($\Delta E/E$), which is typically 0.2 to 0.5%. Electrons with higher or lower energies (velocities) than E either hit the outer or inner cylinders, respectively. They do not exit the analyzer and are not counted. By a sweep of the bias potential on the analyzer, the entire electron spectrum is obtained. Complete AES spectrometers are commercially available and cost about a half-million dollars. Auger electrons are but a part of the total electron yield, $N(E)$, intermediate between low-energy secondary and high-energy elastically scattered electrons. They are barely discernible as small bumps above the background signal. Therefore, to accentuate the energy and magnitude of the Auger peaks, the spectrum is electronically or numerically differentiated, and this gives rise to the common AES spectrum, or $dN(E)/dE$ vs. E response, shown in Fig. 6-15c for Ti. The reader should verify that differentiation of a Gaussian-like peak yields the wiggly narrow double-peak response. By convention, the Auger line energy is taken at the resulting peak minimum.

Two very useful capabilities for thin-film analysis are depth profiling and lateral scanning. The first is accomplished with incorporated ion guns that enable the specimen surface to be continuously sputtered away while Auger electrons are being detected. Multielement composition depth profiles can thus be determined over total film thicknesses of several thousand angstroms by sequentially sampling and analyzing arbitrarily thin layers. Although depth resolution is extremely high, the frequently unknown sputter rates makes precise depth determinations problematical. Through raster or line scanning the electron beam, the AES is converted into an SEM and images of the surface topography are obtained. By modulating the imaging beam with the Auger electron signal, we can achieve lateral composition mapping of the surface distribution of particular elements. Unlike EDX–SEM composition mapping, only the upper few atom layers is probed in this case.

6.4.4.2. Quantification. The determination of the Auger electron yield from which atomic concentrations can be extracted is expressed by a formula similar in form to Eq. 6-19 for the X-ray yield. The use of external standards is very important in quantifying elemental analysis, particularly because standardless computer programs for AES are rather imprecise when compared with those available for EDX analysis. An approximate formula that has been widely used to determine the atomic concentration of a given species A in a matrix of m elements is

$$C_A = \frac{I_A / \bar{S}_A}{\sum_{i=1}^{m} I_i / \bar{S}_i}. \tag{6-20}$$

The quantity I_i represents the intensity of the Auger line and is taken as the peak-to-peak span of the spectral line. The relative Auger sensitivity $\bar{S}_i$ also enters Eq. 6-20. It has values ranging from ~ 0.02 to 1 and depends on the element in question, the particular transition selected, and the electron-beam voltage. Uncertainties in C values so determined are perhaps a few atomic percent at best.

6.4.5. X-Ray Photoelectron Spectroscopy (XPS)

In order to capitalize on the X-ray-induced photoelectron effect, a spectrometer like the one used for AES and shown in Fig. 6-18 is employed. The only difference is the excitation source, which is now a beam of either Mg or Al Kα X-rays. These characteristic X-rays have relatively low energy (e.g., $h\nu_{Mg} = 1254$ eV and $h\nu_{Al} = 1487$ eV) and set an upper bound to the kinetic energy of the detected photoelectrons.

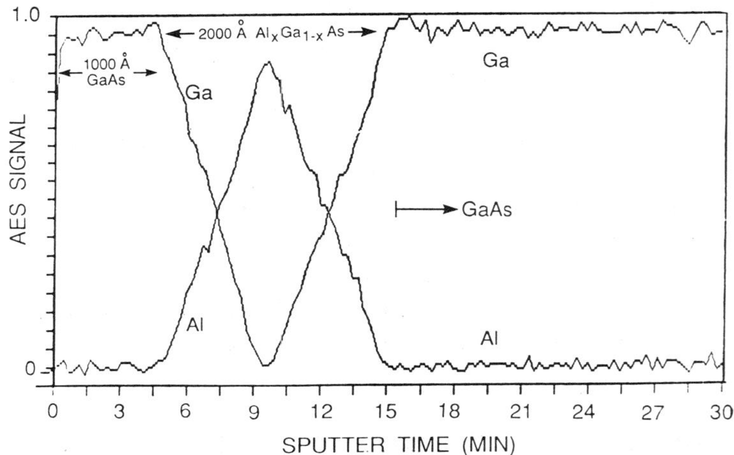

Figure 6-19. AES depth profiles of Al and Ga through GaAs and $Al_xGa_{1-x}As$ films. Signal for As not shown. (Courtesy of R. Kopf, AT&T Bell Laboratories.)

A portion of the XPS spectrum for Ti is shown in Fig. 6-15d, where 2s, $2p_{1/2}$, and $2p_{3/2}$ peaks are evident. Interestingly, characteristic Auger electron transitions (not shown) frequently appear at precisely the energy locations indicated in Fig. 6-15c. The XPS peak positions, however, are shifted slightly by a few electron volts from those predicted by Eq. 6-18 because of work function differences between the specimen and detector.

It is beyond our scope to discuss spectral notation, the chemistry and physics of transitions, and position and width of the lines. What is significant is that linewidths are considerably narrower than those associated with Auger transitions. This fact makes it possible to gain useful chemical bonding information that can also be attained with less resolution by AES, but not by the other surface analytical techniques. It is for this reason that XPS is also known as electron spectroscopy for chemical analysis (ESCA).

Effects due to chemical bonding originate at the valence electrons and ripple beyond them to alter the energies of the core levels in inverse proportion to their proximity to the nucleus. As a result, energy shifts of a few electron volts occur and are resolvable. For example, in the case of pure Ti, the $2p_{3/2}$ line has a binding energy of 454 eV. In compounds this electron is more tightly bound to the Ti nucleus; apparently the electron charge clouds of the neighboring atoms "repel" it. In TiC, TiN, and TiO, the same line is located at $E_B = 455$ eV. Similarly, for the compounds TiO_2, $BaTiO_3$, $PbTiO_3$, $SrTiO_3$, $CaTiO_3$, and $(C_2H_5)_2TiCl_2$ the transition occurs between $E_B = 458$ and 459 eV. Clearly the magnitude of the chemical shift alone is not a sufficient condition to establish the nature of the compound.

Aside from chemical bonding information, XPS has an important advantage relative to AES. X-rays are less prone to damage surfaces than are electrons. For example, electron beams can reduce hydrocarbon contaminants to carbon, destroying the sought-after evidence. For this reason, XPS tends to be preferred in assessing the cleanliness of semiconductor films during MBE growth.

6.4.6. A Couple of Applications in GaAs Films

6.4.6.1. AES. As an example of AES, consider the depth profiles for Ga and Al shown in Fig. 6-19. The structure represents a single-crystal GaAs substrate onto which a 2000-Å thick, compositionally graded film of $Al_xGa_{1-x}As$ was grown by molecular-beam epitaxy methods (Chapter 7). At first, the deposition rate of Al was increased linearly while that of Ga correspondingly decreased until AlAs formed; then the deposition rates were reversed until the GaAs composition was attained after 2000 Å. Finally, a 1000-Å-cap film of GaAs was deposited resulting in a 2000-Å-wide, V-shaped quantum well sandwiched between GaAs layers. Because the lattice constants of GaAs, AlAs, and the intermediate $Al_xGa_{1-x}As$ compositions differ by only 0.16% at most, the entire "lattice-matched" structure is free of crystallographic defects. During AES depth profiling, the film is sputtered away, exposing surface compositions in reverse order to those that were initially deposited. The linearly graded walls of the quantum well reflect the precise control of deposition that is possible. In this case, the values of $\bar{S}_{Al}$ and $\bar{S}_{Ga}$ are apparently independent of composition.

6.4.6.2. XPS. Here we consider an AlGaAs film that has been etched in a $CF_2Cl_2 + O_2$ plasma in order to fabricate devices. It is desired to determine the composition of the film surface relative to that of the underlying material. But both electron impingement and ion bombardment during sputter depth profiling would alter or even remove surface compounds. What is needed is a technique to probe surface layers nondestructively. Angle-resolved XPS is such a method. It is based on altering the takeoff angle for electron detection. If the angle (θ) that exiting photoelectrons make with the surface plane is large, then chemical information on deep surface layers is sampled. However, if electrons exiting at a small grazing angle are detected, then only the top surface layers are probed; photoelectrons generated within deeper layers simply never emerge because their effective range (l_e) is smaller than the now

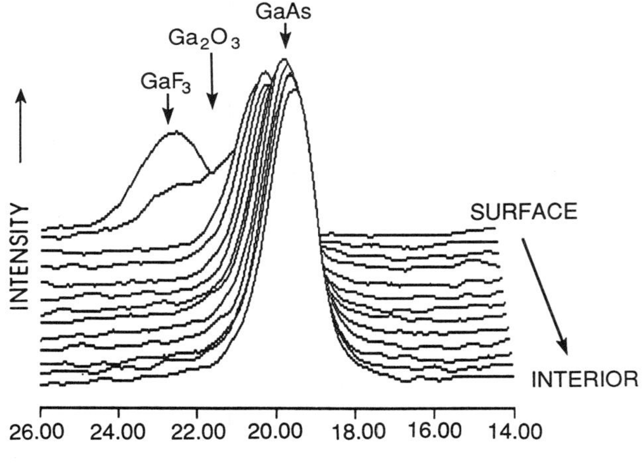

Figure 6-20. Angle-resolved XPS spectra of Ga 3d line as a function of electron detection angle. (Courtesy of M. Vasile, AT&T Bell Laboratories.)

longer geometric escape-path length. In fact, the signal intensity from atoms at depth x varies as $I = I_0 \exp - x / l_e \cos \theta$.

In the analysis shown in Fig. 6-20, θ was varied from 90° to 20° while the Ga 3d peak was scanned. At the surface, bonding associated with GaF_3 and Ga_2O_3 compounds was detected. Deeper within the film only Ga bound within GaAs or AlGaAs is evident.

6.4.7. Rutherford Backscattering (RBS) (Refs. 17, 20)

6.4.7.1. Physical Principles. This popular thin-film characterization technique relies on the use of very high energy (MeV) beams of low mass ions. These have the property of penetrating thousands of angstroms or even microns deep into films or film–substrate combinations. Interestingly, such beams cause negligible sputtering of surface atoms. Rather, the projectile ions lose their energy through electronic excitation and ionization of target atoms. (For further discussion see Section 13.4.2.) These "electronic collisions" are so numerous that the energy loss can be considered to be continuous with depth. Sometimes the fast-moving light ions (usually $^4He^+$) penetrate the atomic electron cloud shield and undergo close-impact collisions with the nuclei of the much heavier stationary target atoms. The resulting scattering from the Coulomb repulsion between ion and nucleus has been long known in nuclear physics as Rutherford scattering. The primary reason that this phenomenon has been so successfully capitalized upon for film analysis is that classical two-body elastic scattering is operative. This, perhaps, makes RBS the easiest of the analytical techniques to understand.

We start by first considering an incident ion of mass (atomic weight) M_0 and energy E_0 incident on a surface as shown in Fig. 6-21. An elastic collision between the ion projectile and a surface atom of mass M occurs such that afterward the ion energy is E_1. The collision is insensitive to the electronic configuration or chemical bonding of target atoms, but depends solely on the masses and energies involved. As a consequence of conserving energy and momentum, it is readily shown that

$$E_1 = \left\{ \frac{\left(M^2 - M_0^2 \sin^2 \theta\right)^{1/2} + M_0 \cos \theta}{M_0 + M} \right\}^2 E_0, \tag{6-21}$$

where θ is the scattering angle. For a particular combination of M_0, M, and θ, the simple formula

$$E_1 = K_M E_0 \tag{6-22}$$

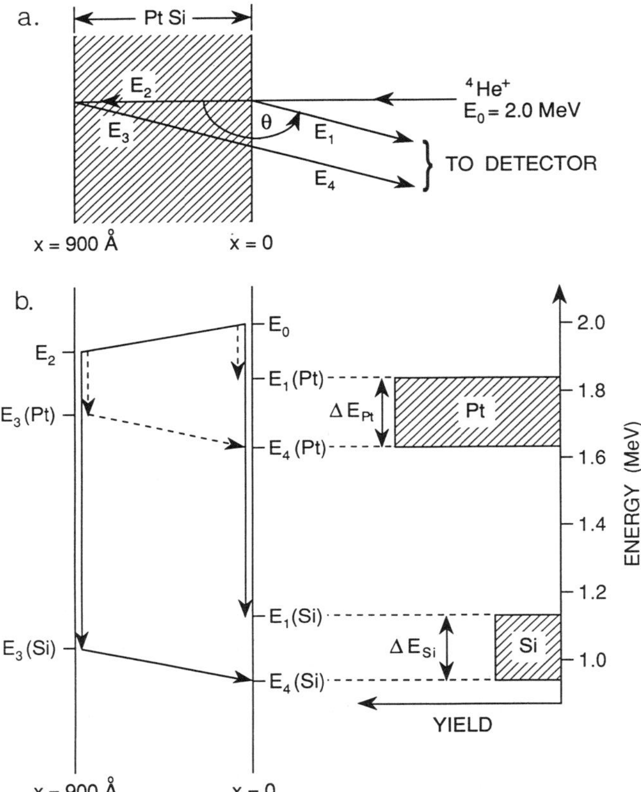

Figure 6-21. (a) Geometry of scattering and notation of energies at the front and back surfaces of a 900-Å-thick PtSi film. (b) $^4He^+$ ion energy as a function of film depth as a result of scattering from Pt and Si. Schematic RBS spectrum shown rotated by 90°. (Reprinted with permission from Elsevier Sequoia S.A., from W. K. Chu, J. W. Mayer, M. A. Nicolet, T. M. Buck, G. Amsel, and F. Eisen, *Thin Solid Films* **17**, 1, 1963).

relates the energy of emergent ions to that of the incident ions. The term K_M is known as the kinematic factor and can be calculated from Eq. 6-21. Once the incident ion, e.g., $^4He^+(M_0 = 4)$ at $E_0 = 2$ MeV, and angular position of the ion detector are selected (θ is typically 170°), K_M just depends on the atomic weight of the target atom. For example, under the conditions just given, $K_{Pt} = 0.922$ and $K_{Si} = 0.565$. This means that if a 2-MeV He ion collides with a Pt atom ($M = 195$) located on the outer surface of a PtSi film, it will backscatter into the detector with an energy of 1.844 MeV; similarly, He ions scattering from surface Si atoms will have their energy reduced to 1.130 MeV.

Consider now what happens when a 2-MeV $^4He^+$ ion beam impinges on a 900-Å platinum–silicide film on a Si wafer substrate. In Fig. 6-21a the film–substrate geometry is shown together with a schematic of the energy changes the ions undergo during nuclear scattering and the corresponding RBS spectrum (Fig. 6-21b). Energy changes ($E_0 \rightarrow E_1$) for scattering from Pt and Si surface atoms are shown on the $x = 0$ axis. The majority of the He ions penetrate below the film surface, however, where they continuously lose energy at a linear rate ($E_0 \rightarrow E_2$) with distance traversed. At any film depth they can suffer an atomic collision. The scattered ion energy is still given by Eq. 6-22, but E_0 is now the incident $^4He^+$ energy at that point in the matrix. Some of the energy-attenuated $^4He^+$ ions can reach the PtSi–Si interface where they may backscatter ($E_3 = K_M E_2$) and again lose energy ($E_3 \rightarrow E_4$) in traversing the film backward until they finally exit. Other $^4He^+$ ions can even penetrate into the Si substrate where they eventually backscatter. It is important to realize that in the course of passage through the film, the ion beam can be thought of as splitting into two elemental components, each spanning a different range of energies. For each broad elemental peak detected, the highest and lowest energies correspond to atoms on the front and back film surfaces, respectively. Statistics largely govern the depth at which scattering occurs and whether Pt or Si atoms participate in the collisions.

Thus, after measurement of the number and energy of backscattered He ions, information on the nature of the elements present, their concentration, and depth distribution can all be simultaneously determined without appreciably damaging the specimen.

6.4.7.2. Equipment. A schematic of the experimental arrangement employed for RBS is shown in Fig. 6-22. Lest the reader be deceived by the figure, it should be appreciated that the actual facility shown in Fig. 6-23 is some 15 m long and occupies more than 100 m^2 of floor space. Ions for analysis (e.g., ^4He, ^{12}C, ^{14}N, etc.) are accelerated by the high voltage generated by the Van De Graaff accelerator. After entering the evacuated ($\sim 10^{-6}$–10^{-7} torr) beam line, the ions are then collimated and focused. Mass selection occurs in the bending magnet, which geometrically disperses ions according to their mass. The resultant ion beam is then raster-scanned across the surface of the specimen. Backscattered ions are analyzed with respect to their energy by a silicon surface barrier detector capable of an energy resolution of $\sim$ 15 keV (peak width at half-maximum amplitude). The electronic pulses are then amplified and sorted according to voltage amplitude (i.e., energy) by a multichannel analyzer to yield the resulting RBS spectrum.

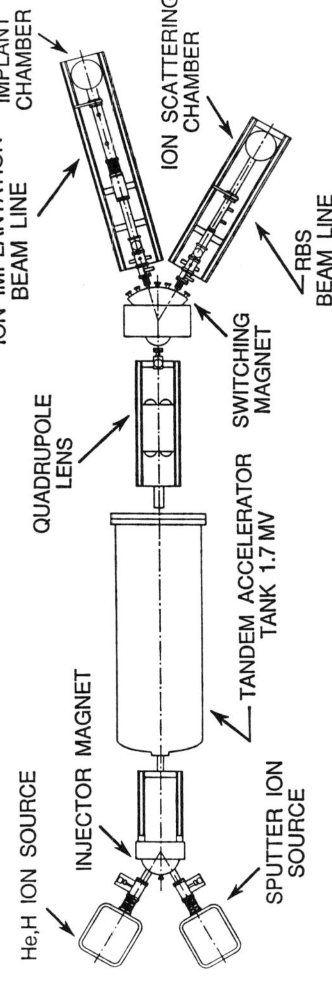

Figure 6-22. Schematic of the 1.7-MeV tandem accelerater, RBS facility at AT&T Bell Laboratories, Murray Hill, NJ.

Figure 6-23. Photograph of 1.7-MeV tandem accelerater–RBS facility. (Courtesy of D. C. Jacobson, AT&T Bell Laboratories.)

6.4.7.3. Capabilities and Limitations

Elemental Information. All elements and their isotopes including Li and those above it in the periodic table are, in principle, detectable with $^4He^+$ ions. The critical test is how well neighboring elements are resolved, and this ultimately depends on the detector resolution. With 2-MeV $^4He^+$, isotopes with $\Delta M = 1$ can generally be separated for M below approximately 40. At values of $M \approx 200$ only atoms for which $\Delta M > 20$ can be resolved. Thus, ^{209}Bi and ^{190}Os would be indistinguishable. The apparent advantage in separating low-Z elements is offset by their low cross sections (σ_i) for scattering. The σ_i are a measure of how efficiently target atoms scatter incoming ions and depend on Z_i, particle energy E, the masses involved, and the angle of scattering. To a good approximation, their dependence on these quantities varies as

$$\sigma = \left(\frac{Z_1 Z_2 q^2}{4E}\right)^2 \left[\left(\sin\frac{\theta}{2}\right)^{-4} - 2\left(\frac{M_0}{M}\right)^2\right], \tag{6-23}$$

where the term in brackets is an important correction for low-mass targets.

Clearly, high-Z elements produce a stronger backscattered signal than low-Z elements.

Consideration of these factors suggests that specimens for RBS analysis should ideally contain elements of widely different atomic weight stacked with the heavy atoms near the surface and the lighter atoms below them. In bilayer film structures it is desirable for the high-M film to be at the outer surface. Otherwise there is the danger that separate elemental peaks will overlap.

In general, when applicable, RBS can detect concentration levels of about 1 at%. The technique is unmatched in determining the stoichiometry of thin-film binary compounds such as metal silicides where accuracies of $\sim \pm 1\%$ or so are achieved.

6.4.7.4. Spatial and Depth Resolution. Since MeV ion beams can only be focused to a spot size of a millimeter or so in diameter, the lateral spatial resolution of RBS is not great. The depth resolution is commonly quoted to be 200 Å. This can be improved to 20 Å by altering the geometry of detection. Grazing exit angles are employed to make the film appear to be effectively thicker. For example, when $\theta = 95°$, corresponding to an exit angle of 5°, ion scattering at a given film depth means that the energy loss path length is an order of magnitude longer. Implicit in the use of RBS is the desirability that the specimen surface and underlying layered structures be precisely planar. Fortunately, polished Si wafers are extraordinarily flat. Films grown or deposited on Si maintain this planarity and are thus excellently suited for RBS analysis. Films with rough surfaces yield broadened RBS peaks.

The maximum film depth that can be probed depends on the ion used, its energy, and the nature of the matrix. Typically, $\sim 1\ \mu m$ is an upper limit for 2-MeV $^4He^+$. On the other hand, $^3H^+$ beams of 2 MeV penetrate $\sim 5\ \mu m$ deep in Si.

6.4.7.5. Quantification. Despite the limitations noted, RBS enjoys the status of being the preferred method of analysis in situations where it is applicable. The basic reasons are that it is quantitative from first principles, does not require elemental standards, and yields simultaneous depth and film thickness information. Clearly, the area under a spectral peak represents the total number of atoms of a given element present within some continuous region or layer. The peak height (H) is directly proportional to the atomic concentration. The peak width (ΔE) depends on the maximum length traversed by projectile ions in the layer. Therefore, ΔE is directly proportional to the layer or film thickness if the ion energy attenuation with distance is known.

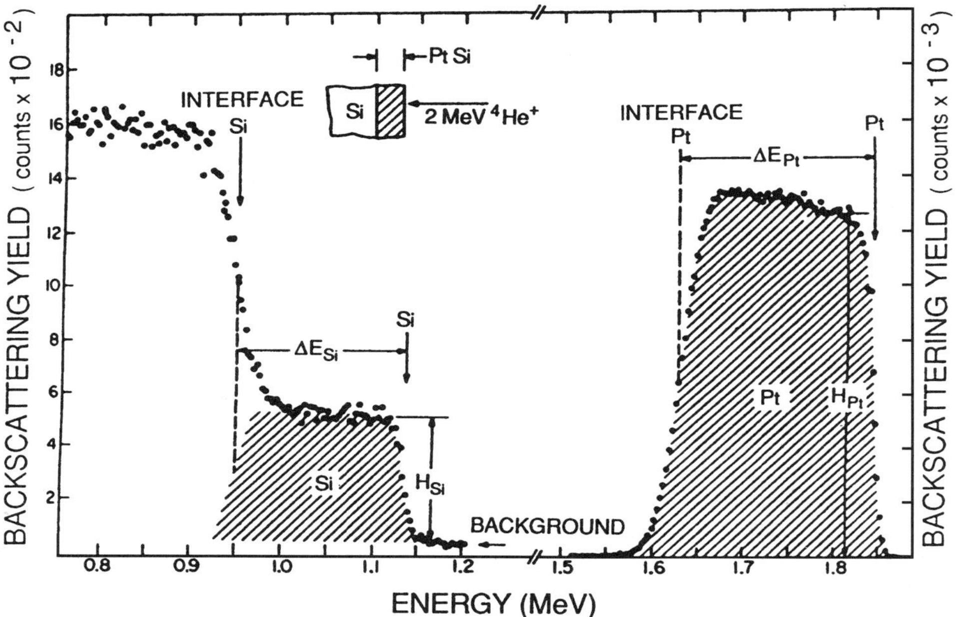

Figure 6-24. Energy spectrum for 2-MeV $^4He^+$ ions backscattered from 900 Å of PtSi on a Si substrate. (Reprinted with permission from Elsevier Sequoia S.A., from W. K. Chu, J. W. Mayer, M. A. Nicolet, T. M. Buck, G. Amsel, and F. Eisen, *Thin Solid Films* **17**, 1, 1963).

Particle range equations derived for nuclear physics applications yield this information to a high degree of accuracy. Hence, RBS is a useful way to determine film thicknesses.

As an exercise in chemical analysis, let us consider the experimental PtSi spectrum (Fig. 6-24) and calculate the relative proportions of Pt and Si present. By analogy to Eq. 6-20, the concentration ratio is given by

$$\frac{C_{\mathrm{Pt}}}{C_{\mathrm{Si}}} = \frac{A_{\mathrm{Pt}}/\sigma_{\mathrm{Pt}}}{A_{\mathrm{Si}}/\sigma_{\mathrm{Si}}}, \tag{6-24}$$

where A_i are the peak areas and the scattering cross sections are, in effect, sensitivity factors. For PtSi, $A_{\mathrm{Pt}}/A_{\mathrm{Si}}$ is measured to be 32, and $\sigma_{\mathrm{Pt}}/\sigma_{\mathrm{Si}}$ is estimated to be equal to $(Z_{\mathrm{Pt}}/Z_{\mathrm{Si}})^2 = 31.8$, so that $C_{\mathrm{Pt}}/C_{\mathrm{Si}} = 1.01$. Alternatively, since $A_i = H_i \Delta E_i$, we have

$$\frac{C_{\mathrm{Pt}}}{C_{\mathrm{Si}}} = \frac{\sigma_{\mathrm{Si}} H_{\mathrm{Pt}} \Delta E_{\mathrm{Pt}}}{\sigma_{\mathrm{Pt}} H_{\mathrm{Si}} \Delta E_{\mathrm{Si}}}. \tag{6-25}$$

Experimental measurement reveals that $H_{Pt}/H_{Si} = 27.5$ and $\Delta E_{Pt}/\Delta E_{Si} = 1.15$, and, therefore, C_{Pt}/C_{Si} is calculated to be 0.99.

Although film thickness can be determined to within 5%, the procedure is somewhat involved. The required ion energy loss with distance data, expressed as stopping cross sections, are tabulated and can be calculated for any matrix (Ref. 13). Film thicknesses can, in principle, also be determined either from known total incident charge and detector solid angle, or through the use of thickness standards. Experimental difficulties, however, limit the utility of these latter methods in practice.

6.4.7.6. Channeling. An interesting effect known as channeling can greatly extend the depth of ion penetration into specially oriented single-crystal film matrices. To understand why, imagine viewing a ball-and-spoke model of a diamond crystal structure along various crystal directions. In many orientations, the model appears impenetrable to impinging ions. But along the [110] direction, a surprisingly large hexagonal tunnel is exposed through which ions can deeply penetrate by undergoing glancing, zig-zag collisions with the tunnel wall atoms. The ion trajectories simply do not bring them close enough to target atoms where they can undergo the nuclear collisions that are particularly effective in slowing them down. Rather, these channeled ions lose energy primarily by electronic excitation of the lattice and therefore range further than if the matrix were, say, amorphous.

Channeling effects can be used advantageously to distinguish RBS spectral features when crystalline, polycrystalline, and amorphous film layers coexist. If, for example, an amorphous Si (a-Si) film covers underlying crystalline Si (c-Si) the $^{4}He^{+}$ yield from Si would abruptly drop to the c-Si value at the a–c interface (see Fig. 13-9). However, during ion implantation doping of semiconductors, channeling effects are unwelcome because the resulting dopant profiles are modified in ways not easily predicted.

6.4.8. Secondary Ion Mass Spectrometry (SIMS) (Ref. 18, 21)

The mass spectrometer, long common in the chemistry laboratory for the analysis of gases has been dramatically transformed in recent years to create SIMS apparatus capable of analyzing the chemical composition of solid surface layers. A critical need to measure thermally diffused and ion-implanted depth profiles of dopants in semiconductor devices spurred the development of SIMS. In typical devices, peak dopant levels are about $10^{20}/cm^3$ while background levels are $10^{15}/cm^3$. These correspond to atomic concentrations in Si of 0.2% to 2×10^{-6}%, respectively. None of the analytical techniques considered thus far has the capability of detecting such low concentration levels. The price paid for this high sensitivity is an extremely complex spectrum of peaks corresponding the the masses of detected ions and ion fragments. This necessitates the use of standards, composed of the specific elements and matrices in question, for quantitative determinations of composition.

In SIMS, a source of ions bombards the surface and sputters neutral atoms, for the most part, but also positive and negative ions from the outermost film layers. Once in the gas phase, the ions are mass-analyzed in order to identify the species present as well as determine their abundance. Since it is the secondary ion emission current that is detected in SIMS, high-sensitivity analysis requires methods for enhancing sputtered-ion yields. Secondary ion emission may be viewed as a special case of (neutral atom) sputtering. However, a comprehensive theory to quantitatively explain all aspects of secondary ion emission (e.g., ion yields S^+ and S^-, escape velocities and angles, dependence on ion projectile and target material, etc.) does not yet exist. Reliable experiments to test proposed theories are difficult to perform. Experimentally, it has been found that different ion beams interact with the specimen surface in profoundly different ways. For example, the positive metal ion yield of an oxidized surface is typically enhanced 10-fold and frequently more relative to a clean surface. This accounts for the common practice of using O_2^- beams to flood the surface when analyzing positive ions. Similarly, the negative ion signals can be enhanced by using Cs^+ primary ion beams.

One of the theories that attempts to explain the opposing effects of O_2^- and Cs^+ beams involves charge transfer by electron tunneling between the target and ions leaving the target surface. Negative ion (O_2^-) bombardment repels charge from the surface, in effect lowering its Fermi energy and raising its effective work function (ϕ). Tunneling is now favored from the surface atom (soon to be ejected positive ion) into the now empty electron states of the target. Similarly, positive ion (Cs^+) bombardment lowers the target work function. Now electrons tunnel from the target into empty levels of surface atoms, enhancing the creation of negative ions. Since these charge transfer processes depend exponentially on ϕ, very large changes in ion yields with small shifts in ϕ are possible.

A schematic depicting the basic elements of a double-focusing SIMS spectrometer is shown in Fig. 6-25. The primary ions most frequently employed

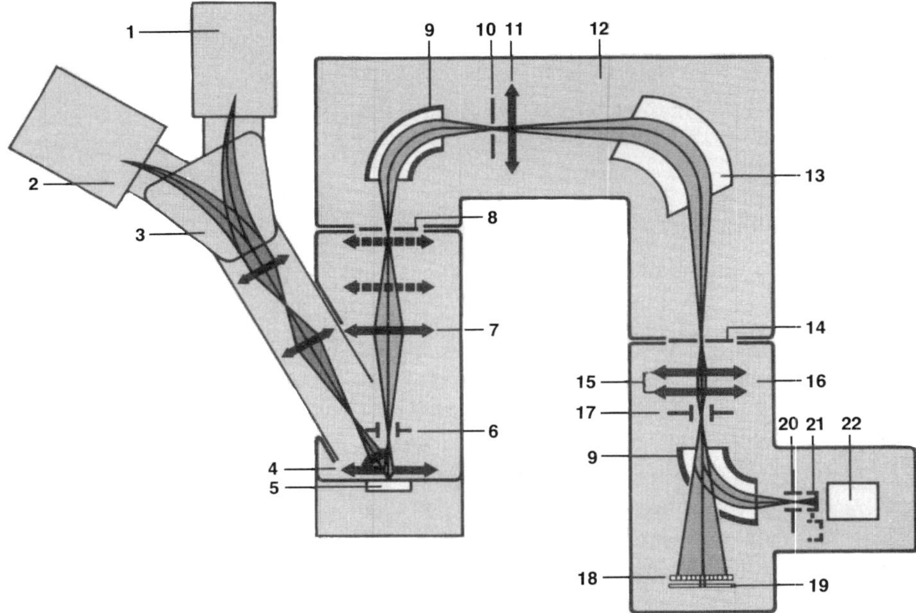

Figure 6-25. Schematic of the ion optical system in the Cameca double-focusing mass spectrometer: 1. Cs ion source; 2. duoplasmatron source; 3. primary beam mass filter; 4. immersion lens; 5. specimen; 6. dynamic transfer system; 7. transfer optical system; 8. entrance slit; 9. electrostatic sector; 10. energy slit; 11. spectrometer lens; 12. spectrometer; 13. electromagnet; 14. exit slit; 15. projection lens; 16. projection display and detection system; 17. deflector; 18. channel plate; 19. fluorescent screen; 20. deflector; 21. Faraday cup; 22. electron multiplier. (Courtesy Cameca Instruments, Inc., Stamford, Connecticut).

are Ar^+, O_2^-, and Cs^+, and these are focused into a beam ranging from 2 to 15 keV in energy. The sputtered charged atoms and compound fragments are extracted and enter an electrostatic energy analyzer similar to the cylindrical and spherical electron energy analyzers employed in AES and XPS work. Those secondary ions that pass now enter a magnetic sector mass filter whose function is to select a particular mass for detection. The desired ion of mass M, charge q, and velocity v traces an arc of radius r in the magnetic field (B) of the electromagnet, given by

$$r = Mv/qB. \tag{6-26}$$

Table 6-4. Comparison between Static and Dynamic SIMS

Variable	Static SIMS	Dynamic SIMS
Residual gas pressure (torr)	10^{-9}–10^{-10}	10^{-7}
Primary ion energy (keV)	0.5–3	3–20
Current density (A/cm^2)	10^{-9}–10^{-6}	10^{-5}–10^{-2}
Area of analysis (cm^2)	0.1	10^{-4}
Sputter rate (sec/monolayer)	10^4	1

Two modes of operation are possible. In the imaging mode, B is fixed, and simultaneous projection of ions of different mass on the channel plate yields mass spectra maps in the form of an ion micrograph (spectro*graphy*). In the mass spectro*metry* mode, B is scanned, and each ion mass is sequentially detected by the electron multiplier detector. Magnetic sector mass filters possess high-resolution capability ($M/\Delta M \approx$ 3000–20,000). Another common version of SIMS employs a quadrupole lens of lower resolution; i.e., $M/\Delta M \approx 500$.

A further distinction is made between what is known as "static" and "dynamic" SIMS. The issue that distinguishes them is the rate of specimen erosion relative to the time necessary to acquire data. Static SIMS requires that data be collected before the surface is appreciably modified by ion bombardment. It is well suited to surface analysis and the detection of contaminants such as hydrocarbons. Dynamic SIMS, on the other hand, implies that high sputtering rates are operative during measurement. This, of course, enables depth profiling of surface layers. Typical operating parameters for both static and dynamic SIMS are given in Table 6-4.

One of the unique features of SIMS is its mass discrimination, which often provides interesting processing information. For example, consider a bipolar transistor that contains both ^{121}Sb and ^{123}Sb in the emitter but only ^{121}Sb in the collector. The emitter doped with both naturally occurring isotopes of antimony must have been diffused while the collector was obviously ion-implanted. Extreme discrimination is required when two species of very similar mass must be distinguished. For example, to separate the ^{31}P (mass = 30.974) signal from that of the interfering ^{30}SiH (mass = 30.982) background level, the resolution required is $M/\Delta M = 3872$. With high-resolution mass spectrometers, these species have been separated and phosphorous doping profiles obtained as shown in Fig. 6-26.

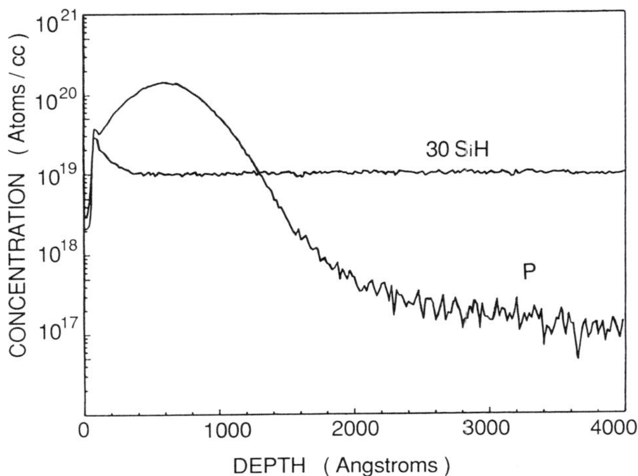

Figure 6-26. SIMS depth profile of P implanted in Si obtained with Cs^+ ions. (Courtesy of H. Luftman, AT&T Bell Laboratories.)

6.4.9. Applications

The chapter closes with the self-explanatory Table 6-5, in which some applications of the surface analytical techniques just considered are listed. The varied phenomena, materials, and structures involved extend into all facets of thin-film science and technology. Ever-expanding applications will ensure future growth of the list.

Exercises

1. Assume you are given samples of the following items:

a. a glass camera lens coated with a purple-colored film.
b. a steel drill coated with a gold-colored metallic layer.
c. a plastic potato chip bag coated with a thin film.

In each case how would you experimentally determine the film or coating composition and thickness?

2. Material deposits onto a quartz crystal thickness monitor from two separate evaporation sources, A and B. Explain possible difference in the

Table 6-5. Applications of Surface Analytical Techniques in Thin Films

Application	Information Obtained, Comments	Technique
1. Nucleation and growth	Distinctions among island, layer and S.K. growth modes (p. 220)	AES
2. Diffusion in metal films	Diffusion coefficients are obtained by sputter sectioning, and surface accumulation methods	AES, SIMS, RBS
3. Doped semiconductors	Diffused and ion-implanted depth profiles in Si and GaAs	SIMS
4. Compound formation	Stoichiometry of intermetallic and silicide compounds, growth kinetics	RBS, AES
5. Investigation of surface residues, stains, haze, and discoloration after processing	Identification of elemental contaminants	AES, SIMS, XPS
	Identification of compound valence and bonding states	XPS
6. Contamination of surfaces by organic materials	Identification of elements	AES, SIMS XPS
	Identification of compounds	XPS
7. Interfacial analysis	Cause of adhesion failure, segregation of impurities at grain boundaries and interfaces	AES, RBS, SIMS, XPS
8. Multilayer films and coatings, superlattices	Stoichiometry, layer thickness, interfacial impurities	AES, RBS, SIMS
9. Determination of crystalline perfection	Channeling spectra distinguish between single-crystal and amorphous Si films	RBS
10. Fracture of coatings	Segregation of impurities at fracture surfaces	AES, SIMS
11. Metal–Semiconductor contacts	Adhesion, contact reactions in Si and GaAs	AES, XPS, SIMS, RBS
12. Dielectric films on metals and semiconductor	Surface contamination, impurity diffusion, and segregation at interfaces (e.g., SiO_2–Si)	AES, SIMS, XPS
13. Molecular-beam epitaxy	Assessment of surface cleanliness prior to deposition, detection of C and O contaminants	AES, XPS

frequency shift if

a. a bilayer film deposits, i.e., first mass m_A from source A, then mass m_B from source B.
b. an alloy film deposits, i.e., m_A and m_B deposit simultaneously. Is the final film thickness the same in each case? Make any assumptions you wish.

3. After monitoring the thickness of a deposited Au film with a 6.0-MHz quartz (AT cut) crystal monitor, a researcher decides to confirm his results employing interferometry. A frequency shift of 10.22 Hz was recorded for the film measuring 1.00 cm^2 in the area. Interferometry with the Hg green line revealed a displacement of 1.75 fringes across the film step. Are these measurements consistent? If not, suggest plausible reasons why not?

[Note: Density of Au is 19.3 g/cm^3.]

4. a. Contrast the optical interference effect when viewing an SiO_2 step gauge in white light and in monochromatic light.
b. Similar step gauges have been prepared for native oxides on GaAs and GaP substrates. How would you account for color differences between these semiconductors coated with oxides of the same thickness?

5. Based on the diffraction pattern of Fig. 6-13, what are the lattice parameters for Ag and Pb? [Note: $\lambda_{CuK\alpha} = 1.54$ Å.]

6. Recommend specific structural and chemical characterization techniques for analyzing the following samples. In each case indicate potential difficulties in the analysis.

a. The surface of a glass slide.
b. A 1000-Å-thick layer of crystalline SiO_2 buried in crystalline Si 3000 Å below the surface.
c. A thin film of a Bi–Th alloy. (Thorium is radioactive emitting high-energy α, β, and γ radiation.)
d. A thin carbon deposit on a Be substrate.
e. A "black" Al film surface consisting of a deeply creviced, rough moundlike topography. (Surface features are typically 1 μm in size.)
f. A liquid Ga film surface at 35 °C.
g. Al_2O_3 powder ~ 0.8 μm in size.

7. Explain how an SEM/EDX facility could be used to measure the thickness of film A without detaching it from substrate B (film thickness standards may be necessary). Can the film thickness be measured this way if film and substrate have the same composition?

8. Auger electrons emanating from A atoms located a depth z below the surface give rise to a signal of intensity

$$I_A = K \int_0^{\infty} C(z) e^{-z/\lambda_e} \, dz,$$

where λ_e is the electron escape depth, $C(z)$ is the concentration of A atoms, and K is a constant.

a. What is the intensity from a pure A material?

b. A pure A substrate is covered by a film of thickness d so that $C_A(z) = 0$ for $d \geq z$, and $C_A(z) = 1$ for $\infty \geq z \geq d$. Show that $I_A = I_A(0)\exp - d/\lambda_e$, where $I_A(0)$ is the signal intensity from a pure A surface. Does this suggest a way to determine d?

9. The surface of a film contained the following elements with the indicated sensitivities, $\bar{S}$, and Auger intensities, I (in arbitrary units)

Element	$\bar{S}$	I
Ga	0.68	6,950
As	0.68	5,100
Al	0.23	3,040
O	0.71	26,900
C	0.30	5,000
F	1.0	40,500
Cl	0.89	3,700

What is the composition of the surface in atomic percent? Are the binding energies of Ga in the compounds GaF_3, Ga_2O_3, and GaAs (Fig. 6-20) consistent with what you know about the chemistry of these materials?

10. a. Sketch the RBS spectrum for a 900-Å-thick Pt film on a Si wafer. How does it differ from the spectrum of 900 Å of PtSi on Si? In both cases 2.0-MeV $^4He^+$ ions are employed.

b. Sketch the RBS spectrum for the case of a 900-Å Si film on a thick Pt substrate.

11. Consider the RBS spectra of Fig. 8-12.

a. Calculate the value of σ_{Au}/σ_{Al} from the initial Au and Al data. How does it compare with the value obtained from Eq. 6-23? Assume $\theta = 170°$.

b. If the initial order of film stacking were reversed, what would the resulting RBS spectrum look like initially and after annealing?

c. Are the stoichiometries of Au_2Al and $AuAl_2$ consistent with the value for $\sigma_{Au} / \sigma_{Al}$?

12. Refer to the RBS spectra of Fig. 13-9.

a. What is the significance of the yield counts of $\sim$ 1000 and 3000 per channel in each Si spectrum?
b. What is the width of the a-Si layer after 15 min at 515 °C?
c. How does the total number of Au atoms partitioned in a-Si change as a function of annealing time?
d. From the width and height of spectral features calculate the solubility of Au in a-Si after 85 min.

13. The implanted P dopant distribution shown in the SIMS spectrum of Fig. 6-26 can be described by the equation

$$C(z) = \frac{\phi}{\sqrt{2\pi}\,\Delta R_p} \exp - \left(\frac{z - R_p}{\sqrt{2}\,\Delta R_p} \right)^2, \qquad (13\text{-}24)$$

with terms defined on p. 613.

a. Show that the distribution appears to be Gaussian by replotting the data in a form suggested by Eq. 13-24; i.e., $\ln C$ vs. $(z - R_p)^2$.
b. What is the value of R_p, the projected range? What is the value for ΔR_p, the longitudinal straggle?
c. What is the dose ϕ?

14. An Ar primary ion beam of 1 keV energy is used to sputter-etch Cu during AES depth profiling. The ion current is 10^{-8} A, and the area sputtered is 0.5 cm × 0.5 cm.

a. Predict the sputter rate of Cu in units of monolayers/min. Assume the (100) surface is exposed, and the lattice parameter of Cu = 3.61 Å.
b. The beam energy is raised to 10 keV where the sputter yield is 6.25. Estimate the rate of Cu removal at a current of 10^{-6} A.

15. Moseley's law for $K\alpha$ X-ray emission lines suggests a correlation between X-ray energy E and atomic number Z; i.e.,

$$E = (3/4)\mathrm{Ry}\, ch(Z - 1)^2,$$

where Ry = Rydberg constant (Ry = 1.0974×10^5 cm^{-1})
c = speed of light
h = Planck constant

a. Calculate the energy of the Kα line for Ti.
b. Make a plot of $\sqrt{E}$ vs. Z (Moseley diagram) utilizing the data of Fig. 6-16. Calculate the slope of the line.
c. Relative to Ti how are the spectral lines of Fe positioned in the energy-level scheme?

References

1.* J. B. Bindell, in *VLSI Technology*, 2nd ed., ed. S. M. Sze, McGraw-Hill, New York (1988).
2.* W. A. Pliskin and S. J. Zanin, in *Handbook of Thin Films Technology*, eds. L. I. Maissel and R. Glang, McGraw-Hill, New York (1970).
3.* K. L. Chopra, *Thin Film Phenomena*, McGraw-Hill, New York (1969).
4.* H. K. Pulker, *Coatings on Glass*, Elsevier, Amsterdam (1984).
5. S. Tolansky, *Multiple Beam Interference Microscopy of Metals*, Academic Press, London (1970).
6.* L. I. Maissel and M. H. Francombe, *An Introduction to Thin Films*, Gordon and Breach, New York (1973).
7. J. B. Theeten and D. E. Aspenes, *Ann. Rev. Mater. Sci.* **11**, 97 (1981).
8. R. Glang, in *Handbook of Thin-Film Technology*, eds. L. I. Maissel and R. Glang, McGraw-Hill, New York (1970).
9.* J. I. Goldstein, D. E. Newbury, P. Echlin, D. C. Joy, C. Fiori, and E. Lifshin, *Scanning Electron Microscopy and X-Ray Microanalysis*, Plenum, New York (1981).
10.* G. Thomas and M. J. Goringe, *Transmission Electron Microscopy of Materials*, Wiley, New York (1979).
11.* R. B. Marcus and T. T. Sheng, *Transmission Electron Microscopy of Silicon VLSI Circuits and Structures*, Wiley, New York (1983).
12. K. N. Tu, *J. Appl. Phys.* **43**, 1303 (1972).
13.* L. C. Feldman and J. W. Mayer, *Fundamentals of Thin-Film Analysis*, North-Holland, New York (1986).
14.* D. Briggs and M. P. Seah, eds., *Practical Surface Analysis by Auger and Photoelectron Spectroscopy*, Wiley, New York (1984).
15. A. W. Czanderna, ed., *Methods of Surface Analysis*, Elsevier, Amsterdam (1975).

*Recommended texts or reviews.

16. H. Windawi and F. F.-L. Ho, eds., *Applied Electron Spectroscopy for Chemical Analysis*, Wiley, New York (1982).
17.* J. R. Bird and J. S. Williams, eds., *Ion Beams for Materials Analysis*, Academic Press, Sydney (1989).
18.* A. W. Benninghoven, F. G. Rudenauer, and H. W. Werner, *Secondary Ion Mass Spectrometry—Basic Concepts, Instrumental Aspects, Applications and Trends*, Wiley, New York (1987).
19.* H. W. Werner and P. P. H. Garten, *Rep. Prog. Phys.* **47**, 221 (1984).
20. W. K. Chu, J. W. Mayer, M. A. Nicolet, T. M. Buck, G. Amsel, and F. Eisen, *Thin Solid Films* **17**, 1 (1963).
21. F. Degreve, N. A. Thorne, and J. M. Lang, *J. Mat. Sci.* **23**, 4181 (1988).

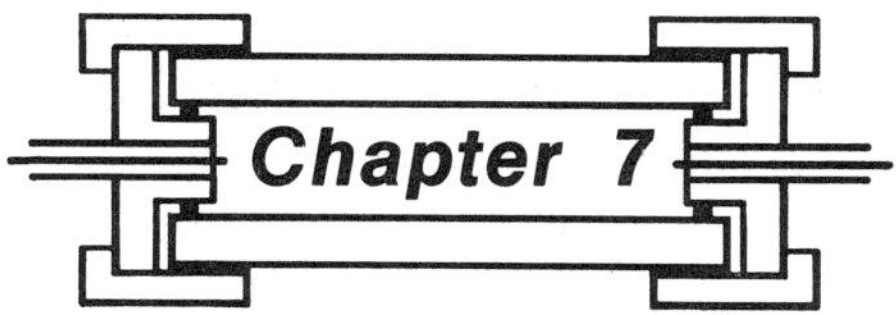

Epitaxy

7.1. Introduction

Two ancient Greek words $\varepsilon\pi\iota$ (*epi*—placed or resting upon) and $\tau\alpha\xi\iota\zeta$ (*taxis*—arrangement) are the root of the modern word *epitaxy*, which describes an extremely important phenomenon exhibited by thin films. Epitaxy refers to extended single-crystal film formation on top of a crystalline substrate. It was probably first observed to occur in alkali halide crystals over a century ago, but the actual word *epitaxy* was apparently introduced into the literature by the French mineralogist L. Royer in 1928 (Ref. 1). For many years the phenomenon of epitaxy continued to be of scientific interest to numerous investigators employing vacuum evaporation, sputtering, and electrodeposition. A sense of much of this early work on island growth systems (e.g., metal films on alkali halide substrates) was given in Chapter 5. Over the past two decades, epitaxy has left the laboratory and assumed crucial importance in solid-state device processing. Interest has centered on epitaxial films exhibiting layer growth. This chapter focuses primarily on such films as well as on several broader issues related to epitaxy.

Two types of epitaxy can be distinguished and each has important scientific and technological implications. *Homoepitaxy* refers to the case where the film and substrate are the same material. Epitaxial (epi) Si deposited on Si wafers is the most significant example of homoepitaxy. In fact, one of the first steps in the fabrication of bipolar and some MOS transistors is the CVD vapor phase epitaxy (VPE) of Si on Si (see Chapter 4). The reader may well ask why the underlying Si wafer is not sufficient; why must the single-crystal Si be extended by means of the epi film layer? The reason is that the epilayer is generally freer of defects, purer than the substrate, and can be doped independently of the wafer. A dramatic improvement in the yield of early bipolar transistors was the result of incorporating the epi-Si deposition step. The second type of epitaxy is known as *heteroepitaxy* and refers to films and substrates composed of different materials, e.g., AlAs deposited on GaAs. Heteroepitaxy is, of course, the more common phenomenon. Optoelectronic devices such as light-emitting diodes and lasers are based on compound semiconductor heteroepitaxial film structures.

The differences between the two basic types of epitaxy are schematically illustrated in Fig. 7-1. When the epilayer and substrate crystal are identical, the lattice parameters are perfectly matched and there is no interfacial bond straining. In heteroepitaxy the lattice parameters are necessarily unmatched, and, depending on the extent of the mismatch, we can envision three distinct epitaxial regimes. If the lattice mismatch is very small, then the heterojunction interfacial structure is essentially like that for homoepitaxy. However, differences in film and substrate chemistry and coefficient of thermal expansion can strongly influence the electronic properties and perfection of the interface. Small lattice mismatch is universally desired and actually achieved in a number of important applications through careful composition control of the materials

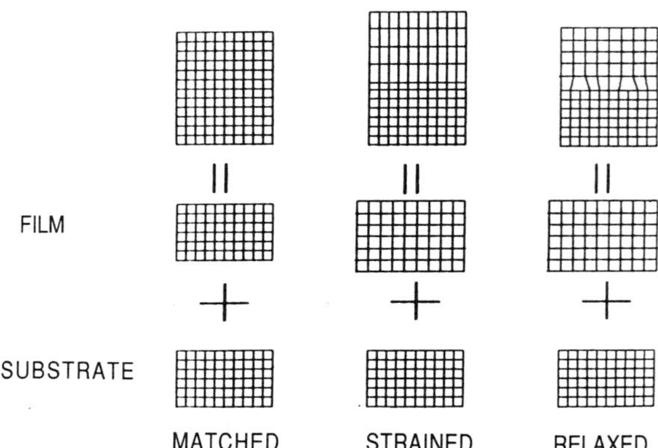

Figure 7-1. Schematic illustration of lattice-matched, strained, and relaxed heteroepitaxial structures. Homoepitaxy is structurally very similar to lattice-matched heteroepitaxy.

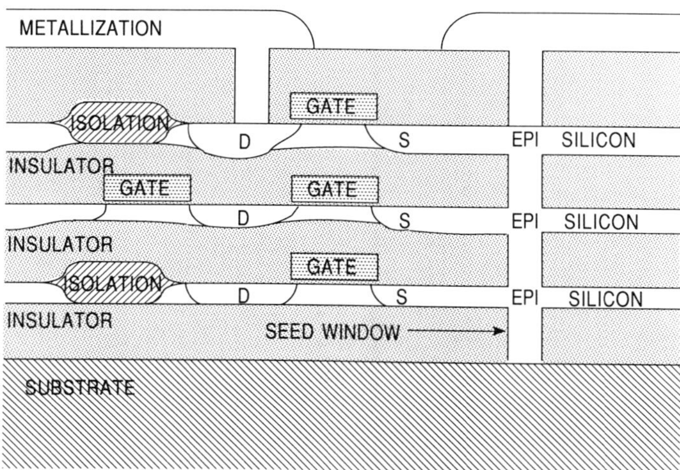

Figure 7-2. Cross-sectional model of a three-dimensional integrated circuit.

involved. Section 7.4 deals with such epitaxial interfaces in compound semiconductors and the devices based on them.

When the film and substrate lattice parameters differ more substantially, we may imagine the other cases in Fig. 7-1. Either edge dislocation defects form at the interface, or the two lattices strain to accommodate their crystallographic differences. The former situation (relaxed epitaxy) generally prevails during later film formation stages irrespective of crystal structure or lattice parameter differences. The latter case is the basis of strained-layer heteroepitaxy. This phenomenon occurs between film–substrate pairs composed of different materials that have the same crystal structure. Lattice parameter differences are an order of magnitude larger than in the case of lattice-matched heteroepitaxy. Structures consisting of Ge_xSi_{1-x} films grown on Si, currently under active research study, are important examples of strained layer epitaxy, and will be discussed further in Section 7.3.

Recently, there has been a great deal of exciting research devoted to both the basic science of epitaxy and its engineering applications. One area has addressed the dream of creating three-dimensional integrated circuits possessing intrinsically high device packing densities. Rather than a single level of processed devices, a vertical multifloor structure can be imagined with each level of devices separated from neighboring ones by insulating films. What is crucial is the ability to grow an epitaxial semiconductor film on top of an amorphous substrate—e.g., Si on SiO_2 as shown in Fig. 7-2 . This will require selective nucleation of epi Si at existing crystalline Si, and nowhere else, followed by lateral growth across surfaces that are ill suited to epitaxy.

Methods to achieve Si on insulator (SOI) epitaxy are mentioned in Section 7.5. A second area involves the fabrication of multilayer heterojunction composites. These remarkable epitaxial film structures include superlattices and quantum wells. Some of their simple properties together with applications involving incorporation into actual devices will be deferred until Chapter 14. The remainder of this chapter is divided into the following major sections:

7.2. Structural Aspects of Epitaxial Films
7.3. Lattice Misfit and Imperfections in Epitaxial Films
7.4. Epitaxy of Compound Semiconductors
7.5. Methods for Depositing Epitaxial Semiconductor Films
7.6. Epitaxial Film Growth and Characterization

7.2. Structural Aspects of Epitaxial Films

7.2.1. Single-Crystal Surfaces

Prior to consideration of epitaxial films, it is instructive to examine the nature of the topmost surface layers of a crystalline solid film. The reason the surface will generally have different properties than the interior of the film can be understood by a schematic cross-sectional view as shown in Fig. 7-3. If the surface structure is the predictable extension of the underlying lattice, we have the case shown in Fig. 7-3a. The loss of periodicity in one direction will tend to alter surface electronic properties and leave dangling bonds to promote chemical reactivity. It is more likely though that the structure shown in Fig. 7-3b will prevail. The absence of bonding forces to underlying atoms results in new equilibrium positions that deviate from those in the bulk lattice. A disturbed surface layer known as the "selvedge" may then be imagined. Within this layer the atoms relax in such a way as to preserve the symmetry of the bulk lattice parallel to the surface but not normal to it. One result of this, for example, could be a surface electric dipole moment in the selvedge. A more extreme structural disturbance is depicted in Fig. 7-3c. Here surface atoms rearrange into a structure with a symmetry that is quite different from the bulk solid. This phenomenon is known as *reconstruction* and can significantly alter many surface structure-sensitive properties, e.g., chemical, atomic vibrations, electrical, optical, and mechanical.

Surface reconstruction is quite common on semiconductor surfaces; perhaps the most famous example occurs on the (111) surface of Si. A cut through the covalent bonds of the bulk (111) plane to create two exposed surfaces leaves

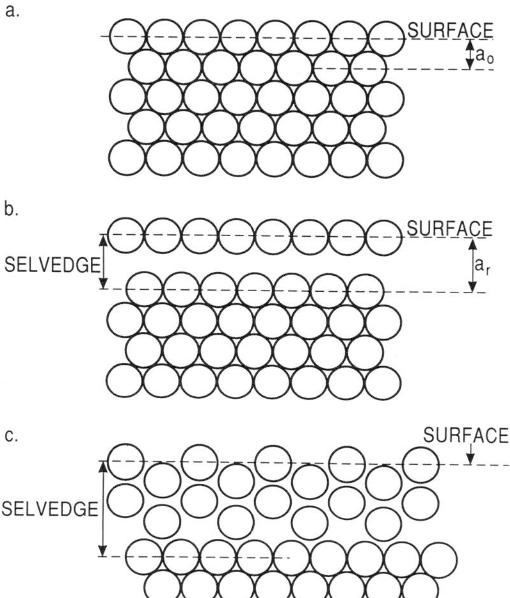

Figure 7-3. Schematic cross-sectional views of close-packed atomic positions at a solid surface: (a) bulk exposed plane; (b) atomic relaxation outward; (c) reconstruction of outer layers. (From Ref. 2, © Oxford University Press, by permission).

covalent bonds dangling normal to the surface into the vacuum. Dangling bonds are energetically unfavorable, and the surface reduces its overall energy by reconstructing in a manner that reduces the number and/or energy of the bonds.

A direct atomic image of the reconstructed surface of (111) Si, obtained by scanning tunneling microscopy (STM), is shown in Fig. 7-4. The Nobel Prize in physics was awarded to the developers of STM, G. Binnig and H. Rohrer, in 1986. The technique involves a highly controlled raster-fashion translation of a metal tip possessing an extremely small radius of curvature (< 1000 Å), over the atomic terrain of a surface. Because the tip is only tens of angstroms from the surface, a tunneling current inversely proportional to this distance, and varying directly with the topography of the surface atoms, flows. This signal is recorded and ultimately converted into an image. STM may be thought or as the quantum mechanical analog of the stylus method for measuring film thickness (see Section 6.2.3).

Two other important techniques for analyzing the structure of crystalline surfaces and epitaxial films in particular are low-energy electron diffraction (LEED) and reflection high-energy electron diffraction (RHEED). Both will be

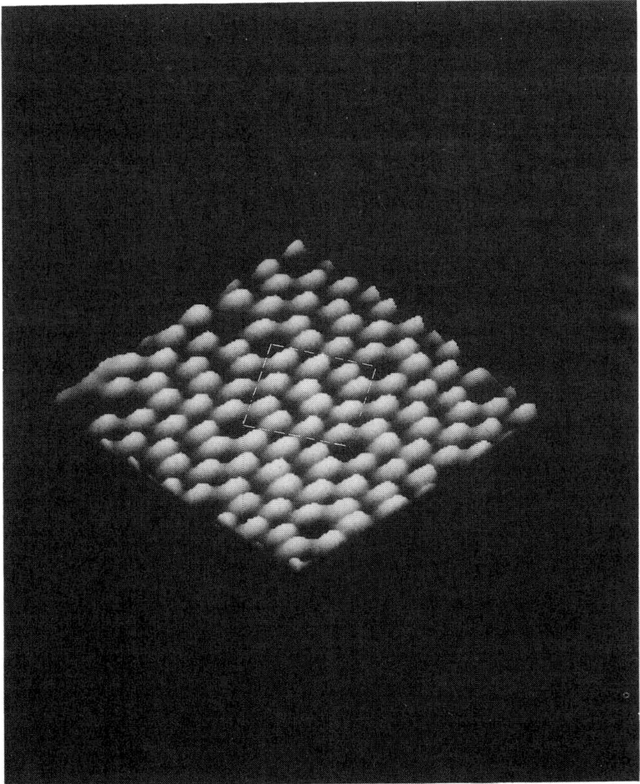

Figure 7-4. Scanning tunneling microscope image of the reconstructed (7 × 7) surface of (111) Si. (Courtesy of Y. Kuk, AT&T Bell Laboratories.)

described in Section 7.6.2 in connection with MBE, but before that we must attend to the two-dimensional geometric arrangements of atoms on a crystalline surface and the notation used to identify them.

7.2.2. Surface Crystallography

Just as there are 14 Bravais lattices in three dimensions, so there are just five unit meshes or nets, corresponding to a two-dimensional surface as shown in Fig. 7-5. Points representing atoms may be arranged to outline (1) squares, (2) rectangles, (3) centered rectangles, (4) hexagons and (5) arbitrary parallelograms. Miller-type indices are used to denote atom coordinates, directions, and distances between lines within the surface.

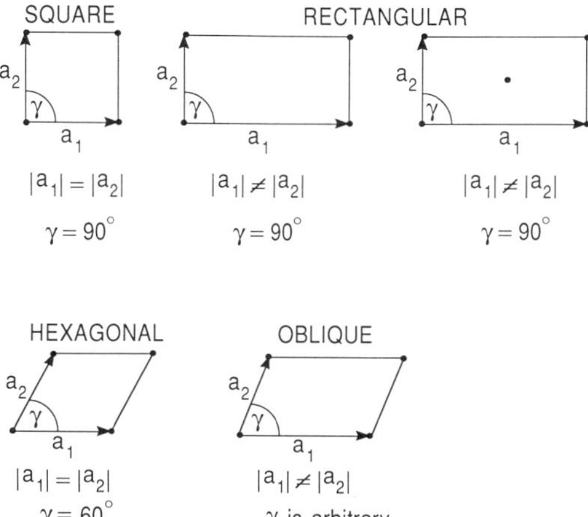

Figure 7-5. The five diperiodic surface nets.

Consider the mesh of substrate atoms in Fig. 7-6a with an array of adatoms situated on the surface as indicated. This combination could, for example, correspond to the early growth of an epitaxial layer on the (100) plane of a BCC crystalline substrate surface. The adsorbate atoms (shaded in) form a rectangular monolayer lattice or overgrowth above the substrate atom positions (dots). Unit dimension vectors describing the monolayer lattice (b_i) are simply related to those of the substrate lattice (a_i). In the x direction $b_1 = 2a_1$, and in the y direction $b_2 = 1a_2$. We, therefore, speak of a $P(2 \times 1)$ overlayer where P indicates that the unit cell is primitive. Similarly, for an overlayer of the same geometry but oriented at 90° with respect to the first case, the notation is $P(1 \times 2)$. Other examples are shown in Fig. 7-6b, c. Note that in Fig. 7-6c the overgrowth layer is rotated with respect to the substrate coordinates and is identified by the letter R. Similarly, C is used to denote the centered lattice.

It is the geometry of the reciprocal lattice, however, and not that of the real lattice, that appears as the visible image in diffraction patterns. As the name implies, the reciprocal lattice has dimensions and features that are "inverse" to those of the real lattice. Long dimensions in real space appear at right angles and are shortened in inverse proportion, within the reciprocal space. The relation between a_1 and a_2 and the reciprocal vectors a_1^* and a_2^* of the unit mesh in the reciprocal lattice, is expressed by

$$a_i \cdot a_j^* = 2\pi\delta_{ij}, \tag{7-1}$$

Similarly,

$$b_i \cdot b_j^* = 2\pi\delta_{ij}. \tag{7-2}$$

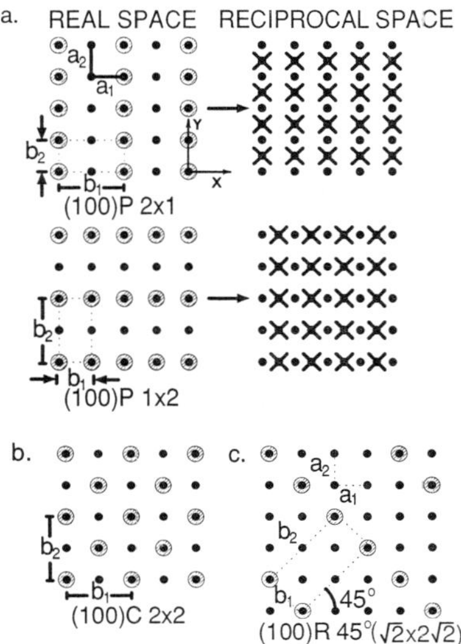

Figure 7-6. (a) Atomic positions of adatoms (shaded) relative to (100) substrate or bulk atoms (dots). Corresponding diffraction patterns are shown on the right. (X refers to adatom lattice.) (b) (100) C(2 × 2) surface structure. (c) (100) R45° ($\sqrt{2} \times 2\sqrt{2}$) surface structure.

where $\delta_{ij} = 0$ if $i \neq j$ and $\delta_{ij} = 1$ if $i = j$. Reciprocal lattices corresponding to the real-space structures of Fig. 7-6a are sketched intuitively in the same figure. The scale is arbitrary, but the lattice periodicity and symmetry are preserved. As a final example, consider what happens when two orientations, $P(2 \times 1)$ and $P(1 \times 2)$, are admixed in roughly equal proportions. The reciprocal lattices from each simply superimpose. It is left as an exercise for the reader to sketch the resultant pattern.

7.2.3. Epitaxial Interface Crystallography

To address issues dealing with the structure of epitaxial interfaces it is first necessary to identify the crystallographic orientation relationships between the film and the substrate. Unlike the notation used to describe the two-dimensional surfaces of the previous section, the traditional (3-D) Miller indices are employed here. For this purpose, the indices of the overgrowth plane are written as (HKL), and those of the parallel substrate plane at the common interface are taken as (hkl). The corresponding parallel directions in the overgrowth and substrate planes, denoted by $[UVW]$ and $[uvw]$, respectively, must also be specified. This tetrad of indices written by convention as

$(HKL) \parallel (hkl)$; $[UVW] \parallel [uvw]$ serves to define the epitaxial geometry. For example, in the case of parallel epitaxy of Ni on cleaved NaCl, the notation would read (001) Ni ‖ (001) NaCl; [100] Ni ‖ [100] NaCl. In this case both planes and directions coincide. For (111) PbTe ‖ (111) $MgAl_2O_4$; $[\bar{2}11]$ PbTe ‖ $[\bar{1}01]$ $MgAl_2O_4$, the interfacial plane is common, but the directions are not. Frequently, the epitaxial relationships can be predicted on the basis of lattice-fitting arguments. Those planes and directions that give the best lattice fit determine the orientation of the film with respect to the substrate.

As an example, consider the growth of a (110) Fe (BCC) film on a (110) GaAs (zinc blende) substrate. The unit cell lattice parameters for Fe and GaAs are 2.866 Å and 5.653 Å, respectively, suggesting that two Fe cells could be accommodated by one of GaAs. A view of the (110) plane of these materials is shown in Fig. 7-7. The resulting epitaxial geometry is denoted by (110) Fe ‖ (110) GaAs; [200] Fe ‖ [100] GaAs.

An important quantity that characterizes epitaxy is the lattice misfit, f, of the film defined as

$$f = (a_0(s) - a_0(f))/a_0(f) = \Delta a_0/a_0 \tag{7-3}$$

where $a_0(f)$ and $a_0(s)$ refer to the *unstrained* lattice parameters of film and substrate, respectively. A positive f implies that the initial layers of the epitaxial film will be stretched in tension, and a negative f means film compression. In the Fe–GaAs system the misfit in the [001] direction is

$$f = \frac{a_0(\mathrm{GaAs}) - 2a_0(\mathrm{Fe})}{2a_0(\mathrm{Fe})} = \frac{5.653 - 2(2.866)}{2(2.866)} = -0.0138.$$

or 1.38%. High-quality epitaxial Fe films have been deposited on GaAs, facilitating fundamental studies of ferromagnetism (Ref. 3).

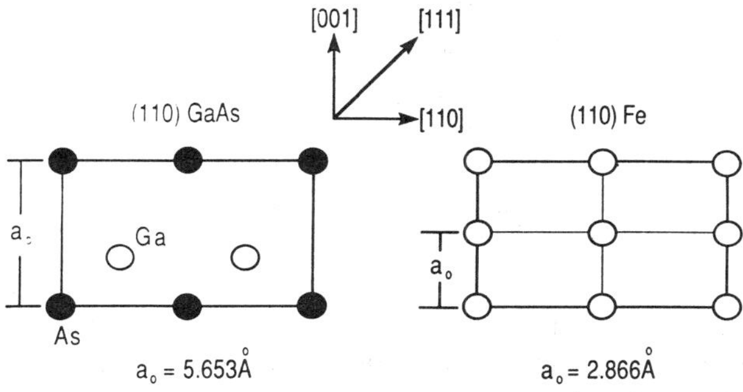

Figure 7-7. Surface nets of (110) planes of GaAs and α-Fe. (From Ref. 3).

7.3. Lattice Misfit and Imperfections in Epitaxial Films

7.3.1. Lattice Misfit

In this section we explore some implications of lattice misfit on the perfection of epitaxial films. The basic theory that accounts for the elastic–plastic changes in the bilayer was introduced by Frank and van der Merwe (Ref. 4). It attempts to account for the accommodation of misfit between two lattices rather than being a theory of epitaxy per se. The theory predicts that any epitaxial layer having a lattice parameter mismatch with the substrate of less than $\sim 9\%$ would grow pseudomorphically; i.e., for very thin films the deposit would be elastically strained to have the same interatomic spacing as the substrate. The interface would, therefore, be coherent with atoms on either side lining up. With increasing film thickness, the total elastic strain energy increases, eventually exceeding the energy associated with a relaxed structure consisting of an array of so-called misfit dislocations separating wide regions of relatively good fit. At this point, the initially strained film would ideally decompose to this relaxed structure where the generated dislocations relieve a portion of the misfit. As the film continues to grow, more misfit is relieved until at infinite thickness the elastic strain is totally eliminated. In the case of epitaxial growth without interdiffuson, pseudomorphism exists only up to some critical film thickness d_c, beyond which misfit dislocations are introduced. Misfit dislocations lie in planes parallel to the interface, and they can be observed by transmission electron microscopy. For example, early epitaxial growth of $CoSi_2$ films on Si is accompanied by the honeycomb array of misfit dislocations shown in the TEM image of Fig. 7-8.

Following Matthews (Refs. 5, 6), an expression for d_c can be calculated by minimizing the sum of the elastic strain energy E_ε (per unit area) and dislocation energy E_d (per unit area) with respect to the film strain ε_f. Assuming that the film and substrate shear moduli, μ, are the same, we have approximate expressions for E_ε and E_d:

$$E_\varepsilon = \frac{2\mu(1+\nu)}{1-\nu}\varepsilon_f^2 d, \tag{7-4}$$

$$E_d = \frac{\mu b(f-\varepsilon_f)}{2\pi(1-\nu)}\ln\left(\frac{R_0}{b}+1\right), \tag{7-5}$$

where d is the film thickness, ν is Poisson's ratio, b is the dislocation Burgers vector, and R_0 is a radius about the dislocation where the strain field terminates. Physically these equations indicate that strain energy is a *volume*

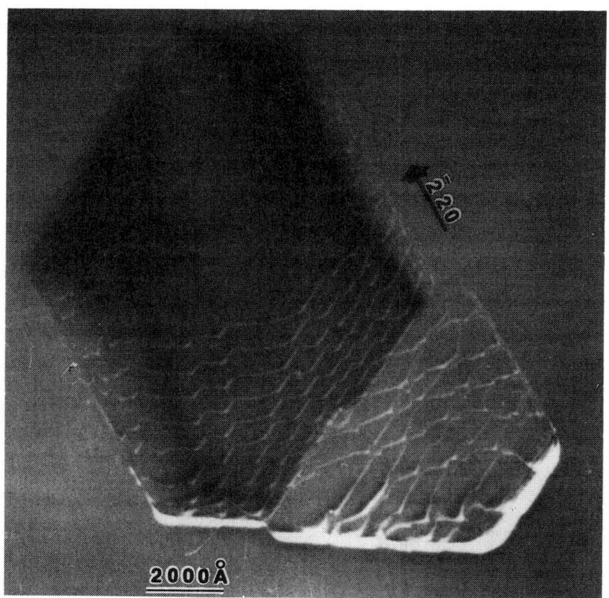

Figure 7-8. Misfit dislocations arising from reaction between 70 Å Co and Si at 880 °C, in ultrahigh vacuum. (TEM (weak beam 220) image courtesy of J. M. Gibson, AT&T Bell Laboratories.)

energy that increases linearly with film thickness. In contrast, dislocation energy is nearly constant with only a weak dependence on d arising from R_0. Therefore, at some value of d, dislocations are favored. Taking the derivative of $E_\varepsilon + E_d$ with respect to ε_f and setting it equal to zero gives for the critical strain a value

$$\varepsilon_f^* = \frac{b}{8\pi(1+\nu)d}\ln\left(\frac{R_0}{b}+1\right). \tag{7-6}$$

The largest possible value for ε_f^* is f. If the value of ε_f^* predicted by Eq. 7-6 is larger than f, the film will strain to match the substrate exactly, in which case ε_f^* will equal f, and E_d will be zero. If $\varepsilon_f^* < f$, a portion of the misfit equal to $f - \varepsilon_f^*$ will be accommodated by misfit dislocations. By assuming $\varepsilon_f^* \approx f$ at $d = d_c$ and that $R_0 \approx d_c$, the critical film thickness prior to misfit dislocation formation is expressed by

$$d_c = \frac{b}{8\pi(1+\nu)f}\ln\left(\frac{d_c}{b}+1\right). \tag{7-7}$$

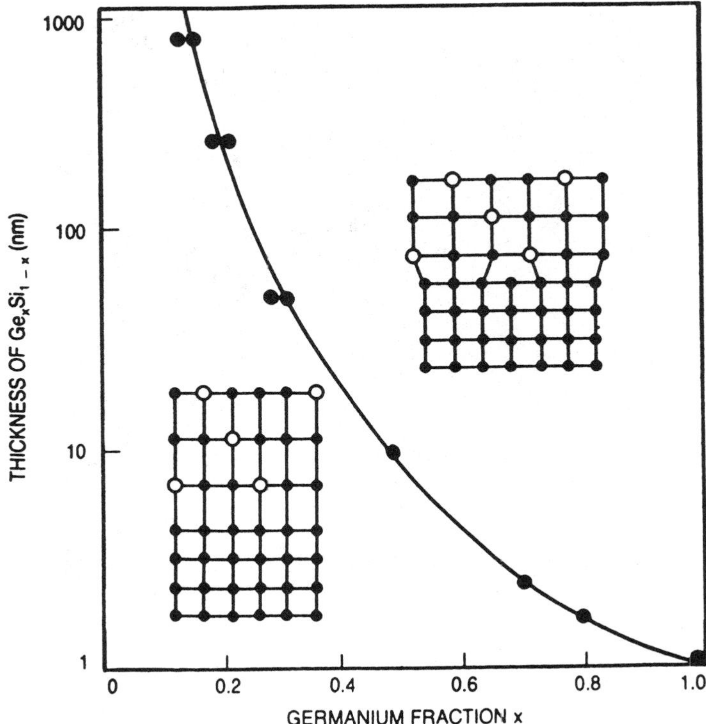

Figure 7-9. Experimentally determined limits for defect-free strained-layer epitaxy of Ge_xSi_{1-x} on Si. Note that f is proportional to Ge fraction. (From Ref. 9).

In the region where d_c is approximately a few thousand angstroms, $d_c \approx b/2f$. This means that the film will be pseudomorphic until the accumulated misfit $d_c f$ exceeds about half the unit cell dimension or $b/2$.

The validity of these ideas has been critically tested on several occasions in Si-base materials. By doping Si wafers with varying amounts of boron, whose atomic size is smaller than that of Si, the lattice parameter of the substrate can be controllably reduced. This affords the opportunity to study defect generation in subsequently deposited epitaxial films under conditions of very small lattice mismatch. As expected, an increase in f resulted in an increase in misfit dislocation density (Ref. 7).

More recently, experimentation on Ge_xSi_{1-x}–Si epitaxial films has extended a test of the theory into the regime of large lattice misfits (Ref. 8). The results are shown in Fig. 7-9 where regions of lattice-strained but defect-free (commensurate) epitaxy are distinguished from those of dislocation-relaxed (discommensurate) epitaxy. Nature is kinder to us than the Matthews theory would suggest, and considerably thicker films than d_c predicted by Eq. 7-7 can be deposited in practice. The reason is that Ge_xSi_{1-x} strained-layer films are not at equilibrium. Extended dislocation arrays do not form instantaneously with well-defined spacings as assumed; rather, dislocations nucleate individually over an area determined by a width w and unit depth, over which atoms above and below the slip plane are displaced by at least $b/2$. For isolated screw dislocations the energy per unit length is equal to $\mu b^2/4\pi \ln R_0/b$. When formed in a film of thickness d, its area energy density is then given by

$$E_d = \frac{\mu b^2}{4\pi(w)} \ln \frac{d}{b}. \tag{7-8}$$

By equating this to E_ε (Eq. 7-4), the energy supplied for nucleation, Bean (Ref. 8) has shown that for $\varepsilon_f = f$

$$d_c \approx \frac{(1-\nu)b^2}{(1+\nu)8\pi w f^2} \ln\left(\frac{d_c}{b}\right). \tag{7-9}$$

The solid line of Fig. 7-9 represents the excellent fit of this equation to the experimental data for the case where w is arbitrarily chosen to be five [110] lattice spacings or ~ 19.6 Å.

When the lattice misfit becomes large, the spacing between misfit dislocations decreases to the order of only a few lattice spacings. In such a case, natural lattice misfit theory breaks down. There are no longer large areas of good fit separated by narrow regions of poor fit. Rather, poor fit occurs everywhere.

7.3.2. Sources of Defects in Epitaxial Films

The issues just raised with respect to misfit dislocations are but a part of a larger concern for defects in epitaxial films. In semiconductors it is well known that such defects as grain boundaries, dislocations, twins, and stacking faults degrade many device properties by altering carrier concentrations and mobilities. By acting as nonradiative recombination centers, they serve to reduce the minority carrier lifetime and quantum efficiency of photonic devices. For these reasons a very considerable effort has been devoted to the study of defects and methods for their elimination. Stringfellow (Ref. 7) has divided the sources of defects in epitaxial films into five categories, and it is instructive to consider them.

7.3.2.1. Propagation of Defects from the Substrate into the Epitaxial Layer. A classic example of defect propagation from the substrate to the

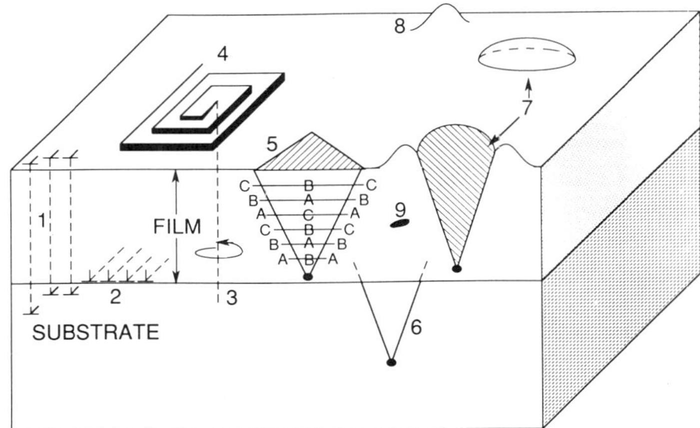

Figure 7-10. Schematic composition of crystal defects in eptiaxial films: (1) threading edge dislocations; (2) interfacial misfit dislocations; (3) threading screw dislocation; (4) growth spiral; (5) stacking fault in film; (6) stacking fault in substrate; (7) oval defect; (8) hillock; (9) precipitate or void.

epitaxial layer results in the extension of an emergent screw dislocation spiral from the substrate surface into the growing film, as shown schematically in Fig. 7-10. Depositing atoms preferentially seek the accommodating ledge sites of the dislocation spiral staircase. Layers of defect-free film then radiate laterally to cover the substrate. Except for the threading screw dislocation, an epitaxial layer of otherwise good fit is possible. Today's semiconductor substrates, including Si and GaAs, are largely "dislocation-free." Occasional substrate dislocations that are present are apparently a source of dislocations observed in homoepitaxial layers. Gross defects such as grain boundaries and twins are, however, never present.

7.3.2.2. Stacking Faults. Stacking faults are crystallographic defects in which the proper order of stacking planes is interrupted. For example, consider the first three atomic planes or layers of a (111) silicon film. Each of these planes may be imagined to be a close-packed array of atomic spheres (Fig. 1-2a), and each successive layer fits into the interstices of the previous layer. Atoms in the second layer (B) have no choice but to nest in one set of interstices of the first (A) layer. It now makes a difference which set of B layer interstices that atoms of the third layer choose to lie in. If they do not lie above the A or B layer atoms, the stacking sequence is ABC, and in a perfect epitaxial film the ABCABCABC, etc., order is preserved. If, however, a plane of atoms is missing from the normal sequence, e.g., ACABCABC, or a plane of atoms has been inserted, e.g., ABCBABCABC, then stacking fault defects are produced. It is established that they propagate from dislocations and oxide precipitates at the substrate interface. Misoriented clusters, or nuclei containing stacking faults, coalesce with normal nuclei and grow into the film in the manner of an inverted pyramid. Continuing growth causes the characteristic closed triangle shown in Fig. 7-10 to become progressively larger. For (100) growth the stacking faults form squares. Appropriate etches are required to reveal these defects. In general, the stacking-fault density of homoepitaxial Si films increases with decreasing growth temperature. In heteroepitaxial films, e.g., GaAs on Si, stacking faults are very common.

Related to stacking faults are the ubiquitous "oval defects" observed during MBE growth (Section 7.5.4) of compound semiconductors. These defects shown schematically in Fig. 7-10 are faceted growth hillocks that nucleate at the film–substrate interface and nest within the epitaxial layer. They usually contain a polycrystalline core bounded by four $\{111\}$ stacking-fault planes. With densities as high as 1000 cm^{-2} or more and sizes ranging from 1 μm^2 to $\sim$ 30 μm^2, these defects are a source of great concern. There are a number of possible sources for oval defects in GaAs, including carbon contamination of the substrate, incomplete desorption of oxides prior to growth, and spitting of Ga and Ga_2O from melts.

7.3.2.3. Formation of Precipitates or Dislocation Loops due to Supersaturation of Impurities or Dopants. This category is self-explanatory. The precipitates and dislocations usually form during cooling and are the result of solid-state reactions subsequent to growth. Films containing high intentional or accidental dopant or impurity levels are susceptible to such defects.

7.3.2.4. Formation of Low-Angle Grain Boundaries and Twins. Low-angle grain boundaries and twins arise from misoriented islands that meet and coalesce. When this happens, small-angle grain boundaries or crystallographic twins result. The lattice stacking is effectively mirrored across a twin plane $\underline{A}$, i.e., . . . CABC$\underline{A}$CBAC . . . . Both types of defects may anneal out by dislocation motion if the temperature is high enough. During heteroepitaxial growth, Matthews (Ref. 5) has suggested that there is some ambiguity in the exact orientation of small nuclei. He has formulated a rule of thumb that the relaxation of elastic misfit strain causes a variation in the orientation of crystal planes (in radians) roughly equal to the magnitude of the lattice misfit f. Such an effect would naturally give rise to a network of small-angle grain boundaries.

7.3.2.5. Formation of Misfit Dislocations. Formation of misfit dislocations has been dealt with in the previous section. The mechanisms by which misfit dislocations form is of interest. Although misfit dislocations lying in the interface between the overgrowth and substrate are the most efficient means of relaxing misfit strain, they are not the only dislocations present. Extending from the substrate and into the epitaxial film (much like the screw dislocation of Fig. 1-6) are so-called threading dislocations. Under the influence of lattice strain, the vertical segments in the substrate and film move in opposite directions, leaving a segment of misfit dislocation lying in the plane of the interface (Fig. 7-10). Therefore, the initial density of threading dislocations should correlate with the final density of misfit dislocations, and this has indeed been observed. Matthews has estimated that misfits of more than 1–2% are necessary to nucleate a typical misfit dislocation, and this is also consistent with experimental observation.

In closing, it is appropriate to comment on the perfection of epitaxial layers. Various levels of perfection can be imagined, depending on methods used to prepare and evaluate films. Early epitaxial semiconductor films were judged to be single crystals based on standard X-ray and electron diffraction techniques that are relatively insensitive to slight crystal misorientations. However, the subsequent electrical characterization of these films yielded significantly poorer electrical properties than anything imaginable in bulk melt-grown single crystals. In general, thinned slices of bulk crystals were more structurally perfect than epitaxial films of equivalent thickness. This is certainly true of epitaxial films exhibiting island and S.K. growth and most of the films reported in the older literature. However, some of today's lattice-matched MBE films, grown under exacting conditions, are indeed structurally perfect when judged by the unambiguous standard of high-resolution TEM lattice imaging. In fact, the crystalline quality now frequently exceeds that attainable in bulk form.

7.4. Epitaxy of Compound Semiconductors

7.4.1. Introduction

Compound semiconductor films have been grown epitaxially on single-crystal insulators (e.g., Al_2O_3, CaF_2) and semiconductor substrates. The latter case

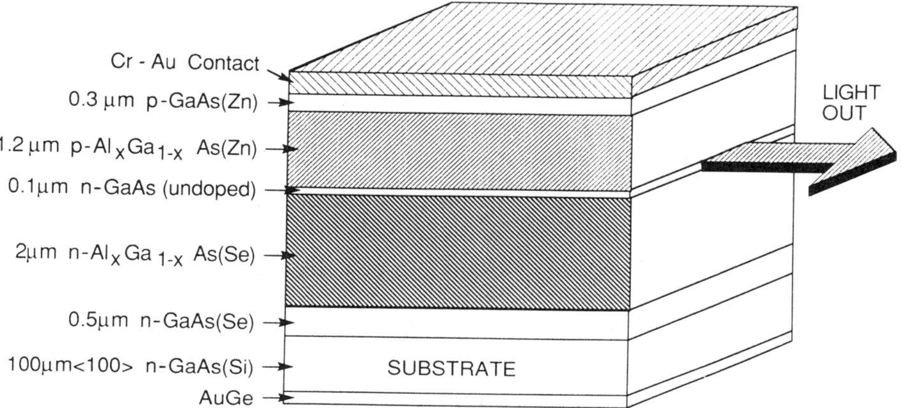

Figure 7-11. Conventional AlGaAs–GaAs double-heterostructure injection laser structure (not to scale.)

has attracted the overwhelming bulk of the attention and, unless indicated otherwise, will be the only systems discussed. The epitaxy of heterojunctions has been crucial in the exploitation of compound semiconductors for optoelectronic device applications. The term *heterojunction* refers to the interface between two single-crystal semiconductors of differing composition and doping level brought into contact. There have been two main thrusts with regard to the epitaxy of heterojunction structures. In the first, a single or generally limited number of different junctions is involved. The intent is to grow suitable binary, ternary, or even quaternary compound layers epitaxially on top of a similar compound substrate, or vice versa. The most common example is the $Al_xGa_{1-x}As$–GaAs combination. As we shall see, good epitaxy is ensured because the two lattices are very well matched to each other (i.e., low misfit), even when the atom fraction x of Al substituted for Ga is high. From a device standpoint, it is significant that the energy band gaps of these materials are different; i.e., GaAs has narrower band gap than any of the $Al_xGa_{1-x}As$ compounds. This means that charge carriers will be confined to the low-energy-gap GaAs film when clad by the wider-gap $Al_xGa_{1-x}As$ heterojunction barriers. This makes carrier population inversion and laser action possible. Such layered structures are utilized in devices such as lasers (Fig. 7-11) and light-emitting diodes (Fig. 7-12), which have served as light sources in fiber optical communication systems. Our scope does not allow for discussion of the operation of these or other thin-film optoelectronic devices. For more information the reader is referred to the many excellent textbooks on solid-state devices (Refs. 10, 11). The second thrust dealing with superlattices will be addressed in Section 14.7.

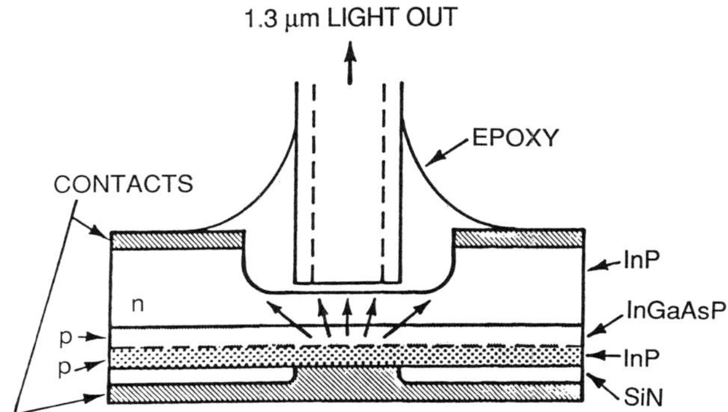

Figure 7-12. Schematic of a surface emitting InGaAsP ($\lambda = 1.3\ \mu m$) light-emitting diode. (Courtesy N. K. Dutta, AT&T Bell Laboratories.)

7.4.2. Compound Semiconductor Materials

Materials employed for epitaxial optoelectronic devices have been drawn largely from a collection of direct-band-gap III-V semiconductors. Although the discussion will be primarily limited to them, the results are applicable to other materials as well (e.g., II-VI compounds). Table 7-1 contains a list of important semiconductors together with some physical properties pertinent to epitaxy. When light is emitted from or absorbed in a semiconductor, energy as well as momentum must be conserved. In a *direct band-gap* semiconductor, the carrier transitions between the valence and conduction bands occur without change in momentum of the two states involved. In the energy–momentum or equivalent energy–wave vector, parabola-like (E vs. k) representation of semiconductor bands (Ref. 10) (Fig. 7-13), emission of light occurs by a vertical electron descent from the minimum conduction-band energy level to the maximum vacant level in the valence band. This is what occurs in the direct band-gap materials GaAs and InP. However, in *indirect band-gap semiconductors* like Ge and Si, the transition occurs with a change in momentum that is essentially accommodated by excitation of lattice vibrations and heating of the lattice. This makes direct electron–hole recombination with photon emission unlikely. But in direct-band-gap semiconductors, such processes are more probable, making them far more efficient (by orders of magnitude) light emitters.

Another implication of the distinction between direct and indirect semiconductors is the variation of the absorption coefficient (α) as a function of photon energy, as shown in Fig. 7-14. The ratio of the photon intensity at a depth x below the surface, $I(x)$, to that incident on it, I_0, is described by

$$I(x) = I_0 \exp - \alpha x. \tag{7-10}$$

Table 7-1. Semiconductor Properties (25 °C)

Material	Lattice Parameter (Å)	Melting Point (K)	Dissociation Pressure (Atm)	Thermal Expansion Coefficient 10^{-6} °C^{-1}	Energy Gap (eV at 25 °C)	Electron Mobility (cm^2/V-s)
Diamond	3.560	~ 4300		1.0	5.4	1800
Si	5.431	1685		2.33	1.12I	1350
Ge	5.657	1231		5.75	0.68I	3600
ZnS	5.409	3200		7.3	3.54	120
ZnSe	5.669	1790		7.0	2.58	530
ZnTe	6.101	1568		8.2	2.26	530
CdTe	6.477	1365		5.0	1.44	700
HgTe	6.460	943		1.9	~ 0.15	
CdS				4.0	2.42D	340
AlAs	5.661	1870	1.4	5.2	2.16I	280
AlSb	6.136	1330	$< 10^{-3}$	3.7	1.60I	900
GaP	5.451	1750	35	5.3	2.24I	300
GaAs	5.653	1510	1	5.8	1.43D	~ 6500
GaSb	6.095	980	$< 10^{-3}$	6.9	0.67D	5000
InP	5.869	1338	25	4.5	1.27D	4500
InAs	6.068	1215	0.3	4.5	0.36D	30000
InSb	6.479	796	$< 10^{-3}$	4.9	0.165D	80000

D = direct; I = indirect.
From Refs. 11, 12.

In all semiconductors, α becomes negligible once the wavelength exceeds the cutoff wavelength. This critical wavelength λ_c is related to the band-gap energy E_g by the well-known relation $E_g = hc/\lambda_c$ or λ_c (μm) $= 1.24/E_g$ (eV). For direct-band-gap semiconductors the value of α becomes large on the short-wavelength side of λ_c, signifying that light is absorbed very close to the surface. For this reason even thin-film layers of GaAs are adequate, for example, in solar cell applications. In Si, on the other hand, α varies more gradually with wavelength less than λ_c because of the necessity for phonon participation in light absorption–carrier generation processes. Therefore, efficient solar cell action necessitates thicker layers if indirect semiconductors are employed.

Device applications require, in addition to a direct-band-gap semiconductor, a specific wavelength or value of E_g at which emission or absorption processes are optimized. A further necessity is close matching of the lattice parameters (a_0) of the heterostructure layers to ensure defect-free interfaces. An extremely handy graphical representation of E_g and a_0 for elemental and major III-V compound semiconductors and their alloys is shown in Fig. 7-15. Through its use, the design and selection of complex semiconductor alloys with

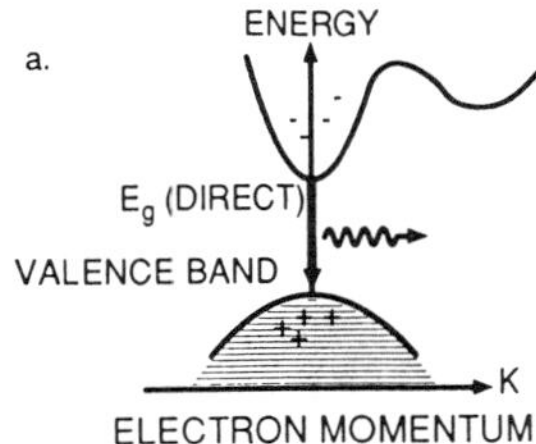

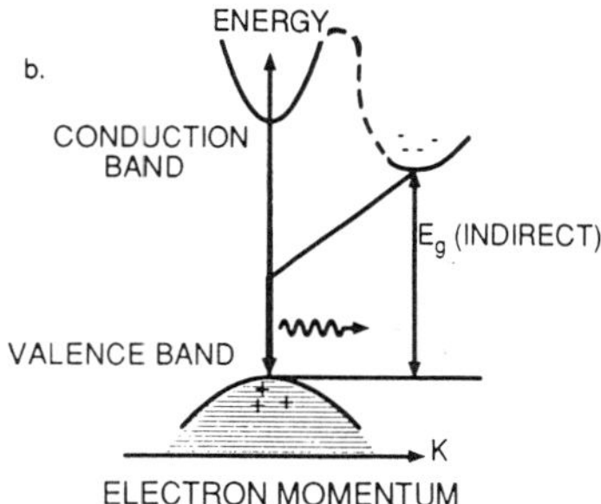

Figure 7-13. Model of electron transitions between conduction and valence bands in (a) direct- and (b) indirect band-gap semiconductors.

the desired properties may be visualized. Elements and binary compounds are represented simply as points. Ternary alloys are denoted by lines between constituent binary compounds. When one of the elements is common to both compounds, a continuous range of solid-solution ternary alloys form when the binaries are alloyed. Thus, the line between InP and InAs represents the collection of $InAs_xP_{1-x}$ ternary solution alloys, with x dependent on the proportions mixed. Within the areas outlined by four binary compounds are the quaternary alloys. Therefore, the $Ga_{1-x}In_xAs_{1-y}P_y$ system may be thought of as arising from suitable combinations of GaAs, GaP, InAs, and InP. There is no need though to start with these four binary compounds when synthesizing a quaternary; it is usually only necessary to control the vapor pressures of the Ga-, In-, As-, and P-bearing species. The solid lines represent the direct-band-gap ternary compounds, and the dashed lines refer to materials with an indirect band gap.

An example will illustrate the use of this important diagram. We assume for simplicity that a linear law of mixtures governs both the resultant lattice constant and energy-gap values. Consider GaAs and AlAs with respective a_0 values of 5.6537 Å and 5.6611 Å. No hint of structural defects at the interface between epitaxial films of these compounds is observed in the lattice image of Fig. 7-16. The lattice mismatch between these compounds is $\Delta a_0 / a_0$ or

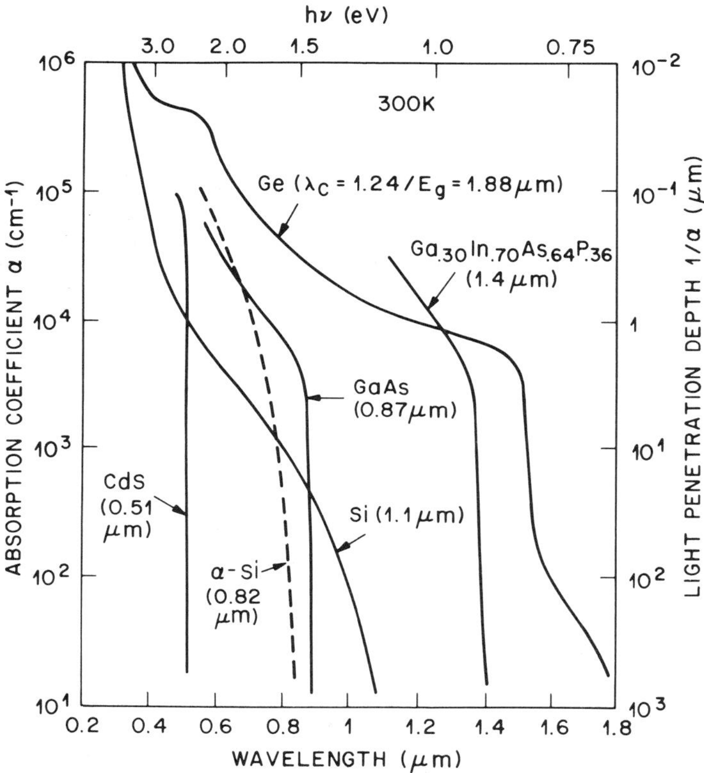

Figure 7-14. Optical absorption coefficients for various semiconductor materials. (Reprinted with permission from J. Wiley and Sons, S. M. Sze, *Semiconductor Devices–Physics and Technology*, 1985).

0.14%. This is also the maximum difference between any GaAs–$Al_xGa_{1-x}As$ heterojunction combination. However, for high device reliability a maximum mismatch of 0.1% is desired. Assuming a $Al_{0.4}Ga_{0.6}As$ cladding layer composition in Fig. 7-11, a_0 is calculated to be 5.6563 Å. The mismatch with GaAs is therefore 0.056% or 0.4 of the maximum mismatch. Similarly, linear interpolation between the energy gaps E_g(GaAs) = 1.435 eV and E_g(AlAs) = 2.16 eV yields a value of $E_g(Al_{0.4}Ga_{0.6}As)$ = 1.73 eV. Vegard's law expressing a linear variation of a_0 with composition is a good assumption for these alloys. However, E_g values can only be linearly interpolated between direct-gap materials and AlAs is an indirect-gap semiconductor. For $Al_xGa_{1-x}As$ it has been found (Ref. 14) that

$$E_g(x) = 1.424 + 1.44x \text{ eV}; \qquad x < 0.45. \tag{7-11}$$

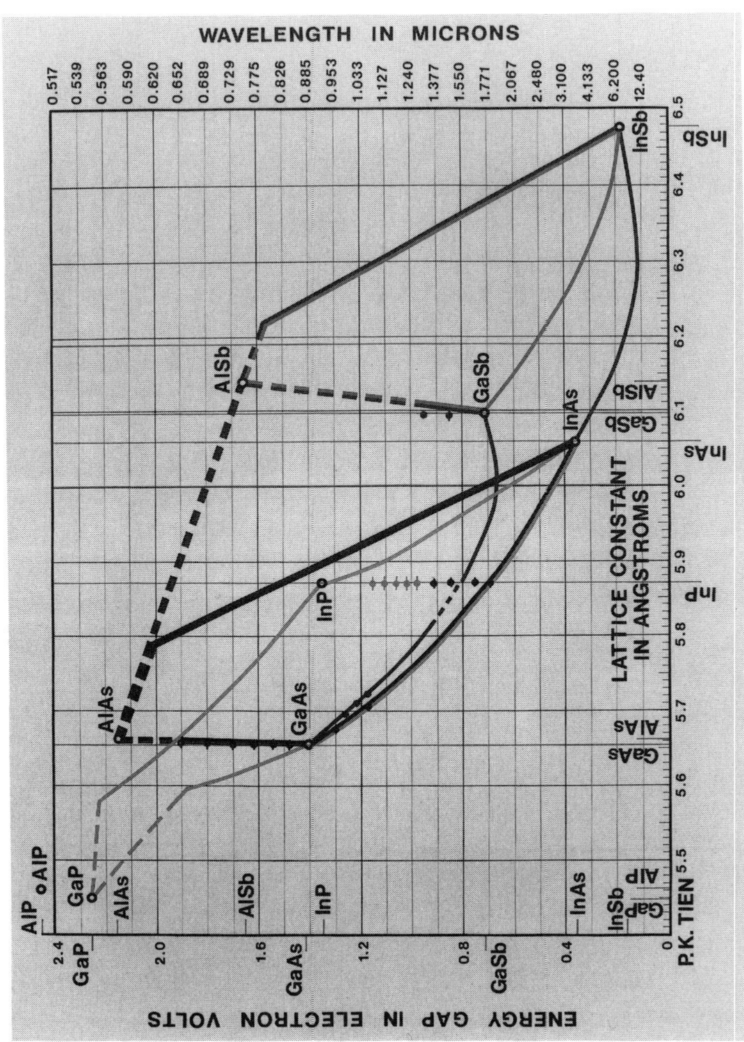

7-15. Energy gaps and corresponding lattice constants for various compound semiconductors. (Courtesy of P. K. Tien, AT&T Bell Laboratories.)

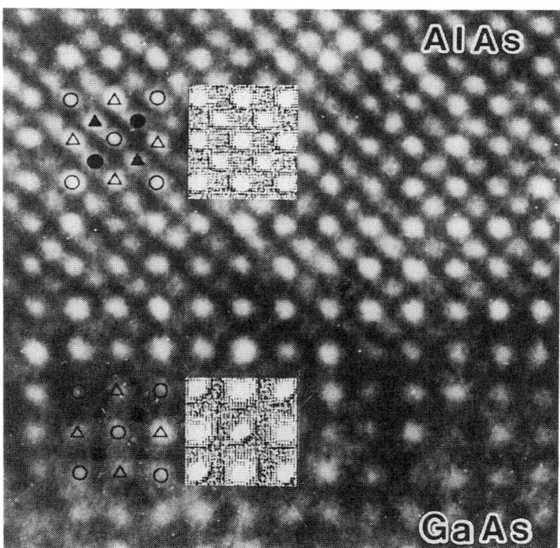

Figure 7-16. Electron microscope lattice image of GaAs–AlAs heterojunction taken with [100] illumination. (From Ref. 13). (Courtesy of JOEL USA, Inc.)

Therefore, the correct value for $E_g(0.4) = 2.00$ eV. In addition, the index of refraction n, required for light-guiding properties, varies as (Ref. 15)

$$n(x) = 3.590 - 0.710x + 0.091x^2. \qquad (7\text{-}12)$$

In summary, it is possible to design ternary alloys with E_g larger than GaAs, with n smaller than GaAs, while maintaining an acceptable lattice match for high-quality heterojunctions. This unique combination of properties has led to the development of a family of injection lasers, light-emitting diodes, and photodetectors based on the GaAs–AlAs system.

7.4.3. Additional Applications

7.4.3.1. Optical Communications. Optical communication systems are used to transmit information optically. This is done by converting the initial electronic signals into light pulses using laser or light-emitting diode light sources. The light is launched at one end of an optical fiber that may extend over long distances (e.g., 40 km). At the other end of the system, the light pulses are detected by photodiodes or phototransistors and converted back into electronic signals that, in telephone applications, finally generate sound. In such a system it is crucial to transmit the light with minimum attenuation or low optical loss. Great efforts have been made to use the lowest-loss fiber possible and minimize loss at the source and detector ends. If optical losses are high, it means that the optical signals must be reamplified and that additional, costly repeater stations will be necessary. The magnitude of the problem can be appreciated when transoceanic communications systems are involved. In silica-based fibers it has been found that minimum transmission losses occur with light of approximately 1.3–1.5 μm wavelength. The necessity to operate within this infrared wavelength window bears directly on the choice of suitable semiconductors and epitaxial deposition technology required to fabricate the required sources and detectors.

Reference to Table 7-1 shows that InP is transparent to 1.3-μm light, and this simplifies the coupling of fibers to devices. A very close lattice match to InP ($a_0 = 5.869$ Å) can be effected by alloying GaAs and InAs. Through the use of Vegard's law, it is easily shown that the necessary composition is $Ga_{0.47}In_{0.53}As$. In the same vein, high-performance lasers based on the lattice-matched GaInAsP–InP system have recently emerged for optical communications use.

7.4.3.2. Silicon Heteroepitaxy (Ref. 8). Since the early 1960s, Si has been the semiconductor of choice. Its dominance cannot, however, be attributed solely to its electronic properties for it has mediocre carrier mobilities and only average breakdown voltage and carrier saturation velocities. The absence of a direct band gap rules out light emission and severely limits its efficiency as a photodetector. Silicon does, however, possess excellent mechanical and chemical properties. The high modulus of elasticity and high hardness enable Si wafers to withstand the rigors of handling and device processing. Its great natural abundance, the ability to readily purify it and the fact that it possesses a highly inert and passivating oxide have all helped to secure the dominant role for Si in solid-state technology. Nevertheless, Si is being increasingly supplanted in high-speed and optical applications by compound semiconductors.

The idea of combining semiconductors that can be epitaxially grown on low-cost Si wafers is very attractive. Monolithic integration of III-V devices with Si-integrated circuits offers the advantages of higher-speed signal processing distributed over larger substrate areas. Furthermore, Si wafers are more robust and dissipate heat more rapidly than GaAs wafers. Unfortunately, there are severe crystallographic, as well as chemical compatibility problems that limit Si-based heteroepitaxy. From data in Table 7-1, it is evident that Si is only closely lattice matched to GaP and ZnS. Furthermore, its small lattice constant limits the possible epitaxial matching to semiconductor alloys. Nevertheless, high-quality, lattice-mismatched (strained-layer) heterostructures of AlGaAs–Si and Ge_xSi_{1-x}–Si have been prepared and show much promise for new device applications.

7.4.3.3. Epitaxy in II-VI Compounds (Ref. 16). Semiconductors based on elements from the second (e.g., Cd, Zn, Hg) and sixth (e.g., S, Se, Te) columns of the periodic table display a rich array of potentially exploitable properties. They have direct energy band gaps ranging from a fraction of an electron volt in Hg compounds to over 3.5 eV in ZnS, and low-temperature carrier mobilities approaching 10^6 cm^2/V-sec are available. Interest in the wide-gap II-VI compounds has been stimulated by the need for electronically addressable flat-panel display devices and for the development of LED and injection lasers operating in the blue portion of the visible spectrum. For these purposes, ZnSe and ZnS have long been the favored candidates. When the group II element is substituted by a magnetic transition ion such as Mn, new classes of materials known as diluted magnetic or semimagnetic semiconductors result. Examples are Cd(Mn)Te or Zn(Mn)Se, and these largely retain the semiconducting properties of the pure compound. But the five electrons in the unfilled 3d shell of Mn give rise to localized magnetic moments. As a result, large magneto-optical effects (e.g., Zeeman splitting in magnetic fields, Faraday rotation, etc.) occur and have been exploited in optical isolator devices. For this, as well as other potential applications in integrated optics, high-quality epitaxial films are essential.

7.5. Methods for Depositing Epitaxial Semiconductor Films

7.5.1. Liquid Phase Epitaxy

In this section an account of the processes used to deposit epitaxial semiconductor films is given. We start with LPE, a process in which melts rather than vapors are in contact with the growing films. Introduced in the early 1960s, LPE is still used to produce heterojunction devices. However, for greater layer uniformity and atomic abruptness, it has been supplanted by CVD and MBE techniques. LPE involves the precipitation of a crystalline film from a supersaturated melt onto the parent substrate, which serves as both the template for epitaxy and the physical support for the heterostructure. The process can be understood by referring to the GaAs binary-phase diagram on p. 31. Consider a Ga-rich melt containing 10 at% As. When heated above 950°C, all of the As dissolves. If the melt is cooled below the liquidus temperature into the two-phase field, it becomes supersaturated with respect to As. Only a melt of lower than the original As content can now be in equilibrium with GaAs. The excess As is, therefore, rejected from solution in the form of GaAs that grows epitaxially on a suitably placed substrate. Many readers will appreciate that the crystals they grew as children from supersaturated aqueous solutions essentially formed by this mechanism.

Through control of the cooling rates, different kinetics of layer growth apply. For example, the melt temperature can either be lowered continuously together with the substrate (equilibrium cooling) or separately reduced some 5–20 °C and then brought into contact with the substrate at the lower temperature (step cooling). Theory backed by experiment has demonstrated that the epitaxial layer thickness increases with time as $t^{3/2}$ for equilibrium cooling and as $t^{1/2}$ for step cooling (Ref. 10). Correspondingly, the growth rates or time derivatives vary as $t^{1/2}$ and $t^{-1/2}$, respectively. These diffusion-controlled kinetics respectively indicate either an increasing or decreasing film growth rate with time depending on mechanism. Typical growth rates range from ~ 0.1 to 1 μm/min. A detailed analysis of LPE is extremely complicated in ternary systems because it requires knowledge of the thermodynamic equilibria between solid and solutions, nucleation and interface attachment

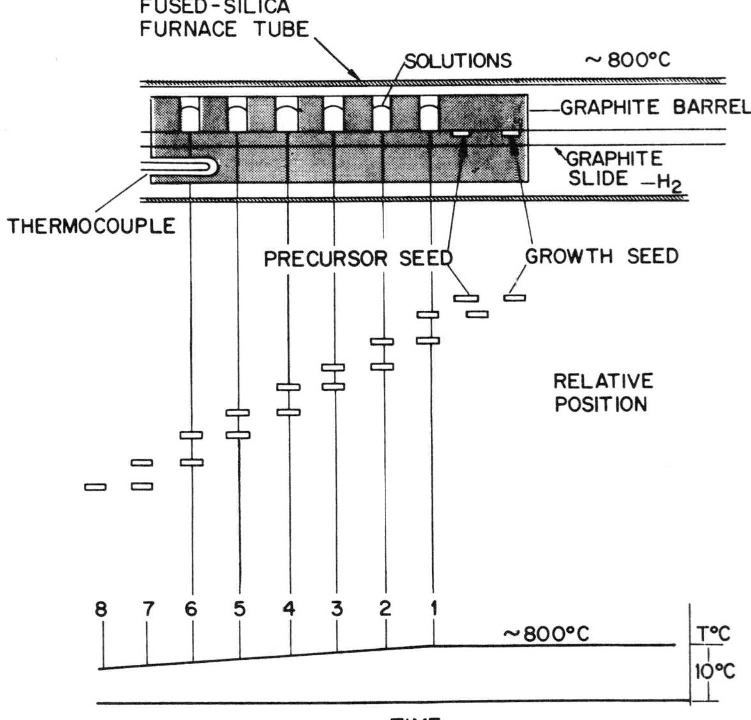

Figure 7-17. Schematic of LPE reactor. (Courtesy of M. B. Panish, AT&T Bell Laboratories.)

kinetics, solute partitioning, diffusion, and heat transfer. LPE offers several advantages over other epitaxial deposition methods, including low-cost apparatus capable of yielding films of controlled composition and thickness, with lower dislocation densities than the parent substrates.

To grow multiple GaAs–AlGaAs heterostructures, one translates the seed substrate sequentially past a series of crucibles holding melts containing various amounts of Ga and As together with such dopants as Zn, Ge, Sn, and Se as shown in Fig. 7-17. Each film grown requires a separate melt. Growth is typically carried out at temperatures of $\sim$ 800 °C with maximum cooling rates of a few degrees Celsius per minute. Limitations of LPE growth include poor thickness uniformity and rough surface morphology particularly in thin layers. The CVD and MBE techniques are distinctly superior to LPE in these regards.

7.5.2. Seeded Lateral Epitaxial Film Growth over Insulators

The methods we describe here briefly have been successfully implemented in Si but not in GaAs or other compound semiconductors. The use of melts suggests the inclusion of this subject at this point. Technological needs for three-dimensional VLSI and isolation of high-voltage devices have spurred the development of techniques to grow epitaxial Si layers over such insulators as SiO_2 or sapphire. In the recently proposed LEGO (lateral epitaxial growth over oxide) process (Ref. 17), the intent is to form isolated tubs of high-quality Si surrounded on all sides by a moat of SiO_2. Devices fabricated within the tubs require the electrical insulation provided by the SiO_2. As a result they are also radiation-hardened or immune from radiation-induced charge effects originating in the underlying bulk substrate. The process shown schematically in Fig. 7-18 starts with patterning and masking a Si wafer to define the tub regions followed by etching of deep-slanted wall troughs. A thick SiO_2 film is grown and seed windows are opened down to the substrate by etching away the SiO_2. Then a thick polycrystalline Si layer ($\sim$ 100 μm thick) is deposited by CVD methods. This surface layer is melted by the unidirectional radiant heat flux from incoherent light emitted by tungsten halogen arc lamps (lamp furnace). The underlying wafer protected by the thermally insulating SiO_2 film does not melt except in the seed windows. Crystalline Si nucleates at each seed, grows vertically, and then laterally across the SiO_2, leaving a single-crystal layer in its wake upon solidification. Lastly, mechanical grinding and lapping of the solidified layer prepares the structure for further microdevice processing. Conventional dielectric isolation processing also employs a thick CVD Si layer. But the latter merely serves as the mechanical handle enabling the bulk of the Si wafer to be ground away.

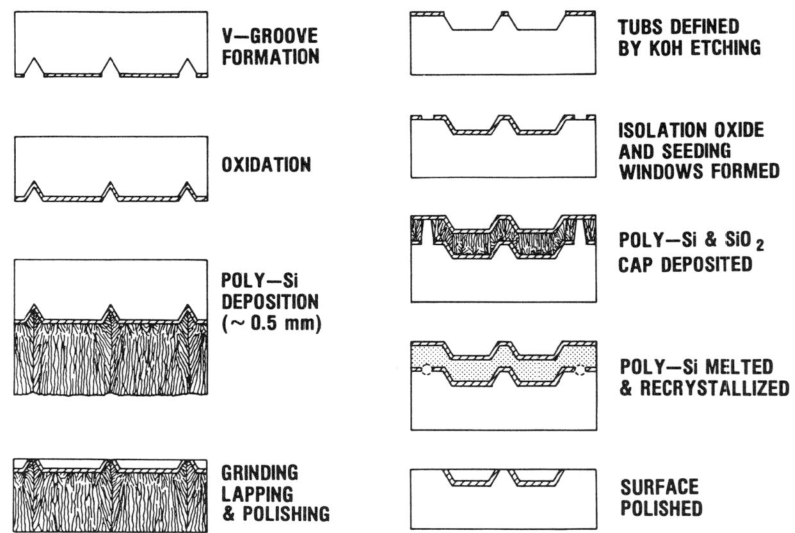

Figure 7-18. Schematics of methods employed to isolate single-crystal Si tubs. (left) conventional dielectric isolation process; (right) LEGO process. (Courtesy of G. K. Celler, AT&T Bell Laboratories.)

An alternative process for broad-area lateral epitaxial growth over SiO_2 employs a strip heat source in the form of a hot graphite or tungsten wire, scanned laser, or electron beam. After patterning the exposed polycrystalline or amorphous Si above the surrounding oxide, the strip sweeps laterally across the wafer surface. Local zones of the surface then successively melt and recrystallize to yield, under ideal conditions, one large epitaxial Si film layer. Analogous processes involving seeded lateral growth and selective deposition from the vapor phase also show much promise.

7.5.3. Vapor Phase Epitaxy (VPE)

An account of the most widely used VPE methods—chloride, hydride, and organometallic CVD processes—has been given in Chapter 4. Here we briefly address a couple of novel VPE concepts that have emerged in recent years. The first is known as vapor levitation epitaxy (VLE), and the geometry is shown in Fig. 7-19. The heated substrate is levitated above a nitrogen track close to a porous frit through which the hot gaseous reactants pass. Upon impingement on the substrate, chemical reactions and film deposition occur while product gases escape into the effluent stream. As a function of radial distance from the center of the circular substrate, the gas velocity increases

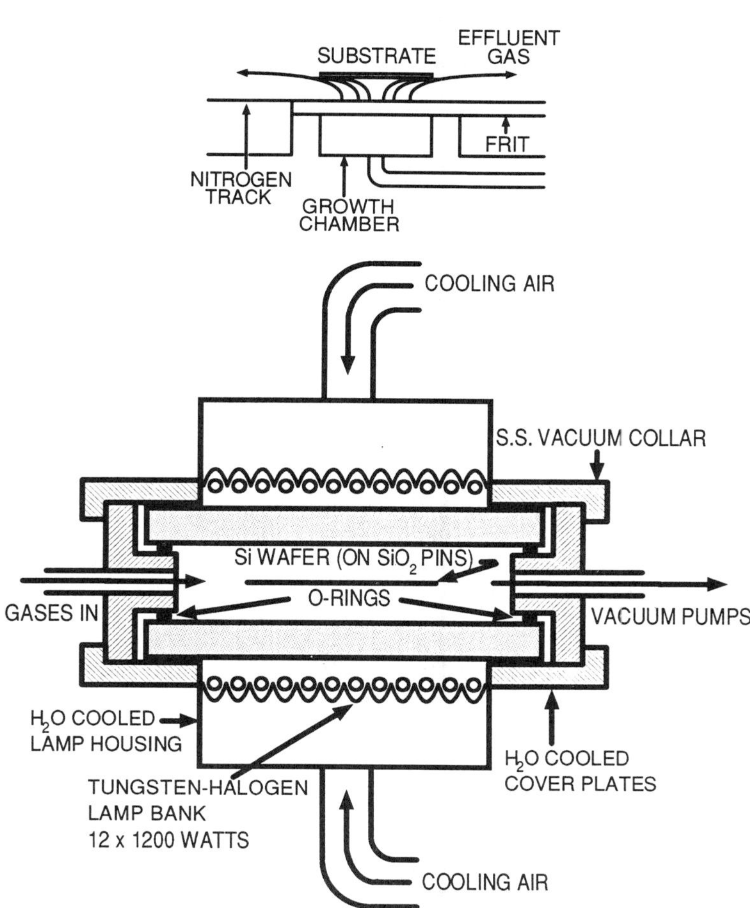

Figure 7-19. (Top) Schematic of VLE process; (bottom) schematic of RTCVD process. (Courtesy of M. L. Green, AT&T Bell Laboratories.)

while the gas concentration profile exhibits depletion. These effects cancel one another, and uniform films are deposited. The VLE process was designed for the growth of epitaxial III-V semiconductor films and has certain advantages worth noting:

1. There is no physical contact between substrate and reactor.
2. Thin layer growth is possible.
3. Sharp transitions can be produced between film layers of multilayer stacks.
4. Commercial scale-up appears to be feasible.

The second method, known as rapid thermal CVD processing (RTCVD), is an elaboration on conventional VPE. Epitaxial deposition is influenced through rapid, controlled variations of substrate temperature. Source gases (e.g., halides, hydrides, metalorganics) react on low-thermal-mass substrates heated by the radiation from external high-intensity lamps (Fig. 7-19). The latter enable rapid temperature excursions, and heating rates of hundreds of degrees Celsius per second are possible. For III-V semiconductors, high-quality epitaxial films have been deposited by first desorbing substrate impurities at elevated temperatures followed by immediate lower temperature growth (Ref. 18).

Very high quality lattice-matched heteroepitaxial films can be grown by CVD methods. This is particularly true of OMVPE techniques where atomically abrupt heterojunction interfaces have been demonstrated in alternating AlAs–GaAs (superlattice) structures. Only molecular-beam epitaxy, which is considered next, can match or exceed these capabilities.

7.5.4. Molecular-Beam Epitaxy (Refs. 19 – 21)

Molecular-beam epitaxy is conceptually a rather simple single-crystal film growth technique that, however, represents the state-of-the-art attainable in deposition processing from the vapor phase. It essentially involves highly controlled evaporation in an ultrahigh-vacuum ($\sim 10^{-10}$ torr) system. Interaction of one or more evaporated beams of atoms or molecules with the single-crystal substrate yields the desired epitaxial film. The clean environment coupled with the slow growth rate and independent control of the beam sources enable the precise fabrication of semiconductor heterostructures at an atomic level. Deposition of thin layers from a fraction of a micron thick down to a single monolayer is possible. In general, MBE growth rates are quite low, and for GaAs materials a value of 1 μm/h is typical.

A modern MBE system is displayed in the photograph of Fig. 7-20. Representing the ultimate in film deposition control, cleanliness and real-time structural and chemical characterization capability, such systems typically cost more than $1 million. The heart of a deposition facility is shown schematically in Fig. 7-21a. Arrayed around the substrate are semiconductor and dopant sources, which usually consist of so-called effusion cells or electron-beam guns. The latter are employed for the high-melting Si and Ge materials. On the other hand, effusion cells consisting of an isothermal cavity with a hole through which the evaporant exits are used for compound semiconductor elements and their dopants. Effusion cells behave like small-area sources and exhibit a cos ϕ emission. Vapor pressures of important compound semiconductor species are displayed in Fig. 3-2.

Figure 7-20. Photograph of multichamber MBE system. (Courtesy of Riber Division, Inc. Instruments SA).

Consider now a substrate positioned a distance l from a source aperture of area A, with $\phi = 0$. An expression for the number of evaporant species striking the substrate is

$$\dot{R} = \frac{3.51 \times 10^{22} PA}{\pi l^2 (MT)^{1/2}} \text{ molecules/cm}^2\text{-sec.} \tag{7-13}$$

As an example, consider a Ga source in a system where $A = 5$ cm^2 and $l = 12$ cm. At $T = 900$ °C the vapor pressure $P_{Ga} \approx 1 \times 10^{-4}$ torr, and substituting $M_{Ga} = 70$, the arrival rate of Ga at the substrate is calculated to be 1.35×10^{14} atoms/cm^2-sec. The As arrival rate is usually much higher, and, therefore, film deposition is controlled by the Ga flux. An average monolayer of GaAs is 2.83 Å thick and contains $\sim 6.3 \times 10^{14}$Ga atoms/cm^2. Hence, the growth rate is calculated to be $(1.35 \times 10^{14})/(6.3 \times 10^{14}) \times 2.83 \times 60 = 36$ Å/min, a rather low rate when compared with VPE.

One of the recent advances in MBE technology incorporates a gas source to supply As and P, as shown in Fig. 7-21b. Organometallics used for this purpose are thermally cracked, releasing the group V element as a molecular beam into the system. Excellent epitaxial film quality has been obtained by this

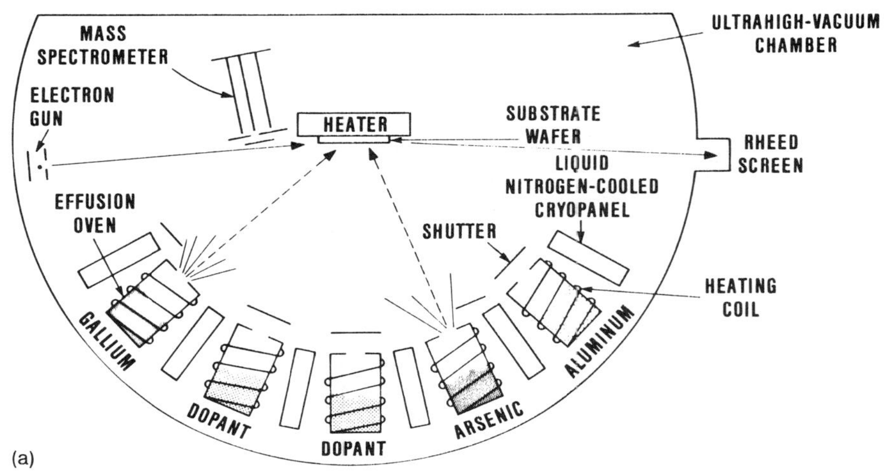

(a)

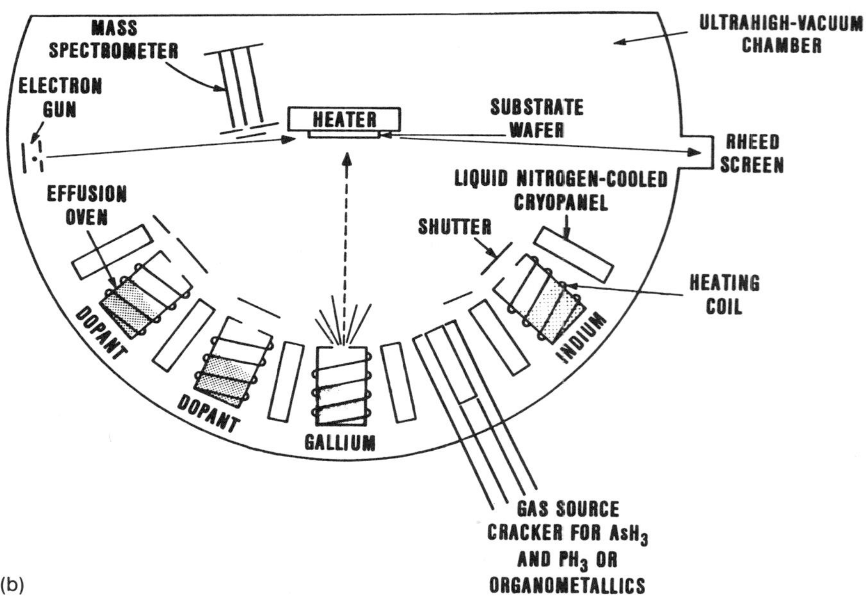

(b)

Figure 7-21. Arrangement of sources and substrate in (a) conventional MBE system, (b) MOMBE system. (Courtesy of M. B. Panish, AT&T Bell Laboratories.)

hybrid MBE–OMVPE process, which is known by the acronym MOMBE. Hydride gas sources (e.g., AsH_3, PH_3) have also been similarly employed in MBE systems.

In many applications, GaAs–$Al_xGa_{1-x}As$ multilayers are required. For this purpose, the Ga and As beams are on continuously, but the Al source is operated intermittently. The actual growth rates are determined by the measured layer thickness divided by the deposition time. The fraction x can be determined from the relation

$$x = \frac{\dot{R}(Al_xGa_{1-x}As) - \dot{R}(GaAs)}{\dot{R}(Al_xGa_{1-x}As)}, \tag{7-14}$$

where the respective deposition rates $\dot{R}$ for GaAs and $Al_xGa_{1-x}As$ must be known. Recommended substrate temperatures for MBE of GaAs range from 500 to 630 °C. Higher temperatures, by about 50 °C, are required for $Al_xGa_{1-x}As$ because AlAs is thermally more stable than GaAs. For InP growth from In and P_2 beams on (100) InP, substrate temperatures of 350–380 °C have been used. Similarly, $In_xGa_{1-x}As$ films, lattice-matched to InP, have been grown between 400 and 430 °C.

7.6. Epitaxial Film Growth and Characterization (Ref. 22)

7.6.1. Film Growth Mechanisms

Irrespective of whether homo- or heteroepitaxy is involved, it is essential to grow atomically smooth and abrupt epitaxial layers. This implies a layer growth mechanism, and thermodynamic approaches to layer growth based on surface energy arguments have been presented in Chapter 5. Ideally, the desired layer-by-layer growth depicted in Fig. 7-22 is achieved through lateral terrace, ledge, and kink extension by adatom attachment or detachment. In this case the new layer does not grow until the prior one is atomically complete. One can also imagine the simultaneous coupled growth of both the new and underlying layers.

In this section we explore the interactions of molecular beams with the surface and the steps leading to the incorporation of atoms into the growing epitaxial film. Although MBE is the focus, the results are, of course, applicable to other epitaxial film growth sequences. The first step involves surface adsorption—the process in which impinging particles enter and interact within the transition region between the gas phase and substrate surface. Two kinds of

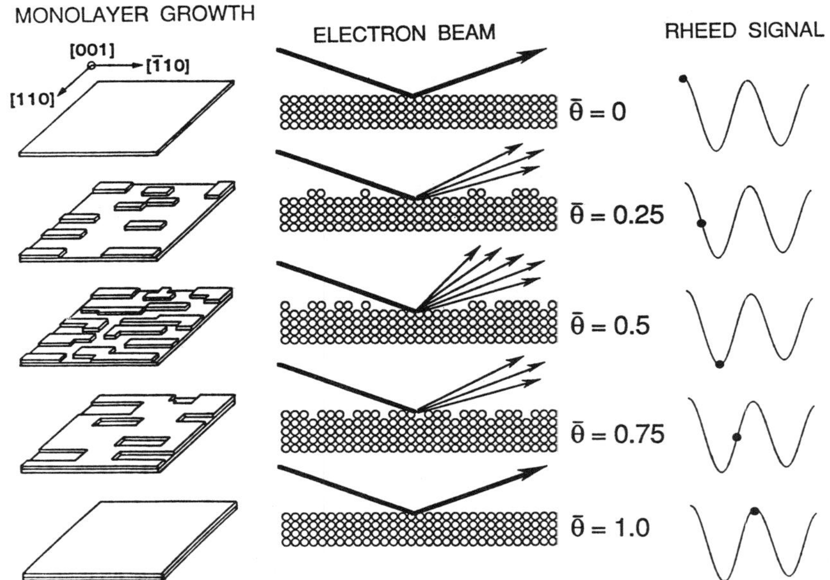

Figure 7-22. Real space representation of the formation of a single complete monolayer; $\bar{\theta}$ is the fractional layer coverage; corresponding RHEED oscillation signal is shown.

adsorption—namely, physical (physisorption) and chemical (chemisorption)—can be distinguished. If the particle (molecule) is stretched or bent but retains its identity, and van der Waals forces bond it to the surface, then we speak of physisorption. If, however, the particle loses its identity through ionic or covalent bonding with substrate atoms, chemisorption is involved. The two can be quantitatively distinguished on the basis of heats of adsorption—H_p and H_c, for physisorption and chemisorption, respectively. Typically, $H_p \sim 0.25$ eV and $H_c \sim 1-10$ eV.

Now consider a beam of Ga atoms incident on a GaAs surface. Below about 480 °C, Ga atoms readily physisorb on the surface, but above this temperature Ga adsorbs as well as desorbs. Time-resolved mass spectroscopy measurements of the magnitude of the atomic flux *desorbing* from the substrate have revealed details of the mechanism of MBE GaAs film growth (Ref. 23). The instantaneous Ga surface concentration, n_{Ga}, is increased by the incident Ga beam flux, $\dot{R}(\mathrm{Ga})$, and simultaneously reduced by a first-order kinetics desorption process. Therefore,

$$\frac{dn_{\mathrm{Ga}}}{dt} = \dot{R}(\mathrm{Ga}) - \frac{n_{\mathrm{Ga}}}{\tau_s(\mathrm{Ga})}, \tag{7-15}$$

where τ_s(Ga) is the Ga adatom lifetime and n_{Ga}/τ_s(Ga) represents the Ga desorption flux $\dot{R}_{\mathrm{des}}$(Ga). Integrating Eq. 7-15 yields

$$\dot{R}_{\mathrm{des}}(\mathrm{Ga}) = \dot{R}(\mathrm{Ga})\left[1 - \exp - t/\tau_s(\mathrm{Ga})\right]. \tag{7-16}$$

For a rectangular pulse of incident Ga atoms, the detected desorption flux closely follows the dependence of Eq. 7-16. Similarly, when the Ga beam is abruptly shut off, the desorption rate decays as $\exp - t/\tau_s$(Ga). The exponential rise and decay of the signal is shown schematically in Fig. 7-23a.

In the case of $\mathrm{As_2}$ molecules incident on a GaAs surface, the lifetime is extremely short ($\tau_s(\mathrm{As_2}) \approx 0$), so the desorption pulse profile essentially mirrors that for deposition (Fig. 7-23b); i.e., $\dot{R}_{\mathrm{des}}(\mathrm{As_2}) = \dot{R}(\mathrm{As_2})$. However, on a Ga-covered GaAs surface, $\tau_s(\mathrm{As_2})$ becomes appreciable, with desorption increasing only as the available Ga is consumed (Fig. 23c). These observations indicate that in order to adsorb $\mathrm{As_2}$ on GaAs at high temperature, Ga adatoms are essential. The detailed model for growth of GaAs requires physisorption of mobile $\mathrm{As_2}$ (or $\mathrm{As_4}$) precursors followed by dissociation and attachment to Ga atoms by chemisorption. Excess As merely re-evaporates, leading to the growth of stoichiometric GaAs. In summary, these adsorption–desorption effects strongly underscore the kinetic rather than thermodynamic nature and control of MBE growth.

The III-V compound semiconductor films are generally grown with a 2- to

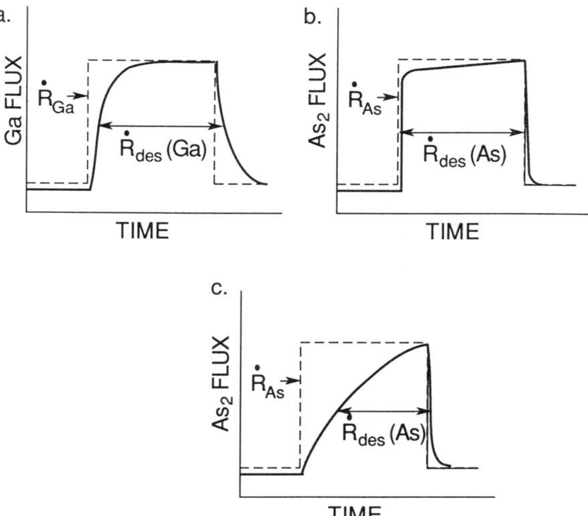

Figure 7-23. Deposition and desorption pulse shapes on (111) GaAs for (a) Ga, (b) $\mathrm{As_2}$, (c) $\mathrm{As_2}$ on a Ga-covered surface. (From Ref. 23)

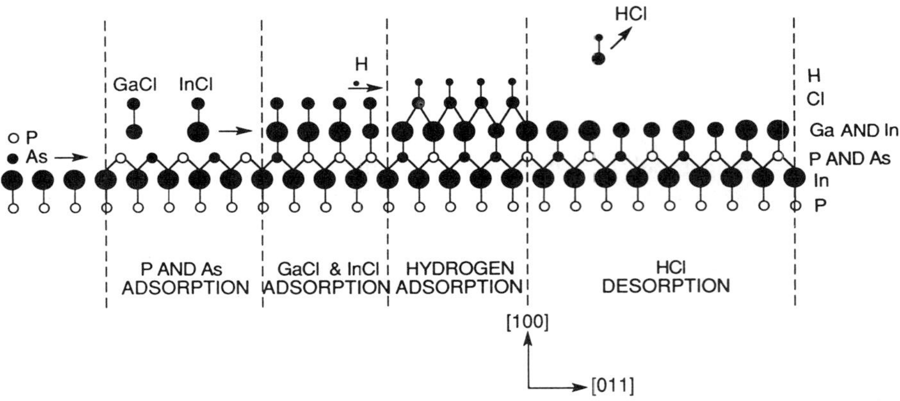

Figure 7-24. Atomic mechanisms involved in the sequential deposition of GaInAsP on InP (Reprinted with permission from John Wiley and Sons, from G. H. Olsen in *GaInAsP Alloy Semiconductors*, ed. by T. P. Pearsall, Copyright © 1982, John Wiley and Sons).

10-fold excess of the group V element. This maintains the elemental V-III impingement flux ratio > 1. In the case of GaAs and $Al_xGa_{1-x}As$, this condition results in stable stoichiometric film growth for long deposition times. In contrast to this so-called As-stabilized growth, there is Ga-stabilized growth, which occurs when the flux ratio is approximately 1. An excess of Ga atoms is to be avoided, though, because it tends to cause clustering into molten droplets. The (100) and (111) surfaces of GaAs and related compounds exhibit a variety of reconstructed surface geometries dependent on growth conditions and subsequent treatments. For As- and Ga-stabilized growth, (2×4) and (4×2) reconstructions, respectively, have generally been observed on (100)GaAs. Other structures (i.e., $C(2 \times 8)$ As and $C(8 \times 2)$ Ga) have also been reported for the indicated stabilized structures. To complicate matters further, intermediate structures, e.g., (3×1), (4×6), (3×6), as well as mixtures also exist within narrow ranges of growth conditions. The complex issues surrounding the existence and behavior of these surface reconstructions are being actively researched.

During epitaxial film deposition of multicomponent semiconductors, the mechanisms of substrate chemical reactions and atomic incorporation can be quite complex. For example, a proposed model for sequential deposition of the first two monolayers of GaInAsP on an InP substrate is depicted in Fig. 7-24 for the hydride process (Ref. 24). The first step is suggested to involve adsorption of P and As atoms. Then GaCl and InCl gas molecules also adsorb in such a way that the Cl atoms dangle outward from the surface. Next, gaseous atomic hydrogen adsorbs and reacts with the Cl atoms to form HCl molecules, which then desorb. Now the process repeats with P and As adsorption, and when the cycle is completed another bilayer of quaternary alloy film deposits. This picture accounts for single-crystal film growth and the development of variable As–P and Ga–In stoichiometries.

7.6.2. In Situ Film Characterization

This section deals with techniques that are capable of monitoring the structure and composition of epitaxial films during in situ growth. Both LEED and RHEED have this ability. They are distinguished in Fig. 7-25. An ultrahigh-vacuum environment is a necessity for both methods because of the sensitivity of diffraction to adsorbed impurities and the need to eliminate electron-beam scattering by gas molecules. In LEED a low-energy electron beam ($\sim$ 10–1000 eV) impinges normally on the film surface and only penetrates a few angstroms below the surface. Bragg's law for both lattice periodicities in the surface plane results in cones of diffracted electrons emanating along forward and backscat-

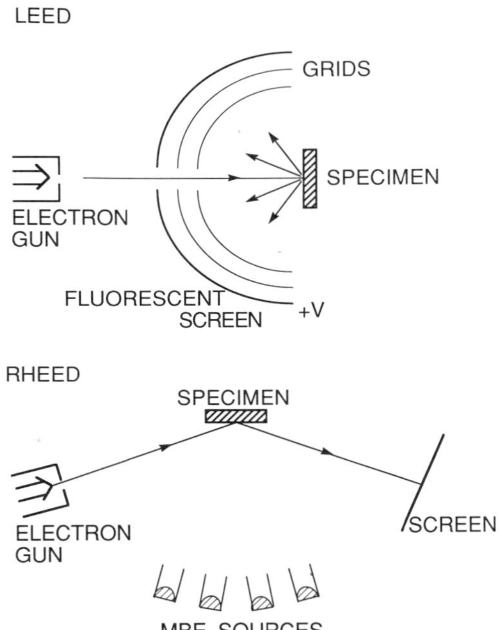

Figure 7-25. Experimental arrangements of LEED and RHEED techniques.

tered directions. Simultaneous satisfaction of the diffraction conditions means that constructive interference occurs where the cones intersect along a set of lines or beams radiating from the surface. These backscattered beams are intercepted by a set of grids raised to different electric potentials. The first grids encountered retard the low-energy inelastic electrons from penetrating. The desired diffracted (elastic) electrons of higher energy pass through and, accelerated by later grids, produce illuminated spots on the fluorescent screen.

In RHEED the electron beam is incident on the film surface at a grazing angle of a few degrees at most. Electron energies are much higher than for LEED and range from 5 to 100 keV. An immediate advantage of RHEED is that the measurement apparatus does not physically interfere with deposition sources in an MBE system the way LEED does. This is one reason why RHEED has become the preferred real-time film characterization accessory in MBE systems.

Both LEED and RHEED patterns of the (7×7) structure of the Si(111) surface are shown in Fig. 7-26. To obtain some feel for the nature of these diffraction patterns, we think in terms of reciprocal space. Arrays of reciprocal lattice points form rods or columns of reciprocal lattice planes shown as vertical lines pointing normal to the real surface. They are indexed as (10), (20), etc., in Fig. 7-27. Consider now an electron wave of magnitude $2\pi/\lambda$ propagating in the direction of the incident radiation and terminating at the origin of the reciprocal lattice. Following Ewald, we draw of sphere of radius $2\pi/\lambda$ about the center. A property of this construction is that the only possible directions of the diffracted rays are those that intersect the reflecting sphere at reciprocal lattice points as shown. To prove this, we note that the normal to the reflecting plane is the vector connecting the ends of the incident and diffracted rays. But this vector is also a reciprocal lattice vector. Its magnitude is $2\pi/a$ (Eq. 7-1), where a is the interplanar spacing for the diffracting plane in question. It is obvious from the geometry that

$$\frac{2\pi}{a} = 2 \times \frac{2\pi}{\lambda} \sin\theta, \tag{7-17}$$

which reduces to Bragg's law, the requisite condition for diffraction. When the electron energies are small as in LEED, the wavelength is relatively large, yielding a small Ewald sphere. A sharp spot diffraction pattern is the result. The intense hexagonal spot array of Fig. 7-26a reflects the sixfold symmetry of the (111) plane, and the six fainter spots in between are the result of the (7×7) surface reconstruction.

In RHEED, on the other hand, the high electron energies lead to a very large Ewald sphere (Fig. 7-27). The reciprocal lattice rods have finite width due to lattice imperfections and thermal vibrations; likewise, the Ewald sphere

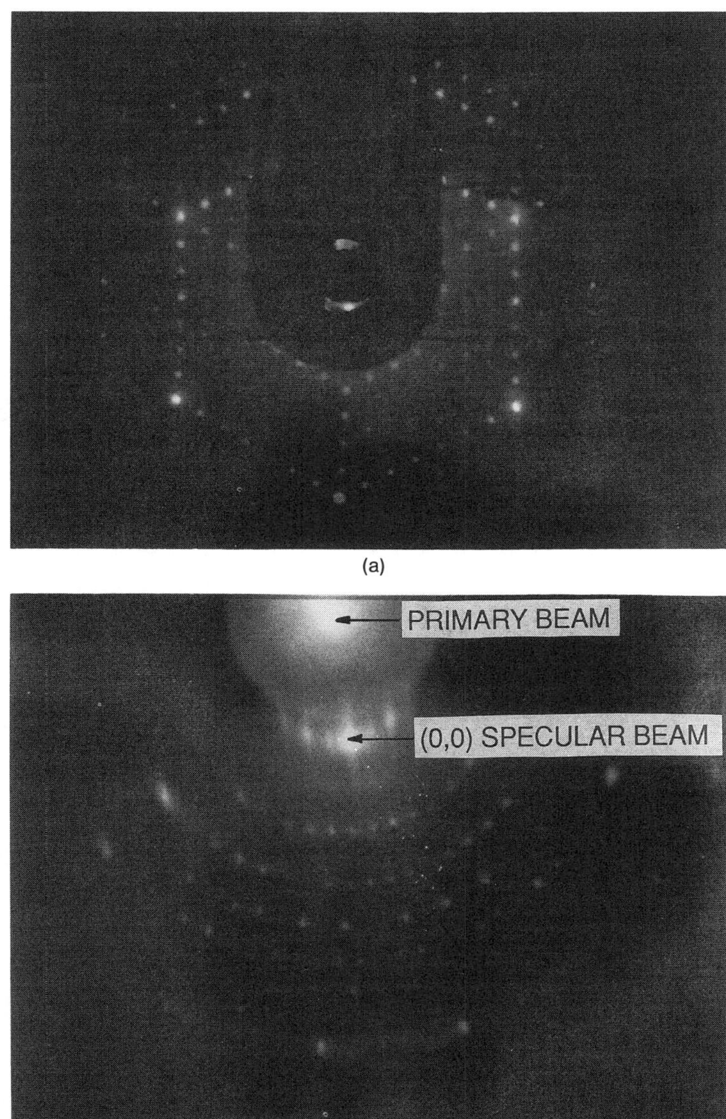

Figure 7-26. (a) LEED pattern of Si surface. (38-eV electron energy, normal incidence) (b) RHEED pattern of Si surface. (5-keV electron energy, along ⟨112⟩ azimuth) (Courtesy H. Gossmann, AT&T Bell Laboratories.)

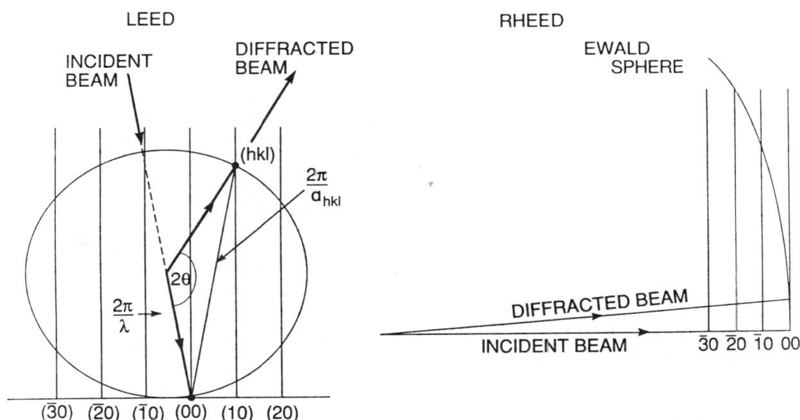

Figure 7-27. Ewald sphere construction for LEED and RHEED methods. The film plane is horizontal, and reciprocal planes are vertical lines.

is of finite width because of the incident electron energy spread. Therefore, the intersection of the Ewald sphere and rods occurs for some distance along their height, resulting in a streaked rather than spotty diffraction pattern. During MBE film growth both spotted and streaked patterns can be observed; spots occur as a result of three-dimensional volume diffraction at islands or surface asperities, whereas streaks characterize smooth layered film growth. These features can be seen in the RHEED patterns obtained from MBE-grown GaAs films (Fig. 7-28).

An important attribute of the RHEED technique is that the diffracted beam intensity is relatively immune to thermal attenuation arising from lattice vibrations. This makes it possible to observe the so-called RHEED oscillations during MBE growth at elevated temperatures. The intensity of the specular RHEED beam undergoes variations that track the step density on the growing surface layer. If we reconsider Fig. 7-22 and associate the maximum beam intensity with the flat surface where the fractional coverage $\bar{\theta} = 0$ (or $\bar{\theta} = 1$), then the minimum intensity corresponds to $\bar{\theta} = 0.5$. During deposition of a complete monolayer, the beam intensity, initially at the crest, falls to a trough and then crests again. Film growth is, therefore, characterized by an attenuated, sinelike wave with a period equal to the monolayer formation time. Under optimal conditions the oscillations persist for many layers and serve to conveniently monitor film growth with atomic resolution.

The temperature above which RHEED oscillations are expected can be easily estimated. The required diffusivity to allow a few atomic jumps to occur and smooth terraces before they are covered by a monolayer (per second) is

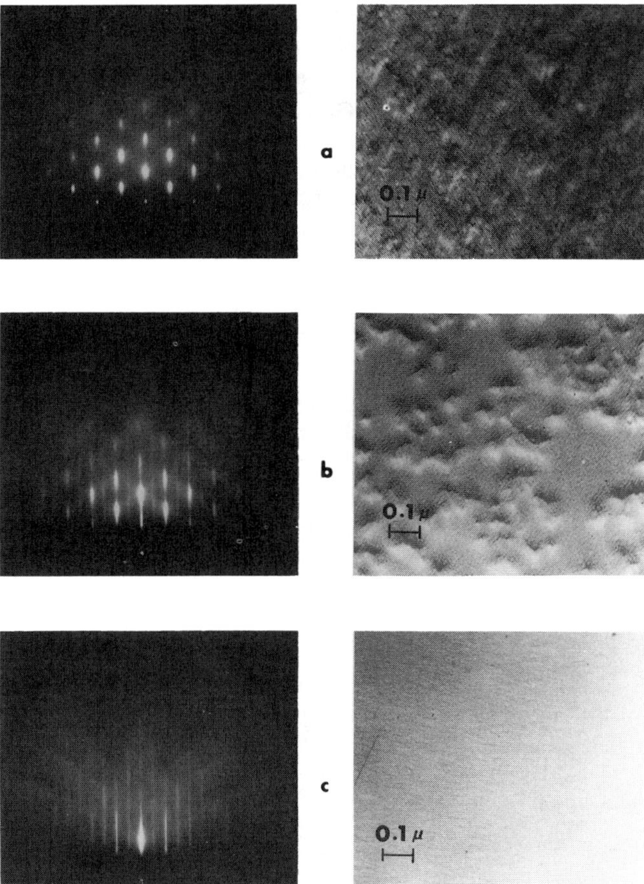

Figure 7-28. RHEED patterns (40 keV, $\langle\bar{1}\,\bar{1}0\rangle$ azimuth) and corresponding electron micrograph replicas (38,400 ×) of same GaAs surface: (a) polished and etched GaAs substrate heated in vacuum to 580 °C for 5 min; (b) 150-Å film of GaAs deposited; (c) 1 μm of GaAs deposited. (Ref. 23), (Courtesy of A. Y. Cho, AT&T Bell Laboratories).

roughly 10^{-15} cm^2/sec. By the example in Section 5.3.1, RHEED oscillations are predicted to occur above $0.2T_M$, $0.12T_M$, and $0.03T_M$ on group IV elements, metal, and alkali halide substrates, respectively, in reasonable agreement with experiment.

7.6.3. X-ray Diffraction Analysis of Epitaxial Films (Refs. 22, 25)

Let us suppose we wish to nondestructively measure the composition of a ternary epitaxial film of $Al_xGa_{1-x}As$ on GaAs to an accuracy of 2%. One way to do this is to use the connection between the lattice parameter a_0 and x.

Vegard's law then suggests that a_0 must be measured to a precision of

$$\frac{\Delta a_0}{a_0} = \frac{2}{100}\left[\frac{a_0(\text{AlAs}) - a_0(\text{GaAs})}{a_0(\text{GaAs})}\right] = 2.8 \times 10^{-5}$$

or 1 part in over 35,000. This is quite a formidable challenge, and neither LEED nor RHEED can even remotely approach such a capability. X-ray diffraction methods can however, but not easily. By Bragg's law (Eq. 7-15), differentiation yields

$$\frac{\Delta a}{a} = \frac{\Delta\lambda}{\lambda} - \frac{\Delta\theta}{\tan\theta}. \tag{7-18}$$

This equation reveals the inadequacy of conventional X-ray diffraction methods in meeting the required measurement precision. For example, typical CuKα ($\lambda = 1.5406$ Å) radiation from an X-ray tube exhibits a so-called spectral dispersion of 0.00046 Å, so $\Delta\lambda/\lambda = 0.0003$. This causes unacceptable diffraction peak broadening. In addition, the angular divergence of the beam must be several seconds of arc, and it is not possible to achieve this with usual slit-type collimation.

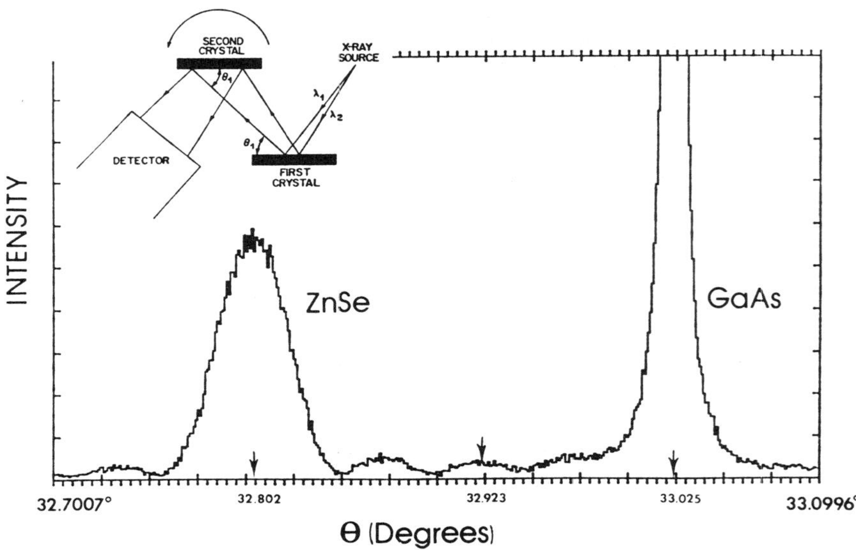

Figure 7-29. Rocking curve for (004) reflection of ZnSe on GaAs. (Courtesy of B. Greenberg, Philips Laboratories, North American Philips Corp.) Inset: Schematic of high-resolution double-crystal diffractometer. (From Ref. 25).

For these reasons, the high-resolution *double-crystal* diffractometer, shown schematically in Fig. 7-29, is indispensible. It has three special features:

1. Very high angular stepping accuracy on the θ axis (i.e., $\sim$ 1 arc second)
2. Very good angular collimation of the incident X-ray beam (i.e., $<$ 10 arc seconds)
3. Elimination of peak broadening due to spectral dispersion

The diffractometer consists of a point-focus X-ray source of monochromatic radiation that falls on a first collimator crystal composed of the same material as the sample epitaxial film (second crystal). When the Bragg condition is satisfied, both crystals are precisely parallel. The Bragg condition is simultaneously satisfied for all source wavelength components—i.e., no wavelength dispersion. During measurement, the sample is rotated or *rocked* through a very small angular range, bringing planes in and out of the Bragg condition. The resulting rocking curve diffraction pattern contains the very intense substrate peak that serves as the internal standard against which the position of the low-intensity epitaxial film peak is measured.

The following example (Ref. 26) involving ZnSe, a potential blue laser material, illustrates the power and importance of the technique. A rocking curve of an 1100-Å film of ZnSe grown epitaxially on a (001) GaAs substrate is shown in Fig. 7-29 for the (004) reflection. In GaAs, a_0 = 5.6537 Å and $a_{(004)}$ = 1.4134 Å. Since λ = 1.5406 Å, Bragg's law yields θ = 33.025°. For ZnSe, a_0 = 5.6690 Å ($a_{(004)}$ = 1.4173), and the expected Bragg angle for *unstrained* ZnSe is 32.923°. But the actual (004) peak appears at 32.802°, which corresponds to $a_{(004)}$ = 1.4219 Å. To interpret these findings, note that the misfit (Eq. 7-3) in this system is -0.27%, and, hence, ZnSe is biaxially compressed in the film plane. Since the film thickness is less than d_c (Eq. 7-7), it has grown pseudomorphically or coherently with GaAs; we can therefore assume that ZnSe has the same lattice constant as GaAs in the interfacial plane. However, normal to this plane the ZnSe lattice expands and assumes a tetragonal distortion. The measured increase in the (004) interplanar spacing of ZnSe is thus consistent with this explanation.

Before leaving the subject of X-ray diffraction, we briefly comment on its application in the characterization of epitaxial superlattices. These structures, discussed further in Section 14.7, contain a synthesized periodicity, associated with numerous alternating layers, superimposed on the crystalline periodicity of each individual layer. Resulting diffraction patterns consist of the intense substrate peak flanked on either side by a satellite structure related to the

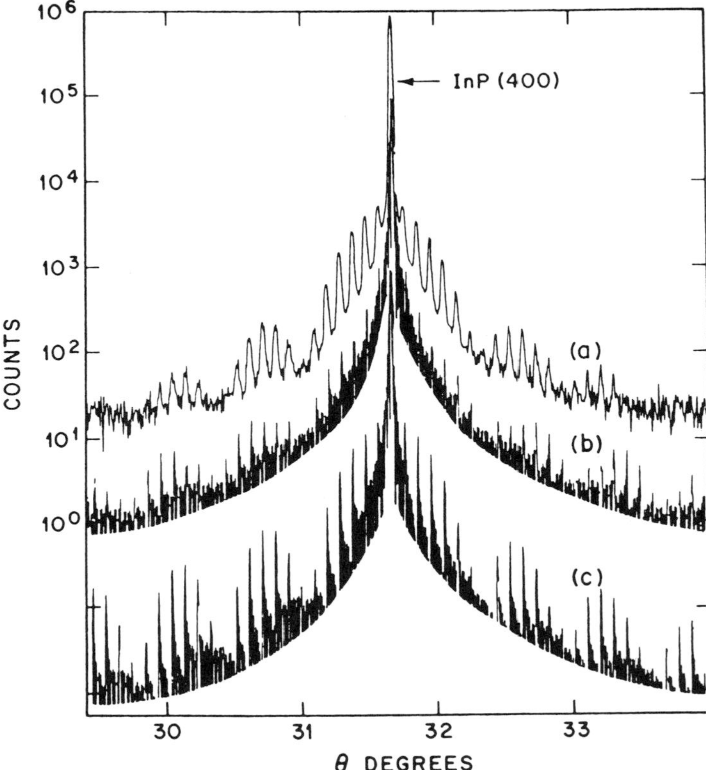

Figure 7-30. High-resolution X-ray rocking curve of a 10-period $Ga_{0.47}In_{0.53}As$–InP superlattice with d_s = 540 Å and 79-Å-thick $Ga_{0.47}In_{0.53}As$ layers (data taken with four-crystal diffractometer): (a) actual data; (b) simulation assuming no interfacial strain; (c) simulation assuming strained layers. (From Ref. 27).

superlattice period d_s. An extension of Bragg's law gives

$$d_s = \frac{(n_i - n_j)\lambda}{2(\sin\theta_i - \sin\theta_j)}, \tag{7-19}$$

where d_s is the thickness of a neighboring pair of film layers, and n_i and n_j are the diffraction orders. As an illustration, the (004) rocking curve for the 10-period $Ga_{0.47}In_{0.53}As$–InP superlattice is shown in Fig. 7-30a. Interestingly, X-ray rocking curves can be computer-modeled to simulate compositional and dimensional information on superlattices containing abrupt interfaces with remarkable precision and sensitivity. Thus, curve *c*, which closely fits the data, models the case where opposite interfaces of each $Ga_{0.47}In_{0.53}As$ layer are strained differently. Curve b, on the other hand, a poorer fit, models the case where interfacial strain is omitted. High-resolution X-ray diffraction methods reveal the excellent microscopic detail with which epitaxial films can be investigated.

7.7. Conclusion

Even after coverage in this as well as Chapter 5, additional references to epitaxial films are scattered in various contexts throughout the remainder of the book. The most extensive treatment, located in Chapter 14, is devoted largely to superlattice structures and emerging electronic devices based on them.

Exercises

1. During a drought, there is frequently enough moisture in the atmosphere to produce clouds but rain does not fall. Comment on the practice of seeding clouds with crystals of AgI to induce ice nucleation and rain formation. [Note: The crystal structure of ice is hexagonal with lattice constants of a = 4.52 Å, c = 7.36 Å; the crystal structure of AgI is also hexagonal with a = 4.58 Å and c = 7.49 Å.]

2. Fe thin films grown on single-crystal Al substrates were found to be essentially dislocation-free to a thickness of 1400 Å, whereas misfit dislocations appeared with thicker films. If the lattice parameters of Fe and Al are 2.867 Å and 4.050 Å, respectively, what are the probable indices describing the epitaxial interface crystallography?

3. A monatomic FCC material has a lattice parameter of 4 Å. For the (110) and (111) surfaces,

   a. sketch the direct crystallographic net indicating the primitive unit cell.
   b. draw the reciprocal net.
   c. compare the patterns you drew with the ball model structures and laser diffraction patterns of Fig. 5-22.
   d. calculate the spacing between rows (hk).
   e. LEED patterns are generated at normal incidence for electron energies of 50 and 200 eV. The crystal surface–screen distance is 200 mm. Index the resulting diffraction patterns that would be seen on the 180° sector screen.

4. a. Calculate the lattice misfit between GaAs and Si.

b. What is the critical thickness for pseudomorphic growth of GaAs films on Si? Is this thickness sufficient for fabrication of devices?

c. Even though GaP films are more closely lattice-matched to Si, what difficulties do you foresee in the high-temperature epitaxial growth of this material on Si substrates?

5. You are asked to suggest II-VI and III-V compounds as heteroepitaxial combinations for potential semiconductor device applications. Mention two such systems that appear promising and indicate the misfit for each.

6. Suppose monolayer formation depicted in Fig. 7-22 corresponds to (2×1) growth. Sketch the next monolayer if growth leads to the (1×2) orientation.

7. After 10 min at 800 °C, a 3-μm-thick layer of GaAs was observed to form for both equilibrium and step-cooling LPE growth mechanisms.

a. How thick were the respective GaAs films after 5 min?

b. At what time will the growth *rates* for equilibrium and step cooling be identical?

8. If the temperature regulation in effusion cells employed in MBE is ± 2 °C, what is the percent variation in the flux of atoms arriving at the substrate for the deposition of GaAs films (Ga evaporated at 1200 K, As_2 evaporated at 510 K.)

9. Sequential layers of GaAs and AlGaAs films were grown by MBE. The GaAs beams were on throughout the deposition, which lasted 1.5 h. The Al beam was alternately on for 0.5 min and off for 1 min during the entire run. Film thickness measurements showed that 1.80 μm of GaAs and 0.35 μm of AlAs were deposited.

a. What are the growth rates of GaAs and $Al_xG_{1-x}As$?

b. What is x, the atom fraction of Al in $Al_xGa_{1-x}As$?

c. What are the thicknesses of the GaAs and $Al_xGa_{1-x}As$ layers?

d. How many layers of each film were deposited? (From A. Gossard, AT&T Bell Laboratories.)

10. It is desired to make diode lasers that emit coherent radiation with a wavelength of 1.24 μm. For this purpose, III-V compounds or ternary solid-solution alloys derived from them can be utilized. At least four possible compound combinations (alloys) will meet the indicated specifications. For each alloy specify the original pair of binary compounds, the composition, and the lattice constant. (Assume linear mixing laws.)

11. The quaternary $Ga_xIn_{1-x}As_yP_{1-y}$ alloy semiconductors have an energy gap and lattice parameter given, respectively, by E_g(eV) = 1.35 − 0.72y + 0.12y^2 and $a_0(x, y)$(Å) = 0.1894y − 0.4184x + 0.0130xy + 5.869. If it is desired to produce light-wave devices operating at 1.32 and 1.55 μm, calculate the values for x and y, assuming perfect lattice matching to InP.

12. If the residence time of Ga adatoms on GaAs is 10 sec at 895 °C and 7 sec at 904 °C, what is the expected residence time at 850 °C?

13. An engineer attached a new cylinder labeled 10% AsH_3 in H_2 to a CVD reactor growing epitaxial GaAs films. Instead of the usual dark gray film, an orange deposit formed on the reactor walls. What went wrong? (From R. Dupuis, AT&T Bell Laboratories.)

14. One strategy for producing GaAs on Si first involves the deposition of a lattice-matched GaP layer. Next, $GaAs_{0.5}P_{0.5}$ is deposited and forms a strained epitaxial layer between GaP and the topmost GaAs film.

a. Indicate the variation in lattice parameter through the structure from Si to GaAs.
b. What is the nature of the lattice deformation at the GaP–$GaAs_{0.5}P_{0.5}$ and $GaAs_{0.5}P_{0.5}$–GaAs interfaces?
c. Very roughly estimate the required thicknesses of both GaP and $GaAs_{0.5}P_{0.5}$ for pseudomorphic growth.

15. X-ray rocking curves for an epitaxial GaInAsP film lattice-matched to InP were recorded for the (002), (006), and (117) reflections using CuKα radiation.

a. What are the Bragg angles for these reflections in InP?
b. Suppose the lattice mismatch between the epilayer and substrate is 1.3×10^{-4}. What Bragg angles correspond to the InGaAsP peaks.
c. For which reflection is the peak separation greatest?

References

1. L. Royer, *Bull. Soc. Fr. Mineral Cristallogr.* **51**, 7 (1928).
2. M. Prutton, *Surface Physics*, 2nd ed., Clarendon Press, Oxford (1983).
3. G. A. Prinz, *MRS Bull.* **XIII(6)**, 28 (1988).
4. F. C. Frank and J. H. van der Merwe, *Proc. Roy. Soc.* **A189**, 205 (1949).

5.* J. W. Matthews, in *Epitaxial Growth*, ed. J. W. Matthews, Academic Press, New York (1975).
6. J. W. Matthews, *J. Vac. Sci. Tech.* **12**, 126 (1975).
7.* G. B. Stringfellow, *Rep. Prog. Phys.* **45**, 469 (1982).
8. J. C. Bean, in *Silicon-Molecular Beam Epitaxy*, eds. E. Kasper and J. C. Bean, CRC Press, Boca Raton, FL (1988).
9. J. C. Bean, *Physics Today* **39(10)**, 2 (1986).
10. S. M. Sze, *Semiconductor Devices—Physics and Technology*, McGraw-Hill, New York (1985).
11.* J. W. Mayer and S. S. Lau, *Electronic Materials Science: For Integrated Circuits in Si and GaAs*, Macmillan, New York (1990).
12. B. R. Pamplin, in *Handbook of Chemistry and Physics*, ed. R. C. Weast, CRC Press, Boca Raton, FL (1980).
13. H. Ichinose, Y. Ishida, and H. Sakaki, *JOEL News* **26E(1)**, 8 (1988).
14. T. F. Kuech, D. J. Wolford, R. Potoemski, J. A. Bradley, K. H. Kelleher, D. Yan, J. P. Farrell, P. M. S. Lesser, and F. H. Pollack, *Appl. Phys. Lett.* **51**, 505 (1987).
15. H. C. Casey, D. D. Sell, and M. B. Panish, *Appl. Phys. Lett.* **24**, 63 (1974).
16. R. L. Gunshor, L. A. Kolodziejski, A. V. Nurmikko, and N. Otsuka, *Ann. Rev. Mater. Sci.* **18**, 325 (1988).
17. G. K. Celler, McD. Robinson, and L. E. Trimble, *J. Electrochem. Soc.* **132**, 211 (1985).
18. M. L. Green, D. Brasen, H. Luftman, and V. C. Kannan, *J. Appl. Phys.* **65**, 2558 (1989).
19.* A. Y. Cho, *Thin Solid Films* **100**, 291 (1983).
20.* K. Ploog, *Ann. Rev. Mater. Sci.* **11**, 171 (1981).
21.* M. B. Panish and H. Temkin, *Ann. Rev. Mater. Sci.* **19**, 209 (1989).
22.* M. A. Herman and H. Sitter, *Molecular Beam Epitaxy—Fundamentals and Current Status*, Springer-Verlag, Berlin (1989).
23.* A. Y. Cho and J. R. Arthur, *Prog. Solid State Chem.* **10**, 157 (1975).
24. G. H. Olsen, in *GaInAsP Alloy Semiconductors*, ed. T. P. Pearsall, Wiley, New York (1982).
25.* A. T. Macrander, *Ann. Rev. Mater. Sci.* **18**, 283 (1988).
26. B. Greenberg, Philips Laboratories, North American Philips Corp. Private communication.
27. J. M. Vandenberg, M. B. Panish, H. Temkin, and R. A. Hamm, *Appl. Phys. Lett.* **53**, 1920 (1988).

*Recommended texts or reviews.

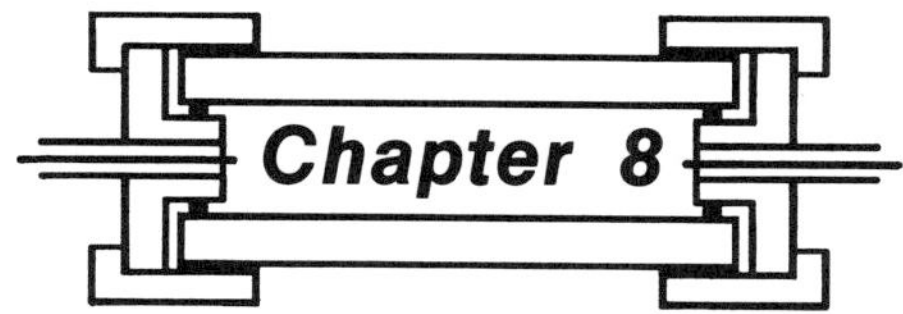

Interdiffusion and Reactions in Thin Films

8.1. Introduction

There is hardly an area related to thin-film formation, properties, and performance that is uninfluenced by mass-transport phenomena. This is especially true of microelectronic applications, where very small lateral as well as depth dimensions of device features and film structures are involved. When these characteristic dimensions (d) become comparable in magnitude to atomic diffusion lengths, then compositional changes can be expected. New phases such as precipitates or layered compounds may form from ensuing reactions, altering the initial film integrity. This, in turn, frequently leads to instabilities in the functioning of components and devices that are manifested by such effects as decrease in conductivity as well as short- or even open-circuiting of conductors, lack of adhesion, and generation of stress. The time it takes for such effects to evolve can be roughly gauged by noting that the diffusion length is given by $\sim 2\sqrt{Dt}$, where D and t are the appropriate diffusivity and time, respectively. Therefore $t \approx d^2/4D$. As we shall see, D values in films are relatively high even at low temperatures, so small film dimensions serve to make these characteristic times uncomfortably short. Such problems frequently surface when neighboring combinations of materials are chemically reactive.

For example, consider the the pitfalls involved in designing a Cu–Ni film couple as part of the contact structure for solar cells (Ref. 1). Readily available high-temperature data in *bulk* metals extrapolated to 300 °C yield a value of 3.8×10^{-24} cm^2/sec for the diffusion coefficient of Cu in Ni. For a 1000Å thick Ni film, the interdiffusion time is thus predicted to be $(10^{-5})^2/4(3.8 \times 10^{-24})$ sec, or over 200,000 years! Experiment, however, revealed that these metals intermixed in less than an hour. When colored metal films are involved, as they are here, the eye can frequently detect the evidence of interdiffusion through color or reflectivity changes. The high density of defects, e.g., grain boundaries and vacancies, causes deposited films to behave differently from bulk metals, and it is a purpose of this chapter to quantitatively define the distinctions. Indeed, a far more realistic estimate of the Cu–Ni reaction time can be made by utilizing the simple concepts developed in Section 8.2. Other examples will be cited involving interdiffusion effects between and among various metal film layer combinations employed in Si chip packaging applications. Practical problems associated with making both stable contacts to semiconductor surfaces and reliable interconnections between devices have been responsible for generating the bulk of the mass-transport-related concerns and studies in thin films. For this reason, issues related to these extremely important subjects will be discussed at length.

While interdiffusion phenomena are driven by chemical concentration gradients, other mass-transport effects take place even in homogeneous films. These rely on other driving forces such as electric fields, thermal gradients, and stress fields, which give rise to respective electromigration, thermomigration, and creep effects that can similarly threaten film integrity. The Nernst–Einstein equation provides an estimate of the characteristic times required for such transport effects to occur. Consider a narrow film stripe that is as wide as it is thick. If it can be assumed that the volume of film affected is $\sim d^3$ and the mass flows through a cross-sectional area d^2, then the appropriate velocity is d/t. By utilizing Eq. 1-35, we conclude that $t \approx RTd/DF$. Large driving forces (F), which sometimes exist in films, can conspire with both small d and high D values to reduce the time to an undesirably short period. As circuit dimensions continue to shrink in the drive toward higher packing densities and faster operating speeds, diffusion lengths will decrease and the surface-area-to-volume ratio will increase. Despite these tendencies, processing temperatures and heat generated during operation are not being proportionately reduced. Therefore, interdiffusion problems are projected to persist and even worsen in the future.

In addition to what may be termed reliability concerns, there are beneficial mass-transport effects that are relied on during processing heat treatments in films. Aspects of both of these broad applications will be discussed in this chapter in a fundamental way within the context of the following subjects:

8.2. Fundamentals of Diffusion
8.3. Interdiffusion in Metal Alloy Films
8.4. Electromigration in Thin Films
8.5. Metal–Semiconductor Reactions
8.6. Silicides and Diffusion Barriers
8.7. Diffusion During Film Growth

Before proceeding, the reader may find the survey of diffusion phenomena given in Chapter 1 useful and wish to review it.

8.2. Fundamentals of Diffusion

8.2.1. Comparative Diffusion Mechanisms

Diffusion mechanisms attempt to describe the details of atomic migration associated with mass transport through materials. The resulting atom movements reflect the marginal properties of materials in that only a very small fraction of the total number of lattice sites, namely, those that are unoccupied, interstitial, or on surfaces, is involved. An illustration of the vacancy mechanism for diffusion was given on p. 36. Similarly, the lattice diffusivity D_L, in terms of previously defined quantities, can be written as

$$D_L = D_0 \exp - E_L / RT, \tag{8-1}$$

where E_L is the energy for atomic diffusion through the lattice on a per-mole basis. In polycrystalline thin films the very fine grain size means that a larger proportion of atom-defect combinations is associated with grain boundaries, dislocations, surfaces, and interfaces, relative to lattice sites, than is the case in bulk solids. Less tightly bound atoms at these nonlattice sites are expected to attract different point-defect populations and be more mobile than lattice atoms. Although the detailed environment may be complex and even varied, the time-averaged atomic transport is characterized by the same type of Boltzmann behavior expressed by Eq. 8-1. Most importantly, the activation energies for grain-boundary, dislocation, and surface diffusion are expected to be smaller than E_L, leading to higher diffusivities. Therefore, such heterogeneities and defects serve as diffusion paths that short-circuit the lattice.

In order to appreciate the consequences of allowing a number of uncoupled transport mechanisms to freely compete, we consider the highly idealized

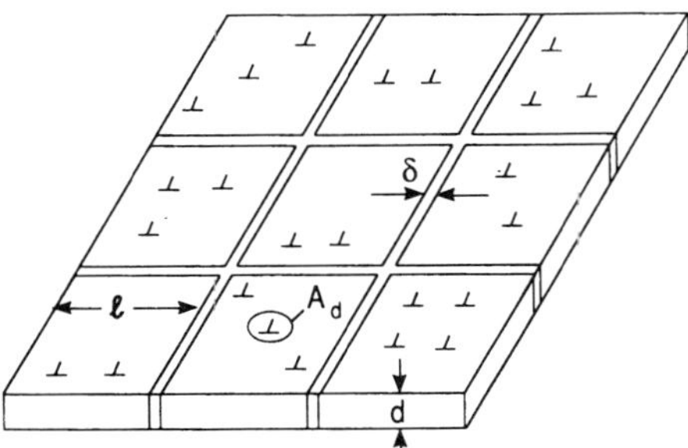

Figure 8-1. Highly idealized polycrystalline film containing square grains, grain boundaries, and dislocations.

polycrystalline film matrix in Fig. 8-1. Grain-boundary slabs of width δ serve as short-circuit diffusion paths even though they may only be 5–10 Å wide. They separate square-shaped grains of side l. Within the grains are dissociated dislocations oriented normal to the film surface. They thread the latter with a density ρ_d per cm^2, and diffusion is assumed to occur through the dislocation core whose cross-sectional area is A_d. Parallel transport processes normal to the film plane are assumed to occur for each mechanism. Under these conditions, the number of atoms ($\dot{n}_i$) that flow per unit time is essentially equal to the product of the appropriate diffusivity (D_i), concentration gradient $(dc/dx)_i$, and transport area involved. Therefore,

Lattice: $$\dot{n}_L = D_L l^2 \left(\frac{dc}{dx} \right)_L, \tag{8-2a}$$

Grain Boundary: $$\dot{n}_b = \delta D_b l \left(\frac{dc}{dx} \right)_b, \tag{8-2b}$$

Dislocation: $$\dot{n}_d = A_d D_d l^2 \rho_d \left(\frac{dc}{dx} \right)_d, \tag{8-2c}$$

where L, b and d refer to lattice, grain-boundary, and dislocation quantities.

The importance of short-circuit mass flow relative to lattice diffusion can be quantitatively understood in the case of face-centered cubic metals where data for the individual mechanisms are available. A convenient summary of resulting diffusion parameters is given by (Ref. 2)

Lattice: $$D_L \approx 0.5 \exp - 17.0 T_M / T \text{ cm}^2/\text{sec}, \tag{8-3a}$$

Grain Boundary: $$\delta D_b \approx 1.5 \times 10^{-8} \exp - 8.9 T_M / T \text{ cm}^3/\text{sec}, \tag{8-3b}$$

Dislocation: $$A_d D_d \approx 5.3 \times 10^{-15} \exp - 12.5 T_M / T \text{ cm}^4/\text{sec}. \tag{8-3c}$$

These approximate expressions represent average data for a variety of FCC metals normalized to the reduced temperature T/T_M, where T_M is the melting point. As an example, the activation energy for lattice self-diffusion in Au is easily estimated through comparison of Eqs. 8-1 and 8-3a, which gives $E_L/RT = 17.0T_M/T$. Therefore, $E_L = 17.0RT_M$ or (17.0)(1.99 cal/mole-K)(1336 K) = 45,200 cal/mole. As a first approximation, the preceding equations can be assumed to be valid for both self- and dilute impurity diffusion. Generalized Arrhenius plots for D_L as a function of T_M/T have already been introduced in Fig. 5-6 for metal, semiconductor, and alkali halide matrices.

Regimes of dominant diffusion behavior, normalized to the same concentration gradient, can be mapped as a function of l and ρ_d by equating the various $\dot{n}_i$ in Eq. 8-2. The equations of the boundary lines separating the operative transport mechanisms are thus

$$\frac{1}{l} = \frac{D_L}{\delta D_b}, \quad \rho_d = \frac{D_L}{A_d D_d}, \quad \text{and} \quad \frac{1}{l} = \frac{A_d D_d \rho_d}{\delta D_b}.$$

These are plotted as $\ln 1/l$ versus ρ_d in Fig. 8-2 at four levels of T/T_M,

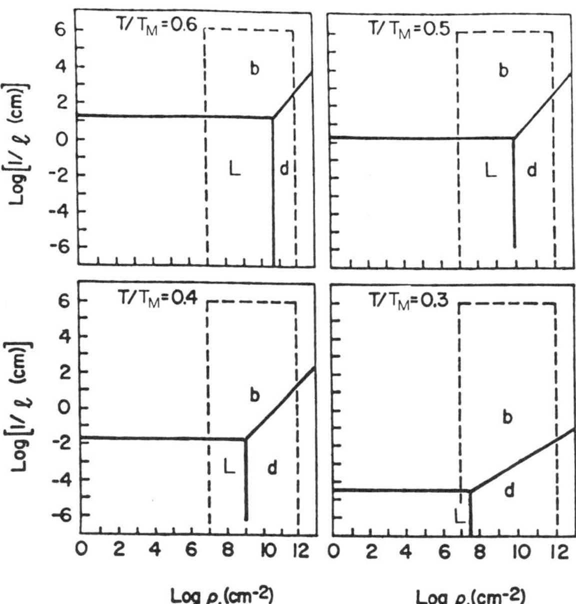

Figure 8-2. Regimes of dominant diffusion mechanisms in FCC metal films as a function of temperature. (Reprinted with permission from Elsevier Sequoia, S.A. from R. W. Balluffi and J. M. Blakely, *Thin Solid Films* **25**, 363, 1975).

employing Eqs. 8-3a, b, c. The broken rectangles represent the range of thin film values for l and ρ_d that occur in practice. For typical metal films with a grain size of 1 μm or less, grain-boundary diffusion dominates at all practical temperatures. Similarly, for dislocation-free epitaxial films where $1/l = 0$, lattice diffusion dominates. Transport at these extremes is intuitively obvious. Where the film structure is such that combinations of mechanisms are operative, different admixtures will occur as a function of temperature. Generally, lower temperatures will favor grain boundary and dislocation short-circuiting relative to lattice diffusion.

Surface diffusion is another transport mechanism of relevance to thin films because of the large ratio of the number of surface-to-bulk atoms. As noted in Chapter 5, this mechanism plays an important role in film nucleation and

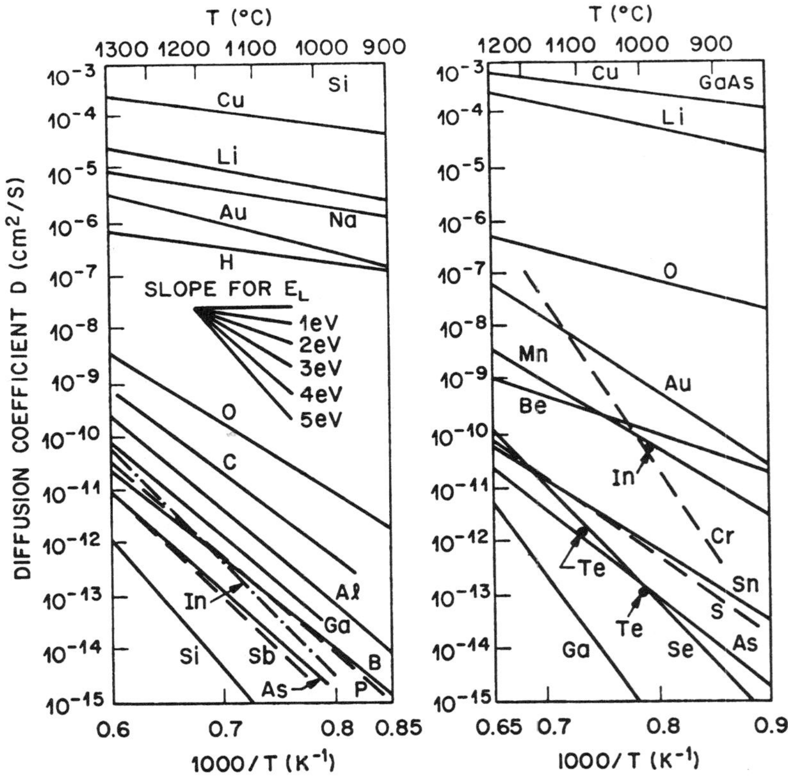

Figure 8-3. Diffusion coefficients of various elements in Si and GaAs as a function of temperature. (Reprinted with permission from John Wiley and Sons, from S. M. Sze, *Semiconductor Devices: Physics and Technology*, Copyright © 1985, John Wiley and Sons).

growth processes. Reduced parameters describing measured surface transport in FCC metals have been suggested (Ref. 3); e.g.,

$$D_s \approx 0.014 \exp - \frac{6.54 T_M}{T} \text{ cm}^2/\text{sec} \qquad \text{for } \frac{T_M}{T} > 1.3.$$

It is well known, however, that surface diffusion varies strongly with ambient conditions, surface crystallography, and the nature and composition of surface and substrate atoms.

Systematics similar to those depicted in Fig. 8-2 also govern diffusion behavior in ionic solids and semiconductors where grain boundaries and dislocations are known to act as short-circuit paths. However, complex space-charge effects in ionic solids make a clear separation of lattice and grain-boundary diffusion difficult in these materials. In semiconductors a great deal of impurity diffusion data exists, and these are used in designing and analyzing doping treatments for devices. This is a specialized field, and complex modeling (Ref. 4) is required to accurately describe diffusion profiles. Due to the importance of Si and GaAs films, preferred lattice dopant diffusion data are presented in Fig. 8-3 (Ref. 5). Some very recent data on diffusion of noble metals Au, Ag, and Cu in amorphous Si films interestingly reveals that the activation energy for diffusion in the disordered matrix is very similar to values obtained for lattice diffusion in crystalline Si. (Ref. 6).

8.2.2. Grain-Boundary Diffusion

Of all the mass-transport mechanisms in films, grain-boundary (GB) diffusion has probably received the greatest attention. This is a consequence of the rather small grain size and high density of boundaries in deposited films. Rapid diffusion within individual GBs coupled with their great profusion make them the pathways through which the major amount of mass is transported. Low diffusional activation energies foster low-temperature transport, creating serious reliability problems whose origins can frequently be traced to GB involvement. This has motivated the modeling of both GB diffusion and phenomena related to film degradation processes.

The first treatment of GB diffusion appeared nearly 40 years ago. The Fisher model (Ref. 7) of GB diffusion considers transport within a semi-infinite bicrystal film initially free of diffusant, as shown in Fig. 8-4. A diffusant whose concentration C_0 is permanently maintained at plane $y = 0$ diffuses into the GB and the two adjoining grains. At low temperatures in typical polycrystalline films, it is easily shown that there is far more transport down

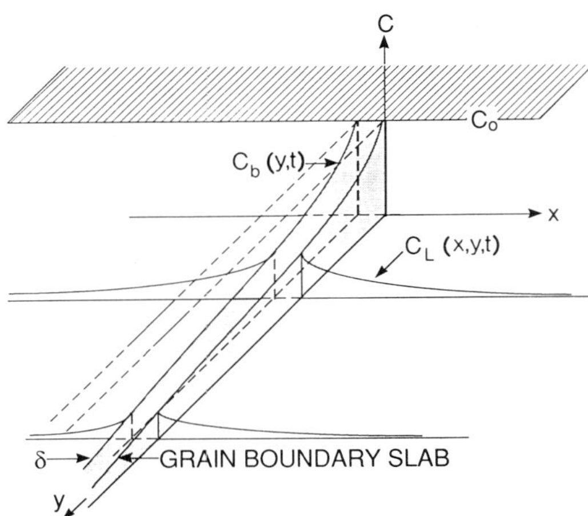

Figure 8-4. Representation of diffusional penetration down a grain boundary (y direction) with simultaneous lateral diffusion into adjoining grains (x direction).

the GB than there is into the matrix of the grains. The ratio of these two fluxes can be estimated through the use of Eqs. 8-2 and 8-3 for FCC metals; i.e.,

$$\frac{\dot{n}_b}{\dot{n}_L} = \frac{\delta D_b}{l D_L} = \frac{3 \times 10^{-8}}{l} \exp \frac{8.1\, T_M}{T}.$$

Assuming $l = 10^{-4}$ cm and $T/T_M = 1/3$, we have $\dot{n}_b/\dot{n}_l = 1.1 \times 10^7$. For this reason, we may envision transport to consist primarily of a deep rapid penetration down the GB from which diffusant subsequently diffuses laterally into the adjoining grains, building up the concentration level there. This is shown schematically in Fig. 8-4 and described mathematically by

$$C_L(x, y, t) = C_0 \exp - \left(\frac{2\sqrt{D_L}}{\delta D_b \sqrt{\pi t}} \right)^{1/2} y \cdot \operatorname{erfc} \frac{x}{2\sqrt{D_L t}}, \qquad (8\text{-}4)$$

where $C_L(x, y, t)$ is the diffusant concentration at any position and time.

The Fisher analysis of the complex, coupled GB–lattice diffusion process yields simplified decoupled solutions—an exponential diffusant profile in the GB and an error function profile within adjoining grains. Experimental verification of Eq. 8-4 is accomplished by measurement of the integrated concentration $\overline{C}$ within incremental slices Δy thick (e.g., by sputtering) normal to the $y = 0$ surface; i.e.,

$$\overline{C} = \int_{-\infty}^{\infty} C_L(x, y, t)\, dx\, \Delta y = \text{const}\; e^{-(2\sqrt{D_L}/\delta D_b \sqrt{\pi t})^{1/2} y}. \qquad (8\text{-}5)$$

The last equation suggests that a plot of $\ln \bar{C}$ versus y is linear. Therefore, the useful result

$$\delta D_b = \frac{1.0}{\sqrt{\pi}} \left(\frac{d \ln \bar{C}}{dy} \right)^{-2} \left(\frac{4D_L}{t} \right)^{1/2} \tag{8-6}$$

emerges. However, in order to obtain δD_b, the value of D_L in the same system must be independently known. This poses no problem usually, since lattice diffusivity data are relatively plentiful. Exact, but far more complicated, integral solutions, that are free of the simplifications of the Fisher analysis, have been obtained by Whipple (Ref. 8) and Suzuoka (Ref. 9). A conclusion, based on these analyses, that has been extensively used is

$$\delta D_b = -0.66 \left(\frac{d \ln \bar{C}}{dy^{6/5}} \right)^{-5/3} \left(\frac{4D_L}{t} \right)^{1/2}. \tag{8-7}$$

Apart from overriding questions of correctness, the difference between Eqs. 8-6 and 8-7 is that $\ln \bar{C}$ is plotted versus y in the former and versus $y^{6/5}$ in the latter. Frequently, however, the experimental concentration profiles are not sufficiently precise to distinguish between these two spatial dependencies. It does not matter that actual films are not composed of bicrystals, but rather polycrystals with GBs of varying type and orientation; the general character of the solutions is preserved despite the geometric complexity. A schematic representation of equiconcentration profiles in a polycrystalline film containing an array of parallel GBs is shown in Fig. 8-5. At elevated temperatures the extensive amount of lattice diffusion masks the penetration through GBs. At the lowest temperatures, virtually all of the diffusant is partitioned to GBs. In between, the admixture of diffusion mechanisms results in an initial rapid penetration down the short-circuit network, which slows down as atoms leak into the lattice. The behaviors indicated in Fig. 8-5 represent the so-called A-, B-, and C-type kinetics (Ref. 10). Polycrystalline film diffusion phenomena have been studied in the B to C range for the most part. Excellent reviews of

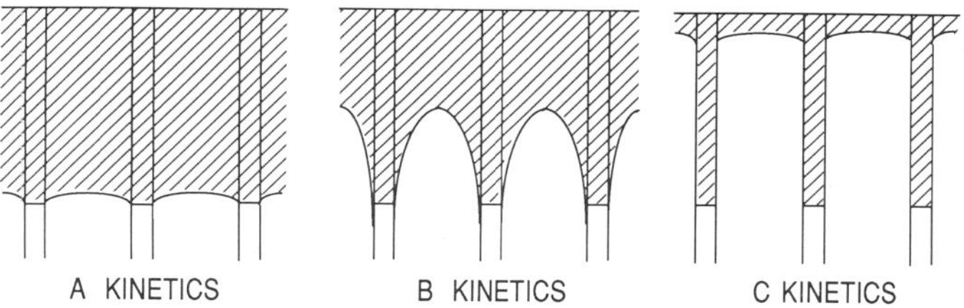

Figure 8-5. Schematic representation of type A (highest-temperature), B, and C (lowest-temperature) diffusion kinetics. (From Ref. 10).

the mathematical theories of GB diffusion including discussions of transport in these different temperature regimes, and applications to thin-film data are available (Refs. 11, 12). The best general source of this information is the volume *Thin Films—Interdiffusion and Reactions*, edited by Poate, Tu, and Mayer (Ref. 12) which also serves as an authoritative reference for much of the material discussed in this chapter. This book also contains a wealth of experimental mass transport data in thin-film systems.

The experimental measurements of the penetration of radioactive ^{195}Au into epitaxial (Fig. 8-6a) and polycrystalline (Fig. 8-6b) Au films provide a test of the above theories. They also importantly illustrate how the spectrum of diffusion behavior can be decomposed into the individual component mechanisms through judicious choice of film temperature and grain size. These data were obtained by incrementally sputter-sectioning the film, collecting the removed material in each section, and then counting its activity level. Very low

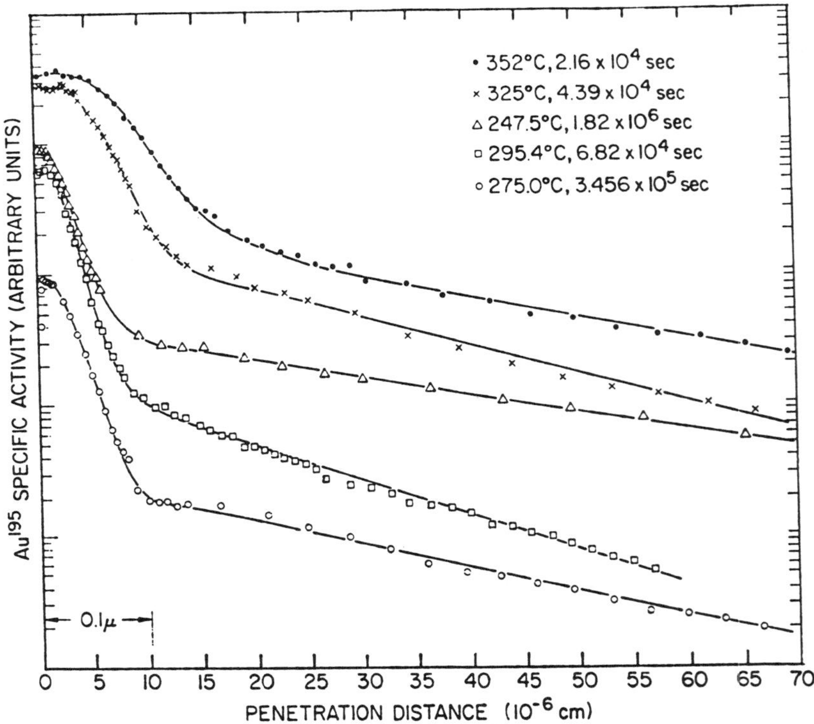

Figure 8-6a. Diffusional penetration profiles of ^{195}Au in (001) epitiaxial Au films at indicated temperatures and times. Lattice and dislocation diffusion dominate. (From Ref. 13).

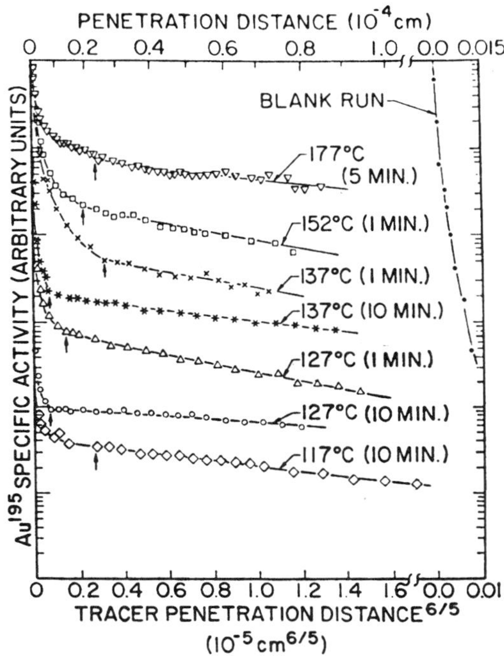

Figure 8-6b. Diffusional penetration profiles of ^{195}Au in polycrystalline films at indicated temperatures and times. Only GB diffusion is evident. (Reprinted with permission from Elsevier Sequoia, S.A., From D. Gupta and K. W. Asai, *Thin Solid Films* **22**, 121, 1974).

concentration levels can be detected because radiation-counting equipment is quite sensitive and highly selective. This makes it possible to measure shallow profiles and detect penetration at very low temperatures. The epitaxial film data display Gaussian-type lattice diffusion for the first 1000 to 1500 Å, followed by a transition to apparent dislocation short-circuit transport beyond this depth. Rather than high-angle boundaries, these films contained a density of some 10^{10} to 10^{11} dissociated dislocations per cm^2. On the other hand, extensive low-temperature GB penetration is evident in fig. 8-6b without much lattice diffusion. The large differences in diffusional penetration between these two sets of data, which are consistent with the systematics illustrated in Fig. 8-2, should be noted. For epitaxial Au a mixture of lattice and dislocation diffusion is expected for $\rho_d \approx 10^{10}/\text{cm}^2$ at temperatures of $\sim 0.4T_M$. Only GB diffusion is expected, however, at temperatures of $\sim 0.3T_M$ for a grain size of 5×10^{-5} cm, and this is precisely what was observed.

8.2.3. Diffusion in Miscible and Compound-Forming Systems

8.2.3.1. Miscible Systems. It is helpful to initiate the discussion on diffusion in miscible systems by excluding the complicating effects of grain boundaries. Bulk materials contain large enough grains so that the influence of GBs is frequently minimal. For thin films a couple where both layers are single crystals (e.g., a heteroepitaxial system) must be imagined. Under such conditions, the well-established macroscopic diffusion analyses hold. Upon interdiffusion in miscible systems, there is no crystallographic change, for this would imply new phases. Rather, each composition will be accessed at some point or depth within the film as a continuous range of solid solutions is formed. When the intrinsic atomic diffusivities are equal, i.e., $D_A = D_B$, the profile is symmetric and Eq. 1-27a governs the resultant diffusion. On the other hand, it is more common that $D_A \neq D_B$, so A and B atoms actually migrate with unequal velocities because they exchange with vacancies at different rates.

As an example of a miscible system, consider the much-studied Au–Pd polycrystalline thin-film couple in Fig. 8-7 (Ref. 15). Both AES sputter sectioning and RBS methods were employed to obtain the indicated profiles, whose apparent symmetry probably reflects the lack of a strong diffusivity dependence on concentration. It is very tempting to analyze these data by fitting them to an error-function-type solution. Effective diffusivity values could be obtained, but they would tend to have limited applicability because of the heterogeneous character of the film matrix. It must not be forgotten that

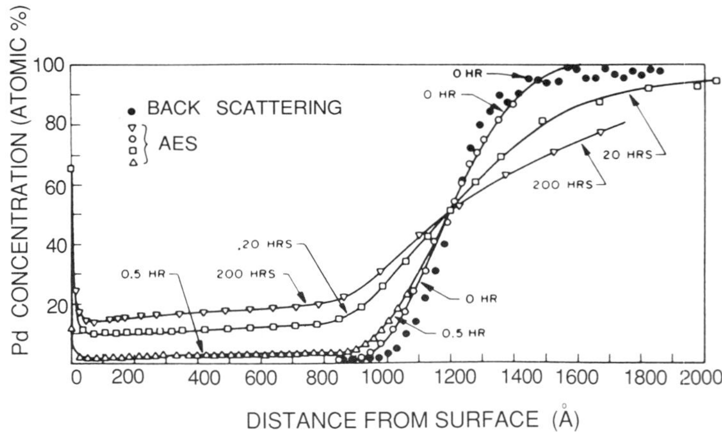

Figure 8-7. Palladium concentration profiles in a Au–Pd thin-film diffusion couple measured by RBS and AES techniques. (Reprinted with permission from Elsevier Sequoia, S.A., from P. M. Hall, J. M. Morabito and J. M. Poate, *Thin Solid Films* **33**, 107, 1976).

GB diffusion is the dominant mechanism in this couple. Therefore, the appropriate GB analysis is required in order to extract fundamental transport parameters. With this approach, it was found that a defect-enhanced admixture of GB and lattice diffusion was probably responsible for the large changes in the overall composition of the original films. Diffusional activation energies obtained for this film system are typically 0.4 times that for bulk diffusion, in accord with the systematics for GB diffusion.

8.2.3.2. Compound-Forming Systems. Many of the interesting binary combinations employed in thin-film technology react to form compounds. Since the usual configuration is a planar composite structure composed of elemental films on a flat substrate, layered compound growth occurs. The concentration–position profile in such systems is schematically indicated in Fig. 8-8. Each of the terminal phases is assumed to be in equilibrium with the intermediate compound. The compound shown is stable over a narrow rather than broad concentration range. Both types of compound stoichiometries are observed to form. With time, the compound thickens as it consumes the α phase at one interface and the β phase at the other interface.

It is instructive to begin with the simplified analysis of the kinetics of compound growth based on Fig. 8-8. Only the γ phase (compound) interface in equilibrium with α is considered. Since A atoms lost from α are incorporated into γ, the shaded areas shown are equal. With respect to the interface

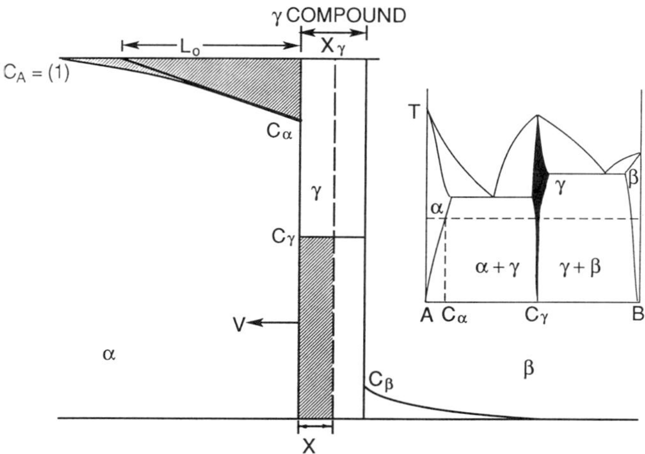

Figure 8-8. Depiction of intermediate compound formation in an A–B diffusion couple. Reaction temperature is dotted in on phase diagram.

moving with velocity V, the following mass fluxes of A must be considered:

$$\text{flux into interface} = C_\alpha V - D_\alpha \left(\frac{dC_\alpha}{dx} \right)_{\text{int}}.$$

$$\text{flux away from interface} = C_\gamma V.$$

These fluxes remain balanced for all times, so by equating them we have

$$V = \frac{dX}{dt} = \frac{D_\alpha (dC_\alpha / dx)}{C_\alpha - C_\gamma}, \tag{8-8}$$

where X is the compound layer thickness. From the simple geometric construction shown, dC_α / dx can be approximated by $(C_A - C_\alpha)/L_0$. Therefore as growth proceeds, L_0 increases while V decreases. If the shaded area within the α phase can be approximated by $(1/2)L_0(C_A - C_\alpha)$, and this is set equal to $C_\gamma X$, then $L_0 = 2C_\gamma X/(C_A - C_\alpha)$. Substituting for dC_α / dx in Eq. 8-8 leads to

$$\frac{dX}{dt} = \frac{D_\alpha}{2X} \frac{(C_A - C_\alpha)^2}{(C_\alpha - C_\gamma)C_\gamma}. \tag{8-9}$$

Upon integrating, the α-γ portion of the compound layer thickness is obtained as

$$X = \frac{\left[D_\alpha (C_A - C_\alpha)^2 \right]^{1/2} t^{1/2}}{\left[(C_\alpha - C_\gamma) C_\gamma \right]^{1/2}}. \tag{8-10}$$

A similar expression holds for the β–γ interface, and both solutions can be added together to yield the final compound layer thickness X_γ; i.e.,

$$X_\gamma = \text{const } t^{1/2}. \tag{8-11}$$

The important feature to note is that parabolic growth kinetics is predicted. Thermally activated growth is also anticipated, but with an effective activation energy dependent on an admixture of diffusion parameters from both α and β phases.

Among the important and extensively studied thin-film compound-forming systems are Al–Au (used in interconnection/contact metallurgy) and metal–silicon (used as contacts to Si and SiO_2); they will be treated later in the chapter. Parabolic growth kinetics is almost always observed in these systems. When diffusion is sufficiently rapid; however, growth may be limited by the speed of interfacial reaction. Linear kinetics varying simply as t then ensue, but only for short times. For longer times linear growth gives way to diffusion-controlled parabolic growth.

8.2.4. The Kirkendall Effect

The Kirkendall effect has served to illuminate a number of issues concerning solid-state diffusion. One of its great successes is the unambiguous identification of vacancy motion as the operative atomic transport mechanism during interdiffusion in binary alloy systems. The Kirkendall experiment requires a diffusion couple with small inert markers located within the diffusion zone between the two involved migrating atomic species. An illustration of what happens to the marker during thin-film silicide formation is shown in Fig. 8-9. Assuming that metal (M) atoms exchange sites more readily with vacancies than do Si atoms, more M than Si atoms will sweep past the marker. In effect, more of the lattice will move toward the left! To avoid lattice stress or void generation, the marker responds by shifting as a whole toward the right. The reverse is true if Si is the dominant migrating specie. Such marker motion has indeed been observed in an elegant experiment (Ref. 16) employing RBS methods to analyze the reaction between a thin Ni film and a Si wafer. Implanted Xe, which served as the inert maker, moved toward the surface of

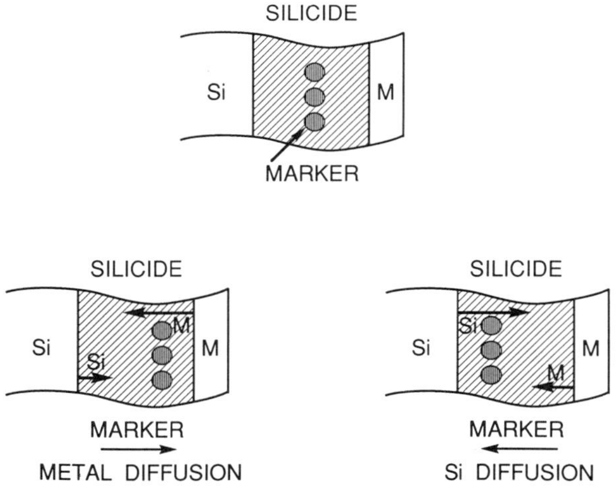

Figure 8-9. Schematic of Kirkendall marker motion during silicide formation. (Reprinted with permission from John Wiley and Sons, from J. M. Poate, K. N. Tu and J. W. Mayer, eds., *Thin Films: Interdiffusion and Reactions*, Copyright © 1978, John Wiley and Sons).

the couple during formation of Ni_2Si. The interpretation, therefore, is that Ni is the dominant diffusing specie.

8.2.5. Diffusion Size Effects (Ref. 17)

A linear theory of diffusion has been utilized to describe the various transport effects we have considered to this point and, except for this section, will be assumed for the remainder of the thin-film applications in this book. The macroscopic Fick diffusion equations defined by Eqs. 1-21 and 1-24 suffice as an operating definition of what is meant by linear diffusion theory. There are, however, nonlinear diffusion effects that may arise in thin-film structures when relatively large composition changes occur over very small distances (e.g., superlattices). To understand nonlinear effects, we reconsider atomic diffusion between neighboring planes in the presence of a free-energy gradient driving force. As a convenient starting point, two pertinent equations (1-33 and 1-35) describing this motion are reproduced here:

$$r_N = 2\nu \exp - \frac{G_D}{RT} \sinh \frac{\Delta G}{RT}, \tag{8-12}$$

$$v = DF/RT. \tag{8-13}$$

An expansion of $\sinh(\Delta G/RT)$ yields

$$\sinh \frac{\Delta G}{RT} = \frac{\Delta G}{RT} + \frac{1}{3!}\left(\frac{\Delta G}{RT}\right)^3 + \frac{1}{5!}\left(\frac{\Delta G}{RT}\right)^5 + \cdots. \tag{8-14}$$

Under conditions where $\Delta G/RT < 1$, the higher-order terms are small compared with the first, which is the source of the linear effects expressed by the Nernst–Einstein equation. Linear behavior is common because the lattice cannot normally support large energy gradients.

Now consider nonideal concentrated alloys where the free energies per atom or chemical potentials, μ_i, at nearby planes 1 and 2 are defined by (Eq. 1-9)

$$\mu_1 = \mu^\circ + kT \ln a_1 = \mu^\circ + kT \ln \gamma_1 C_1,$$

$$\mu_2 = \mu^\circ + kT \ln a_2 = \mu^\circ + kT \ln \gamma_2 C_2. \tag{8-15}$$

Here $\mu = G/N_A$, N_A is Avogadro's number, the activity a is defined by the product of the activity coefficient γ and concentration C, and μ° is the chemical potential of the specie in the standard state. The force on an atom (f) is defined by the negative spatial derivative of μ:

$$f = -\frac{d\mu}{dx} \approx \frac{\mu_2 - \mu_1}{N_l a_0},$$

and the force on a mole of atoms is

$$F = N_A f.$$

If F is also defined as $2\,\Delta G/a_0$ (Eq. 1-35), then

$$\frac{\Delta G}{RT} = \frac{a_0 F}{2RT} = \frac{\ln(\gamma_2 C_2/\gamma_1 C_1)}{2N_l}, \tag{8-16}$$

where N_l is the number of lattice spacings (a_0) included between planes 1 and 2. The ratio $\gamma_2 C_2/\gamma_1 C_1$ typically ranges from 10 to 10^3. In conventional thin films, $N_l > 100$, so $\Delta G/RT$ is small compared with unity. Only the first term in the expansion of $\sinh(\Delta G/RT)$ need be retained, which, as noted earlier, defines diffusion in the linear range.

Imagine now what happens when N_l is about 5 to 10 so that film dimensions of only 10–20 Å are involved. At the highest values of $\gamma_2 C_2/\gamma_1 C_1$, the quantity $\Delta G/RT$ is approximately unity. The higher-order terms in Eq. 8-14 can no longer be neglected now, and this leads to the nonlinear region of chemical diffusion. If only the cubic term is retained, a combination of Eqs. 8-12 to 8-14 and 1-33 to 1-34 yields

$$v = \frac{D}{RT}\left(F + \frac{a_0^2 F^3}{24(RT)^2}\right) \tag{8-17}$$

rather than the Nernst–Einstein relation. For an ideal solution, $\mu = kT \ln C$, and therefore $f = (-kT/C)(dC/dx)$. The macroscopic mass flux J is given by the product of atomic concentration and velocity, and accordingly

$$J = Cv = -D\frac{dC}{dx} - \frac{Da_0^2}{24C^2}\left(\frac{dC}{dx}\right)^3. \tag{8-18}$$

The first term is the readily recognizable Fickian flux, which is the source of linear diffusion effects. By taking the negative divergence of the flux (Eqs. 1-23 and 1-24), we obtain

$$\frac{\partial C}{\partial t} = D\frac{\partial^2 C}{\partial x^2} + \frac{Da_0^2}{8C^2}\left(\frac{dC}{dx}\right)^2\frac{\partial^2 C}{\partial x^2}. \tag{8-19}$$

This fifth-degree nonlinear differential equation was developed by Tu (Ref. 17), who termed it the "kinetic nonlinear" equation. It should not be confused with a similar "thermodynamic nonlinear" equation containing a $\partial^4 C/\partial x^4$ term used to describe diffusion in compositionally modulated films of small layer thickness.

As a simple example illustrating diffusion size effects, consider a composite film structure with layers d thick. Initially, the solute concentration varies sinusoidally above an average level C_0 as

$$C(x,0) = C_0 + C_i \sin(\pi x/d) \qquad (C_i = \text{constant}). \tag{8-20}$$

If the film layer is thick, linear diffusion is expected. The solution that satisfies the Fick diffusion equation under the given conditions is

$$C(x,t) = C_0 + C_i \sin\frac{\pi x}{d} \exp - \frac{\pi^2 Dt}{d^2}. \tag{8-21}$$

When heated, compositional gradients throughout the film structure are reduced through interdiffusion. A measure of the extent of homogenization is

$$(C(d/2,t) - C_0)/C_i = e^{-\pi^2 Dt/d^2}. \tag{8-22}$$

As the film thickness shrinks, homogenization occurs more rapidly since diffusion distances are reduced. On further size reduction, a point is reached where large free-energy gradients cause entrance into the regime of nonlinear diffusion. Now Eq. 8-19 applies, and the reader can easily show that Eq. 8-21 is no longer a solution. Initially, compositional smoothing will be governed by the nonlinear term, and the actual kinetics and spatial concentration distribution will therefore be rather complex. But after some homogenization, $\ln(\gamma_2 C_2/\gamma_1 C_1)$ diminishes to the point where the linear diffusion term dominates the interdiffusion. An exponential decay of the profile amplitude should then characterize the long-time kinetics.

8.3. Interdiffusion in Metal Alloy Films

8.3.1. Reactions at a Solder Joint

One of the best ways to appreciate the importance of interdiffusion effects in metal films is to consider the interfacial region between a fabricated Si chip and the solder joint that connects it to the outside world. This is shown in Fig. 8-10, and, following Tu (Ref. 17), we note that the two levels of Al metalization interconnections, which contact the Si devices above, must also be bonded to the solder ball below. Anyone who has tried to solder Al is acquainted with the difficulties involved. In this case, they are overcome by using an evaporated Cr–Cu–Au thin-film structure. Since the surface of Al is easily oxidized, it is difficult to solder with the Pb–Sn alloy, so a Cu layer is introduced. The intention is to utilize the fast Cu–Sn reaction to form intermetallic compounds. However, Cu adheres poorly to oxidized Al and SiO_2 surfaces. Moreover, when molten, the relatively massive Pb–Sn consumes the

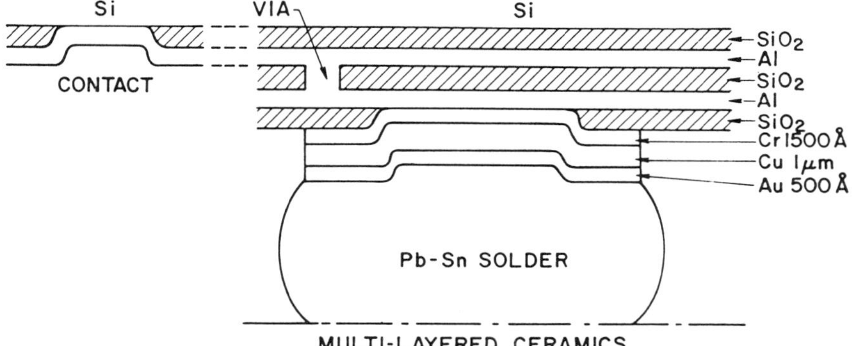

Figure 8-10. Schematic diagram of solder contact to Al by means of the trimetal Cr–Cu–Au film metallization scheme. (Reprinted with permission from Ref. 17).

Cu and then tends to dewet on the Al surface. Therefore, Cr is introduced as a glue layer between the Al and SiO_2 and to prevent the molten solder from dewetting. The Cu surface needs to be protected against atmospheric corrosion because corroded Cu surfaces do not solder well. Therefore, a thin film of Au is introduced to passivate the Cu surface. Since Au dissolves rapidly in Pb and Sn, a solder richer in Pb than the eutectic composition is used. This allows for enough Pb to dissolve Au and sufficient Sn to react with Cu. The reaction between the excess Pb and Cu is limited because these elements do not form extended solid solutions or intermetallic compounds. Thus, in order to fulfill the functions of adhesion, soldering, and passivation, this elaborate trimetal film structure is required.

There are additional solid-state diffusion effects between metal layers to contend with. At temperatures close to 200 °C, Cu can diffuse rapidly through GBs of Cr even though these metals are basically immiscible. When this happens, the interfacial adhesion at the Cr–SiO_2 interface is adversely affected. Moreover, Cu can diffuse outward through the Au in which it is miscible. At the Au surface it forms an oxide that interferes with soldering. This one method of joining (the flip-chip technology) has generated a host of mass-transport phenomena. For this reason, there has been considerable interest in thin-film interdiffusion studies of Cr–Cu, Au–Cu, and Cu–Sn systems. Other binary combinations from this group of involved elements, such as Al–Au and Al–Cu, have received even more attention. Miscible, immiscible, as well as compound-forming systems are represented. For the most part, vacancy diffusion is the accepted diffusion mechanism. However, in the case of the noble metals Cu and Au in the group IV matrices of Sn and Pb, anomalously rapid migration through interstitial sites is believed to occur.

8.3.2. GB Diffusion in Alloy Films

There have been a considerable number of fundamental interdiffusion studies in metal alloy films that have been interpreted in terms of GB transport models. In order to exclude the complicating effects of compounds and precipitates, the systems selected were primarily limited to those displaying solid solubility. Results from a representative group of investigations are entered in Table 8-1, where the diffusivities are expressed in terms of the pre-exponential factor (δD_b) and activation energy.

Two broad categories of experimental techniques were employed in gathering these data. Sputter sectioning through the film is the basis of the first technique. In the commonly employed AES depth-profiling method, the film is analyzed continuously as it is simultaneously thinned by sputter etching. Profiles appear similar to those shown in Fig. 8-6a, b.

The second category is based on permeation and surface accumulation techniques. An example shown in Fig. 8-11a utilizes AES signal-sensing methods to detect Ag penetration through a Au film and subsequent spreading along the exit surface (Ref. 19). The signal reflects this by building slowly at first after an incubation period, then changing more rapidly and finally saturating at long times. Diffusivities are then unfolded by fitting data to an assumed kinetics (shown in Fig. 8-11a) of concentration buildup with time. When type C kinetics (Fig. 8-5) prevails, D_b values may be simply estimated by equating the film thickness to the GB diffusion length $\sim 2\sqrt{D_b t}$, where t

Table 8-1. Grain-Boundary Diffusion in Thin Metal Films

Diffusant	Matrix	δD_b (cm^3/sec)	E_b (eV)	Remarks
Al	Cu	5.1×10^{-7}	0.94	Polycrystalline films
^{195}Au	Au	9.0×10^{-10}	1.00	Polycrystalline films
^{195}Au	Au	1.9×10^{-10}	1.16	(100) epitaxial
Cu	Al	4.5×10^{-8}	1.00	Polycrystalline
		1.8×10^{-10}	0.87	films
Ag	Au	4.5×10^{-12}	1.20	(111) epitaxial
^{195}Au	Ni–0.5%Co	1.4×10^{-10}	1.60	Polycrystalline films
Sn	Pb	6.9×10^{-10}	0.62	Polycrystalline films
Cr	Au	5.0×10^{-11}	1.09	Textured film
Ag	Cu	1.5×10^{-13}	0.75	Polycrystalline films
Pt	Cr	5.0×10^{-10}	1.69	Polycrystalline films
Cu	Al–0.2%Co	2.0×10^{-7}	0.56	Polycrystalline films
Ag	Au	5.0×10^{-14}	1.10	(100) epitaxial
^{119}Sn	Sn	5.0×10^{-8}	0.42	Polycrystalline films

From Ref. 18.

is the time required for the AES signal to appear. Interestingly, a grain-boundary Kirkendall effect shown in Fig. 8-11b has been suggested to account for the unequal GB diffusion rates of Ag and Au.

The high sensitivity inherent in detecting a signal rise from the low background enables very small atomic fluxes to be measured. This makes it possible to monitor transport at quite low temperatures. On the other hand, disadvantages include the electron-beam heating of films during measurement, and the need to maintain ultrahigh vacuum conditions because of the sensitivity of surface diffusion to ambient contamination.

Results for the sectioning methods suggest two GB diffusion regimes:

1. $D_b = 0.3 \exp - 7.5T_M/T$ cm^2/sec for large-angle boundaries where $E_b = 0.5E_L$.
2. $D_b = 2 \exp - 12.5T_M/T$ cm^2/sec for epitaxial films containing subboundaries or dissociated dislocations where $E_b = 0.7E_L$.

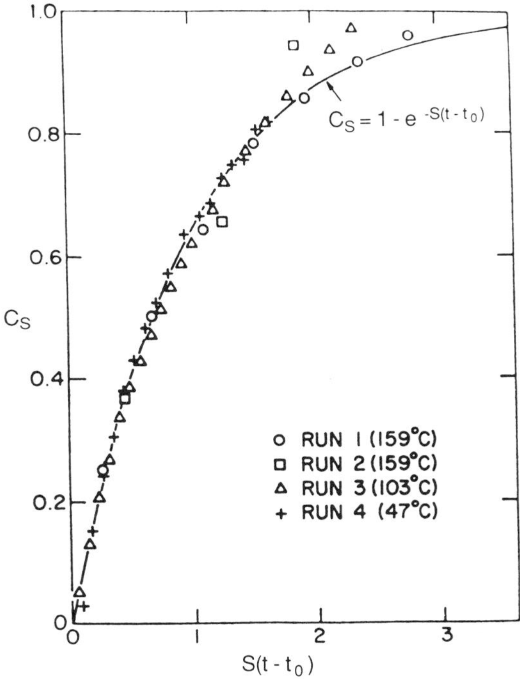

Figure 8-11a. Accumulation (C_s) of Ag on Au film surfaces as a function of normalized (reduced) time for different diffusional annealing temperatures. Constants S and t_0 (incubation time) depend on the run involved. (From Ref. 19).

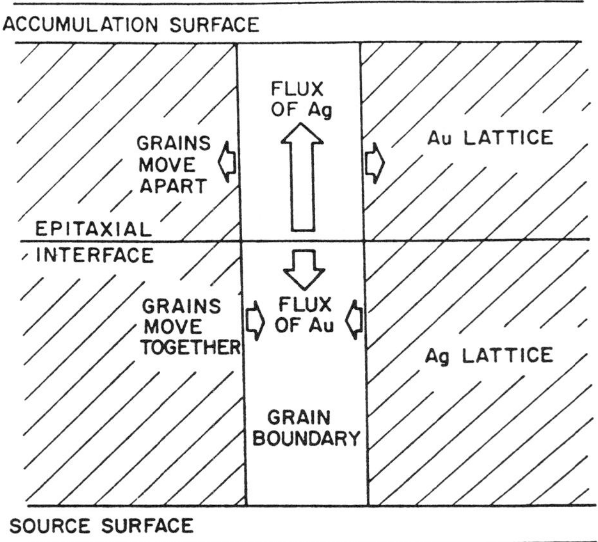

Figure 8-11b. Model of grain-boundary Kirkendall effect. (From Ref. 19).

For the permeation and surface accumulation methods, the activation energies appear to be somewhat lower. Data derived from these techniques are more susceptible to GB structure, since diffusant is transported through the most highly disordered boundaries first.

8.3.3. Formation of Intermetallic Compound Films

Perhaps the best-known example of intermetallic compound formation between metal films involves the interaction between Al and Au. For at least 20 years the combination of Al films and Au wires or balls served to satisfy bonding requirements in the semiconductor industry, but not without a significant number of reliability problems. When devices are heated to 250–300 °C for a few days, an Al–Au reaction proceeds to form a porous intermetallic phase around the Au ball bond accompanied by a lacy network of missing Al. One of the alloy phases, $AuAl_2$, is purple, which accounts for the appellation "purple plague." Even though the evidence is not definitive that the presence of $AuAl_2$ correlates with bond embrittlement, lack of strength, or degradation, formation of this compound is viewed with concern.

An extensive RBS study (Ref. 20) of compound formation in the Al–Au thin-film system is worth reviewing since it clarifies the role of film thickness,

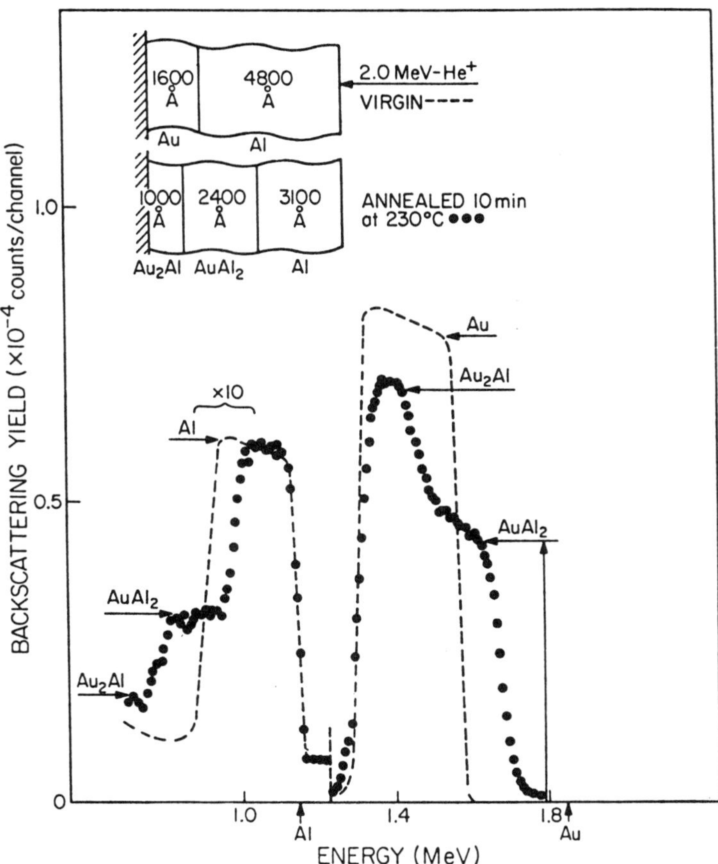

Figure 8-12. RBS spectrum showing formation of $AuAl_2$ and Au_2Al phases at 230 °C. (From Ref. 20).

temperature and time in influencing the reaction. Film couples with Al on top of Au were annealed, yielding RBS spectra such as shown in Fig. 8-12. Because there is such a large mass difference between Al and Au, there is no peak overlap despite the unfavorable spatial ordering of the metal layers. The stoichiometry of the peak shoulders in the annealed films corresponds to the compounds Au_2Al and $AuAl_2$, and the high-energy tail is indicative of Au transport through the Al film to the surface. From the fact that the $AuAl_2$ always appears at a higher energy than the Au_2Al, we know that it is closer to the air surface of the couple. By the methods outlined in Section 6.4.7, the compound thickness can be determined, and Fig. 8-13a reveals that both compounds grow with parabolic kinetics. The slopes of the compound thick-

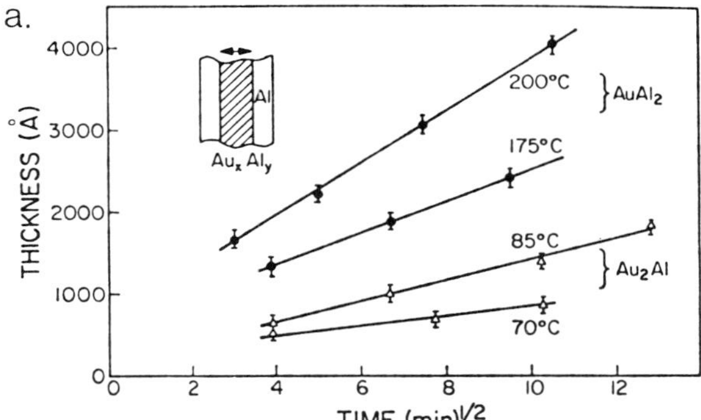

Figure 8-13a. Compound film thickness versus square root of time for $AuAl_2$ and Au_2Al at different annealing temperatures.

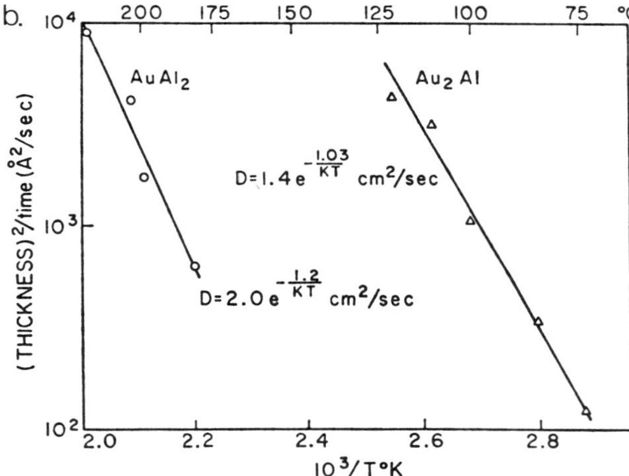

Figure 8-13b. Arrhenius plots for the kinetics of formation of $AuAl_2$ and Au_2Al compounds. (From Ref. 20).

ness–time$^{1/2}$ curves are proportional to the ubiquitous Boltzmann factor. Therefore, by plotting these slopes (actually the logs of the square of the slope in this case) versus $1/T$ K in the usual Arrhenius manner (Fig. 8-13b), we obtain activation energies for compound growth. The values of 1.03 and 1.2 eV can be roughly compared with the systematics given for FCC metals to elicit some clue as to the mass-transport mechanism for compound formation. Based on Au, these energies translate into equivalent Boltzmann factors of $\exp - 8.9T_M/T$ and $\exp - 10.4T_M/T$, respectively, suggesting a GB-assisted diffusion mechanism. Lastly, it is interesting to note how the sequence of

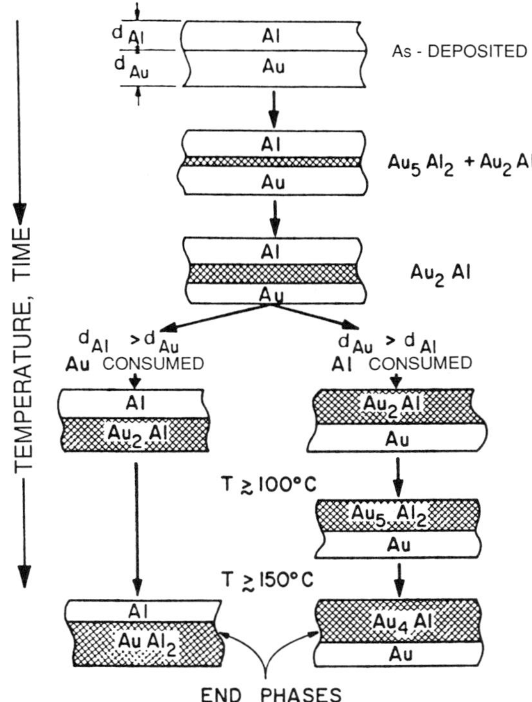

Figure 8-14. Schematic diagrams illustrating compound formation sequence in Al–Au thin film couples. End phases depend on whether $d_{\mathrm{Al}} > d_{\mathrm{Au}}$ or $d_{\mathrm{Au}} > d_{\mathrm{Al}}$.

compound formation (Fig. 8-14) correlates with the equilibrium phase diagram (not shown). When the film thickness of Al exceeds that of Au, then the latter will be totally consumed, leaving excess Al. The observed equilibrium between Al and $AuAl_2$ layers is consistent with the phase diagram. Similarly, excess Au is predicted to finally equilibrate with the Au_4Al phase, as observed.

8.4. Electromigration in Thin Films

Electromigration, a phenomenon not unlike electrolysis, involves the migration of metal atoms along the length of metallic conductors carrying large direct current densities. It was observed in liquid metal alloys well over a century ago and is a mechanism responsible for failure of tungsten light-bulb filaments. Bulk metals approach the melting point when powered with current densities (J) of about 10^4 A/cm^2. On the other hand, thin films can tolerate densities of

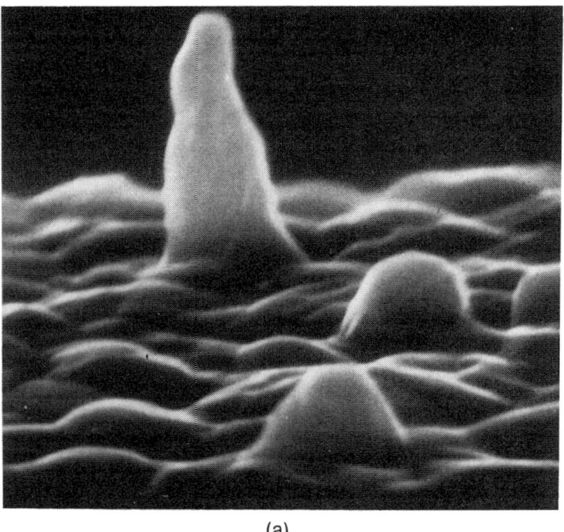

(a)

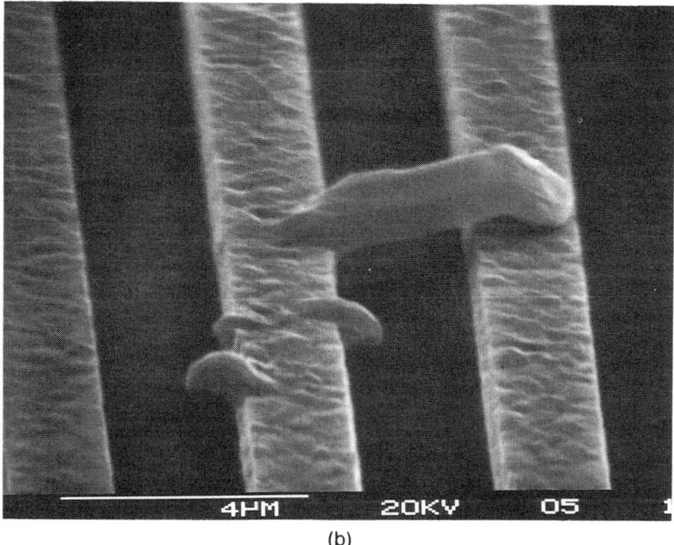

(b)

Figure 8-15. Manifestations of electromigration damage in Al films: (a) hillock growth, (from Ref. 21, courtesy of L. Berenbaum); (b) whisker bridging two conductors (courtesy of R. Knoell, AT & T Bell Laboratories); (c) nearby mass accumulation and depletion (courtesy S. Vaidya, AT & T Bell Laboratories).

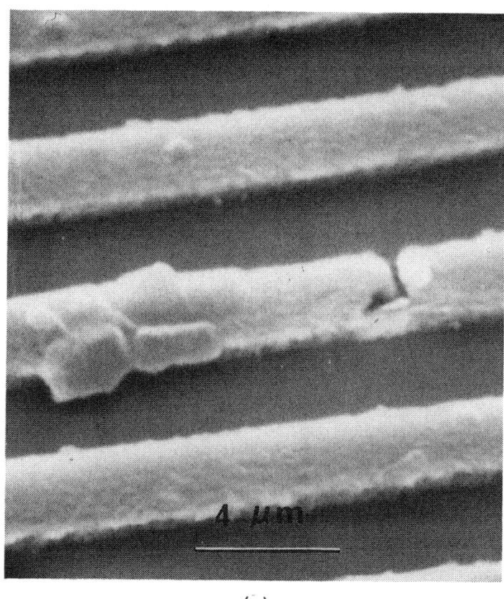

(c)

Figure 8-15. *Continued.*

10^6 A/cm^2 without immediate melting or open-circuiting because the Joule heat is effectively conducted away by the substrate, which behaves as a massive heat sink. In a circuit chip containing some 100,000 devices, there is a total of several meters of polycrystalline Al alloy interconnect stripes that are typically less than 1.5 μm wide and 1 μm thick. Under powering, at high current densities, mass-transport effects are manifested by void formation, mass pileups and hillocks, cracked dielectric film overlayers, grain-boundary grooving, localized heating, and thinning along the conductor stripe and near contacts. Several examples of such film degradation processes are shown in Fig. 8-15. In bootstrap fashion the damage accelerates to the point where open-circuiting terminates the life of the conductor. It is for these reasons that electromigration has been recognized as a major reliability problem in integrated circuit metallizations for the past quarter century. Indeed, there is some truth to a corollary of one of Murphy's laws—"A million-dollar computer will protect a 25-cent fuse by blowing first." Analysis of the extensive accelerated testing that has been performed on interconnections has led to a general relationship between film mean time to failure (MTF) and J given by

$$\mathrm{MTF}^{-1} = K(\exp - E_e/kT)J^n. \tag{8-23}$$

As with virtually all mass-transport-related reliability problems, damage is thermally activated. For Al conductors, n is typically 2 to 3, and E_e, the

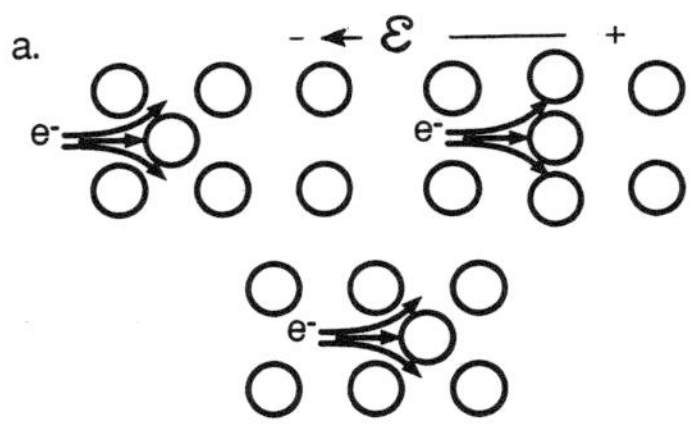

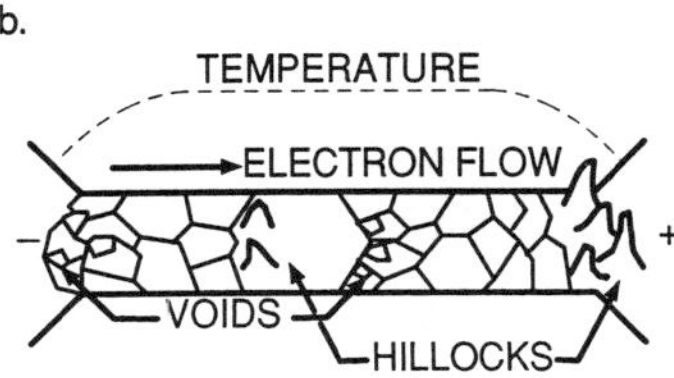

Figure 8-16. (a) Atomic model of electromigration involving electron momentum transfer to metal ion cores during current flow. (b) Model of electromigration damage in a powered film stripe. Mass flux divergences arise from nonuniform grain structure and temperature gradients.

activation energy for electromigration failure, ranges from 0.5 to 0.8 eV, depending on grain size. In contrast, an energy of 1.4 eV is associated with bulk lattice diffusion so that low-temperature electromigration in films is clearly dominated by GB transport. The constant K depends on film structure and processing. Current design rules recommend no more than 10^5 A/cm^2 for stripe widths of $\sim$ 1.5 μm. Although Eq. 8-23 is useful in designing metalizations, it provides little insight into the atomistic processes involved.

The mechanism of the interaction between the current carriers and migrating atoms is not entirely understood, but it is generally accepted that electrons streaming through the conductor are continuously scattered by lattice defects. At high enough current densities, sufficient electron momentum is imparted to atoms to physically propel them into activated configurations and then toward the anode as shown in Fig. 8-16. This electron "wind" force is oppositely directed to and normally exceeds the well-shielded electrostatic force on atom cores arising from the applied electric field $\mathscr{E}$. Therefore, a net force F acts on the ions, given by

$$F = Z^* q \mathscr{E} = Z^* q \rho J, \tag{8-24}$$

where q is the electronic charge and $\mathscr{E}$ is, in turn, given by the product of the electrical resistivity of the metal, ρ, and J. An "effective" ion valence Z^* may be defined, and for electron conductors it is negative in sign with a magnitude usually measured to be far in excess of typical chemical valences.

On a macroscopic level, the observed mass-transport flux, J_m, for an element of concentration C is given by

$$J_m = Cv = CDZ^* q \rho J / RT, \tag{8-25}$$

where use has, once again, been made of the Nernst–Einstein relation. Electromigration is thus characterized at a fundamental level by the terms Z^* and D. Although considerable variation in Z^* exists, values of the activation energy for electrotransport in films usually reflect a grain-boundary diffusion mechanism.

Film damage is caused by a depletion or accumulation of atoms, which is defined by either a negative or positive value of dC/dt, respectively. By Eq. 1-23,

$$\frac{\partial C}{\partial t} = -\frac{\partial}{\partial x}\left(\frac{CDZ^* q \rho J}{RT}\right) - \frac{\partial}{\partial T}\left(\frac{CDZ^* q \rho J}{RT}\right)\frac{\partial T}{\partial x}. \tag{8-26}$$

The first term on the right-hand side reflects the isothermal, *structurally induced* mass flux divergence, and the second term represents mass transport in the presence of a *temperature gradient*. The resulting transport under these distinct conditions can be qualitatively understood with reference to Fig. 8-16b, assuming that atom migration is solely confined to GBs and directed toward the anode. Let us first consider electromigration under isothermal conditions. Because of varying grain size and orientation distributions, local mass flux divergences exist throughout the film. Each cross section of the stripe contains a lesser or greater number of effective GB transport channels. If more atoms enter a region such as a junction of grains than leave it, a mass pileup or growth can be expected. A void develops when the reverse is true. At highly heterogeneous sites where, for example, a single grain extends across the stripe width and abuts numerous smaller grains, the mass accumulations and depletions are exaggerated. For this reason, a uniform distribution of grain size is desirable. Of course, single-crystal films would make ideal interconnections because the source of damage sites is eliminated, but it is not practical to deposit them.

Electromigration frequently occurs in the presence of nonuniform temperature distributions that develop at various sites within device structures—e.g., at locations of poor film adhesion, in regions of different thermal conductivity, such as metal–semiconductor contacts or interconnect-dielectric crossovers, at nonuniformly covered steps, and at terminals of increased cross section. In addition to the influence of microstructure, there is the added complication of the temperature gradient. The resulting damage pattern can be understood by considering the second term on the right-hand side of Eq. 8-26. For the polarity shown, all terms in parentheses are positive and $Cq\rho J/RT$ is roughly temperature independent, whereas DZ^* increases with temperature. Therefore, dC/dt varies as $-dT/dx$. Voids will thus form at the negative electrode, where $dT/dx > 0$, and hillocks will grow at the positive electrode, where $dT/dx < 0$. Physically, the drift velocity of atoms at the cathode increases as they experience a rising temperature. More atoms then exit the region than flow into it. At the anode the atoms decelerate in experiencing lower temperatures and thus pile up there. An analogy to this situation is a narrow strip of road leading into a wide highway (at the cathode). The bottleneck is relieved and the intercar spacing increases. If further down the highway it again narrows to a road, a new bottleneck reforms and cars will pile up (at the anode).

Despite considerable efforts to develop alternative interconnect materials, Al-base alloys are still universally employed in the industry. Their high conductivity, good adhesion, ease of deposition, etchability, and compatibility with other processing steps offset the disadvantages of being prone to corrosion and electromigration degradation. Nevertheless, attempts to improve the quality of Al metallizations have prompted the use of alternative deposition methods as well as the development of more electromigration-resistant alloys. With regard to the latter, it has been observed that Al alloyed with a few percent Cu can extend the electromigration life by perhaps an order of magnitude relative to pure Al. Reasons for this are not completely understood, but it appears that Cu reduces the GB migration of the solvent Al. The higher values for E_b which are observed are consistent with such an interpretation. Other schemes proposed for minimizing electromigration damage have included

1. Dielectric film encapsulation to suppress free surface growths
2. Incorporation of oxygen to generally strengthen the matrix through dispersion of deformation-resistant Al_2O_3 particles
3. Deposition of intervening thin metal layers in a sandwichlike structure that can shunt the Al in case it fails

The future may hold some surprises with respect to electromigration lifetime. Experimental results shown in Fig. 8-17 reveal reduction of film life as the linewidth decreases from 4 to 2 μm in accord with intuitive expectations. However, an encouraging increase in lifetime is surprisingly observed for submicron-wide stripes. The reason for this is the development of a bamboo-like grain structure generated in electron-beam evaporated films. Because the GBs are oriented normal to the current flow, the stripe effectively behaves as a single crystal. Similar benefits are not as pronounced in sputtered films.

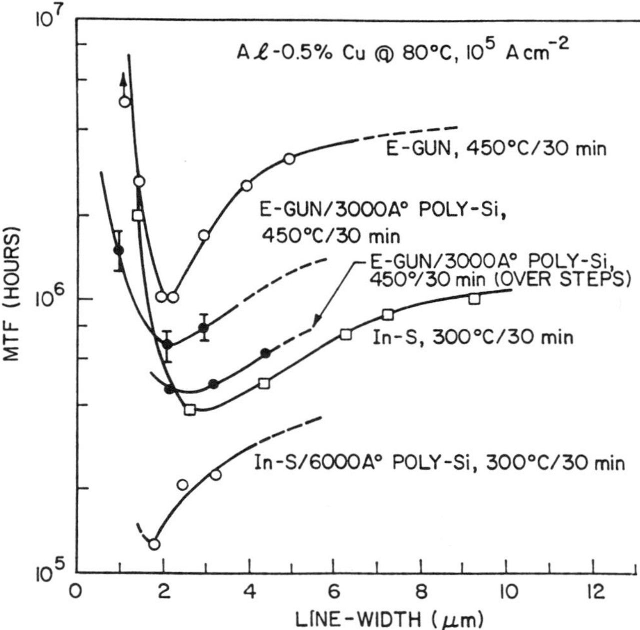

Figure 8-17. Mean time to failure as a function of stripe linewidth for evaporated (E-gun) and sputtered (S-gun, In–S) Al films. (From Ref. 22).

8.5. Metal - Semiconductor Reactions

8.5.1. Introduction to Contacts

All semiconductor devices and integrated circuits require contacts to connect them to other devices and components. When a metal contacts a semiconductor surface, two types of electrical behavior can be distinguished in response to an applied voltage. In the first type, the contact behaves like a *P–N* junction and rectifies current. The ohmic contact, on the other hand, passes current equally as a function of voltage polarity. In Section 10.4 the electrical properties of metal–semiconductor contacts will be treated in more detail.

Contact technology has dramatically evolved since the first practical semiconductor device, the point-contact rectifier, which employed a metal whisker that was physically pressed into the semiconductor surface. Today, deposited thin films of metals and metal compounds are used, and the choice is dictated by complex considerations; not the least of these is the problem of contact

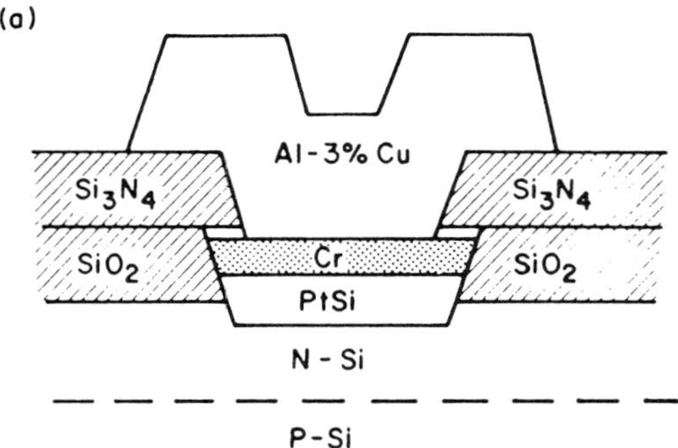

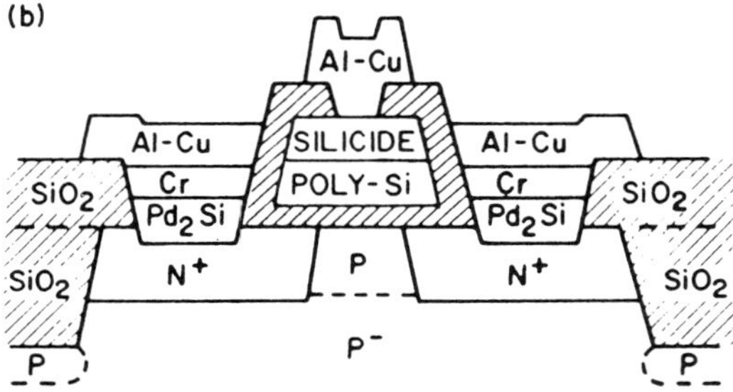

Figure 8-18. Schematic diagrams of silicide contacts in (a) bipolar and (b) MOS field effect transistor configurations. (Reprinted with permission from Ref. 17, © 1985 Annual Reviews Inc.).

instability during processing caused by mass-transport effects. For this reason, elaborate film structures are required to fulfill the electrical specifications and simultaneously defend against contact degradation. The extent of the problem can be appreciated with reference to Fig. 8-18, where both bipolar and MOS field effect transistors are schematically depicted. The operation of these devices need not concern us. What is of interest are the reasons for the Cr and metal silicide films that serve to electrically connect the Si below to the Al–Cu metal interconnections above. These bilayer structures have replaced the more obvious direct Al–Si contact, which, however, continues to be used in other applications. Contact reactions between Al and Si are interesting metallurgically and provide a good pedagogical vehicle for applying previously developed concepts of mass transport. A discussion of this follows. Means of minimizing Al–Si reactions through intervening metal silicide and diffusion-barrier films will then be reviewed.

8.5.2. Al - Si Reactions

Nature has endowed us with two remarkable elements: Al and Si. Together with oxygen, they are the most abundant elements on earth. It was their destiny to be brought together in the minutest of quantities to make the computer age possible. Individually, each element is uniquely suited to perform its intended function in a device, but together they combine to form unstable contacts. In addition to creating either a rectifying barrier or ohmic contact, they form a diffusion couple where the extent of reaction is determined by the phase diagram and mass-transport kinetics. The processing of deposited Al films for contacts typically includes a 400 °C heat treatment. This enables the Al to reduce the very thin native insulating SiO_2 film and ''sinter'' to Si, thereby lowering the contact resistance. Reference to the Al–Si phase diagram (Fig. 1-13) shows that at this temperature Si dissolves in Al to the extent of about 0.3 wt%. During sintering, Si from the substrate diffuses into the Al via GB paths in order to satisfy the solubility requirement. Simultaneously, Al migrates into the Si by diffusion in the opposite direction. As shown by the sequence of events in Fig. 8-19, local diffusion couples are first activated at several sites within the contact area. When enough Al penetrates at one point, the underlying *P–N* junction is shorted by a conducting metal filament, and junction ''spiking'' or ''spearing'' is said to occur.

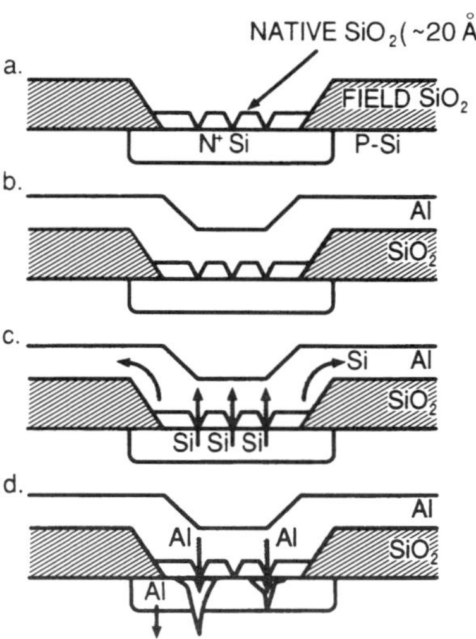

Figure 8-19. Schematic sequence of Al–Si interdiffusion reactions leading to junction spiking.

The remedy for the problem seems simple enough. By presaturating the Al with Si the driving force for interdiffusion disappears. Usually a 1 wt% Si–Al alloy film is sputtered for this purpose. However, with processing another complication arises. During the heating and cooling cycle Si is first held in solid solution but then precipitates out into the GBs of the Al as the latter becomes supersaturated with Si at low temperatures. The irregularly shaped Si precipitate particles, saturated with Al, grow epitaxially on the Si substrate. Electrically these particles are P type and alter the intended electrical characteristics of the contact. Thus, despite ease in processing, Al contact metallurgy is too unreliable in the VLSI regime of very shallow junction depths. For this reason, noble metal silicides such as Pd–Si have largely replaced Al at contacts.

There is yet another example of Al–Si reaction that occurs in field effect transistors. In this case, however, the contact to the gate oxide (SiO_2), rather than to the semiconductor source and drain regions, is involved. Historically, Al films were first used as gate electrodes, but, as noted on p. 24, they tend to reduce SiO_2, which is undesirable. Other metals are also problematical

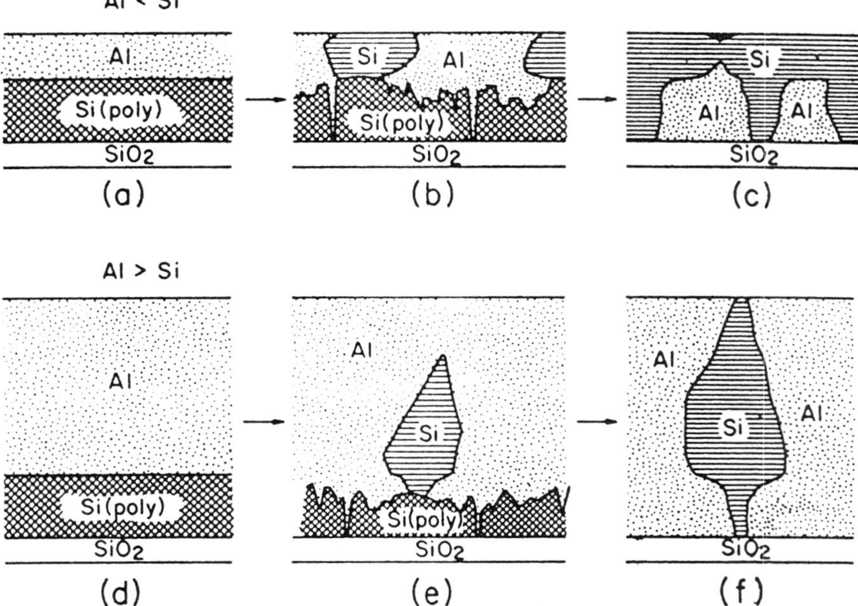

Figure 8-20. Depiction of reactions between Al and polysilicon films during annealing. Figures a, b, c refer to the case where $d_{Si} > d_{Al}$. Figures d, e, f refer to case where $d_{Al} > d_{Si}$. (From Ref. 23).

because of the potential reaction to form a silicide as well as oxide; an example is $3Ti + 2SiO_2 \rightarrow TiSi_2 + 2TiO_2$. For reliable device performance, the foregoing considerations have led to the adoption of poly-Si films as the gate electrode. Although there is now no driving force promoting reaction between Si and SiO_2, the chronic problem of Si–Al interdiffusion has re-emerged. The Al interconnections must still make contact to the gate electrode. To make matters worse, reaction of Al with poly Si is even more rapid than with single-crystal Si because of the presence of GBs. The dramatic alteration in the structure and composition in the Al–poly-Si–layered films following thermal treatment is shown schematically in Fig. 8-20. Reactions similar to those previously described for the Al–Si contact occur, and resultant changes are sensitive to the ratio of film thicknesses. It is easy to see why electrical properties would also be affected. Therefore, intervening silicide films and diffusion barriers must once again be relied on to separate Al from Si.

8.6. Silicides and Diffusion Barriers

8.6.1. Metal Silicides

In the course of developing silicides for use in contact applications, a great deal of fundamental research has been conducted on the reactions between thin metal films and single-crystal Si. Among the issues and questions addressed by these investigations are the following:

1. Which silicide compounds form?
2. What is the time and temperature dependence of metal silicide formation?
3. What atomic mass-transport mechanisms are operative during silicide formation? Which of the two diffusing species migrates more rapidly?
4. When the phase diagram indicates a number of different stable silicide compounds, which form preferentially and in what reaction sequence?

Virtually all thin-film characterization and measurement tools have been employed at one time or another in studying these aspects of silicide formation. In particular, RBS methods have probably played the major role in shaping our understanding of metal–silicon reactions by revealing compound stoichiometries, layer thicknesses, and the moving specie. Examples of the spectra obtained and their interpretation have been discussed previously. (See Section 6.4.7).

A summary of kinetic data obtained in silicide compounds formed with near-noble, transition, and refractory metals is contained in Table 8-2. This

Table 8-2. Silicide Formation

Silicide	Formation Temperature (°C)	Activation Energy (eV)	Growth Rate	Moving Specie	Formation Energy at 298 K (kcal/mole)
Au_2Si	100				
Co_2Si	350–500	1.5	$t^{1/2}$	Co	−27.6
Ni_2Si	200–350	1.5	$t^{1/2}$	Ni	−33.5
NiSi	350–700	1.4	t		−20.5
Pt_2Si	200–500	1.5	$t^{1/2}$	Pt	−20.7
PtSi	300	1.6	$t^{1/2}$		−15.8
FeSi	450–550	1.7	$t^{1/2}$	Si	−19.2
RhSi	350–425	1.95	$t^{1/2}$	Si	−16.2
HfSi	550–700	2.5	$t^{1/2}$	Si	−34
IrSi	400–500	1.9	$t^{1/2}$		−16.2
$CrSi_2$	450	1.7	t		−28.8
$MoSi_2$	525	3.2	t	Si	−31.4
WSi_2	650	3.0	t, $t^{1/2}$	Si	−22.2

From Refs. 12 and 24.

large body of work can be summarized in the following way:

Silicide	Formation Temp. (°C)	Growth Rate	Activation Energy (eV)
M_2Si	200	$t^{1/2}$	1.5
MSi	400	$t^{1/2}$	1.6–2.5
MSi_2	600	$t^{1/2}$	1.7–3.2

Three broad classes of silicides are observed to form: the metal-rich silicide (e.g., M_2Si), the monosilicide (MSi), and the silicon-rich silicide (e.g., MSi_2). As a rough rule of thumb, the formation temperature ranges from one third to one half the melting point (in K) of the corresponding silicide. Since fine-grained metal films are involved, it is not surprising that this rule is consistent with the GB diffusion regime. The activation energies roughly correlate with the melting point of the silicide, in agreement with general trends noted earlier.

In the metal-rich silicides, the metal is observed to be the dominant mobile specie, whereas in the mono- and disilicides Si is the diffusing specie. The crucial step in silicide formation requires the continual supply of Si atoms through the breaking of bonds in the substrate. In the case of disilicides, high temperatures are available to free the Si for reaction. At lower temperatures there is insufficient thermal energy to cause breaking of Si bonds, and the metal-rich silicides thus probably form by a different mechanism. It has been suggested that rapid interstitial migration of metal through the Si lattice assists bond breaking and thus controls the formation of such silicides.

The sequence of phase formation has only been established in a few silicide systems. Perhaps the most extensively studied of these is the Ni–Si system, for which the phase diagram and compound formation map are provided in Fig. 8-21. The map shows that Ni_2Si is always the first phase to form during low-temperature annealing. Clearly, Ni_2Si is not in thermodynamic equilibrium with either Ni or Si, according to the phase diagram. What happens next

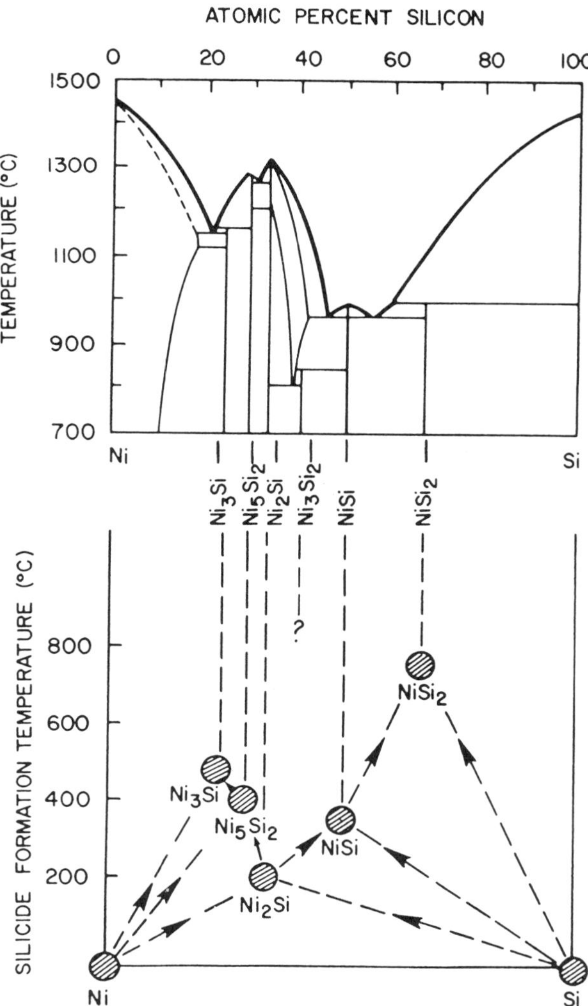

Figure 8-21. Map of thin-film Ni silicide formation sequence. Phase diagram of Ni–Si system shown on top. (Reprinted with permission from Ref. 17, © 1985 Annual Reviews Inc.).

depends on whether Si or Ni is present in excess. In the usual former case, where a Ni thin film is deposited on a massive Si wafer, the sequence proceeds first to NiSi and then to $NiSi_2$ at elevated temperatures. However, when a film of Si is deposited on a thicker Ni substrate, then the second and third compounds become Ni_5Si_2 and Ni_3Si. At elevated temperatures the resultant two-phase equilibrium (i.e., Si–$NiSi_2$ or Ni–Ni_3Si) conforms to the phase diagram. The question of the first silicide to form is a more complicated issue. It may be related to the ability to vapor-quench alloys to nucleate very thin, prior amorphous film layers. It is well known that bulk amorphous phases are readily formed by quenching metal-silicon eutectic melts. Therefore, it is suggested that silicide compounds located close to low-temperature eutectic compositions are the first to form.

Interestingly, in bulk diffusion couples all compounds appear to grow simultaneously at elevated temperatures. This does not seem to happen in films (at low temperature), but more sensitive analytical techniques may be required to clarify this issue.

8.6.2. Diffusion Barriers

Diffusion barriers are thin-film layers used to prevent two materials from coming into direct contact in order to avoid reactions between them. Paint and electrodeposited layers are everyday examples of practical barriers employed to protect the underlying materials from atmospheric attack. In a similar vein, diffusion barriers are used in thin-film metallization systems, and the discussion will be limited to these applications. We have already noted the use of silicides to prevent direct Al–Si contact. Ideally, a barrier layer X sandwiched between A and B should possess the following attributes (Ref. 25):

1. It should constitute a kinetic barrier to the traffic of A and B across it. In other words, the diffusivity of A and B in X should be small.
2. It should be thermodynamically stable with respect to A and B at the highest temperature of use. Further, the solubility of X in A and B should be small.
3. It should adhere well to and have low contact resistance with A and B and possess high electrical and thermal conductivity. Practical considerations also require low stress, ease of deposition, and compatibility with other processing.

Some of these requirements are difficult to achieve and even mutually exclusive so that it is necessary to make compromises.

A large number of materials have been investigated for use as barrier layers between silicon semiconductor devices and Al interconnections. These include

Table 8-3. Aluminum–Diffusion-Barrier–Silicon Contact Reactions

Diffusion Barrier	Reaction Temperature (°C)	Reaction Products	Failure Mechansims
Cr	300	Al_7Cr	C (E_C = 1.9 eV)
V	450	Al_3V, Al–V–Si	C (E_C = 1.7 eV)
Ti	400	Al_3Ti	C (E_C = 1.8 eV)
Ti–W	500		D
ZrN	550	Al–Zr–Si	C
PtSi	350	Al_2Pt, Si	C
Pd_2Si	400	Al_3Pd, Si	C
NiSi	400	Al_3Ni, Si	C
$CoSi_2$	400	Al_9Co_2, Si	C
$TiSi_2$	550	Al–Ti–Si	D
$MoSi_2$	535	$Al_{12}Mo$, Si	D
Ti–Pd_2Si	435	Al_3Ti	C
W–$CoSi_2$	500	$Al_{12}W$	C
TiN–PtSi	600	AlN, Al_3Ti	C
TiC–PtSi	600	Al_4C_3, Al_3Ti	C
TaN–NiSi	600	AlN, Al_3Ta	C

C = Compound formation; D = Diffusion
From Ref. 26.

silicides, refractory metals, transition metal alloys, transition metal compounds, as well as dual-layer barriers such as refractory metal–silicide, transition metal–silicide and transition metal compound–silicide combinations (Ref. 26). A compilation of these materials and reaction products is given in Table 8-3. Stringent physical requirements and the complexity of low-temperature interdiffusion and reactions have frequently necessitated the use of "diffusion barriers" to protect diffusion barriers. In order to gain a complete picture of the effectiveness of diffusion barriers, we need analytical techniques to reveal metallurgical interactions and their effect on the electrical properties of devices. For this reason, RBS measurements and, to a lesser extent, SIMS and AES depth profiling have been complemented by various methods for determining barrier heights (Φ_B) of contacts (Section 10.4). Changes in Φ_B are a sensitive indicator of low-temperature reactions at the metal–Si interface.

To appreciate the choice of barrier materials, we first distinguish among three models that have been proposed for successful diffusion-barrier behavior (Ref. 25).

1. Stuffed Barriers. Stuffed barriers rely on the segregation of impurities along otherwise rapid diffusion paths such as GBs to block further passage of two-way atomic traffic there. The marked improvement of sputtered Mo and Ti–W alloys as diffusion barriers when they contain small quantities of intentionally added N or O impurities is apparently due to this mechanism. Impurity concentrations of $\sim 10^{-1}$ to 10^{-3} at% are typically required to decorate GBs and induce stuffed-barrier protection. In extending the electromigation life of Al, Cu may in effect "stuff" the conductor GBs.

*2. **Passive Compound Barriers.*** Ideal barrier behavior exhibiting chemical inertness and negligible mutual solubility and diffusivity is sometimes approximated by compounds. Although there are numerous possibilities among the carbides, nitrides, borides, and even the more conductive oxides, only the transition metal nitrides, such as TiN, have been extensively explored for device applications. TiN has proved effective in solar cells as a diffusion barrier between N–Si and Ti–Ag, but contact resistances are higher than desired in high-current-density circuits.

*3. **Sacrificial Barriers.*** A sacrificial barrier maintains the separation of A and B only for a limited duration. As shown in Fig. 8-22, sacrificial barriers exploit the fact that reactions between adjacent films in turn produce uniform layered compounds AX and BX that continue to be separated by a narrowing X barrier film. So long as X remains and compounds AX and BX possess adequate conductivity, this barrier is effective. The first recognized application of a sacrificial barrier involved Ti, which reacted with Si to form Ti_2Si and with Al to form $TiAl_3$. Judging from the many metal aluminide and occasional Al–metal–silicon compounds in Table 8-3, sacrificial barrier reactions appear to be quite common.

If the reaction rate kinetics of both compounds, i.e., AX, BX, are known, then either the effective lifetime or the minimum thickness of barrier required may be predicted. The following example is particularly instructive (Ref. 1). Suppose we consider a Ti diffusion barrier between Si and Al. Without imposition of Ti, the Al–Si combination is unstable. The question is, how

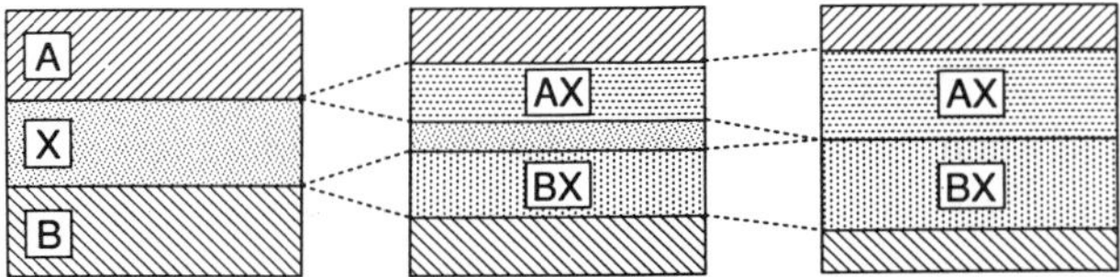

Figure 8-22. Model of sacrificial barrier behavior. A and B films react with barrier film X to form AX and BX compounds. Protection is afforded as long as X is not consumed. (Reprinted with permission from Elsevier Sequoia, S.A., from M.-A. Nicolet, *Thin Solid Films* **52**, 415, 1978).

much Ti should be deposited to withstand a thermal anneal at 500 °C for 15 min? At the Al interface, $TiAl_3$ forms with parabolic kinetics given by

$$d^2_{TiAl_3} = (1.5 \times 10^{15}) e^{-1.85\,eV/kT} t \; (\text{Å}^2), \tag{8-27}$$

where d_{TiAl_3} is the thickness of the $TiAl_3$ layer and t is the time in seconds. Similarly, the reaction of Ti with Si results in the formation of $TiSi_2$ with a kinetics governed by

$$d^2_{TiSi_2} = (5.74 \times 10^{9}) e^{-1.3\,eV/kT} t \; (\text{Å}^2). \tag{8-28}$$

For the specified annealing conditions, $d_{TiAl_3} = 1100$ Å and $d_{TiSi_2} = 130$ Å. An insignificant amount of Ti is consumed under ambient operating conditions. Therefore, the minimum thickness of Ti required is the sum of these two values, or 1230 Å.

In conclusion, we note that semiconductor contacts are thermodynamically unstable because they are not in a state of minimum free energy. The imposition of a diffusion barrier slows down the equilibration process, but the instability is never actually removed. Enhanced reliability is bought with diffusion barriers, but at the cost of increasing structural complexity and added processing expense.

8.7. Diffusion During Film Growth

We close the chapter by considering diffusion effects in films growing within a gas-phase ambient. In addition to the diffusional exchange between gas atoms and growing film, or the redistribution of atoms between film and substrate, there is the added complexity of transport across a moving boundary. Such effects are important in high-temperature oxidation of Si, one of the most-studied film growth processes. The resulting amorphous SiO_2 films find extensive use in microelectronic applications as an insulator, and as a mask used to pattern and expose some regions for processing while shielding other areas. In contrast to film deposition, where the atoms of the deposit originate totally from the vapor phase (as in CVD of SiO_2), oxidation relies on the reaction between Si and oxygen to sustain oxide film growth. This means that for every 1000 Å of SiO_2 growth, 440 Å (i.e., $1000\rho_{SiO_2}M_{Si}/\rho_{Si}M_{SiO_2}$) of Si substrate is consumed. The now-classic analysis of oxidation due to Grove (Ref. 27) has a simple elegance and yet accurately predicts the kinetics of thermal oxidation. In this treatment of the model, we assume a flow of gas containing oxygen parallel to the plane of the Si surface. In order to form oxide at the Si–SiO_2 interface, the following sequential steps are assumed to occur:

1. Oxygen is transported from the bulk of the gas phase to the gas–oxide interface.

2. Oxygen diffuses through the growing solid oxide film of thickness d_0.

3. When oxygen reaches the Si–SiO_2 interface, it chemically reacts with Si and forms oxide.

The respective mass fluxes corresponding to these steps can be expressed by

$$J_1 = h_G(C_G - C_0), \tag{8-29}$$

$$J_2 = D(C_0 - C_i)/d_0, \tag{8-30}$$

$$J_3 = K_S C_i, \tag{8-31}$$

where the concentrations of oxygen in the bulk of the gas, at the gas–SiO_2 interface, and at the SiO_2–Si interface are respectively, C_G, C_0, and C_i. The quantities h_G, D, and K_S represent the gas mass-transport coefficient, the diffusion coefficient of oxygen in SiO_2, and the chemical reaction rate constant, respectively. Constants D and K_S display the usual Boltzmann behavior but with different activation energies, and h_G has a weak temperature dependence.

By assuming steady-state growth implying $J_1 = J_2 = J_3$, we easily solve for C_i and C_0 in terms of C_G:

$$C_0 = \frac{C_G(1 + K_S d_0/D)}{1 + K_S/h_G + K_S d_0/D}, \tag{8-32a}$$

$$C_i = \frac{C_G}{1 + K_S/h_G + K_S d_0/D}. \tag{8-32b}$$

Clearly, the grown SiO_2 has a well-fixed stoichiometry so that C_0 and C_i differ only slightly in magnitude, but sufficiently to establish the concentration gradient required for diffusion. In fact, $C_0 = C_i = C_G/(1 + K_S/h_G)$ in the so-called reaction-limited case where $D \gg K_S d_0$. Here, diffusion is assumed to be very rapid through the SiO_2, but the bottleneck for growth is the interfacial chemical reaction. On the other hand, under diffusion control, D is small so that $C_0 \approx C_G$ and $C_i \approx 0$. In this case the chemical reaction is sufficiently rapid, but the supply of oxygen is rate-limiting. The actual oxide growth rate is related to the flux, say J_3, and therefore the thickness of oxide at any time is expressed by

$$d(d_0)/dt = K_S C_i/N_0, \tag{8-33}$$

where N_0 is the number of oxidant molecules incorporated into a unit volume of film. For oxidation in dry O_2 gas, $N_0 = 2.2 \times 10^{22}$ cm^{-3}, whereas for steam (wet) oxidation $N_0 = 4.4 \times 10^{22}$ cm^{-3}, because half as much oxygen is contained per molecule.

Substitution of Eq. 8-32b into 8-33 and direct integration of the resulting differential equation yields

$$d_0^2 + Ad_0 = B(t + \tau), \tag{8-34}$$

where

$$A = 2D\left(\frac{1}{h_G} + \frac{1}{K_S}\right), \quad B = \frac{2DC_G}{N_0}, \quad \text{and} \quad \tau = \frac{d_i^2 + Ad_i}{B}.$$

The constant of integration τ arises only if there is an initial oxide film of thickness d_i present prior to oxidation, and therefore Eq. 8-34 is useful in describing sequential oxidations. A solution to this quadratic equation is

$$\frac{d_0}{A/2} = \sqrt{1 + \frac{t + \tau}{A^2/4B}} - 1, \tag{8-35}$$

from which the limiting long- as well as short-time growth kinetics relationships are easily shown to be

$$d_0^2 \approx Bt \qquad \text{for } t \gg A^2/4B, \tag{8-36}$$

$$d_0 \approx \frac{B}{A}(t + \tau) \qquad \text{for } t + \tau \ll \frac{A^2}{4B}. \tag{8-37}$$

The reader will recall similar parabolic and linear growth in metal compound and silicide films. All film growth is probably linear to begin with because parabolic growth implies an infinite initial thickening rate.

Values for the parabolic and linear rate constants for SiO_2, grown from (111) Si, are approximately (Ref. 28)

$$B = 186 \exp -\frac{0.71 \text{ eV}}{kT} \ \mu\text{m}^2/\text{hr} \qquad (\text{wet } O_2), \tag{8-38a}$$

$$B = 950 \exp -\frac{1.24 \text{ eV}}{kT} \ \mu\text{m}^2/\text{hr} \qquad (\text{dry } O_2), \tag{8-38b}$$

$$\frac{B}{A} = 7.31 \times 10^7 \exp -\frac{1.96 \text{ eV}}{kT} \ \mu\text{m/hr} \qquad (\text{wet } O_2), \tag{8-39a}$$

$$\frac{B}{A} = 5.89 \times 10^6 \exp -\frac{2.0 \text{ eV}}{kT} \ \mu\text{m/hr} \qquad (\text{dry } O_2), \tag{8-39b}$$

where all constants are normalized to 760 torr. Equations 8-36 and 8-37 serve as an aid in designing oxidation treatments. Different activation energies for B are obtained in wet and dry O_2 because the migrating species in each case is

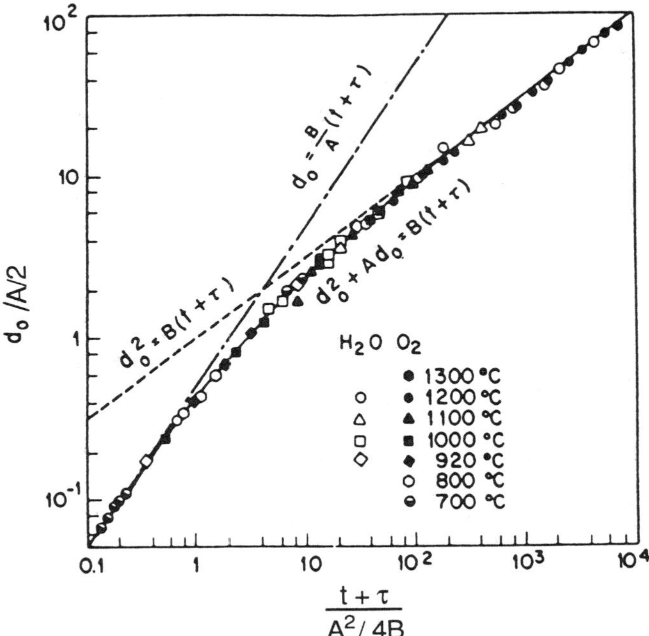

Figure 8-23. Oxidation kinetics behavior of Si in terms of dimensionless oxide thickness and time. The two limiting forms of the kinetics are shown. From A. S. Grove, *Physics and Technology of Semiconductor Devices*, Copyright © 1967, John Wiley and Sons. (Reprinted with permission.)

different, e.g., H_2O and OH as opposed to O_2 or O. However, the activation energy for B/A reflects the chemical reaction at the Si–SiO_2 interface and is the same regardless of the nature of the oxygen-bearing diffusant. A single dimensionless thickness–time plot shown in Fig. 8-23 very neatly summarizes Si oxidation behavior. The limiting linear and parabolic growth kinetics regimes are clearly identified.

Not all oxidation processes, however, display linear or parabolic growth kinetics. Some examples are presented in Chapter 12 in connection with protective oxide coatings.

EXERCISES

1. a. Establish generalized expressions for the lattice diffusivity as a function of temperature for semiconductors and alkali halides, using Fig. 5-6.

b. How does your expression for D_L compare with the diffusivity values for Si in Si (self-diffusion) in Fig. 8-3?

2. A P–N junction is produced by diffusing B from a continuous source ($C_0 = 10^{19}$ cm^{-3}) into an eptiaxial Si film with a background N level of 10^{15} cm^{-3}. Diffusion is carried out at 1100 °C for 30 min.

a. How far beneath the Si surface is the junction (i.e., where $C_N = C_P$). Use Eq. 1-27a.
b. If there is a 1% error in temperature, what is the percent change in junction depth?
c. By what percent will the junction depth change for a 1% change in diffusion time?

3. At what temperature will the number of Au atoms transported through grain boundaries equal that which diffuses through the lattice if the grain size is 2 μm? 20 μm?

4. The equation for transport of atoms down a single grain boundary where there is simultaneous diffusion into the adjoining grains is

$$\frac{\partial C_b}{\partial t} = D_b \frac{\partial^2 C_b}{\partial y^2} - \frac{2D_L}{\delta}\left(\frac{dC_L}{dx}\right)_{x=\delta/2}$$

Derive this equation by considering diffusional transport into and out of an element of grain boundary δ wide and dy long.

5. In the Pd–Au thin-film diffusion couple an approximate fit to the data of Fig. 8-7 can be made employing the equation

$$C(x, t) = \frac{C_0}{2}\operatorname{erfc}\frac{x}{2\sqrt{Dt}}.$$

a. Plot $C(x, t)$ vs. x.
b. From values of $dC/dx\,|_{x=0}$, estimate the values of D for the 0-h, 20-h, and 200-h data. Are these D values the same?
c. What accounts for the apparent interdiffusion between Au and Pd at 0 h?

6. a. Calculate the activation energy for dislocation pipe diffusion of Au in epitaxial Au films from the data of Fig. 8-6a.
b. Calculate the activation energy for grain-boundary diffusion of Au in polycrystalline Au films from the data of Fig. 8-6b.

In both cases make Arrhenius plots of the diffusivity data. Assume

$$D_L = 0.091 \exp - \frac{41.7 \text{ (kcal/mole)}}{RT} \text{ cm}^2/\text{sec}.$$

7. An N-type dopant from a continuous source of concentration C_0 is diffused into a P-type semiconductor film containing a single grain boundary oriented normal to the surface. If the background dopant level in the film is C_B, write an expression for the resulting P–N junction profile (y vs. x) after diffusion.

8. A 1-μm-thick film of Ni was deposited on a Si wafer. After a 1-h anneal at 300 °C, 600 Å of Ni_2Si formed.

a. Predict the thickness of Ni_2Si that would form if the Ni–Si couple were heated to 350 °C for 2 h.
b. In forming 600 Å of Ni_2Si, how much Si was consumed? [Note: The atomic density of Ni is 9×10^{22} atoms/cm^3.]
c. The lattice parameters of cubic Ni_2Si and Si are 5.406 Å and 5.431 Å, respectively. Comment on the probable nature of the compound–substrate interface.

9. The thermal stability of a thin-film superlattice consisting of an alternating stack of 100-Å-thick layers of epitaxial GaAs and AlAs is of concern.

a. If chemical homogenization of the layers is limited by the diffusion of Ga in GaAs, estimate how long it will take Ga to diffuse 50 Å at 25 °C?
b. *Roughly* estimate the temperature required to produce layers of composition $Ga_{0.75}Al_{0.25}As$–$Ga_{0.25}Al_{0.75}As$ after a 1-h anneal.

10. For electromigration in Al stripes assume $E_e = 0.7$ eV and $n = 2.5$ in Eq. 8-23. By what factor is MTF shortened (or extended) at 40 °C by

a. a change in E_e to 0.6 eV?
b. no change in E_e but a temperature increase to 85 °C?
c. a decrease in stripe thickness at a step from 1.0 to 0.75 μm?
d. an increase in current from 1 to 1.5 mA?

11. a. When there are simultaneous electromigration and diffusional fluxes of atoms, show that

$$\frac{\partial C}{\partial t} = D\frac{\partial^2 C}{\partial x^2} - v\frac{\partial C}{\partial x},$$

with v defined by Eq. 8-25.

b. For a diffusion couple ($C = C_0$ for $x < 0$ and $C = 0$ for $x > 0$) show that

$$C(x, t) = \frac{C_0}{2}\left[\exp\frac{vx}{D}\,\text{erfc}\frac{x + vt}{\sqrt{4\,Dt}} + \text{erfc}\frac{x - vt}{\sqrt{4\,Dt}}\right]$$

is a solution to the equation in part (a) and satisfies the boundary conditions.

c. A homogeneous Al film stripe is alloyed with a cross stripe of Cu creating two interfaces; (+) Al–Cu/Al and (−) Al/Al–Cu. Show, using the preceding solution, that the concentration profiles that develop at these interfaces obey the relation.

$$\ln\frac{C_+(x, t)}{C_-(x, t)} = \frac{vx}{D}.$$

12. The surface accumulation interdiffusion data of Fig. 8-11a can be fitted to the normalized equation $C_S = 1 - \exp - S(t - t_0)$. For

run 1: $S = 7.1 \times 10^{-1}\ \text{sec}^{-1}$,
run 2: $S = 1.4 \times 10^{-2}\ \text{sec}^{-1}$,
run 3: $S = 7.4 \times 10^{-4}\ \text{sec}^{-1}$,
run 4: $S = 2.6 \times 10^{-5}\ \text{sec}^{-1}$.

a. If S is thermally activated, i.e., $S = S_0 \exp - E_S/kT$ (S_0 = constant), make an Arrhenius plot and determine the activation energy for diffusion of Ag in Au films.

b. What diffusion mechanism is suggested by the value of E_S?

13. a. Compare the time required to grow a 3500 Å thick SiO_2 film in dry as opposed to wet O_2 at 1100 °C. Assume the native oxide thickness is 30 Å.

b. A window in a 3500 Å SiO_2 film is opened down to the Si substrate in order to grow a gate oxide at 1000 °C for 30 minutes in dry O_2. Find the resulting thickness of both the gate and surrounding (field) oxide films.

References

1. M.-A. Nicolet and M. Bartur, *J. Vac. Sci. Tech* **19**, 786 (1981).
2. R. W. Balluffi and J. M. Blakely, *Thin Solid Films* **25**, 363 (1975).
3. N. A. Gjostein, *Diffusion*, American Society for Metals, Metals Park, Ohio (1973).

4. J. C. C. Tsai, in *VLSI Technology*. 2nd ed., ed. S. M. Sze, McGraw-Hill, New York (1988).
5. S. M. Sze, *Semiconductor Devices–Physics and Technology*, Wiley, New York (1985).
6. D. C. Jacobson, *Ion-Beam Studies of Noble Metal Diffusion in Amorphous Silicon Layers*, Ph.D. Thesis, Stevens Institute of Technology (1989).
7. J. C. Fisher, *J. Appl. Phys.* **22**, 74 (1951).
8. R. T. Whipple, *Phil. Mag.* **45**, 1225 (1954).
9. T. Suzuoka, *Trans. Jap. Inst. Met.* **2**, 25 (1961).
10. L. G. Harrison, *Trans. Faraday. Soc.* **57**, 1191 (1961).
11.* A Gangulee, P. S. Ho, and K. N. Tu, eds., *Low Temperature Diffusion and Applications to Thin Films*, Elsevier, Lausanne (1975); also *Thin Solid Films* **25** (1975).
12.* J. M. Poate, K. N. Tu, and J. W. Mayer, eds., *Thin Films—Interdiffusion and Reactions*, Wiley, New York (1978).
13. D. Gupta, *Phys. Rev.* **7**, 586 (1973).
14. D. Gupta and K. W. Asai, *Thin Solid Films* **22**, 121 (1974).
15. P. M. Hall, J. M. Morabito, and J. M. Poate, *Thin Solid Films* **33**, 107 (1976).
16. K. N. Tu, W. K. Chu, and J. W. Mayer, *Thin Solids Films* **25**, 403 (1975).
17.* K. N. Tu, *Ann. Rev. Mater. Sci.* **15**, 147 (1985).
18.* D. Gupta and P. S. Ho, *Thin Solid Films* **72**, 399 (1985).
19. J. C. M. Hwang, J. D. Pan, and R. W. Balluffi, *J. Appl. Phys.* **50**, 1349 (1979).
20. S. U. Campisano, G. Foti, R. Rimini, and J. W. Mayer, *Phil. Mag.* **31**, 903 (1975).
21. M. Ohring and R. Rosenberg, *J. Appl. Phys.* **42**, 5671 (1971).
22. S. Vaidya, T. T. Sheng, and A. K. Sinha, *Appl. Phys. Lett.* **36**, 464 (1980).
23. K. Nakamura, M.-A. Nicolet, J. W. Mayer, R. J. Blattner, and C. A. Evans, *J. Appl. Phys.* **46**, 4678 (1975).
24. G. Ottavio, *J. Vac. Sci. Tech.* **16**, 1112 (1979).
25. M.-A. Nicolet, *Thin Solid Films* **52**, 415 (1978).
26. M. Wittmer, *J. Vac. Sci. Tech.* **A2**, 273 (1984).
27. A. S. Grove, *Physics and Technology of Semiconductor Devices*, Wiley, New York (1967).
28. L. E. Katz, in *VLSI Technology*, 2nd ed., ed. S. M. Sze, McGraw-Hill, New York (1988).

*Recommended texts or reviews.

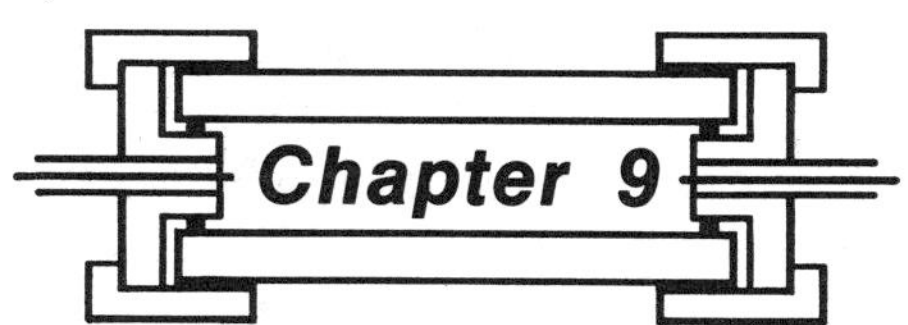

Mechanical Properties of Thin Films

9.1. Introduction

Interest in mechanical-property effects in thin films has focused on two major issues. The primary concern has been with the deleterious effects that stress causes in films. This has prompted much research to determine the type, magnitude, and origin of stress as well as means of minimizing or controlling stresses. A second important concern is related to enhancing the mechanical properties of hardness and wear resistance in assorted coating applications. The topic of stress in films has historically generated the greatest attention and will be our major interest in this chapter. A discussion of the mechanical properties of metallurgical and protective coatings is the focus of Chapter 12.

It is virtually always the case that stresses are present in thin films. What must be appreciated is that stresses exist even though films are not externally loaded. They directly affect a variety of phenomena, including adhesion, generation of crystalline defects, perfection of epitaxial deposits, and formation of film surface growths such as hillocks and whiskers. Film stresses that tend to increase with thickness are a prime limitation to the growth of very thick films because they promote film peeling. In addition, film stresses influence band-gap shifts in semiconductors, transition temperatures in superconductors, and magnetic anisotropy. Substrate deformation and distortion also necessarily arise from stresses in the overlying films. In most applications, this is not a troublesome issue because substrates are usually relatively massive compared to films. In integrated circuit technology, however, even slight bowing of silicon wafers presents significant problems with regard to maintaining precise tolerances in the definition of device features.

The existence of stresses in thin electrodeposited films has been known since 1858, when the English chemist Gore noted (Ref. 1): "In electrodeposits generally the inner and outer surfaces are in unequal states of cohesive tension frequently in so great a degree as to rend the deposit extensively and raise it from the cathode in the form of a curved sheet with its concave side to the anode." The concave bending of the cantilevered cathodic electrode implies, as we shall see, a tensile stress in the deposit. Despite the passage of years, the origins of stress in electrodeposited films are still not completely understood. A similar state of affairs exists in both physical and chemical vapor-deposited thin films; stresses exist, they can be measured, but their origins are not known with certainty.

There is a great body of information on mechanical effects in bulk materials that provides a context for understanding film behavior. At one extreme is the elastic regime, rooted in the theory of elasticity, which forms the basis of structural mechanics and much engineering design. Here the material elongations (i.e., strains) are linearly proportional to the applied forces (i.e., stresses). Upon unloading, the material snaps back and regains its original shape. At the other extreme are the irreversible plastic effects induced at stress levels above the limit of the elastic response (i.e., the yield stress). All sorts of mechanical forming operations in materials (e.g., rolling, extrusion, drawing) as well as failure phenomena (e.g., creep, fatigue, fracture) are manifestations of plastic-deformation effects. Plasticity, unlike elasticity, is difficult to model mathematically because plastic behavior is nonlinear and strongly dependent on the past thermomechanical processing and treatment history of the material.

In the packaging and attachment of semiconductor chips to circuit modules and boards, a new collection of structural-mechanics applications has recently emerged (Ref. 2). The involved components are small, thicker than thin films, but vastly smaller than conventional engineering structures. Nevertheless, our understanding of the mechanical behavior of electronic packaging materials used—metals (e.g., Pb–Sn, Al), ceramics (e.g., Al_2O_3, SiO_2 glasses), semiconductors (e.g., Si, GaAs), and polymers (e.g., epoxies, polyimide)—and the structural mechanics of combinations of involved components (e.g., solder bumps and joints, chip bonding pads, die supports, encapsulants, etc.) has evolved from well-established bulk elastic and plastic phenomena and analyses.

It comes as no surprise that the varied mechanical properties of thin films also span both the elastic and plastic realms of behavior. For this reason we begin with an abbreviated review of relevant topics dealing with these classic subjects prior to the consideration of mechanical effects in films. The chapter outline follows:

9.2. Introduction to Elasticity, Plasticity, and Mechanical Behavior
9.3. Internal Stresses and Their Analyses
9.4. Stress in Thin Films
9.5. Relaxation Effects in Stressed Films
9.6. Adhesion

9.2. Introduction to Elasticity, Plasticity, and Mechanical Behavior

9.2.1. Elastic Regime

An appropriate way to start a discussion of mechanical properties is to consider what is meant by stress. When forces are applied to the surface of a body, they act directly on the surface atoms. The forces are also indirectly transmitted to the internal atoms via the network of bonds that are distorted by the internally developed stress field. If the plate in Fig. 9-1a is stretched by equal and opposite axial tensile forces F, then it is both in mechanical equilibrium as well as in a state of stress. Since the plate is in static equilibrium, it can be cut as shown in Fig. 9-1b, revealing that internal forces must act on the exposed surface to keep the isolated section from moving. Regardless of where and at what orientation the plate is cut, balancing forces are required to sustain equilibrium. These internal forces distributed throughout the plate constitute a state of stress. In the example shown, the normal force F divided by the area A defines the tensile stress σ_x. Similarly, normal stresses in the remaining two coordinate directions, σ_y and σ_z, can be imagined under more complex loading conditions. If the force is directed into the surface, a compressive stress arises. Convention assigns it a negative sign, in contrast to the positive sign for a tensile stress. In addition, mechanical equilibrium on internal surfaces cut at an arbitrary angle will generally necessitate forces and stresses resolved in the plane itself. These are the so-called shear stresses. The tensile force of Fig. 9-1a produces maximum shear stresses on planes inclined at 45° with respect to the plate axis (Fig. 9-1a). If the normal tensile stress $\sigma_x = F/A$, then the force resolved on these shear planes is $F \cos 45° = \sqrt{2}/2\ F$. The area of the shear planes is $A/\cos 45° = 2/\sqrt{2}\ A$. Therefore, the shear stress $\tau = F/2A$

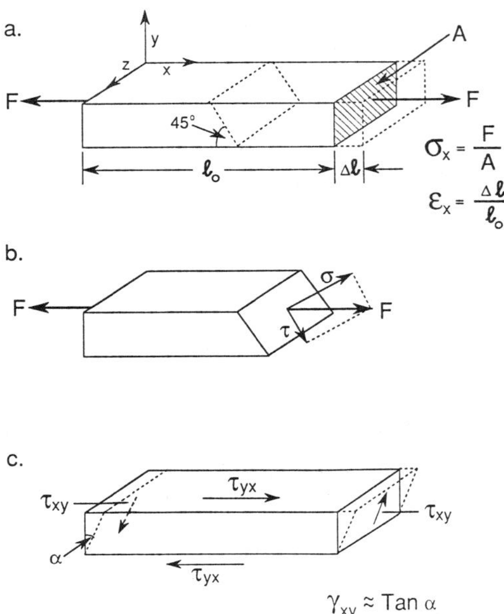

Figure 9-1. (a) Tensile force applied to plate. (b) Arbitrary free-body section revealing spatial distribution of stress through plate. Both tensile and shear stresses exist on exposed plane. (c) Distortion in plate due to applied shear stress.

and is half that of the tensile stress. Shear stresses are extremely important because they are essentially responsible for the plastic deformation of crystalline materials. Two subscripts are generally required to specify a shear stress: the first to denote the plane in which shear occurs, and the second to identify the direction of the force in this plane. If, for example, a shear force were applied to the top surface of the plate, it distorts into a prism. For all forces and moments to balance, a tetrad of shear stresses must act on the horizontal as well as vertical faces. It is left as an exercise to show that $\tau_{xy} = \tau_{yx}$ in equilibrium, and similarly, for τ_{yz} and τ_{xz}.

The application of tensile forces extends the plate of Fig. 9-1 by an amount Δl. This results in a normal strain ε_x, defined by $\varepsilon_x = \Delta l / l_0$, where l_0 is the original length. similarly, in other directions the normal strains are ε_y and ε_z. In the example given, the plate also contracts laterally in both the y and z directions in concert with the longitudinal extension in the x direction. Therefore, even though there is no stress in the y direction, there is a strain ε_y given by $\varepsilon_y = -\nu\varepsilon_x$; similarly for ε_z. The quantity ν, a measure of this lateral contraction, is Poisson's ratio, and for many materials it has a value of about 0.3. Under the action of shear stresses, shear strains (γ) are induced;

these are essentially defined by the tangent of the shear distortion angle α in Fig. 9-1c.

In the elastic regime, all strains are small, and Hooke's law dominates the response of the system; i.e.,

$$\sigma_x = E\varepsilon_x, \tag{9-1a}$$

where E is Young's modulus. (Values of E are entered in Tables 9-2 and 12-1 for a variety of materials of interest.) When a three-dimensional state of stress exists,

$$\varepsilon_x = (1/E)\left[\sigma_x - \nu(\sigma_y + \sigma_z)\right] \tag{9-1b}$$

(similarly for ε_y and ε_z). This formula simplifies to Eq. 9-1a in the absence of σ_y and σ_z. For shear stresses, Hooke's law also applies in the form

$$\tau_{xy} = \mu\gamma_{xy} \tag{9-2}$$

(similarly for τ_{xz} and τ_{yz}), where μ is the shear modulus. These equations only strictly apply to isotropic materials where of the three elastic constants E, μ, and ν, only two are independent; they are connected by the relation $\mu = E/2(1 + \nu)$.

In anisotropic media, such as the single-crystal quartz plate used to monitor film thickness during deposition (Chapter 6), the elastic constants reflect the noncubic symmetry of the crystal structure. Although there are more elastic constants to contend with, Hooke's law is still valid. In addition to describing stress–strain relationships, elasticity theory is concerned with specifying the stress and strain distribution throughout the volume of shaped bodies subjected to arbitrary loading conditions. In Section 9.3, we employ some of the simpler concepts to derive basic formulas used to determine the stress in films.

9.2.2. Tensile Properties of Thin Metal Films

Tensile tests are widely used to evaluate both the elastic and plastic response of bulk materials. Although direct tensile tests in films are not conducted with great frequency today, past measurements are interesting because of the basic information they conveyed on the nature of deformation processes in metal films. Many of the results and their interpretations can be found in the old but still useful review by Hoffman (Ref. 3). Unlike its bulk counterpart, tensile testing of thin films is far from a routine task and has all the earmarks of a research effort. The extreme delicacy required in the handling of thin films has posed a great experimental challenge and stimulus to the ingenuity of investigators. Detachment of films from substrates and methods for gripping and aligning them, applying loads, and measuring the resultant mechanical response are some of the experimental problems. Loading is commonly achieved by electromagnetic force transducers, and strains are typically measured by optical methods. The most novel microtensile testing devices have been incorporated within electron microscopes, enabling direct observation of defects and recording of diffraction patterns during straining.

A typical stress–strain curve for gold is shown in Fig. 9-2a. The maximum tensile strength, or stress at fracture in this case, is only somewhat higher than typical intrinsic or residual stress levels about which we will speak later. The strain at fracture is only ~ 0.8%, which is much more than an order of magnitude smaller than that observed in bulk Au. Loading and unloading curves of varying slope (or modulus of elasticity) have raised questions of whether E differs from the bulk value and whether it is film-thickness-depen-

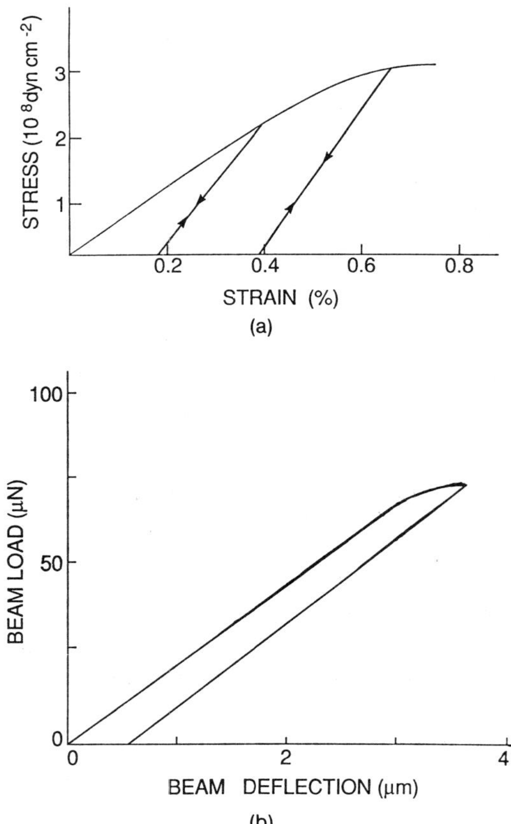

Figure 9-2. (a) Stress-strain behavior for a Au film. (From Ref. 4). (b) Load-deflection behavior for a 0.87-μm-thick Au cantilever film. (From Ref. 7).

dent. On both accounts experimental data indicate no abnormal effects. Above the elastic limit there is considerable evidence for plasticity in the form of observed dislocation motion, stress relaxation, and creep effects as well as regions of localized thinning. Unlike bulk metals, there is no regime of easy dislocation glide in films; rather, polycrystalline metal films sometimes contain initial dislocation densities of 10^{10} to 10^{12} cm^{-2}, which are higher by an order of magnitude or so than those found in heavily worked and strain-hardened bulk metals.

The surprisingly high tensile strengths of metal films have been the subject of much interest. Typical strengths exceed those for hard-drawn metals by a factor of 2 to 10, and may be as much as 100 times that of annealed bulk metals. Polycrystal films are usually stronger than single-crystal films, reflecting the role of grain boundaries as obstacles to dislocation motion. Reported values for the maximum tensile stresses or strengths of FCC metal films span a range from $\mu/40$ to $\mu/120$ with a wide degree of scatter, particularly when results from different laboratories are compared. The highest strengths are close to those theoretically predicted, assuming deformation occurs by rigid lattice displacements. This means that the high dislocation densities in films leave few avenues available for either generation or motion of new dislocations.

Specific microscopic mechanisms for the strength of films are quite detailed. Some insight into what is involved, however, can be gained by considering two relationships borrowed from theories of mechanical behavior of bulk materials. The first simply estimates the shear stress required to cause dislocations to effectively bypass obstacles situated a distance l apart and thereby produce plastic deformation. Thus,

$$\tau = \mu b / l, \tag{9-3}$$

where b is the Burgers vector. The distance l is usually taken as the spacing between dislocation pinning points such as precipitates, grain boundaries, or other dislocations. If l is imagined to be the film thickness (d), then it is clear that film strength is predicted to vary inversely with d. The second relationship is the Hall–Petch equation, which connects the yield stress to grain size l_g. Variants of this equation relate ultimate tensile strength σ_{TS} to grain size as well. Therefore,

$$\sigma_{TS} = \sigma_a + K l_g^{-1/2}. \tag{9-4}$$

In a modified version, the film thickness replaces the mean grain size. This is a justifiable substitution since the two quantities are indirectly related. In Eq. 9-4, σ_a is an intrinsic stress level and K is a constant. Again on the basis of this formula, thinner films are expected to be stronger. Although this is certainly true in the scaling down from (bulk) micron-thick foils to submicron-thick films, it is not certain whether there is a thinness beyond which no further strengthening occurs. Experimental data are contradictory with respect to this issue.

9.2.3. Bulge Testing of Films

Bulge testing is widely used to determine the mechanical properties of thin films and membranes. In this test the film–substrate assembly is sealed to the end of a hollow cylindrical tube so that it can be pressurized with gas. The maximum height of the resulting hemispherical bulge in the film is then measured optically with a microscope or interferometer and converted to strain. A relationship between the dome height h produced by the differential pressure P has been determined to be (Ref. 3)

$$P = \frac{4\,dh}{r^2}\left(\sigma_0 + \frac{2}{3}\frac{E}{1-\nu}\frac{h^2}{r^2}\right). \tag{9-5}$$

The film thickness and specimen radius are d and r, respectively, and σ_0 is the residual stress in the film under a zero pressure differential. To illustrate bulge testing and how data are analyzed, we consider a recent study conducted on both epitaxial and polycrystalline Si membranes (Ref. 5). These materials are candidates for X-ray lithography mask substrates, an application requiring high fracture strength and excellent dimensional stability. Freestanding membranes 1 μm thick and 38 mm in diameter, supported by a thicker substrate ring of either Si or SiO_2, were prepared by selective masking and etching methods.

The pressure-deflection response for the polycrystalline membrane is shown in Fig. 9-3a. An excellent fit of the data to Eq. 9-5 is evident, enabling values of σ_0 and $E/(1-\nu)$ to be determined. Despite the nonlinear membrane deflection with pressure, the response is actually elastic and not plastic. Repeated pressurization cycles did not result in any appreciable deterioration of reproducibility in mechanical response. In Fig. 9-3b the membrane is stressed to failure at higher pressure levels. The fracture stress is given by $\sigma_f = P_f r^2/4\,dh_f$, where P_f and h_f are the values at fracture. The following results compare the properties of the two Si film materials.

	E (10^{12} dynes/cm^2)	σ_0 (10^9 dynes/cm^2)	σ_f (10^9 dynes/cm^2)
Epitaxial	1.8	0.52	1.9
Polycrystalline	2.5	1.7	4.2

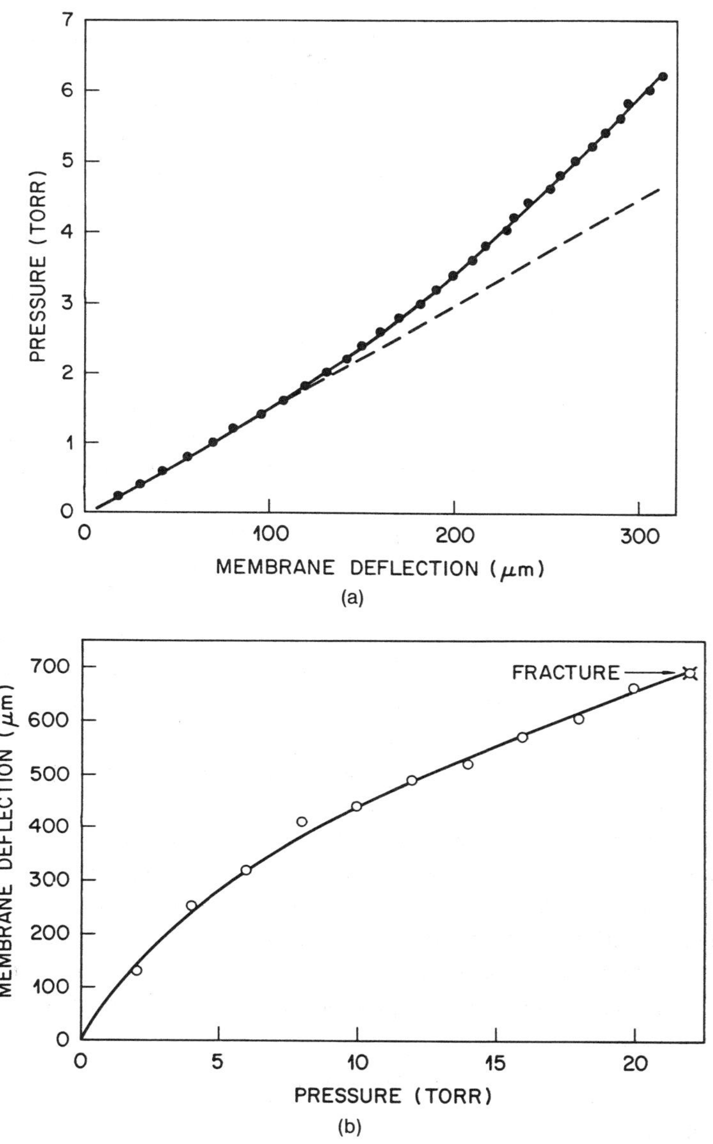

Figure 9-3. (a) Pressure-membrane deflection characteristics of a 1-μm-thick poly-Si membrane. (Solid curve is Eq. 9-5; dashed curve represents $P = 4\,dh\sigma_0/r_0^2$.) (b) Membrane deflection vs. differential pressure measured to the point of fracture. (From Ref. 5).

9.2.4. Other Testing Methods

Recent years have witnessed the development of new techniques to measure the mechanical properties of films under the application of minute loads. Simultaneously, very small displacements are detected so that a continuous "stress–strain"-like curve is obtained. These techniques are based on the use of the Nanoindenter, a load-controlled submicron indentation instrument that is commercially available (Ref. 6). Its chief application has been indentation hardness testing, a subject more fully treated in Chapter 12. In addition to hardness, indentation tests have been used to indirectly measure a wide variety of mechanical properties in bulk materials, such as flow stress, creep resistance, stress relaxation, fracture, toughness, elastic modulus, and fatigue behavior.

Similar future tests on films and coatings with the Nanoindenter are certain. In this instrument the typical resolution of the displacment-sensing system is 2 Å, and that of the loading system is 0.3 μN. Due to the very small volume ($\sim$ 1 μm^3) sampled, the technique may be regarded as a mechanical properties microprobe by analogy to chemical microprobes (e.g., AES, SIMS). In one application, the Nanoindenter loading mechanism was used to load and measure the deflection of cantilever microbeams (Ref. 7). The latter were fabricated from Si wafers by employing micromachining techniques (Section 14.1.2). These involve standard photolithographic and etching processes borrowed from microelectronics technology to generate a variety of geometric shapes. Dimensions are larger than those employed in VLSI technology, but significantly smaller than what can be machined or fabricated by traditional methods. Thus freestanding microbeams $\sim$ 1 μm thick, 20 μm wide, and 20 μm long have been fabricated and tested by employing the experimental arrangement depicted schematically in Fig. 9-4. By evaporating or growing films on the Si substrate, then etching the latter away, we can extend the technique to different film materials. A typical beam load-deflection curve for a Au microbeam is shown in Fig. 9-2b. During loading, the linear elastic, as well as nonlinear, strain-hardening plastic regimes are observed. Formulas for the elastic deflection

$$\delta = 4F(1 - \nu^2)l^3 / wd^3E, \tag{9-6}$$

and yield stress

$$\sigma_y = 6lF_y / wd^2 \tag{9-7}$$

of the beam enable E and σ_y to be extracted from the data. Here l, w, and d are the beam length, width, and thickness, and F_y is the load marking the deviation from linearity in the loading response.

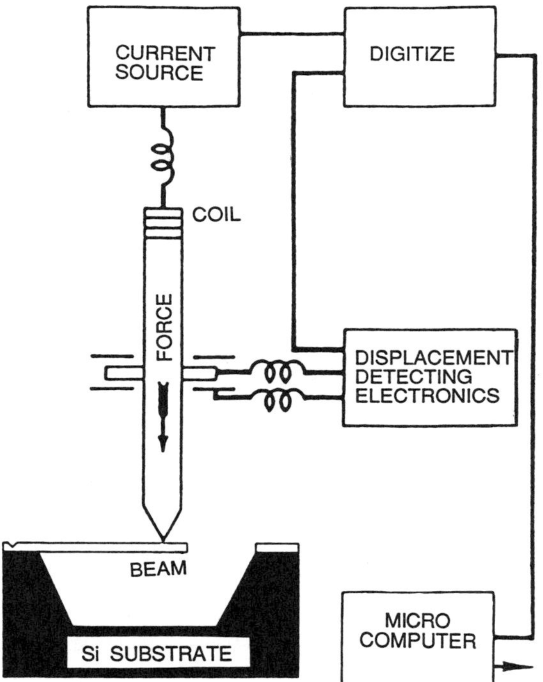

Figure 9-4. Schematic of Nanoindenter loading mechanism and a cantilever microbeam. (From Refs. 6 and 7, with permission from the Materials Research Society).

Further information on microbeam deflection and related submicron indentation testing techniques for films and coatings is provided in the recent review by Nix (Ref. 7). This reference is highly recommended for its treatment of misfit dislocations in epitaxial films (a subject already discussed in Chapter 7), as well as coverage of elastic and plastic phenomena including internal stress, strength, and relaxation effects in films. It is to these topics that we now turn our attention.

9.3. Internal Stresses and Their Analysis

9.3.1. Internal Stress

Implicit in the discussion to this point is that the stresses and the effects they produce are the result of *externally* applied forces. After the load is removed, the stresses are expected to vanish. On the other hand, thin films are stressed even without the application of external loading and are said to possess internal or residual stresses. The origin and nature of these internal residual stresses are the sources of many mechanical effects in films and a primary concern of this chapter.

Residual stresses are, of course, not restricted to composite film–substrate structures, but occur universally in all classes of homogeneous materials under special circumstances. A state of nonuniform plastic deformation is required, and this frequently occurs during mechanical or thermal processing. For example, when a metal strip is reduced slightly by rolling between cylindrical rolls, the surface fibers are extended more than the interior bulk. The latter restrains the fiber extension and places the surface in compression while the interior is stressed in tension. This residual stress distribution, is locked into the metal, but can be released like a jack-in-the-box. Machining a thin surface layer from the rolled metal will upset the mechanical equilibrium and cause the remaining material to bow. Residual stresses arise in casting, welds, machined and ground materials, and heat-treated glass. The presence of residual stresses is usually undesirable, but there are cases where they are beneficial. Tempered glass and shot-peened metal surfaces rely on residual compressive stresses to counteract harmful tensile stresses applied in service.

A model for the generation of internal stress during the deposition of films is illustrated in Fig. 9-5. Regardless of the stress distribution that prevails, maintenance of mechanical equilibrium requires that the *net* force (F) and

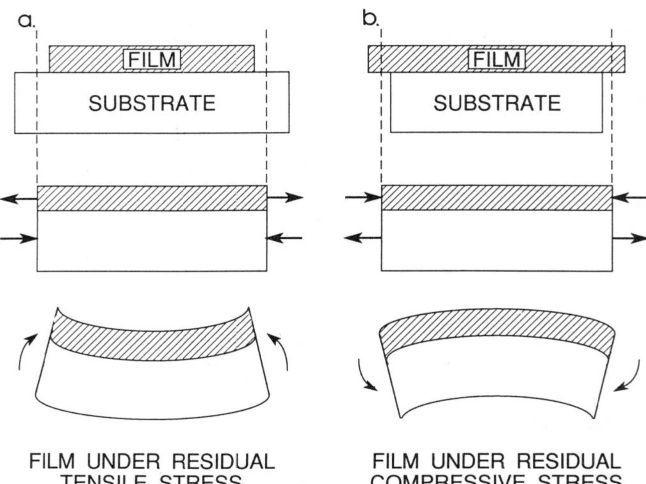

Figure 9-5. Sequence of events leading to (a) residual tensile stress in film; (b) residual compressive stress in film.

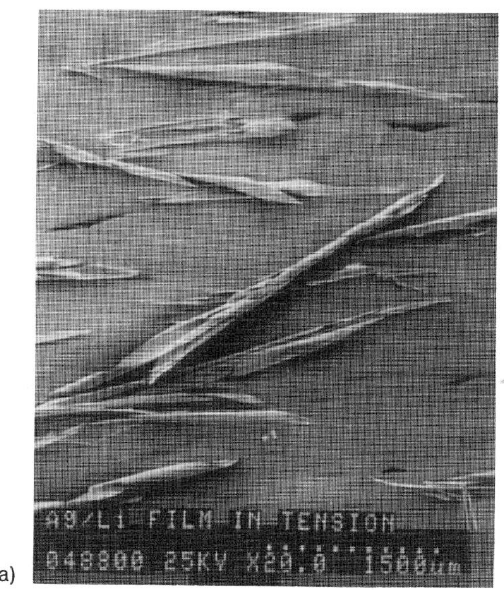

(a)

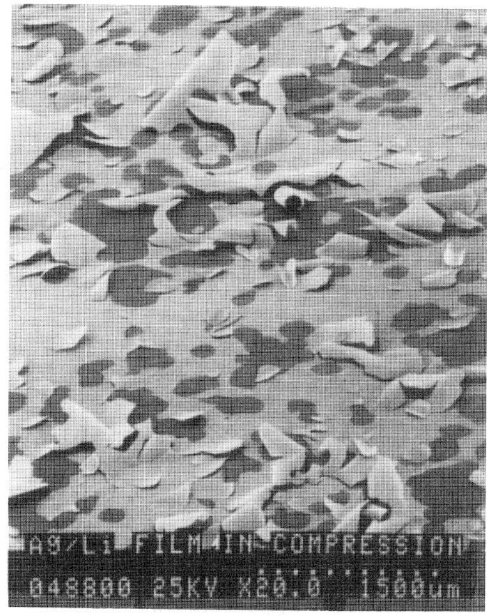

(b)

Figure 9-6. Stresses in silver lithium thin films: (a) tensile film failures during deposition; (b) compressive film failures during aging in Ar. (Courtesy of R. E. Cuthrell, Sandia National Lab).

bending moment (M) vanish on the film–substrate cross section. Thus

$$F = \int \sigma \, dA = 0, \tag{9-8a}$$

$$M = \int \sigma y \, dA = 0, \tag{9-8b}$$

where A is the sectional area and y is the moment lever arm. Intuitive use will be made of these basic equations, and they are applied analytically in deriving the Stoney formula in the next section. In the first type of behavior shown in Fig. 9-5a, the growing film initially shrinks relative to the substrate. Surface tension forces are one reason why this might happen; the misfit accompanying epitaxial growth is another. Compatibility, however, requires that both the film and substrate have the same length. Therefore, the film is constrained and stretches, and the substrate accordingly contracts. The tensile forces developed in the film are balanced by the compressive forces in the substrate. However, the combination is still not in mechanical equilibrium because of the uncompensated end moments. If the film–substrate pair is not restrained from moving, it will elastically bend as shown, to counteract the unbalanced moments. Thus, films containing internal tensile stresses bend the substrate concavely upward. In an entirely similar fashion, compressive stresses develop in films that tend to initially expand relative to the substrate (Fig. 9-5b). Internal compressive film stresses, therefore, bend the substrate convexly outward. These results are perfectly general regardless of the specific mechanisms that cause films to stretch or shrink relative to substrates. Sometimes the tensile stresses are sufficiently large to cause film fracture. Similarly, excessively high compressive stresses can cause film wrinkling and local loss of adhesion to the substrate. Examples of both effects are shown in Fig. 9-6. A discussion of the causes of internal stress will be deferred until Section 9.4. We now turn our attention to a quantitative calculation of film stress as a function of substrate bending.

9.3.2. The Stoney Formula

The formulas that have been used in virtually all experimental determinations of film stress are variants of an equation first given by Stoney in 1909 (Ref. 8). This equation can be derived with reference to Fig. 9-7, which shows a composite film–substrate combination of width w. The film thickness and Young's modulus are d_f and E_f, respectively, and the corresponding substrate values are d_s and E_s. Due to lattice misfit, differential thermal expansion, film growth effects, etc., mismatch forces arise at the film–sub-

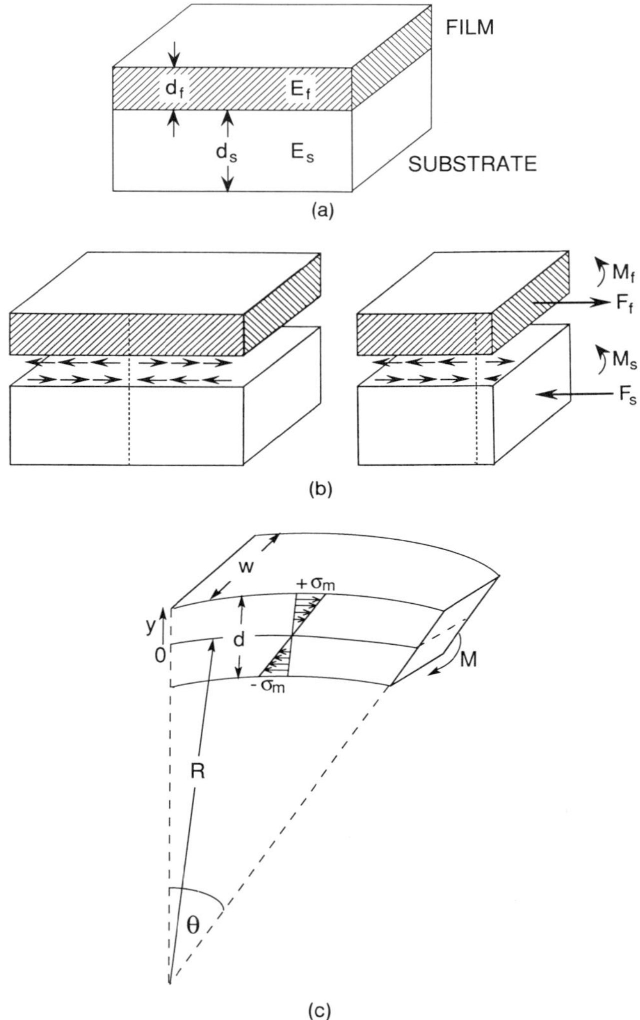

Figure 9-7. Stress analysis of film–substrate combination: (a) composite structure; (b) free-body diagrams of film and substrate with indicated interfacial forces and end moments; (c) elastic bending of beam under applied end moment.

strate interface. In the free-body diagrams of Fig. 9-7b each set of interfacial forces can be replaced by the statically equivalent combination of a force and moment: F_f and M_f in the film, F_s and M_s in the substrate, where $F_f = F_s$. Force F_f can be imagined to act uniformly over the film cross section $(d_f w)$ giving rise to the film stress. The moments are responsible for the bowing of the film–substrate composite. Equation 9-8b requires equality of the clockwise and counterclockwise moments, a condition expressed by

$$((d_f + d_s)/2)F_f = M_f + M_s. \tag{9-9}$$

Consider now an isolated beam bent by moment M, as indicated in Fig. 9-7c. In this case, the deformation is assumed to consist entirely of the extension or contraction of longitudinal beam fibers by an amount proportional to their distance from the central or neutral axis, which remains unstrained in the process. The stress distribution reflects this by varying linearly across the section from maximum tension $(+\sigma_m)$ to maximum compression $(-\sigma_m)$ at the outer beam fibers. In terms of the beam radius of curvature R, and angle θ subtended, Hooke's law yields

$$\sigma_m = E\left\{\frac{(R \pm d/2)\theta - R\theta}{R\theta}\right\} = \pm\frac{Ed}{2R}. \tag{9-10}$$

Corresponding to this stress distribution is the bending moment across the beam section:

$$M = 2\int_0^{d/2} \sigma_m w\left(\frac{y}{d/2}\right) y\,dy = \frac{\sigma_m d^2 w}{6} = \frac{Ed^3 w}{12R}. \tag{9-11}$$

By extension of this result, we have

$$M_f = E_f d_f^3 w/12R \qquad \text{and} \qquad M_s = E_s d_s^3 w/12R.$$

Lastly, in order to account for biaxial stress conditions, it is necessary to replace E_f by $E_f/(1 - \nu_f)$, and similarly for E_s. Substitution of these terms in Eq. 9-9 yields

$$\left(\frac{d_f + d_s}{2}\right)F_f = \frac{w}{12R}\left[\left(\frac{E_f}{1 - \nu_f}\right)d_f^3 + \left(\frac{E_s}{1 - \nu_s}\right)d_s^3\right]. \tag{9-12}$$

Since d_s is normally much larger than d_f, the film stress σ_f is, to a good approximation, given by

$$\sigma_f = \frac{F_f}{d_f w} = \frac{1}{6R}\frac{E_s d_s^2}{(1 - \nu_s)d_f}. \tag{9-13}$$

Equation 9-13 is the Stoney formula. Values of σ_f are determined through measurement of R. The reader should be wary of σ_f values in the literature, since the $(1 - \nu_s)$ correction is frequently omitted.

9.3.3. Thermal Stress

Thermal effects provide important contributions to film stress (Ref. 9). Films and coatings prepared at elevated temperatures and then cooled to room temperature will be thermally stressed, as will films that are thermally cycled or cooled from the ambient to cryogenic temperatures. To see what the magnitude of the thermal stress is, consider a rod of length l_0 and modulus E, clamped at both ends. If its temperature is reduced from T_0 to T, it would tend to shrink in length by an amount equal to $\alpha(T - T_0)l_0$, where α is the coefficient of linear expansion ($\alpha = \Delta l / l_0 \, \Delta T(K^{-1})$). But the rod is constrained and is, therefore, effectively elongated in tension. The tensile strain is simply $\varepsilon = \alpha(T - T_0)l_0 / l_0 = \alpha(T - T_0)$, and the corresponding thermal stress, by Hooke's law, is

$$\sigma = E\alpha(T - T_0). \tag{9-14}$$

Consider now the film–substrate combination of Fig. 9-7 subjected to a temperature differential ΔT. The film and substrate strains are, respectively,

$$\varepsilon_f = \alpha_f \, \Delta T + F_f(1 - \nu_f)/E_f d_f w, \tag{9-15a}$$

$$\varepsilon_s = \alpha_s \, \Delta T - F_f(1 - \nu_s)/E_s d_s w. \tag{9-15b}$$

But strain compatibility requires that $\varepsilon_f = \varepsilon_s$; therefore, the thermal mismatch force is

$$F_f = \frac{w(\alpha_s - \alpha_f)\,\Delta T}{((1 - \nu_f)/d_f E_f) + ((1 - \nu_s)/d_s E_s)}. \tag{9-16}$$

If $d_s E_s/(1 - \nu_s) \gg d_f E_f/(1 - \nu_f)$, the thermal stress in the film is

$$\sigma_f(T) = F_f / d_f w = (\alpha_s - \alpha_f)\,\Delta T E_f/(1 - \nu_f). \tag{9-17}$$

Note that the signs are consistent with dimensional changes in the film–substrate. Films prepared at elevated temperatures will be residually compressed when measured at room temperature ($\Delta T < 0$) if $\alpha_s > \alpha_f$. In this case the substrate shrinks more than the film. Overall, the system must contract a fixed amount, but a compromise is struck; the substrate is not allowed to contract fully and is, therefore, placed in tension, and the film hindered from shrinking, is consequently forced into compression. An example of this occurs in TiC coatings on steel. At the CVD deposition temperature of 1000 °C, it is assumed the coating and substrate are unstressed. For the values $\alpha_{\text{steel}} = 11 \times 10^{-6}$ K^{-1}, $\alpha_{\text{TiC}} = 8 \times 10^{-6}$ K^{-1}, $\text{E}_{\text{TiC}} = 4.5 \times 10^{12}$ dynes/cm^2 (450 kN/mm^2), and $\nu_f = 0.19$, the compressive stress calculated for TiC, using Eq. 9-17, is 1.67×10^{10} dynes/cm^2 (1.67 kN/mm^2) at 0 °C.

Lastly, by equating Eqs. 9-13 and 9-17, we obtain

$$\frac{1}{R} = \frac{6E_f(1 - \nu_s)}{E_s(1 - \nu_f)} \frac{d_f}{d_s^2}(\alpha_s - \alpha_f)\,\Delta T. \tag{9-18}$$

This modification of Stoney's equation represents the extent of bowing when differential thermal expansion effects cause the stress. In order to generally convert the measured deflection into film stress, we note that the curvature is related to the second derivative of the beam displacement; i.e., $1/R = d^2y(x)/dx^2$. After integration, $y(x) = x^2/2R$. For a cantilever of length l, if the free-end displacement is δ, then $\delta = l^2/2R$, and Stoney's formula yields

$$\sigma_f = \delta E_s d_s^2/3l^2(1 - \nu_s)d_f. \tag{9-19}$$

9.4. Stress in Thin Films

9.4.1. Stress Measurement Techniques

Two film–substrate configurations have been primarily used to determine internal stresses in films. In the first the substrate is fashioned into the shape of a cantilever beam. The film is deposited on one surface, and the deflection of the free end of the bent beam is then determined (Fig. 9-8a). A common measure of the sensitivity of the measurement is the smallest detectable force

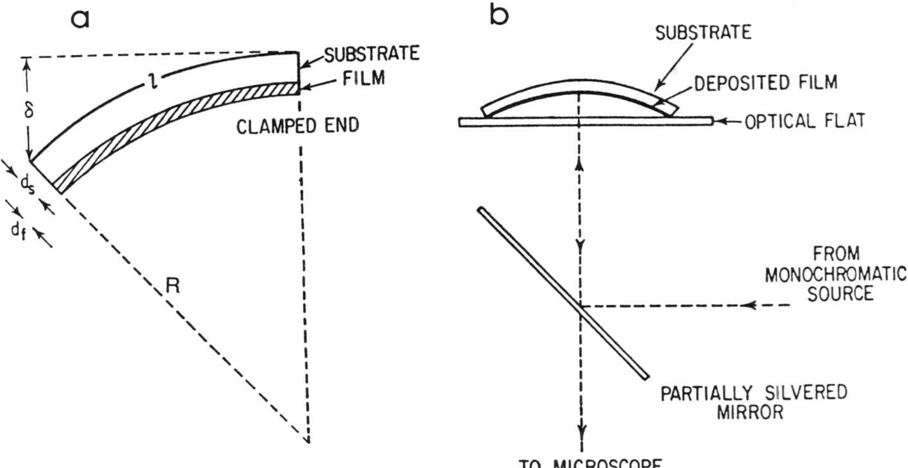

Figure 9-8. Schematic diagrams of film stress measurement techniques. (From Ref. 10): (a) bending of cantilever beam; (b) bowing of circular plate.

per unit width S, defined by the product of the film stress and thickness. By Eq. 9-19,

$$S = \sigma_f d_f = \delta E_s d_s^2 / 3l^2(1 - \nu_s). \tag{9-20}$$

This definition circumvents difficulties in the magnitude of σ_f in very thin films since d_f is no longer in the denominator; S is simply proportional to the deflection. To obtain σ_f, d_f must be independently known. For very thin films, measurement sensitivity is an issue of concern. The minimum value of S detected is frequently taken as a measure of the sensitivity. Typical values of S for the different experimental techniques are given in Table 9-1.

Measurements of stress are frequently made in real time during film formation and growth. Sensitive electromechanical or magnetic restoration of the null, or undeflected position of the beam, in combination with measurement of the restoring force enables continuous monitoring. Null methods have a couple of advantages in film stress determinations relative to techniques in which beam (or plate) deflections are measured. One is the lack of stress relaxation in the film because the substrate is effectively restrained from deflecting. The second is that the frequently unknown value of Young's modulus is, surprisingly, *not* required to evaluate the stress in the film. Since there is no deflection, Stoney's formula is inappropriate. Rather, the restoring force establishes a moment, and from an equation of the type $M = \sigma_m d^2 w/6$ (Eq. 9-11) the stress can be evaluated.

Measurements for a number of metals evaporated onto glass cantilever substrates at room temperature are shown in Fig. 9-9. In addition to σ_f, which can be determined at any value of d_f, the slope at any point of the S–d_f plot

Table 9-1. Sensitivity of Film Stress Measurement Techniques

Method	Substrate Configuration	Detectable Force per Unit Width (dynes/cm)
Optical	B	800
Optical	C	250
Capacitance	C	500
Magnetic restoration	C	250
Electromagnetic restoration	C	150
Mechanical	C	1
Electromechanical	C	1
Interferometric	C	0.5
Interferometric	P	15
X-ray		500

B = beam supported on both ends; C = cantilever beam; P = circular plate.
From Refs. 3 and 10.

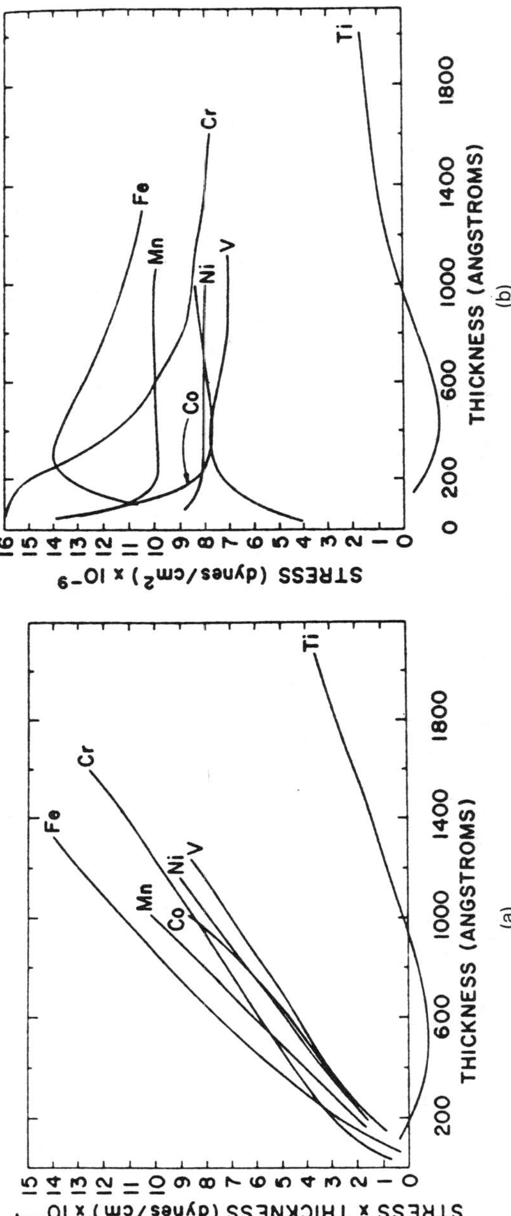

Figure 9-9. Internal stress values in a number of evaporated metal thin films: (a) $S = \sigma_f d_f$ vs. film thickness; (b) σ_f vs. film thickness. (From Ref. 11, reprinted by permission of the Electrochemical Society).

yields a quantity known as the incremental or instantaneous stress $\sigma_f(z)$. The latter is useful in characterizing dynamic changes in film stress. Thus,

$$S = \int_0^{d_f} \sigma_f(z)\, dz \qquad \text{or} \qquad \sigma_f(z) = \frac{dS}{dz}, \tag{9-21}$$

where $\sigma_f(z)$ is the stress present in a layer thickness dz, at a distance z from the film–substrate interface. The reader should be alert to the fact that S, σ_f, and $\sigma_f(z)$ have all been used to report stress values in the literature. Unless otherwise noted, the single term σ_f will be used for the remainder of the chapter.

The second type of substrate configuration frequently employed is the circular plate. In this case, Eq. 9-19 is assumed to describe the resultant stress, where l is now the plate radius and δ represents the center deflection. Not only is the plate an important test geometry, it has obvious applications in optical components. As a practical example, consider a circular glass plate window with a diameter-to-thickness ratio of 16:1 having the elastic properties $E = 6.37 \times 10^{11}$ dynes/cm^2 and $\nu = 0.25$. If one face is coated with a 1000-Å-thick single-layer antireflection film of MgF_2, a stress of 3.33×10^9 dynes/cm^2 develops after deposition. The depression of the center relative to the circumference is calculated to be 7.5×10^{-6} cm. For a wavelength $\lambda = 5000$ Å the extent of bowing corresponds to 0.15λ. This change from a planar surface to a paraboloid of revolution is significant in the case of high-precision optical surfaces where the tolerance may be taken to be 0.05λ. In the case of thin lenses, stress distortions can also change critical spacings in multi-element optical systems. A typical arrangement for stress measurement of plates is shown in Fig. 9-8b, where the change in the optical fringe pattern between the film substrate and an optical flat is used to measure the deformation. Interference patterns illustrating different biaxial stress states are shown in Fig. 9-10. Alternatively, a calibrated optical microscope can be used to measure the extent of bowing.

There are a number of issues relevant to the experimental determination of accurate values of film stress. Some deal with the validity of elastic theory itself in treating the film–substrate composite properties, geometries, and deflections involved. In general, if deflections are small compared with the substrate beam or plate thickness, the simple theory suffices. Substrate elastic constant values should be chosen with care and checked to determine whether they are isotropic. Special care must be exercised when determining the stress in epitaxial films because elastic constants of both films and substrates are anisotropic, complicating the stress analysis.

A number of useful methods for determining stress in films, including

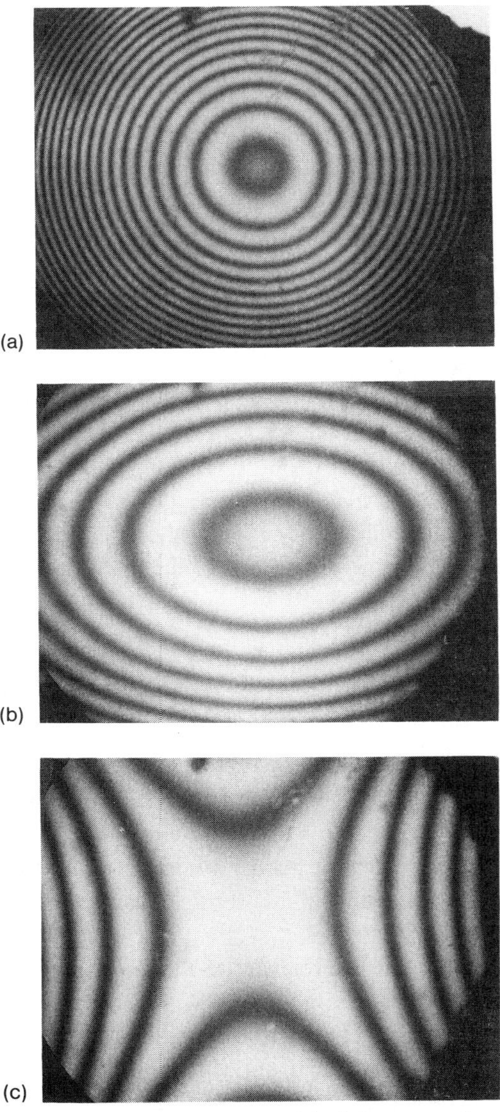

Figure 9-10. Interference fringe patterns in biaxially stressed, sputtered Mo films. (From Ref. 12): (a) balanced biaxial tension (or compression) ($\sigma_x = \sigma_y$); (b) unbalanced biaxial tension (or compression) ($\sigma_x \neq \sigma_y$); (c) one component tensile, one component compressive.

epitaxial films, are based on X-ray diffraction methods. As an example, consider a polycrystalline film containing an isotropic biaxial tensile stress distribution in the xy plane ($\sigma_z = 0$). The film contracts in the z direction by an amount (see Eq. 9-1b, also application to epitaxial films, p. 349)

$$\varepsilon_z = (-\nu/E)(\sigma_x + \sigma_y) = -\nu(\varepsilon_x + \varepsilon_y). \tag{9-22}$$

By measuring the lattice spacing in the stressed film $\bar{a}_0$ as well as unstressed bulk lattice (a_0) with X-rays, we can determine ε_z directly; i.e., $\varepsilon_z = -(\bar{a}_0 - a_0)/a_0$. Since $\sigma_x = \sigma_y = \sigma_f$,

$$\sigma_f = -\frac{E}{2\nu}\left(\frac{\bar{a}_0 - a_0}{a_0}\right). \tag{9-23}$$

The accuracy of the X-ray technique is considerably extended with high-precision lattice parameter determinations. Precise determination of $\bar{a}_0$ and a_0 is complicated by line broadening due to small grain size, dislocations, twins, stacking faults, and nonuniform microstrains.

9.4.2. Measured Intrinsic Stress Behavior

9.4.2.1. Evaporated Films. Despite the apparent simplicity of the experimental techniques and corresponding defining stress equations, the measured values of the intrinsic contribution to σ_f display bewildering variations as a function of deposition variables, nature of film–substrate combination, and film thickness. Some of the variety is evident in the measured stress values for 1000-Å-thick metal and nonmetal films (evaporated on room-temperature glass or silica substrates), which are tabulated in Table 9-2. These values for evaporated films should be considered representative rather than precise. Even though data published by different investigators employing similar and different measurement techniques are frequently inconsistent, the following trends can be discerned from published results in this field:

a. In metals the film stress is invariably tensile with a magnitude ranging from 10^8 to 10^{10} dynes/cm^2.
b. There is no apparent strong dependence of stress on the nature of the substrate.
c. In dielectric films, compressive and tensile stresses arise.
d. The magnitude of the stress in nonmetallic films is frequently small.

A simple way to rationalize the difference in behavior between metals and nonmetals is to note that metals are strong in tension, but offer little resistance

Table 9-2. Intrinsic Stress in Evaporated Films

Metals*	σ_f (10^9 dynes/cm^2)	E (10^{12} dynes/cm^2)
Ag	0.2	0.76
Al	−0.74	0.69
Au	2.6	0.80
Co	8.4	2.06
Cu	0.6	1.17
Cr	8.5	2.48
Fe	11.0	2.0
In	~0	0.11
Mn	9.8	1.58
Mo	10.8	3.24
Pd	6	1.12
Ti	~0	1.15
Zr	7	0.94
Nonmetals**	**σ_f (10^9 dynes/cm^2)**	**E (10^{12} dynes/cm^2)**
C (graphite)	−4	0.4
Ge	2.3	1.58
Si	3	2.0
Te	0.6	
ZnS	−1.9	0.54
MgF_2	3 to 7	1.17
PbF_2	−0.2	
Cryolite	0.2	
CaF_2	0.2	
CeF_3	2.2	
SiO	0.1	
$PbCl_2$	0.8	
CdTe	−1.4	
Tl (I, Br)	−0.07	
$TlCl_3$	−0.3	
$ThOF_2$	1.5	

Note: 1 dyne/cm^2 = 10^{-7} MN/m^2 = 10^{-1} Pa.
*From Refs. 3 and 11.
**From Refs. 3, 13, and 14.

to compression, whereas insulators and semiconductors are strong in compression and weak in tension. Therefore, metals prefer to be in tension, and nonmetals are best deposited in a state of compression.

9.4.2.2. Thickness Dependence. The data of Fig. 9-9 provide a valuable means for evaluating distinctions and similarities among different metals because of the common in situ measurement technique, the high vacuum (10^{-6} to 10^{-7} torr) maintained during deposition and measurement, the similar film deposition rates, and the absence of a thermal stress contribution. With few exceptions, the film stress is always tensile. Hard refractory metals and metals with high melting points generally tend to exhibit higher residual stresses than softer, more easily melted metals. Appreciable film stress arises after only 100 Å or so of deposition, after which large stresses continue to develop up to a thickness of roughly 600 Å. This thickness increment range is, not surprisingly, coincident with the typical coalescence and channel stages of growth leading to continuous film formation. There is thus good reason for the use of the term *growth stress* to denote intrinsic stress (σ_I). With further film growth the stress does not change appreciably.

9.4.2.3. Temperature Effects. When film deposition occurs at temperatures different from that at which the stress is measured, the previously considered differential thermal expansion contribution (Eq. 9-17) superimposes on the growth stress; i.e.,

$$\sigma_f = \sigma_f(T) + \sigma_I, \tag{9-24}$$

Precise determinations of σ_I require subtraction of the $\sigma_f(T)$ correction. In addition, heated substrates alter the intrinsic stresses largely by promoting defect annealing and the processes of recrystallization or even grain growth if the temperature is high enough. The resultant softening relaxes the growth stresses, which fall rapidly with temperature. Diffusion of impurities into and out of the film is also accelerated, and this can give rise to substantial stress change. The total film stress may then show a minimum or even reverse in sign. Both kinds of behavior have been observed in practice. One implication of stress reversal is the possibility of depositing films with low or even near-zero stress levels. This is frequently a desirable feature, e.g., for magnetic thin films. For the required critical properties to be achieved, film composition, deposition rates, and substrate temperature must be optimally adjusted.

9.4.2.4. Sputtered Films. Unlike evaporated films, generalizations with respect to stress are difficult to make for sputtered metal films because of the complexity of the plasma environment and the effect of the working gas. However, at low substrate temperatures, compressive film stresses are often observed. The fact that the extent of compression varies directly with the amount of trapped gas has pointed to the latter as the source of stress. But this is too simplistic a view. Sputtered films display a rich variety of effects, including tensile-to-compressive stress transitions as a function of process variables. For example, in rf-diode-sputtered tungsten films a stress reversal from tension to compression was achieved in no less than three ways (Ref. 15):

a. By raising the power level about 30 W at zero substrate bias
b. By reversing the dc bias from positive to negative
c. By reducing the argon pressure

Oxygen incorporation in the film favored tension, whereas argon was apparently responsible for the observed compression.

The results of extensive studies by Hoffman and Thornton (Ref. 16) on magnetron-sputtered metal films are particularly instructive since the internal stress correlates directly with microstructural features and physical properties. Magnetron sputtering sources have made it possible to deposit films over a wide range of pressures and deposition rates in the absence of plasma bombardment and substrate heating. It was found that two distinct regimes,

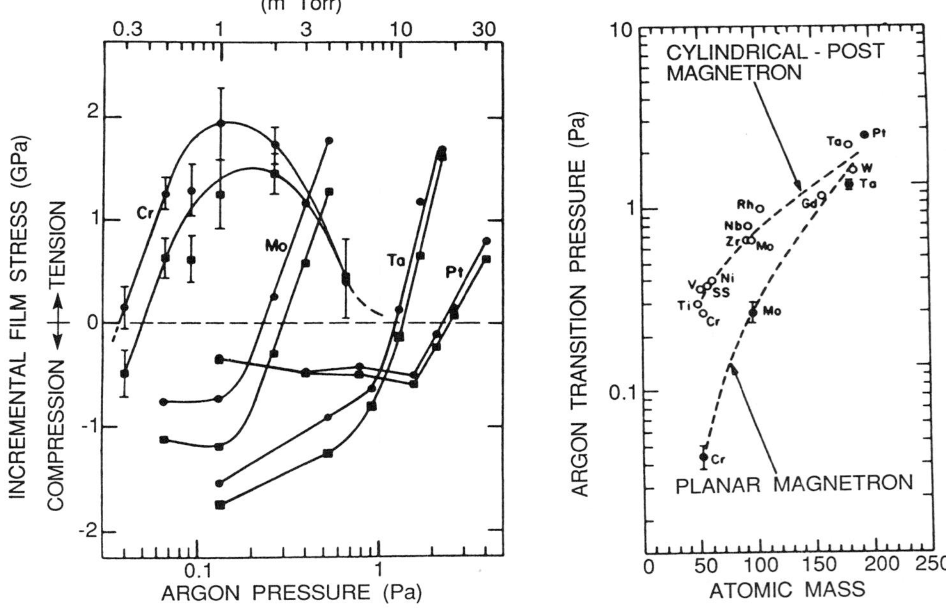

Figure 9-11. (a) Biaxial internal stresses as a function of Ar pressure for Cr, Mo, Ta, and Pt films sputtered onto glass substrates: ● parallel and ■ perpendicular to long axis of planar cathode. (From Ref. 16). (b) Ar transition pressure vs. atomic mass of sputtered metals for tensile to compressive stress reversal. (From Ref. 16).

separated by a relatively sharp boundary, exist where the change in film properties is almost discontinuous. The transition boundary can be thought of as a multidimensional space of the materials and processing variables involved. On one side of the boundary, the films contain compressive intrinsic stresses and entrapped gases, but exhibit near-bulklike values of electrical resistivity and optical reflectance. This side of the boundary occurs at low sputtering pressures, with light sputtering gases, high-mass targets, and low deposition rates. On the other hand, elevated sputtering pressures, more massive sputtering gases, light target metals, and oblique incidence of the depositing flux favor the generation of films possessing tensile stresses containing lesser amounts of entrapped gases. Internal stress as a function of the Ar pressure is shown in Fig. 9-11a for planar magnetron-deposited Cr, Mo, Ta, and Pt. The pressure at which the stress reversal occurs is plotted in Fig. 9-11b versus the atomic mass of the metal.

Comparison with the zone structure of sputtered films introduced in Chapter 5 reveals that elevated working pressures are conducive to development of columnar grains with intercrystalline voids (zone 1). Such a structure exhibits high resistivity, low optical reflectivity, and tensile stresses. At lower pressures the development of the zone 1 structure is suppressed. Energetic particle bombardment, mainly by sputtered atoms, apparently induces compressive film stress by an atomic peening mechanism.

9.4.3. Some Theories of Intrinsic Stress

Over the years, many investigators have sought universal explanations for the origin of the constrained shrinkage that is responsible for the intrinsic stress. Buckel (Ref. 17) classified the conditions and processes conducive to internal stress generation into the following categories, some of which have already been discussed:

1. Differences in the expansion coefficients of film and substrate
2. Incorporation of atoms (e.g., residual gases) or chemical reactions
3. Differences in the lattice spacing of monocrystalline substrates and the film during epitaxial growth
4. Variation of the interatomic spacing with the crystal size
5. Recrystallization processes
6. Microscopic voids and special arrangements of dislocations
7. Phase transformations

One of the mechanisms that explains the large intrinsic tensile stresses observed in metal films is related to item 5. The model by Klokholm and Berry

(Ref. 11) suggests that the stress arises from the annealing and shrinkage of disordered material buried behind the advancing surface of the growing film. The magnitude of the stress reflects the amount of disorder present on the surface layer before it is covered by successive condensing layers. If the film is assumed to grow at a steady-state rate of $\dot{G}$ monolayers/sec, the atoms will on average remain on the surface for a time $\dot{G}^{-1}$. In this time interval, thermally activated atom movements occur to improve the crystalline order (recrystallization) of the film surface. These processes occur at a rate r described by an Arrhenius behavior,

$$r = \nu_0 e^{-E_r/RT_S}, \tag{9-25}$$

where ν_0 is a vibrational frequency factor, E_r is an appropriate activation energy, and T_S is the substrate temperature. On this basis it is apparent that high-growth stresses correspond to the condition $\dot{G} > r$, low-growth stresses to the reverse case. At the transition between these two stress regimes, $\dot{G} = r$ and $E_r/RT_S = 32$, if $\dot{G}$ is 1 sec^{-1} and ν_0 is taken to be 10^{14} sec^{-1}. Experimental data in metal films generally show a steep decline in stress when $T_M/T_S = 4.5$, where T_M is the melting point. Therefore, $E_r = 32RT_M/4.5 = 14.2T_M$. In Chapter 8 it was shown that for FCC metals the self-transport activation energies are proportional to T_M as $34T_M$, $25T_M$, $17.8T_M$, and $13T_M$ for lattice, dislocation, grain-boundary, and surface diffusion mechanisms, respectively. The apparent conclusion is that either surface or grain-boundary diffusion of vacancies governs the temperature dependence of film growth stresses by removing the structural disorder at the surface of film crystallites.

Hoffman (Ref. 18) has addressed stress development due to coalescence of isolated crystallites when forming a grain boundary. Through deposition neighboring crystallites enlarge until a small gap exists between them. The interatomic forces acting across this gap cause a constrained relaxation of the top layer of each surface as the grain boundary forms. The relaxation is constrained because the crystallites adhere to the substrate, and the result of the deformation is manifested macroscopically as observed stress.

We can assume an energy of interaction between crystallites shown in Fig. 9-12 in much the same fashion as between atoms (Fig. 1-8b). At the equilibrium distance a, two surfaces of energy γ_s are eliminated and replaced by a grain boundary of energy γ_{gb}. For large-angle grain boundaries $\gamma_{gb} \approx (1/3)\gamma_s$, so that the energy difference $2\gamma_s - \gamma_{gb} \approx (5/3)\gamma_s$ represents the depth of the potential at a. As the film grows, atoms are imagined to individually occupy positions ranging from r (a hard-core radius) to $2a$ (the

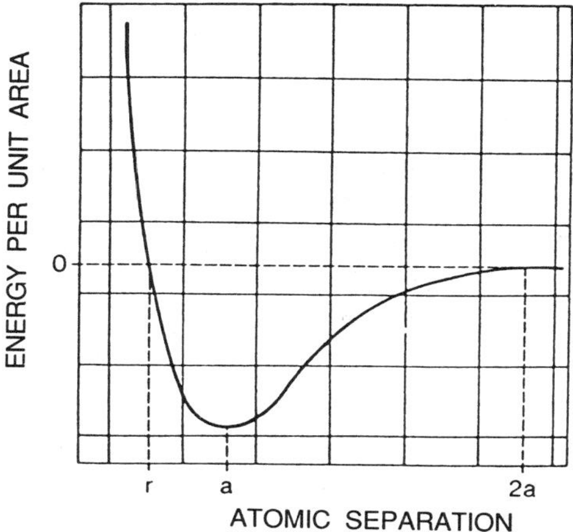

Figure 9-12. Grain-boundary potential. (Reprinted with permission from Elsevier Sequoia, S.A., from R. W. Hoffman, *Thin Solid Films* **34**, 185, 1976).

nearest-neighbor separation) with equal probability. Between these positions the system energy is lowered. If an atom occupies a place between r and a, it would expand the film in an effort to settle in the most favored position—a.

Similarly, atoms deposited between a and $2a$ cause a film contraction. Because the potential is asymmetric, contraction relative to the substrate dominates leading to tensile film stresses. An estimate of the magnitude of the stress is

$$\sigma = \frac{E\Delta P}{1 - \nu \cdot \bar{d}_c}, \tag{9-26}$$

where $\bar{d}_c$ is the mean crystallite diameter and P is the packing density of the film. The quantity Δ is the constrained relaxation length and can be calculated from the interaction potential between atoms. When divided by $\bar{d}_c$, $\Delta/\bar{d}_c$ represents an "effective" strain. In Cr films, for example, where $E/(1 - \nu) = 3.89 \times 10^{12}$, $\bar{d}_c = 130$ Å, $\Delta = 0.89$ Å, and $P = 0.96$, the film stress is calculated to be 2.56×10^{10} dynes/cm^2. Employing this approach, Pulker and Mäser (Ref. 19) have calculated values of the tensile stress in MgF_2 and compressive stress in ZnS in good agreement with measured values.

A truly quantitative theory for film stress has yet to be developed, and it is doubtful that one will emerge that is valid for different film materials and methods of deposition. Uncertain atomic compositions, structural arrangements and interactions in crystallites and at the film–substrate interface are not easily amenable to a description in terms of macroscopic stress–strain concepts.

9.5. Relaxation Effects in Stressed Films

Until now, we have only considered stresses arising during film formation processes. During subsequent use, the grown-in elastic–plastic state of stress in the film may remain relatively unchanged with time. However, when films are exposed to elevated temperatures or undergo relatively large temperature excursions, they frequently display a number of interesting time-dependent deformation processes characterized by the thermally activated motion of atoms and defects. As a result, local changes in the film topography can occur and stress levels may be reduced. In this section we explore some of these phenomena that are exemplified in materials ranging from lead alloy films employed in superconducting Josephson junction devices to thermally grown SiO_2 films in integrated circuits.

9.5.1. Stress Relaxation in Thermally Grown SiO_2

As noted previously (page 395), a volume change of some 220% occurs when Si is converted into SiO_2. This expansion is constrained by the adhesion in the plane of the Si wafer surface. Large intrinsic compressive stresses are, therefore, expected to develop in SiO_2 films in the absence of any stress relaxation. A value of 3×10^{11} dynes/cm^2 has, in fact, been estimated (Ref. 20), but such a stress level would cause mechanical fracture of both the Si and SiO_2. Not only does oxidation of Si occur without catastrophic failure, but virtually no intrinsic stress is measured in SiO_2 grown above 1000 °C. To explain the paradoxical lack of stress, let us consider the viscous flow model depicted in Fig. 9-13. For simplicity, only uniaxial compressive stresses are assumed to act on a slab of SiO_2, which is free to flow vertically. The SiO_2 film is modeled as a viscoelastic solid whose overall mechanical response reflects that of a series combination of an elastic spring and a viscous dashpot (Fig. 13b). Under loading, the spring instantaneously deforms elastically, whereas the dashpot strains in a time-dependent viscous fashion. If ε_1 and ε_2 represent the strains in the spring and dashpot, respectively, then the total strain is

$$\varepsilon_T = \varepsilon_1 + \varepsilon_2 . \tag{9-27}$$

The same compressive stress σ_x acts on both the spring and dashpot so that $\varepsilon_1 = \sigma_x / E$ and $\dot{\varepsilon}_2 = \sigma_x / \eta$, where $\dot{\varepsilon}_2 = d\varepsilon_2 / dt$, and η is the coefficient of viscosity. Here we recognize that the *rate* of deformation of glassy materials, including SiO_2, is directly proportional to stress. Assuming ε_T is constant,

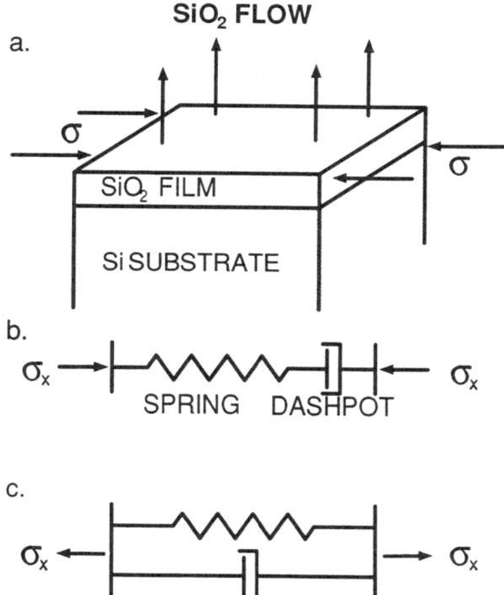

Figure 9-13. (a) Viscous flow model of stress relaxation in SiO_2 films. (From Ref. 20); (b) spring–dashpot model for stress relaxation; (c) spring–dashpot model for strain relaxation.

$\dot{\varepsilon}_1 = -\dot{\varepsilon}_2$ or $(1/E)d\sigma_x/dt = -\sigma_x/\eta$. Upon integration, we obtain

$$\sigma_x = \sigma_0 e^{-Et/\eta}. \tag{9-28}$$

The initial stress in the film, σ_0, therefore relaxes by decaying exponentially with time. With $E = 6.6 \times 10^{11}$ dynes/cm^2 and $\eta = 2.8 \times 10^{12}$ dynes-sec/cm^2 at 1100 °C, the time it takes for the initial stress to decay to σ_0/e is a mere 4.3 sec. Oxides grown at this temperature are, therefore, expected to be unstressed. Since η is thermally activated, oxides grown at lower temperatures will generally possess intrinsic stress. The lack of viscous flow in a time comparable to that of oxide growth limits stress relief in such a case. Typically, intrinsic compressive stresses of 7×10^9 dynes/cm^2 have been measured in such cases.

9.5.2. Strain Relaxation in Films

It is worthwhile to note the distinction between stress and strain relaxation. Stress relaxation in the SiO_2 films just described occurred at a constant total strain or extension in much the same way that tightened bolts lose their tension with time. Strain relaxation, on the other hand, is generally caused by a constant load or stress and results in an irreversible time-dependent stretching (or contraction) of the material. The latter can be modeled by a spring and dashpot connected in parallel combination (Fig. 13c). Under the application of a tensile stress the spring wishes to instantaneously extend, but is restrained from doing so by the viscous response of the dashpot. It is left as an exercise for the reader to show that the strain relaxation in this case has a time dependence given by

$$\varepsilon = \frac{\sigma_0}{E}\left(1 - \exp - \frac{Et}{\eta}\right). \tag{9-29}$$

In actual materials complex admixtures of stress and strain relaxation effects may occur simultaneously.

Film strains can be relaxed by several possible deformation or strain relaxation mechanisms. The rate of relaxation for each mechanism is generally strongly dependent on the film stress and temperature, and the operative or dominant mechanism is the one that relaxes strain the fastest. A useful way to represent the operative regime for a given deformation mechanism is through the use of a map first developed for bulk materials (Ref. 21), and then extended to thin films by Murakami *et al.* (Ref. 22). Such a map for a Pb–In–Au film is shown in Fig. 9-14 where the following four strain relaxation mechanisms are taken into account:

1. Defectless Flow. When the stresses are very high, slip planes can be rigidly displaced over neighboring planes. The theoretical shear stress of magnitude $\sim \mu/20$ is required for such flow. Stresses in excess of this value essentially cause very large strain rates. Below the theoretical shear stress limit the plastic strain rate is zero. Defectless flow is dominant when the normalized tensile stress (σ/μ) is greater than $\sim 9 \times 10^{-2}$, or above the horizontal dotted line. This regime of flow will not normally be accessed in films.

2. Dislocation Glide. Under stresses sufficiently high to cause plastic deformation, dislocation glide is the dominant mechanism in ductile materials. Dislocation motion is impeded by the presence of obstacles such as impurity atoms, precipitates, and other dislocations. In thin films, additional obstacles to dislocation motion such as the native oxide, the substrate, and grain boundaries are present. Thus, the film thickness d and grain size, l_g, may be thought of as obstacle spacings in Eq. 9-3. An empirical law for the dislocation glide strain rate $\dot{\varepsilon}_2$ as a function of stress and temperature is

$$\dot{\varepsilon}_2 = \dot{\varepsilon}_0(\sigma/\sigma_0)\exp - \Delta G/kT, \tag{9-30}$$

where σ_0 is the flow stress at absolute zero temperature, ΔG is the free energy required to overcome obstacles, $\dot{\varepsilon}_0$ is a pre-exponential factor, and kT has the usual meaning.

3. Dislocation Climb. When the temperature is raised sufficiently, dislocations can acquire a new degree of motional freedom. Rather than be impeded by obstacles in the slip plane, dislocations can circumvent them by climbing vertically and then gliding. This sequence can be repeated at new obstacles. The resulting strain rate of this so-called climb controlled creep depends on temperature and is given by

$$\text{at } T > 0.3T_M; \qquad \dot{\varepsilon}_3 = A_3 \frac{\mu b}{kT} D_b \left(\frac{\sigma}{\mu}\right)^5, \tag{9-31}$$

$$\text{at } T > 0.6T_M; \qquad \dot{\varepsilon}_4 = A_4 \frac{\mu b}{kT} D_L \left(\frac{\sigma}{\mu}\right)^7. \tag{9-32}$$

Here, D_b and D_L are the thermally activated grain-boundary and lattice diffusion coefficients, respectively, and A_3 and A_4 are constants.

4. Diffusional Creep. Viscous creep in polycrystalline films can occur by diffusion of atoms within grains (Nabarro–Herring creep) or by atomic transport through grain boundaries (Coble creep). The respective strain rates are given by

$$\dot{\varepsilon}_5 = A_5 \frac{\mu}{kT} \frac{\Omega}{l_g d} D_L \left(\frac{\sigma}{\mu}\right), \tag{9-33}$$

$$\dot{\varepsilon}_6 = A_6 \frac{\mu}{kT} \frac{\Omega\, \delta D_b}{l_g d^2} \left(\frac{\sigma}{\mu}\right), \tag{9-34}$$

where in addition to constants A_5 and A_6, Ω is the atomic volume and δ is the grain-boundary width. It is instructive to think of the last two equations as variations on the theme of the Nernst–Einstein equation (Eq. 1-35). The difference is that in the present context the applied stress (force) is coupled to the resultant rate of straining (velocity). Rather than the linear coupling of $\dot{\varepsilon}$ and σ in diffusional creep, a stronger nonlinear dependence on stress is observed for dislocation climb processes.

In constructing the deformation mechanism map, the process exhibiting the largest strain relaxation rate is calculated at each point in the field of the normalized stress–temperature space. The field boundaries are determined by equating pairs of rate equations for the dominant mechanisms and solving for the resulting stress dependence on temperature.

9.5.3. Relaxation Effects in Metal Films during Thermal Cycling

An interesting application of strain relaxation effects is found in Josephson superconducting tunnel–junction devices (Ref. 23) (These are further discussed

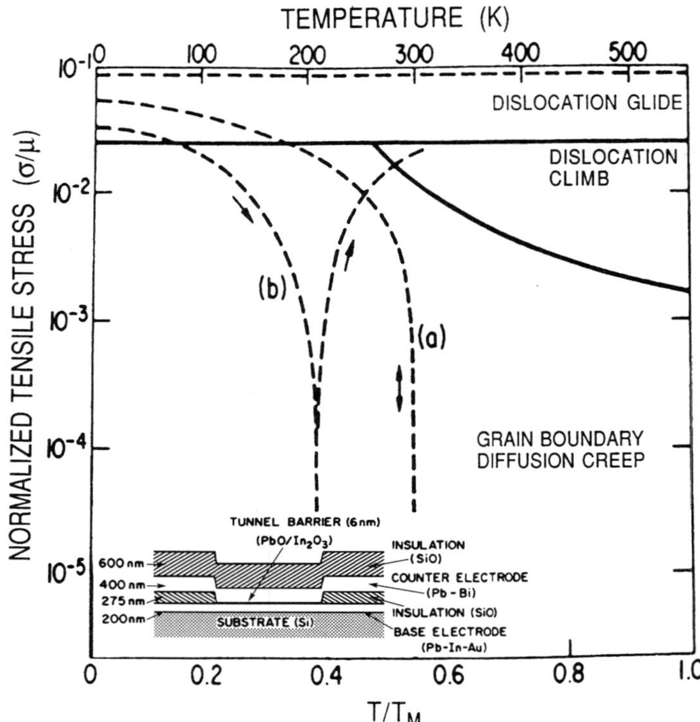

Figure 9-14. Deformation mechanism map for Pb–In–Au thin films. (From Ref. 23). Inset: Schematic cross section of Pb alloy Josephson junction device. (From Ref. 22).

in Chapter 14.) A schematic cross section of such a device is shown in the inset of Fig. 9-14. The mechanism of operation need not concern us, but their very fast switching speeds (e.g., $\sim 10^{-11}$ sec) combined with low-power dissipation levels (e.g., $\sim 10^{-6}$ W/device) offer the exciting potential of building ultrahigh speed computers based on these devices. The junction basically consists of two superconducting electrodes separated by an ultrathin 60-Å-thick tunnel barrier. Lead alloy films serve as the electrode materials primarily because they have a relatively high superconducting transition temperature* and are easy to deposit and pattern. The thickness of the tunnel barrier oxide is critical and can be controlled to within one atomic layer through oxidation of

*The application described here predates the explosion of activity in $YBa_2Cu_3O_7$ ceramic superconductors (see Chapter 14).

Pb alloy films. Fast switching and resetting times are ensured by the low dielectric constant of the $PbO-In_2O_3$ barrier film. A serious materials-related concern with this junction structure is the reliability of the device during thermal cycling between room temperature and liquid helium temperature (4.2 K) where the device is operated. The failure of some devices is caused by the rupture of the ultrathin tunnel barrier due to the mismatch in thermal expansion between Pb alloys and the Si substrate on which the device is built. During temperature cycling the thermal strains are relaxed by the plastic deformation processes just considered resulting in harmful dimensional changes.

Let us now trace the mechanical history of an initially unstressed Pb film as it is cooled to 4.2 K. Assuming no strain relaxation, path a in Fig. 9-14 indicates that the grain-boundary creep field is traversed from 300 to 200 K, followed by dislocation glide at lower temperatures. Because cooling rates are high at 300 K, there is insufficient thermal energy to cause diffusional creep. Therefore, dislocation glide within film grains is expected to be the dominant deformation mechanism on cooling. If, however, no strain relaxation occurs, the film could then be rewarmed and the $\sigma-T$ path would be reversibly traversed if, again, no diffusional creep occurs. Under these conditions the film could be thermally cycled without apparent alteration of the state of stress and strain. If, however, a relaxation of the thermal strain by dislocation glide did occur upon cooling, then the path followed during rewarming would be along b. Because the coefficient of thermal expansion for Pb exceeds that of Si, a large tensile stress initially develops in the film at 4.2 K. As the temperature is raised, dislocation glide rapidly relaxes the stress so that at 200 K the tensile stress effectively vanishes. Further warming from 200 to 300 K induces compressive film stresses. These provide the driving force to produce micron-sized protrusions or so-called hillock or stunted whisker growths from the film surface. This manifestation of strain relaxation is encouraged because grain-boundary diffusional creep is operative in Pb over the subroom temperature range.

It is clear that in order to prevent the troublesome hillocks from forming, it is necessary to strengthen the electrode film. This will minimize the dislocation glide that originally set in motion the train of events leading to hillock formation. Practical methods for strengthening bulk metals include alloying and reducing the grain size in order to create impediments to dislocation motion. Indeed, the alloying of Pb with In and Au caused fine intermetallic compounds to form, which hardened the films and refined the grain size. The result was a suppression of strain relaxation effects and the elimination of hillock formation. Overall, a dramatic reduction in device failure due to thermal cycling was realized. Nevertheless, for these and other reasons, Nb, a much harder material than Pb, has replaced the latter in Josephson junction computer devices.

9.5.4. Hillock Formation

In multilayer integrated devices hillocks are detrimental because their penetration of insulating films can lead to electrical short circuits. Hillocks and whiskers have been observed to sprout during electromigration (see Section 8.4 and Fig. 8-15a). Where glass films overlay interconnections, they serve to conformally constrain the powered metal conductors. The situation is much like a glass film vessel pressurized by an electromigration mass flux. Compressive stresses in the conductor induced by electrotransport can be relieved by extrusion of hillocks or whiskers, which sometimes leads to cracking of the insulating dielectric overlayer. Interestingly, processes that reduce the compression or create tensile stresses, such as current reversal during electromigration or thermal cycling, sometimes cause the hillocks to shrink in size.

From the foregoing examples it is clear that the rate of relieval of compressive stress governs hillock growth. Dislocation flow mechanisms cannot generally relax stress because the intrinsic stress level present in soft polycrystalline metal films is insufficient to activate dislocation sources within grains, at grain boundaries, or at the film surface. However, diffusional creep processes can relieve the stress. We close this section with the suggestion that diffusional creep relaxation of the compressive stress in a film is analogous to the outdiffusion of a supersaturated specie from a solid, e.g., outgassing of a strip. The rate of stress change is then governed by

$$\frac{\partial \sigma(x,t)}{\partial t} = D\frac{\partial^2 \sigma(x,t)}{\partial x^2}, \tag{9-35}$$

where compressive stress simply substitutes for excess concentration in the diffusion equation. If, for example, a film of thickness d contains an initial internal compressive stress $\sigma(0)$ and stress-free surfaces at $x = 0$ and $x = d$, i.e., $\sigma(0, t) = \sigma(d, t) = 0$, then the stress relaxes according to the equation

$$\sigma(x,t) = \frac{4\sigma(0)}{\pi}\sum_{n=0}^{\infty}\left\{\frac{1}{2n+1}\sin\frac{(2n+1)\pi x}{d}\exp -\frac{(2n+1)^2\pi^2 Dt}{d^2}\right\}. \tag{9-36}$$

Boundary value problems of this kind have been treated in the literature to account for hillock growth kinetics, and the reader is referred to original sources for details (Ref. 24).

9.6. ADHESION

9.6.1. Introduction

The term *adhesion* refers to the interaction between the closely contiguous surfaces of adjacent bodies, i.e., a film and substrate. According to the American Society for Testing and Materials (ASTM), adhesion is defined as the condition in which two surfaces are held together by valence forces or by mechanical anchoring or by both together. Adhesion to the substrate is certainly the first attribute a film must possess before any of its other properties can be further successfully exploited. Even though it is of critical importance adhesion is one of the least understood properties. The lack of a broadly applicable method for quantitatively measuring "adhesion" makes it virtually impossible to test any of the proposed theories for it. This state of affairs has persisted for years and has essentially spawned two attitudes with respect to the subject (Ref. 25). The "academic" approach is concerned with the nature of bonding and the microscopic details of the electronic and chemical interactions at the film–substrate interface. Clearly, a detailed understanding of this interface is essential to better predict the behavior of the macrosystem, but atomistic models of the former have thus far been unsuccessfully extrapolated to describe the continuum behavior of the latter. For this reason the "pragmatic" approach to adhesion by the thin-film technologist has naturally evolved. The primary focus here is to view the effect of adhesion on film quality, durability, and environmental stability. Whereas the atomic binding energy may be taken as a significant measure of adhesion for the academic, the pragmatist favors the use of large-area mechanical tests to measure the force or energy required to separate the film from the substrate. Both approaches are, of course, valuable in dealing with this difficult subject, and we shall adopt aspects of these contrasting viewpoints in the ensuing discussion of adhesion mechanisms, measurement methods, and ways of influencing adhesion.

9.6.2. Energetics of Adhesion

From a thermodynamic standpoint the work W_A required to separate a unit area of two phases forming an interface is expressed by

$$W_A = \gamma_f + \gamma_s - \gamma_{fs}\,. \tag{9-37}$$

The quantities γ_f and γ_s are the specific surface energies of film and substrate, and γ_{fs} is the interfacial energy. A positive W_A denotes attraction (adhesion), and a negative W_A implies repulsion (de-adhesion). The work W_A is largest when materials of high surface energy come into contact such as metals with high melting points. Conversely, W_A is smallest when low-surface-energy materials such as polymers are brought into contact. When f and s are identical, then an interfacial grain boundary forms where $\gamma_f + \gamma_s > \gamma_{fs}$. Under these circumstances, $\gamma_f = \gamma_s$ and γ_{fs} is relatively small; e.g., $\gamma_{fs} \approx (1/3)\gamma_s$ in metals. If, however, a homoepitaxial film is involved, then $\gamma_{fs} = 0$ by definition, and $W_A = 2\gamma_s$. Attempts to separate an epitaxial film from its substrate will likely cause a *cohesion* failure through the bulk rather than an *adhesion* failure at the interface. When the film–substrate combination is composed of different materials, γ_{fs} may be appreciable, thus reducing the magnitude of W_A. Interfacial adhesion failures tend to be more common under such circumstances. In general, the magnitude of W_A increases in the order (a) immiscible materials with different types of bonding, e.g., metal–polymer, (b) solid–solution formers, and (c) same materials. Measured values of adhesion will differ from intrinsic W_A values because of contributions from chemical interactions, interdiffusional effects, internal film stresses, interfacial impurities, imperfect contact, etc.

9.6.3. Film - Substrate Interfaces

The type of interfacial region formed during deposition will depend not only on W_A but also on the substrate morphology, chemical interactions, diffusion rates and nucleation processes. At least four types of interfaces can be distinguished, and these are depicted in Fig. 9-15.

1. The abrupt interface is characterized by a sudden change from the film to the substrate material within a distance of the order of the atomic spacing (1–5 Å). Concurrently, abrupt changes in materials properties occur due to the lack of interaction between film and substrate atoms, and low interdiffusion rates. In this type of interface, stresses and defects are confined to a narrow planar region where stress gradients are high. Film adhesion in this case will be low because of easy interfacial fracture modes. Roughening of the substrate surface will tend to promote better adhesion.

2. The compound interface is characterized by a layer or multilayer structure many atomic dimensions thick that is created by chemical reaction and diffusion between film and substrate atoms. The compounds formed are frequently brittle because of high stresses generated by volumetric changes accompanying reaction. Such interfaces arise in oxygen-active metal films on

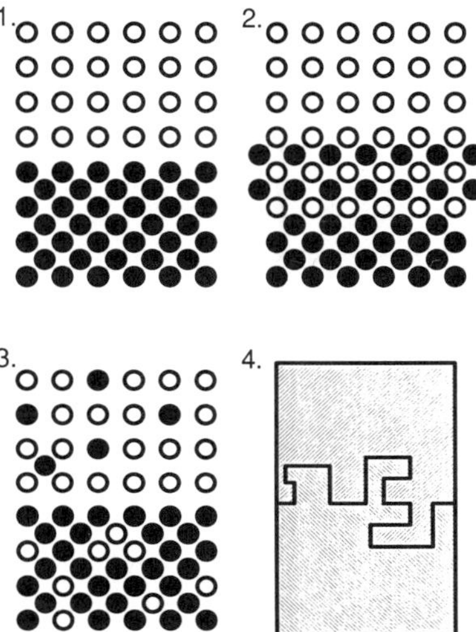

Figure 9-15. Different interfacial layers formed between film and substrate: (1) abrupt interface; (2) compound interface; (3) diffusion interface; (4) mechanical anchoring at interface.

oxide substrates or between intermetallic compounds and metals. Adhesion is generally good if the interfacial layer is thin, but is poor if thicker layers form.

3. The diffusion interface is characterized by a gradual change in composition between film and substrate. The mutual solubility of film and substrate precludes the formation of interfacial compounds. Differing atomic mobilities may cause void formation due to the Kirkendall effect (Chapter 8). This effect tends to weaken the interface. Usually, however, interdiffusion results in good adhesion. A related type of transition zone which can strongly promote adhesion is the interfacial "pseudodiffusion" layer. Such layers are formed when film deposition occurs under the simultaneous ion bombardment present during sputtering or ion plating. In this way backscattered atoms sputtered from the substrate efficiently mix with the incoming vapor atoms of the film to be deposited. The resulting condensate may be thought of as a metastable phase in which the solubility of the components involved exceed equilibrium limits. The generally high concentration of point defects and structural disorder introduced by these processes greatly enhance "diffusion" between materials that do not naturally mix or adhere.

Important examples of interdiffusion adhesion are to be found in polymer systems that are widely used as adhesives. In view of the above, it is not surprising that interdiffusion of polymer chains across an interface requires that the adhesive and substrate be mutually soluble and that the macromolecules or segments be sufficiently mobile. Such conditions are easily met in the autoadhesion of elastomers and in the solvent bonding of compatible amorphous plastics.

4. The mechanical interface is characterized by interlocking of the depositing material with a rough substrate surface. The adhesion strength depends primarily on the mechanical properties of film and substrate and on the interfacial geometry. A tortuous fracture path induced by rough surfaces and mechanical anchoring leads to high adhesion. Mechanical interlocking is relied upon during both electroplating and vacuum metalization of polymers.

9.6.4. Theories of Adhesion (Ref. 25)

The adsorption theory is most generally accepted and suggests that when sufficiently intimate contact is achieved at the interface between film and substrate, the surfaces will adhere because of the pairwise interaction of the involved atoms or molecules. There is no reason to believe that the forces that act in adhesion are any different from those that are functional within bulk matter. Therefore, the interaction energy typically follows the behavior depicted in Fig. 1-8b as a function of separation distance regardless of the type of materials or surface forces involved. It is believed that the largest contribution to the overall adhesion energy is provided by van der Waals forces (physiosorption). These are classified into London, Debye, and Keesom types depending, respectively, on whether neither, one, or both of the paired atoms possess electric dipoles. Interaction energies between film and substrate atoms typically fall off as the sixth power of the separation distance. The resulting forces are weak and secondary bonding is said to exist with energies of 0.1 eV per atomic pair. In addition to van der Waals forces, chemical interactions (chemisorption) also contribute to adhesion. Stronger primary covalent, ionic, and metallic binding forces are involved now, and bond energies of 1 to 10 eV can be expected.

For a typical interface containing some 10^{15} primary bonds/cm^2 at 1 eV per bond, the total energy is 10^{15} eV/cm^2 or 1600 ergs/cm^2. This corresponds to typical surface energies of metals. The bonding force can be obtained from the bond energy if its variation with separation distance is known. If, for example, the adhesion energy drops to zero when the surfaces are parted by some 5 Å, then the specific adhesion force is $F_A = (1600 \text{ ergs/cm}^2)/5 \times 10^{-8}$ cm or 3.2×10^{10} dynes/cm^2. In contrast, van der Waals adhesion forces are expected to be an order of magnitude less or roughly 10^9 dynes/cm^2. Secondary bonding forces alone may result in adequate adhesion, but the presence of primary bonds can considerably increase the joint strength. Surface-specific analytical techniques such as laser-Raman scattering, X-ray photoelectron spectroscopy, and SIMS have yielded definitive evidence that primary interfacial bonding contributes significantly to the intrinsic adhesion.

Exchange of charge across film–substrate interfaces also contributes to adhesion. As a result, electrical double layers consisting of oppositely charged sheets develop and exert adhesive forces. The latter, however, are generally small compared with physiosorption forces. The situation is like that of a parallel-plate capacitor. Chapman (Ref. 25) has estimated that the attractive force is $Q^2/2\varepsilon_0$ per unit area, where Q is the charge density/cm^2 and ε_0 is the permittivity of free space. If $Q = 10^{11}$–10^{13} electronic charges/cm^2 then the resulting attractive forces are 10^4–10^8 dynes/cm^2. These are small compared with other force contributions to adhesion.

Theories do not always provide guidelines on how to practically achieve good film adhesion in practice. Conventional wisdom, for example, suggests using very clean substrates. This is not necessarily true for the deposition of metals on glass substrates because optimum adhesion appears to occur only when the metal contacts the substrate through an oxide bond. Thus Al adheres better when there is some Al_2O_3 present between it and the glass substrate. It is not surprising that strong oxide formers adhere well to glass. Intermediate oxide layers can be produced by depositing metals with large heats of oxide formation such as Cr, Ti, Mo, and Ta. Reactions of the type given by Eq. 1-16 proceed at the interface resulting in good adhesion. Conversely, the noble metals such as Au and Ag do not form oxides readily and, accordingly, adhere poorly to glass, a fact reflected in low film stresses (Table 9-2). To promote adhesion, it is common practice, therefore, to first deposit a few hundred angstroms of an intermediate oxygen-active metal to serve as the "glue" between the film and substrate. This is the basis of several multilayer metallization contact systems such as Ti–Au, Ti–Pd–Au and Ti–Pt–Au. After deposition of the intermediate glue layer, the second film should be deposited without delay, for otherwise the glue metal may oxidize and impede adhesion of the covering metal film.

9.6.5. Adhesion Tests

Although there are no ways to directly measure interfacial atomic bond strengths, numerous tests characterize adhesion practically. These tests have been recently reviewed by Steinmann and Hintermann (Ref. 26), and Valli (Ref. 27). Essentially two types of tests are distinguished by whether tensile or shear stresses are generated at the interface during testing.

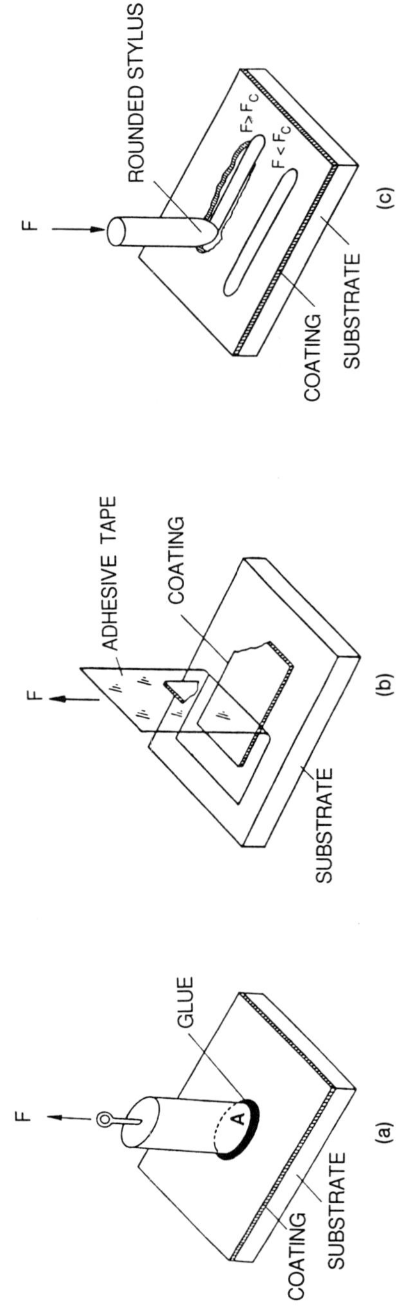

Figure 9-16. Adhesion test methods: (a) pull-off test; (b) adhesive tape test; (c) scratch test. (a. and b. from Ref. 26, c. from Ref. 14).

9.6.5.1. Tensile-Type Tests. The simplest of these include direct pull-off as well as so-called topple tests and both are used primarily for coatings. As Fig. 9-16a indicates, force is applied to a member glued or soldered to the coating, and the resultant load to cause interfacial separation is then measured. Misalignment problems associated with normal pulling are partially overcome by applying a torque in the topple tests. The value of F_A is equal to F, the applied force at separation divided by the contact area A.

Acceleration tests also generate tensile stresses in the coating but without the disadvantage of glues and mechanical linkages. In the ultracentrifugal method a coated cylinder is levitated electromagnetically and spun at ever-increasing speed until the coating debonds from the substrate.

Pulsed lasers have also been used to measure adhesion forces. When the back of the substrate is exposed to the laser pulse, successive compressive and tensile shock waves rapidly flex the substrate backward and then forward, detaching the coating in the process. Adhesion is characterized by the energy absorbed per unit area.

9.6.5.2. Shear-Type Tests. The adhesive tape test developed over a half century ago provides the simplest and quickest qualitative measure of film or coating adhesion. Schematically indicated in Fig. 9-16b, the test can distinguish between complete lifting, partial lifting, or complete adhesion with a little bit of discrimination. The test can also be made semiquantitative by controlling the angle of pull and the rate of pulling. With improved adhesives the force required to peel the tape is measured as a function of angle; the force extrapolated to zero angle is a measure of the adhesion. In such tests it is necessary that the tape–film bond be stronger than the film–substrate bond.

9.6.5.3. Scratch Tests. The scratch test shown schematically in Fig. 9-16c is a widely used means of evaluating the adhesion of films. The test consists of drawing a stylus or indenter of known radius of curvature over a film or coating under increasing vertical loads. Resultant scratches are observed under an optical or scanning electron microscope in order to estimate the minimum or critical load required to scribe away the film and leave a clear channel or visible substrate behind. The elastoplastic deformation is complicated, however, and films can be thinned and appear translucent while still adhering to the substrate. Alternatively, films can remain opaque when detached. Commercial equipment is available to enable the critical load to be determined on the basis of a single scratch. This is accomplished by ramping the indenting load between set limits, followed by visual examination of the scratch to determine the critical load F_c that just causes adhesion failure. The scratching process is also accompanied by the emission of acoustic signals that are small in magnitude when the film adheres at low loads. The onset of large acoustic emission caused by shearing or fracture at the film–substrate interface has been taken as a measure of the critical de-adhesion load, thus obviating the need for microscopic examination. Theoretical analyses relating the critical load, stylus geometry, and scratch dimensions to the specific adhesion force have been made. One such relation is

$$F_A = KH_v F_c / \pi R^2, \tag{9-38}$$

where the magnitude of coefficient K depends on the model details (K can range from 0.2 to 1), H_v is the Vickers hardness (see page 562), and R is the radius of the stylus tip.

At present there is little quantitative agreement in F_A values obtained from different adhesion test methods. Rather, individual tests are well suited to internal comparisons of the same film–substrate combination prepared in different ways.

EXERCISES

1. Identical metal films of equal thickness, deposited on both sides of a thin substrate strip are found to possess a residual tensile stress. One of the films is completely removed by sputter-etching. Qualitatively describe how the remaining film-substrate combination deforms or bows.

2. Stress fields exists around dislocations resulting in matrix distortions shown in Fig. 1-6.

a. A row of edge misfit dislocations of the same sign (orientation) lies within a thin film close to and parallel to the substrate interface. Comment on the internal stress in the film.
b. How would the film stress differ if the dislocations were screw type?
c. Due to annealing, some dislocations climb vertically and some disappear. How does this affect internal stress?

3. It is desired to grow epitaxial films of GaSb on AlSb substrates by deposition at 500 °C. Refer to Table 7-1.

a. What is the expected lattice mismatch at 500 °C?
b. What thermal stress can be expected in the film at 20°C if $E_{GaSb} = 91.6$ GPa and $\nu_{GaSb} = 0.3$?

4. Suppose $S = Kd^n$ describes the behavior of the stress $(\sigma_f) \times$ thickness (d) of a film as a function of d. (K and n are constants.) Contrast the variation of film stress and instantaneous stress versus d.

5. a. Consider the *strain* relaxation of a parallel spring–dashpot combination under constant loading and derive Eq. 9-29.

b. The intrinsic stress in a SiO_2 film is 10^{10} dynes/cm^2. If the coefficient of viscosity of SiO_2 film is $\eta(T) = 1.5 \times 10^{-8} \exp E_v/RT$ ($E_v = 137$ kcal/mole) over the temperature range 900–1500 °C, how long will it take the film to reach half of its final strain at 1000 °C. Assume $E = 6.6 \times 10^{11}$ dynes/cm^2, and the units of η are Poise.

6. An engineer wishes to determine whether there will be more bow at 20 °C in a Si wafer with a 1-μm-thick SiO_2 film, or with a 1-μm-thick Si_3N_4 film. Both films are deposited at 500 °C on a 0.5 mm/Si wafer. At the deposition temperature the intrinsic stresses are -3×10^9 dynes/cm^2 for SiO_2 and -6×10^9 dynes/cm^2 for Si_3N_4. If the respective moduli are $E_{SiO_2} = 7.3 \times 10^{11}$ dynes/cm^2, $E_{Si_3N_4} = 15 \times 10^{11}$ dynes/cm^2, and the thermal expansion coefficients are $\alpha_{SiO_2} = 0.55 \times 10^{-6}$ °C^{-1}, $\alpha_{Si_3N_4} = 3 \times 10^{-6}$ °C^{-1}, calculate the radius of curvature for each wafer. [Note: $E_{Si} = 16 \times 10^{11}$ dynes/cm^2, $\alpha_{Si} = 4 \times 10^{-6}$ °C^{-1}.]

Assume Poisson's ratio for film and substrate is 0.3. What would the radii of curvature be in a 15-cm-diameter wafer? What is the difference in height between the edge and center of the wafer?

7. When sequentially deposited films are all very thin compared with the substrate, each film imposes a separate bending moment and separate curvature. Since moments are additive, so are the curvatures.

a. Show that

$$\frac{1}{R_1} + \frac{1}{R_2} + \cdots + \frac{1}{R_n} = \frac{1 - \nu_s}{E_s}\frac{6}{d_s^2}(\sigma_1 d_1 + \sigma_2 d_2 + \cdots + \sigma_n d_n),$$

where $1, 2, \ldots, n$ denotes the film layer, and σ_n and d_n the film stress and thickness.

b. A 5000-Å-thick Al film is deposited stress free on a 12.5-cm-diameter Si wafer (0.5 mm thick) at 250 °C such that there is no stress relaxation on cooling to 20 °C. Next, the Al–Si combination is heated to 500 °C where Al completely relaxes. A 2-μm-thick Si_3N_4 film is then deposited with an intrinsic compressive stress of 700 MPa. What is the final radius of curvature after cooling to 20 °C? Note the following materials properties.

	Si	Al	Si_3Ni_4
E (GPa)	160	66	150
α °C^{-1}	4×10^{-6}	23×10^{-6}	3×10^{-6}

Assume $\nu = 0.3$ for all materials.

8. Unlike the usual *thin*-film–*thick*-substrate combination treated in this chapter, consider thin-film multilayers. For adjacent films 1 and 2 the corresponding film thicknesses, moduli, and *un*strained lattice parameters are d_1, E_1, $a_0(1)$ and d_2, E_2, a_0 (2). There is a common lattice parameter, $\bar{a}_0$, at the interface between films.

a. What is the strain in each film?
b. What are the corresponding stresses?
c. If the forces are equilibrated, show that

$$\bar{a}_0 = a_0(1)a_0(2)\left(1 + \frac{E_2}{E_1}\frac{d_2}{d_1}\right) \Bigg/ a_0(2) + \frac{E_2 d_2}{E_1 d_1} a_0(1).$$

9. In Fig. 14-17 the structure of the (250 Å) Si–(75 Å) $Ge_{0.4}Si_{0.6}$ superlattice is shown. The [100] moduli for Ge and Si are $E_{Ge} = 141$ GPa, $E_{Si} = 181$ GPa and $a_0(\mathrm{Ge}) = 5.66$ Å, $a_0(\mathrm{Si}) = 5.43$ Å are the corresponding lattice parameters. If the properties of $Ge_{0.4}Si_{0.6}$ are assumed to be derived from weighted composition averages of pure component properties, find

a. the common interfacial (in-plane) lattice parameter using the results of the previous problem.
b. the strains and stresses in the Si and $Ge_{0.4}Si_{0.6}$ layers.
c. the *strained* lattice parameters *normal* to the film layers.

Assume Poisson's ratio is 0.37.

10. Consider a substrate of thickness d_s containing deposited films of thickness d_f on either side that are uniformly stressed in tension to a level of σ_f. The substrate is assumed to be uniformly compressed. Film and substrate have the same elastic constants.

a. Determine the substrate stress, assuming force equilibrium prevails.
b. Show that under the foregoing conditions the net moment with respect to an axis at the center of the substrate vanishes.
c. One film is totally annealed so that its stress vanishes. The substrate and other film are unaffected in the process. What is the net force imbalance or resultant force? What is the net moment imbalance or resultant moment?

d. In the absence of external constraints the film–substrate will elastically deform to find a new equilibrium stress distribution with zero resultant force and moment. A uniform force as well as a moment (arising from a linear force distribution through the film–substrate cross section) are required to counter the mechanical imbalance of part (c). What is the stress contribution to the remaining film from the uniform force? What is the maximum stress contribution to the remaining film from the moment?

e. What is the *new* maximum stress in the remaining film and what sign is it?

11. Voids and porosity are sometimes observed in abrupt, compound, diffusion, and mechanical interfaces between films and substrates. Distinguish the sources of these defects at these interfaces. Which interfaces are likely to contain microcracks? Why?

References

1. G. Gore, *Trans. Roy. Soc.* (*London*), Part 1, 185 (1858).
2. D. P. Seraphim, R. Lasky, and C. Y. Li, eds. *Principles of Electronic Packaging*, McGraw-Hill, New York (1989).
3.* R. W. Hoffman, in *Physics of Thin Films*, Vol. 3, eds. G. Hass and R. E. Thun, Academic Press, New York (1966).
4. C. A. Neugebauer, *J. Appl. Phys.* **32**, 1096 (1960).
5. L. E. Trimble and G. K. Celler, *J. Vac. Sci. Tech.* **B7**, 1675 (1989).
6. W. C. Oliver, *MRS Bull.* **XII(5)**, 15 (1986).
7.* W. D. Nix, *Met. Trans.* **20A**, 2217 (1989).
8. G. G. Stoney, *Proc. Roy. Soc. London* **A82**, 172 (1909).
9. E. Suhir and Y.-C. Lee, in *Handbook of Electronic Materials*, Vol. 1, ed. C. A. Dostal, ASM International, Metals Park, Ohio (1989).
10.* D. S. Campbell, in *Handbook of Thin Film Technology*, eds. L. I. Maissel and R. Glang, McGraw-Hill, New York (1970).
11. E. Klokholm and B. S. Berry, *J. Electrochem. Soc.* **115**, 823 (1968).
12. R. E. Cuthrell, D. M. Mattox, C. R. Peeples, P. L. Dreike, and K. P. Lamppa, *J. Vac. Sci. Tech.* **A6**(5), 2914 (1988).
13. A. E. Ennos, *Appl. Opt.* **5**, 51 (1966).

*Recommended texts or reviews.

14.* H. K. Pulker, *Coatings on Glass*, Elsevier, Amsterdam (1984).
15. R. W. Wagner, A. K. Sinha, T. T. Sheng, H. J. Levinstein, and F. B. Alexander, *J. Vac. Sci. Tech.* **11**, 582 (1974).
16. D. W. Hoffman and J. A. Thornton, *J. Vac. Sci. Tech.* **20**, 355 (1982).
17. W. Buckel, *J. Vac. Sci. Tech.* **6**, 606 (1969).
18. R. W. Hoffman, *Thin Solid Films* **34**, 185 (1976).
19. H. K. Pulker and J. Mäser, *Thin Solid Films* **59**, 65 (1979).
20. E. A. Irene, E. Tierney, and J. Angilello, *J. Electrochem. Soc.* **129**, 2594 (1982).
21. M. F. Ashby, *Acta Met.* **20,** 887 (1972).
22.* M. Murakami, T. S. Kuan, and I. A. Blech, *Treatise on Materials Science and Technology*, Vol. 24, eds. K. N. Tu and R. Rosenberg, Academic Press, New York (1982).
23. C. J. Kircher and M. Murakami, *Science* **208** 944 (1980).
24. P. Chaudhari, *J. Appl. Phys.* **45**, 4339 (1974).
25.* B. N. Chapman, *J. Vac. Sci. Tech.* **11**, 106 (1974).
26. P. A. Steinmann and H. E. Hintermann, *J. Vac. Sci. Tech.* **A7**, 2267 (1989).
27. J. Valli, *J. Vac. Sci. Tech.* **A4**, 3007 (1986).

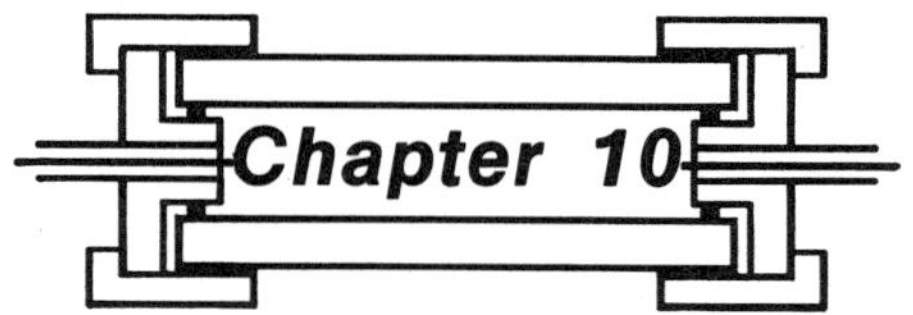

Electrical and Magnetic Properties of Thin Films

10.1. Introduction to Electrical Properties of Thin Films

10.1.1. General Considerations

Electrical properties of thin films have long been of practical importance and theoretical interest. The solid-state revolution has created important new roles for thin film electrical conductors, insulators, and devices. What was once accomplished with large discrete electrical components and systems is now more efficiently and reliably achieved with microscopic thin-film-based integrated circuit chips. Regardless of the class of material involved, its physical state or whether it is in bulk or film form, an electric current of density J (amps/cm^2) is said to flow when a concentration of carriers n (number/cm^3) with charge q moves with velocity v (cm/sec) past a given reference plane in response to an applied electric field $\mathscr{E}$ (V/cm). The magnitude of the current flow is expressed by the simple relation

$$J = nqv. \tag{10-1}$$

For most materials, especially at small electric fields the carrier velocity is proportional to $\mathscr{E}$ so that

$$v = \mu\mathscr{E}. \tag{10-2}$$

The proportionality constant or velocity per unit field is known as the mobility μ. Therefore,

$$J = nq\mu\mathscr{E}, \tag{10-3}$$

and by Ohm's law ($J = \sigma\mathscr{E}$) the conductivity σ or reciprocal of the resistivity ρ is given by

$$\sigma = 1/\rho = nq\mu. \tag{10-4}$$

Quantitative theories of electrical conductivity seek to define the nature, magnitude, and attributes of the material constants in these equations. Corollary questions revolve about how n and v or μ vary as a function of temperature, composition, defect structure, and electric field. An alternative complementary approach to understanding the response of materials to electrical fields involves electronic band structure considerations that, as noted in Chapter 1, have successfully modeled property differences. Comprehensive descriptions of conduction integrate what might be termed the "charge carrier dynamics" approach with the band structure viewpoint. The former is more intuitive and will be adopted here for the most part, but resort will also be made to band diagrams and concepts.

This chapter focuses primarily on the electrical conduction properties of thin metal, insulating, and superconducting films. Almost half of the classic *Handbook of Thin Film Technology*, edited by Maissel and Glang, is devoted to a treatment of electrical and magnetic properties of thin films. Though dated, this handbook remains a useful general reference for this chapter. Much of what is already known about bulk conduction provides a good basis for understanding thin-film behavior. But there are important differences that give thin films unique characteristics and these are enumerated here:

1. *Size effects or phenomena that arise because of the physically small dimensions involved*—Examples include surface scattering and quantum mechanical tunneling of charge carriers.
2. *Method of film preparation*—It cannot be sufficiently stressed that the electrical properties of metal and insulator films are a function of the way they are deposited or grown. Depending on conditions employed, varying degrees of crystal perfection, structural and electronic defect concentrations, dislocation densities, void or porosity content, density, grain morphology, chemical composition and stoichiometry, electron trap densities, eventual contact reactions, etc., result with dramatic property implications. Insulators (e.g., oxides, nitrides) are particularly prone to these effects and metals are less affected.

3. *Electrode effects*—Frequently the substrate and a subsequently deposited conducting film become the electrodes for the film in question that is sandwiched in between. In general, insulating films cannot be considered apart from the electrodes that contact their surfaces. The electrical response of structures containing insulator (I) or oxide (O) films between metals (MIM), semiconductors (SOS), and mixed electrodes (MIS, MOS) is strongly influenced by the specific metal or semiconductor electrode materials employed. Interfacial adhesion, stress, interdiffusion, incorporated or adsorbed impurities, are some of the factors that can alter the character of charge transport at an interface.
4. *Degree of film continuity*—conduction mechanisms in discontinuous, island structure films differ from those in continuous films.
5. *Existence of high electric field conduction phenomena*—Moderate voltages applied across very small dimensions conspire to make high field effects readily accessible in films.
6. *High chemical reactivity*—films are susceptible to aging or time-dependent changes in electrical properties due to corrosion, absorption of water vapor, atmospheric oxidation and sulfidation, and low-temperature solid-state reactions.

In virtually all cases, thin metal films are more resistive, while insulating films are more conductive than their respective bulk counterparts. For metals the differences between film and bulk electrical properties are relatively small; in insulators the differences can be huge. Why this is so will unfold in the ensuing pages.

10.1.2. Measurement of Film Resistivity

A number of techniques have been employed to measure electrical properties of thin films. Some are adaptations of well-known methods utilized in bulk materials. For insulating films, where current flows through the film thickness, electrodes are situated on opposite film surfaces. Small evaporated or sputtered circular electrodes frequently serve as a set of equivalent contacts; the substrate is usually the other contact. If charge leaks along the surface from contact to contact, circumventing through-film conduction, then a guard electrode is required.

For more conductive metal and semiconductor films, it is common to place all electrodes on the same film surface. Such measurements employ four terminals–two to pass current and two to sense voltage. Several contact configurations shown in Fig. 10-1 are suitable for this purpose. Through the

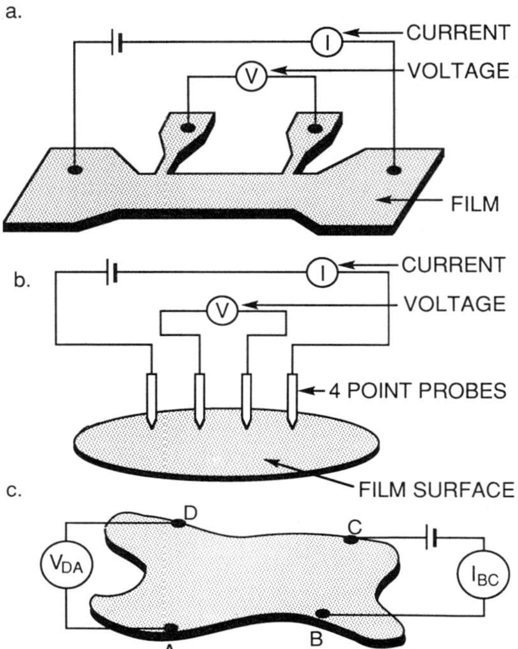

Figure 10-1. Techniques for measuring thin-film electrical resistance. (a) Four-terminal method for conducting stripes. Current is passed through outer terminals, and voltage is measured across inner terminals. (b) Four-point probe method for measuring sheet resistance. (c) Van der Pauw method for measuring resistivity of arbitrarily shaped film.

use of lithographic patterning methods (Chapter 14), long stripes can be configured with outer current and inner voltage leads that are extensions of the film itself (Fig. 10-1a). In this way contact resistance problems are eliminated. Independent measurements of voltage and current yield the film resistance through Ohm's law, and the resistivity if film dimensions are known.

A very common way to report values of thin-film resistivity is in terms of sheet resistance with units "ohms per square." To understand this property and the units involved, consider the film of length l, width w, and thickness d in Fig. 10-2. If the film resistivity is ρ, the film resistance is $R = \rho l/wd$. Furthermore, in the special case of a square film ($l = w$),

$$R = R_s = \rho/d \quad \text{ohms}/\square, \tag{10-5}$$

where R_s is independent of film dimensions other than thickness. Any square, irrespective of size, would have the same resistance. As an example, consider

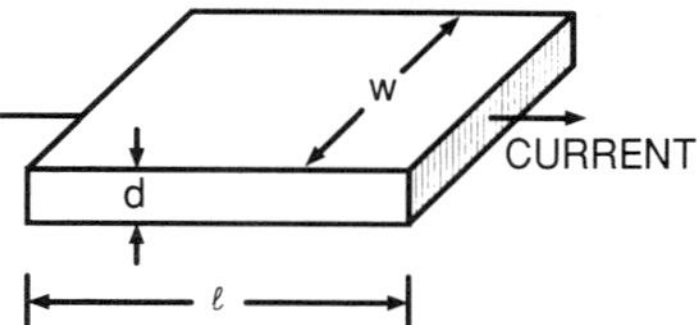

Figure 10-2. Thin-film conductor with length l, width w, and thickness d.

a film stripe measuring 3 μm × 30 μm with $R_s = 15\ \Omega/\square$. The overall stripe resistance is $10 \times 15 = 150\ \Omega$ because there are 10 squares (3 μm × 3 μm) in series.

A very convenient way to measure the sheet resistance of a film is to lightly press a four-point metal-tip probe assembly into the surface as shown in Fig. 10-1b. The outer probes are connected to the current source, and the inner probes detect the voltage drop. Electrostatic analysis of the electric potential and field distributions within the film yields

$$R_s = KV/I, \tag{10-6}$$

where K is a constant dependent on the configuration and spacing of the contacts. If the film is large in extent compared with the probe assembly and the probe spacing large compared with the film thickness, $K = \pi/\ln 2 = 4.53$. Otherwise correction factors must be applied (Ref. 1). Four-point probe assemblies are available commercially with square as well as the more common linear contact arrays.

Van der Pauw (Ref. 2) has devised a perfectly general method to determine the resistivity of a film using four probes located arbitrarily at points A, B, C, D on the surface (Fig. 10-1c). Two sets of current–voltage measurements using alternate contacts are required. The value of ρ can be extracted from the formula

$$\exp - \frac{\pi d}{\rho} \frac{V_{CD}}{I_{AB}} + \exp - \frac{\pi d}{\rho} \frac{V_{DA}}{I_{BC}} = 1. \tag{10-7}$$

10.2. Conduction in Metal Films

10.2.1. Matthiessen's Rule

Much can be learned about electrical conduction in metals from Matthiessen's rule. Originally suggested for bulk metals, it is also valid for thin metal films. It simply states that the various electron scattering processes that contribute to

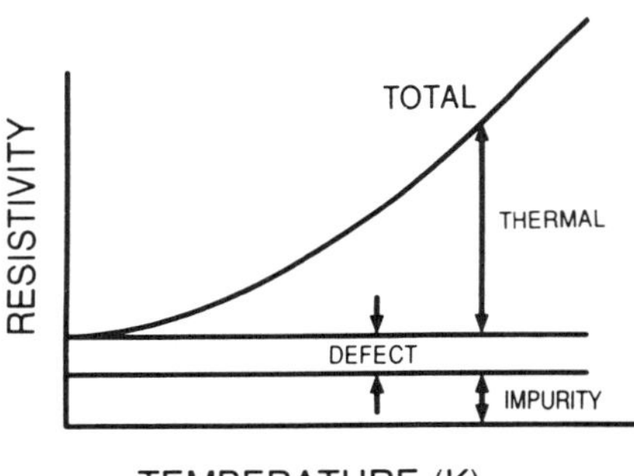

Figure 10-3. Schematic electrical resistivity dependence as a function of temperature for a metal. Thermal, impurity, and defect contributions are shown.

the total resistivity (ρ_T) of a metal do so independently and additively; i.e.,

$$\rho_T = \rho_{\text{Th}} + \rho_I + \rho_D \tag{10-8}$$

The individual thermal, impurity, and defect resistivities are ρ_{Th}, ρ_I, and ρ_D, respectively. Electron collisions with vibrating atoms (phonons) displaced from their equilibrium lattice positions are the source of the thermal or phonon (ρ_{Th}) contribution, which increases linearly with temperature. This causes the well-known positive temperature coefficient of resistivity, an unambiguous sign of metallic behavior. Impurity atoms, defects such as vacancies, and grain boundaries locally disrupt the periodic electric potential of the lattice. Due to atomic valence and size differences at these singularities, electrons are effectively scattered. However, the contributions from ρ_I and ρ_D are temperature independent, as shown in Fig. 10-3 if the concentration of impurities and defects is low.

As an application of Matthiessen's rule, consider the problem of assessing the purity and defect content of "pure" metal films. A measure of these properties is the residual resistivity ratio or RRR. The latter is defined by RRR $= \rho_T(300\text{ K})/\rho_T(4.2\text{ K})$ and is readily determined experimentally by measuring the resistivity at the two indicated temperatures. For relatively pure annealed metal films at 300 K, $\rho_{\text{Th}} \gg \rho_I + \rho_D$. At 4.2 K, however, lattice vibrations are frozen out and thermal scattering is dramatically reduced compared to impurity and defect scattering; i.e., $\rho_I + \rho_D > \rho_{\text{Th}}$. Therefore, to a good approximation,

$$\text{RRR} \sim \rho_{\text{Th}}(300\text{ K})/\rho_I + \rho_D(4.2\text{ K}), \tag{10-9}$$

and it is clear that the purer and more defect-free the film is, the higher the ratio is. Resistivity ratio determinations of sputtered films as well as sputtering target metals have been a useful way to assess their quality. Measured values of RRR broadly range from below 10 to several thousand. For extremely pure metal films, chemical analysis methods fail because of limited sensitivity, and the value of RRR is the only practical means available for determining purity.

An alternative formulation of Matthiessen's rule in terms of the respective mean free-electron lengths (λ) between collisions has the form

$$1/\lambda_T = 1/\lambda_{\text{Th}} + 1/\lambda_I + 1/\lambda_D . \tag{10-10}$$

This result stems from the intuitive direct proportionality between σ and λ, so that $\rho_i \propto 1/\lambda_i$. Theory has suggested that λ_T can be hundreds to thousands of angstroms—lengths difficult to reconcile with classical models of vibrating atoms. Clearly quantum effects are involved here. In films of thickness d that are sufficiently thin so that $d < \lambda_T$, the possibility arises that a new source of scattering—that due to the film surfaces—can increase the measured film resistivity. It is to this subject that we now turn our attention.

10.2.2. Electron Scattering from Film Surfaces

The problem of electron scattering from thin metal film surfaces has been of interest for almost a century; even today the subject is not free of controversy. Critical issues revolve about the nature of the film surface and what happens when an electron fails to execute the full mean-free path because its motion is interrupted through collision with the film surface. The electrons then undergo either specular or diffuse scattering. In specular or elastic scattering the electron reflects in much the same way a photon does from a mirror. In this case one can imagine removing the surface and doubling the film thickness (or tripling it for two surface reflections, etc.). The electron now continues on an imaginary straight path to complete the path it would have in bulk, finally scattering at a point interior to the extended surface. If this happens, the film resistivity is the same as in the bulk, and there is no film thickness effect on resistivity. When scattering is totally nonspecular, or diffuse (inelastic), the electron mean-free path is terminated by impinging on the film surface. After a surface collision the electron trajectory is independent of the impingement direction, and the subsequent scattering angle is random. The resistivity rises because fewer electrons flow through the reference plane registering current flow.

Reducing the geometry in one dimension creates an additional film contribution, i.e., λ_f, to be added to the other λ in Eq. 10-10; if λ_f is smallest, then it

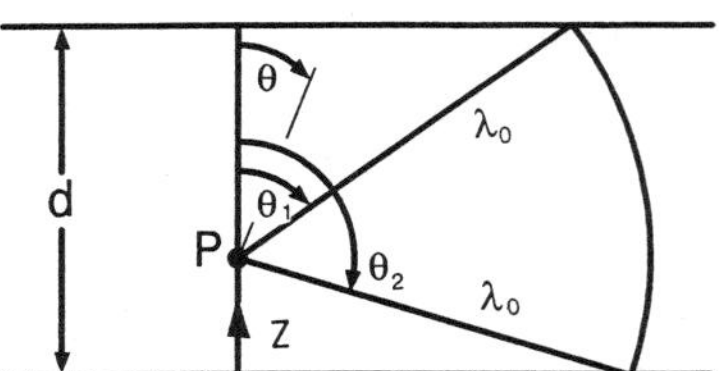

Figure 10-4. Diffuse scattering of electrons from film surface in Thomson model.

will dominate the effective film resistivity. Thomson (Ref. 3), in 1901, provided a simple way to visualize this size effect in terms of classical physics. Consider a metal whose bulk conductivity is characterized by λ_0, which is larger than the film thickness d, as shown in Fig. 10-4. For an electron starting at point P, a distance z from the film surface moving in a direction making an angle θ with the vertical, the effective mean-free path is variable, depending on θ, and is given by

$$\lambda = \begin{cases} \dfrac{d - z}{\cos\theta}, & 0 \le \theta \le \theta_1, \\ \lambda_0, & \theta_1 \le \theta \le \theta_2 \qquad \cos\theta_1 = \dfrac{d - z}{\lambda_0}, \\ \dfrac{-z}{\cos\theta}, & \theta_2 \le \theta \le \pi \qquad \cos\theta_2 = -\dfrac{z}{\lambda_0}. \end{cases} \tag{10-11}$$

The mean value of λ (i.e., λ_f) is obtained by averaging λ over all angles for all distances z. Therefore,

$$\lambda_f = \frac{1}{2d}\int_0^d dz \int_0^\pi \lambda \sin\theta \, d\theta . \tag{10-12}$$

Breaking the total limits of integration into the three angular wedges dictated by Eq. 10-11, we obtain, after integration,

$$\lambda_f = \frac{d}{2}\left(\ln\frac{\lambda_0}{d} + \frac{3}{2}\right). \tag{10-13}$$

Finally, the film conductivity σ_f and resistivity ρ_f relative to bulk values are given by

$$\frac{\sigma_f}{\sigma_0} = \frac{\rho_0}{\rho_f} = \frac{\kappa}{2}\left(\ln\frac{1}{\kappa} + \frac{3}{2}\right), \tag{10-14}$$

where $\kappa = d/\lambda_0$. Clearly, as d or κ shrinks, λ_f is diminished and ρ_f rises; a size effect is exhibited.

A more accurate quantum theory of thin-film conductivity was developed by Fuchs in 1938 (Ref. 4) and elaborated upon in the ensuing half-century by other investigators, most notably Sondheimer (Ref. 5). The Fuchs–Sondheimer (F–S) theory removed the shortcomings in the Thomson development by considering the quantum behavior of the free electrons, the statistical distribution of their λ values in bulk, and the fact that many electron mean-free paths originate at the film surface. The specific details of the calculation for the film resistivity are beyond our scope here, but the exact solution to the film conductivity (resistivity) according to Fuchs for pure *diffuse* scattering is

$$\frac{\sigma_f}{\sigma_0} = \frac{\rho_0}{\rho_f} = 1 - \frac{3}{4}\int_0^{\pi} \sin^3\phi \cos\phi \frac{\lambda_0}{d}\left[1 - \exp - \left(\frac{d}{\lambda_0 \cos\phi}\right)\right] d\phi. \tag{10-15}$$

For very thin films ($\lambda_0 \gg d$ or $\kappa \ll 1$) a useful approximation is

$$\frac{\sigma_f}{\sigma_0} = \frac{\rho_0}{\rho_f} \approx \frac{3}{4}\kappa\left(\ln\frac{1}{\kappa} + 0.423\right), \tag{10-16}$$

a result that preserves the character of the Thomson equation. In thicker films or coatings ($\kappa \gg 1$),

$$\sigma_f/\sigma_0 = \rho_0/\rho_f \approx 1 - 3/8\kappa. \tag{10-17}$$

A more realistic model of scattering envisions an admixture of both specular and diffuse contributions, with P being the fraction of surface scattering events that are specular. Under such circumstances approximate formulas for thin and thick films are, respectively (Ref. 6),

$$\frac{\sigma_f}{\sigma_0} = \frac{\rho_0}{\rho_f} \approx \frac{3}{4}\kappa(1 + 2P)\left(\ln\frac{1}{\kappa} + 0.423\right), \tag{10-18}$$

$$\frac{\sigma_f}{\sigma_0} = \frac{\rho_0}{\rho_f} \approx 1 - \frac{3}{8}\left(\frac{1 - P}{\kappa}\right). \tag{10-19}$$

The results are plotted as ρ_f/ρ_0 vs. log κ in Fig. 10-5a for various values of P. For purely specular scattering ($P = 1$), $\rho_f = \rho_0$ irrespective of film thickness or κ. In contrast, for diffuse scattering ($P = 0$), the extent that ρ_f exceeds ρ_0 is solely dependent on κ. Maximum resistivity enhancements are attained for the largest λ_0 values (achieved at the lowest temperatures) and the smallest d values. Experimentally, however, it is difficult to maintain continuous films below about $\kappa = 0.05$.

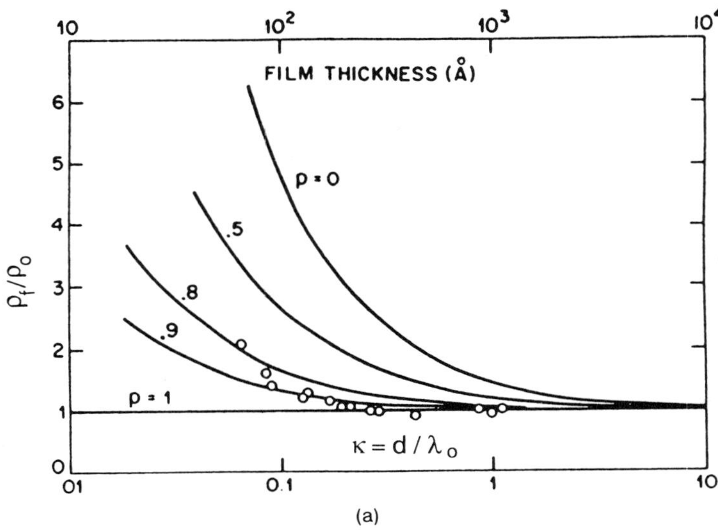

(a)

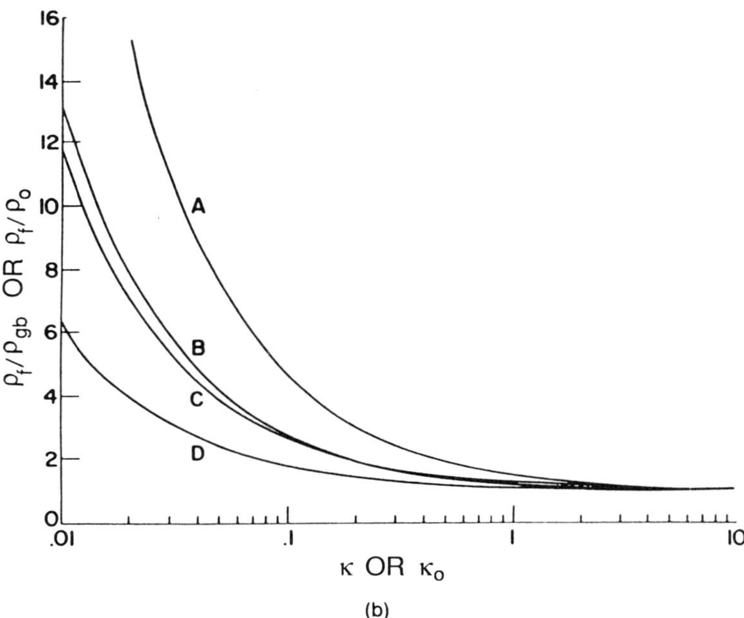

(b)

Figure 10-5. (a) ρ_f/ρ_0 vs. log κ. Data points refer to $CoSi_2$. (From Ref. 10). (b) Comparison between F–S and M–S theories: (A) F–S theory $P = 0$; (B) M–S theory $\beta = 1$, $P = 0$; (C) F–S theory $P = 0.5$; (D) M–S theory $\beta = 1$, $P = 0.5$. (From Ref. 8). (Note: $\kappa_0 = R/1 - R$)

Similar expressions have been developed for other related transport properties in thin films such as the temperature coefficient of resistivity, magnetoresistance, Hall effect, and assorted thermoelectric effects (Refs. 5–7).

10.2.3. Grain-Boundary (GB) Scattering

Grain boundaries, across which crystal orientations change, are expected to be effective electron scatterers and behave like film interfaces in this regard. As the crystallite size becomes smaller than the electron mean-free path, a contribution to the film resistivity from GB scattering arises. Mayadas and Shatzkes (M–S) (Ref. 8) have treated this effect, assuming the average grain diameter is equal to the film thickness and that only GBs whose normals lie in the film plane contribute to the scattering. The theory predicts that the GB resistivity ρ_{gb} is essentially given by

$$\frac{\rho_0}{\rho_{gb}} = 3\left[\frac{1}{3} - \frac{1}{2}\beta + \beta^2 - \beta^3 \ln\left(1 + \frac{1}{\beta}\right)\right]. \tag{10-20}$$

The new parameter, β, related to the relaxation time for GB scattering, depends on the grain geometry and its scattering power. Explicitly,

$$\beta = \frac{\lambda_0}{D}\frac{R}{1 - R},$$

where D is the average grain diameter and R $(0 < R < 1)$ is a GB electron reflection parameter. In the M–S theory, R plays a similar role to P in the F–S equations. If $R = 1$, the film resistivity is infinite because electrons are reflected back to where they originated. In addition, R depends on GB type, impurity adsorption at GBs, and the intergrain geometry and radius of curvature. Analysis of experiments have revealed that values for R are 0.24 for Cu and 0.17 for Al.

The complex picture that emerges is that the film resistivity arises from three scattering mechanisms: one due to bulk scattering (phonons, impurities, and point defects), a second due to the film surfaces, and a third due to GBs. Comparison of the F–S and M–S theories is made in Fig. 10-5b when $D = d$. A significant conclusion is that it is difficult to separate surface from GB scattering effects. Film-thickness-dependent resistivity effects in polycrystalline films could very well be due to GB rather than surface scattering.

10.2.4. Comparison with Experiment

A detailed review of the experimental data to date led one critic in 1983 to conclude "... there are virtually no studies on the resistivity of thin metal films from which useful values of $\rho_0\lambda_0$ (or ρ_f and P) may be deduced" (Ref.

9). Among the reasons for this pessimistic assessment are the following:

1. The structure and morphology of the films have only rarely been characterized.
2. Each film interface possesses a different surface roughness and degree of specularity. These factors have been taken into account in more sophisticated theories by adding new parameters which are difficult, if not impossible, to compare with experiment.
3. The role of scattering from lattice defects has not been established.
4. Electron scattering from grain boundaries has not always been taken into account (especially in the early work) despite the fact that polycrystalline rather than epitaxial films are usually involved.

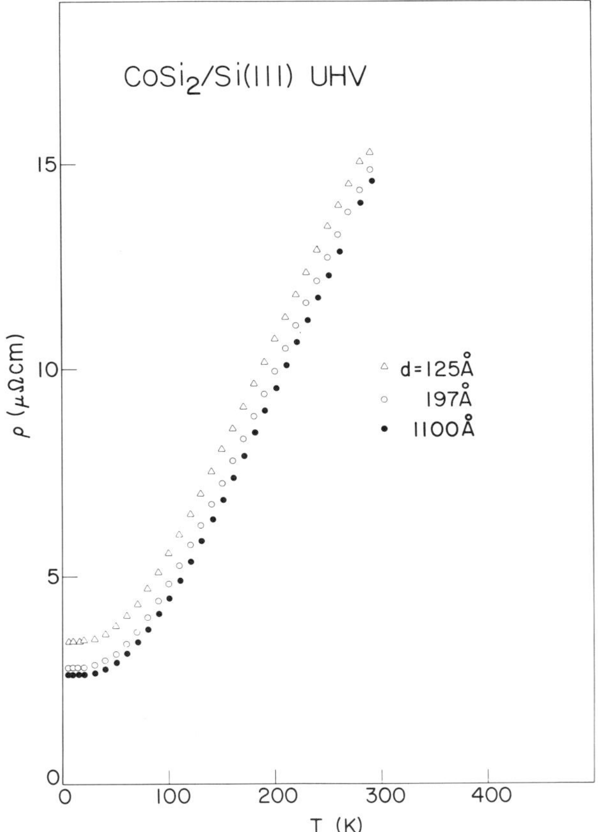

Figure 10-6. Temperature dependence of resistivity of $CoSi_2$ films. The 125-Å and 197-Å films are epitaxial. The 1100-Å film is polycrystalline. (From Ref. 10).

The critic might have been more charitable in assessing a more recent study (Ref. 10) designed to critically test the extent of specular scattering in metallic $CoSi_2$ films. Epitaxial films were prepared by first depositing Co on atomically clean Si wafers under ultrahigh-vacuum conditions. This was followed by vacuum annealing to promote silicide formation. Film thickness and stoichiometry were determined to an accuracy of $\sim 5\%$ by RBS measurements. Cross-sectional TEM methods revealed that the $CoSi_2$–Si interface was nearly atomically perfect (Fig. 1-4), whereas the outer surface was extremely smooth. Independent low-temperature magnetoresistance measurements indicated that λ_0 was 970 Å. Resistivity data recorded as a function of thickness and temperature are shown in Figs. 10-5a and 10-6. They show little dependence on thickness down to 60 Å. An average degree of specularity of about 90% from both the free surface and $CoSi_2$–Si interface is suggested. Few data on metal films exhibit such a small size effect or high specularity for the d/λ_0 range explored. Although these results cannot be generalized to other metal film systems, the experimental approach, and film preparation and characterization techniques are worthy of emulation in future studies of this kind.

10.2.5. Thin-Film Resistors

Applications for thin- and thick-film resistors are numerous and varied. They include microelectronic circuits of all kinds, hybrid circuits consisting of combinations of integrated circuits and discrete components, electromechanical sensors (e.g., strain gauges), chemical sensors, heating elements, etc. To meet these needs, a broad range of R_s (e.g., 10–1000 $\Omega/\square$) corresponding roughly to ρ ranging from 50 to 2000 $\mu\Omega$-cm, is required. In resistor design, film length, width, thickness, and R_s values must be optimized. Clearly, high resistance values necessitate long, narrow, and thin geometries fabricated from high-R_s materials. On the other hand, applications such as interconnections and semiconductor contacts require very low R_s values. In addition, resistors must frequently possess a low temperature coefficient of resistivity (TCR) (< 100 ppm/°C) and be stable with respect to environmental degradation. Some of the common materials together with their electrical properties are listed in Table 10-1. The negative values of TCR mean that the materials, e.g., Ta, Cr-Si, are not truly metallic when sufficient O or N is incorporated during processing, such as reactive evaporation or sputtering. Among the more noteworthy materials in this table are Ta and In_2O_3:Sn. Together with Ta_2O_5, a common capacitor dielectric, it has been possible to completely base a thin-film RC circuit technology on Ta. Indium–tin oxide (ITO) films have the

Table 10-1. Electrical Properties of Thin- and Thick-Film Resistor Materials

Material	Resistivity	TCR	TCR Range
	$\mu\Omega$-cm	ppm/°C	°C
Pd–Ag	38	±50	0 to 100
Ni(80)Cr(20)	110	±85	−55 to 100
Ni(76)Cr(20)Al(2)Fe(2)	133	± 5	−65 to 250
Ta (α-BCC)	25–50	+500 to +1800	
TaN*	~ 250	~ −100	
Cr–SiO(10)	~ 400	−300	200 °C
In_2O_3:Sn	~ 10^2–10^4		
Cr–SiO(40)	~ 3500	−300	200 °C
SnO_2 (undoped)	~ 10^4	−400 to −900	
SnO_2 (doped)	~ 10^2–10^4		

*Values depend strongly on composition.
From Ref. 11.

interesting property of being transparent conductors. They have a number of important uses and are discussed more fully in Section 11.2.5.

Metal oxide mixtures are known as cermets and can have widely varying ρ and TCR values, depending on metal content (Ref. 12). To a large extent, the laws of mixtures govern whether more conducting or insulating behavior dominates. An important class of these materials has found wide use in *thick*-film microcircuitry applications. Thick-film resistor compositions consist of a resistive component, a glassy phase that acts as the binder and an organic suspension medium. These materials in the form of inks or pastes are applied by a silk-screen-like process to ceramic substrates and sintered at high temperatures, thus driving off the organics. The active resistor ingredient is usually one of the following: Pd–PdO–Ag, precious metals other than Pd and Ag, RuO, Tl_2O_3, In_2O_3, SnO_2 as well as W–Ta carbides, and Ta–Ta nitrides. Compositions are available having fired sheet resistivities ranging from 1 $\Omega/\square$ to 10 M$\Omega/\square$ and TCR values from +500 ppm/°C to −800 ppm/°C. Typical resistors are 20–30 μm thick and have somewhat wider widths.

10.3. Electrical Transport in Insulating Films

10.3.1. Introduction

Insulating films are required to electrically isolate conducting components or devices as well as serve as the dielectric within capacitors. Examples of such materials include oxides such as SiO_2, Al_2O_3, and Ta_2O_5 and nitrides such as Si_3N_4 and AlN. The traditional picture of an insulator is a material that possesses very few *free* charge carriers at practical temperatures. This is a consequence of the large energy band gap. In practice, however, insulating films are generally amorphous and the usual model of sharply defined energy bands can not be readily applied; rather, fuzzy tails arise both at the top of the valence band and bottom of the conduction band. These extend into the gap and overlap to form a continuous electron state distribution. When this happens insulators assume semiconductor-like properties. Due to structural imperfections there may actually be a relatively high density of charge carriers. They tend to be localized or trapped at these centers for long times, and the insulating behavior stems from such carriers having very low mobilities.

A rich array of electrical conduction phenomena is displayed by insulating films when sandwiched between combinations of metal and semiconductor electrodes. Under the influence of an applied electric field, electrons, holes, and ions may migrate giving rise to measurable currents. The identification of the dominant transport mechanism—the one that is rate-limiting—is the key to understanding the relationships between the material parameters of the insulator and contacts, and the resulting current–voltage characteristics. Two broad categories of rate-limiting transport mechanisms have been identified (Refs. 13–15).

1. ***Barrier-Limited.*** These mechanisms are operative in the vicinity of the contact/insulator interface. Conduction is limited by the transfer of charge from the contact to the insulator; once charge is injected it has little difficulty migrating to the other electrode. Schottky emission and tunneling are the most important examples of barrier-limited conduction.

2. ***Bulk-Limited.*** In this case sufficient numbers of charge carriers are injected into the insulator conduction band by Schottky emission or tunneling. However they experience difficulty in reaching the other electrode because of bulk transport limitations. Examples include space charge-limited, intrinsic, and Poole–Frenkel conduction mechanisms.

Table 10-2 enumerates some barrier- and bulk-limited conduction mechanisms. The specific theoretical current (J)–voltage or field $(\mathscr{E})$ characteristics, together with explicit material constants that can be extracted from measurement, are listed. There are two outstanding features of current transport in insulating films. The J–$\mathscr{E}$ characteristics are frequently nonohmic, and, with the exception of tunneling, the conduction mechanisms are thermally activated. Band diagram and lattice potential models have provided useful ways

Table 10-2. Conduction Mechanisms in Insulators

Mechanism	J-$\mathscr{E}$ Characteristics		Experimentally Derivable Material Constants
1. Schottky Emission	$J_S = AT^2 \exp - \frac{q\Phi_B}{kT} \exp\left[\frac{1}{kT}\left(\frac{q^3\mathscr{E}}{4\pi\varepsilon_i}\right)^{1/2}\right]$	(10-21)	Φ_B
2. Tunneling	$J_T = \frac{q^2\mathscr{E}^2}{8\pi h\Phi_B} \exp - \left[\frac{8\pi(2m)^{1/2}}{3hq\mathscr{E}}(q\Phi_B)^{3/2}\right]$	(10-22)	Φ_B
3. Space Charge Limited	$J_{SCL} = \frac{9\mu\varepsilon_i}{8}\frac{\mathscr{E}^2}{d}$	(10-23)	—
4. Ionic Conduction	$J_I = \frac{a\mathscr{E}}{kT} \exp - \frac{E_I}{kT}$	(10-24)	E_I
5. Intrinsic Conduction	$J_{In} = bT^{3/2} \exp - \frac{E_g}{2kT} \cdot \mathscr{E}$	(10-25)	E_g
6. Poole-Frenkel Emission	$J_{PF} = c\mathscr{E} \exp - \frac{E_i}{kT} \exp\left[\frac{1}{kT}\left(\frac{q^3\mathscr{E}}{\pi\varepsilon_i}\right)^{1/2}\right]$	(10-26)	E_i

a, b, c = constant.
ε_i = insulator dielectric constant.

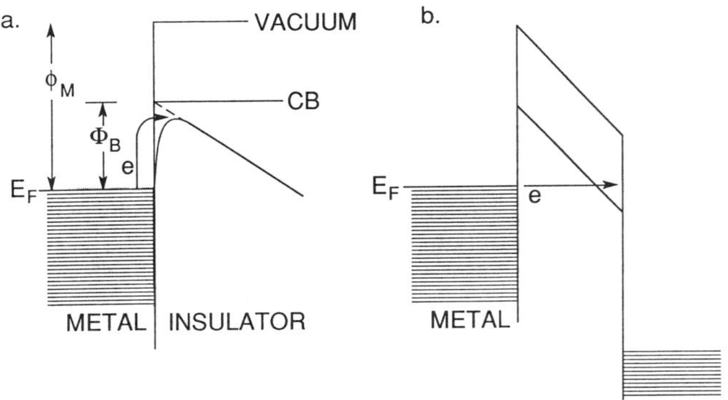

Figure 10-7. Barrier limited conduction mechanisms. (a) Schottky emission; (b) tunneling.

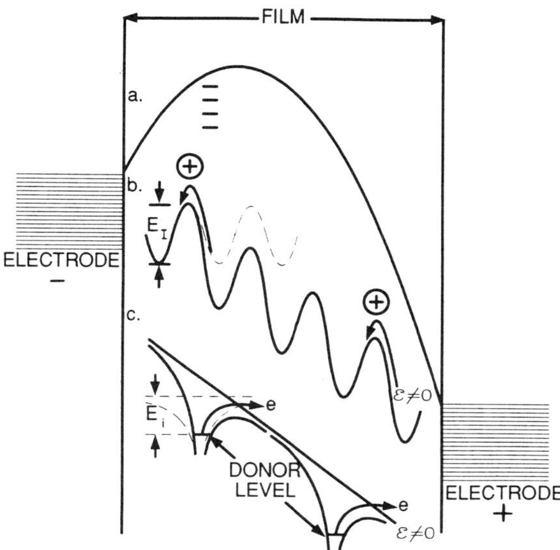

Figure 10-8. Bulk-limited conduction mechanisms. (Dotted lines refer to $\mathcal{E} = 0$) (a) space-charge-limited; (b) ionic conduction of cations ⊕; (c) Poole–Frenkel.

to picture the transport processes. These are depicted in Fig. 10-7 for barrier-limited and in Fig. 10-8 for bulk-limited transport mechanisms.

10.3.2. Specific Electrical Transport Mechanisms

It is worthwhile to comment briefly on each mechanism separately.

1. Schottky Emission. The resemblance of this mechanism to thermionic emission of electrons from a heated metal into vacuum should not go unnoticed. Thermionic emission is described by

$$J_{TE} = AT^2 \exp - q\phi_M/kT, \tag{10-27}$$

where ϕ_M is the work function and A is the Richardson constant equal to 120 A/cm^2-K^2. Typically, $q\phi_M$ is equal to 4–5 eV. Rather than acquire this energy, electrons need only surmount the smaller energy barrier $q\Phi_B$ to access the empty conduction-band states of the insulator. Values of Φ_B depend on the particular metal–insulator or metal–semiconductor contact in question as shown in Fig. 10-9.

The origin of the barrier lowering or Schottky effect in high electric fields is due to the continuously varying rather than abrupt electron potential energy

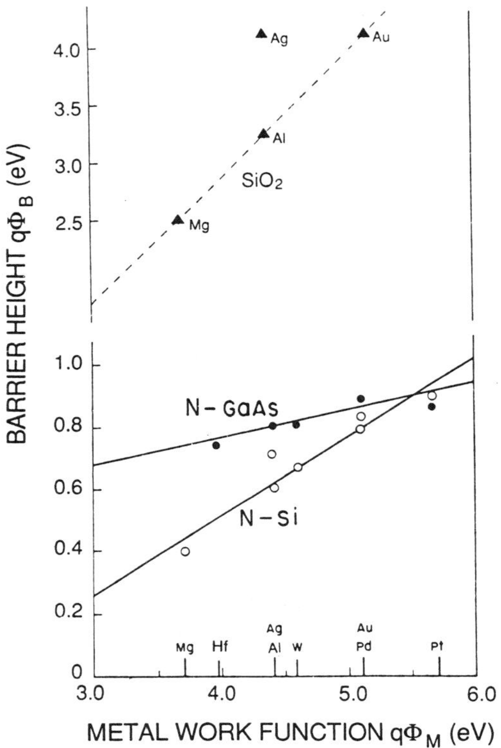

Figure 10-9. Barrier height energies of various metals on SiO_2, GaAs, and Si vs. metal work function. (From Refs. 10, 16, and 17).

distribution at the metal–insulator junction. When the electron enters the insulator, charge is redistributed on the electrode to maintain equipotential surfaces. The result is to produce a so-called image field that adds to the applied field and helps to round and lower the potential barrier (see Fig 10-7a).

2. Tunneling. Tunneling involves charge transport through an insulating medium separating two conductors that are extremely closely spaced. The phenomenon need not involve planar electrodes. Electrical conduction in discontinuous metal films occurs by electron tunneling between islands when the interparticle spacing is small enough. At sufficiently high values of $\mathscr{E}$, electrons penetrate or burrow horizontally (at constant energy) into the vacant insulator conduction-band states rather than surmount Φ_B vertically as in Schottky emission. Tunneling cannot occur classically but is rather a quantum mechanical effect. Because the metal wave functions extend into the insulator, there is a probability that electrons can exist there. If the insulator is thin enough, say ~ 30 Å, electron wave functions can even extend into the opposing metal. The phenomenon of tunneling is important in other applications discussed in this book (e.g., in the scanning tunneling microscope p. 311, in superconductors p. 483, and in tunneling transistors p. 672. In addition, both Schottky emission and tunneling mechanisms are operative at metal–semiconductor contacts (Section 10.4.1).

3. Space-Charge-Limited (SCL) Conduction. This bulk-limited mechanism occurs because the rate of carrier injection from the contact exceeds the rate at which charge can be transported through the film. A space-charge cloud develops that discourages further charge injection and leads to nonlinear conduction effects. At low levels of carrier injection Ohm's law is obeyed. If traps are present, the SCL current will be reduced because carriers are removed by empty traps. For a uniform distribution of trapping states with energies distributed throughout the band gap, a thermally activated current flows proportional to $\mathscr{E}^2/d \cdot \exp - \alpha\mathscr{E}/kT$, where α is a constant. This type of conduction has been observed in thin polymer films.

4. Ionic Conduction. High-temperature conduction in thick-film and bulk insulators frequently occurs by ionic rather than electronic motion. Ions are large cumbersome carriers of low mobility. They require high activation energies (i.e., ~ 1–3 eV) to execute nearest-neighbor diffusive jumps, and when ions diffuse they also transport charge. Since ionic transport is basically solid-state electrolysis, atomic or molecular species will necessarily be discharged at the contacts. An important example of ionic conduction is the migration of sodium ions in thin SiO_2 films. This phenomenon leads to operating instabilities in field effect transistors.

5. Intrinsic Conduction. This mechanism involves direct electronic excitation from the valence to conduction band. Because E_g is large in insulators, intrinsic conduction is negligible at low temperatures.

6. Poole–Frenkel Emission. Bulk charge trapped by impurity levels in the insulator band gap can be transferred to the conduction band by *internal* emission processes. In Poole–Frenkel emission the impurity is viewed as an ionized donor with a hydrogen-like Coulomb potential. When filled by an electron, the defect center is neutral. In an applied field the potential well distorts asymmetrically, as shown in Fig. 10-8. Trapped electrons can escape by thermal activation over the reduced energy barrier corresponding to that required to ionize the donor. Net current flows due to successive electron jumps between Poole–Frenkel traps. Note the factor of 2 difference between

Poole–Frenkel and Schottky emission that arises due to consideration of a fixed rather than image charge.

A variant of Poole–Frenkel emission involves quantum mechanical tunneling into the insulator conduction band by electrons trapped within the spherical donor potential well. The J–$\mathscr{E}$ characteristics resemble those of tunneling for this case.

10.3.3. Comparison with Experiment

The varied conduction mechanisms insulating films exhibit lead to a rich assortment of J-V characteristics. Identification of the dominant type of

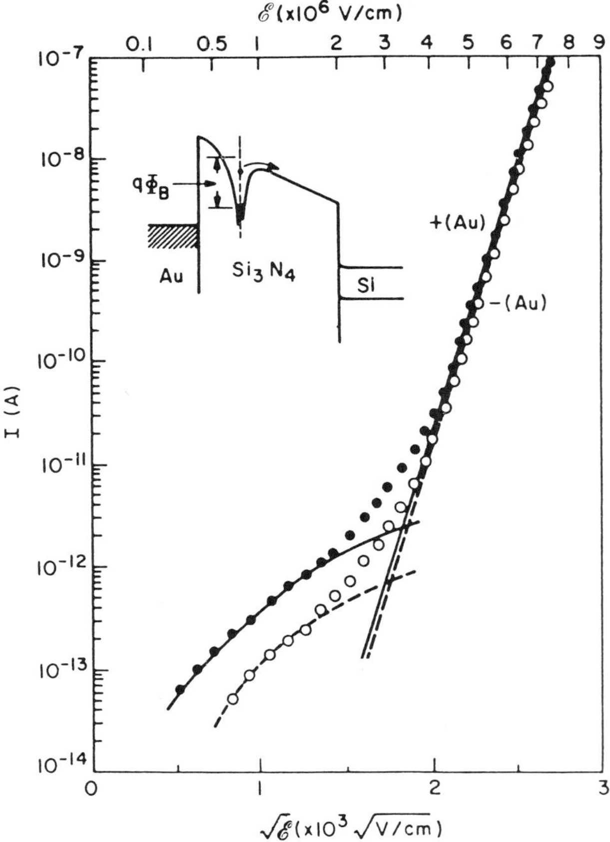

Figure 10-10a. Current (I)–electric field (E) characteristics for Au–Si_3N_4–Si (N–0.0005 Ω cm) structure at 298 K. Si_3N_4 film thickness is 500Å and contact area is 1.6×10^{-4} cm^2.

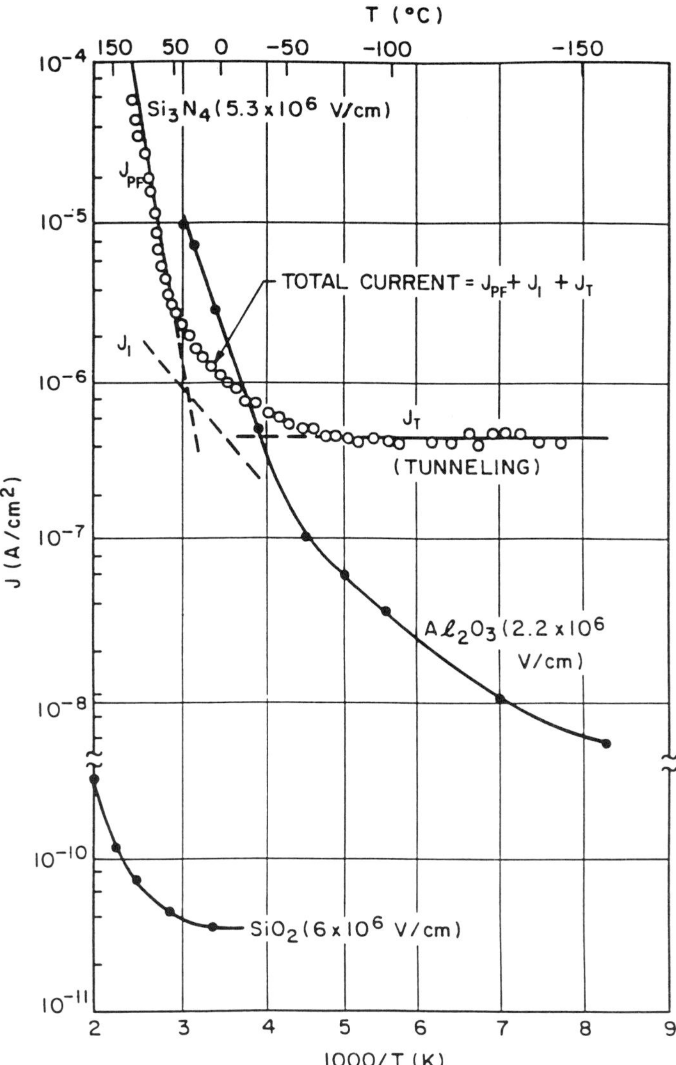

Figure 10-10b. Arrhenius plots of current density versus $1/T$ for Si_3N_4, Al_2O_3, and SiO_2 films at the indicated electric fields (Ref. 1).

transport generally necessitates measurement over a broad range of experimental variables, e.g., $\mathscr{E}$, T and film thickness. Even so, it is not always easy to unambiguously specify the operative conduction mechanism. For example, temperature effects are generally dominated by the Boltzmann factor, which frequently masks the pre-exponential temperature dependence. Thus, ln J_S / T^2, ln $J_I T$, ln $J_{In} / T^{3/2}$, and ln J_{PF} each plotted versus $1/T$ yields an Arrhenius curve of similar appearance. Schottky barrier lowering effects and Poole–Frenkel conduction are difficult to distinguish. Probable admixtures of conduction mechanisms operating in series or parallel further complicate the issue.

Despite these impediments, careful measurements in specific systems have yielded characteristics in agreement with theory. For example, consider the data in Fig. 10-10 for Si_3N_4 films sandwiched between Au and highly doped Si electrodes. The high field dc response (Fig. 10-10a) is consistent with Poole–Frenkel conduction while at low values of $\mathscr{E}$ ohmic (intrinsic) behavior (dotted in) is observed. Polarity reversal results in a different response, an effect attributable to the difference in Φ_B at Au/Si_3N_4 and Si/Si_3N_4 interfaces. The spectrum of conduction regimes can also be distinguished on the basis of temperature as shown in Fig. 10-10b. As the temperature is lowered, Poole–Frenkel conduction gives way to ohmic behavior and then to tunneling. Simultaneously, the effective activation energy for conduction declines in stages until it reaches zero, a certain sign that tunneling dominates. Except for the extent of conduction, Al_2O_3 and SiO_2 films display similar characteristics as a function of temperature. Conduction behavior in insulators is often investigated as a function of film thickness. It is left to the reader to suggest how resultant current–voltage characteristics can be used to identify specific conduction mechanisms in this case.

10.4. Semiconductor Contacts and MOS Structures

10.4.1. The Metal – Semiconductor Contact

Good electrical contacts to semiconductors are crucial for the proper functioning of circuits because signals enter and leave devices through them. One requirement for contacts is chemical stability, a subject already treated in Section 8.5.2 in connection with Al–Si. Further, it is essential that the contact does not introduce unwanted electrical characteristics such as rectification of signals or high resistance into the circuit. Band diagrams provide a convenient way to model the electrical response of contacts. As an example, consider the metal–N-type semiconductor contact of Fig. 10-11 where the respective work function and Fermi level magnitudes are given by $\phi_S < \phi_M$ and $E_{F_S} > E_{F_M}$. When both isolated materials are brought into contact, equilibrium of E_F requires transfer of electrons to the metal. But this depletes the semiconductor of electrons in the vicinity of the interface. As a result the semiconductor

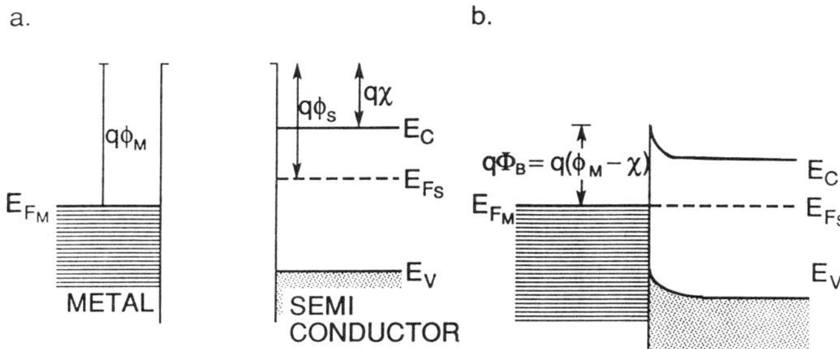

Figure 10-11. (a) Band diagrams for isolated metal and *N*-type semiconductor. (b) Schottky barrier formation at metal–semiconductor contact.

acquires less of an *N*-type character, and this is why the valence and conduction-band edges bend up at the interface. Beyond ~ 1 μm or so, depending on doping level, the original *N*-type semiconductor bands are unaltered.

An important consequence of forming the contact is that a potential energy barrier of magnitude $q(\phi_M - \phi_S)$ is built in at the junction. Readers familiar with elementary semiconductor diode behavior will recognize that the contact in question will function like a *P–N* junction that also has a built-in potential. Furthermore, for low semiconductor doping levels the dominant electron transport mechanism is Schottky emission. The barrier that electrons must surmount by thermal excitation is $\Phi_B = \phi_M - \chi$, where χ is the electron affinity. A synthesis of the above characteristics suggests that the contact—alias Schottky diode—preserves and integrates both attributes. Under the application of bias voltage V, the well-known junction diode equation describing the current (J) response is operative

$$J = J_0(\exp qV/kT - 1), \tag{10-28}$$

For this case it is appropriate to let $J_0 = AT^2 \exp - q\Phi_B/kT$.

Ohmic contacts ideally have linear J–$\mathscr{E}$ characteristics and pass current equally with either voltage polarity. They are required to have negligible resistance and exhibit small voltage drops in comparison to bulk semiconductor behavior. A convenient measure of the contact quality is the specific contact resistance R_c (Ω-cm^2). It is defined through Ohm's law by

$$R_c = (\partial J/\partial V)^{-1}_{V=0} \tag{10-29}$$

Explicit differentiation of Eq. 10-28 yields

$$R_c = \frac{k}{qAT} \exp \frac{q\Phi_B}{kT}. \tag{10-30}$$

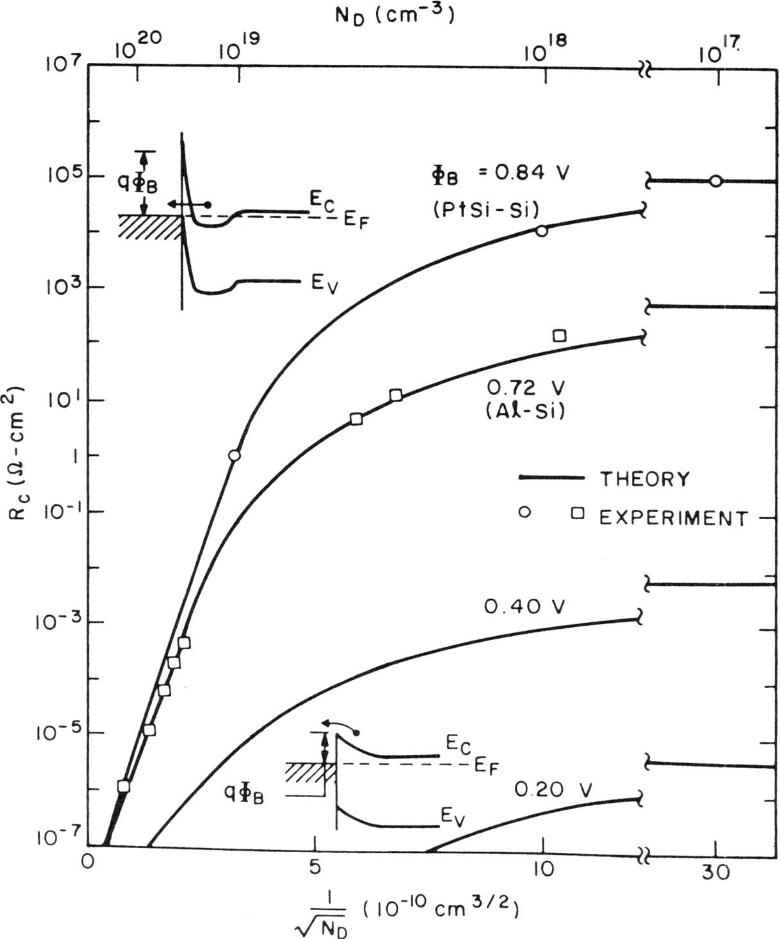

Figure 10-12. Calculated and measured values of contact resistance as a function of doping level N_D. Upper insert shows tunneling. Lower insert shows Schottky emission. (Reprinted with permission from John Wiley and Sons, from S. M. Sze, *Semiconductor Devices: Physics and Technology*, Copyright © 1985, John Wiley and Sons).

This equation is valid for low semiconductor doping levels. When, however, the doping concentration increases, the width of the depletion region shrinks. Correspondingly, the probability of electron tunneling from the contact into the semiconductor increases, reducing R_c in the process. The dependence of R_c on doping level is shown in Fig. 10-12. A target value of $R_c \sim 10^{-7}$ Ω-cm^2 is desired for device applications. Doping semiconductors heavily to levels of

10^{20} cm^{-3} or more is the only practical way to achieve this. The alternative of using metals with low Φ_B values is not feasible. In fact, Φ_B for a large number of otherwise suitable metal and silicide films on *N*-type Si ranges from 0.5 to 0.9 V (Figs. 10-9 and 10-12). For *P*-type Si, Φ_B is somewhat lower (0.4 to 0.6V), but still too high to yield low resistance contacts in the absence of heavy doping.

10.4.2. MOS Structures

By interposing a thin oxide film between planar metal and semiconductor electrodes, we create the technologically important MOS structure. This structure is an essential building block of silicon field effect transistors. Many excellent texts on semiconductor device physics (Refs. 17, 18) offer a detailed account of the electrical behavior of devices containing this structure. Because MOS capacitor properties have proven to be valuable in diagnosing the quality of the oxide and gate electrode films employed, it is worthwhile to qualitatively understand how this structure responds to applied voltages. Let us first consider an ''ideal'' MOS structure shown in Fig. 10-13. It is ideal because both the *P*-type semiconductor and metal gate have identical work functions. Further it is assumed there is no charge at the semiconductor–oxide (S–O) interface or within the oxide film proper. This means that with no voltage applied Fermi level equilibration leaves all bands horizontal or flat. When a negative potential V is applied to the metal, the energy of all of its conduction electrons is raised uniformly by an amount qV relative to the semiconductor. Holes are attracted to the S–O interface making the semiconductor appear to be even more *P*-type than originally. This causes the band bending at the interface shown in Fig. 10-13b, which graphically describes a semiconductor state known as *accumulation*. The high interfacial hole density serves to make the semiconductor an effective capacitor plate. For high V values the MOS capacitance is solely due to the oxide, and it reaches a maximum value given by $C_0 = \varepsilon_0 A_0 / d_0$, where ε_0, d_0, and A_0 are the oxide dielectric constant, thickness, and area, respectively.

As the polarity of V is reversed the overall capacitance decreases. The reason is that holes are repelled from the S–O interface. In this *depletion* regime the semiconductor acts like a dielectric with a capacitance governed by the width of the depletion region (Fig. 10-13c). There are now two capacitors in series with a lower overall capacitance than that of either semiconductor or oxide alone. (Recall that the overall capacitance is the product of the two individual (oxide and semiconductor) capacitances divided by their sum.) Note that now the bands bend the other way—the slope of the oxide conduction-band

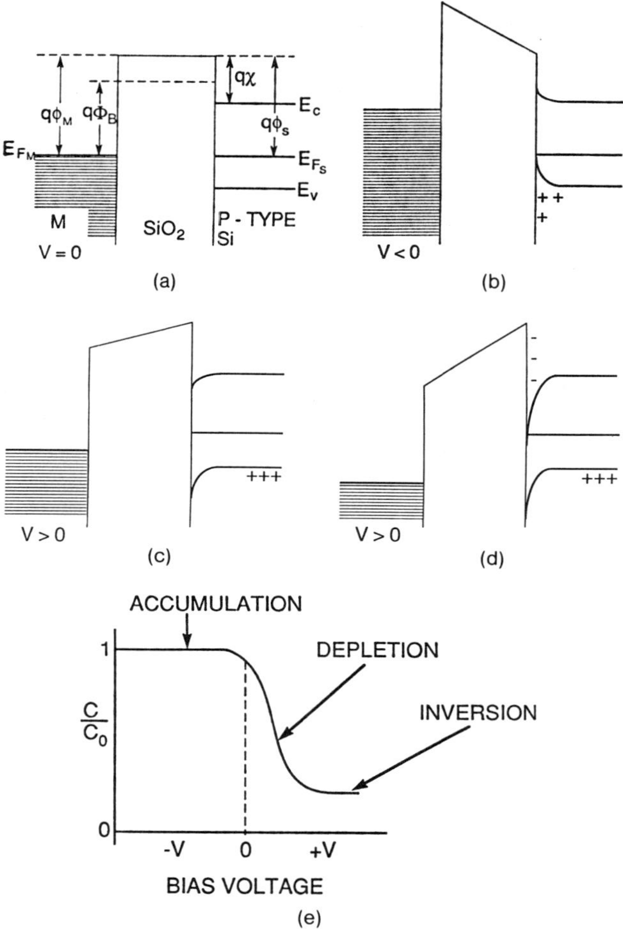

Figure 10-13. Band diagrams for ideal metal oxide P-type Si structure under different voltage bias conditions: (a) $V = 0$; (b) $V < 0$, accumulation; (c) $V > 0$, depletion; (d) $V > 0$, inversion; (e) $C-V$ characteristics.

edge (or electric field) reverses sign, and near the interface the semiconductor is, interestingly, acquiring an N-like character. At still higher positive voltages electrons are drawn to the interfacial region where the semiconductor is said to be locally inverted. Extreme band bending is characteristic of the *inversion* regime (Fig. 10-13d). None of the material properties, i.e., ϕ_M, ϕ_S, Φ_B, or E_g, are altered by the application of voltage. With inversion, the overall capacitance reaches a minimum because the width of the semiconductor region, depleted of holes, is maximum; the oxide capacitance of course remains constant throughout. The ideal capacitance–voltage ($C-V$) characteristics are shown in Fig. 10-13e for a high-frequency test signal. In this case all incremental charge appears at the edge of the depletion region. At low frequencies, however, the minority carriers (electrons) can follow the ac gate voltage variation and exchange charge with the inversion layer. The capacitance rises since only C_0 is now involved. Similar but opposite polarity considerations apply to N-type MOS structures.

In silicon-based MOS structures the gate oxide is amorphous SiO_2 grown by thermal oxidation in dry O_2. Oxides as thin as 100 Å are currently used in devices. Polysilicon, metal silicides, and combinations known as polycides are increasingly employed today in gate electrode structures supplanting the Al films of yesteryear. Real MOS structures are nonideal; now there are work function differences between the gate and Si, and both mobile oxide and fixed interfacial surface charge may exist. The net effect of these contributions is to cause initial band bending even with no applied bias. Also, the ideal $C-V$ characteristics are shifted along the V axis by an amount equal to the flat–band voltage V_{FB} given by (Ref. 17)

$$V_{FB} = \phi_M - \phi_S - \frac{Q_{ss}}{C_0} - \frac{1}{C_0}\int_0^{d_0} \frac{x}{d_0}\rho(x)\,dx. \qquad (10\text{-}31)$$

In this formula Q_{ss} is the S–O interfacial charge density and $\rho(x)$ is the spatial density of oxide charge with x, the distance into the oxide, measured from the metal interface.

10.4.3. Dielectric Breakdown

Application of very high electric fields across insulating films can sometimes lead to local destruction of the material, a phenomenon known as dielectric breakdown. This effect has long been studied in bulk as well as thin-film insulators and occurs at fields considerably in excess of those causing the ohmic or field-dependent, nonohmic conduction considered previously. Breakdown appears to be the result of local injection of excess charge from an electrode which then triggers a lightning-like current discharge. In its wake are breakdown channels, holes, or melted surface craters, and sometimes even metal whiskers or dendrites that bridge and short the electrodes. Studies of dielectric breakdown phenomena in films have been performed on a large number of oxides (e.g., SiO_2, Al_2O_3, Ta_2O_5, ZrO_2, HfO_2, etc.) as well as other insulators, such as Si_3N_4, diamond, mylar, polymers etc. Typical breakdown fields range from 10^5 to 10^7 V/cm so that relatively low applied voltages ($\sim$ 1–100 V) can damage thin ($\sim$ 1000 Å) insulators. It is for this reason that technological interest has centered on insulating films employed in microelectronics, notably the gate oxide, where dielectric breakdown is a serious reliability concern. The remainder of this section will therefore be devoted to SiO_2 films (Ref. 19).

It is generally agreed that electron impact ionization is responsible for intrinsic breakdown in SiO_2 films. In this process, electrons colliding with lattice atoms break valence bonds, creating electron–hole pairs. These new electrons accelerate in the field and through repeated impacts generate more electrons. Ultimately a current avalanche develops that rapidly and uncontrollably leads to excessive local heating and dielectric failure. Typical of theoretical modeling (Refs. 19, 20) of breakdown is the consideration of three interdependent issues;

1. Electrode charge injection into the insulator, e.g., by tunneling (Eq. 10-22). This formula connects current and applied field.
2. The local electric field that is controlled by the relatively immobile hole density through the Poisson equation.
3. A time-dependent change in hole density that increases with extent of impact ionization, but decreases with amount of hole recombination or drift away. The resulting rate equation depends on both current and field.

Simultaneous satisfaction of these coupled relationships leads to the prediction that the current–voltage characteristics display negative resistance. This appears as a knee in the response above a critical applied voltage and reflects a current runaway instability. Another prediction is that the average breakdown field rises sharply as the film thickness decreases.

Reliability concerns for thin SiO_2 films in MOS transistors have fostered much statistical analysis of life-testing results and some typical experimental findings include:

1. The histogram of the number of breakdown failures due to intrinsic causes peaks sharply at about 1.1×10^7 V/cm (Fig. 10-14a). The failure probability is nil below 7×10^6 V/cm.
2. Time to failure (TTF) accelerated testing has revealed that

$$\mathrm{TTF} \propto \exp \frac{0.33 \text{ eV}}{kT} \cdot \exp - 2.47 \text{ V}, \tag{10-32}$$

where the exponentials represent individual temperature and voltage acceleration factors (Ref. 21).

3. Contrary to theory, thinner oxides present a greater failure risk. The lifetime dependence on oxide film thickness is shown in Fig. 10-14b. Note

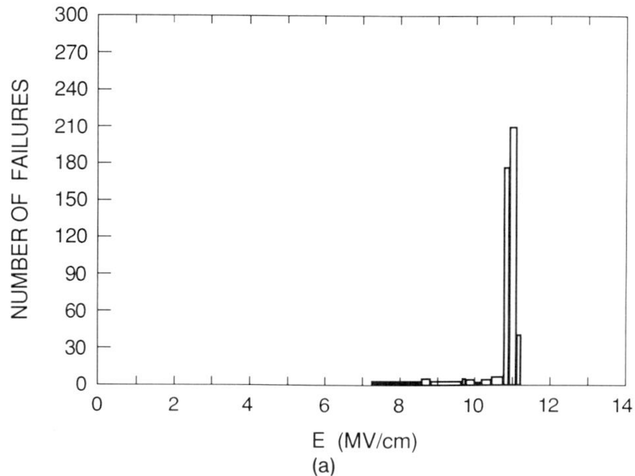

(a)

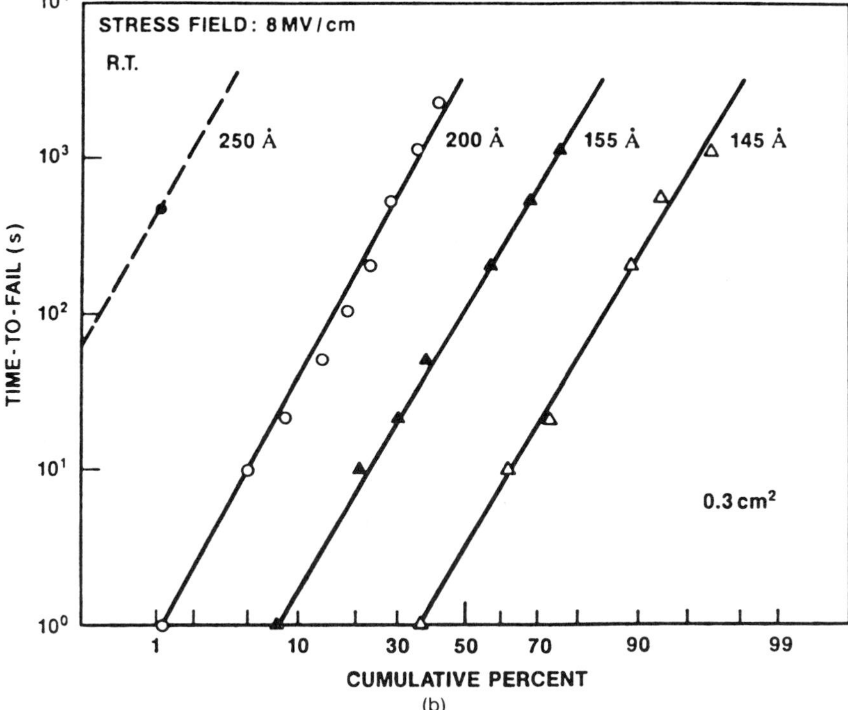

(b)

Figure 10-14. (a) Histogram of number of failures versus applied electric field in thin SiO_2 films. (From D. R. Wolters and J. J. van der Schoot, *Philips J. Res.* **40**, 115, 1985). (b) Time-dependent dielectric breakdown in SiO_2. (Courtesy of A. M-R Lin, AT&T Bell Laboratories.)

that a 100-Å difference in film thickness signifies a four-order-of-magnitude change in failure time.

4. Film defects cause breakdown to occur at smaller than intrinsic fields and in correspondingly shorter times.

As an example illustrating the use of Eq. 10-32 assume that the lifetime of SiO_2 films is 100 h during accelerated testing at 125 °C and 9 V. What lifetime can be expected at 25 °C and 8 V? Clearly, TTF = $100 \exp 0.33/k(1/T_2 - 1/T_1) \exp - 2.47(V_2 - V_1)$, where $T_2 = 298$, $T_1 = 398$, $V_2 = 8$, $V_1 = 9$, and $k = 8.63 \times 10^{-5}$ eV/K. Substitution yields a value for TTF = 29,700 h.

10.5. Superconductivity in Thin Films

10.5.1. Overview

The discovery in 1986–1987 that superconductivity is exhibited by oxide materials at temperatures above the boiling point of liquid nitrogen ignited intense worldwide research devoted to understanding and exploiting the phenomenon. For a perspective of superconducting effects in these new materials and prospects for thin-film uses, it is worthwhile to view the subject against the 75-year backdrop of prior activity. This pre-1988 ''classical'' experience with superconductivity will therefore be surveyed first in this chapter (Ref. 22); in Chapter 14 high-temperature superconductivity will be discussed.

Superconductivity was discovered by Kamerlingh Onnes, who, in 1911, found that the electrical resistance of Hg vanished below 4.15 K. Actually it was estimated from the time decay of (nearly) persistent supercurrents in a toroid that the resistivity of the superconducting state does not exceed $\sim 10^{-20}$ Ω-cm, some 14 orders of magnitude below that for Cu. Some basic attributes possessed by superconductors have been experimentally verified and theoretically addressed over the years. These are briefly enumerated here.

1. Occurrence of Superconductivity. The phenomenon of superconductivity has been observed to occur in at least 26 elements and in hundreds and perhaps thousands of metallic alloys and compounds. It is favored by a large atomic volume or lattice parameter, and when there are between two and eight valence electrons per atom.

2. Critical Temperature and Magnetic Field. The superconducting state only exists in a specific range of temperature (T) and magnetic field strength

Table 10-3. Values of T_c and H_s for Superconducting Materials

Element	T_c (K)	H_s (Oe)	Alloy or Compound	T_c (K)	H_s (Oe)**
Al	1.19	98.8	V_3Ga	14.8	25×10^4
In	3.41	285	V_3Si	16.9	24×10^4
La(β)	5.9	1000	Nb_3Sn	18.3	28×10^4
Nb	9.2	2,000*, 3,000**	Nb_3Ga	20.2	34×10^4
Pb	7.18	800	Nb_3Ge	22.5	38×10^4
Re	1.70	200	$PbMo_5S_8$	14.4	60×10^4
Sn	3.72	308	NbN	15.7	15×10^4
Ta	4.48	825	$YBa_2Cu_3O_7$	93	—
Tc	8.22	—	BiSrCaCuO	107	—
Th	1.37	161	TlBaCaCuO	120	—
Tl	2.39	170			
V	5.13	1290*, 7000**			

*$H_s(l)$
**$H_s(u)$
From Refs. 22 and 23.

(H). Critical values of these quantities are experimentally found to be closely described by

$$H_s = H_0(1 - (T/T_c)^2), \tag{10-33}$$

where H_s is the critical field, H_0 is the maximum field at $T = 0$ K, and T_c is the highest temperature at which superconductivity is observed. On an H vs. T plot the division between superconducting and normal conduction regimes is defined by Eq. 10-33. A temperature spread of only $\sim 10^{-3}$ K (10^{-4} K in pure metals, 10^{-2} K in alloys) about the value of T_c characterizes the sharpness of the transition between the two states. Superconductivity can be extinguished by exposure to a field greater than H_s or by passing a supercurrent that induces a magnetic field in excess of H_s. Values of T_c and H_s are listed in Table 10-3 for a number of superconducting materials.

3. Meissner Effect. One of the remarkable features of the superconducting state is the Meissner effect. It is characterized by the exclusion of magnetic flux and, hence, electrical currents from the bulk of the superconductor. The exclusion is not total, however, and both flux and current are confined to a surface layer known as the penetration depth λ_s. The London theory of superconductivity indicates that

$$\lambda_s(T) = \lambda_s(0)\left[1 - (T/T_c)^4\right]^{-1/2}, \tag{10-34}$$

where $\lambda_s(0)$ is the penetration depth at 0 K. Typically, λ_s is 500–1000 Å. The Meissner effect means that if a superconductor is approached by an H field, screening currents are set up on its surface. This screening current establishes an equal and opposite H field so that the net field vanishes in the superconductor interior. The now common displays of permanent magnets levitated over chilled high-T_c superconductors is visual evidence of the Meissner effect.

4. Type I and Type II Superconductors. There are two types of superconductors: type I (or soft) and type II (or hard). With the exception of Nb and V the elements are type I superconductors. In such materials the superconducting transition is abrupt, and flux penetrates only for fields larger than H_s. In type II superconductors (exemplified by Nb, V, alloys (e.g., Mo–Re, Nb–Ti), and A-15 compounds (e.g., Nb_3Sn, Nb_3Ge)), there are two critical fields $H_s(l)$ and $H_s(u)$, the lower and upper values. If the applied field is below $H_s(l)$, type II behavior is the same as that displayed by type I superconductors. For fields above $H_s(l)$ but below $H_s(u)$, there is a mixed superconducting state, whereas for $H > H_s(u)$, normal conductivity is observed. Importantly, type II superconductors can survive in the mixed state up to extremely high H values (e.g., in excess of 10^5 gauss). This property has earmarked their use in commercial superconducting magnets. In the mixed state, just above $H_s(l)$, flux starts to penetrate the material in microscopic tubular filaments (~ 1000 Å in diameter), known as fluxoids or vortices, that lie parallel to the field direction. The core of the fluxoid is normal while the sheath is superconducting; the circulating supercurrent of the latter establishes the field that keeps the core normal. Fluxoids, which are usually arranged in a lattice array, grow in size as the field is increased with progressively more flux penetration. Above $H_s(u)$, flux penetrates everywhere. The current flow is not entirely lossless in the mixed state, however, because a small amount of power is dissipated by viscous fluxoid motion. Fluxoid pinning due to introduction of alloying elements or defects is a practical way to minimize this energy loss.

5. The BCS Theory. The theory by Bardeen, Cooper, and Schrieffer (BCS) (Ref. 24) in 1957 provided the basis for understanding superconductivity at a microscopic level, superceding previous phenomenological approaches. Central to the BCS theory is the complex coupling between a pair of electrons of opposite spin and momentum through an interaction with lattice phonons. The electrons that normally repel each other develop a mutual attraction, forming Cooper pairs. A measure of the average maximum length at which the phonon coupled attraction can occur is known as the coherence length ξ. Schrieffer described the theoretical issue as ''how to choreograph a dance for more than a million, million, million couples'' so that they condense into a single state that moves in step or flows like a frictionless fluid (Ref. 24). Since the electron coupling is weak, the energy difference between normal and superconducting states is small with the latter lying a distance 2Δ below the former. Thus, a forbidden energy gap of width

$$2\Delta = 3.5kT_c \tag{10-35}$$

appears in the density of states centered about the Fermi energy at 0 K. This predicted relationship has been verified in many superconductors by tunneling measurements, which are described in the next section.

When the temperature is raised, the amplitude and frequency of lattice atomic motion increase, interfering with the propagation of phonons between correlated Cooper pairs. The attraction between electrons is diminished and 2Δ decreases. At $T = T_c$, $\Delta = 0$. Any perturbation in structure or composition extending over the coherence length can alter T_c or 2Δ, placing a practical limit on useful superconducting behavior.

10.5.2. Superconductivity in Thin Films; Tunneling

Thin films have traditionally played a critical role in testing theories of superconductivity and in establishing new effects. Superconductivity apparently persists to film thicknesses of ~ 10 Å. Lower limits are difficult to establish because films of such thickness are generally discontinuous. The dependence of T_c on deposition conditions and film thickness has been studied for a long time, and interesting, though not easily predictable or explainable, effects have been reported. When either λ_s or ξ becomes comparable to the thin film thickness, deviation from bulk superconducting properties may be expected. For example, enhanced superconductivity has been reported in vapor-quenched, amorphous Bi and Be films where T_c values of 6 and 8 K were obtained, even though these metals are not superconducting in bulk. Higher T_c values with decreasing film thickness have been observed by several investigators. The size of these effects ranges from fractions to several degrees K and depends on the magnitude and sign of the film stress, impurities, lattice imperfections, and grain size in generally inexplicable ways. A link between T_c and the fundamental nature of the material is suggested by the BCS formula

$$T_c = \frac{1.14h\nu}{k} \exp - \frac{1}{N(E_F)U} \tag{10-36}$$

(for $N(E_F)U \ll 1$). The quantity $N(E_F)$ is the density of states at the Fermi level, U is the magnitude of the attractive electron–lattice interaction, and ν corresponds to the lattice (Debye) frequency. Normally $N(E_F)U$ is weakly sensitive to lattice dimensions and has a value between 0.1 to 0.5. Furthermore if $N(E_F)$, U, or ν increase, so does T_c. However, connections between these quantities, on the one hand, and film composition and structure, on the other, are uncertain at best.

The most extensive experimentation in thin films has involved tunneling phenomena. Unlike the tunneling between normal metals considered earlier (Section 10.3.1), a superconducting tunnel junction consists of two metal films, one or both being a superconductor, separated by an ultrathin oxide or insulator film. Tunneling currents generally flow when electrons emerge from one metal to occupy allowable empty electron states of the same energy in the opposite metal. Through application of voltage bias, relative shifts of the entire electron distribution of both metals occur, either permitting or disallowing tunneling transitions. Thus, if electrons at the Fermi level of a normal metal lie opposite the forbidden energy gap at the Fermi level of the superconductor, no tunnel current flows. Translation of band states by a voltage Δ/q, or half the energy gap, causes occupied energy levels of the former to line up with unoccupied levels of the latter resulting in current flow. If both electrodes are the same superconductor, a voltage corresponding to the whole energy gap must be applied before tunnel current flows. Current–voltage characteristics corresponding to these two cases are shown in Fig. 10-15. A more complicated behavior is exhibited when two different superconductors with energy gaps

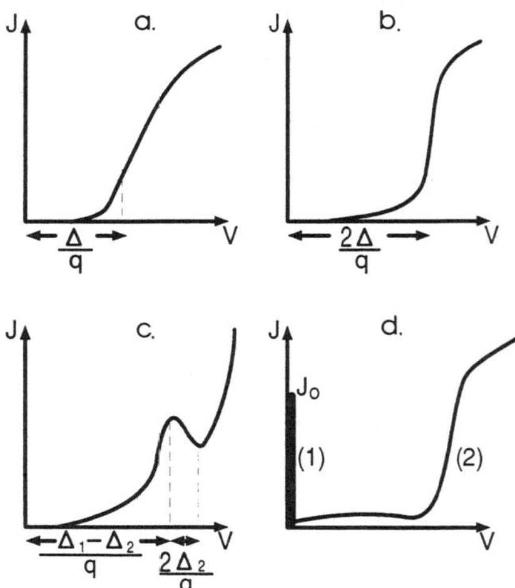

Figure 10-15. Current–voltage characteristics of tunnel junctions: tunnel junctions (a) one metal normal—one metal superconducting. (b) both metals identical superconductors. (c) both metals superconducting but with different energy gaps. (d) Josephson tunneling branch (1) and normal superconducting tunneling branch (2). J_0 is the critical junction current density.

$2\Delta_1$ and $2\Delta_2$ are paired. By yielding precise values of 2Δ, such measurements have provided direct experimental verification of the BCS theory.

One of the very important advances in superconductivity was the remarkable discovery by Josephson (Ref. 25) that supercurrents can tunnel through a junction. Thus tunneling of Cooper pairs and not only electrons is possible. Two superconducting electrodes sandwiching an ultrathin insulator ~ 50 Å thick are required. The current–voltage characteristic has two branches (Fig. 10-15d). The normal tunneling branch is similar to Fig. 10-15c but with a reduced negative resistance feature. The Josephson tunneling current branch consists of a current spike; no voltage develops across the superconducting junction in this case. Because the Josephson current is extremely sensitive to H fields, the junction can be easily switched from one branch to the other.

Josephson devices known as SQUIDs (superconducting quantum interference devices) capitalize on these effects to detect very small H fields or to switch currents at ultrahigh speed in computer logic circuits. These applications will be described in more detail in Section 14.8.3.

10.6. Introduction to Ferromagnetism

The remainder of this chapter is devoted to some of the ferromagnetic properties of thin films (Refs. 26, 27). We start with the idea that magnetic phenomena have quantum mechanical origins stemming from the quantized angular momentum of orbiting and spinning atomic electrons. These circulating charges effectively establish the equivalent of microscopic bar magnets or magnetic moments. When neighboring moments due to spin spontaneously and cooperatively order in parallel alignment over macroscopic dimensions in a material to yield a large moment of magnetization (M), then we speak of ferromagnetism. The quantity M is clearly a vector with a magnitude equal to the vector sum of magnetic moments per unit volume. In an external magnetic field (H) the interaction with M yields a field energy density (E_H) given by

$$E_H = -\mathbf{H} \cdot \mathbf{M}. \tag{10-37}$$

However, no external field need be applied to induce the ferromagnetic state. The phenomenon of ferromagnetism has a number of characteristics and properties worth noting at the outset.

1. Elements (e.g., Fe, Ni, Co), alloys (e.g., Fe–Ni, Co–Ni), oxide insulators (e.g., nickel–zinc ferrite, strontium ferrite) and ionic compounds (e.g., $CrBr_3$, EuS, EuI_2) all exhibit ferromagnetism. Not only are all crystal

structures and bonding mechanisms represented, but amorphous ferromagnets have also been synthesized (e.g., melt-quenched $Fe_{80}B_{20}$ ribbons and vapor-deposited Co–Gd films).

2. Quantum mechanical exchange interactions cause the parallel spin alignments that result in ferromagnetism. It requires an increase in system energy to disorient spin pairings and cause deviations from the parallel alignment direction. This energy, known as the exchange energy (E_{ex}), is given by

$$E_{ex} = A_x(\nabla\phi)^2 \tag{10-38}$$

and is a measure of the "stiffness" of M or how strongly neighboring spins are coupled. The exchange constant A_x is a property of the material and equal to $\sim 10^{-6}$ ergs/cm in Ni–Fe. Avoidance of sharp gradients in ϕ, the angle between **M** and the easy axis of magnetization, leads to small values of E_{ex}.

3. Absorbed thermal energy serves to randomize the orientation of the spin moments μ_s. At the Curie temperature (T_C) the collective alignment collapses, and the ferromagnetism is destroyed. By equating the thermal energy absorbed to the internal field energy ($\mu_s H_i$), i.e., $kT_C = \mu_s H_i$, values of H_i can be estimated. The internal field H_i permeating the matrix is established by exchange interactions. Typically, H_i is predicted to be in excess of 10^6 Oe, an extremely high field.

4. Magnetic anisotropy phenomena play a dominant role in determining the magnetic properties of ferromagnetic films. By anisotropy we mean the tendency of **M** to lie along certain directions in a material rather than be isotropically distributed. In single crystals, **M** prefers to lie in the so-called easy direction, say [100] in Fe and [111] in Ni. To turn M into other orientations, or harder directions, requires energy (i.e., magnetocrystalline anisotropy energy (E_K)). Consider now a fine-grained polycrystalline ferromagnetic film of Permalloy ($\sim$ 80 Ni–20 Fe). Surprisingly, it also exhibits anisotropy with M lying in the film plane. In such a case E_K is a function of the orientation of M with respect to film coordinates. For uniaxial anisotropy

$$E_K = K_u \sin^2\theta, \tag{10-39}$$

where K_u is a constant with units of energy/volume, and θ is the angle between the in-plane saturation magnetization and the easy axis. The source of the anisotropy is not due to crystallographic geometry but rather to the anisotropy arising from shape effects (i.e., shape anisotropy).

When ferromagnetic bodies are magnetized, magnetic poles are created on the surface. These poles establish a demagnetizing field (H_d) proportional and antiparallel to M, i.e., $H_d = -NM$, where N is known as the demagnetizing factor and depends on the shape of the body. For a thin film, $N \approx 4\pi$ in the direction normal to the film plane. Therefore, $H_d = -4\pi M$. In evaporated Permalloy films H_d can be as large as $\sim 10^4$ Oe. However, in the film plane H_d is much smaller so that M prefers to lie in this plane. There are other magnetic films of great technological importance—garnets for magnetic bubble devices (Section 10.8.4) and Co–Cr for perpendicular magnetic recording applications (Section 14.4.3)—where M is *perpendicular* to the film plane. Associated with H_d is magnetostatic energy (E_M) of amount

$$E_M = (1/2)H_dM \tag{10-40}$$

per unit volume. The 1/2 arises because self-energy is involved, i.e., H_d is created from the distribution of M in the film. In the hard direction the energy density is therefore

$$E_M = 2\pi M^2. \tag{10-41}$$

The origins of anisotropy are complex and apparently involve directional ordering of magnetic atom pairs, e.g., Fe–Fe. Film anisotropy is affected by film deposition method and variables, impingement angle of the incident vapor flux, applied magnetic fields during deposition, composition, internal stress,

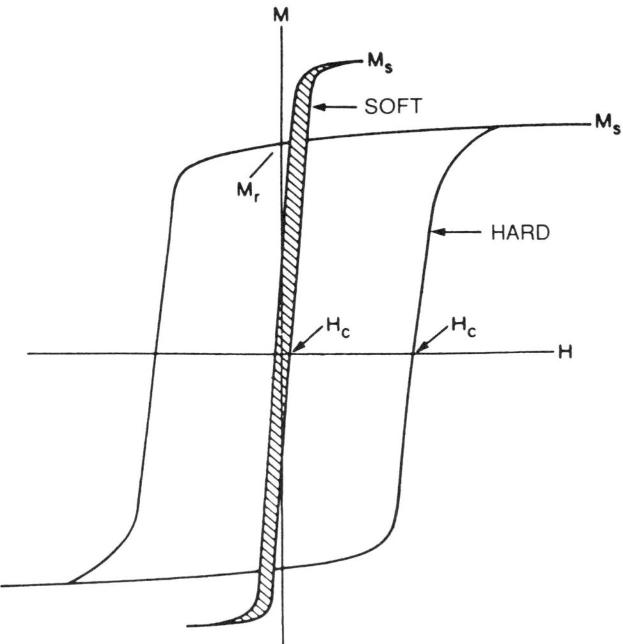

Figure 10-16. Schematic hysteresis loops for soft and hard magnetic materials. For soft magnets $H_c \leq 0.05$ Oe. For hard magnets $H_c \geq 300$ Oe. (From Ref. 28).

and columnar grain morphology, in not readily understood ways. Even amorphous ferromagnetic films exhibit magnetic anisotropy.

5. Not all ferromagnetic materials are magnets or have the ability to attract other ferromagnetic objects. The reason is that the matrix decomposes into an array of domains, each of which has a constant M but is differently oriented. Hence, $\sum M = 0$ over macroscopic dimensions. Domains facilitate magnetic flux closure and reduce stray external magnetic fields—effects that minimize

Table 10-4. Properties of Soft and Hard Magnetic Thin Films

Soft Magnetic Materials	$4\pi M_s$ (kG)	H_c (Oe)	H_k (Oe)	Application
Permalloy	10	0.5	5	Computer memory, magnetoresistance detectors, recording heads
CoZr (amorphous)	14	< 0.5	2-5	—
$Fe_{80}B_{20}$ (amorphous)	15	0.04	7	—
$Fe_{72}Si_{28}$ (amorphous)	12	0.2	4	—
$Fe_{72}Si_{18}C_{10}$ (amorphous)	16	0.2	7	—
$Y_3Fe_5O_{12}$	1	~ 0	1000	Magnetic bubble memory devices
Hard Magnetic Materials				
Co-Re	6-9	700		Longitudinal magnetic recording media
Co-Pt	10-18	1100-1800		"
Co-Ni	10-15	1000-1300		"
Co-Ni-W	5.5	650		"
Fe_3O_4	5	300		"
γFe_2O_3(Co)	3	700		"
γFe_2O_3(Os)	3	2100		"
Cr-Co*	4-7	500-2000 H_c (perpendicular)		Perpendicular magnetic recording media
TbFe		2000-10000		Magneto-optic recording media
GdCo	~ 10	1000-2000		"
GdTbFe		1000-3000		"

*Values for M_s and H_c depend strongly on composition and method of deposition. H_c (parallel) values are typically $0.5H_c$ (perpendicular).

Note: 1 Oe = 80 A/m; 1 G = 10^{-4} T; H_k (anisotropy field) $= 2K_u/M_s$

From Refs. 28-30.

E_M. In bulk materials, domains are frequently smaller than the grain size, whereas the reverse is true in films.

The response of a ferromagnet to an externally applied magnetic field is its most important engineering characteristic. Initially, M increases with H and eventually levels off at the saturation magnetization value M_s. The application of H causes favorably oriented domains to grow at the expense of unfavorably oriented ones by domain boundary migration. At high enough fields, M even rotates. Further cycling of H in both positive and negative senses yields the well-known hysteresis loop. Depending on its size and shape ferromagnets are subdivided into two types—hard and soft, as indicated schematically in Fig. 10-16. The distinction is based on the magnitude of the coercive force H_c or field required to reduce M to zero. Hard magnetic materials with large H_c are hard to magnetize and hard to demagnetize. That is why they make good permanent magnets and are used for magnetic recording media. Soft magnetic materials, on the other hand, both magnetize and switch magnetization directions easily (small H_c). These are just the properties required for computer memory or recording head applications. In Table 10-4 the properties of both soft and hard magnetic thin-film materials are listed. How these ferromagnetic properties are manifested in applications is explored in the remainder of this chapter as well as in Chapter 14.

10.7. Magnetic Film Size Effects — M_S vs. Thickness and Temperature

10.7.1. Theory

Magnetic property size effects are expected simply because the electron spin in an atom on the surface of a uniformly magnetized ferromagnetic film is less tightly constrained than spins on interior atoms. Fewer exchange coupling bonds on the surface than in the interior is the reason. Therefore, the question has been raised of how thin films can be and still retain ferromagnetic properties. At least four decades of both theoretical and experimental research have been conducted on the many aspects of this fundamental issue and related ones. The two theoretical approaches—spin wave and molecular field—both predict that a two-dimensional network of atoms of ferromagnetic elements should not be ferromagnetic but rather paramagnetic at absolute zero. At low temperatures a ferromagnet has very nearly its maximum magnetization. The deviation from complete saturation (ΔM_s) is due to waves of reversed spin propagating through the material. By summing the spin waves according to the

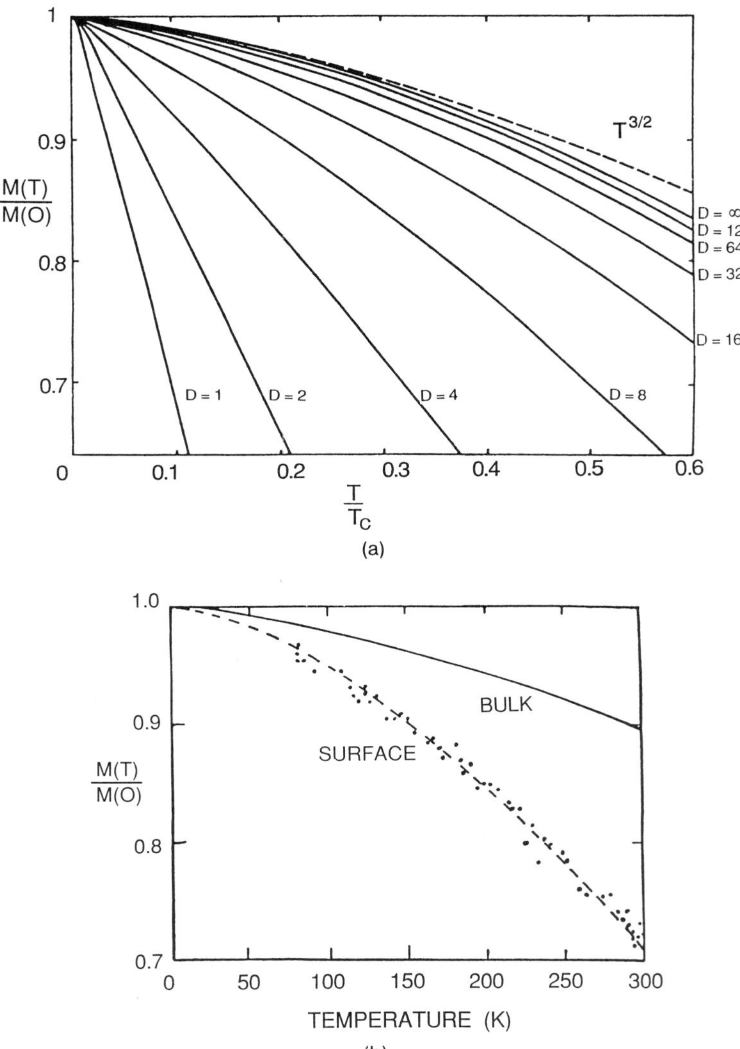

Figure 10-17. (a) Calculated temperature dependence of normalized saturation magnetization for different numbers of film layers (D). (From Ref. 31). (b) Temperature dependence of relative bulk and surface magnetization in the ferromagnetic glass $Fe_{40}Ni_{40}B_{20}$ (From Ref. 33).

rules of quantum statistical mechanics, we obtain the magnitude of the deviation at any temperature ($\Delta M_s(T)$) relative to absolute zero $M_s(0)$. In bulk materials it is generally accepted that

$$\Delta M_s(T)/M_s(0) = BT^{3/2}, \tag{10-42}$$

where B is the spin wave parameter. Its value at the surface has been calculated to be twice that in the bulk. In thin films, theoretical treatments of magnetic size effects are a subject of controversy. Early calculations show a relative decrease in M vs. T for films of varying thickness as indicated in Fig. 10-17. For films thinner than four atomic layers M_s varies linearly over a wide temperature range.

The molecular field approach replaces the exchange interaction between neighboring spins around a particular atom by an effective molecular field. A statistical accounting of the number of interactions between an atom in the jth layer of a film with other atoms in the same as well as $j - 1$ and $j + 1$ layers is the approach taken. Such calculations typically reveal that M begins to decrease below the value in bulk when the film thickness is less than some number of lattice spacings (e.g., 10), corresponding to a film thickness of perhaps 30 Å.

In recent years, quantum calculations of ferromagnetic films have achieved a high level of sophistication (Ref. 32). Spin densities in the ground state of Fe and Ni films consisting of a few atomic layers have, interestingly, been predicted to lead to an *enhancement* of the magnetic moment per atom in the outermost layer (e.g., by 20% in (001)Ni and 34% in (001)Fe compared to the bulk value found four layers away). Such surprising results rule out the existence of magnetically "dead" layers reported in the literature.

Importantly, none of the aforementioned theories explicitly takes into account such surface effects as lattice relaxation or distortion, surface reconstruction, and pseudomorphic growth at real surfaces and interfaces. Rather, perfectly planar surfaces are assumed.

10.7.2. Experiment

Many experimental methods have evolved to yield direct or indirect evidence of ferromagnetic order in thin films. They broadly fall into three categories, depending on whether the

1. spin polarization of electrons,
2. net magnetic moment of the sample, or
3. internal magnetic (hyperfine) field

is measured. The first relies on extracting electrons from the conduction bands of ferromagnetic solids by photoelectron emission and analyzing their energies by methods similar to those employed in Auger spectroscopy. Assuming no spin flips occur during emission, the number of majority and minority spins relative to the direction of magnetization of the surface can be determined. For example, the net surface magnetization of $Fe_{40}Ni_{40}B_{20}$, an amorphous ferromagnet ($T_C = 700$ K), was measured by detecting elastically backscattered spin polarized electrons (Ref. 33). The results are depicted in Fig. 10-17b for 90 eV electrons which are estimated to probe only the topmost one or two atomic layers (~ 2.5 Å). Method 2 relies on very sensitive magnetometers to directly yield macroscopic M vs. H behavior. Although relatively free from interpretation problems associated with indirect methods 1 and 3, the measurements are not surface selective. In the third method the hyperfine magnetic field H_{eff}, which is to a good approximation proportional to the local atomic moment, is measured. Nuclear physics techniques such as nuclear magnetic resonance and Mössbauer effect are used; only the latter will be discussed here at any length.

The Mössbauer effect is based on the spectroscopy of specific low-energy nuclear γ-rays that are emitted (without recoil) from excited radioactive atoms embedded in a source. These γ-rays are absorbed by similar nonradioactive ground state atoms contained within an absorber matrix. In the most famous Mössbauer transition, ^{57}Co nuclei emit 14.4-keV γ-rays and decay to the ^{57}Fe ground state. An absorber containing ^{57}Fe, an isotope present in natural Fe with an atomic abundance of 2.2%, can absorb the γ-ray if its nuclear levels are very precisely tuned to this exact energy. Otherwise there is no absorption, and the γ-ray will simply pass through the absorber and be counted by a γ-ray detector. This is usually the case because differences in the local electromagnetic environment of Fe^{57} in both the source and absorber alter the 14.4-keV level slightly, destroying the resonance. Fortunately very small, easily produced Doppler effect energy shifts, caused by relative source–absorber velocities of only $\sim \pm 1$ cm/sec, can increase or decrease the energy sufficiently to restore the resonance. Mössbauer spectra thus reveal relative energy differences in γ-ray transitions of ^{57}Fe as they are affected by atomic surroundings. In a ferromagnetic absorber the ^{57}Fe nucleus is immersed in the internal magnetic field that splits the nuclear levels, an effect that is the counterpart to Zeeman splitting of atomic electron levels. Six transitions can now be accessed by using an appropriate (unsplit) source.

Mössbauer spectra of Fe film absorbers of varying thickness grown epitaxially on Ag are shown in Fig. 10-18. The detection of γ-ray-induced conversion electrons plus the use of enriched ^{57}Fe layers provide the necessary sensitivity

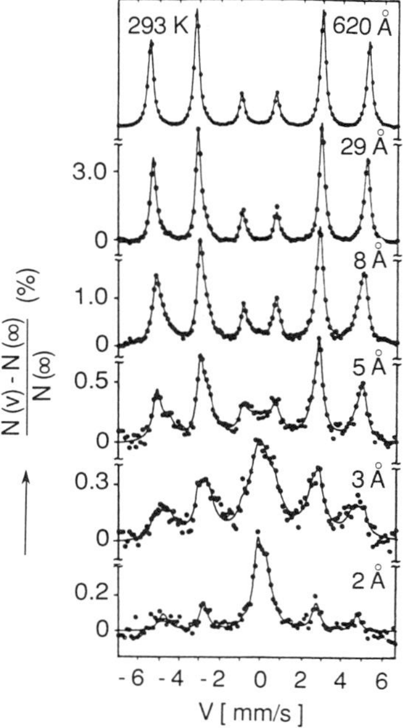

Figure 10-18. Mössbauer spectra of ultrathin epitaxial (110) Fe films on (111) Ag at room temperature. (From Ref. 32 with permission from Elsevier Science Publishers).

to probe ultrathin films. The six spectral lines characteristic of ferromagnetic behavior are clearly discernible at film thicknesses of 5 Å. In the 620-Å film the velocity or energy span between outer lines is essentially equivalent to the bulk H_{eff} of 330 kOe. Peak broadening plus a decreased span in the very thinnest films signify a distribution of somewhat reduced field strengths.

10.8. Magnetic Thin Films for Memory Applications

10.8.1. Introduction

Interest in magnetic films arose primarily because of their potential as computer memory elements. Although semiconductor memory is firmly established today, a quarter of a century ago small bulk ferromagnetic ferrite cores were employed for this purpose. Even earlier it was discovered that magnetic films deposited in the presence of a magnetic field exhibited square hysteresis loops. This meant that magnetic films could be used as a bistable element capable of switching from one state to another (e.g. from 0 to 1). Switching times were about 10^{-9} sec, a factor of 100 shorter than that for ferrite cores. The promise of higher-speed computer memory and new devices fueled a huge research and development effort focused primarily on Permalloy films. Despite initial enthusiasm it was found that careful control of magnetic properties produced formidable difficulties. The metallurgical nightmare of film impurities, imperfections, and stress was among the reasons that actual performance of these films fell short of originally anticipated standards (Ref. 34).

In the mid-1960s an entirely new concept for computer memory and data storage applications was introduced by investigators at Bell Laboratories (Ref. 35). It employed special magnetic thin films (e.g., garnets) possessing cylindrical domains known as bubbles. Unlike the switching of M in Permalloy films, information is processed through the generation, translation, and detection of these bubbles (Section 10.8.4). Domain behavior is critical to both approaches, and we therefore turn our attention to this subject now.

10.8.2. Domains in Thin Films — *M* in Film Plane

When Permalloy and other soft magnetic materials are vapor-deposited in a magnetic field of ~ 100 Oe, M lies in the film plane and in the field direction. Uniaxial anisotropy develops such that a 180° rotation of M occurs across the boundary or wall separating adjacent domains. Schematic illustrations of two ways the rotation can be accommodated are shown in Fig. 10-19. In the Bloch wall, which is common in bulk ferromagnets, the spins undergo a rotation about an axis parallel to the hard direction. There are two types of Bloch walls: one in which the magnetization in the wall center points upward, and one in which it points down. Therefore on the film surface there are free-magnetic poles just above the wall region. These establish stray fields that increase the magnetostatic energy of the system. With decreasing film thickness E_M increases since more free poles exist. In very thin films there is another type of wall with a much lower value of E_M. It is shown in Fig. 10-19b, and is known as the Néel wall. In Néel walls the direction of magnetization turns about an axis perpendicular to the film plane; there are no free poles in this case.

In both types of domain walls E_K is smallest when the change in M is abrupt—i.e., when the wall is as narrow as possible; but this serves to increase E_{ex}, which is minimized when spin pairings remain tightly aligned—i.e., when the wall is as wide as possible. A compromise is struck when the total magnetic energy, $E_T = E_K + E_{ex} + E_M$, is minimized with respect to number of wall spins. Typically, domain walls are 1000 Å wide and have an effective surface energy of several ergs/cm^2.

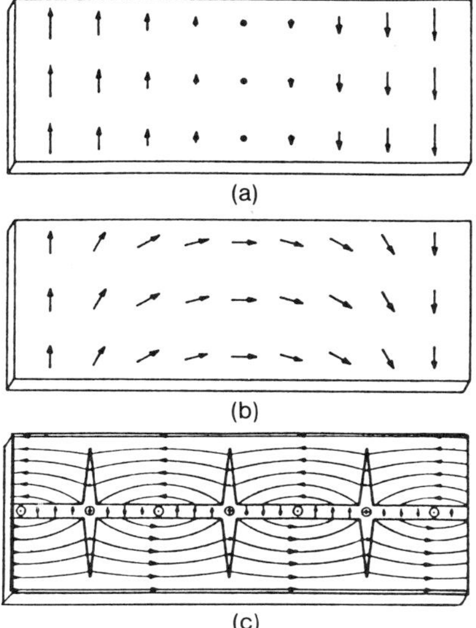

Figure 10-19. Schematic illustrations of magnetization directions at domain walls. (a) 180° Bloch wall; (b) 180° Néel wall; (c) cross-tie wall. (From Ref. 34).

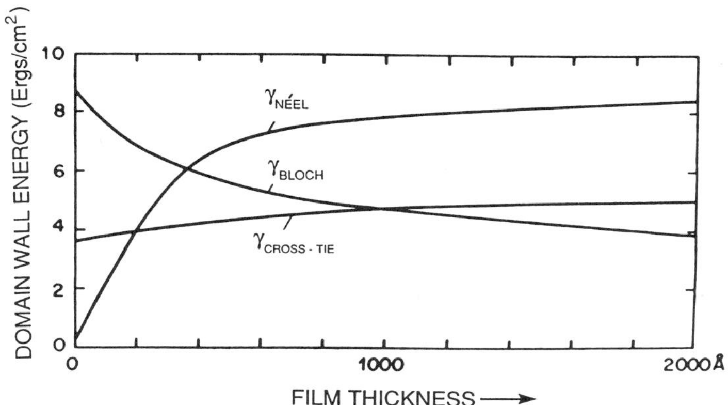

Figure 10-20. Calculated surface energy of a Bloch wall, a Néel wall, and a cross-tie wall as a function of film thickness. ($A_x = 10^{-6}$ erg/cm, $M_s = 800$ gauss, $K_u = 1000$ ergs/cm^3). (From Ref. 27 with permission from McGraw-Hill, Inc.)

The results of a calculation of the surface energy of Bloch, Néel, and cross-tie walls as a function of film thickness are shown in Fig. 10-20. Cross-tie walls are essentially variants of unipolar Néel walls and are shown in Fig. 10-19c. They consist of tapered Bloch wall lines jutting out in both directions from the Néel wall spine. This configuration promotes magnetic flux closure and possesses a lower overall energy than the simple Néel wall. The

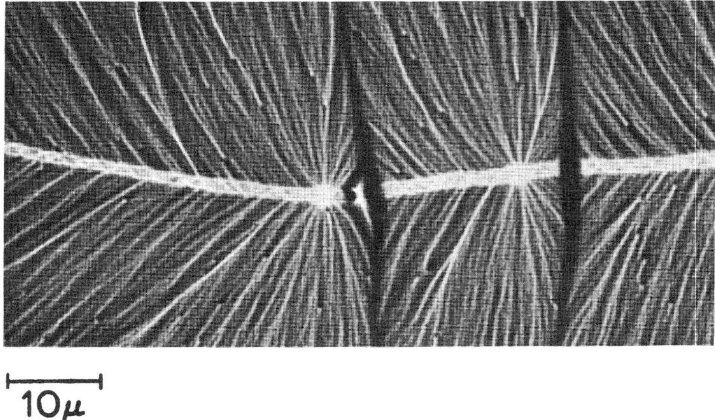

Figure 10-21. Lorentz micrograph of a cross-tie Néel wall transition in a 300-Å-thick Permalloy film. (From Ref. 27 with permission from McGraw-Hill, Inc.)

very thinnest films are predicted to contain Néel walls and films thicker than 1000 Å, Bloch walls. In between cross-tie walls are stable; they have been observed in Permalloy films within the predicted film thickness range, as shown in the Lorentz micrograph of Fig. 10-21. In this technique the film is mounted slightly above (or below) the focal plane of the objective lens in a transmission electron microscope. Electrons passing through the film are deflected owing to the Lorentz forces, and produce a kind of shadow image of the magnetization. The resolution of the technique is sufficient to detect magnetic ripple. The latter is a fine wrinkling substructure within domains where M undergoes slight periodic misalignments from the uniaxial direction. Ripple is apparently due to complex coupling effects between exchange and magnetostatic forces in neighboring crystallites and extends over tens to hundreds of angstroms.

10.8.3. Single-Domain Behavior

When films with uniaxial anisotropy are exposed to magnetic fields, the magnetization direction switches the way single domain particles do. Because of the importance of switching phenomena in memory applications, it is of interest to consider a simple model for this behavior (Ref. 36). A central assumption is that Eq. 10-39 holds. Equilibrium states of minimum energy exist for $\theta = 0$, the easy magnetization direction, as well as for $\theta = \pi$. But for $\theta = \pi/2$, $3\pi/2$, the hard directions, there are energy maxima. The total energy of the film in an applied magnetic field with components H_x and H_y in the easy and hard directions is

$$E_T = -M_s H_x \cos\theta - M_s H_y \sin\theta + K_u \sin^2\theta. \tag{10-43}$$

In *stable* equilibrium the magnetization angle is determined by the conditions that

$$\partial E_T/\partial\theta = 0, \tag{10-44}$$

which amounts to a vanishing torque, and

$$\partial^2 E_T/\partial\theta^2 > 0. \tag{10-45}$$

If $\partial^2 E_T/\partial\theta^2 < 0$, the equilibrium is *unstable*; transitions from unstable to stable states occur at critical fields for which $\partial^2 E_T/\partial\theta^2 = 0$. Successive differentiations yield the conditions for stable equilibrium, respectively,

$$\partial E_T/\partial\theta = M_s H_x \sin\theta - M_s H_y \cos\theta + 2K_u \sin\theta\cos\theta = 0, \tag{10-46a}$$

$$\partial^2 E_T/\partial\theta^2 = M_s H_x \cos\theta + M_s H_y \sin\theta + 2K_u(\cos^2\theta - \sin^2\theta) > 0. \tag{10-46b}$$

We are now in a position to calculate hysteresis loops. The two simplest cases occur when the fields are applied in the easy and hard directions.

1. Easy direction ($H_y = 0$). From Eq. 10-46 the two solutions are

$$M_s H_x = -2K_u \cos\theta; \qquad \sin\theta = 0.$$

For the first solution, $\partial^2 E_T/\partial\theta^2 = -2K_u \sin^2\theta$ is always negative, (except for $\theta = 0$ and $\theta = \pi$), and hence the equilibrium is not stable. However, the second solution yields $\theta = 0, \pi$; it is easily shown that $\theta = 0$ is stable for $H_x > -2K_u/M_s$, and $\theta = \pi$ is stable when $H_x < 2K_u/M_s$. These values of H_x are defined as the anisotropy field H_k. The magnetization in the easy direction is $M_x = M_s \cos\theta$, so that $M_x = \pm M_s$. In Fig. 10-22a the resulting square-loop hysteresis curve is drawn, where H_k is equivalent to H_c.

2. Hard direction ($H_x = 0$). In this case M_y reaches $+M_s$ when H_y exceeds $2K_u/M_s$, and has the value $-M_s$ when $H_y < -2K_u/M_s$. For applied H_y fields between these values, M_y varies linearly; i.e., $M_y = M_s^2 H_y/2K_u$. The multivalued character of the loop vanishes as shown in the hysteresis curve of Fig. 10-22b.

Loops observed experimentally in films differ from the calculated ones. Whereas square loops in the easy direction have been measured, coercive fields are usually much smaller than $2K_u/M_s$. The reason is that magnetiza-

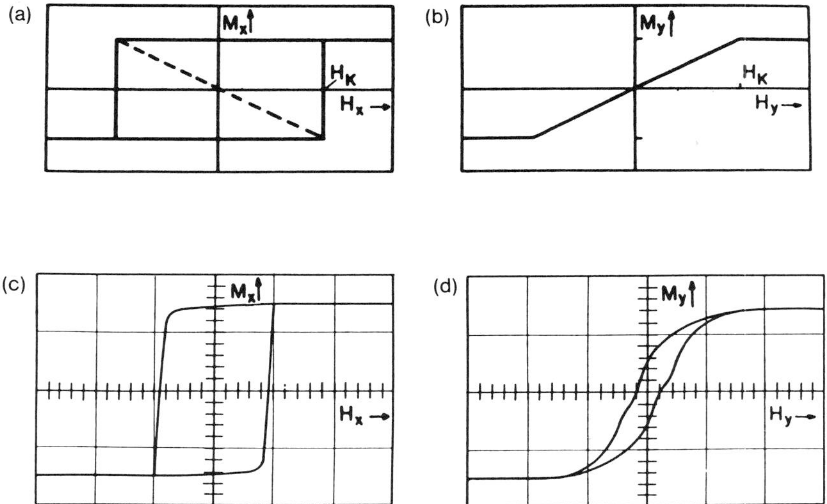

Figure 10-22. Theoretical hysteresis loops: (a) In the easy direction; (b) in the hard direction. Experimental hysteresis loops (Permalloy, 300 Å thick); (c) in the easy direction; (d) in the hard direction. (From Ref. 27 with permission from McGraw-Hill, Inc.).

tion reversal does not occur by uniform rotation, but by domain translation and rotation.

10.8.4. *M* Perpendicular to Film Plane

Imagine a bar magnet compressed axially to thin film dimensions. It would have numerous north poles on one surface opposed by the same number of south poles on the other surface giving rise to a large value of E_M. Unlike films that develop an in-plane magnetization in such a case, there are materials where M points *normal* to the film surface (Ref. 37). Examples are single-crystal films of magnetoplumbite ($PbO{-}6Fe_3O_4$), *ortho*-ferrites (RE FeO_3, with RE a rare earth element), and, most importantly, magnetic garnets (e.g., $Y_3Fe_5O_{12}$). Competition between E_M and E_K results in the formation of domains essentially possessing a uniaxial *perpendicular* anisotropy. The domain structure in such films is striped, displaying the fingerprint pattern of Fig. 10-23. Dark and light domains have oppositely pointed magnetization vectors. By viewing these transparent films through crossed polarizers, one notes an optical contrast between oppositely magnetized domains. The Faraday effect, a magneto-optical phenomenon, is responsible. It causes the plane of

Figure 10-23. Domain pattern in yttrium iron garnet film viewed in transmitted polarized light (200 ×).

transmitted polarized light to rotate, depending on the direction of M in individual domains.

What makes these materials remarkable is that for certain applied (strip-out) fields, the unfavored stripe domains will shrink into stable right-cylindrical domains called *bubbles*. An important property of the bubbles is their ability to be moved laterally through the film. This is the basis for their use in commercial bubble devices for the computer memory and data recording markets. In these applications a thin-film array of conductors and Permalloy films patterned in various shapes (chevrons, I and T bars, etc.) are deposited on top of the bubble film. Bubbles can then be generated, moved, switched, counted, and annihilated in a very animated way in response to the driving fields. The stability, size and speed of bubbles are the key design parameters affecting device reliability, memory capacity, and data rate, respectively.

1. Stability. Isolated bubble domain stability occurs when the ratio E_K/E_M (or, equivalently, $K_u/2\pi M_s^2$) is greater than unity. A high value of K_u, or in-plane anisotropy, encourages the 90° rotation into the perpendicular orientation. Unless the ratio is sufficiently large, in-plane drive fields can strip out bubbles into stripe domains.

2. Size. The bubble diameter is predicted to be about $8\sqrt{A_x K_u}/\pi M_s^2$ in size. It is left to the reader to show that $\sqrt{A_x K_u}$ is proportional to the domain wall energy; a small value of the latter fosters small bubbles. A large value of E_M (or M_s^2) also favors small bubbles by reducing the surface density of free poles through domain formation.

3. Speed. Bubble speed is determined by the product of the drive field minus the threshold field for movement (coercive field), and the bubble mobility μ_m (velocity per drive field gradient). The latter is proportional to $(1/\alpha)\sqrt{A_x/K_u}$, where α is the Gilbert damping or magnetic viscosity parameter.

Selection of optimum properties clearly involves trade-offs. Garnets possess the best combination of properties and have been most widely employed for magnetic bubble devices. Specifications currently call for: bubble size = 0.5–1.0 μm, $4\pi M_s$ = 500–1000 gauss, H_k = 1000 Oe, coercive field ~ 0, and $\mu_m > 300$ cm/sec-Oe. A considerable number and range of possible chemical substitutions are available to modify the basic garnet composition—$\{Y^{3+}\}_3[Fe^{3+}]_2(Fe^{3+})_3O_{12}^{-2}$. For example {Y, La to Lu, Bi, Ca, Pb}, [Fe, Mn, Sc, Ga, Al], and (Fe, B, Ga, Al, Ge) ions are used to facilitate growth of solid solution films with required device properties. Epitaxial garnet films are usually grown by liquid phase methods employing single-crystal $Gd_3Ga_5O_{12}$ substrates. Since bubble motion is adversely affected by film defects elimination of the latter is essential.

10.8.5. Memory Device Configurations (Ref. 27, 38)

The chapter closes by briefly conveying some notion of the two basic thin-film magnetic memory schemes that have been devised. In Fig. 10-24 a portion of a memory system based on in-plane magnetization film elements is shown. The latter are small isolated rectangles crisscrossed by three sets of conducting stripes. Word (W) drive lines run parallel to the easy axis which points either to the left (0) or to the right (1) when there is no applied magnetic field. A pulse field is then applied in the W line in the hard direction; M rotates toward this direction, and either a positive or negative signal is induced in the sense (S) line. If the W line signal exceeds H_K, then M rotates fully into the hard direction. When the W signal is reduced, the direction of M wavers. To avoid this instability, we apply a bit (B) field parallel to the easy axis to store the desired 0 or 1 state. In order to write, we must have the W pulse field large enough to drive all the bits into the hard direction. Similarly, the bit

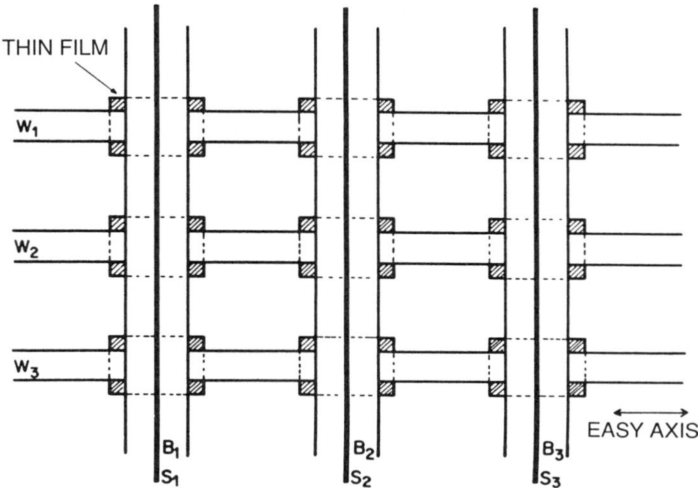

Figure 10-24. A memory plan with word lines W_1, W_2, and W_3, bit lines B_1, B_2, and B_3 and sense lines S_1, S_2, and S_3. (From Ref. 27 with permission from McGraw-Hill, Inc.).

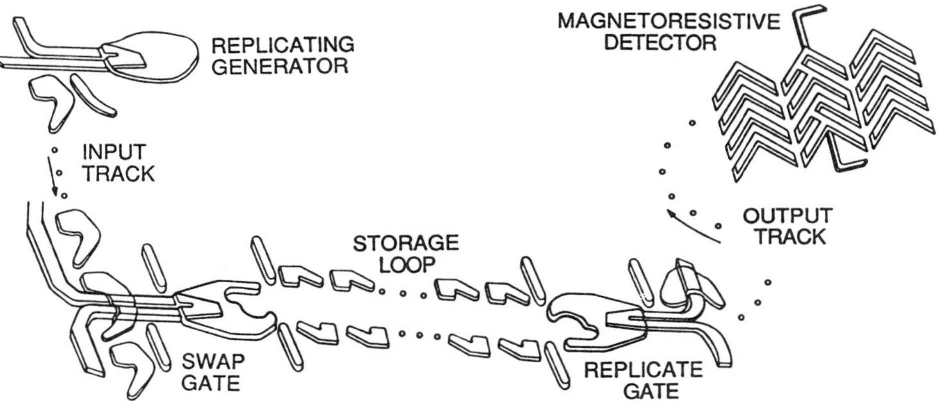

Figure 10-25. Schematic arrangement of a complete bubble memory device. (From Ref. 38).

pulses must be large enough to ensure complete rotation either to the right or left without disturbing bits on other W lines.

The schematic arrangement of a complete magnetic bubble memory device is shown in Fig. 10-25. Bubble creation occurs in the replicating generator where the magnetic field from a current pulse cuts a seed bubble in two. The seed bubble remains under a large Permalloy film patch for further bubble generation while the freshly nucleated bubble enters the input track. There the bubble moves within constraining Permalloy film elements under the influence of current-induced, rotating magnetic fields. The swap gate enables bubbles to be transferred out of a storage loop and another bubble (or no bubble) to be simultaneously transferred to replace it. (Only one of a block of storage loops is shown.) Replication is similar to replicate generation; the difference is that bubbles in the storage loop, rather than form a seed bubble, are replicated. Finally bubbles are detected by the Permalloy magnetoresistance detector. After coming off the output track the chevron stretcher expands the bubble into a long stripe to maximize the detector signal. As the stripe passes under the interconnected column of chevrons (the detector), it changes the resistance of the Permalloy and gives rise to the output signal.

Bubble memory devices with capacities in excess of a megabit are commercially available and offer advantages relative to disks and tapes. These include high storage capacities ($\sim 10^9$ bits/cm^2) with 0.5-μm bubbles, absence of mechanical wear, nonvolatile memory, and a wide-temperature-range read–write memory.

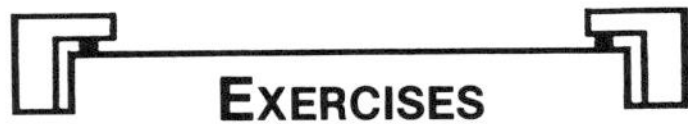

Exercises

1. During four-point probe resistance measurements it is desired to limit currents to less than 0.050 A to prevent overheating the probe tips. The typical digital voltmeter available for measurement of the potential drop has a range of 10 mV to 100 V. Which of the following 5000-Å-thick film materials have sheet resistance that are readily measurable by this method.

a. Cu; $\rho = 1.73 \times 10^{-6}$ Ω-cm?
b. Si; $\rho = 2$ Ω-cm?
c. ZrO_2; $\rho = 10^{14}$ Ω-cm?
d. $CoSi_2$; $\rho = 15 \times 10^{-6}$ Ω-cm?
e. TiN; $\rho = 100 \times 10^{-6}$ Ω-cm?

2. A thin-film window de-icer resistor meanders over a length of 5 m and is 1 mm wide. It is designed to deliver a total power of 5 W, employing a 12-V power source. For a 5000-Å-thick film, what sheet resistance is required?

3. Derive an expression for the ratio of the electrical resistance of a metal film stripe of length $2l$ evaporated from a surface source a distance h away to that of a uniformly thick-film stripe having the same number of atoms. Assume the film conductor is piecewise straight.

4. The resistivity of an evaporated metal thin film measured at low temperature is higher than the bulk resistivity value extrapolated to the same temperature. Reasons postulated for this behavior are (1) the size effect due to surface scattering, (2) a high concentration of quenched-in vacancies, and (3) generation of dislocation defects during cooling. Design a course of experimentation to resolve the cause.

5. An insulating film is characterized by $\Phi_B = 1$ eV. How much does the Schottky emission current change at 100 °C for

a. a 1% change in temperature?
b. a 1% change in electric field?
c. a 1% change in film contact area?

6. Conductor stripes deposited over insulators create resistor–capacitor structures whose *RC* time constants serve to limit the speed of circuit response.

a. Derive a general expression for *RC* in terms of the length, width, thickness, and resistivity of the conductor, and the corresponding area, thickness, and dielectric constant (ε_i) of the insulator.
b. For 1-μm-wide, 1-μm-thick, 1-cm-long Al stripes ($\rho = 2.7 \times 10^{-6}$ Ω-cm) across a 0.5-μm-thick SiO_2 film ($\varepsilon_i = 3.8$), what is *RC* (sec/cm)? [Note: The permittivity of free space is 8.85×10^{-14} F/cm.]
c. What is *RC* (sec/cm) if $TaSi_2$ ($\rho = 55 \times 10^{-6}$ Ω-cm) and poly-Si ($\rho = 1000 \times 10^{-6}$ Ω-cm) are substituted for Al?

7. Two metal contact pads 1 cm apart on a bare NaCl substrate are used to monitor electrical resistance changes in a metal film that is deposited between them. Schematically plot the expected resistance–time behavior from the start of evaporation until the film is completely continuous. What are the operative conduction mechanisms during the various stages of film formation? Contrast the resistance response for high and low substrate temperatures. Also contrast the resistance response for island and layer growth mechanisms.

8. Using the approach of Thomson, calculate the mean-free electron path (λ_f) in a thin metal film if specular scattering occurs at the air interface while the scattering at the substrate interface is diffuse.

9. Consider an $M/SiO_2/N{-}Si$ capacitor structure (assume no oxide or interfacial charge).

a. Sketch band diagram representations for accumulation, depletion, and inversion regimes.
b. Sketch the $C-V$ characteristic.
c. Sketch a series of $C-V$ characteristics as a function of SiO_2 film thickness.
d. Qualitatively indicate how the $C-V$ characteristics change as a function of Si doping level.

10. A bulk plate of SiO_2 and a 3000-Å film of SiO_2 are electrically contacted on both surfaces by Al. Assume the bulk resistivity of SiO_2 is 10^{15} Ω-cm and that an electric field of 10^5 V/cm is applied in both cases at 300 K.

a. What current density flows through the bulk plate?
b. What current density flows in the film if Schottky emission controls conduction?

11. A 1000-Å SiO_2 film underwent dielectric breakdown when an electric field of 1.2×10^7 V/cm was applied. It was observed that a 1-μm^2 damage channel through the SiO_2 shorted the Au electrodes and that failure occurred in 10^{-4} sec. If current flow is controlled by Schottky emission, estimate the maximum temperature rise in the film if its heat capacity is 0.24 cal/g °C.

12. a. Derive a general expression for the specific contact resistance if conduction occurs via tunneling.
b. Do the same for Poole-Frenkel conduction.
c. Estimate a value for R_c given the J-$\mathscr{E}$ characteristics shown in Fig. 10-10a.

13. A sample of a 1-μm-thick (magnetic)film of $Y_3Fe_5O_{12}$ on a nonmagnetic $Gd_3Ga_5O_{12}$ substrate disk (2-cm diameter, 0.75-mm thick) fell from a microscope stage. Attempts to lift the film–substrate with a common bar magnet proved unsuccessful. Why? When suspended from a slender thread, however, the film–substrate is attracted to the magnet. Why?

14. a. At zero magnetic field the total energy (E_T) of a magnetic film is increased when the number of domains is increased. Why? At the same time E_M decreases. Why? Schematically sketch these contributions to E_T and graphically indicate the optimum number of domains.
b. The film in part (a) is magnetized, and cylindrical bubble domains interact with the applied field. Assume the only contributions to E_T are domain wall energy (E_W), and E_M. Making appropriate assump-

tions, sketch the dependence on bubble radius of the individual energy contributions and graphically indicate the equilibrium domain radius.

15. a. Use Fig 10-17 to construct a plot of the magnetization as a function of film thickness at $T/T_C = 0.3$.

b. The difference in velocity between the outermost Mössbauer lines is proportional to the magnetization of Fe in Fig. 10-18. By measuring this velocity span, plot the variation of film magnetization as a function of film thickness. Compare results of (a) and (b).

16. For the single-domain film show that

$$H_x^{2/3} + H_y^{2/3} = (2K_u/M_s)^{2/3}.$$

This equation describes an astroid. To see what it looks like, make a plot in H_x, H_y coordinates.

References

1. S. M. Sze, *Physics of Semiconductor Devices*, 2nd Ed., J. Wiley and Sons, New York (1981).
2. L. J. Van der Pauw, *Philips Res. Repts.* **13**, 1 (1958).
3. J. J. Thomson, *Proc. Cambridge Phil. Soc.* **11**, 120 (1901).
4. K. Fuchs, *Proc. Cambridge Phil. Soc.* **34**, 100 (1938).
5. E. H. Sondheimer, *Advan. Phys.* **1**, 1 (1951).
6.* L. I. Maissel, in *Handbook of Thin Film Technology*, eds. L. I. Maissel and R. Glang, McGraw-Hill, New York (1970), p. 13-1.
7. M. A. Angadi, *J. Mat. Sci.* **20**, 761 (1985).
8. A. F. Mayadas and M. Shatzkes, *Phys. Rev.*, *B***1**, 1382 (1970).
9. J. R. Sambles, *Thin Solid Films* **106**, 321 (1983).
10. J. C. Hensel, R. T. Tung, J. M. Poate, and F. C. Unterwald, *Phys. Rev. Lett.* **54**, 1840 (1985).
11. L. I. Maissel, in *Handbook of Thin Film Technology*, eds. L. I. Maissel and R. Glang, McGraw-Hill, New York (1970), p. 18-3.
12. B. Abeles, in *Applied Solid State Science*, Vol. 6, ed. R. Wolfe, Academic Press, New York (1976).
13. S. R. Pollack and J. A. Seitchik, in *Applied Solid State Science*, Vol. 1, ed. by R. Wolfe, Academic Press, New York (1969).
14.* A. K. Jonscher and R. M. Hill, in *Physics of Thin Films*, Vol. 8, eds.

*Recommended texts or reviews.

M. H. Francombe and R. W. Hoffman, Academic Press, New York (1975).

15.* J. G. Simmons, in *Handbook of Thin Film Technology*, eds. L. I. Maissel and R. Glang, McGraw-Hill, New York (1970).

16. T. C. McGill, *J. Vac. Sci. Tech.* **11**, 935 (1974).

17.* S. M. Sze, *Semiconductor Devices—Physics and Technology*, Wiley, New York (1985).

18. B. G. Streetman, *Solid State Electronic Devices*, 2nd ed., Prentice-Hall, Englewood Cliffs, NJ (1980).

19. D. R. Wolters and J. J. van der Schoot, *Philips J. Res.* **40**, 115–136, 137–163, 164–192 (1985).

20. T. H. DiStefano and M. Shatzkes, *Appl. Phys. Lett.* **25**, 685 (1974).

21. A. M-R Lin, AT&T Bell Laboratories, private communication.

22. J. E. C. Williams, *Superconductivity and Its Applications*, Pion Limited, London (1970).

23. D. D. Hughes, in *Treatise on Materials Science and Technology*, Vol. 14, eds. T. Luhman and D. D. Hughes, Academic Press, New York (1979).

24. J. Bardeen, L. N. Cooper, and J. R. Schrieffer, *Phys. Rev.* **108**, 1175 (1957).

25. B. D. Josephson, *Rev. Mod. Phys.* **36**, 216 (1964).

26.* K. I. Chopra, *Thin Film Phenomena*, McGraw-Hill, New York (1969).

27.* S. Middelhoek, in *Magnetic Properties of Materials*, ed. J. Smit, McGraw-Hill, New York (1971).

28.* J. K. Howard, *J. Vac. Sci. Tech.* **A4**, 1 (1986).

29. T. C. Arnoldussen, *Proc. IEEE* **74**, 1528 (1986).

30. W. H. Meiklejohn, *Proc. IEEE* **74**, 1570 (1986).

31. W. Doring, *Z. Naturf.* **16a**, 1146 (1961).

32. G. Bayreuther, *J. Mag. Mag. Mat.* **38**, 273 (1983).

33. D. T. Pierce, R. J. Celotta, J. Urguris and H. C. Siegman, *Phys. Rev.* **B26**, 2566 (1982).

34.* E. W. Pugh, in *Physics of Thin Films*, Vol. 1, ed. G. Hass, Academic Press, New York (1963).

35. A. H. Bobeck and E. Della Torre, *Magnetic Bubbles*, North-Holland, Amsterdam (1975).

36. E. C. Stoner and E. P. Wohlfarth, *Phil. Trans. Roy. Soc.* (*London*) **A240**, 599 (1948).

37. E. A. Giess, *Science* **208**, 938 (1980).

38. D. K. Rose, P. J. Silverman, and H. A. Washburn, in *VLSI Electronics: Microstructure Science*, Vol. 4, ed. N. G. Einspruch, Academic Press, New York (1982).

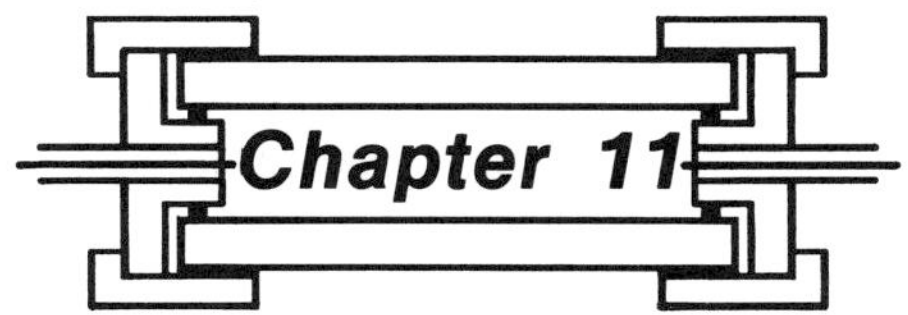

Optical Properties of Thin Films

11.1. Introduction

Thin films were first exploited practically for their optical properties. In the latter part of the nineteenth and first half of the twentieth centuries, the reflecting properties of metal films were utilized in assorted components of precision optical equipment. A noteworthy example was the Fabry–Perot interferometer developed in 1899, which required mirrors of very high reflectance or "finesse." This instrument enabled impressive accuracy to be attained in spectroscopy, thereby greatly advancing research in this field. The utility of dielectric films in optical applications was interestingly recognized as a result of observations by early spectroscopists and microscopists, notably Lord Rayleigh and Fraunhofer. They noticed that atmospheric corrosion of the lens surfaces of their instruments actually resulted in an *enhanced* overall transmission rather than a deterioration of performance. Interference effects of a surface layer was quickly discovered as the cause and it was not long before this damaging effect was capitalized upon in the form of antireflection (AR) coatings. These were first produced commercially by chemical etching, a process which persisted until the 1950s. In the mid-1930s, however, AR coatings were first produced by vacuum evaporation techniques and eventually proved to be more versatile and reliable than those made by etching. Coated lenses then found rapid application in optical imaging equipment such as cameras, telescopes, binoculars, and microscopes. Similar coatings were subsequently employed in dielectric mirrors, optical filters, and selective absorbers (Ref. 1).

Current interest in optical films still centers around traditional optical components. But, in addition, a variety of new applications has emerged—end mirrors for lasers, antireflection coatings for solar cells, and films for energy conservation systems, to mention a few. The portion of the electromagnetic spectrum involved in virtually all optical film applications falls within the span from the ultraviolet (UV) to the infrared (IR) with particular emphasis focused in the visible region. This chapter will therefore be limited to the effects in this broad spectral domain. A common optical theory governs the phenomena exhibited by the many thin-film coating applications irrespective of operating wavelength range. Specific metal, dielectric, and semiconductor films have been deposited, frequently in layered combinations, to produce the necessary optical characteristics. For these reasons the broad topical outline of the chapter includes:

11.2. Properties of Optical Film Materials
11.3. Thin-Film Optics
11.4. Multilayer Optical Film Applications

Additional discussion of thin films employed in optoelectronic devices and optical communications (Chapter 7), integrated optics (Chapter 14), and optical recording (Chapter 14) can be found in the indicated chapters.

Ample measure of the overall importance of optical film properties and components is found in the relatively large number of books devoted to the subject (Ref. 2–6). With regard to these the excellent treatment by Anders is recommended for its lucid presentation of the analysis, design, and production of coatings. Similarly the more recent and accessible book by Pulker is recommended for its wealth of useful data on optical coatings, and how they are processed and used. Readers seeking a more specialized treatment of particular topics should consult the widely acclaimed series *Physics of Thin Films*. Of the 67 review articles published in 13 volumes until 1987, fully 19 deal with optical properties and applications of metal and dielectric thin films.

11.2. Properties of Optical Film Materials

11.2.1. General Considerations

In order to understand the optical behavior of films and film systems, one must become familiar with the optical constants of materials, their origins, magnitudes, and how they depend on the way films are processed. The purpose of Section 11.2 is to provide a brief survey of these issues. The unifying concept that embraces all optical properties is the interaction of electromagnetic radiation with the electrons of the material. On this basis, optical properties are interpretable from what we know of the electronic structure and how it is affected by atomic structure, bonding, impurities, and defects. Quantum mechanics is required for a detailed description, but this is well beyond the scope intended here. Discussións will stress meanings and implications rather than mathematical rigor.

Electromagnetic radiation propagates differently in materials than in free space because of the presence of charge. As a result, there is a change in the wave velocity and intensity of the radiation described by the *complex* index of refraction

$$N = n - ik. \tag{11-1}$$

The quantity n is the real index of refraction, and k is the index of absorption, which is also known as the extinction coefficient. These are the two material optical constants that we are concerned with here. The spatially dependent portion of the electric field of a wave propagating in the x direction is then expressed by

$$\mathscr{E} = \mathscr{E}_0 \exp - \left(\frac{i2\pi Nx}{\lambda} \right) = \mathscr{E}_0 \exp - \frac{2\pi kx}{\lambda} \exp - \frac{i2\pi nx}{\lambda}, \tag{11-2}$$

where $\mathscr{E}_0$ is the field amplitude and λ is the wavelength. In free space the index of refraction is unity and the wave velocity is c, the speed of light; in the medium these respective quantities are n and c/n. The real function $\exp - 2\pi kx/\lambda$ represents an exponential damping or attenuation of the wave due to some absorption process within the material. On the other hand, the imaginary exponential portion of Eq. 11-2 contains n and reflects propagation without absorption. All materials exhibit varying proportions of these two linked attributes of N. In the highly absorbing metals, for example, n is usually small compared with k. On the other hand, the dielectric films used for optical purposes are highly nonabsorbing and k is vanishingly small compared with n. Multiplying $\mathscr{E}$ by its complex conjugate (i.e., $i = \sqrt{-1}$ replaced by $-i$) leads to the well-known expression for the attenuation of the intensity, $I \propto \mathscr{E}\mathscr{E}^*$ or $I \propto \mathscr{E}_0^2 \exp - 4\pi kx/\lambda$. Therefore,

$$I = I_0 \exp - \alpha x, \tag{11-3}$$

where the absorption constant α is defined as $4\pi k/\lambda$, and I_0 is the intensity of incident radiation. A common unit for I_0 is W/cm^2. An important application of this formula occurs in optical communications systems. A measure of the quality of the waveguide frequently quoted is attenuation in decibels (dB); the number of dB $= 10 \log_{10} I_0/I$. By comparison with Eq. 11-3, 1 dB/cm is equal to 4.34α. Extremely low absorption losses of $\sim$ 1 dB/km are common in silica-based optical fibers. In contrast, a 1-dB loss occurs within a few angstroms of penetration beneath the surface of a metal. This corresponds to a difference of some 12 orders of magnitude in absorption behavior on the part of materials.

Of the total radiation energy incident on an object, a fraction R is reflected from the top surface and a fraction T is transmitted through the bottom surface. The remaining fraction is lost through electronic absorption (A) processes and by scattering (S) at surface and volume imperfections. Surface roughness, internal boundaries, and density fluctuations arising from porosity, pinholes, microcracks, particulate incorporation, and impurities are sources of scattering. Adding the various contributions gives

$$R + T + A + S = 1. \tag{11-4}$$

Of the terms, R is of the greatest interest to us. For light passing through a medium of refractive index n_0, impinging normally on a transparent film of index n_1,

$$R = \left(\frac{n_1 - n_0}{n_1 + n_0}\right)^2. \tag{11-5}$$

If, however, the film is absorbing with index of absorption k_1

$$R = \frac{(n_0 - n_1)^2 + k_1^2}{(n_0 + n_1)^2 + k_1^2}. \tag{11-6}$$

The origins of both formulas will be indicated subsequently.

11.2.2. Optical Properties of Metals and Mirrors

The large density of empty, closely spaced electron energy states above the Fermi level in a metal plays an important role in influencing its optical properties. Incident photons over a wide wavelength range are readily absorbed by conduction-band electrons. These excited electrons move to higher energy levels where they undergo collisions with lattice ions, and the extra energy is dissipated in the form of phonons. The lattice is thus heated, and we speak of absorption. Alternatively, if the probability of colliding with an ion is small, the electron will emit a photon as it drops back to a lower energy level. This results in the strongly reflected beam exhibited by metals in the visible and infrared region. The characteristic color of some metals is due to the preferential absorption of some portion of the visible spectrum. In gold, for example, the green portion is absorbed, and the metal assumes the coloration

Table 11-1. Optical Constants of Metal Films Employed for Mirrors

	Silver		Gold		Copper		Aluminum	
Wavelengths, μ	*n*	*k*	*n*	*k*	*n*	*k*	*n*	*k*
0.40	0.075	1.93	1.45		0.85		0.40	3.92
0.45	0.055	2.42	1.40	1.88	0.87	2.20	0.49	4.32
0.50	0.050	2.87	0.84	1.84	0.88	2.42	0.62	4.80
0.55	0.055	3.32	0.34	2.37	0.72	2.42	0.76	5.32
0.60	0.060	3.75	0.23	2.97	0.17	3.07	0.97	6.00
0.65	0.070	4.20	0.19	3.50	0.13	3.65	1.24	6.60
0.70	0.075	4.62	0.17	3.97	0.12	4.17	1.55	7.00
0.75	0.080	5.05	0.16	4.42	0.12	4.62	1.80	7.12
0.80	0.090	5.45	0.16	4.84	0.12	5.07	1.99	7.05
0.85	0.100	5.85	0.17	5.30	0.12	5.47	2.08	7.15
0.90	0.105	6.22	0.18	5.72	0.13	5.86	1.96	7.70
0.95	0.110	6.56	0.19	6.10	0.13	6.22	1.75	8.50
2.0	0.48	14.4	0.54	11.2			2.30	16.5
3.0					1.22	17.1		
4.0	1.89	28.7	1.49	22.2			5.97	30.3
6.0	4.15	42.6	3.01	33.0			11.0	42.4
7.0					5.25	40.7		
8.0	7.14	56.1	5.05	43.5			17.0	55.0
10.0	10.69	69.0	7.41	53.4			25.4	67.3

From Ref. 7.

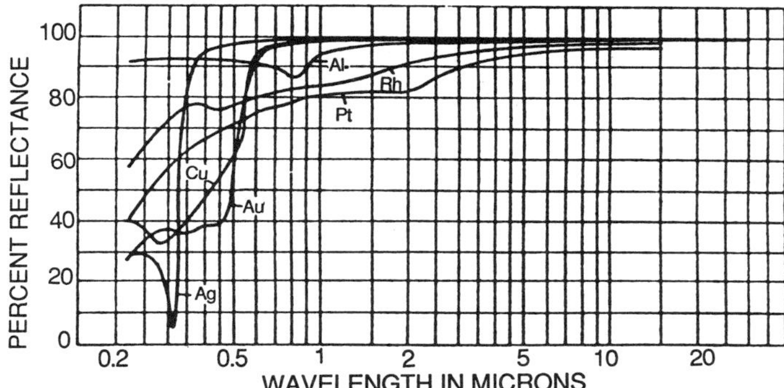

Figure 11-1. Reflectance of Al, Ag, Au, Cu, Rh, and Pt mirror coatings from the UV to the IR. (From Ref. 8).

of the reflected red and yellow light. Silver and aluminum reflect all portions of the visible spectrum and therefore appear to have a white color.

Optical constants for the most important mirror-coating metals, Ag, Au, Cu, and Al, are given in Table 11-1. The corresponding reflectivities are plotted as a function of wavelength in Fig. 11-1. A noteworthy feature of the optical response is the decreased reflectivity in the ultraviolet and, in particular, the abrupt absorption edge exhibited by the noble metals. The absorption is due to interband electron transitions (e.g., 3d → 4s in Cu). In aluminum, however, and rhodium, to a lesser extent, high reflectance extends into the ultraviolet. In the infrared all metals are very highly reflective. This fact is explained by the Hagen–Rubens relation, which holds when the electron relaxation time is short compared with the period of the incident wave. Physically, electrons can then respond sufficiently rapidly to prevent the external electric field from penetrating the metal.

The optical properties of thin films are somewhat different from those of bulk metals. In ultrathin films (< 100 Å thick) variations in film continuity make the concept of optical constants problematical. In thicker continuous films, n and k values tend to be slightly smaller and larger, respectively, than comparable bulk values. The differences stem largely from variables encountered in evaporation or sputtering, the almost universally employed processes for depositing metal film mirrors. In metals such as Al, which readily getter gases, the reflectivity is sensitive to deposition conditions. Lower operating pressures and higher deposition rates result in purer films with enhanced reflectivities as shown in Fig. 11-2. The reason is due in part to the absence of surface oxides, nitrides, etc., which serve to increase absorption. More noble metals like Rh and Pt are, on the other hand, relatively immune to deposition conditions. Other variables affecting reflectivity include incident angle of

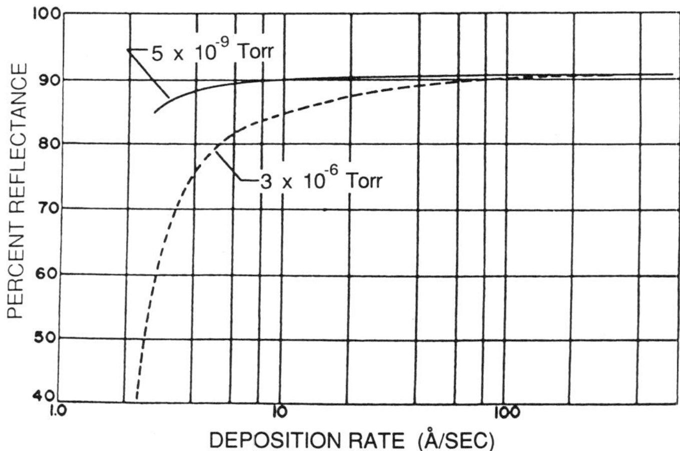

Figure 11-2. Reflectance of Al as a function of deposition rate in high (3×10^{-6} torr) and ultrahigh (5×10^{-9} torr) vacuum. Films were aged 24 h in air and measured at a wavelength of 2000 Å. (From Ref. 7).

vapor flux, surface topography, and substrate temperature (Refs. 7, 8). In metals such as Al, Ag, and Au, high substrate temperatures result in coarser-grained films with rough surfaces that tend to diffusely scatter light. A contrary behavior is exhibited by Rh whose reflectivity is enhanced by some 2–6% over the spectral range from 0.4–2.2 μm, when deposited at 400°C rather than 40°C.

The reader should appreciate that even small reflectance enhancements are very significant in mirror applications. Therefore, much care is exercised during deposition, and steps are frequently taken subsequently to prevent degradation of reflectance due to oxidation (Al), sulfide tarnishing (Ag), and mechanical scratching. Mirrors are therefore frequently coated with hard transparent protective overlayers. Thus films of SiO, SiO_2, and Al_2O_3 are commonly used to protect evaporated Al mirrors, but usually at the cost of increasing absorbance. The superior reflectance of Ag is offset by its poor adhesion and susceptibility to atmospheric attack. Various substrate film glue layers (e.g., Al_2O_3, nichrome) and coatings (e.g., Al_2O_3, SiO) have been used to improve adhesion, but Ag film use in mirrors remains restricted. Films of Rh are ideally suited as front surface mirrors because they are hard and chemically very durable giving them long-term stability. Adhesion to fused silica based substrates is often poor, however. Despite its relatively low reflectivity, evaporated Rh has found application in telescope mirrors, optical reflectivity standards and in mirrors for medical applications.

11.2.3. Optical Effects in Dielectrics

Dielectric materials employed in optical coating applications include fluorides (e.g., MgF_2, CeF_3) oxides (e.g., Al_2O_3, TiO_2, SiO_2) sulfides (e.g., ZnS, CdS) and assorted compounds (e.g., ZnSe, ZnTe). Bonding characteristics ranging from ionic to covalent are represented in these materials. An essential common feature of dielectric optical materials in their very low absorption ($\alpha < 10^3$/cm) in some relevant portion of the spectrum; in this region they are essentially transparent (e.g., fluorides and oxides in the visible and infrared, chalcogenides in the infrared).

The refractive index n is basically the only optical constant of interest insofar as optical coating design is concerned. In different materials the refracted beam travels at different velocities because incident and forward-scattered beams interact and produce a phase shift. This phase shift can be interpreted in terms of a difference in velocity between incident and refracted beams. As might be expected the magnitude of n depends on the strength of the refracted beam, which in turn depends on the density of electrons. In polar materials such as oxides, glasses, and compound semiconductors the lattice is essentially a collection of *permanent* electric dipole moments (p_e). The magnitude of p_e is equal to the product of the effective charge (nuclear, electronic) and the distance between charge centers. Wave retardation occurs because of the interaction between the electromagnetic radiation ($\mathscr{E}$ field) and the stretched permanent dipoles. In nonpolar solids like diamond and Si, the incident radiation creates *induced* electric dipole moments by displacing atomic electrons relative to the nuclear charge. In this case, $p_e = \alpha_p \mathscr{E}$, where α_p is known as the polarizability. It is beyond our scope to derive the link between the microscopic polarization and the macroscopic index of refraction known as the Lorenz–Lorentz equation, namely

$$\alpha_p = \frac{3}{4\pi N_A} \frac{(n^2 - 1)}{(n^2 + 2)} \frac{M}{\rho}, \tag{11-7}$$

where N_A is Avogadro's number, M is the molecular weight, and ρ is the density (Ref. 9). Thus, high refractive indices are associated with large ionic polarizabilities, which increase with the size of the ion and with the degree of negative charge on isoelectric ions. In glasses and cubic crystals, n is independent of crystallographic directions. In other crystal systems, n is large in close-packed directions. Similarly more open structures of high-temperature polymorphic phases have lower refractive indices than crystalline forms of the same composition. In SiO_2, for example, in order of increasing density $n_{glass} = 1.46$, $n_{tridymite} = 1.47$, $n_{cristobalite} = 1.49$, and $n_{quartz} = 1.55$. Compounds with predominant covalent bonding have higher refractive indices than ionically bound solids. For example, in order of increasing degree of covalent bonding $n_{ZnCl_2} = 1.68$, $n_{ZnO} = 2.08$, $n_{ZnS} = 2.37$, $n_{ZnSe} = 2.57$, and $n_{ZnTe} = 3.56$.

The excellent transmission of dielectric materials in the visible region of the spectrum is terminated at short wavelengths with the onset of the ultraviolet absorption edge. The critical radiation wavelength λ_c at which this occurs is given by the familiar equation

$$\lambda_c = hc/E_g \qquad \text{or} \qquad \lambda_c\,(\mu\text{m}) = 1.24/E_g\,(\text{eV}). \tag{11-8}$$

Values physically correspond to electronic transitions from the filled valence-band levels across the energy gap E_g to the unfilled conduction-band states. Multiple peaks near the UV absorption edge indicate the complexity of these processes. At long wavelength the high optical transmission is once again limited, this time by absorption due to the vibration of lattice ions in resonance with the incident radiation. The frequency of maximum absorption is related to the force constant and masses of vibrating anions and cations.

11.2.4. Optical Effects in Semiconductors

Lastly, some elementary aspects of the optical properties of semiconductors will be briefly considered. The absorption coefficient behavior of a number of important semiconductor materials was previously noted in Fig. 7-13. The most prominent feature is the rapid decrease in absorption at the critical or cutoff wavelength λ_c. Here as in dielectrics the magnitude of the energy gap between the valence and conduction band governs the value of λ_c. For wavelengths larger than λ_c, semiconductors are essentially transparent because no mechanisms exist to excite electron transitions. However, for wavelengths shorter than λ_c, electrons can be stimulated into the conduction band. In addition, the generated free carriers can now absorb quanta of energy and occupy excited levels in the continuum of conduction-band states. Semiconductors then behave like metals and are highly reflective.

An interesting phenomenon occurs in this regime of free-carrier light absorption. This effect has relevance to semiconducting In_2O_3:Sn films that are widely used because they have the unusual property of being transparent conductors (see Section 11.2.5). In semiconductors (and metals), values of n and k are generally not independent of each other but are connected by electromagnetic theory through so-called dispersion relations (Ref. 10)

$$n^2 - k^2 = \varepsilon_1 - n_c q^2 / m^* \varepsilon_0 (\omega^2 + \gamma^2), \tag{11-9}$$

$$2nk = \frac{\gamma n_c q^2}{m^* \varepsilon_0 \omega (\omega^2 + \gamma^2)}. \tag{11-10}$$

In these formulas, ε_1 and ε_0 are the permitivities of the solid and free space, respectively, n_c is the carrier concentration in the conduction band, m^* is the effective mass of the charge carrier, q is the electronic charge, ω is the angular frequency of the incident radiation of wavelength λ ($\omega = 2\pi c/\lambda$), and γ is the reciprocal of the carrier relaxation time. Basically, γ is inversely related to the electrical conductivity. When $n = k$, the optical properties change radically. The critical value of ω for which this happens is called the plasma frequency ω_p, and corresponding to it is the plasma wavelength λ_p. At the optical transition the conductivity is high, and γ^2 may be neglected. It is then easy to show that

$$\lambda_p = \frac{2\pi c}{q} \left(\frac{\varepsilon_0 \varepsilon_1 m^*}{n_c} \right)^{1/2}, \tag{11-11}$$

which only depends on the carrier concentration. When $\lambda > \lambda_p$, the film exhibits metal-like reflection, whereas if $\lambda < \lambda_p$ the high transmittance of a

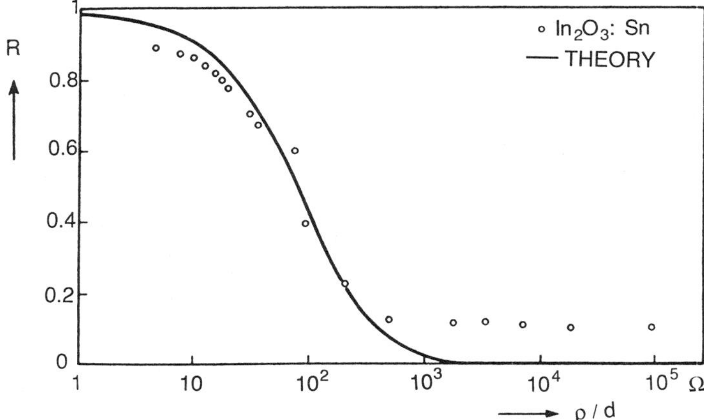

Figure 11-3. Reflectance versus surface resistivity (sheet resistance) for In_2O_3:Sn layers. Variations in film thickness and carrier concentration are responsible for the wide range of ρ/d values. (From Ref. 11).

dielectric occurs. This means that In_2O_3:Sn films can be prepared to possess the desired, though not independent, admixture of optical reflectivity and electrical resistivity as shown in Fig. 11-3.

In photonic devices the operating wavelength is usually either λ_c (light-emitting diodes, lasers) or less than λ_c (photodetectors, solar cells). It is essential in the latter applications that light be absorbed for the devices to operate. The opposite is required of semiconductors in optical coating applications where transparency is essential. Another difference relates to structure. In photonic devices either bulk or thin-film single-crystal semiconductors are essential for operation. On the other hand, polycrystalline or amorphous Ge and Si are deposited in optical coatings. The extension of the spectral regime of transparency down to $\lambda_c = 0.82$ μm for amorphous relative to polycrystalline Si, is an advantage in optical coatings; in solar cell use however, the constricted range of absorption is a disadvantage, other things being equal. Films possessing tailor-made energy band gaps and refractive indices, prepared from compound semiconductor mixtures, have been discussed previously (Chapter 7).

11.2.5. Transparent Conducting Films (Ref. 12)

Somewhere between metals and dielectrics is an interesting class of materials known as transparent conductors. According to electromagnetic theory, high conductivity and optical transparency are mutually exclusive properties since photons are strongly absorbed by the high density of charge carriers. Although there are materials that separately are far more conductive or transparent, the transparent conductors dealt with here exhibit a useful compromise of both desirable properties. Broadly speaking, transparent conducting films consist either of very thin metals or semiconducting oxides. The first widespread use of such films was for transparent electrical heaters in aircraft windshield de-icing applications during World War II. Today they are used for automobile and airplane window defrosters, liquid crystal and gas-discharge displays, front electrodes for solar cells, antistatic coatings, heating stages for optical microscopes, IR reflectors, photoconductors in television camera vidicons, and Pockel cells for laser Q-switches.

Metals that have been used as transparent conductors include Au, Pt, Rh, Ag, Cu, Fe, and Ni. Simultaneous optimization of conductivity and transparency presents a considerable challenge in film deposition. At one extreme are discontinuous islands of considerable transparency but high resistivity; at the other are films that coalesce early and are continuous, possessing high conductivity but low transparency. For these reasons, the semiconducting oxides SnO_2, In_2O_3, CdO, and, more commonly, their alloys (e.g., indium–tin oxide, also known as ITO) doped In_2O_3 (with Sn, Sb), and doped SnO_2 (with F, Cl, etc.) are used.

Deposition processes have included chemical and physical methods. Hydrolysis of chlorides and pyrolysis of metalorganic compounds are examples of the former; reactive evaporation and sputtering in an oxygen environment are examples of the latter. Optimum film properties require maintenance of tight stoichiometry. Since glass substrates, which are often heated close to the softening temperature, are commonly employed, care must be taken to prevent stresses and warpage. Generally films of comparable quality can be produced by chemical or physical deposition processes.

The inverse relation between film reflectivity and sheet resistance should be noted in Fig. 11-3.

11.2.6. Optical Properties of Thin Dielectric and Semiconductor Films

The optical properties of dielectric and semiconductor films are collected in Table 11-2. Practically all of the data are for evaporated films, but some results for sputtered films are included. Only a portion of the exhaustive information on these films, provided by Pulker (Ref. 5) in his excellent treatment of this subject, is considered here. Among the important issues concerning optical films are the magnitudes of n and k, the spectral range of transmission, the dependence of properties on film structure and deposition processes, and environmental effects. These issues will now be addressed in turn.

11.2.6.1. Magnitude of n and k — Spectral Range. For the most part, n in fluoride and oxide films has a value less than 2 at the reference wavelength of 0.55 μm. For many applications, however, it is important to have films with higher refractive index in the visible range. To meet these needs, materials like ZnS and ZnSe are employed. High transmittance is an essential requirement in optical films, and as an arbitrary criterion only materials with an absorption constant less than $\alpha = 10^3$/cm are entered in Table 11-2. In practice, however, only films with significantly lower absorption can be tolerated. For

Table 11-2. Optical Properties of Films

Film Composition	Structure (packing density P)	Transmittance Range (μm)	Index of Refraction
NaF	C	0.2–	1.29–1.30
LiF	C	0.11–7	1.3
CaF_2	C; $P = 0.57–1$	0.15–12	1.23–1.46
Na_3AlF_6	C; $P = 0.88$	0.2–14	1.32–1.35
AlF_3	A; $P = 0.64$	0.2–	1.23
MgF_2	C; $P = 0.72$	0.11–4	1.32–1.39
ThF_4	A	0.2–15	1.50
LaF_3	C; $P = 0.80$	0.25–2	1.55
CeF_3	C; $P = 0.80$	0.3–5	1.63
SiO_2	A; $P = 0.9$	0.2–9	1.45–1.46
Al_2O_3	A; $P = 1$	0.2–7	1.54
MgO	C	0.2–8	1.7
Y_2O_3	A	0.3–12	1.89
La_2O_3	A	0.3–	1.98
CeO_2	C	0.4–12	2.2
ZrO_2	A; $P = 0.67$	0.34–12	1.97
SiO	A	0.7–9	2.0
ZnO	C	0.4–	2.1
TiO_2	A	0.4–3	1.9
ZnS	C; $P \geq 0.94$	0.4–14	2.3
CdS	C	0.55–7	2.5
ZnSe	C	0.55–15	2.57
PbTe		3.5–20	5.6
Si	A	1–9	3.4
Ge	A	2–23	4.4

C = crystalline; A = amorphous.
From Ref. 5.

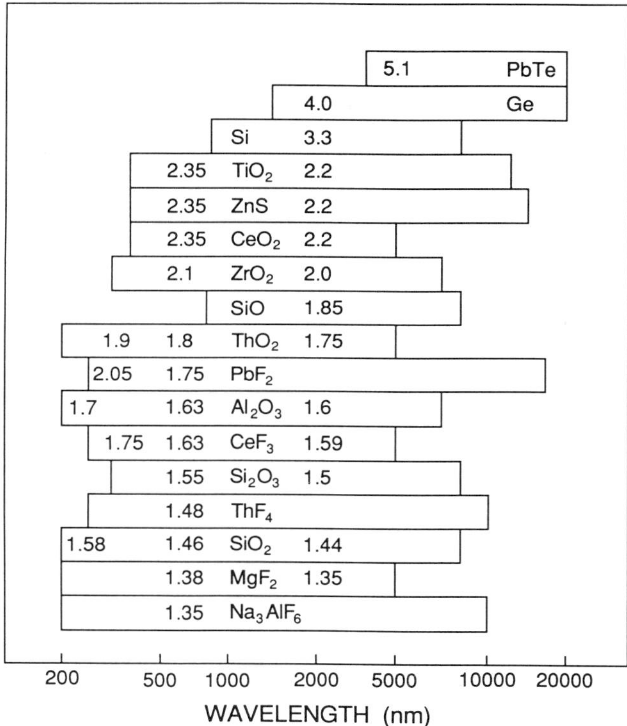

Figure 11-4. Spectral region of high transparency of dielectric films. Indices of refraction are given at the indicated wavelengths. (From Ref. 8).

example, in laser AR coatings losses must be kept to less than 0.01%, corresponding to $k \approx 4 \times 10^{-5}$ or $\alpha = 10/\text{cm}$ at $\lambda = 5500$ Å. The spectral regions of high transparency for the most commonly employed dielectric films are depicted in Fig. 11-4 together with corresponding n values.

11.2.6.2. Film Structure and Its Effect on n. Generally n values in films are lower than those in bulk. For example, $n = 2.35$ for bulk cubic ZnS at 6330 Å, whereas $n = 2.27$ in a 1000-Å-thick film. Thinner films show even lower values, and thicker film behavior approaches that of the bulk. Film structure apparently has an important influence on optical properties. The connection is through Eq. 11-7, which indicates that n decreases with decreasing film density, a fact commonly observed in porous films.

Oxides like Al_2O_3, SiO_2, ZrO_2, and TiO_2 have a tendency to be amorphous when deposited as films. In common with metals and semiconductors,

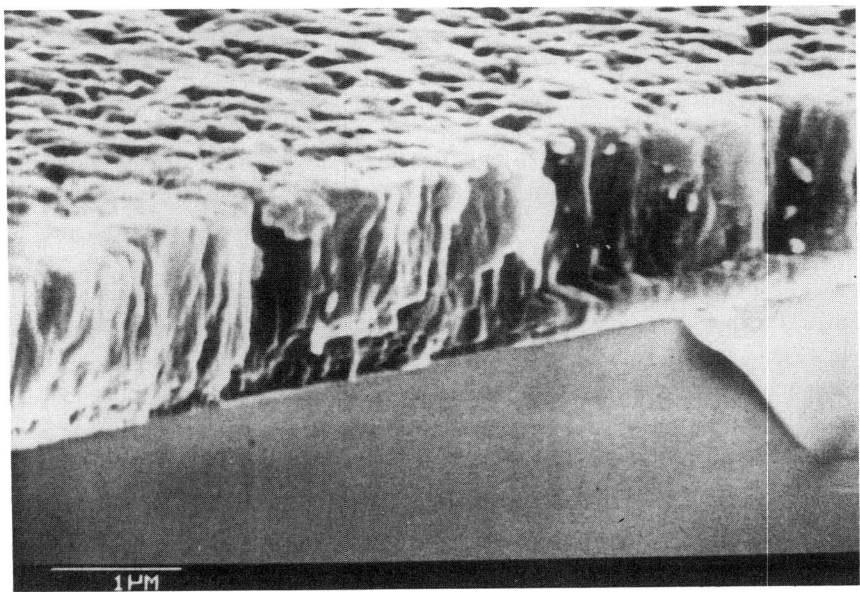

Figure 11-5. SEM micrograph of columnar structure of a ZnS film layer deposited by diode ion plating. (Courtesy of H. A. Macleod).

however, some dielectric films also exhibit a zone structure. As an example, the columnar structure of ZnS is shown in Fig. 11-5. Depending on the proportions of grains and voided boundaries present, differing film densities result. The packing density P (defined by Eq. 5-44) for optical films lies in the range 0.7–1.0. The simplest model of a columnar grain structure consists of an array of close-packed cylindrical columns of identical diameter, and it is a simple matter to show that $P = 0.907$ in this case. Other structural models are possible. For example, unevenly contracting columns will form truncated cones with $P < 0.907$; columns expanding into a hexagonal shape will be more densely packed than cylinders and $P > 0.907$.

Various relationships have been proposed to relate refractive index to microstructure, or n or P (Ref. 13). The simplest is a linear law of mixing; i.e.,

$$n = n_s P + n_v(1 - P), \qquad (11\text{-}12)$$

where n_s and n_v are the refractive indices associated with the solid film and voids (or pores) respectively.

11.2.6.3. Deposition Processes and Their Effect on n and k. In order to produce films with low absorption losses, we must carefully control various parameters in the traditionally employed evaporation processes. First of all, the materials selected for deposition should be carefully selected as to purity. Substrates must be scrupulously cleaned. It is recommended that vacuum-sintered and outgassed materials, or even pieces of dense single crystals, be used to avoid gas outbursts and spitting from the melt during evaporation. Reactions of fluorides or oxides with resistively heated crucible materials lead to contamination and absorption in the UV. Electron-beam evaporation has overcome this difficulty, and virtually all of the oxides can be deposited as highly transparent, undecomposed films this way. To ensure a high and constant evaporation rate and avoid decomposition through overheating, a defocused electron beam (e.g., $\sim$ 2 cm in diameter) is recommended. Another technique for producing highly transparent oxides, even when lower oxides or metals are used as starting materials, is reactive evaporation. In such a case, evaporant and oxygen impingement rates must be carefully controlled.

The substrate temperature during deposition has an important effect on refractive index. In general, n increases as the substrate temperature is raised. In the range 40–300 °C, the difference is small in Al_2O_3 because the films are amorphous at both substrate temperatures. In CeO_2, however, an increase in n occurs over the same temperature range (Ref. 8). These films were crystalline, and the grain size increased substantially with increasing substrate temperature. The rise in n apparently stemmed from pore elimination during sintering.

Improvement in the preparation of optical thin films over the years can be appreciated in the case of ZnS. A steady decrease in k from 1.1×10^{-2} to 2.0×10^{-3} and then to 9×10^{-5} was reported in 1952, 1956, and 1974, respectively, over the spectral range 5100 to 5500 Å. No doubt, general property improvement and film reproducibility were the result of more careful attention to deposition process parameters. The relationship of the latter to film properties is conveniently summarized in Table 11-3. Strong, established, and possible dependencies are noted. The complexity of the interplay between film properties and process variables is evident, particularly when one parameter may alter several properties.

The quest for improved and more reproducible coatings is unceasing as the needs of modern precision optics extends to such applications as high-power lasers and laser gyroscopes, where extreme stability and high reflectance are required. In recent years, the use of sputter deposition and ion-assisted technology in particular, has resulted in significant improvements in optical properties of films. Although still largely in the research and development stage, the results are promising enough to warrant some discussion here. The deposition processes employed include ion-assisted deposition (IAD), ion plating, conventional sputtering as well as the variants employing ion sources

Table 11-3. Influence of Process Variables on Optical Properties of Thin Films

Film Property	Substrate Material	Substrate Cleaning	Starting Material	Glow Discharge	Evaporation Method	Rate	Pressure	Vapor	Substrate Temperature
Refractive index			E	E		S	S	S	S
Transmission			E	E	E	P	P		P
Scattering	S	E	P	P	E	P	S	E	S
Geometric thickness		E		E		E	E	E	S
Stress	S	P	S			E	E	E	E
Adherence	E	S		S	E	E	E	E	E
Hardness	E	E		E		E	S	E	S
Temperature stability	E					E	E	E	S
Insolubility	P	E	E	E		E	S	E	S
Resistance to laser radiation	P	E	E	E	E	E			E
Defects	S	E	E	E	S	E	P		E

S = strong effect; E = established effect; P = possible effect.
From Ref. 14.

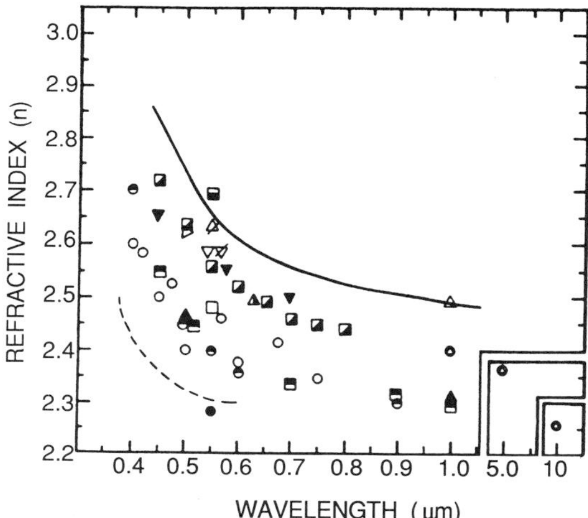

Figure 11-6. Summary of the variation of n in TiO_2 films deposited by ion-based and conventional methods. Solid line is the bulk TiO_2 (rutile) data. IAD—○●◓; sputtering—△▽▲▼; reactive evaporation, □; dotted line represents data from conventionally deposited films. (From Ref. 14 © 1986 Chapman and Hall).

discussed in Chapter 3. As a result columnar growth is interrupted, the crystallite size decreases, the packing factor increases, and the film density approaches bulk values with net improvement in optical film properties. A good example is provided by amorphous TiO_2 where the spread of the refractive index obtained by different investigators, employing various ion-based techniques, is presented in Fig. 11-6. Although very few results approach the optical properties of bulk rutile, in virtually all cases at all wavelengths, conventionally deposited TiO_2 has a lower refractive index than films prepared employing ion bombardment. Convincing evidence for the benefits of ion bombardment was obtained through real-time monitoring of light transmission through substrates during deposition. Switching the ion beam on and off caused corresponding variations in the transmission to occur in concert. They were interpreted in terms of changes in n and k. Experiments on ZrO_2 films also clearly revealed an increase in n and decrease in k as a result of ion bombardment (Ref. 14).

11.2.6.4. Environmental Effects.

Exposure of optical films to elevated temperatures and corrosive and humid environments has deleterious effects on their properties. Temperature variations produce reversible and irreversible property changes. Typically, $dn/dT \sim 10^{-5}$ to 10^{-4} per degree Celsius, and positive as well as negative refractive index changes are observed. Absorption coefficient variations in semiconducting films are also reversible over small temperature ranges, but for large temperature excursions irreversible changes occur. Films may densify with an increase in packing and refractive index. Reactions between film layers may form compounds and even free metal, with attendant increases in absorption. Thermally induced stresses having harmful implications with respect to film adhesion and cracking are also a possibility. Resistance of optical films to etching and chemical attack of saltwater environments, SO_2 and H_2S gases, cleaning agents, acids, and sometimes even fungus, is highly desirable.

The subject of water absorption by optical films, particularly those of low refractive index, has been of concern for many years. Water vapor penetrates through the microvoid network and gaps between columnar grains of low-packing-density films by capillary action. The net effects are swelling, changes in refractive index, and overall shifts in the spectral transmittance of optical coatings.

11.3. Thin-Film Optics

11.3.1. Nonabsorbing Films (One Interface)

In this section we present the theory of the optical behavior of a single transparent (nonabsorbing) film on a nonabsorbing substrate. Transparent films are of great practical importance because applications of all kinds are based on one or, more commonly, multiple layers of such films. Crucial to understanding these applications is how radiation is reflected and transmitted at a single interface between two media of differing index of refraction.

Let us consider Fig. 11-7, which depicts an interface between an optically isotropic medium with an index of refraction n_1 and a vacuum region of index n_0 $(= 1)$. Electromagnetic radiation is incident on the n_0/n_1 interface at an angle ϕ_0 and refracted in the medium at an angle ϕ_1 with respect to the normal. The electric field vector of the incident wave, $\mathscr{E}_i$, is decomposed into two components, $\mathscr{E}_i^{\perp}$ and $\mathscr{E}_i^{\parallel}$, respectively perpendicular and parallel, to the plane of incidence. Through the use of the Maxwell equations with appropriate interfacial boundary conditions, the amplitudes of the components of the transmitted $(\mathscr{E}_1^{\perp}, \mathscr{E}_1^{\parallel})$ and reflected $(\mathscr{E}_0^{\perp}, \mathscr{E}_0^{\parallel})$ waves in the medium and vacuum can be determined. The results of the calculation are the celebrated Fresnel coefficients for reflection (r) and transmission (t) at the n_0/n_1

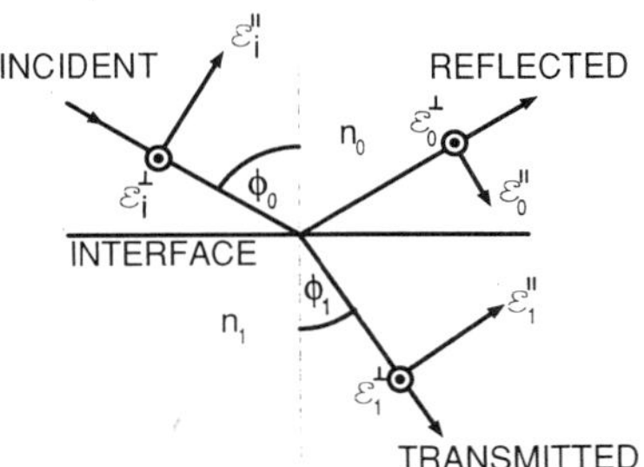

Figure 11-7. Light incident on an interface at an oblique angle undergoing both reflection and refraction.

interface given by

$$r_1^{\perp} = \frac{\mathscr{E}_0^{\perp}}{\mathscr{E}_i^{\perp}} = \frac{n_0 \cos \phi_0 - n_1 \cos \phi_1}{n_0 \cos \phi_0 + n_1 \cos \phi_1}, \tag{11-13a}$$

$$r_1^{\parallel} = \frac{\mathscr{E}_0^{\parallel}}{\mathscr{E}_i^{\parallel}} = \frac{n_0 \cos \phi_1 - n_1 \cos \phi_0}{n_0 \cos \phi_1 + n_1 \cos \phi_0}, \tag{11-13b}$$

$$t_1^{\perp} = \frac{\mathscr{E}_1^{\perp}}{\mathscr{E}_i^{\perp}} = \frac{2 n_0 \cos \phi_0}{n_0 \cos \phi_0 + n_1 \cos \phi_1}, \tag{11-13c}$$

$$t_1^{\parallel} = \frac{\mathscr{E}_1^{\parallel}}{\mathscr{E}_i^{\parallel}} = \frac{2 n_0 \cos \phi_0}{n_0 \cos \phi_1 + n_1 \cos \phi_0}. \tag{11-13d}$$

Extensions of these equations are the basis for the mathematical analysis of the optical behavior of single films and multilayer film structures. Changing the subscripts from 0, 1 to 1, 2 makes the results applicable to the film (n_1)–substrate (n_2) interface in a single film layer.

For normal incidence, where $\phi_0 = \phi_1 = 0$ the distinction between the perpendicular and parallel electric field components vanishes and the reflection and transmission coefficients are given by

$$r = r_1^{\perp} = r_1^{\parallel} = \frac{n_0 - n_1}{n_1 + n_0}, \tag{11-14}$$

$$t = t_1^{\perp} = t_1^{\parallel} = \frac{2 n_0}{n_1 + n_0}. \tag{11-15}$$

The equation for r contains information on the amplitude and phase of the reflected wave. For example, if $n_0 = 1$ and $n_1 = 1.52$, $r = (1 - 1.51)/(1 + 1.52) = -0.206$, which is equal to $+0.206 e^{\pm i\pi}$. This represents the well-known phase jump of π during reflection at an optically denser medium. If, however, the wave propagates in the reverse direction, at the n_1/n_0 interface $n_0 = 1.5$ and $n_1 = 1$ (for consistency) and $r = (1.52 - 1)/(1.52 + 1) = 0.206$. There is no phase jump in this case. For nonabsorbing media the ratio of the energy reflected to that incident is r^2 or R, the reflectance; its magnitude is given by Eq. 11-5. Similarly the ratio of the energy transmitted to that incident is $(n_1/n_0)^2 t^2$ or T, the transmittance, given by

$$T = 4n_1 n_0/(n_1 + n_0)^2. \tag{11-16}$$

Clearly, $R + T = 1$.

11.3.2. Nonabsorbing Film (Two Interfaces)

We now consider reflection of light at both film interfaces in Fig. 11-8 where radiation may be partially reflected or transmitted. The separation between the boundaries, d, is of the order of magnitude of the wavelength λ so that interference effects occur. It is assumed that the film is plane parallel and homogeneous. The optical properties of the film will now depend on n_0, n_1, n_2, d, and λ. If a plane wave of unit amplitude is normally incident on the n_0/n_1 interface at A, the fraction reflected is $r_1 = (n_0 - n_1)/(n_1 + n_0)$, and the wave transmitted has amplitude $\sqrt{1 - r_1^2}$. At point B of the film–substrate (n_1/n_2) interface, $r_2 = (n_1 - n_2)/(n_2 + n_1)$ so that a wave of amplitude $r_2\sqrt{1 - r_1^2}$ is reflected back to point C on the first interface. Here the wave is reflected again into the film with amplitude $-r_1 r_2 \sqrt{1 - r_1^2}\, e^{i\delta}$. Compared with the first reflection at A, the emerging wave at C has twice traversed the optical thickness, $n_1 d$, of the film. The total path difference, $2n_1 d$, implies a phase δ of amount $\delta = (2\pi/\lambda)$(path difference), and therefore $\delta = 4\pi n_1 d/\lambda$. The portion of the wave transmitted into free space at C is attenuated by the factor $\sqrt{1 - r_1^2}$ so that its amplitude is given by $r_2(1 - r_1^2)$. Adding the two emergent waves in amplitude and phase yields

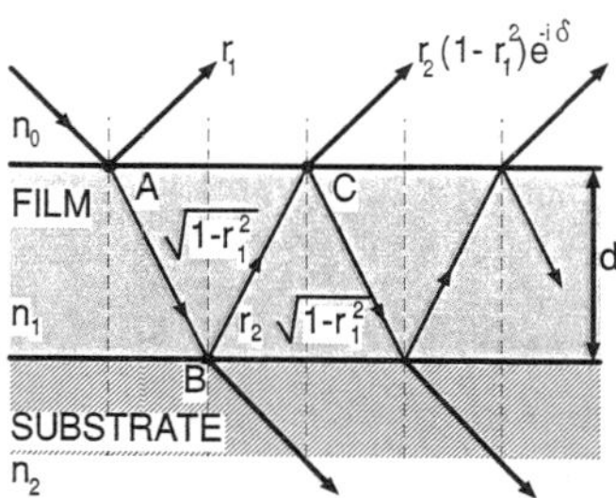

Figure 11-8. Reflections at both surfaces of a nonabsorbing, homogeneous thin film at normal incidence.

$r_1 + r_2(1 - r_1^2)e^{-i\delta}$. Continuing the process and summing the multiply reflected beams whose amplitudes decline with repeated interfacial impingement, the result is

$$r = \frac{r_1 + r_2 \exp - i\delta}{1 + r_2 r_1 \exp - i\delta}. \tag{11-17}$$

By squaring r according to the rules of complex number algebra, the reflectance R is expressed by

$$R = \frac{r_1^2 + r_2^2 + 2r_1 r_2 \cos\delta}{1 + r_1^2 r_2^2 + 2r_1 r_2 \cos\delta}. \tag{11-18}$$

Similarly, the transmittance is given by

$$T = \frac{n_2}{n_0} \frac{t_1 t_2}{1 + r_1^2 r_2^2 + 2r_1 r_2 \cos\delta} \tag{11-19}$$

where t_1, t_2 are extensions of Eq. 11-15.

The first of these equations is an extremely important one with many implications for the design and analysis of optical films. First, R is plotted as a function of phase or optical thickness in Fig. 11-9 for a number of different film refractive indices. As a reference, the refractive index of the uncoated,

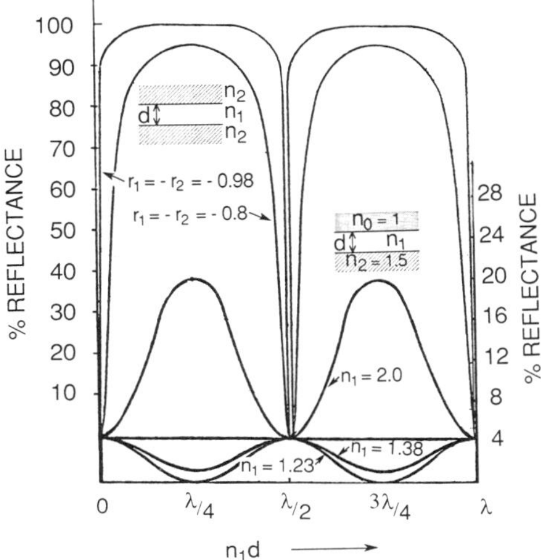

Figure 11-9. (Lower portion) Reflectance as a function of $n_1 d$ for films of various n_1 values on a glass substrate of $n_2 = 1.5$ (also see Fig. 6-2); (upper portion) reflectance at a nonabsorbing uniform film between two media with same index of refraction. (From Ref. 3).

nonabsorbing substrate is $n_2 = 1.5$, the value for plate glass. The oscillatory nature of the reflected light intensity, caused by interference effects, has a periodicity related to the film thickness and index of refraction. This is the basis for experimentally determining the thickness of transparent films if the index of refraction is known (see Chapter 6). Conversely, the optical properties of the film can be determined at a particular wavelength if its thickness is known. Maxima or minima in the reflected intensity occur at specific film thicknesses, for given wavelengths, depending on whether the refractive index of the film is greater or less than that of the substrate. In the former case the reflectivity is enhanced, whereas in the latter case reflectivity is diminished. Optimization of these two effects has led to the development of dielectric mirrors and antireflection coatings, respectively.

To quantify the issues related to antireflectivity, Eq. 11-17 reveals that r vanishes when $r_1 + r_2 \exp - i\delta = 0$ or when the denominator goes to infinity. The latter is an impossibility, since r_1 and r_2 are less than or equal to 1. The remaining condition can be decomposed into two real transcendental equations: (a) $r_1 + r_2 \cos \delta = 0$ and (b) $r_2 \sin \delta = 0$. Equation (b) implies that $\delta = 0$, $\pm\pi$, $\pm 2\pi$, $\pm 3\pi$, etc., but the simultaneous satisfaction of equation (a) requires the selection of $\delta = \pm\pi$, $\pm 3\pi$, $\pm 5\pi$, etc. Under these conditions, $r_1 - r_2 = 0$ or $(n_0 - n_1)/(n_0 + n_1) = (n_1 - n_2)/(n_1 + n_2)$. Therefore,

$$n_1 = \sqrt{n_0 n_2}\,. \tag{11-20}$$

Since $\delta = 4\pi n_1 d_1/\lambda = \pi, 3\pi, 5\pi$, etc,

$$n_1 d = \frac{\lambda}{4}, \frac{3\lambda}{4}, \frac{5\lambda}{4}, \text{etc.} \tag{11-21}$$

Equations 11-20 and 11-21 represent the amplitude and phase conditions for zero reflectance, respectively. In the design of a one-layer antireflection coating, the film index of refraction should be the geometric mean of the refractive indices of adjacent media. This is only strictly true for the wavelength λ for which the optical thickness of the film is $\lambda/4$, $3\lambda/4$, etc.

To coat a glass lens ($n_2 = 1.52$), a film with $n_1 = \sqrt{1.52} = 1.23$ is optimal for antireflection purposes. Clearly, this is only one consideration among many, including availability, ease of deposition, hardness, and environmental stability, which must be taken into account when choosing the film layer. The most widely used AR coating is a $\lambda/4$-thick film of MgF_2 with $n_1 = 1.38$. It can be used to coat either glass or acrylic substrates. In the absence of an AR coating, glass will exhibit a reflectance of $((1.0 - 1.52)/(1.0 + 1.52))^2 = 0.043$. Suppose it is desired to reduce the reflectance at a wavelength of 5500 Å. Then the film thickness required is $\lambda/4n_1$ or

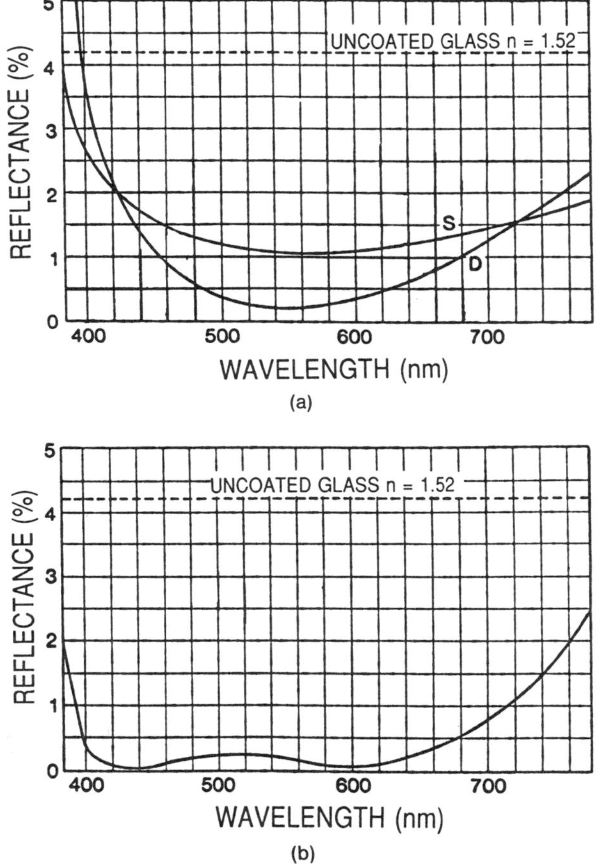

Figure 11-10. (a) Single (*S*) and double (*D*) layer antireflection coating characteristics. (b) Broadband antireflection coating characteristics. (From Ref. 5).

5500/4(1.38) = 996 Å. Under these conditions the reflectance is given by Eq. 11-18 with $r_1 = (1.0 - 1.38)/(1.0 + 1.38) = -0.160$, $r_2 = (1.38 - 1.52)/(1.38 + 1.52) = -0.0483$, and $\delta = \pi$. Substitution in Eq. 11-18 yields a value of $R = 0.0126$, indicating an almost fourfold decrease in reflectivity. Greater improvements occur for higher n_2 values of the underlying substrate. As an example, for an uncoated glass with $n_2 = 1.75$ the reflectance is 0.074. With a quarter-wave-thick MgF_2 coating, R is reduced to 0.0025. At other wavelengths, but the same optical thickness, R will be different because n_1 varies with λ (i.e., dispersion) and because of changes in δ. The reflectance reduction with a single-layer AR coating as a function of wavelength is shown in Fig. 11-10.

It is instructive to end the discussion with several observations made by Anders. (Ref. 3)

1. There is a more rapid variation of δ and hence R with λ for a $3\lambda/4$ film than for a $\lambda/4$ film. Therefore, R will be less dependent on wavelength with a $\lambda/4$ coating.
2. It is not always true that films of high refractive index give a high reflectance, whereas those with low refractive index yield AR coatings. The rule is that if the reflected amplitudes r_1 and r_2 are of the same sign, antireflection behavior is observed; if they are of opposite sign, then reflection from the surface is enhanced.
3. For very large amplitude values of $r_1 = -r_2$, R approaches 100% and the reflection becomes zero only in narrow wavelength bands at $\lambda/2$, λ, $3\lambda/2, \ldots$. This occurs physically when a film is sandwiched between two media of the same refractive index, i.e., cemented film $(n_2/n_1/n_2)$, as shown in Fig. 11-9.

11.3.3. Absorbing Films

The mechanisms by which materials absorb radiation were treated earlier. Absorption effects can be formally incorporated into the Fresnel equations by replacing the refractive index n by the complex refractive index; i.e., $N = n - ik$. For the case of reflection due to normal incidence of light at an interface between nonabsorbing and absorbing media of refractive indices n_0 and $n_1 - ik_1$, respectively,

$$r_1 = \frac{n_0 - n_1 + ik_1}{n_0 + n_1 - ik_1}. \tag{11-22}$$

By evaluating $|r_1|^2$, we have the reflectance formula Eq. 11-6.

As an example, consider the reflectance of Al front surface mirrors produced some 15 years apart. Hass (Ref. 15) measured the optical constants of Al to be $N_{Al} = 0.76 - i5.5$ in 1946 and $N_{Al} = 0.81 - i5.99$ in 1961. Substitution in Eq. 11-6 with $n_0 = 1$ yields respective reflectances of 0.909 and 0.916. Improved deposition technology including higher and cleaner vacua, purer metal, and higher evaporation rates were probably the cause of the enhanced reflectance. An R value of 0.91 could be achieved with a hypothetical absorption-free material with $n = 43$. This extremely high value can be thought of as the effective refractive index for aluminum.

A frequently asked question regarding thin films is, how thick must a metal film (on a transparent substrate) be before it is continuous? By this is meant the thickness at which it can no longer be seen through. A simple estimate can be obtained by arbitrarily assuming that a drop in transmitted intensity by a factor of $1/e$ occurs when the film is continuous. Therefore, the use of Eq. 11-3 with $I/I_0 = 1/e$ yields $4\pi kd/\lambda = 1$, or $d = \lambda/4\pi k$. It is clear that the answer to the question not only depends on the type of metal but also on the wavelength of light used to view it. The critical thickness for Al films at 5500 Å is 82 Å, whereas for Au films it is 185 Å. It is common experience, however, that films that are considerably thicker exhibit some transparency. The reason is that ultrathin films condense in an island structure of discrete clusters rather than as planar, continuous, homogeneous layers assumed in the optical theory. An alternative approach to this problem, which is left to the reader as a lengthy but healthy exercise in the use of complex numbers, is to consider the optical structure $n_0/n_1 - ik_1/n_2$ corresponding to free-space/metal film/substrate. From Eq. 11-19, T can be calculated for different film thicknesses.

By inverting the order of the last two optical components, i.e., $n_0/n_1/n_2 - ik_2$, we have the case of the back surface or protected mirror. It is commonly believed that the reflection properties of the mirror are unaffected by the protective layer. In reality, the latter actually reduces the reflectance in the visible, and, particularly, in the UV and IR ranges.

The remaining case, $n_0/n_1 - ik_1/n_2 - ik_2$, models the optical behavior of an absorbing film on an absorbing substrate. This structure was recently used to determine the real-time kinetics of regrowth of epitaxial Si into an amorphous Si (surface) layer. By bouncing a He–Ne laser beam off the surface and monitoring the reflected beam intensity, the instantaneous position of the epitaxial–amorphous interface could be unfolded from the attenuated periodic signal (Ref. 16). Such a measurement is possible because the optical constants of crystalline and amorphous Si differ. (See Problem 9, p. 543.)

11.4. Multilayer Optical Film Applications

11.4.1. Introduction

Once the basic principles governing the applications of single dielectric films and their deposition methods were firmly established, extension to multilayer systems was naturally driven by several factors (Ref. 17):

1. By suitable variations in design, it is possible to obtain improved AR properties over a broader spectral range.
2. Systems with a vast variety of optical filtering properties can be achieved

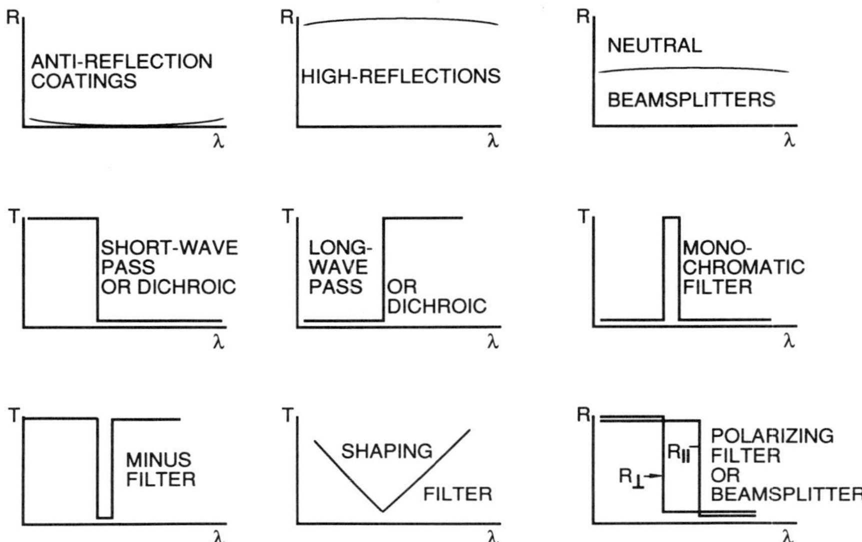

Figure 11-11. Typical applications of thin films and film systems in optics. (From Ref. 5).

usually with the use of many film layers (sometimes a dozen or more) but with only a very limited number of materials (e.g., MgF_2 and ZnS).

3. Multilayer optical filters have advantages over other types of filters. The reason is that there is very little absorption loss in dielectric film layers, since they rely on the effects of interference.
4. The principles of design of optical systems applicable to one region of the electromagnetic spectrum (e.g., visible) are also valid in other regions (e.g., UV and IR).

Various types of thin-film optical component characteristics are shown in Fig 11-11 where the desired reflectance and transmittance properties are schematically indicated as a function of wavelength.

11.4.2. AR Coatings

Antireflection coatings constitute the overwhelming majority of all optical coatings produced. They are used on the lenses of virtually all optical equipment, including cameras, microscopes, binoculars, range finders, telescopes, and on opthalmic glasses. Because of the reflection at each air–glass interface, intolerably large light losses can rapidly mount in complex lens systems. Neglecting absorption effects, the transmission of an optical system is given by

$$T = (1 - R_1)(1 - R_2)(1 - R_3)\ldots, \tag{11-23}$$

where the R_i are the reflectances (Eq. 11-5) at the individual optical interfaces. For example, in a system with uncoated lenses consisting of 20 interfaces, each with $R = 0.05$, the value of $T = (0.95)^{20} = 0.358$. If, however, R is reduced to 0.01 by means of AR coatings, then $T = 0.818$. The measured transmission is actually somewhat higher than these estimates because light is backreflected at internal air–glass interfaces. The improvement is impressive indeed. In addition to enhancing light transmission, AR coatings reduce glare. The so-called veiling glare causes a reduction in image contrast by illuminating regions of the image that should normally be dark. Lastly, since lens surfaces fortuitously act as mirrors in addition to refractors, spurious ghost images are frequently generated. These are also reduced through the use of AR coatings. Other optical systems that derive benefit from the use of such coatings to maximize the capture of light include solar cells, infrared detectors, and magneto-optical devices.

In the case of the double-layer coating where the indices of refraction vary successively as n_0 $(= 1)/n_1/n_2/n_3$ from free space to the substrate, the complex reflectivity amplitude is given by

$$r = \frac{r_1 + r_2 e^{-i\delta_1} + r_3 e^{i(\delta_1+\delta_2)} + r_1 r_2 r_3 e^{-i\delta_2}}{1 + r_1 r_2 e^{-i\delta_1} + r_1 r_3 e^{-i(\delta_1+\delta_2)} + r_2 r_3 e^{-i\delta_2}} \tag{11-24}$$

by analogy with Eq. 11-17. For normal incidence the indicated r for each of the three interfaces is given by

$$r_i = \frac{n_{i-1} - n_i}{n_{i-1} + n_i} \quad \text{and} \quad \delta_1 = \frac{4\pi n_1 d_1}{\lambda}; \quad \delta_2 = \frac{4\pi n_2 d_2}{\lambda},$$

where d_1 and d_2 are the thicknesses of the coating layers. Zero reflectance at one wavelength will obtain when the condition $n_2 = \mathrm{n}_1\sqrt{n_3}/n_0$ is fulfilled. Once film n_1 has been selected, this condition serves to specify the optimal value of n_2. The improvement of a double-layer AR coating relative to the single-film coating is shown in Fig. 11-10a. Interestingly, although the double layer results in a considerable reflectance reduction at wavelengths centered about 5500 Å, the response is worse at the spectral extremes due to the high curvature of the R vs. λ dependence. Greater care is required in controlling the film thickness in bilayer coatings than in single layers. In the latter a film thickness error simply means that the reflectance minimum is shifted to another wavelength. In contrast, an error in double-layer thicknesses can not only eliminate reflection minima but even increase reflectance. Multilayer film thicknesses must be even more stringently controlled.

The extension of the analysis to a multilayer stack of dielectric films of various thicknesses and n values is straightforward, though cumbersome. Exact formulas exist for three and more layers. Modern broadband AR coatings generally consist of three to seven film layers. An example of the reflectance characteristics of such a multilayer coating is shown in Fig. 11-10b.

11.4.3. Multilayer Dielectric Stacks

Since the high reflectance of a single $\lambda/4$ film is due to the constructive interference of the beams reflected at both surfaces, the effect can be enhanced by phase agreement in the reflected beams from multiple film layers. What is required is a stack of alternating high (H) and low (L) index $\lambda/4$ films. Next to the substrate is the usual high index layer so that the stacking order is HLHLHLHL. . . . For z layers it has been calculated that the maximum reflectance is given by (Refs. 4, 17)

$$R_{\lambda/4} = \left(\frac{n_{\mathrm{H}}^{z+1} - n_{\mathrm{L}}^{z-1} n_2}{n_{\mathrm{H}}^{z+1} + n_{\mathrm{L}}^{z-1} n_2} \right)^2 \tag{11-25}$$

where n_{H}, n_{L}, and n_2 are the high, low, and substrate indices. An expansion of Eq. 11–18 for $n_1 > n_2$ shows that the z layers are equivalent to a single layer whose effective refractive index is equal to $\sqrt{n_{\mathrm{H}}^{z+1}/n_{\mathrm{L}}^{z-1}}$.

The spectral characteristics of such a multilayer stack are shown in Fig. 11-12 for the case of a variable number of alternating layers of ZnS and MgF_2. Also shown is a portion of the microstructure of a multifilm stack composed of these materials. It is clear that the magnitude of the reflectance increases with the number of layers. The number of sideband oscillations outside the high-reflectance zone also increases with number of layers. The spectral width of the high reflectance zone is a function of the ratio of the refractive indices of the involved films, and there are a couple of practical ways to extend it. One is to select materials with n_{H} and n_{L} that are higher and lower, respectively, than those of ZnS and MgF_2. Another is to broaden the basis of design to include several wavelengths. In such a case the dielectric stack would be composed of staggered layer thicknesses so that consecutive maxima would overlap. In this way 15 layers of ZnS and Na_3AlF_6 with different optical thicknesses can be used to span the visible range. By similar methods dielectric mirrors are designed to operate in the infrared or ultraviolet with very small residual

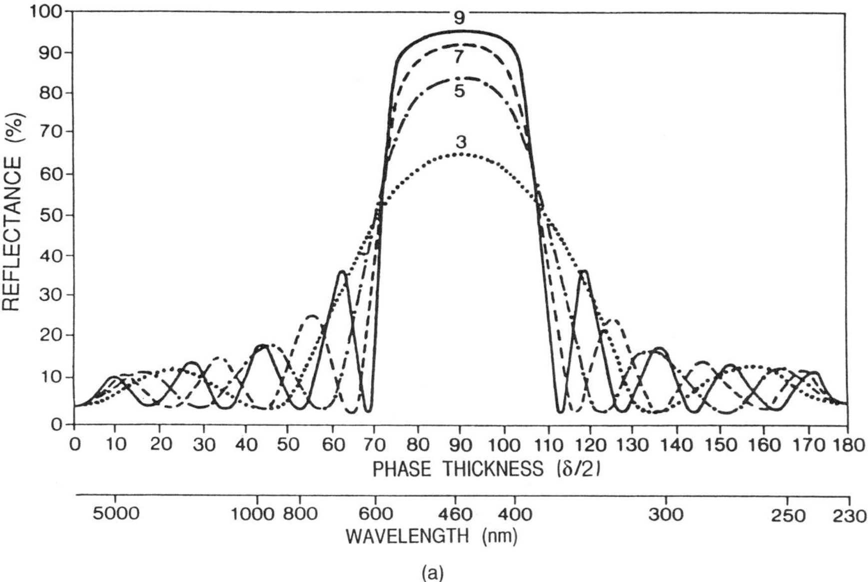

(a)

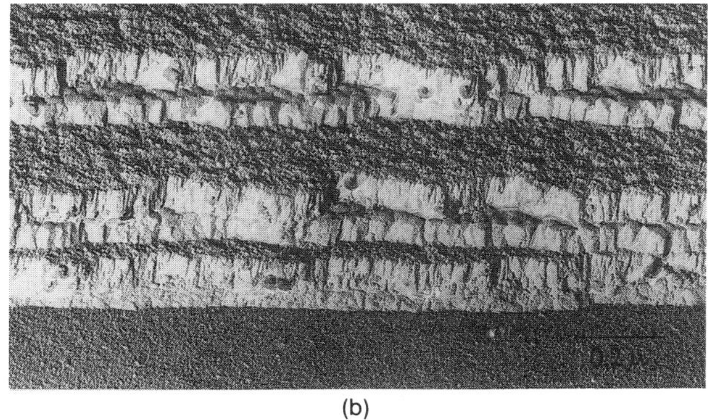

(b)

Figure 11-12. (a) Spectral characteristics of multilayer stacks formed of alternating $\lambda/4$ layers of ZnS and MgF_2 on glass ($n_2 = 1.52$) as a function of $2\pi nd/\lambda$. Normally incident light with $\lambda = 4600$ Å assumed. Number of layers in each stack is indicated. (From Ref. 18). (b) Transmission electron micrograph of a replica of the ZnS/MgF_2 multilayer cross section. (Courtesy of K. H. Guenther).

absorption. In reducing the difference between n_H and n_L, a narrow-band reflection filter, the minus filter of Fig. 11-11, can be generated.

Multilayer dielectric interference systems are ideally suited as reflection coatings for fully reflecting and partially transmitting laser mirrors. Negligible absorption means that reflectances of almost 100% can be achieved. Typical material combinations have included $ZnS-ThF_4$, TiO_2-SiO_2, and other oxide combinations in either broad or narrow spectral-band mirror configurations. Much attention must be paid to substrates employed where low light scattering and good film adhesion are critical requirements.

11.4.4. Cold Light and Heat Mirrors

There are two noteworthy practical variants of dielectric mirrors—cold light and heat mirrors. The cold light mirror spectral characteristics are shown in Fig. 11-13. It has high reflectivity for visible light but a high transmission for IR radiation. These characteristics are particularly suited to motion picture or slide projectors in order to avoid overheating the photographic emulsion. Intense light sources (e.g., carbon arc, xenon lamps) emit IR radiation in addition to visible light and the heat generated by the former must be dissipated. A cold mirror is thus placed at 45° in front of the light source. The heating infrared radiation passes through it while the nonheating visible light reflects off to illuminate the object. Metals cannot be used because they are good reflectors of the IR. Interference films are required and these must have low absorption in the IR. In addition the first film on the glass should be material having high reflectance in the visible and transmitting in the IR (e.g., Ge or Si). A few alternating $\lambda/4$ amplifying film layers on top of this help achieve the high reflectance over a suitably wide visible bandwidth.

Heat or dark mirrors have characteristics that are inverse to those of cold mirrors (Fig. 11-14). There are two approaches to achieving high visual transmittance simultaneously with high IR reflectance. The first is to employ

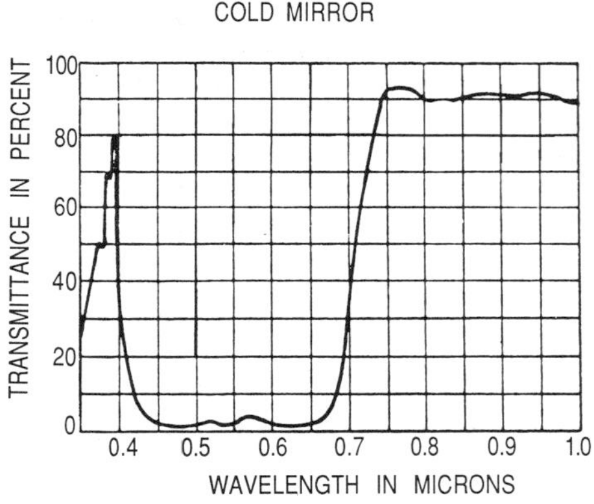

Figure 11-13. Spectral characteristics of a cold light mirror. (From Ref. 19 © Laurin Publishing Co. Inc.).

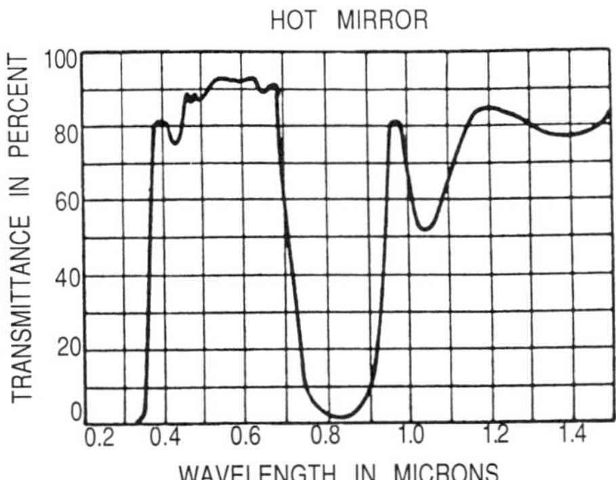

Figure 11-14. Spectral characteristics of a heat or dark mirror. (From Ref. 19 © Laurin Publishing Co. Inc.).

interference phenomena in an all-dielectric film stack. The second makes use of the properties of transparent conducting films. Consider the application to a low-pressure sodium vapor lamp, which consists of a Na-filled discharge tube within an evacuated glass envelope. For optimum Na pressure, the discharge tube must be kept at a temperature of about 260 °C. The necessary power for this is supplied by the gas discharge. However, the tube loses heat through radiation of energy in the far IR. Therefore, to conserve energy the inside of the envelope is coated so as to enable the (cold) yellow light to emerge while reflecting the IR back to the discharge tube.

In another energy-saving application, home window panes coated with heat mirrors would reflect heat back into the house in the winter. In the summer the window could be reversed so that the coating could reflect the IR from the sun and help provide interior cooling.

11.4.5. Photothermal Coatings

The direct conversion of solar radiation into energy for heating or cooling applications is a vital component of energy supply and conservation strategies. Coatings play an important role in photothermal conversion, and it is appropriate to briefly consider them because of their outward resemblance to the above mirrors. They differ because the substrate is usually a heat-absorbing metal panel. In addition, they are designed for optimal response to the spectral characteristics of sunlight. The situation can be modeled by noting that $A + R = 1$, where A is the coating absorbance. Strong absorption of sunlight in the range of 0.3–2.0 μm is required to heat the substrate. However, a portion of the heat will be lost by reradiation from the surface, reducing the overall conversion efficiency. Therefore, a second requirement of the coating surface is a low emittance or high reflectivity in the spectral region of reradiation—2–10 μm. Emittance ε is defined by the ratio of power emitted by a given surface to that of a blackbody. Clearly, higher values of A/ε result in desired higher equilibrium temperatures reached by the coating. (The similarity to the radiation limited temperature reached during sputtering should be noted. See p. 117.)

Solar absorbing coatings have been produced by physical and chemical vapor deposition techniques as well as by electroplating, anodization, acid dipping, painting, and spraying. Compositions include NiS–ZnS (black Ni), Cr–Cr oxide (black Cr), Al_2O_3–metal, SiO–metal, PbS, and Zn to name a few. Typical absorptances range from 0.90 to 0.98, and emittances of 0.1 are common.

11.4.6. Optical Filters

Filters are optical components that selectively change either the intensity or spectral distribution of light emitted by a source. They can be designed to change spectral characteristics over the total, a substantial fraction of the total, or only over an extremely narrow portion of the total wavelength range. Respective examples of these are shown in Fig. 11-11 and include

1. Neutral or gray filters, which reduce the light intensity equally for all wavelengths.
2. Broadband, short- or long-wave pass filters. The cold light and heat mirrors just described are specific examples.
3. Narrow bandpass or monochromatic filters.

Thin film coatings to achieve these ends consist of thin metal films, dielectric films, multilayer metal and dielectric film combinations, and all dielectric film stacks. These can be deposited on clear and colored glass substrates to produce the desired effects. In the very broadest usage of the term, filters can be thought to include mirrors and antireflection coatings but these optical devices are usually considered separate categories. Since the subject is a large one, discussion will be limited.

11.4.6.1. Neutral Filters. Neutral density filters consist of single metallic films of varying thicknesses on glass. They produce the desired uniform attenuation of light by reflection and absorption effects. Metals such as Cr, Pd, Rh, and Ni–Cr alloys are used for this purpose. The filter is usually characterized by its optical density, which is defined by log I/I_0 (see Eq. 11-3). Important applications of neutral filters can be found in spectroscopy equipment, color photography, and microscopy. They can be fabricated to span the visible as well as IR and UV portions of the spectrum.

11.4.6.2. Broadband Filters. Low- and high-pass edge filters fall into the category of broadband filters. They are characterized by an abrupt change between a region of high transmission and a region where light is rejected. Such an edge band filter is shown in Fig. 11-15, and is used to block out UV radiation from a mercury light source. Similar filters can create distinctions in light transmission and rejection between the visible and IR and well as across a narrow wavelength range entirely within the visible, IR or UV. Filters manufactured for the near-IR and visible employ Ag films, whereas Al is used for those operating in the UV. These metals are coated with dielectrics such as MgF_2, PbF_2, cryolite, and ThF_4. In the IR, Ge, Si, and Te layers find common use. All dielectric multilayer mirror systems can also be used as the basis for the design of edge filters, particularly those that require a sharp transmission between the pass and stop portions of the transmittance curves. The way to sharpen the transition is to increase the number of layers in the stack. Unfortunately, the amplitude and frequency of the sideband oscillations in the passband also increase when this is done. Suppression of these oscillations or "ripple" is one of the major concerns of filter designers.

Other common broadband filters consist of colored absorbing glasses in combination with interference edge filters or a pair of interference edge filters.

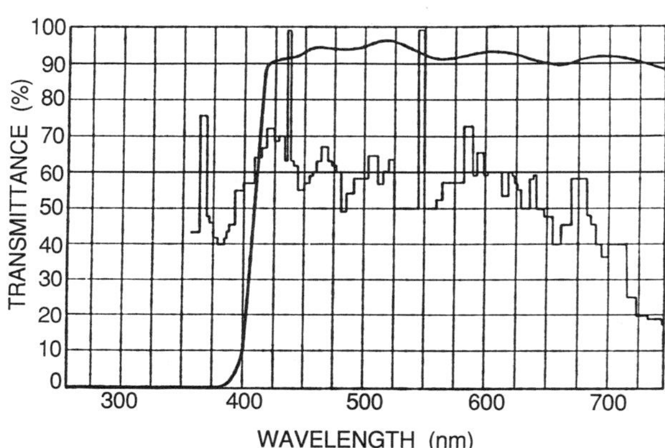

Figure 11-15. Mercury lamp light source spectrum and UV blocking filter characteristics. (From Ref. 5).

The latter can be made to have the inverse characteristics of the all-dielectric mirror stack—i.e., with high transmission instead of high reflectance, and vice versa. Wide ranges of the visible or IR can be selectively filtered this way.

There are many applications of wide-band filters in color photography, TV cameras, color separation schemes, studio illumination, microscopy, etc. We close with an additional pair of applications. The first involves using a filter to minimize heating of Si solar cells by eliminating the IR component from sunlight. Electron-hole pairs are only generated for wavelengths less than 1 μm and the cell is more efficient when cool. An edge filter with a cutoff beyond this wavelength would be called for. Such a filter can be combined with an antireflection coating to optimize efficiency.

A second interesting example involves filters employed in fluorescence microscopy (Ref. 5). Sometimes the excitation and emission wavelength bands used are so closely spaced that, unless precautions are taken, the two overlap, resulting in swamping of the fluorescent light output by the strong source light. This happens for example with FTIC, a fluorochrome employed in immunofluorescence. For excitation, maximum absorption occurs at 0.490 μm, and the emission maximum occurs at 0.520–0.525 μm. An edge filter with an exceedingly high steepness at ~ 0.500 μm is required. A filter with no less than 31 TiO_2–SiO_2 layers is required to suppress unwanted source radiation to levels of ~ 0.1% in the region where excitation occurs.

11.4.6.3. Narrow-Band Filters. These filters can be traced back to the use of the Fabry–Perot interferometer. The optical arrangement involved consists of two parallel facing, partially transmitting silver film mirrors separated by an air or dielectric layer. Light incident normally on this pair of mirrors is strongly transmitted only in a very narrow spectral range. This is a very surprising result, since one would expect the mirrors to reflect and filter the light; what little light the first allowed to be transmitted would be reflected back by the second mirror so that none would get through. This does not happen, however. Assume, for example, that the mirrors transmit 2% of the light and that 1 W of monochromatic light is incident. If the distance or cavity between mirrors is not an integral number of wavelengths long, the light waves that penetrate the first mirror will bounce to and fro and soon be out of phase. Of the 0.02 W incident on the second mirror, 0.0004 W will eventually be transmitted. If the cavity is, however, resonant, all waves will be in phase and their amplitudes will add, so that perhaps 50 W will circulate between the mirrors. Then 2%, or approximately 1 W, will be transmitted. This effect is relied upon in laser operation. The transmission maxima occur for $\lambda_0 = 2nd$, where nd is the effective optical thickness of the spacer layer.

Narrow bandpass filters can be fabricated in virtually any region of the spectrum. Figure 11-9 gives us a clue as to what is required. As the refractive index of the deposited interference film increases, not only does the reflectance increase at $\lambda/4$ but the region of high transmittance at $\lambda/2$ narrows considerably. The case where $r_1 = -r_2 = -0.98$ combines the high transmittance over a narrow range. Two conditions must be fulfilled to achieve this. First the optical structure must be symmetric about the spacer layer so that $|r_1| = |r_2|$. Second, high reflectance is required at each layer–substrate interface. Metal film mirrors can accomplish this but at the expense of some absorption losses. A desirable alternative when low loss is essential is to employ an all-dielectric film stack. The role of the stratified dielectric structure is to increase the reflectance by essentially raising the effective index of refraction as noted earlier.

11.4.7. Conclusion

In virtually all of the applications in this chapter the individual dielectric films have traditionally been modeled solely in terms of two parameters—thickness and refractive index. This simple approach will be inadequate in the future because of the steadily increasing performance requirements of advanced precision optical systems. The gap between theoretically predicted characteristics and performance attained in practice can be narrowed only by modifying the basic theory to include second-order effects. These include

1. Dispersion or the variation of refractive index with wavelength
2. Small amounts of absorption
3. Inhomogeneities resulting in the variation of refractive index throughout single films
4. Anisotropy in the refractive index with direction of radiation
5. Departures from perfectly planar boundaries

Concurrently, great strides have been made in improving the quality of optical materials and in controlling deposition processes. Likewise, characterization techniques have reached such high degrees of precision that measurements have exposed weaknesses in the theory and design of multilayer film systems. Computer-aided interactive feedback integrating theory, design, processing and performance of multilayer coatings is essential. In these ways, experience and art, which have so long and so well served the optical coating field, are being supplanted by more exact scientific approaches.

EXERCISES

1. Schematically sketch the optical absorption of two semiconductor films as a function of wavelength if one film is doped more heavily than the other. Is there a difference in absorption at the wavelength corresponding to E_g?

2. a. If the index of refraction of a GaInAsP semiconductor laser is $n_1 = 3.52$, what is the reflectance at the air interface?

b. To reduce R at the laser exit window a single-layer AR coating is required. What index of refraction and film thickness would you recommend for a 1.3-μm device?

3. Prove that

a. without AR coatings, surfaces of higher refracting glasses produce higher values of R than those of lower refracting glasses.

b. higher refracting glasses increase the effectiveness of a single $\lambda/4$ AR layer.

4. Compare the spectral response of the single AR layer ($n_0 = 1/n_1 = 1.38(\lambda/4)/n_2 = 1.52$) and the two-layer AR coating

$$\left(n_0 = 1/n_1 = 1.38(\lambda/4)/n_2 = 1.70(\lambda/4)/n_3 = 1.52\right)$$

by calculating R at $\lambda = 500$, 550, and 600 nm for each. [Note: The $\lambda/4$ layers are selected for $\lambda = 550$ nm, and n is assumed to be independent of λ.]

5. A 7.5-cm-long glass slide substrate of index of refraction $n_2 = 1.5$ is coated with ZnS for which $n_1 = 2.3$. Graph the expected percent reflectance at 0.55 μm as a function of position along the slide if

a. a uniform 2000-Å film is deposited.

b. a wedge-shaped film (zero thickness at one end, 2000-Å thick 10 cm away) is deposited.

c. an evaporated film is deposited from a surface source 10 cm directly below the center of the slide. The maximum film thickness is 2000 Å.

6. For an additional charge lenses on eyeglasses are coated. How does this enhance wearer personal appearance and vision or extend lens life?

7. A protected Al mirror is characterized by $n_0/n_1/n_2 - ik_2$, with $n_0 = 1$, $n_1 = 1.52$, $n_2 = 0.81$, and $k_2 = 5.99$ at $\lambda = 0.55$ μm.

a. Calculate r_1, r_2, and R.
b. What is the reflectance of the mirror?
c. How does R for an *unprotected* mirror compare with the answer to part (b)?

8. A step gauge consisting of thermal SiO_2 films on a Si substrate, varying in thickness from 200 to 5000 Å is viewed with a HeNe laser ($\lambda = 6328$ Å). If the refractive index of Si is $N = 4.16 - i0.018$, plot the reflectance versus SiO_2 film thickness.

9. A thin amorphous Si (a-Si) film ($n_a - ik_a$) on a (100) Si (c-Si) substrate ($n_c - ik_c$) shrinks in thickness during solid-phase epitaxial regrowth at elevated temperature. A He–Ne laser ($\lambda = 6328$ Å) probe beam reflects from both the surface and a–c interface establishing interference effects in the backscattered optical signal.

a. Show that the reflectivity for any given a-Si film thickness, d, is given by

$$R = \left[\frac{r_1 + r_2 e^{-4\pi k_a d/\lambda} e^{-i4\pi n_a d/\lambda}}{1 + r_1 r_2 e^{-4\pi k_a d} e^{-i4\pi n_a d/\lambda}}\right]^2.$$

b. If $N_a = 4.85 - i0.61$ and $N_c = 4.16 - i0.018$, calculate R as a function of d over the range 4000 to 0 Å. Characterize the resultant reflectivity oscillations. [Note: When $d > 4000$ Å, there is essentially no contribution from c-Si and $R = 0.438$. At $d = 0$, $R = 0.375$.]
c. Suppose the layers are reversed and a film of c-Si is at the surface on top of a thicker a-Si substrate layer beneath. How does the R vs. d(c-Si) dependence differ from the R vs. d(a-Si) dependence of the previous case?

10. Ion bombardment deposition of a (HLH, etc.) multilayer dielectric stack of nine layers of ZnS and MgF_2 on glass ($n_2 = 1.52$) raises each of the respective refractive indices by 4%. What change in reflectivity can be expected for such a structure relative to a traditionally evaporated stack? What if there were only five layers?

11. The inner surface of an incandescent lamp bulb is coated with a thin-film sandwich consisting of

ZnS (0.03 μm thick)
Ag (0.02 μm thick)
ZnS (0.03 μm thick)
Explain the function of these layers and the overall behavior of the lamp.

12. Explain why the thin-film coating consisting of air/SiO(λ/4)/Ge(λ/4)/opaque Al/glass substrate has a reflectance-wavelength response as follows:

Wavelength	%R
Visible	~ 2
1 μm	80
1.2 μm	90
> 2 μm	> 95

13. Calculate R for the following dielectric stacks.

System	No. of Layers	n Substrate	H.L.	λ (μm)
SH	1	1	CeO_2 ,Na_3AlF_6	0.55
SHLH	3	1.52	CeO_2 ,Na_3AlF_6	0.55
SHLHLH	5	1.52	CeO_2 ,Na_3AlF_6	0.55
SHLHLH	5	1.45	Ge, Na_3AlF_6	2.0
SHLHLHLH	7	1.50	ZnS, Na_3AlF_6	0.59

References

1.* H. A. Macleod, in *Applied Optics and Optical Engineering*, Vol. X, eds. R. R. Shannon and J. C. Wyant, Academic Press, New York (1987).
2.* O. S. Heavens, *Optical Properties of Thin Solid Films*, Dover, New York (1965).
3.* H. Anders, *Thin Films in Optics*, Focal Press, London (1967).
4.* K. I. Chopra, *Thin Film Phenomena*, McGraw-Hill, New York (1969).
5.* H. K. Pulker, *Coatings on Glass*, Elsevier, Amsterdam (1984).
6.* H. A. Macleod, *Thin-Film Optical Filters*, Adam Hilger, London and Macmillan, New York (1987).

*Recommended texts or reviews.

7. G. Hass, J. B. Heaney, and W. R. Hunter, in *Physics of Thin Films*, Vol. 12, eds. G. Hass, M. H. Francombe, and J. L. Vossen, Academic Press, New York (1982).
8. G. Hass and E. Ritter, *J. Vac.Sci. Tech.* **4**, 71 (1967).
9. C. Kittel, *Introduction to Solid State Physics*, 4th ed., Wiley, New York (1971).
10. N. F. Mott and H. Jones, *The Theory and Properties of Metals and Alloys*, Clarendon Press, Oxford, (1936).
11. H. Köstlin and G. Frank, *Philips Tech. Rev.* **41**, 225 (1983/4).
12. J. L. Vossen, in *Physics of Thin Films*, Vol. 9, eds. G. Hass, M. H. Francombe, and R. W. Hoffman, Academic Press, New York (1977).
13. M. Harris, H. A. Macleod, S. Ogura, E. Pelletier, and B. Vidal, *Thin Solid Films* **57**, 173 (1979).
14. P. J. Martin, *J. Mater. Sci.* **21**, 1 (1986).
15. G. Hass, *Optik* **1**, 8 (1946); G. Hass and M. Waylonis, *J. Opt. Soc. Am.* **51**, 719 (1961).
16. G. L. Olsen and J. A. Roth, *Mat. Sci. Repts.* **3**, 1 (1988).
17.* P. H. Lissberger, *Rep. Prog. Phys.* **33**, 197 (1970).
18. S. Penselin and A. Steudel, *Z. Phys.* **142**, 21 (1955).
19. The Optical Industry and Systems Purchasing Directory-Encyclopdia (1979).

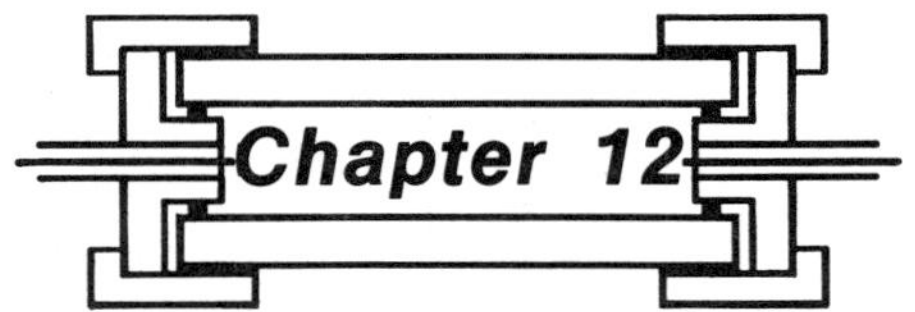

Metallurgical and Protective Coatings

12.1. Introduction

Paralleling the dramatic development of thin-film technology in microelectronics have been the no less than remarkable advances in what may be conveniently called metallurgical and protective coatings. The unusual materials which comprise these coatings are drawn from several classes of solids and include ionic ceramic oxides (e.g., Al_2O_3, ZrO_2, TiO_2), covalent materials (e.g., SiC, BC, diamond), transition metal compounds (e.g., TiC, TiN, WC) and metal alloys (e.g., CoCrAlY, NiAl, NiCrBSi). As a whole they are characterized by extremely high hardness, very high melting points, and resistance to chemical attack, attributes that have earmarked their use in critical applications where one or more of these properties is required; correspondingly the respective categories of hard, thermal, and protective coatings denote the functions to which they are put. Hard coatings of TiN and TiC, for example, are used to extend the life of cutting tools, dies, punches, and in applications such as ball bearings to minimize wear. The collection of coated cutting tools and dies shown in Fig. 12-1 is representative of the widespread commercial use of this technology in machining and forming operations. Thermal coatings find extensive use in gas turbine engines where they help to

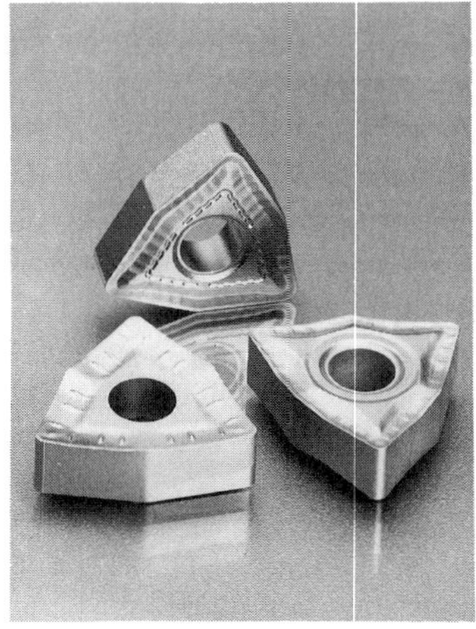

Figure 12-1. (Left) Assorted cutting and forming tools coated with TiN and multilayer coatings. (Courtesy Multi-Arc Scientific Coatings). (Top right) HSS forming and sheet metal dies coated with TiN and TiC. (Courtesy Ti Coating Inc.) (Lower right) multilayer coated cutting tool inserts. (Courtesy of S. Wertheimer, ISCAR Ltd.)

improve the performance and extend the life of compressor and turbine components. As the name implies, protective coatings are intended to defend the underlying materials, usually metals, from harsh gaseous or aqueous environments that cause corrosive attack. Such coatings have found applications in chemical and petroleum industries, coal gasification plants, as well as in nuclear reactors.

Employing coatings represents a significant departure from traditional engineering design and manufacturing practices. Processing components beyond the primary manufacturing steps of casting, forging, extrusion, machining and grinding, pressing and sintering, etc., has generally been resisted. This is due in part to a reluctance to tamper with the product, and to leave well enough alone. A compelling case was not made for the cost effectiveness of additional treatments. However, more recently several important factors have combined to firmly establish the practice of modifying the surface properties of engineering materials and components.

1. In many critical applications the design specifications call for properties that are simply beyond the capabilities of the commonly available and routinely processed materials. The new limits of behavior demanded can be met by the use of the unusually hard, temperature- and degradation resistant materials noted earlier. However, these materials are extremely difficult to fabricate in bulk form.
2. Concerns of limited availability of strategic materials, the thrust toward energy efficiency and independence, and an increasingly competitive world economy have exerted a strong impetus to considerably tighten engineering design, improve performance, and economize on materials utilization.
3. High-quality coatings possessing fewer surface imperfections than comparable pressed and sintered bulk parts made from powder, can now be reproducibly deposited. This is due to the advances made in our basic understanding of the deposition processes and the development of improved coating and deposition techniques.
4. The commercial availability of the necessary deposition chambers or reactors, hardware, computer-controlled processing equipment, and high-purity sources of precursor gases, powders and sputtering targets has facilitated the option of employing coatings.

Various combinations of the above factors have then resulted in the marriage of coatings to the underlying base materials, each with their particular set of desirable and complementary properties. For example, many structural materials with adequate high-temperature mechanical properties simply do not have the ability to withstand high-temperature oxidation, corrosion, particle erosion, and wear. On the other hand, the materials that do possess the environmental resistance either do not qualify as structural materials because of low toughness or, if they do, are prohibitively expensive to fashion in bulk form.

Before we turn to the main subjects of the chapter, it is worth noting some of the similarities and differences between the present mechanically and environmentally functional *coatings*, and the *thin films* of prior book chapters. In common, many coatings are deposited by the same type of physical (PVD) and chemical (CVD) vapor deposition techniques. Adhesion to the substrate, development of desirable structure and properties, and meeting performance standards are universal concerns. Among the differences are the following:

1. The coatings we will be considering are far thicker than thin films. Whereas a couple of microns, at most, is the arbitrary upper limit to what we have

called films, coatings typically range from several to tens and even hundreds of microns in thickness.

2. The maintenance of precise coating thickness and uniformity is not usually a major concern. There is generally a broad range of acceptable coating thicknesses. This is in contrast to the critical thickness tolerances and uniform coverage that must be achieved in optical and microelectronic films.
3. The substrate is frequently an integral part of the coating system. In diffusion coatings, for example, metalloid as well as metal elements are diffused into the substrate, creating thick, solute-enriched layers beneath the surface.
4. For the most part, the substrates employed for hard and protective coatings are rather special metals and alloys such as tool, high-speed, and stainless steels; iron-, cobalt-, and nickel-base superalloys; sintered tungsten carbide; titanium, etc. The use of the term *metallurgical coating* is based in part on this fact.
5. There are many methods for producing coatings. In addition to the vapor phase atomistic deposition processes (PVD, CVD) for films, coatings are also formed by

   a. Deposition of particulates (e.g., by thermal spraying of metal or oxide powders either through a hot flame or an even hotter plasma)
   b. Immersion of substrates in molten baths or heated solid packs
   c. Electrolytic processes such as electroplating, fused salt electrolysis, and electroless plating
   d. Miscellaneous processes, e.g., welding and enameling

6. Except for epitaxial semiconductor films, most thin-film depositions are carried out at relatively low-substrate temperatures. Metallurgical and protective coatings, however, are frequently deposited at elevated temperatures. Certainly this is true of the CVD coatings, and, therefore, atomic interdiffusion and reactions generally occur at the interface between coating and substrate. Compositional change can either be beneficial or detrimental to adhesion and coating properties depending on the materials involved.

The bulk of the chapter will be concerned with hard coatings and issues related to them. Somewhat lesser emphasis is placed on thermal and environmental coatings. To limit the treatment to manageable proportions, we deal with properties and the phenomena they influence in a fundamental way. The more widely used vapor deposition processes will be primarily discussed to maintain a consistency with prior chapters. Electrodeposition, for example, will not be mentioned again, since there is already a huge and accessible literature on the subject. Case histories and examples are always interesting and will be interspersed where appropriate. The specific topical outline of the rest of the chapter is

12.2. Hard Coating Materials
12.3. Hardness and Fracture
12.4. Tribology of Films and Coatings
12.5. Diffusional, Protective, and Thermal Coatings

12.2. Hard Coating Materials

12.2.1. Compounds and Properties

Hard coating materials can be divided into three categories, depending on the nature of the bonding. The first includes the *ionic* hard oxides of Al, Zr, Ti, etc. Next are the *covalent* hard materials exemplified by the borides, carbides, and nitrides of Al, Si, and B, as well as diamond. (See Section 14.2.) Finally, there are the *metallic* hard compounds consisting of the transition metal borides, carbides, and nitrides. Typical mechanical and thermal property values for important representatives of these three groups of hard materials are listed in Table 12-1. The reader should be aware that these data were gathered from many sources (Refs. 1–6) and that there is wide scatter in virtually all reported property values. Differences in processing (e.g., CVD, PVD, and sintering of powders), variations in structure (e.g., grain size, porosity, density, defects) and composition (e.g., metal–nonmetal ratio, purity), together with statistical error in measurement, contribute to the uncertainties. Perusal of this tabulated information leads to the following broad conclusions:

1. All of these compounds have extremely high hardnesses. This can be appreciated by noting that heat-treated tool steel has a hardness of about $H_v = 850$. Hardness is the most often quoted material property of hard coatings. Therefore, Section 12.3 has been specially reserved for an extensive discussion of the concept of hardness, the technique of measurement, and the significance of its magnitude in coatings.
2. These compounds have very high melting points and decomposition temperatures. For example, the decomposition temperatures of TaC, HfC, and diamond exceed the melting point for tungsten (MP = 3410 °C).
3. The modulus of elasticity is lowest for the ionic solids. In comparison, only

Table 12-1. Mechanical and Thermal Properties of Coating Materials

Material	Melting or Decomposition Temperature (°C)	Hardness ($kg\text{-}mm^{-2}$)	$H = H_0e^{-aT}$ (Eq. 12-4) H_0 ($kg\text{-}mm^{-2}$)	a (10^{-4} C^{-1})	Density ($g\text{-}cm^{-3}$)	Young's Modulus ($kN\text{-}mm^{-2}$)	Thermal Expansion Coefficient (10^{-6} K^{-1})	Thermal Conductivity (Wm^{-1} K^{-1})	Fracture Toughness ($MPa\text{-}m^{1/2}$)
Ionic									
Al_2O_3	2047	2100	2300	7.85	3.98	400	6.5	~ 25	3.5
TiO_2	1867	1100	1250	5.99	4.25	200	9.0	9	
ZrO_2	2710	1200			5.76	200	8.0	1.5	4–12
SiO_2	1700	1100			2.27	151	0.55	2	< 1
Covalent									
C (Diamond)	3800	~ 8000			3.52	1050	1	1100	
B_4N	2450	~ 4000			2.52	660	5		
BN	2730	~ 5000			3.48	440			
SiC	2760	2600	2800	0.90	3.22	480	5.3	84	3
Si_3N_4	1900	1700	1900	2.79	3.19	310	2.5	17	4
AlN	2250	1200			3.26	350	5.7		
Metal Compounds									
TiB_2	3225	3000	3500	18.9	4.5	560	7.8	30	
TiC	3067	2800	3300	18.3	4.9	460	8.3	34	0.46
TiN	2950	2100	2100	23.5	5.4	590	9.3	30	
HfN			2000	8.57			6.9	13	
HfC	3928	2700	3000	14.7	12.3	460	6.6		
TaC	3985	1600	1800	6.75	14.5	560	7.1	23	
WC	2776	2300	2350	3.62	15.7	720	4.0	35	
Substrate Materials									
High-Speed Steel	1400	900			7.8	250	14	30	50–170
WC-6%Co		1500				640	5.4	80	11.4
Ti	1667	250			4.5	120	11	13	80
Ni Superalloys	1280				7.9	214	12	62	> 100

the stiffest metals have modulus values overlapping those of the oxides listed.

4. The linear thermal expansion coefficient generally increases in going from the covalent to metallic to ionic hard compounds. Metals tend to have thermal expansion coefficients that are higher by approximately a factor of two or more than these hard compounds.
5. The thermal conductivity of the hard metallic and covalent compounds is comparable to that of the transition metals and their alloys. Good metallic electrical conductors have proportionally higher thermal conductivities. Ceramic oxides are the poorest thermal conductors.

The last two properties have important implications for the properties and performance of coatings. An important source of coating residual stress is the thermal contribution generated by the difference in expansion between coating and substrate. The illustrative problem dealing with TiC on steel (p. 419) is worth reviewing and indicates the magnitude of possible effects.

The susceptability to cracking of coatings subjected to varying temperature histories is an important limitation to the performance of thermal coatings. To see how thermal cracking can occur, consider the rapid cooling of a high-temperature component. The surface coating contracts more than the interior, which is still relatively hot. As a result, the surface forces the interior into compression and is itself stretched in tension. The coatings we have been discussing are weak in tension and prone to fracture at incipient crack flaws. A measure of this fracture susceptibility is given by the thermal shock parameter S_T. This composite quantity is derivable from heat transfer and thermal stress considerations that are defined respectively by the equations

$$Q = -\kappa(\Delta T/\Delta x) \tag{12-1}$$

and

$$\sigma = (E\Delta\alpha\,\Delta T)/(1-\nu). \tag{12-2}$$

In the first equation the heat flux Q is given by the product of the thermal conductivity κ and the temperature gradient. The second equation was considered previously (Eq. 9-17), where $\Delta\alpha$ is the difference in thermal expansion coefficient between coating and substrate. Elimination of ΔT yields

$$S_T^{-1} = \frac{\sigma}{Q\Delta x} = \frac{E\Delta\alpha}{(1-\nu)\kappa}. \tag{12-3}$$

Therefore, for a given thermal flux through a coating of thickness Δx, the stress that develops is inversely proportional to the thermal shock parameter. Ceramic oxides with the lowest S_T values are most prone to fracture by thermal shock, and metals with the largest S_T values are least susceptible. The hard metal and covalent compounds are intermediate in this regard.

12.2.2. Deposition Processes (Ref. 7)

Hard coatings have now been commercially deposited on perhaps hundreds of millions of cemented tungsten carbide cutting tools by CVD methods since the mid-1960s. However, advances in physical vapor deposition methods in the past decade have led to the fruitful competition in many applications between these two broad deposition processes. It is instructive to briefly review and compare several aspects of these processes as they relate to hard coatings. The more extensive discussion of CVD and PVD in Chapters 3 and 4 provide the appropriate background.

Since hard coatings are deposited on varied substrates (e.g., carbides, high-speed tool steels, ball-bearing steels, watch cases, jewelry, etc.), a range of substrate temperatures is involved, necessitating different deposition techniques and conditions. For example, TiN can be deposited by CVD over a broad range of temperatures using the same $TiCl_4$ precursor. Some typical reactions leading to the attractive, lustrous golden TiN deposits are

1. High temperature: 1200 °C > T > 850 °C

$$2TiCl_4 + N_2 + 4H_2 \rightarrow 2TiN + 8HCl.$$

2. Moderate temperature: 850 °C > T > 700 °C

$$TiCl_4 + CH_3CN + (5/2)H_2 \rightarrow Ti(C,N) + CH_4 + 4HCl.$$

3. Low temperature: 700 °C > T > 300 °C

$$2TiCl_4 + 2N_2 + 7H_2 + \text{Ar plasma} \rightarrow 2TiN + 8HCl + 2NH_3.$$

Analagous reactions involving gaseous carbon compounds, such as CH_4 result in TiC deposits. Precursor gases and operating temperatures for these as well as other coatings have been given previously in Chapter 4.

One of the important advantages of CVD methods is the ability to batch-coat large numbers of small tools at one time. The schematic of such a commercial system capable of individual or sequential TiC, TiN, and Al_2O_3 depositions is shown in Fig. 12-2a. Typical gas feed proportions for these materials are

TiC–$TiCl_4$ (1–2%);	CH_4 (3–6%);	H_2 (96–92%)
TiN–$TiCl_4$ (1–2%);	N_2 (22–26%);	H_2 (77–72%)
Al_2O_3–$AlCl_3$ (0.18–0.3%);	CO (0.2–0.5%);	H_2 (99%)

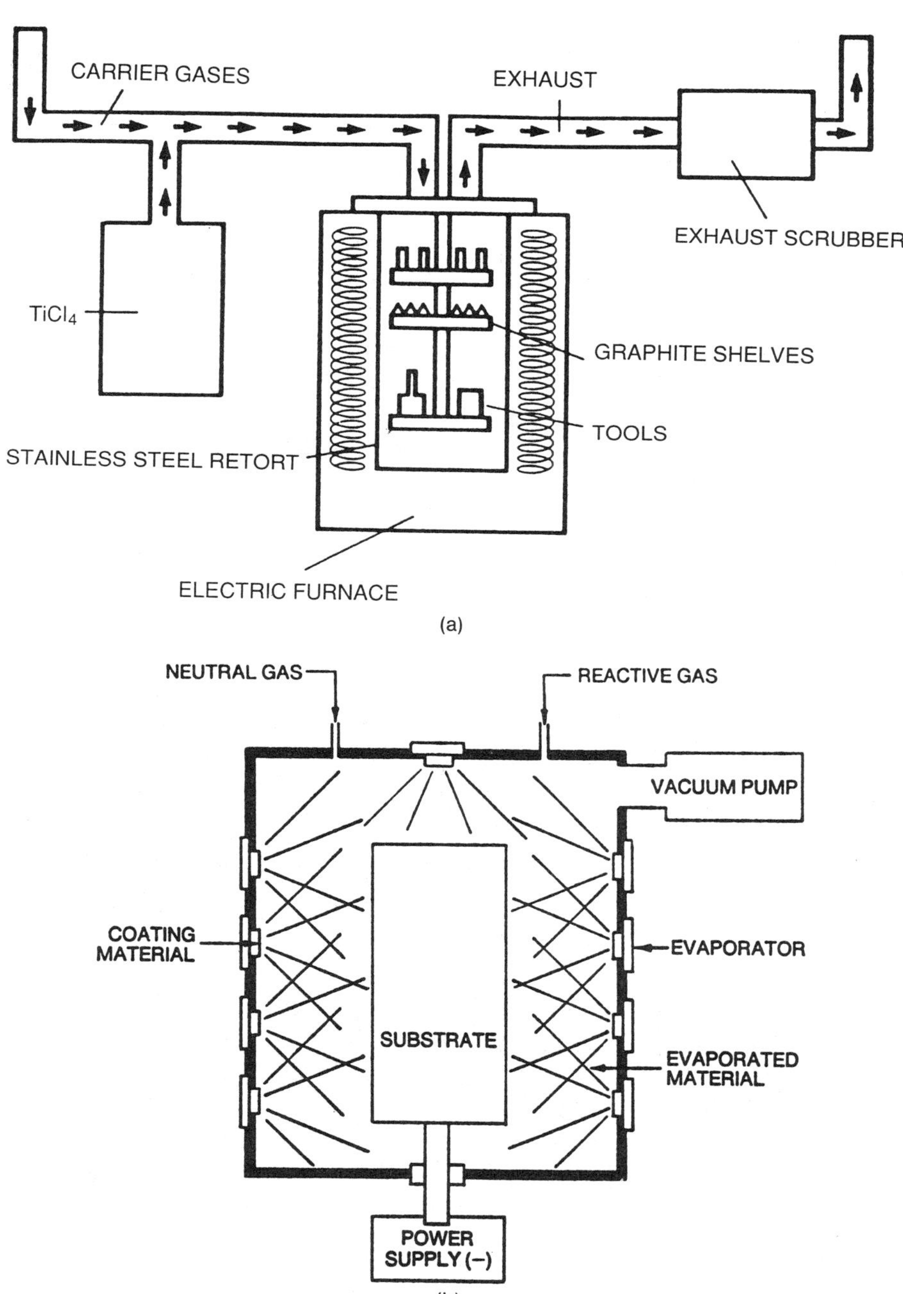

Figure 12-2. (a) Schematic view of commercial CVD reactor for deposition of TiC, TiN and Al_2O_3 on carbide cutting tools. (b) Cathodic-arc process for reactive evaporation of TiN coatings. (Courtesy of A. Gates, Multi-Arc Scientific Coatings Inc.)

Temperatures within the range 700–1100 °C are maintained to ±2 °C. As many as 10–20,000 WC tool inserts (1/2 in square) can be simultaneously given a ~ 6-μm-thick coating during a several-hour treatment cycle. The fracture surface morphologies of single-layer CVD TiC, TiN and TiCN as well as HfN coatings deposited on cemented carbide substrates are shown in Fig. 12-3. Rough transverse fracture surfaces containing columnar-like grains rising from a sharply defined substrate interface, typically characterize these coating structures. In the coating plane a mountainous relief of jagged crystals is evident.

The three basic PVD coating processes are evaporation, sputtering, and ion plating together with reactive variants incorporating chemical reactions (Refs.

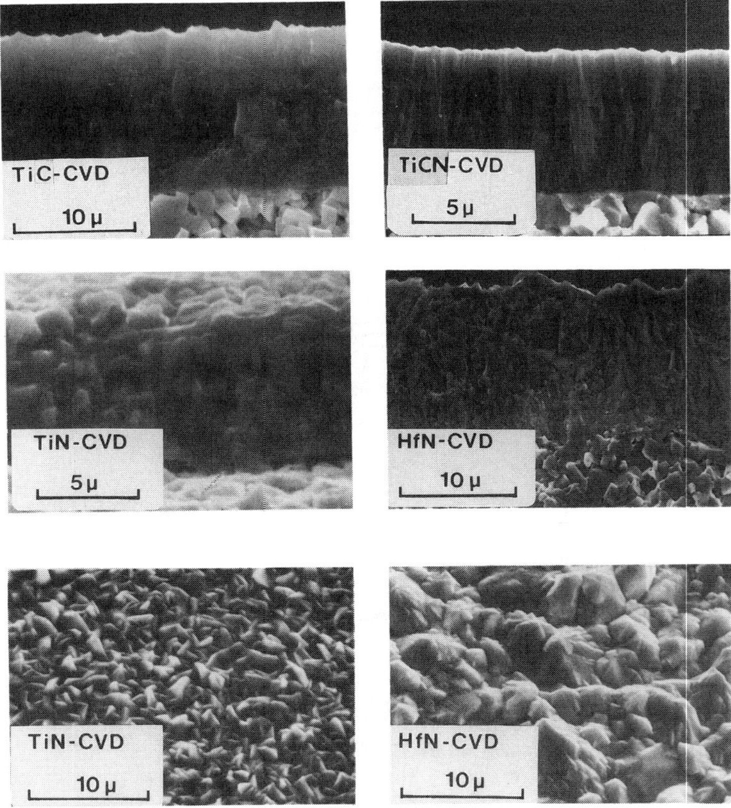

Figure 12-3. Morphology of representative CVD coatings viewed in the SEM. Top–Transverse fracture surfaces of TiC and TiCN. Middle–Transverse fracture surfaces of TiN and HfN. Bottom–Lateral surface structures of TiN and HfN. Substrate is cemented carbide. (Courtesy of D. T. Quinto, Kennametal, Inc., from Ref. 8).

8, 9). An important advantage of PVD is that substrate temperatures in the range of 200–600 °C, rather than 800–1200 °C for CVD, are utilized. This leads to somewhat finer coating grain sizes possessing slightly higher hardnesses. Because of the high deposition temperatures in CVD, substrate steels are sometimes softened, requiring subsequent hardening and tempering treatments. Undesirable part distortion occasionally results from this sequence of thermal treatments. In addition, substrate elements such as iron, chromium, and carbon diffuse into the coating altering adhesion as well as substrate integrity. These difficulties do not arise in the low-temperature PVD coatings. An added advantage of commercial PVD processes is the generally higher deposition rates of 4–6 μm/hour relative to typical CVD rates of 1–2 μm/hour. Nevertheless, throughput of small parts is substantially greater for CVD because substrate coverage is not limited by line-of-sight gas impingement as in PVD. On the other hand, the hazard of handling toxic and corrosive gases is considered a disadvantage of CVD processes. For larger tools such as drills and milling cutters, the PVD techniques of reactive evaporation and sputtering are widely employed. Commercially, uniform coating of tools and dies is accomplished through multiple evaporators arranged to surround the work as shown in Fig. 12-2b. While single (e.g., TiC), double (e.g., TiC–TiN)

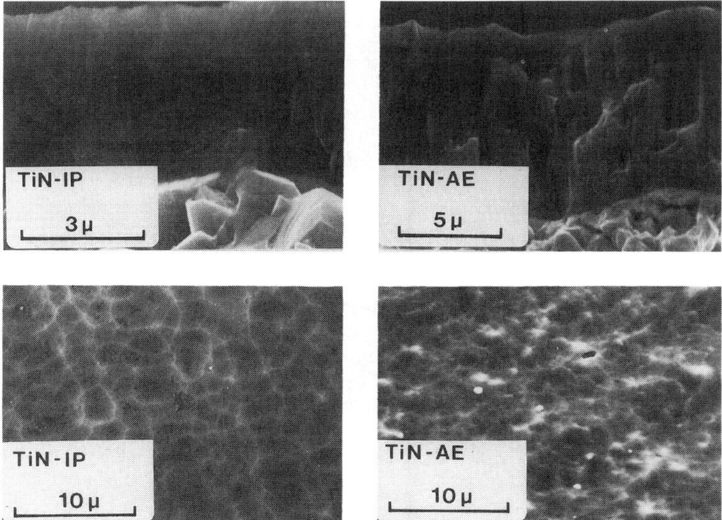

Figure 12-4. Morphology of PVD coatings of TiN prepared by ion plating (IP) and arc evaporation (AE) viewed in the SEM. Top–Transverse fracture surfaces. Bottom–Lateral surface structures. Substrate is cemented carbide. (Courtesy of D. T. Quinto, Kennametal, Inc., from Ref. 8).

and even triple (e.g., TiC–Al_2O_3–TiN) hard layers are commonly produced by CVD, commercial PVD coatings are largely limited to single TiN layers. Transverse fracture and lateral surface morphologies of TiN prepared by two different PVD processes are reproduced in Fig. 12-4. The smooth fracture surfaces composed of densely packed, fine columnar grains should be compared with the microstructures of CVD TiN (Fig. 12-3). Differences in grain morphology can be attributed to varying conditions of deposition temperature and energetic bombardment by vapor phase atoms.

12.2.3. Alloy Compounds (Refs. 3, 10)

Just as metallic alloys display properties superior to each of the component elements, so there has been considerable interest in improving the properties of the transition metal compounds through alloying. Fortunately, this is possible because the binary carbides and nitrides tend to be miscible in the solid state. For example, at elevated temperature, TiC reacts with VC, ZrC, NbC, HfC,

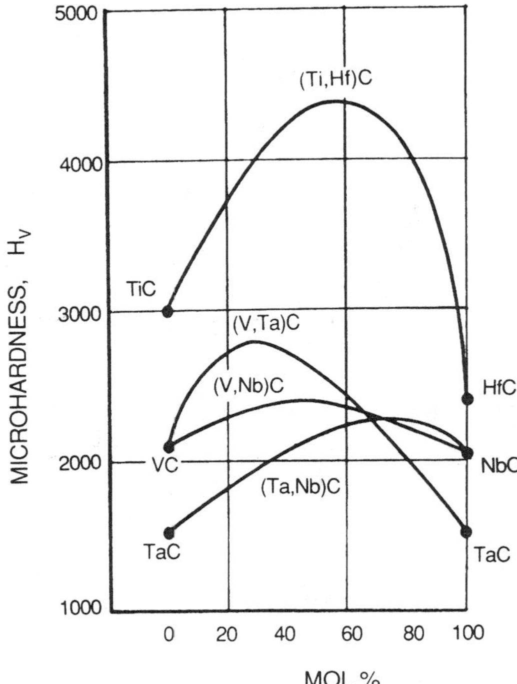

Figure 12-5. Microhardness of mixed carbides due to solid solution and precipitation hardening (From Ref. 3).

and TaC to produce mixed carbide solid solutions, such as (Ti, V)C, (Ti, Zr)C, (Ti, Nb)C. Since these carbides have the same crystal structures, TiC–MC (M = V, Zr, etc.) phase diagrams resemble the one for Ge–Si shown in Fig. 1-12, at least at high temperatures. Similarly, TiN can be alloyed with VN, ZrN, NbN, and HfN to produce solid solutions of mixed nitrides. The increase in hardness for the mixed carbides can be quite extensive, as shown in Fig. 12-5. Substitution on the metal sublattice results in maximum hardening at specific alloy concentrations. Similar increases in hardness have been observed in the mixed metal nitride solid solutions as well as in metal carbonitrides, e.g., Ti(C, N), Hf(C, N). Property benefits also extend to oxide alloys. For example, aluminum oxynitride has greater chemical and thermal stability than Al_2O_3 alone and is used to coat tools.

Quaternary alloys arising from mixtures of carbides and nitrides of different metals (e.g., Ti–V–C–N) have also been investigated and appear to provide a promising basis for the selection of yet other new composite coating materials. Subsequent heat treatment of the alloyed carbides, nitrides, and carbonitrides can lead to complex low-temperature phase transformations and precipitation reactions. Dispersions of hard phases such as TiB_2 or B_4C in TiC or TiN matrices have produced coatings that show better wear resistance and adhesion as well as higher toughness than single-phase layers.

12.2.4. Multilayer Coatings (Refs. 3, 11)

Distinct from *multicomponent single* layers is the *multilayer* concept. Several strategies have been adopted in creating composite multilayers:

1. Mutually soluble layers, e.g., TiC–TiN or Al_2O_3–AlN, are sequentially deposited. Alloys with enhanced properties are produced this way.
2. Compounds possessing coherent interfaces such as TiC–TiB_2 or TiN–TiB_2 are sequentially deposited. Adhesion between layers is promoted by low energy bonding between common elements.
3. Separate layers of largely noninteracting compounds are sequentially deposited. Each layer has a specific purpose, such as interfacial adhesion, diffusion barrier, abrasion resistance, etc.

The last approach is one that has attained considerable commercial success. Cross sections of such coatings on WC inserts for assorted cutting tools are shown in Fig. 12-6. They contain some combination of TiC, TiN, and Al_2O_3 layers deposited by CVD and are typically 6–9 μm thick. The multilayer of Fig. 12-6a is tough and suitable for rough, semifinishing, and interrupted

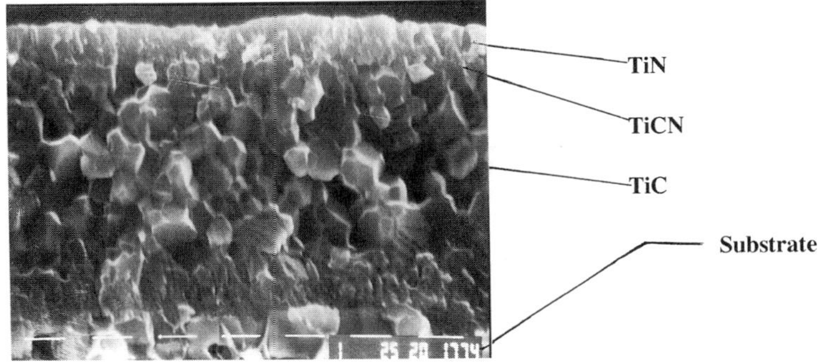

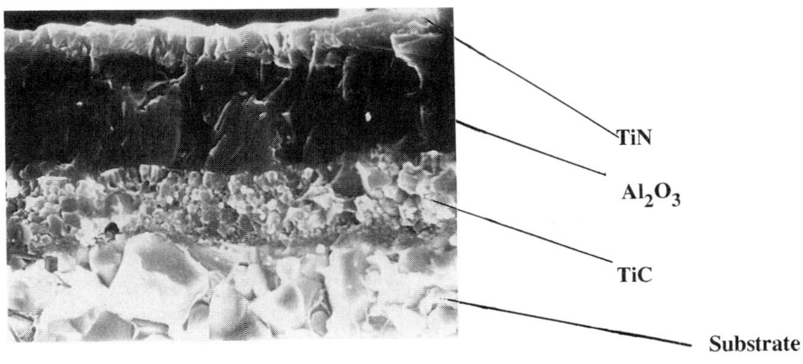

Figure 12-6. SEM images of CVD multilayer coatings for cutting tool inserts. (a) Carbide substrate/TiC/TiCN/TiN (5500 ×). (b) Carbide substrate–TiC–Al_2O_3–TiN (3500 ×). (Courtesy of S. Wertheimer, ISCAR Ltd.)

machining of plain carbon, alloy, tool, and stainless steels. The multilayer of Fig. 12-6b containing Al_2O_3 has many of the same characteristics but is capable of cutting at higher speeds without tool failure. Because of its low thermal conductivity, Al_2O_3 shields the cemented carbide substrate and enables the generated heat to be diverted to the chip instead. Among the reasons for the use of TiN as the outer coating layer are its high lubricity, chemical stability, and the lack of adherence of abrasive oxide particles to it. As indicated in Section 12.4.3 tools coated with multilayers suffer considerably less wear than tools with only single-layer TiC or TiN coatings.

12.3. Hardness and Fracture

12.3.1. Hardness

Hardness is an important material property of concern in films utilized for electronic and optical as well as mechanically functional applications. It affects wear resistance and plays an important role in the friction and lubrication of surface films in contact. Hardness, a complex property related to the strength of interatomic forces, apparently depends on more than one variable. Hard materials are generally modeled by a deep potential energy well (Fig. 1-8b) with steep walls. These characteristics imply a combination of high cohesive energy and short bond length. Measures of the cohesive energy are the heat of sublimation and enthalpy of compound formation (ΔH_{298}). If some account is

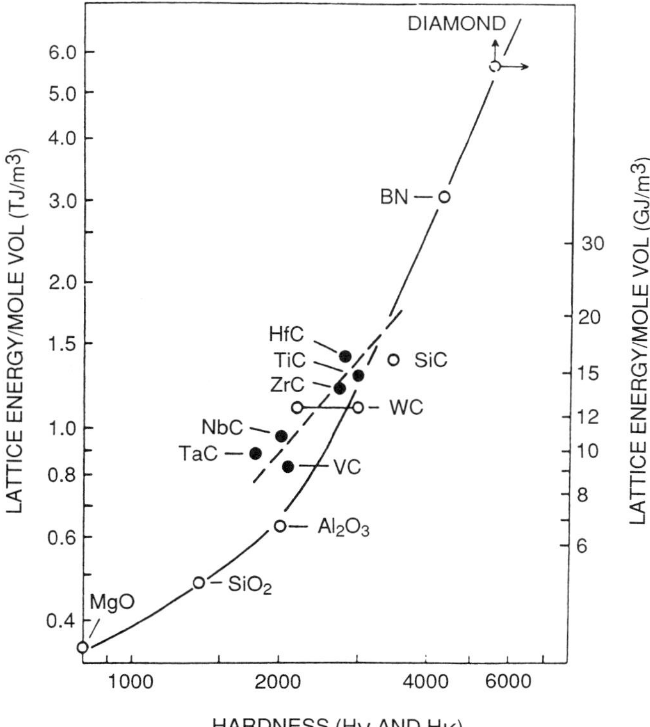

Figure 12-7. Relation between hardness and cohesive energy per molar volume expressed in terms of heat of sublimation (○-Left axis) and heat of formation (●-Right axis).

taken of the bond length by dividing the cohesive energy by the molar volume of the material, then the correlations with hardness shown in Fig. 12-7 result. It is apparent that the hardest materials are covalently bonded, and that increasing the ionic character of the bond leads to reduced hardness. At a microscopic level, directional bonds more readily resist distortion and rupture by concentrated loads than do ionic bonds.

The hardness of a material is usually defined as its resistance to local plastic deformation. Since the hardness test consists of pressing a hard indenter into the surface, it is equivalent to performing a highly localized compression test. A correlation between hardness H and yield strength σ_y is therefore expected. Typically $H \approx 3\sigma_y$ in bulk metals, but the correlation has not been directly verified in films or hard coatings. It is well known that σ_y for bulk materials decreases with temperature, a fact that facilitates hot mechanical forming operations. Therefore, it can be anticipated that the hot hardness of coatings will be lower than that at ambient temperatures for which values are usually quoted. It is these hot hardness values that are important in thermal coatings and in applications such as machining. A functional relationship between H and T given by

$$H = H_0 e^{-aT} \tag{12-4}$$

has been proposed (Ref. 6). Values for the constants H_0 and a, determined from experimental data, are given in Table 12-1.

12.3.2. Hardness Testing

Hardness testing of films and coatings is a relatively simple (though deceptively so) measurement to perform. The most frequently employed methods are modifications of techniques having long standing in the metallurgical community. The Vickers hardness test, also known as the diamond pyramid hardness (DPH) test, employs an indenter consisting of a square-based diamond pyramid ground to have a face angle of 136°. H_v, the Vickers hardness number, is obtained as the ratio of the applied load L to the surface area of the resulting indentation. It is given by

$$H_v = 2\cos 22^\circ \frac{L}{l_v^2} = \frac{1.854L}{l_v^2}\ \text{kg/mm}^2, \tag{12-5}$$

where l_v is the indentation diagonal. A related test is the Knoop microhardness test, which employs a rhombic-based diamond pyramid indenter where the length ratio of the major to minor indentation diagonal is 7.11; furthermore, the minor diagonal is four times the penetration depth. The Knoop hardness is given by

$$H_k = 14.22\,L/l_k^2 \text{ kg/mm}^2 \tag{12-6}$$

where l_k is the length of the major indentation diagonal. Care must be taken in distinguishing H_v and H_k values, which are sometimes used interchangeably, even though the former are some 10–15% lower than the latter.

Testing loads applied to the indenter are selected to produce an indent whose lateral dimensions are large enough to be measured with an optical microscope or SEM, and whose depth is small compared with the film thickness. Caution must be exercised so that the plastically deformed zone under the indenter does not extend to the substrate where stress interactions may be reflected in erroneous hardness readings. Film-thickness-to-indentation-depth ratios ranging from 5 to 10 appear to be safe in most situations. For example, suppose that in TiN coatings, a value of 7 has been established as safe. In the Vickers test this means that the indent geometry requires that the film thickness must be at least larger than l_v. For a TiN coating where $H_v = 2000$ kg/mm^2, a 10-g load generates an l_v value of ~ 3 μm. This would be the minimum acceptable film thickness for hardness testing in this case. Unusually small indents of less than a few microns in size should be avoided because of the difficulty in resolving dimensions. In TiN films, for example, the same indent measured by the same investigator optically and with an SEM gave values of 3238 and 2359 kg/mm^2, respectively (Ref. 12). Hardness comparisons should, therefore, only be made when testing methods and film thicknesses are maintained approximately constant.

Further complications arise in interpreting hardness values in hard materials. First, indentation-induced cracking phenomena make it difficult to measure indent dimensions precisely. Second, and perhaps more importantly, the quoted hardness values are load-dependent. For small loads, it is observed that the indentation dimension l varies with load as $L = K_1 l^m$, where K_1 is a constant and $1 < m \leq 2$. Elimination of l_v (or l_k) in Eq. 12-5 (or 12-6) yields

$$\frac{dH}{dL} = \frac{m-2}{m} K_2 \left(\frac{K_1}{L} \right)^{2/m}, \tag{12-7}$$

where K_2 is either 1.854 or 14.22. A drop in hardness with increasing indenter load is thus predicted because $m - 2$ is negative. In polycrystalline SiC, WC–Co cemented carbide, and bearing steels, m equals 1.70, 1.88, and 1.92, respectively (Ref. 13). It is this load dependence of hardness that may be partially responsible for claims of unusually outstanding hardness values in films and coatings.

It is appropriate to end this section with brief mention of *nano*, rather than *micro*, hardness indentation measurements. The Nanoindenter, which was introduced previously (Section 9.2.4) is capable of reproducibly making indentations as shallow as 500 Å and as small as 5000 Å across. Unlike the Vickers and Knoop tests, the Nanoindenter produces a triangular pyramidal indentation. Small load ranges of 0–20 mN and 0–0.120 mN, make it possible to space indents $\sim 1\ \mu m$ apart when determining the hardness profile across a coating surface.

12.3.3. Effect of Microstructure on Hardness

A fundamental cornerstone of materials science is the relationship between microstructure and properties. Following Sundgren and Hentzell (Ref. 4) connections between the hardness of films and coatings and various microstructural characteristics will be discussed.

12.3.3.1. Grain-Size and Grain-Boundary Structure. In metal and alloy films the hardness correlates well with the Hall–Petch relation (Eq. 9-4), which may be written in the variant form

$$H = H_i + K_H l_g^{-1/2}, \tag{12-8}$$

where H is the hardness, H_i is the intrinsic hardness of a single crystal, l_g is the grain size, and K_H is a material constant. In refractory compound coatings, however, Eq. 12-8 appears to be invalid because the hardness of fine-grained films is frequently close to the bulk material value.

What does apparently affect the hardness and strength of refractory compounds is the perfection of the grain boundaries. Porosity and fine microcracks are very deleterious to such coatings and lower their strength and hardness significantly. Metal films are, however, somewhat more tolerant of such defects, whose influence can be blunted by plastic deformation effects. Increasing the substrate temperature is the simplest and most common way to reduce the grain-boundary defect structure and enhance the hardness of these compounds. The effect can be rather dramatic. For example, an increase in substrate temperature from 100 to 600 °C raised the hardness of magnetron-sputtered TiN from 1300 to 3500 kg/mm^2 (Ref. 14). Elevated temperatures eliminate void networks and evidently promote the strengthening of grain boundaries. The grain size increases as does hardness, contrary to the Hall–Petch prediction. With further grain growth at still higher temperatures, the hardness drops and reaches bulk levels. Metal films display a rather different character with respect to the influence of substrate temperature. Raising the latter increases thermal stresses and grain size, but reduces the intrinsic stress and hardness.

12.3.3.2. Metastable Structures. Metastable phases are frequently observed in refractory compound films. As in the case with metastable metal alloy films, high deposition rates and low substrate temperatures are conducive to the formation of nonequilibrium structures and a fine grain size. Manifestations of the metastability are the incorporation of C and N in interstitial lattice sites and the generation of supersaturated solid solutions. This generally occurs during PVD rather than CVD, which is usually carried out under conditions closer to thermodynamic equilibrium. The incorporated interstitials tend to distort the lattice and the subsequent difficulty in initiating dislocation motion is reflected in increased hardness. The effect can be quite large. Hardnesses of 2500 to 3500 kg/mm^2 are generally found in reactively sputtered HfN films compared to 1600 in CVD-grown films. Apparently the bombardment of the growing HfN film by the sputtering gas forces N into tetrahedral interstitial positions. This creates a high compressive stress in the plane of the film and thus a higher hardness.

In general, metastable phases and structures can be frozen in up to temperatures of $0.3T_M$ (T_M is the melting point). As a consequence, metastable hard coating systems can be used at temperatures up to 550–800 °C (for $T_M =$ 2500–3300 °C) (Ref. 10).

12.3.3.3. Impurities. Since hard PVD coatings are generally grown in medium vacuum or under even higher pressure ambients, the incorporation of noble gases, C, N, and O from residual gases, and impurities from chamber hardware and walls is not uncommon. The deposit impurities are located in both substitutional, interstitial as well as grain-boundary sites at total levels up to a few atomic percent. Even at such low concentrations, the effect on hardening can be pronounced. The mechanism of hardening due to impurities apparently involves the electrostatic attachment of the latter to charged dislocations in ionic materials and to dangling dislocation bonds in covalent compounds. Such interactions limit dislocation mobility by pinning effects.

12.3.3.4. Film Texture. By texture we mean the preponderance of one (or more) crystallographic planes oriented parallel to the film surface compared with the case of randomly oriented planes. In the latter, isotropic behavior may be expected. Films grown by PVD and CVD techniques usually display a preferred orientation, however, with low index planes lying parallel to the substrate surface, creating a texture that is strongly dependent on virtually all deposition and process variables. Factors of as much as 2 in hardness have been observed in cubic coatings (e.g., TiC, TiN, ZrC) as a function of the preferred orientation plane [e.g., (111), (100) and (110)]; similar hardness anisotropy on basal (0001) and prismatic (1100) planes of hexagonal materials (e.g., WC, SiC) has been reported.

12.3.4. Fracture

It is a fact of nature that materials that are extremely hard are simultaneously brittle and prone to fracture. The phenomenon of fracture is of cardinal concern in a great many materials engineering applications; once fracture of a component or structure occurs, other issues quickly assume secondary importance. Films and coatings are no exception. For example, with the exception of oxidation wear, all wear mechanisms are based on some sort of crack development that creates new surfaces, from which particles can be detached by fracture processes. In a similar vein, high-temperature fracture or spalling of coatings leaves the underlying substrate unprotected and at the mercy of harsh corrosive atmospheres.

At the outset it is important to distinguish between brittle and ductile fracture. The latter occurs after the material has undergone some plastic deformation. Metals tend to undergo ductile fracture upon overloading. Brittle fracture, on the other hand, occurs rapidly, without warning and in such a way that the broken pieces can usually be neatly fitted together. Generally, brittle fracture occurs more readily in materials having small tensile strengths compared with their compressive strengths. Thus, glasses, ceramic oxides, and covalent, as well as hard metal compounds, to a lesser extent, are particularly prone to brittle fracture.

Two approaches to fracture of coatings will be presented next. The paramount assumption is that the materials involved possess an intrinsic collection of structural flaws distributed laterally as well as through the coating thickness. Voids, porosity, interconnected voids or porosity, voided or grooved grain boundaries, local regions of de-adhesion, etc. may be viewed, in a broader context, as flaws and even incipient cracks. Under either external or residual internal stressing, each flaw will locally concentrate stress and the surrounding material will tend to deform. If the stresses are sufficiently large, they can ultimately destroy the coating by crack propagation if tensile, and by wrinkling or buckling if compressive.

12.3.5. Statistical Strength of Brittle Materials

A useful concept for interpreting the strength of brittle solids is that of the "weakest link." Under load the worst flaw is assumed to govern the strength and useful life of the object. Since the flaw size geometry and distribution varies from sample to sample, there will be a distribution in fracture strength. The Swedish engineer Weibull suggested a way of handling the statistics of strength. He defined the survival probability $P(V_0)$ as the fraction of identical samples each of volume V_0 that survive loading to a tensile strength σ,

$$P(V_0) = \exp\left[-\left(\sigma/\sigma_f\right)^m\right], \tag{12-9}$$

where σ_f and m are constants. When $\sigma = 0$, then $P(V_0) = 1$ and all samples survive. At large stress levels all samples fail and $P(V_0) \to 0$. When $\sigma = \sigma_f$, then $P(V_0) = 1/e = 0.37$ and 37% of the samples survive. The constant m, known as the Weibull modulus, reflects how rapidly the strength falls as σ_f is approached. The greater the m value, the greater is the probability that the material will fail at very nearly the same stress level. In a material like steel, $m = 100$ and the reader can show that a plot of P vs. σ resembles a step function. Thus, for $\sigma < \sigma_f$, $P = 1$ and for $\sigma > \sigma_f$, $P = 0$. This expresses the fact that steel has a rather well-defined fracture stress, σ_f. For brittle ceramics, however, m may be 5–10, and there is a much greater variability in strength, as evidenced by a statistically large spread in hardness values. These facts are displayed in Fig. 12-8 on the Weibull probability axes. If the sample volume

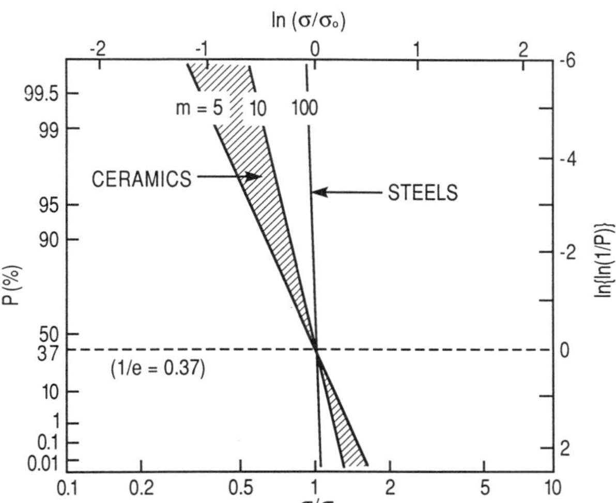

Figure 12-8. Survival probability of various materials plotted on Weibull probability axes. (Reprinted with permission. From Ref. 1; © 1986; Pergamon Press).

(V) is n times larger than V_0, i.e., $V = nV_0$, the probability of incorporating a fatal defect is enhanced leading to a smaller value of P. Under these circumstances, $P(V) = [P(V_0)]^n$ and

$$P(V) = \exp\left[-\frac{V}{V_0}\left(\frac{\sigma}{\sigma_f}\right)^m\right]. \tag{12-10}$$

Equation 12-10 is more strictly valid for bulk solids, but it has been applied to the study of macrocrack distributions in TiC coatings 2–10 μm thick (Ref. 15). As expected, more cracks are observed in thicker deposits, a reason for restricting coating thickness in practice. To analyze such situations, we can substitute film thickness for volume in Eq. 12-10.

12.3.6. Griffith Crack Theory and Fracture Toughness

We now turn to the issue of the stability of a *single* crack under load. Fracture mechanics provides a basis for addressing this situation. The Griffith theory considers a thin flat crack of length $2l$ and depth d in a plate (coating) that is assumed to be elastically isotropic. Tensile stresses act to open the crack, as shown in Fig. 12-9a. As the applied stress increases, more and more elastic strain energy E_ε is released from the coating. At any stress level this energy has a magnitude of $(1/2)\sigma\varepsilon$ (ε is the strain) per unit volume and represents

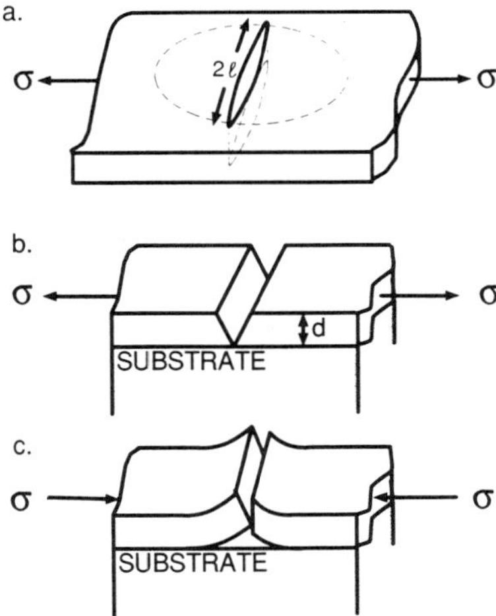

Figure 12-9. (a) Flat elliptical crack of length $2l$ in a coating loaded uniaxially in tension. (b) Crack leading to fracture of coating. (c) Crack leading to delamination of coating.

the area under the elastic stress–strain curve. By Hooke's law, $\sigma = E\varepsilon$ and

$$E_\varepsilon = \sigma^2/2E. \tag{12-11}$$

The elastic forces are resisted by interatomic bonds that must be broken for further crack extension. At any instant the total energy E_T associated with these opposing tendencies is given by

$$E_T = -\frac{\sigma^2}{2E}\pi l^2 d + 4\gamma l d. \tag{12-12}$$

The elastic strain energy is assumed to be roughly concentrated in a disk of volume $\pi l^2 d$ surrounding the crack. The second term represents the effective energy associated with broken bonds, where γ is the surface energy per unit area of crack. Rapid extension of the crack occurs when the rate of strain energy release exceeds the rate at which energy is absorbed by bonds at the crack tip. The instability occurs when $dE_T/dl = 0$, and simple differentiation then yields a critical stress for crack propagation given by

$$\sigma_c = \sqrt{4\gamma E/\pi l}. \tag{12-13}$$

This important formula connects *Design*, *Processing and Materials*; σ_c can be thought of as a *design* stress, l as the flaw size in the coating, an unavoidable consequence of *processing* (deposition) and γE as the intrinsic *material* property. It is common to refer to $\sqrt{4\gamma E}$ as K_c, a material property known as the fracture toughness. Fast fracture will then occur when

$$\sigma\sqrt{\pi l} \geq K_c. \tag{12-14}$$

As a simple application of Eq. 12-14, consider a coating of thickness d with a through crack to the substrate. Either fracture (Fig. 12-9b) or delamination of the coating (Fig. 12-9c) could occur under stress application. The condition for the former possibility is $\sigma_c d^{1/2} > K_c(c)/\sqrt{\pi}$, where $K_c(c)$ is the fracture toughness of the coating. Similarly, coating delamination will occur when the stress normal to the surface reaches a value such that $\sigma_c d^{1/2} > K_c(i)/\sqrt{\pi}$. In this expression $K_c(i)$ is equal to $\sqrt{4\gamma E}$, with γ interpreted as an interfacial debonding energy.

In microhardness tests on brittle materials and coatings, cracks usually spread out from the indentation crater along low-index cleavage planes. Through measurement of the indentation crack dimensions and hardness, values of $K_c(c)$ and $K_c(i)$ have been estimated through complex, semiquantitative relationships (Ref. 16). K_c has the peculiar units of psi-in$^{1/2}$ or MPa-$\sqrt{m}$, and values for several important coating materials and substrates are entered in Table 12-1. Metals are typically more than an order of magnitude tougher than either ceramic oxides or covalent materials in resisting fracture.

As an example, let us estimate the critical stress required to propagate microdefect cracks 10^4 Å in size within an Al_2O_3 coating. For Al_2O_3, $K_c(c) \approx 3.5$ MPa-$\sqrt{m}$ and therefore by Eq. 12-14, $\sigma_c = 3.5/\sqrt{\pi 10^4 \times 10^{-10}} = 1.97 \times 10^3$ MPa. This value is equivalent to 200 kg/mm^2 or 2×10^{10} dynes/cm^2, which is an order of magnitude larger than expected internal stresses. Therefore, it can be concluded that flaws of this size are subcritical. The benevolence of nature in generally providing for internal compressive (growth) stresses in brittle coatings should be appreciated. However, through-thickness cracks in thicker coatings are much more susceptible to propagation leading to eventual coating fracture, particularly in the presence of tensile stresses. These are commonly generated in situations where coatings are stretched during contact with other material surfaces. We now turn our attention to such effects.

12.4. Tribology of Films and Coatings

Tribology is the study of phenomena associated with interacting surfaces in relative motion. Adhesion, hardness, friction, wear, erosion, and lubrication are among some of the scientific and technological concerns of this important subject. From an engineering standpoint three separate categories of behavior can be distinguished based on the relative magnitudes of friction and wear (Ref. 17):

1. Friction and wear are both low. This is the case in bearings, gears, cams, and slideways.
2. Friction is high but wear is low. This combination of properties is desired in devices that use friction to transmit power, e.g., clutches, belt drives, tires.
3. Friction is low and wear of one contacting body is high. This is the situation that prevails during material removal processes such as machining, cutting, and grinding. In these operations plastic deformation is involved, whereas in the first two categories the contacting bodies are generally elastically stressed.

We now explore the concepts of friction and wear in more detail.

12.4.1. Friction

All engineering surfaces are rough and characterized by a density of projections or asperities with some distribution of heights. Surfaces may be thought

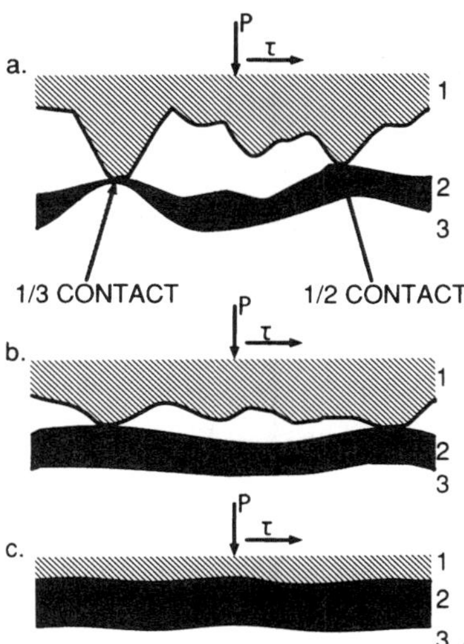

Figure 12-10. Types of contact between rough surfaces (1 and 3) through film layer (2). (Reprinted with permission from Elsevier Sequoia, S.A., from J. Halling, *Thin Solid Films* **108**, 103, 1983). (a) Film thickness is less than roughness. (b) Film thickness is of the same order as roughness. (c) Film thickness is much greater than roughness.

of as producing contact with each other at the summits of the asperities. The sum of these local contact areas represents the *real* area of contact, which may only be but a fraction of the apparent geometric contact area. In the majority of contacts surface films are present with properties different from the underlying bulk material. Surface films may be gases, fluids (e.g., oil), or deposited solid layers such as graphite, metals, or ceramic coatings. In Fig. 12-10 we consider three cases of contact between rough surfaces with an intervening film layer (Ref. 17). In the first, the film thickness is less than the roughness so that some asperities pierce the film enabling contact between bulk bodies (i.e., 1/3 contact). In Fig. 12-10b the film (2) is still thin but of the order of roughness dimensions. All contacts are now between the upper body and the film (i.e., 1/2 contact). In Fig. 12-10c the film is thicker, and all contacts are also of 1/2 type. Surface roughness is not expected to play a marked role here.

If the subscripts a and b are associated with 1/3 and 1/2 contacts, respectively, then standard notions of friction yield

$$f_a = \frac{\tau_a}{P_a}; \qquad f_b = \frac{\tau_b}{P_b}. \tag{12-15}$$

In these definitions f is a nondimensional term representing the ratio of the interfacial shear stress τ to the applied normal pressure P. The overall coefficient of friction f is given by the ratio of the total real shear-to-normal forces or

$$f = \frac{\tau_a A_a + \tau_b A_b}{P_a A_a + P_b A_b}, \tag{12-16}$$

where note is taken of the total real area A for each type of contact. Furthermore, if the respective forces borne by each contact are $F_a = P_a A_a$ and $F_b = P_b A_b$, then

$$f = f_b\left(\frac{\bar{f}\bar{F} + 1}{\bar{F} + 1}\right), \tag{12-17}$$

where $\bar{f} = f_a / f_b$ and $\bar{F} = F_a / F_b$. No assumption has been made as to the nature of the film and therefore Eq. 12-17 is generally applicable to all types of contacts. For example, in the absence of a film, $F_b = 0$ and $f = f_a$ as in the simple theory of friction. In the limit of large $\bar{F}$, the value of f also tends to f_a. At the other extreme where $F_a \rightarrow 0$, $f \rightarrow f_b$ and frictional effects are governed by the film characteristics.

The dependence of frictional effects on the film thickness is of considerable technological importance. Consider a soft solid film such as lead on steel. As the film thickness increases, f decreases from f_a (steel–steel) to f_b (lead–steel), reflecting the changing nature of contact. Beyond this thickness, τ essentially stays the same but P, which is related to the effective film hardness changes. Thin lead films are harder than thicker ones because of the support provided by the steel. Therefore, P decreases as the film thickness increases, leading to an increase in f_b or f. This causes a friction variation in film thickness that passes through a minimum. The same effect occurs with thin fluid films used to lubricate metal surfaces in contact. The well-known Stribeck curve then displays a minimum in the coefficient of friction as a function of thickness or equivalent quantity (viscosity · speed/pressure). As the film thickness increases, boundary, quasi-hydrodynamic and hydrodynamic lubrication regimes are successively operative.

An important set of applications involves the dry lubrication of moving parts where fluid lubrication is not possible, e.g., at high and low temperatures or pressures. Thin, low-friction solid films composed of chalcogenides, oxides, fluorides, or carbon are used instead. In particular, MoS_2 is a favored lubricant and has been deposited by reactively sputtering MoS_2 targets in an H_2S ambient.

12.4.2. Wear Mechanisms

Wear may be defined as the progressive removal of material from surfaces that are under load in relative motion. Several different mechanisms have been identified to characterize contact wear, and these are briefly described below (Ref. 18).

12.4.2.1. Adhesive Wear. Adhesive wear occurs when applied tangential forces cause fracture between surfaces bonded at asperities. One option is for the fracture path to follow the original microwelded interface. Other paths lie above or below the interface when the strength of the bonds between asperities exceeds the cohesive strength of the bodies in contact. The result is material transfer, usually from the softer to harder body. During subsequent surface motion cycles, these particles may eventually be removed by fatigue fracture. In more severe cases of adhesive wear, smearing, galling, and seizure of surfaces may occur.

12.4.2.2. Abrasive Wear. Abrasive wear is a form of cutting wear where the material is removed by hard wear particles, by hard asperities, or by hard particles entering the interface from the environment.

12.4.2.3. Fatigue Wear. Fatigue wear occurs in situations where there is repeated loading and unloading of surfaces in contact. Failure may initiate at both surface flaws or cracks or at subsurface inhomogeneities. Crack growth eventually results in detached wear particles.

12.4.2.4. Fretting Wear. Fretting wear may be viewed as a type of fatigue wear that occurs under conditions of oscillatory movement of small amplitude (in the range of 1–200 μm), but relatively high frequency. Many sequential damage processes occur during fretting, including breakup of protective films, adhesion and transfer of material, oxidation of metal wear particles, and nucleation of surface cracks.

12.4.2.5. Delamination Wear. Delamination wear takes the form of regular detachments of thin platelike particles from wearing surfaces due to the influence of high tangential (friction) forces in the surface contact zone. During cyclic loading the cracks that develop propagate parallel to the surface at a depth governed by the material properties and coefficient of friction.

12.4.2.6. Oxidation Wear. Oxidation wear arises from the continuous rubbing and removal of surface films produced by reaction with the environment. Wear damage is modest in this case because oxide regrows soon after it is lost. However, at high temperatures where chemical reactions are accelerated oxidation wear is aggravated.

Quantifying the various wear mechanisms is a challenge that continues to engage the efforts of workers in the field. However, expressions for adhesive wear have been developed based on the probability that a junction between the two sliding surfaces produces a wear particle. A commonly employed phenomenological formula is

$$V = K_W Fl/H, \tag{12-18}$$

where V is the volume of material removed, F is the normal force, l is the length of travel, H is the indentation hardness of the softer body, and K_W is a dimensionless wear coefficient. A similar formula holds for wear of films attached to substrates. The value of K_W depends on a number of factors. Most important is the nature of the materials in contact, but the extent and type of lubrication, surface temperature, and nature of the chemical environment also play a significant role. In order of increasing wear are the material pairs: nonmetal–metal, nonmetal–nonmetal, dissimilar metals, and similar metals.

12.4.3. Tool Wear

In the important application of hard coatings for tools, wear under cutting conditions appears to be governed by two processes: abrasive wear and chemical dissolution. The latter process is important at tool temperatures above 1000 °C, which are typical for the high-speed machining of steel with coated WC tools. Dissolution of the coating atoms in the hot chips softens the cutting edge and leads to tool degradation. For the case of abrasive wear, a simple semiquantitative model to predict the relative performance of tool coatings in machining metals has been proposed and is worth presenting here. The results should be applicable to machining at low speeds or at low temperatures. The basic assumption is that three-body abrasion describes abrasive wear. In this model the abrasive particles are free to roll between two solid bodies translating relative to each other. One can imagine that abrasive inclusions at the (two-body) chip-tool interface approximate this condition.

In three-body abrasion Rabinowicz (Ref. 19) has found that the wear behavior can be divided into three regions. These depend on the ratio of the hardness of the abrasive inclusions (e.g., oxides, compounds, precipitates, hard phases, etc.) in the workpiece to that of the surface being abraded (i.e., the tool coating). Defining the former by H_a and the latter by H_t, the wear volume (V_t) removed from the tool is essentially expressed in each hardness regime by

$$V_t = \frac{K_W Fl}{3H_t}; \qquad \frac{H_t}{H_a} < 0.8, \tag{12.19a}$$

$$V_t = \frac{K_W Fl}{5.3H_t}\left(\frac{H_t}{H_a}\right)^{-2.5}; \qquad 1.25 > \frac{H_t}{H_a} > 0.8, \tag{12.19b}$$

$$V_t = \frac{K_W Fl}{2.43H_t}\left(\frac{H_t}{H_a}\right)^{-6}; \qquad \frac{H_t}{H_a} > 1.25. \tag{12.19c}$$

When the coating is soft relative to the workpiece inclusions, then Eqs. 12-18 and 12-19a are basically the same and tool wear is rapid. With harder, more wear-resistant coatings, V_t decreases dramatically as an inverse power of H_t/H_a.

As an example, let us try to predict the relative abrasive wear suffered by TiN and high-speed (HSS) tools when machining steel containing hard abrasive particles of Fe_3C. Under cutting conditions it is assumed that the tool–work interface reaches 200 °C. To apply Eqs. 12-19, we need hot hardness values, which can be calculated from Eq. 12-4 and the data in Table 12-1. The estimated hardnesses are

TiN = 1310 kg/mm^2

HSS = 900 kg/mm^2

Fe_3C = 865 kg/mm^2

For TiN, $H_{TiN}/H_{Fe_3C} = 1.51$ (region c), while for HSS, $H_{HSS}/H_{Fe_3C} = 1.04$ (region b). Therefore, substitution in Eq. 12-19 yields

$$\frac{V_{HSS}}{V_{TiN}} = \frac{2.43}{5.3}\,\frac{1310}{900}\,\frac{(1.04)^{-2.5}}{(1.51)^{-6}} = 7.2$$

Therefore, TiN-coated tools are predicted to outlast HSS tools by a factor of over 7 in this case.

There have been numerous comparisons between the performance of coated vs. uncoated tools. In one study (Ref. 20) the ratio of the number of acceptable holes drilled by TiN-coated twist drills vs. uncoated HSS drills, prior to regrinding, was determined. The ratios ranged from 3 when drilling plain carbon steel to over 5 for both spring steel and gray cast iron. Reactive ion plating was used to deposit the TiN coatings in order to minimize drill

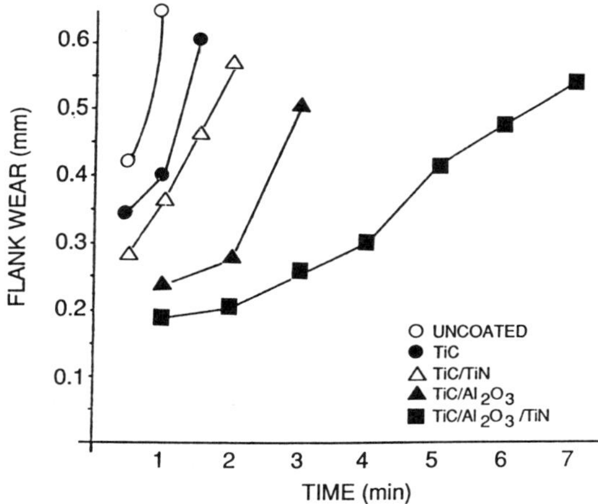

Figure 12-11. Flank wear of uncoated and coated carbide lathe tools as a function of time in machining cast iron. (Courtesy of A. Gates, Multi-Arc Scientific Coatings.)

distortion. The results of similar benefits derived from single-layer TiC coatings relative to uncoated tools are shown in Fig. 12-11. Comparisons are made for lathe tools turning cast iron. Significantly, bilayer TiC–TiN and trilayer TiC–Al_2O_3–TiN coatings are considerably longer-lived for some of the reasons noted earlier. Enhanced thermal shock and mechanical impact resistance enable these coated tools to maintain high cutting edge strength and more effectively cope with tool face, edge, and flank wear.

12.4.4. Erosion (Ref. 21)

Erosion is a type of abrasive wear in which material is lost from a surface that is in relative motion with a fluid (gas or liquid) containing solid particles. Unlike three-body wear, which was just considered, erosion is a two-body wear phenomenon. It occurs in earth-moving and agricultural equipment as well as in machinery handling slurries. We will, however, only briefly consider erosion and erosion-resistant coatings as they relate to gas-turbine engines (Ref. 22). Ingestion of particulate matter in the front end of the engine results in erosion problems downstream. Large particles at the front end typically impinge normally and wear down the leading edge of the airfoil. Further downstream in the compressor, the particles tend to be smaller and are centrifuged toward the rotor periphery. The lower impingement angle affects the midchord and even trailing edge of the airfoil. Factors such as particle

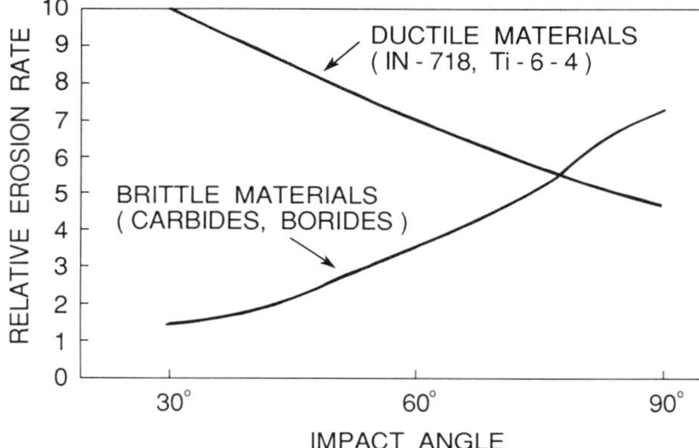

Figure 12-12. Erosion behavior of ductile and brittle materials. (From Ref. 22).

velocity, size, shape, hardness, and concentration all directly affect the magnitude of erosion. However, it is the angle of impingement of the erodent that appears to have the greatest influence on erosion. Ductile materials exhibit maximum erosion at impingement angles of 15° to 30° and erode less at higher angles. Apparently, low-angle impingement results in a tensile component that stretches the material to fracture, whereas compression and hardening of the surface by peening are the result of normal impact. Conversely, brittle materials exhibit a continuously increasing erosion rate with impingement angle and erode the maximum rate at 90°. These opposing trends are illustrated in Fig. 12-12.

Coatings have been used to counter erosion problems, but the selection of materials that can provide protection over the entire angular range has proven elusive. For example, compressor blade materials such as Ti–6V–4Al or Inconel 718 are ductile and can withstand high-angle impingement. On the other hand, carbide or boride coatings, which may be perhaps, 10 times more resistant to erosion at low angles, provide little or no protection at high angles. Composite coatings have been proposed as a solution.

12.4.5. Ball-Bearing Coatings

One of the most critical applications of hard coatings has been in ball bearings. Used in virtually all precision rotating machinery, the four basic-bearing components are the inner race, the outer race, the balls, and the cage as shown in Fig. 12-13a. The surface quality of the metallic components in rolling and sliding contact is of great importance. In particular, the roughness of the

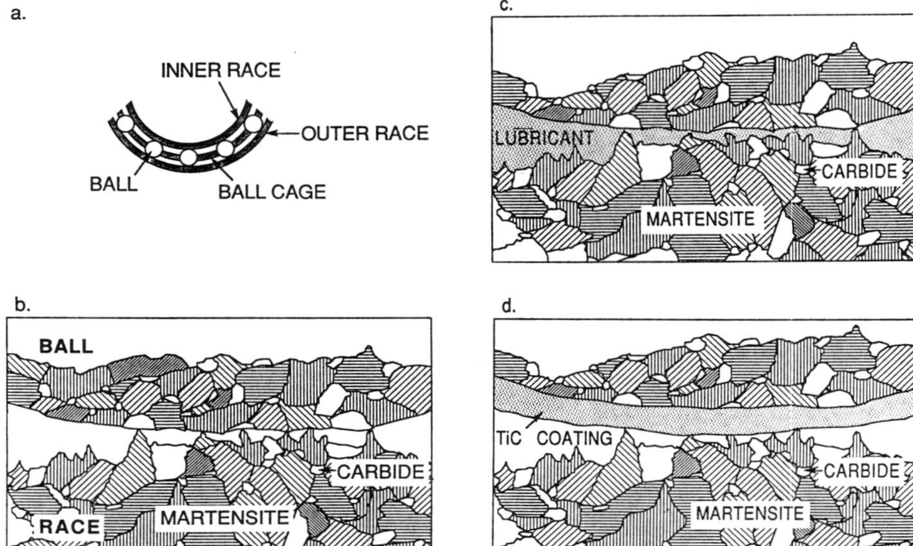

Figure 12-13. (a) Ball-bearing components: inner race or ring; outer race or ring; ball; cage or ball retainer. (b) Schematic representation of unlubricated steel ball-to-steel race contact. (c) Schematic representation of lubricated ball-to-race contact. (d) Schematic representation of unlubricated TiC coated ball-to-race contact. (Reprinted with permission from Elsevier Sequoia, S.A., H. J. Boving and H. E. Hintermann, *Thin Solid Films* **153**, 253, 1987).

raceways and balls strongly influences the effective life of the bearing. When the roughness is very low, then the oil film carries the load as desired, and minimal metal-to-metal contact occurs. However, when the roughness is high, the surface asperities occasionally impinge to cause local contact and micro-welding. This is shown in Figs. 12-13b and c, representations of the steel ball–steel race contact in the absence and presence of oil–grease lubrication. Even though the microwelds rupture almost instantaneously, the bearing interface roughens. The more rapid deterioration of the bearing when no lubricant is present is evident. Many more microwelds form, and their fracture releases hard abrasive metal particles. Such a situation arises in bearing applications where no lubricant is permitted because of environmental restrictions. For such demands, bearings coated with several microns of hard compounds such as TiC and TiN exhibit dramatically improved behavior as illustrated in Figure 12-13d. In this case a TiC-coated ball contacts an uncoated race. Longer bearing life, accompanied by lower noise and vibration, occurs because of several beneficial effects. When the steel impinges on TiC almost no microwelding or adhesion occurs between these dissimilar materials. Upon contact, the harder TiC tends to flatten the raceway asperities by plastic deformation. This process, accompanied by a smaller temperature increase than during microweld fracture, leads to lower wear and slower lubricant degradation.

There are several issues related to bearing coatings that deserve further comment. First is the question of what to coat—the races, the balls, or both. The 1000 °C CVD TiC coating process represents a severe thermal treatment for high-precision bearing components. Furthermore, final hardening and quenching treatments are required. The geometric and size distortions accompanying these thermal cycles are decided disadvantages, especially for large bearing rings and raceways. Tribology considerations require that only one partner of the contact couple be coated. The logical alternative to coated races is to use coated balls, an increasingly accepted option. Second, there is the challenge to high-temperature CVD by low-temperature PVD processes. Even so, CVD has significant advantages. Since the coating treatment lasts for several hours, a significant amount of diffusion between the substrate and coating occurs. This results in better adhesion, progressively graded mechanical properties across the interface and improved fatigue resistance. Finally, the as-deposited TiC coating surface is too rough for use in bearings; however, with high-precision lapping and polishing the coated balls become extremely smooth with a surface roughness considerably lower than attainable with uncoated steel.

The benefits of coated bearings in specific unusual applications have been noted in the literature (Ref. 24). One example involves the orbiting European Meteosat telescope. The positioning mechanism of this telescope contains ball bearings with TiC-coated races and steel balls operating under a vacuum of 7×10^{-9} torr at temperatures between -80 to 120 °C. The bearings have functioned perfectly for several years. In other coated-bearing applications involving nuclear reactors and navigation gyroscope motors, performance at elevated temperatures (i.e., 300 °C) and high rotational speeds (i.e., 24,000 rpm) were respectively evaluated. In both cases TiC-coated bearings considerably outperformed uncoated bearings.

From these and other testimonials on bearing behavior, it is clear that the 5 millenia evolution of ways to support and move loads from sliding to rolling friction, from single contact wheels to multiple contact rollers has entered a modern phase of coating utilization. One might say that it is a whole new "ball" game.

12.5. Diffusional, Protective, and Thermal Coatings

12.5.1. Diffusion Coatings

Diffusion coatings are not coatings in the sense normally meant in this chapter. They are produced by a type of CVD reaction in which the element of interest (e.g., C, N, B, Si, Al, or Cr) is deposited on and diffused into a metal substrate (usually steel), in which it is soluble. The corresponding carburizing, nitriding, boronizing, siliciding, aluminizing, and chromizing processes yield surfaces that are considerably harder or more resistant to environmental attack than the base metal. Doping of semiconductors in which infinitesimal levels of solute are involved should be distinguished from diffusional coating processes. Through diffusion, the surface layers are frequently enriched beyond the matrix solubility limit, and when this happens, compounds (e.g., Fe_3C, Fe_4N, Fe_2B) or intermediate phases (e.g., iron and nickel aluminides) precipitate, usually in a finely dispersed form. Sometimes, however, a continuous subsurface compound layer forms. Since these compounds and phases are frequently harder than the matrix, they strengthen the surface to a depth determined by the diffusional penetration. The lack of a readily identifiable planar interface between different materials means that there is no need to be concerned about adhesion in such diffused layers.

Carburization of steel is easily the most well-known and widely used diffusional surface treatment. Carbon-rich gases such as methane are made to flow over low-to-medium carbon steels (0.1 to 0.4 wt% C) maintained at temperatures of ~ 900 °C. Pyrolysis at the metal surface releases elemental carbon that diffuses into austenite or γ-Fe, a high-temperature, face-centered cubic phase of Fe capable of dissolving about 1.25 wt% C at 920 °C. After sufficient carbon enrichment, γ-Fe can be subsequently transformed to the hard tetragonal martensite phase simply by rapidly quenching the hot steel to ambient temperature. A hard, wear-resistant case or layer of martensite containing roughly 1 wt% C then surrounds the softer mild steel core. Many automotive parts, machine components and tools such as gears, shafts, and chisels are carburized. The hard-wearing surface is backed by the softer, but tougher matrix that is required to absorb impact loading.

In order to design practical diffusional coating treatments, we must have phase compositions and solubilities, available from phase diagrams, together with diffusivity data. For example, the subsurface carbon concentration $C(x, t)$ during carburization of mild steel of composition C_0 is given by

$$C(x, t) - C_0 = (C_S - C_0)\operatorname{erfc}\frac{x}{2\sqrt{Dt}}, \tag{12-20}$$

where C_S, the surface carbon concentration, depends on the solubility of carbon in the steel at the particular temperature. Other terms in Eq. 12-20 have been previously defined (*cf*. p. 35); the value for the diffusivity of C in Fe is given by $D = 0.02 \exp[-(20.1 \text{ kcal/mole})/RT] \text{ cm}^2/\text{sec}$. For typical temperatures (~ 920 °C) and times (~ 1 h) case depths of the order of 1000 μm are produced. Even harder steel surfaces on steel can be produced by nitriding. Ammonia pyrolysis at 525 °C provides the N, which then penetrates the steel with a diffusivity given by $D = 0.003 \exp[-(18.2 \text{ kcal/mole})/RT] \text{ cm}^2/\text{sec}$. After two days case layers possessing a hardness of H_v 900–1200 extending about 300 μm deep can be expected. Conventional nitriding should be compared with ion-implantation methods for introducing nitrogen into steels. This technology, discussed in Chapter 13, only modifies layers several thousand angstroms deep.

As a final, but nevertheless important, example of a diffusion-coating process we consider aluminizing. Coatings based on Al have been used for several decades to enhance the environmental resistance of materials to high-temperature oxidation, hot corrosion, particle erosion, and wear. Aluminized components find use in diverse applications—nuclear reactors, aircraft, and chemical processing and coal gasification equipment.

Metals subjected to aluminizing treatments include Ni-base as well as Fe-base superalloys, heat-resistant alloys, and a variety of stainless steels. In common these alloys all contain substantial amounts of Ni, which is required for reaction with Al. Parts to be coated are packed in a retort containing Al salts, activators, and gases capable of reacting and transporting the Al to the surface being treated, in a CVD-like process. Upon solid-state diffusion, the intermetallic compound NiAl forms on the surface. This layer is hard and lacks ductility, but exhibits low wear and friction as well as impressive high-temperature corrosion resistance to both sulfur-containing gases and liquid sodium. Beyond the outer NiAl layer is a region containing a fine dispersion of Ni_3Al precipitates that serve to strengthen and toughen the matrix. Typically both regions combined do not extend deeper than ~ 100 μm from the surface.

12.5.2. Oxidation and Oxide Films

The universal response of metal surfaces exposed to oxygen-bearing atmospheres is to oxidize. The oxidation product may be a thin adherent film that protects the underlying metal from further attack, or a thicker porous layer that may flake off and offer no protection. In this section, discussion is limited to oxidation via high-temperature exposure; aqueous corrosion oxidation phenomena are already the subject of a broad and accessible literature. From the

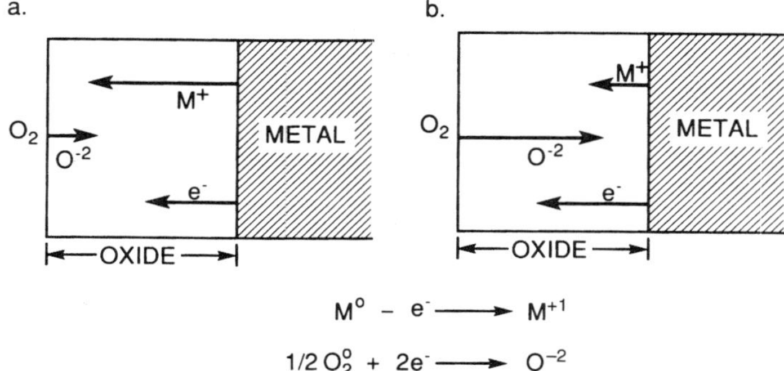

Figure 12-14. Mechanisms of oxidation: (a) oxide growth at oxide–ambient interface. (b) oxide growth at oxide–metal interface.

standpoint of thermodynamics all of the structural metals exhibit a tendency to oxidize. As noted in Chapter 1, the driving force for oxidation of a given metal depends on the free-energy change for oxide formation. What thickness of oxide will form and at what rate are questions dependent on complex kinetics and microstructural considerations, and not on thermodynamics. As shown in Fig. 12-14, two simultaneous processes occur during oxidation. At the metal–oxide interface neutral metal atoms lose electrons and become ions that migrate through the oxide to the oxide–ambient interface. The released electrons also travel through the oxide and serve to reduce oxygen molecules to oxygen ions at the surface. If metal cations migrate more rapidly than oxygen anions (e.g., Fe, Cu, Cr, Co), oxide grows at the oxide–ambient interface. On the other hand, oxide forms at the metal–oxide interface when metal ions diffuse more slowly than oxygen ions (e.g., Ti, Zr, Si). An important implication is that highly insulating oxides, such as Al_2O_3, SiO_2, do not grow readily because electron mobility, so central to the process, is low. This is what limits their growth and results in ultrathin protective native oxide films.

The model of growth kinetics developed for oxidation of Si, and Eq. 8-34 in particular, is applicable to other systems. Both parabolic oxide growth under diffusion-controlled conditions, as well as linear oxide growth when interfacial reactions limit oxidation, are frequently observed. However, not all oxidation processes fit the aforementioned categories, and other growth rate laws have been experimentally observed in various temperature and oxygen pressure regimes (Ref. 25). Specific formulas for the oxide thickness d_0, with constants $C_1, C_2, \ldots, C_7$, include

Cubic rate law

$$d_0^3 = C_1 t + C_2 \qquad \text{(e.g., Ti-400 °C)}. \tag{12-21}$$

Logarithmic

$$d_0 = C_3 \ln(C_4 t + C_5) \qquad (\text{e.g., Mg-100 °C}). \tag{12-22}$$

Inverse Logarithmic

$$1/d_0 = C_6 - C_7 \ln t \qquad (\text{e.g., Al-100 °C}). \tag{12-23}$$

In fact, careful plotting of data reveals that many metals and alloys apparently exhibit a number of different rates, depending on temperature. Most metals gain weight during oxidation, but, interestingly, metals like Mo and W lose weight during oxidation. The reason is that the oxide films that form (MoO_3 and WO_3) are volatile and evaporate as soon as they form.

The physical integrity of the oxide coating is the key issue that determines its ability to protect the underlying metal. If the oxide that forms is dense and thin, then it can generally be tolerated. If it is porous and continues to grow and spall off, the exposed underlying metal will undergo further deterioration. Whether the oxide formed is dense or porous can frequently be related to the ratio of oxide volume produced to the metal consumed. The quotient, known as the Pilling–Bedworth ratio, is given by

$$\frac{\text{Volume of oxide}}{\text{Volume of metal}} = \frac{M_{\text{o}} \rho_{\text{m}}}{x M_{\text{m}} \rho_{\text{o}}}, \tag{12-24}$$

where M and ρ are the molecular weight and density, respectively, of the metal (m) and oxide (o), and x is the number of metal atoms per molecule of oxide M_xO. If the ratio is less than unity, then compatability with the metal will create residual tensile stresses in the oxide. This will generally split it, much like dried wood, and make it porous, affording little protection to the underlying metal. If the ratio is close to or greater than unity, there is a good chance the oxide will not be porous; it may even be protective. On the other hand, if the ratio is much larger than unity, the oxide will acquire a residual compressive stress. Wrinkling and buckling of the oxide may cause pieces of it to spall off. For example, in the case of Al_2O_3 the Pilling–Bedworth ratio is calculated to be 1.36, whereas for MgO the ratio is 0.82. The lack of a protective oxide in the case of Mg has limited its use in structural applications.

What we have said of oxidation applies as well to the sulfidation of metals in SO_2 or H_2S ambients. Metal sulfides are particularly deleterious because of their low melting temperatures. Liquid sulfide films tend to wet grain boundaries and penetrate deeply, causing extension of intergranular cracks. Whether the elevated temperature atmosphere is oxidizing or sulfidizing, structural metals must be generally shielded by protective or thermal coatings.

12.5.3. Thermal Coatings (Refs. 26, 27)

Ever-increasing demands for improved fuel efficiency in both civilian and military jet aircraft has continually raised operating temperatures of turbine engine components. Among those requiring protection are turbine blades, stators, and gas seals. The metals employed for these critical applications are Co-, Ni-, and Fe-base superalloys, which possess excellent bulk strength and ductility properties at elevated temperatures. A widely used cost-effective way to achieve yet higher temperature resistance to degradation in the hot gas environment is to employ an additional thermal barrier coating (TBC) system. This consists of a metallic bond coat and a top layer composed primarily of ZrO_2. The bond coating, as the name implies, is the glue layer between the base metal and the outer protective oxide. Its function is not unlike that of a bond or primer coating used to prepare surfaces for painting. Typical bond coatings consist of MCrAlY or MCrAlYb, where M = Ni, Co, Fe. Original bond coating compositions such as Ni–26Cr–6Al–0.15Y (in wt%) have been continually modified in an effort to squeeze more performance from them. The role of Y or other rare earth substitutes is critical. These elements apparently protect the bond coat from oxidation and shift the site of failure from the base metal and coat interface to within the outer thermal barrier oxide. Just why is not known with certainty; it appears that these reactive metals easily diffuse along the boundaries of the plasma-sprayed particles of the bond coating, oxidize there, and limit further oxygen penetration.

The use of ZrO_2 is based on a desirable combination of properties: melting point = 2710 °C, thermal conductivity = 1.7 W/m-K, and thermal expansion coefficient = $9 \times 10^{-6}\ K^{-1}$ (Ref. 28). However, the crystal structure undergoes transformation—from monoclinic to tetragonal to cubic—as the temperature increases, and vice versa, as the temperature decreases. A rapid, diffusionless martensitic transformation of the structure occurs in the temperature range of 950–1400 °C accompanied by a volume contraction of 3–12%. The thermal stresses so generated lead to fatigue cracking, which signifies that ZrO_2 alone is unsuitable as a TBC. The ZrO_2 overlayers are generally stabilized with 2–15 wt% CaO, MgO, and Y_2O_3. Through alloying with these oxides, a partially stable cubic structure is maintained from 25 °C to 2000 °C. Actually the tetragonal and monoclinic phases coexist together with the cubic phase, whose stabilization depends on the amount of added oxide. Cubic phase stabilization results in stress-induced transformation toughening, which can be understood as follows. If a crack front meets a tetragonal particle, the latter will transform to the monoclinic phase a process that results in a volume increase. The resultant compressive stresses blunt the advance of cracks, toughening the matrix.

Both bond and thermal barrier coatings are usually deposited by means of plasma spraying. This process is carried out in air and utilizes a plasma torch, commonly fashioned in the form of a handheld gun. An arc emanates from the gun electrodes and is directed toward the workpiece. Powders of the coating material are introduced into the plasma by carrier gases that drive them into the arc flame. There they melt and are propelled to the workpiece surface where they splat and help to build up the coating thickness. Typical bond and thermal barrier coating thicknesses are 200 and 400 μm, respectively. Exposures to temperatures of 1100 to 1200 °C, to thermal cycling, and to stresses are common in the use of TBC systems.

EXERCISES

1. a. If the potential energy of interaction between neighboring atoms in hard compounds is

$$V(r) = -A/r^m + B/r^n$$

(see problem 1-5), show that the modulus of elasticity is given by

$$E = m(n-m)A/a_0^{m+3}$$

(a_0 is the equilibrium lattice spacing).

b. Show that E is proportional to the *binding energy density* or

$$E = -mnV(r = a_0)/a_0^3 .$$

c. How well does this correlation fit the data of Fig. 12-7?

2. Why are epitaxial hard coatings of TiC or TiN not practical or of particular interest?

3. A hardness indenter makes indentations in the shape of a tetrahedron whose base is an equilateral triangle that lies in the film plane.

a. If the side of this triangle has length l_t, what is the depth of penetration of the indenter?

b. What is the hardness value in terms of applied load and l_t?

4. Compare the relative abrasive wear of TiC-coated tools vs. HSS tools when machining steel containing Fe_3C particles. Assume the same wear model and machining characteristics as in the illustrative problem on p. 575.

5. Molten steel at 1500 °C can be poured into a quartz crucible resting on a block of ice without cracking it. Why? Calculate the stress generated.

6. A coating has small cracks of size l that grow by fretting fatigue. Assume the crack extension rate is given by

$$\frac{dl}{dN} = A\,\Delta K^m \qquad \text{with } \Delta K = \sqrt{\pi l}\,\Delta\sigma,$$

where $\Delta\sigma$ the range of cyclic stress, N is the number of stress cycles, and A and m are constants. By integrating this equation, derive an expression for the number of cycles required to extend a crack from l_i to l_f.

7. Speculate on some of the implications of the Weibull distribution for hard coating materials if

a. the volume can be replaced by the coating thickness.
b. the tensile strength is proportional to coating hardness (this is true for some metals).
c. the Weibull modulus is 10.

8. Oxidation rates are observed to vary as

a. $\dfrac{d(d_0)}{dt} = \dfrac{A}{d_0}$

b. $\dfrac{d(d_0)}{dt} = B\exp - Cd_0$

c. $\dfrac{d(d_0)}{Dt} = D\exp\dfrac{E}{d_0}$

A, B, C, D, E are constants. Derive explicit equations for the oxide thickness (d_0) vs. time (t) for each case.

9. Contrast the materials and processes used to coat sintered tungsten carbide lathe tool inserts and high-speed steel end-mill cutters.

10. By accident a very thin *discontinuous* rather than continuous film of TiC was deposited on the steel races of ball bearings. How do you expect this to affect bearing life?

11. The Taylor formula $Vt_l^n = c$, widely used in machining, relates the lifetime (t_l) of a cutting tool to the cutting velocity (V). Constants n and c depend on the nature of the tool, work, and cutting conditions. A TiN-coated cutting tool failed in 40 min when turning a 10-cm-diameter steel shaft at 300 rpm. At 250 rpm the tool failed after 2 h. What will the tool life be at 400 rpm?

12. Assume that K_w for adhesive wear is 10^{-10} for 52100 steel balls on 52100 steel races and an order of magnitude lower for TiC-coated bearings.

a. Approximately how many revolutions are required to generate a wear volume of 10^{-5} cm^3 in an all-steel, 2-cm-diameter bearing if $H = 900$ kg/mm^2 and $F = 200$ kg?
b. At 1000 rpm how long will it take to produce this amount of wear in the coated bearing?

13. A problem arising during the CVD deposition of TiC on cemented carbides is the loss of C from the substrate due to reaction with $TiCl_4$. This leads to a brittle *decarburized* layer (the η phase) between substrate and coating. Assuming that interstitial diffusion of C in W is responsible for the effect, sketch the expected C profile in the substrate after a 2-h exposure to 1000 °C, where the diffusivity is 10^{-9} cm^2/sec.

14. The surface of a HSS drill is exposed to a flux of depositing Ti atoms and a N_2 plasma during reactive ion plating at 450 °C. It is assumed that an effective surface concentration of 50 at% N is maintained that can diffuse into the substrate. Under typical deposition conditions roughly estimate the ratio of layer thicknesses of TiN to Fe_4N formed as a function of time.

REFERENCES

1. M. F. Ashby and D. R. H. Jones, *Engineering Materials 1 and 2*, Pergamon Press, Oxford (1980 and 1986).
2.* E. A. Almond, *Vacuum* **34**, 835 (1984).
3. H. Holleck, *J. Vac. Sci. Tech.* **A4**, 2661 (1986).
4.* J. E. Sundgren and H. T. G. Hentzell, *J. Vac. Sci. Tech.* **A4**, 2259 (1986).
5. M. Rühle, *J. Vac. Sci. Tech.* **A3**, 749 (1985).
6. B. M. Kramer and P. K. Judd, *J. Vac. Sci. Tech.* **A3**, 2439 (1985).
7.* R. F. Bunshah, ed., *Deposition Technologies For Films and Coatings*, Noyes, Park Ridge, NJ (1982).
8. D. T. Quinto, G. J. Wolfe and P. C. Jindal, *Thin Solid Films* **153**, 19 (1987).

*Recommended texts or reviews.

9.* R. F. Bunshah and C. Deshpandey, in *Physics of Thin Films*, Vol. 13, eds. M. H. Francombe and J. L. Vossen, Academic Press, New York (1987).
10. D. Holleck and H. Schulz, *Thin Solid Films* **153**, 11 (1987).
11. A. Layyous and R. Wertheim, *J. de Phys. Colloque* C5, 423 (1989).
12. W. D. Sproul, *Thin Solid Films* **107**, 141 (1983).
13. C. A. Brookes, *Science of Hard Materials*, Plenum Press, New York (1983).
14. W. D. Münz and G. Hessberger, *Vak. Tech.* **30**, 78 (1981).
15. G. Gille, *Thin Solid Films* **11**, 201 (1984).
16. P. K. Mehrotra and D. T. Quinto, *J. Vac. Sci. Tech.* **A3**, 2401 (1985).
17. J. Halling, *Thin Solid Films*, **108**, 103 (1983).
18. M. Antler, *Thin Solid Films*, **84**, 245 (1981).
19. E. Rabinowicz, *Lubr. Eng.*, **33**, 378 (1977).
20. R. Buhl, H. K. Pulker, and E. Moll, *Thin Solid Films*, **80**, 265 (1981).
21. R. C. Tucker, in *Metals Handbook*, Vol. 11, 9th ed., American Society for Metals (1986).
22. R. V. Hillery, *J. Vac. Sci. Tech.* **A4**, 2624 (1986).
23. H. J. Boving and H. E. Hintermann, *Thin Solid Films* **153**, 253 (1987).
24. H. E. Hintermann, *Thin Solid Films* **84**, 215 (1981).
25. S. Mrowec and T. Werber, *Gas Corrosion of Metals* U.S. Dept. of Commerce, NTIS TT 76-54038, Springfield, VA (1978).
26. S. Stecura, *Thin Solid Films* **136**, 241 (1986).
27. R. A. Miller and C. C. Berndt, *Thin Solid Films* **119**, 195 (1984).
28. G. Johner and K. K. Schweitzer, *Thin Solid Films* **119**, 301 (1984).

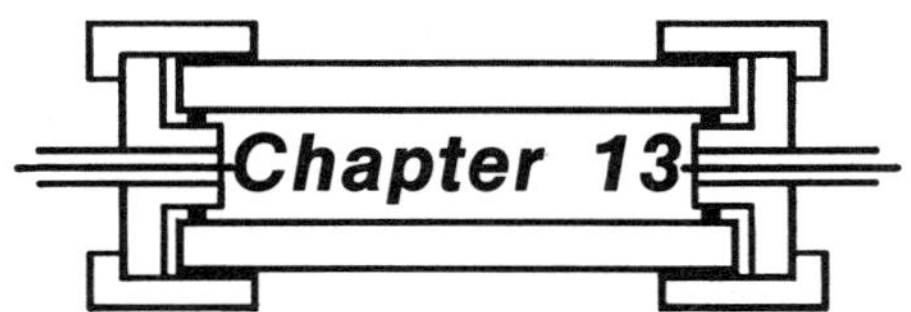

Modification of Surfaces and Films

13.1. Introduction

Two main approaches to improving or altering the surface properties of solids have evolved over the years. The more traditional one involves the deposition of films and coatings from solid, liquid, and vapor sources. Processing utilizing these methods has totally dominated our attention in the book until this point. But there is another approach of more recent origin based on modifying existing surfaces through the use of directed-energy sources. These include photon, electron, and ion beams, and it is their interaction with surfaces that will be the focus of this chapter.

Coherent (laser) and incoherent light sources, as well as electron beams, modify surface layers by heating them to induce melting, high-temperature solid-state annealing or phase transformations, and, occasionally, vaporization. In the case of lasers, the relation between the required power density and irradiation time is depicted in Fig. 13-1 for a number of important commercial processing applications. However, the focus of this book is thin films and in the applications shown much thicker layers of material are modified. These materials processing techniques will, therefore, not be discussed in any detail, nor will there by any additional mention of electron beams. Their heating effects are basically equivalent to those produced by lasers of comparable power. Furthermore, the great depth of the heat-affected zone is more typical of bulk rather than surface processing. The thin-film or layer-modification regime we shall be concerned with is characterized by approximate laser energies of ~ 0.1–$2\ \text{J/cm}^2$, interaction times of $\sim 10^{-9}$ to 10^{-6} sec, and power densities of $\sim 10^6$ to $10^8\ \text{W/cm}^2$. These conditions prevail in the indicated region of Fig. 13-1. Surface layers ranging from 0.1 to 10 μm in thickness are correspondingly modified by melting under such conditions. The melting–solidification cycle frequently does not restore the surface structure and properties to their original states. Rather, interesting irreversible changes may occur. For example, one consequence of laser processing can be an ultrahigh quench rate with the retention of extended solid solutions, metastable crystalline phases, and, in some cases, amorphous materials. Directed thermal energy sources have also been employed to effect annealing, surface alloying, solid-state transformations and homogenization. The controlled epitaxial regrowth of molten Si layers over SiO_2 or insulators, discussed in Chapter 7, is an important example of the great potential of such processing.

Like photon and electron beams, ion beams play an indispensible role in surface analytical methods and have also achieved considerable commercial success in surface processing. In the very important ion-implantation process, ion beams have totally revolutionized the way semiconductors are doped. Depending on the specific ion projectile and matrix combination, dopants can be driven below the semiconductor surface to readily predictable depths through control of the ion energy. Unlike traditional diffusional doping where the highest concentration always occurs at the surface, ion-implanted distributions peak beneath it. The reduction of the threshold voltage required to trigger

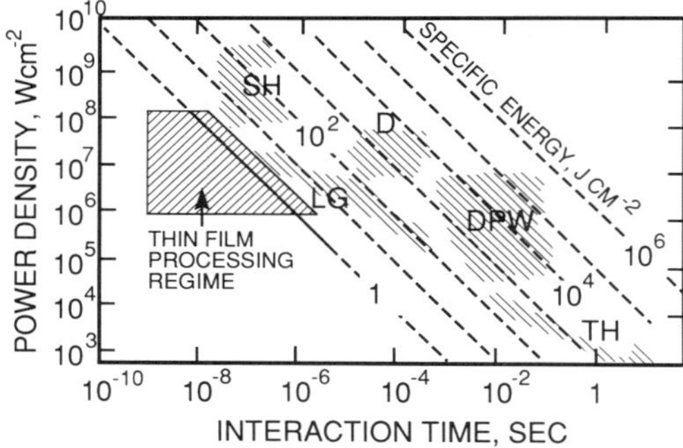

Figure 13-1. Laser processing regimes illustrating relationships between power density, interaction times and specific energy. D-drilling; SH-shock hardening; LG-laser glazing; DPW-deep penetration welding; TH-transformation hardening.

current flow in MOS transistors, by means of ion implantation, ushered in battery-operated, handheld calculators and digital watches. Today ion-implantation doping is practiced in MOS as well as bipolar transistors, diodes, high-frequency devices, optoelectronic devices, etc., fabricated from silicon and compound semiconductors. Achievements in microelectronics encouraged broader use of ion implantation to harden mechanically functional surfaces, improve their wear and fatigue resistance, and make them more corrosion resistant. Critical components such as aircraft bearings and surgical implant prostheses have been given added value by these treatments. In addition, there are other novel ion-beam-induced surface-modification phenomena such as ion-beam mixing, or subsurface epitaxial growth, that may emerge from their current research status into future commercial processes.

The purpose of this chapter is to present the underlying principles of the interaction of directed-energy beams with surfaces, together with a description of the changes which occur and why they occur. Accordingly, the subject matter is broadly subdivided into the following sections:

13.2. Lasers and Their Interaction with Surfaces
13.3. Laser Modification Effects and Applications
13.4. Ion-Implantation Effects in Solids
13.5. Ion-Beam Modification Phenomena and Applications

13.2. Lasers and Their Interactions With Surfaces

13.2.1. Laser Sources

The intense scientific and engineering research associated with the development of lasers has resulted in much innovation and rapid growth of applications. Space limitations preclude any discussion of the details of the theory of laser construction, operation, and applications, which are all covered admirably in other textbooks (Ref. 1). Suffice it to say, that all lasers contain three essential components: the lasing medium, the means of excitation, and the optical feedback resonator. The most common lasers employed in materials processing contain either gaseous or solid-state lasing media (Ref. 2). Gas lasers include the carbon dioxide (CO_2:N_2:He), argon ion and xenon fluoride excimer types. The solid-state varieties used are primarily the chromium-doped ruby, the neodymium-doped yttrium–aluminum–garnet and neodymium-doped glass laser. These solid-state lasers are excited through pumping by incoherent light derived from flash lamps. Gas lasers, on the other hand, are excited by

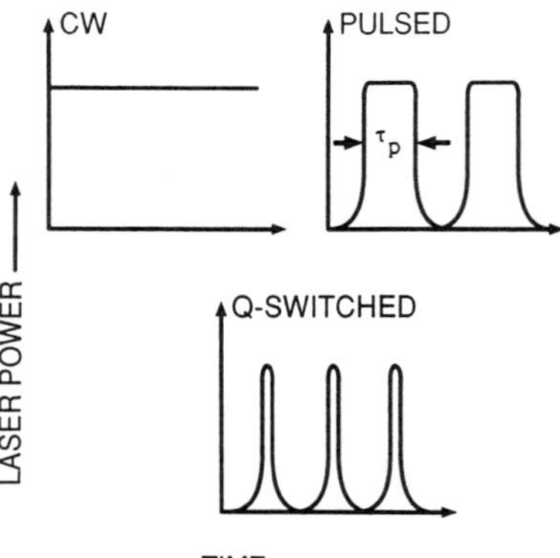

Figure 13-2. Output power for three modes of laser operation. (From Ref. 2).

means of electrical discharges. Laser excitation may be continuous or cw, pulsed, or Q-switched to provide the different output powers shown schematically in Fig. 13-2. The distinctions in these power-time characteristics are important in the various materials processing applications. In the welding and drilling of metals, for example, advantage is taken of the power–time profile in the pulsed and Q-switched lasers. Both the reflectance and the thermal diffusivity of metals decrease with increasing temperature. Therefore, the high-power leading edge of these lasers is used to preheat the metal and enhance the efficiency of the photon–lattice phonon energy transfer.

In Table 13-1 the common lasers employed in surface processing together with their pertinent operating characteristics are listed. Among the important laser properties are spatial intensity distribution, the pulse width, and pulse repetition rate. The spatial distribution of emitted light depends on the cavity configuration with Gaussian (TEM_{00}) intensity profiles common. Because a uniform laser flux is desirable in surface processing, methods have been developed to convert emission modes into the "top-hat" spatial profile. The dwell time or pulse length, τ_p, ranges from less than 10 nsec to 200 nsec for Q-switched lasers, and many orders of magnitude longer for other types of lasers. Repetition rates for pulsed and switched lasers range from one in several seconds to many thousands per second. Although the low repetition rates of Q-switched lasers may not be practical in industrial processing applications because the duty cycle (i.e., time on/time off $\approx 10^{-8}$), is low, they are useful for laboratory research.

It is the magnitudes of both the absorbed radiant power and τ_p that determine the effective depth of the surface layers modified through melting or redistribution of atoms. Generally, the smaller values of τ_p result in submicron

Table 13-1. Typical Parameters for Lasers Used in Surface Modification

Laser	Wavelength (μm)	Focused Average Power (W*)	Focused optical spot (mm)	Power density (MW cm^{-2})	Dwell/pulse Length (ns)	Repetition Rate (Hz)	Spatial Distribution
CW-CO_2	10.6	200–15000	0.1–10	1–10	10^7–10^{10}	—	Multimode, top hat, or Gaussian
Q-switched Nd:YAG	1.06	0.5–3	20–40	80–200	130	11000	Gaussian
Frequency-doubled *Q*-switched Nd:YAG	0.53	0.05–0.3	10–30	30–100	120	5000	Gaussian
Q-switched ruby	0.69	1–10*	6	10–200	2–80	0.03	Top hat
Pulsed ruby, Nd:YAG or Nd:glass	0.69, 1.06	2–50*	0.5	0.5–10	2×10^6–10^7	0.1–10	Multimode

*For low repetition rate, laser output is quoted in J per pulse.
From Ref. 2.

melt depths. Melting and extensive interdiffusion over tens to hundreds of microns occur with the longer irradiation times possible with cw lasers. No single laser spans the total range of accessible melt depths. Section 13.2.3 is devoted to the quantitative modeling of the temperature–distance–time interrelationships in the heat-affected zone.

13.2.2. Laser Scanning Methods

Practical modification of large surface areas with narrowly focused laser beams necessarily implies some sort of scanning operation as shown in Fig. 13-3a. For cw lasers the surface generally rotates past the stationary beam in a manner reminiscent of a phonograph record past a needle. Through additional *x-y* motion, radial positioning and choice of rotational speed, a great latitude in transverse velocities (v) is possible. This also means a wide selection of interaction or dwell times, t_d, given by $t_d = d_m / v$, where d_m is the effective

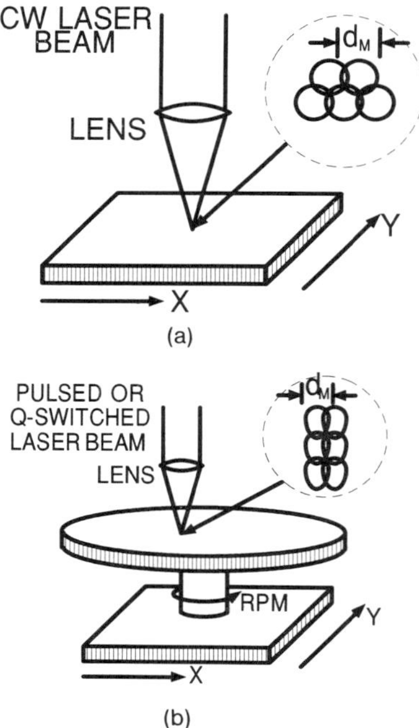

Figure 13-3. Schemes for broad area modification of laser-heated surfaces. (a) Surface translated past stationary laser beam in X and Y directions. (b) Surface rotated relative to laser beam at low speed. Inserts show enlarged views of melt trails. (From Ref. 2).

melt trail diameter. Typically, t_d ranges between tens of microseconds to hundreds of milliseconds. In this case the surface-modified region appears to consist of a chain of overlapping elliptical melt puddles.

The experimental arrangement for processing using pulsed or Q-switched lasers is shown in Fig. 13-3b. In this case, discrete, overlapping circular-modified (melted) regions are generated by a train of laser pulses. For area coverage larger than the individual melt spots, a mechanism for raster scanning must be provided. This is usually accomplished by computer-controlled x-y stepping of substrates.

13.2.3. Thermal Analysis of Laser Annealing (Refs. 4, 5)

The substrate heating caused by an incident laser pulse is due to electronic excitation processes accompanying the absorption of light. Typical pulse durations of 1 nsec or longer far exceed the relaxation time for electronic

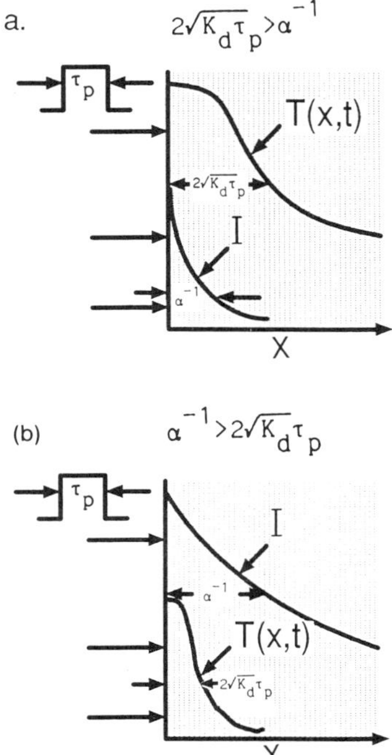

Figure 13-4. Schematic laser pulse intensity and temperature profiles for a penetration depth α^{-1} that is (a) small and (b) large compared with the thermal diffusion length, $2\sqrt{K_d \tau_p}$. (From Ref. 3).

transitions ($\sim 10^{-12}$ sec). Therefore, it is permissible to assume that the thermal history of the irradiated sample can be modeled by continuum non-steady-state heat-conduction theory. The fundamental equation for the temperature $T(x, t)$ that has to be solved is

$$\rho c \frac{\partial T(x,t)}{\partial t} - \kappa \frac{\partial^2 T(x,t)}{\partial x^2} - A(x,t) = 0; \qquad 0 < x < \infty, \quad (13\text{-}1)$$

where the first two terms representing conventional one-dimensional heat conduction should be familiar to readers. The term $A(x, t)$, in units of W/cm^3, is the spatial and time-dependent power density absorbed from the incident laser pulse. Other quantities which appear are ρ, the density, c, the heat capacity, κ, the thermal conductivity, and x and t, the distance measured from the surface into the interior and time, respectively. Depending on the relative value of the absorption length, α^{-1} cm, of the laser light within the specimen surface, two limiting regimes of thermal response can be distinguished, as shown in Fig. 13-4.

13.2.3.1. Strong Thermal Diffusion ($2\sqrt{K_d \tau_p} \gg \alpha^{-1}$). When the thermal diffusion length $2\sqrt{K_d \tau_p}$ (where K_d the thermal diffusivity $= \kappa/\rho c$) is much larger than α^{-1}, then the heat source is essentially a surface source. This is the situation for metal surfaces where light penetration is extremely limited. Hence, we assume that $A(x, t)$ can be written as

$$A(x,t) = I_0(1 - R)\delta(x - 0)H(t - \tau_p). \qquad (13\text{-}2)$$

The individual factors physically express that a rectangular laser pulse of power density I_0 is incident for time τ_p, after which the pulse amplitude is zero. The Heaviside function $H(t - \tau_p)$ mathematically describes this time dependence. A fraction of the incident radiant energy is reflected from the surface whose reflectivity is R. The remainder is concentrated only at the surface; hence, the use of the delta function, $\delta(x - 0)$.

The boundary value problem that models laser heating in the semi-infinite medium is then expressed by the following conditions. Initially,

$$T(x,0) = T_0; \qquad 0 \le x < \infty, \qquad (13\text{-}3a)$$

where T_0 is the ambient temperature. The first set of boundary conditions concerns heat transfer through the surface. Thus,

$$\frac{\partial T(0,t)}{\partial x} = -\frac{I_0(1 - R)}{\kappa}; \qquad x = 0, \qquad (13\text{-}3b)$$

assumes a constant heat flux during heating, i.e., for $\tau_p > t > 0$. For time $t > \tau_p$ corresponding to cooling,

$$\partial T(0, t)/\partial x = 0. \tag{13-3c}$$

The second boundary condition specifies T far from the surface

$$T(\infty, t) = T_0. \tag{13-3d}$$

Closed-form solutions for both transient heating and cooling can be obtained by Laplace transform methods. During heating ($t < \tau_p$)

$$T(x, t) = \frac{I_0(1 - R)}{\kappa} \sqrt{4K_d t}\, \mathrm{ierfc} \frac{x}{2\sqrt{K_d t}} + T_0, \tag{13-4}$$

where

$$\mathrm{ierfc}\, z = \int_z^\infty \mathrm{erfc}\, y\, dy = \frac{1}{\sqrt{\pi}} e^{-z^2} - z\, \mathrm{erfc}\, z. \tag{13-5}$$

Explicitly,

$$T(x, t) = T_h(x, t) = \frac{I_0(1 - R)}{\kappa} \left[\sqrt{\frac{4K_d t}{\pi}} \exp \frac{-x^2}{4K_d t} - x\, \mathrm{erfc} \frac{x}{\sqrt{4K_d t}} \right] + T_0. \tag{13-6a}$$

During cooling the temperature drops for all $t > \tau_p$ and

$$T(x, t) = T_c(x, t) = T_h(x, t) - \frac{I_0(1 - R)}{\kappa} \left[\sqrt{\frac{4K_d(t - \tau_p)}{\pi}} \exp - \frac{x^2}{4K_d(t - \tau_p)} - x\, \mathrm{erfc} \frac{x}{\sqrt{4K_d(t - \tau_p)}} \right]. \tag{13-6b}$$

In this development it has been tacitly assumed that K_d is temperature-independent, and that the laser-beam diameter is larger than $2\sqrt{K_d t_p}$. These assumptions have considerably simplified the analysis, the first by linearizing Eq. 13-1, and the second by justifying one-dimensional heat diffusion through neglect of the otherwise lateral heat flow.

At the surface of the material ($x = 0$) Eq. 13-6a reduces to

$$T(0, t) = \frac{2I_0(1 - R)}{\kappa} \sqrt{\frac{K_d t}{\pi}} + T_0, \tag{13-7a}$$

and Eq. 13-6b similarly becomes

$$T(0, t) = \frac{2I_0(1 - R)}{\kappa}\left[\sqrt{\frac{K_d t}{\pi}} - \sqrt{\frac{K_d(t - \tau_p)}{\pi}}\right] + T_0. \quad (13\text{-}7b)$$

Through differentiation of these equations with respect to time, the surface heating and quenching rates are calculated to be

$$\frac{dT(t)}{dt} = \frac{I_0(1 - R)}{\kappa}\sqrt{\frac{K_d}{\pi t}} \quad (13\text{-}8a)$$

and

$$\frac{dT(t)}{dt} = \frac{I_0(1 - R)}{\kappa}\left[\sqrt{\frac{K_d}{\pi t}} - \sqrt{\frac{K_d}{\pi(t - \tau_p)}}\right], \quad (13\text{-}8b)$$

respectively.

As an example, consider a 1.8-J/cm^2 ruby laser pulse of 10-nsec width incident on a Ni surface at $T_0 = 0$ °C, for which $\kappa = 0.92$ W/cm-°C and $K_d = 0.98$ cm^2/sec. In this case $I_0 = 1.8 \times 10^8$ W/cm^2, and if $R = 0.9$, the

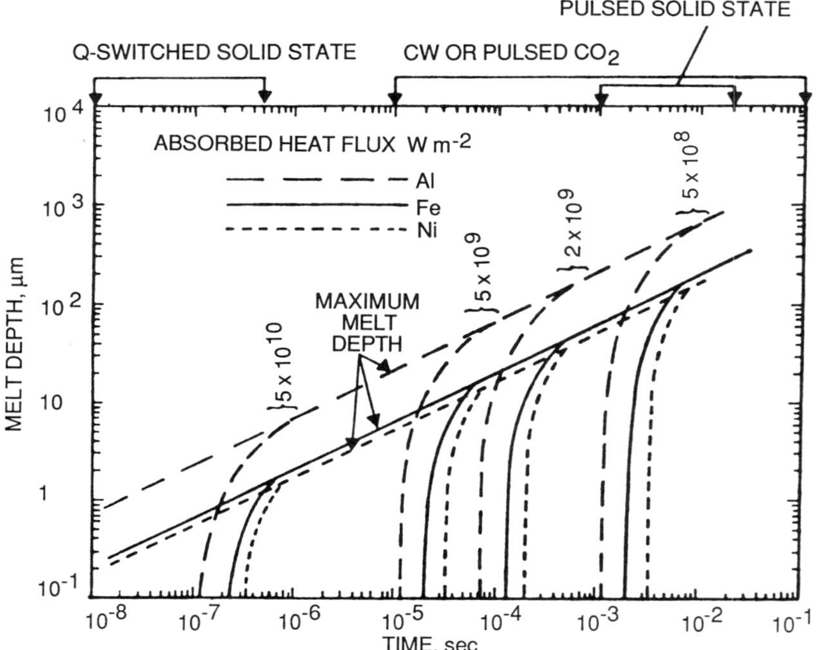

Figure 13-5. Calculated melt depths vs. irradiation time for Al, Fe and Ni based on one-dimensional computer heat flow model. (From Ref. 6).

time it will take the surface to reach the melting point (1455 °C) is calculated to be 4.4 nsec, using Eq. 13-7a. In another 5.6 nsec a maximum surface temperature of 2190 °C is attained. The thickness of Ni that has totally melted can be estimated with the use of Eq. 13-6a. Substituting $T = 1455$ °C and $t = 1 \times 10^{-8}$ sec, trial-and-error solution yields a value $x \sim 4300$ Å. Due to neglect of (1) radiation and convection heat losses, (2) latent heat absorption during melting and liberation upon solidification, and (3) temperature dependence of thermal constants—these calculated effects have been considerably overestimated. More precisely determined melt depths vs. irradiation time and input power are depicted in Fig. 13-5 for Al, Fe, and Ni. Submicron melt depths for Q-switched laser pulses are typical.

Lastly, it is instructive to estimate the melt quenching rate. The instantaneous value is time-dependent (Eq. 13-8) and for the example given above, calculation at 20 nsec yields a rate of -3.2×10^{10} °C/sec (for $T = 896$ °C). Similarly, the quench rate at 50 nsec is -8.37×10^{9} °C/sec (for $T = 580$ °C). Such ultrahigh quench rates from the liquid phase are sufficient to freeze in a variety of metastable chemical and structural states in many alloy systems.

13.2.3.2. Adiabatic Heating ($2\sqrt{K_d \tau_p} < \alpha^{-1}$, Ref. 7). In this regime the temperature of the surface is largely determined by the initial distribution of the energy absorbed from the laser beam. Light penetrates within the material and the thermal evolution during the pulse duration overshadows heat diffusion effects that can be neglected; adiabatic heating prevails then. Such a situation is applicable to the laser modification of semiconductor surfaces where the distribution of light intensity is given by

$$I = I_0(1 - R)\exp - \alpha x \quad \mathrm{W/cm^2}. \tag{13-9}$$

The heat generation rate is equal to $-dI/dx$, so

$$A(x, t) = \alpha I_0(1 - R)\exp - \alpha x \quad \mathrm{W/cm^3}. \tag{13-10}$$

Upon substitution in Eq. 13-1 and neglecting $\kappa(\partial^2 T/\partial x^2)$, we obtain a temperature rise

$$\Delta T(x, \tau_p) = \frac{\alpha \tau_p I_0(1 - R)\exp - \alpha x}{\rho c} \tag{13-11}$$

at $t = \tau_p$ after integration. The threshold incident power required to just initiate surface melting is therefore

$$I_0(\text{threshold}) = c\rho(T_M - T_0)/(1 - R)\tau_p \alpha, \tag{13-12}$$

where T_M is the melting point. Cooling rates can be estimated if it is assumed that heat transfer occurs by conduction over a distance α^{-1}. Associated with α^{-1} is the heat diffusion (quenching) time, $t_q = 1/(2\alpha^2 K_d)$. Therefore, a surface quench rate of

$$\Delta T/t_q = -2\alpha^3 I_0(1 - R)\tau_p K_d/\rho c \qquad (13\text{-}13)$$

is predicted which, interestingly, depends on α^3.

As an application of these equations let us consider Si for which $\rho = 2.33$ g/cm^3, $c = 0.7$ J/g-°C, $K_d = 0.14$ cm^2/sec, $R = 0.35$ (at 0.69 μm), and $\alpha = 2.5 \times 10^3$ cm^{-1}. For the 10-nsec, 1.8-J/cm^2 ruby laser pulse previously

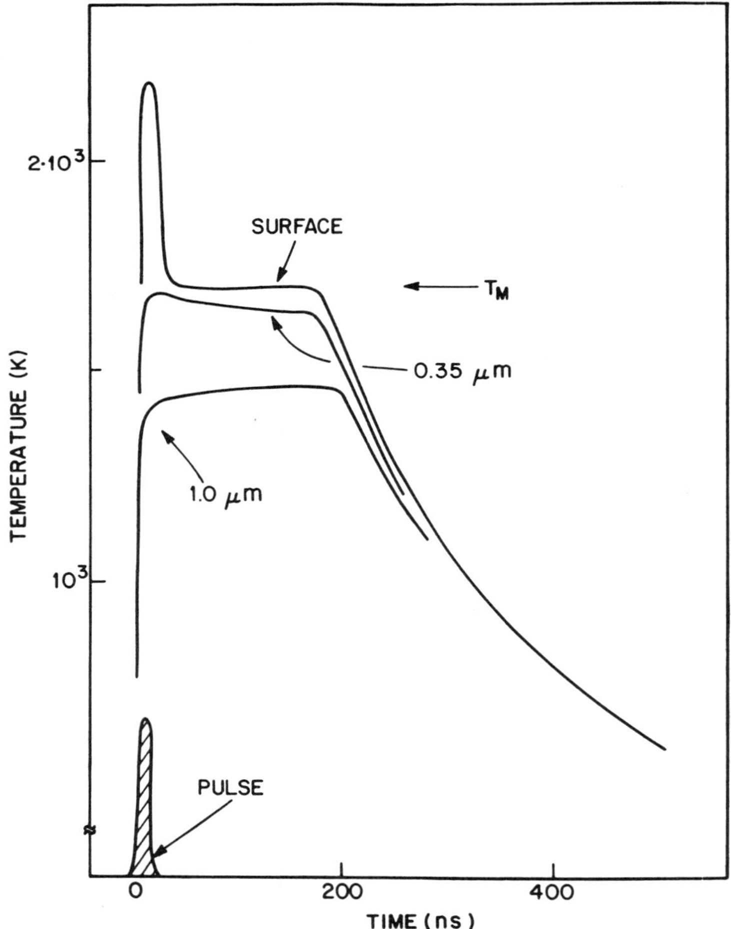

Figure 13-6. Calculated temperature vs. time at different depths below Si surface irradiated with a 10-nsec ruby laser pulse of 1.7-J/cm^2 energy density. (From Ref. 8).

considered, ΔT ($x = 0$, $t = 10$ nsec) is estimated to be ~ 1800 °C (Eq. 13-11), and the quench rate is calculated to be $\sim -4.8 \times 10^9$ °C/sec (Eq. 13-13). These values are only approximate, and more exact computer analyses that account for thermal losses, latent heat effects and temperature (and position) dependent material constants exist. One such calculation for the temperature history at different depths beneath an irradiated single-crystal Si surface is shown in Fig. 13-6.

13.2.4. Solidification Rate

If directional solidification is assumed, then an estimate of the solidification rate, defined as the rate of movement of the melt interface, may be obtained by evaluating dx/dt directly from the prior heat flow analysis. Since

$$\frac{dx}{dt} = \left(\frac{dT}{dt}\right)_{T_M}\left(\frac{dT}{dx}\right)_{T_M}^{-1}, \tag{13-14}$$

the direct dependence on cooling rate and inverse dependence on temperature gradient should be noted. Brute-force calculation of the involved factors is tedious and best left to the computer. Instead, it is instructive to take an intuitive approach based on heat transfer considerations. During solidification, the latent heat of fusion, H_f, liberated at the advancing solid–liquid interface, is conducted into the substrate. Therefore, the thermal power balance per unit area of interface, which limits the solidification velocity, is given by

$$\rho H_f \frac{dx}{dt} = \kappa_S\left(\frac{dT}{dx}\right)_S - \kappa_L\left(\frac{dT}{dx}\right)_L. \tag{13-15}$$

Here ρ is the density, S and L refer to solid and liquid, respectively, and both derivatives must be evaluated at the melt interface. Because the molten film is roughly at constant temperature during solidification, $\kappa_L(dT/dx)_L$ may be neglected compared to $\kappa_S(dT/dx)_S$. Note that in this formulation dx/dt is directly proportional to $dT/dx|_S$. For the case when $2\sqrt{K_d t_M} > \alpha^{-1}$, the temperature gradient is roughly $(T_M - T_0)/2\sqrt{K_d t_M}$, where the denominator is a measure of the thermal diffusion length for melt time t_M. Therefore, an estimate of the solidification rate is

$$\frac{dx}{dt} = \kappa_S(T_M - T_0)/2\rho H_f\sqrt{K_d t_M}. \tag{13-16}$$

Again for the case of Ni where $\rho = 8.9$ g/cm^3 and $H_f = 151$ cal/g, with other constants previously given, $dx/dt = 5 \times 10^3$ cm/sec, assuming $t_M =$ 10 nsec. In actuality, melt times are generally longer serving to reduce the value of dx/dt.

Alternatively, for the case where $\alpha^{-1} > 2\sqrt{K_d t_M}$, direct differentiation of Eq. 13-11 yields

$$dT/dx = -\alpha\,\Delta T = -\alpha(T_M - T_0). \tag{13-17}$$

After substitution in Eq. 13-15, we have

$$dx/dt = \kappa_S \alpha(T_M - T_0)/\rho H_f. \tag{13-18}$$

Evaluating this equation for Si where $H_f = 264$ cal/g, and other constants were previously given, we obtain $dx/dt = 2.1 \times 10^3$ cm/sec.

The estimates of dx/dt are strongly dependent on the laser pulse power and duration, and are probably too high by a factor of ~ 2 to 5. Nevertheless, these ultrahigh solidification rates of several meters/sec are many orders of magnitude larger than conventional solidification rates in bulk materials. For example, single crystals of Si are typically pulled at rates of only 3×10^{-3} cm/sec.

13.3. Laser Modification Effects and Applications

13.3.1. Regrowth Phenomena in Silicon (Ref. 9)

In this section attention is directed to the structural and compositional property changes produced in silicon surface layers as a result of laser processing. Silicon has been singled out as the vehicle for discussion because of the large volume of study devoted to this important material. Furthermore, many of the phenomena observed in Si can be readily understood in the context of traditional solidification and recrystallization theories that have evolved over the past four decades.

13.3.1.1. Impurity-Free Si. Laser melting of single-crystal Si wafer surfaces results in *liquid phase epitaxial* (LPE) regrowth. However, when ultrashort picosecond pulses are applied, crystalline → liquid → amorphous Si transitions can be sequentially induced.

The phenomenon of *solid phase epitaxial* (SPE) regrowth of amorphous silicon layers upon surface annealing is worth noting. As we shall see later, ion implantation methods can be used to "amorphize," or make amorphous, surface layers of Si. The latter can be recrystallized by laser annealing, in what amounts to a second surface modification treatment. Better control over SPE can be exercised, however, by means of simple furnace annealing. The result is a well-defined constant planar regrowth velocity with thermally activated kinetics given by

$$\frac{dx}{dt} = a\nu \exp - \frac{E}{kT}, \tag{13-19}$$

where a is the atomic spacing and ν is the lattice frequency. The activation energy (E) for impurity-free SPE is 2.35 eV, a value that corresponds to cooperative bond breaking and rearrangement at the moving amorphous–crystalline interface. In its wake, dangling bonds are eliminated and interfacial bond straining and distortion are minimized. In contrast to the several m/sec laser-induced solidification rates, SPE regrowth proceeds at a velocity of only ~ 1 Å/sec at 500 °C—a difference spanning some 10 orders of magnitude. Epitaxial regrowth rates of Si-implanted amorphous Si are strongly dependent on substrate orientation with (100) and (111) exhibiting the highest and lowest magnitudes, respectively.

13.3.1.2. Doped Si

Epitaxial Regrowth. Depending on the type and concentration of impurity atoms, and the thermal processing parameters, a rich assortment of regrowth effects is possible. Consider what happens to an implanted distribution of Bi atoms in Si after irradiation by a 100-nsec ruby laser pulse with an energy density of 2 J/cm^2. As indicated in Fig. 13-7, the distribution, originally centered 1500 Å deep, is swept toward the surface during solidification. The reason for this has to do with two facts. The first is that when a liquid and solid are in equilibrium at some temperature, solute generally has greater solubility in the liquid phase; second, solute atoms diffuse rapidly in the liquid phase but are essentially immobile in the solid. When a planar liquid–solid front now passes by the implanted solute, successive partitioning occurs between the two phases in an attempt to maintain a fixed (equilibrium) solute concentration ratio; i.e., $k_0 = C_S/C_L$, at the interface. Also variably known as the segregation, partition or distribution coefficient, k_0 values depend on the dopant in question; it can, for example, range from 2.5×10^{-5} for Au to 0.8 for B in Si. A computer simulation of the solidification sequence of events is shown in Fig. 13-8 for the case of a solute with $k_0 = 0.1$, where regrowth proceeds at a rate of 2 m/sec. The process resembles zone refining, a technique employed to purify rods of material by directionally sweeping impurities and concentrating them at one end by means of a moving molten zone.

A recent intriguing result is the demonstration of the *solid-state* analog of

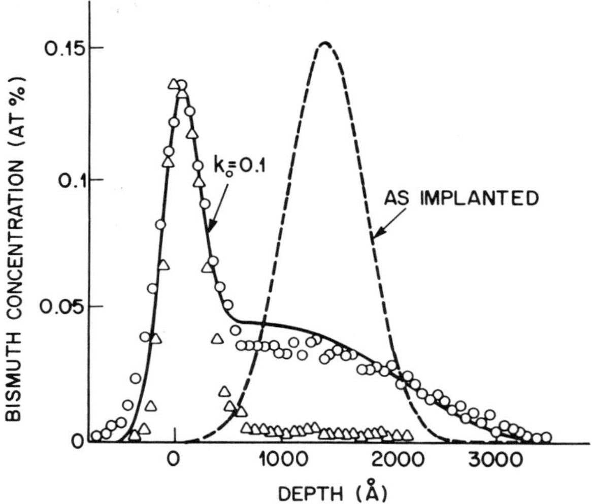

Figure 13-7. RBS depth profile of 250-keV Bi implanted in (100) Si (dashed line). After irradiation with 100-nsec ruby laser pulses of 2 J/cm^2 the full line profile results. Circles (triangles) represent random (aligned) RBS spectra. (From Ref. 10).

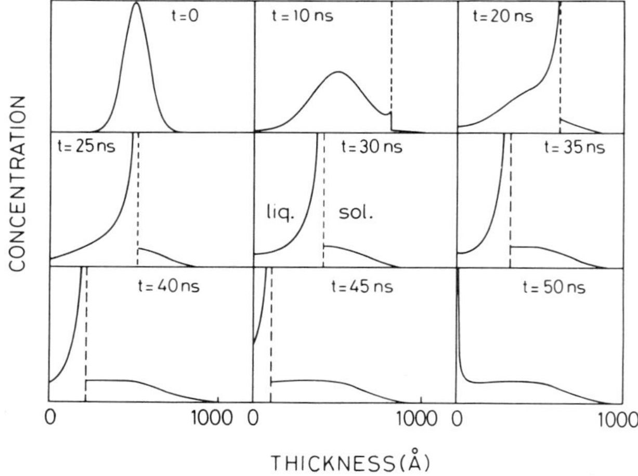

Figure 13-8. Computed solute profiles in the liquid and solid at the indicated times during solidification. (From Ref. 8).

zone refining in an amorphous Si layer doped with implanted Au. During epitaxial regrowth of Si, the amorphous Si layer shrinks in extent as the interface moves toward the surface (Fig. 13-9). Simultaneously, Au preferentially partitions into the amorphous phase and is concentrated there leaving a virtually Au-free epitaxial Si region behind. Interestingly, Au solubilities in

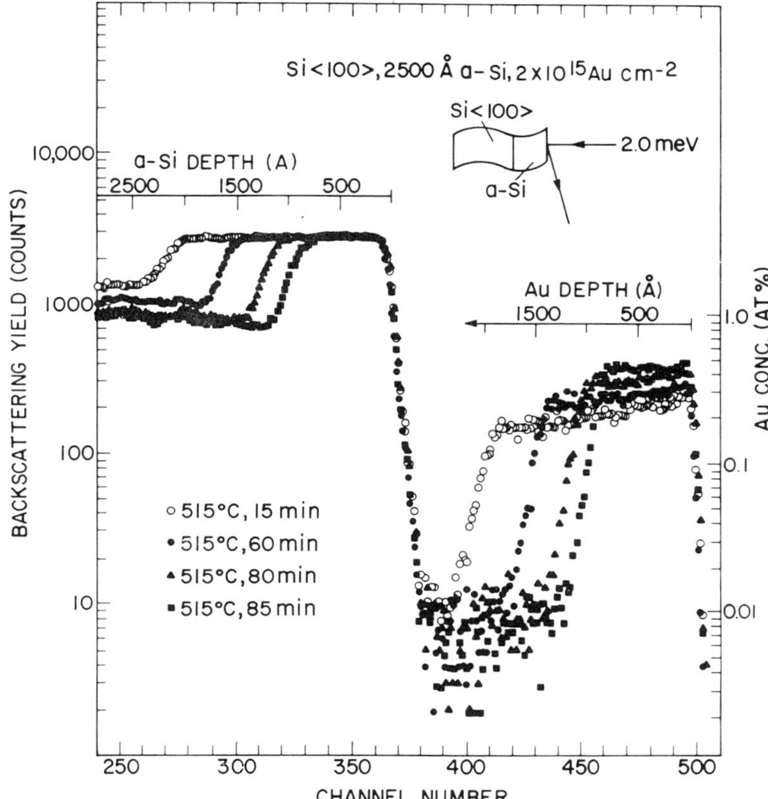

Figure 13-9. Channeling RBS spectra showing amorphous Si thickness and Au depth profiles as a function of annealing time. (From Ref. 11).

amorphous Si many orders of magnitude greater than in crystalline Si, were measured (Ref. 11).

13.3.1.3. Solute Trapping Effects (Ref. 12). One of the consequences of rapid solidification is the trapping of solute within the regrown matrix, causing equilibrium solid solubility limits to be exceeded. The effects can be quite appreciable. For example, whereas the equilibrium solid solubility limit of Bi in Si is $8 \times 10^{17}/\mathrm{cm}^3$ values as high as $4 \times 10^{20}/\mathrm{cm}^3$ were measured after regrowth at velocities of 4.5 m/sec. Similarly, for As, Ga, Sb, and In, solubility enhancements of 4, 10, 30, and 200, respectively, have been reported. One explanation of the trapping process involves a competition between interface and impurity kinetics. The time (t_i) required to regrow a layer equal to the interface thickness (a_i) is $t_i = a_i / v_s$, where v_s is the solidification velocity dx/dt. Meanwhile, the average time the dopant spends in this region is $t_d \approx a_i^2/2D_i$, where D_i is the diffusivity in the interface region. Trapping occurs if the dopant resides in the interface region longer than the time required to regrow it, i.e., when $t_d > t_i$. On the other hand, when the reverse is true, dopant is rejected into the high-solubility phase, i.e., the liquid. The critical velocity is then $v_{\text{crit}} = 2D_i/a_i$; values of $a_i = 10^{-7}$ cm and $D_i \approx 10^{-5}$ cm^2/sec lead to $v_{\text{crit}} = 2$ m/sec, which is in reasonable agreement with experience.

Another explanation (Ref. 12) suggests that solute trapping is *not* due to atoms being unable to get out of the way of the rapidly advancing interface. Rather, atoms are incorporated into the crystalline lattice in greater concentrations than predicted by equilibrium thermodynamics. Evidence for this assertion is the much larger k_0 value of 0.1 (instead of the equilibrium value of 7×10^{-4}) required to fit the experimental Bi profile in Fig. 13-7. Also the lower RBS spectrum (triangles) exhibiting channeling indicates that the Bi dopant primarily resides on lattice, not interstitial sites.

One manifestation of small solute additions is the very marked effect on epitaxial regrowth behavior. For example, less than 1 at% of B, P, and As increase the regrowth velocity while O, N, and C reduce it. It is well known that virtually all elemental additions impede the processes of recrystallization and grain growth in bulk metal alloys by impeding dislocation motion and acting as a drag on mobile grain boundaries. Evidently additional complex electronic effects of dopants at the amorphous–crystal (or liquid–epitaxial) interface are at play in semiconductors.

13.3.1.4. Constitutional Supercooling. One of the interesting phenomena that can occur when the temperature increases into the growth medium is the existence of a region ahead of the interface that is effectively *supercooled*. The reason is due to the buildup of rejected solute into the liquid ahead of the solidification front, and the effect it produces is known as *constitutional supercooling*. It is so named because the *effective* melt temperature varies with solute concentration in a manner predicted by the binary phase or constitution diagram. If this *constitutional* temperature varies less rapidly with distance into the melt than the *actual* temperature, the growth interface is planar (no constitutional supercooling); if the reverse is true, local instabilities occur and the planar solidification front breaks down through growth of feathery projections known as dendrites, or through formation of a cellular substructure. The criterion for unstable interfacial motion is

$$\frac{mC_0}{D}\left(\frac{1}{k_0} - 1\right)\frac{dx}{dt} > \frac{dT}{dx}, \tag{13-20}$$

where m is the slope of the liquidus line on the phase diagram, C_0 is the composition of the liquid far from the interface, and the other quantities have been previously introduced.

In the regime of laser processing, the inequality is frequently exceeded and interfacial instabilities develop. The recrystallization breakdown occurs on the scale of $D/(dx/dt) = (10^{-4}\ \mathrm{cm^2/sec})/(100\ \mathrm{cm/sec})$ or about 100 Å.

13.3.2. Laser Surface Alloying (LSA)

When a thin metal film A on a metal substrate B is exposed to laser radiation, the combination can be alloyed through melting to yield a new modified surface layer. This LSA process can be understood with reference to the schematic cross-sectional views shown in Fig. 13-10. A laser pulse causes film A to melt and the resulting liquid/solid front sweeps past the original A–B interface; interdiffusion of film and substrate atoms occurs as irradiation terminates. The maximum melt depth is reached in Fig. 13-10d, where the atomic mixing is quite vigorous. Resolidification then begins and the solid–liquid interfacial velocity, which is initially zero, increases very rapidly. Interdiffusion continues within the liquid but the resolidified metal cools so rapidly that atoms are immobilized and frozen in place. The final result is an alloy of nominal composition A_xB_{1-x} which is not necessarily homogeneous. As an example, consider the surface alloying of a thin Au film on a Ni substrate. Typical Q-switched laser pulses generate total melt times (t_M) ranging from 50 to 500 nsec and produce atomic mixing over the diffusion distance $\sim 2\sqrt{Dt_M}$. The diffusivity of atoms in liquid metals is rarely outside the range of 10^{-5} to 10^{-4} cm^2/sec so that compositional change can be expected over a distance of 140–1400 Å. The higher estimate roughly agrees with the Au concentration profile data of Fig. 13-11. In this example the total melt depth is about 4500 Å and considerably exceeds the initial Au thickness. If, on the other hand, t_M is lengthened to $\sim 50\ \mu$sec, a situation arising when using a cw–CO_2 laser pulse, then the molten pool is deeper and the diffusional length exceeds 4500 Å. Complete homogenization clearly occurs in this case.

Considerable LSA experimentation has been conducted on assorted binary alloy systems (Ref. 2). They include (1) those in which the two components are mutually soluble in both liquid and solid states, e.g., Cr–Fe, Au–Pd, W–V; (2) those where there is appreciable liquid solubility (miscibility) but limited solid solubility, e.g., Cu–Ag, Au–Ni, Cu–Zr; and (3) those that exhibit both liquid and solid phase immiscibility, e.g., Pb–Cu, Ag–Ni, Cu–Mo. Category 1 lends itself to thermodynamically favored interdiffusional mixing. Except for rare or expensive alloying elements, however, laser processing in these

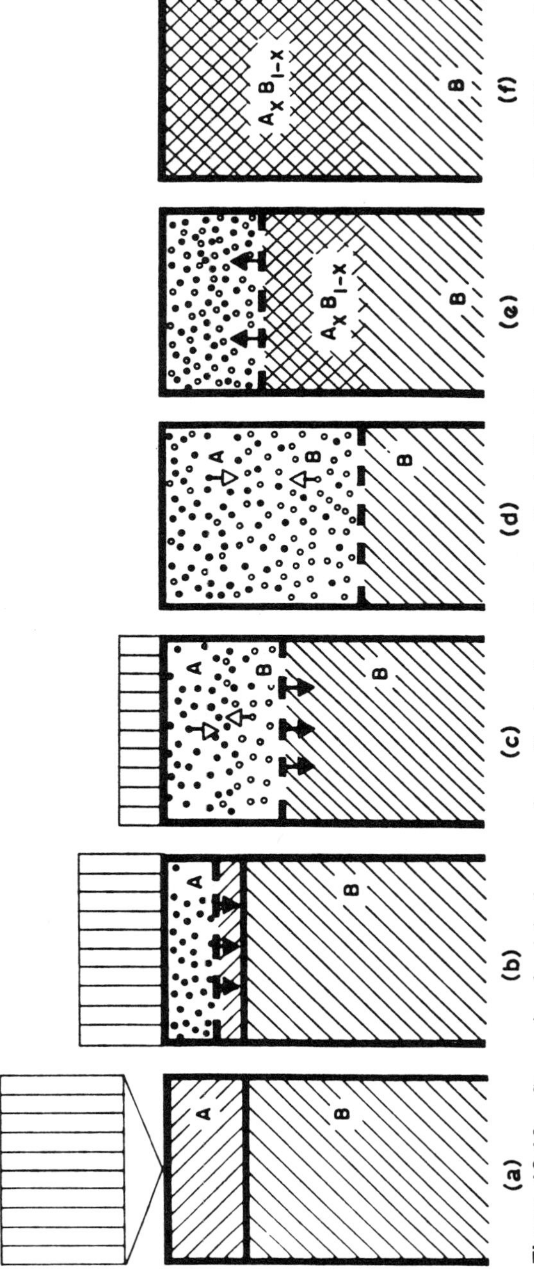

Figure 13-10. Stages involved in laser surface alloying from the time the incident laser pulse arrives until solidification is complete. (From Ref. 2).

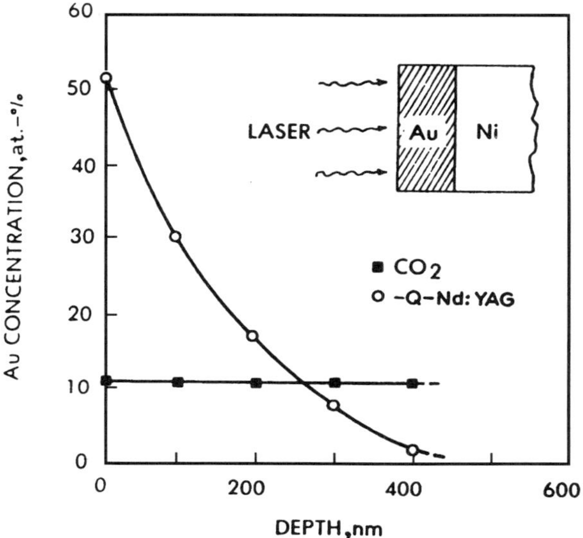

Figure 13-11. Surface alloying of Au film with a Ni substrate after irradiation with CO_2 and Nd:YAG lasers. (From Ref. 2).

systems offers few advantages over conventional bulk alloying processes. Thermodynamic obstacles to mixing in category 3 are not easily overcome even with LSA methods. Only when the films are thin enough and the melt temperature high enough is there the chance that a single-phase liquid will form, which then can be quenched to retain metastable phases. Otherwise, predictable phase separation will occur. Intermediate category 2 offers the greatest potential for quenching in metastable and amorphous phases. It is this class of binary systems which had been previously studied by vapor-quenching methods (Chapter 5) over a decade earlier. The objective was the same as LSA —to extend solubility of terminal phases, freeze in metastable phases and produce amorphous phases by suppressing crystallization. Laser scanning methods offer the best means of achieving these ends over large surface areas.

13.4. Ion-Implantation Effects in Solids

13.4.1. Introduction

Ion-surface interactions have already been discussed within several contexts in this book. Sputtering for film deposition and Rutherford backscattering for microanalysis are the most important examples; typical ion energies involved are 5 keV and 2 MeV, respectively. At ion energies between these extremes, i.e., tens to hundreds of keV, the probability is great that projectile ions will be implanted hundreds to thousands of angstroms deep beneath the surface.

As a surface modification technique, ion implantation has a number of important advantages as well as disadvantages (Ref. 13). Among the advantages are controllable and reproducible subsurface depth concentrations, no sacrifice of bulk properties, low-temperature processing, no significant dimensional change in implanted objects, extension of solid solubility limits, formation of metastable phases, and vacuum cleanliness. Significant limitations include line-of-sight processing, shallow penetration of ions, lattice damage, and, of course, very high capital equipment and processing costs. Despite the latter drawbacks, ion implantation is not only indispensible in VLSI processing, but its use has been explored as a means of beneficially modifying virtually every surface property of interest.

Modification of surfaces occurs because the newly implanted distributions of chemical species are accompanied by considerable structural disorder. Sometimes it is possible to induce compositional change without appreciable structural modification, e.g., in the case of low-dosage implants followed by thermal annealing. Alternatively, crystalline targets can be disordered structurally and even made amorphous by implanting ions that are identical to matrix atoms. In this case no chemical change is effected. Frequently, however, both compositional and structural changes are inseparably linked and serve to broaden the number of possible ways surfaces can be modified. Some choice over the extent of modification can be exercised through control of processing variables.

13.4.2. Energy Loss and Structural Modification

The collisions of an individual energetic ion in a solid cause the motion of atoms and the excitation of electronic states. At the outset the ions primarily induce "gentle" electronic transitions that cause relatively little structural damage. Nevertheless, the electronic structure is excited by Coulomb interactions with moving ions. A relatively narrow lightning-like trail surrounds the ion track and defines a region of intense electronic excitation, e.g., ionization, secondary and Auger electron production, electron–hole pair formation, luminescence, etc. Through these mechanisms of energy loss the ion slows sufficiently until it begins to set in motion violent nuclear collision cascades along its trajectory. These cascades are the result of displaced atoms that dislodge yet other atoms so that a jagged branched trail is produced.

Through these interactions the ion energy (E) continually decreases with distance (z) traversed in a very complicated way. For simplicity the energy loss is expressed by

$$-\frac{dE}{dz} = N[S_e(E) + S_n(E)], \tag{13-21}$$

where $S_e(E)$ and $S_n(E)$ are the respective electronic and nuclear stopping powers (in units of eV-cm^2), and N is the target atom density (in units of atoms/cm^3). The magnitudes of both stopping powers depend on the atomic numbers and masses of the ions as well as target atoms. Typically, electronic stopping results in energy losses of 5–10 eV/Å as opposed to the higher losses of 10–100 eV/Å for nuclear stopping. The Lindhard–Scharff–Schiott (LSS) theory (Ref. 14) has quantified the nature of these ion–solid interactions and predicted the extent of partitioning of the energy loss. It is well established that $S_e(E)$ varies as $E^{1/2}$. The energy dependence of $S_n(E)$ varies widely with ion–matrix system; $S_n(E)$ is large at low E but relatively small and constant at high E. An explicit expression for $S_n(E)$ is

$$S_n(E) = \frac{1}{N}\frac{dE}{dz} = \frac{\pi}{2} a Z_1 Z_2 q^2 \frac{M_1}{M_1 + M_2}, \tag{13-22}$$

where the atomic numbers and masses of the projectile ion and target nucleus are Z_1, Z_2, and M_1, M_2, respectively, and q is the electronic charge. Values of a, the screening radius for the collision, lie between 0.1 and 0.2 Å. In Chapter 3 it was noted that the sputter yield, which is related to $S_n(E)$, rises at first, reaches a maximum, and then falls with increasing energy. At low energies $S_n > S_e$, whereas the reverse is true at higher energy. The critical energy E_c for which $S_e(E_c) = S_n(E_c)$, depends on the nature of the ion–target combination. For example, in amorphous GaAs, $E_c = 15$ keV for Be, and 800 keV for Se.

From the standpoint of structural modification, matrix damage and atomic relocation following nuclear collisions are cardinal issues (Ref. 15). In addition to the cascades referred to earlier, which can affect a considerable portion of the matrix, depending on the extent of the branch overlap, there are also "spikes." When the density of energy deposition in either electronic or nuclear cascades exceeds a certain level, the result is a thermal spike. These are launched when bombarding particles transfer energies of a few hundred eV to lattice atoms with the virtually instantaneous liberation of heat. In Cu, calculation has shown that the spike will heat a sphere of approximately 20 Å in diameter to the melting point within 5×10^{-12} sec. In another 3×10^{-11} sec, the temperature decays to 500 °C, quenching the motion of some 1000 atoms in the process (Ref. 16). Small, highly energized cascade regions melt and quench too rapidly to grow epitaxially within the surrounding crystalline material. As a result, amorphous zones form. In silicon these have been found (Ref. 17) to be more resistant to formation of defect-free material on low-temperature annealing than thin amorphous–Si surface layers or films formed by laser melting or by physical vapor deposition. In this sense the amorphous cascade or spike regions have a different defective nature.

The dose of ions required to amorphize a crystalline matrix can be roughly estimated by assuming that the energy density is essentially the same as that needed for melting. In the case of Si this amounts to about 20 eV/atom, or

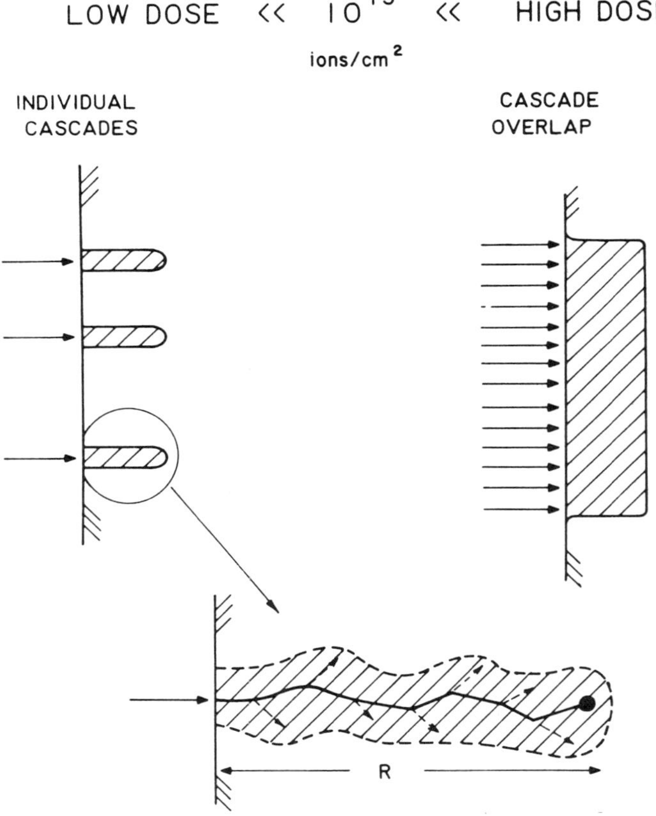

Figure 13-12. Schematic of individual cascades at low doses and overlapping cascades at high doses. R is typically 100 Å, the collision cascade time is $\sim 10^{-14}$ sec and the thermal quench time is 10^{-12} sec. (Reprinted with permission from Plenum Publishing Corp., from Ref. 15).

~ 10^{24} eV/cm^3. For ions of energy E_0 the dose ϕ is

$$\phi = 10^{24} R_p / E_0 \quad \text{ions/cm}^2 \tag{13-23}$$

where R_p (soon to be discussed in the next section) is the ion range. In the case of 100-keV B ions, for which $R_p = 3200$ Å, $\phi = 3.2 \times 10^{14}$ ions/cm^2. In practice a dose greater than 10^{16} ions/cm^2 is required, indicating that damage and recrystallization effects occur simultaneously during implantation.

A montage of assorted electronic and nuclear cascades is depicted in Fig. 13-12. Typical picosecond time scales rule out real-time monitoring of these events. Dimensions of several nanometers mean that high-resolution TEM techniques are required to image damage details.

13.4.3. Compositional Modification

From the previous discussion it is clear that no two ions will execute identical trajectories, but will participate in some admixture of nuclear and electronic collision events. Furthermore, the collective damage and zigzag motion of ions within the matrix cause them to deviate laterally from the surface entry point. When summed over the huge number of participating ions, these factors lead to the statistical distribution of ions as a function of position shown in Fig. 13-13. The concentration of implanted ions ideally has a Gaussian depth dependence given by

$$C(z) = \frac{\phi}{\sqrt{2\pi}\,\Delta R_p} \exp - \left(\frac{z - R_p}{\sqrt{2}\,\Delta R_p} \right)^2 \tag{13-24}$$

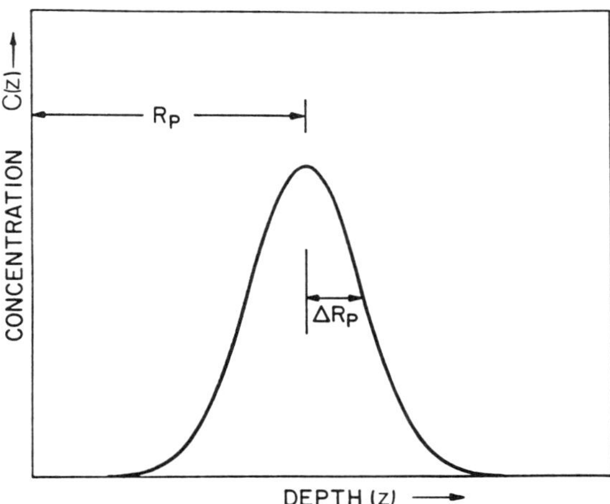

Figure 13-13. Distribution of implanted ions as a function of depth.

with a peak magnitude varying directly as the fluence or dose ϕ of incident ions. Dose has units of number (of ions) per cm^2 and is related to the measured time integrated current or charge Q deposited per unit surface area A. Specifically,

$$\phi = Q/nqA, \tag{13-25}$$

where n is the number of electronic charges, q, per ion. The projected range R_p is the depth most ions are likely to come to rest at, yielding a peak concentration of $\phi/\sqrt{2\pi}\,\Delta R_p$. Note that the actual distance an ion travels is greater than the component projected normal to the target surface. This is analogous to the total distance executed in atomic random walk jumps exceeding the *net* diffusional displacement. Experimental spread in the ion range is accounted for by the term ΔR_p, the standard deviation or longitudinal "straggle" of the distribution. Similarly ΔR_l, the lateral ion straggle, is a measure of the spread in the transverse direction.

Values of R_p and ΔR_p have been calculated to accuracies of a few percent for many ion-target atom combinations as a function of energy (Ref. 18). Results for N and Cr in Fe are shown in Fig. 13-14. For the same energy the lighter N ions penetrate further as intuitively expected. Typically R_p is about 0.1 μm. Similar information for the common dopants B, P and As in Si, SiO_2 and photoresist matrices, is commonly used in designing VLSI implantation doping processes, and can be found in standard texts on the subject (Refs. 19, 20). Our primary concern, however, is not with semiconductors and it is,

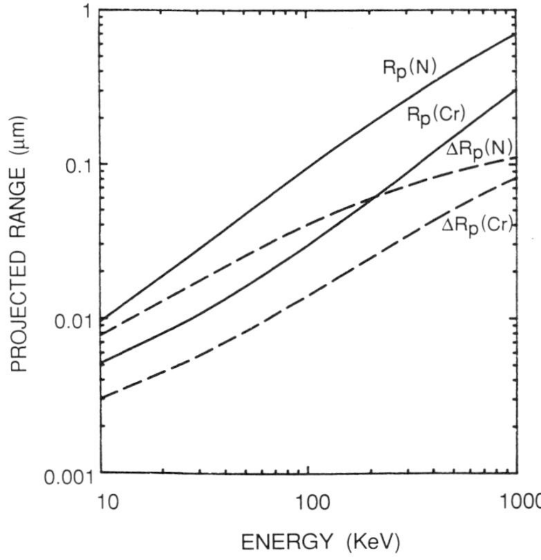

Figure 13-14. Calculated values of R_p and ΔR_p for N and Cr in Fe as a function of energy. (From Ref. 18).

therefore, instructive to mention a number of factors that distinguish these two kinds of ion-implantation applications.

1. The chief one from which many differences stem is the dose employed. In semiconductor technology typical doses of 10^{12} to $10^{15}/\text{cm}^2$, or about a monolayer at most, lead to dopant volume concentrations of less than 0.1 at%. In nonsemiconductor applications fluences of 10^{15}–$10^{17}/\text{cm}^2$ are common and result in atomic concentrations of several percent.
2. A consequence of high implanted concentrations is the deviation from predicted Gaussian profiles which strictly prevail in the low-dose regime. Concentration-dependent chemical affinities, radiation-enhanced diffusion, and higher target temperatures because of beam heating, contribute to altering the implanted profile in the high-dose regime.
3. Greater surface sputtering arises from high-dose implants. This is an obvious consequence of the higher ion flux bombardment and causes the target surface to recede. The effect can be appreciable as shown in Fig. 13-15 for the case of 150-keV Ni ions implanted into 430 stainless steel at a

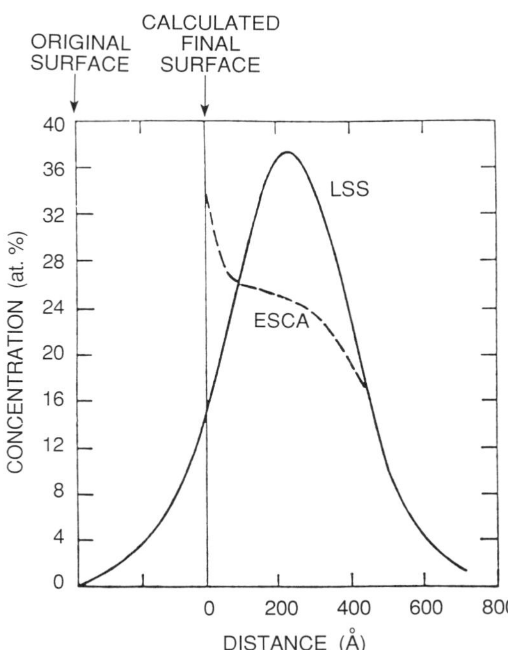

Figure 13-15. Comparison of calculated (LSS) and measured (ESCA) distribution of Ni implanted into stainless steel. (Reprinted with permission from Elsevier Sequoia, S. A., from S. B. Agarwal, Y. F. Yang, C. R. Clayton, H. Herman and J. K. Hirvonen, *Thin Solid Films* **63**, 19, 1979).

dose of $5 \times 10^{17}/\text{cm}^2$. Together with radiation enhanced diffusion, the apparent preferential sputtering of Cr results in the buildup of Ni at the surface and the gross alteration of the Gaussian profile.

4. In semiconductor applications it is essential that the structurally damaged matrix be restored to crystalline perfection and that the dopants be electrically activated. A high-temperature ($\sim$ 900 °C) anneal fosters both defect annihilation and crystal regrowth, where dopant incorporation at desired lattice sites occurs. No such post-implantation treatment is normally carried out in the nonsemiconductor applications. On the contrary, the reaction and retention of nonequilibrium surface phases and structures, not possible via conventional thermodynamic and diffusional routes are frequently desiderata.

13.5. Ion-Beam Modification Phenomena and Applications

13.5.1. Ion-Beam Mixing (Ref. 22)

Ion-beam mixing phenomena deal with compositional and structural changes in a two- or multiple-component system under the influence of ion radiation. The effect commonly occurs during sputtering and results in changes in surface composition during depth profiling analysis by SIMS and AES techniques. For example, consider a thin film of A on substrate B bombarded by a beam of inert-gas ions. Typically, the ion range (R) exceeds the escape depth of the sputtered A atoms. If R does not exceed the thickness of A, then only A atoms sputter. If, after some sputtering, R now extends into the substrate region, the atomic displacements and interdiffusion that occur within collision cascades will cause A and B to intermix. The mixing occurs locally at the interface and eventually links with other similarly intermixed zones to create a continuous ion-beam mixed layer. Now B atoms also enter the stream of sputtered atoms because the combination of continued surface erosion and interfacial broadening, due to ion mixing, has brought them closer to the surface.

It is of interest to estimate the extent of interfacial broadening. To do this, we set R equal to the half-width of the broadened ion-mixed layer and assume that nuclear energy loss dominates with a stopping power independent of energy (Ref. 23). Therefore, from Eq. 13-21,

$$R = \int_0^R dz = \int_{E_0}^0 \frac{dE}{NS_n(E)} = \frac{E_0}{(dE/dz)_n}, \tag{13-26}$$

where $(dE/dz)_n = -NS_n(E)$. Direct substitution in Eq. 13-22 for the case of Ar in Cu shows that $(dE/dz)_n \approx 125$ eV/Å (for $a = 10^{-9}$ cm). Thus, for every keV of ion energy the altered layer extends about 8 Å.

Through the use of high-energy ion beams, mixing reactions occur over substantial dimensions. Films can, therefore, be effectively alloyed with substrates and layered, but normally immiscible films can be homogenized with the assistance of ion implantation. As an example, consider the multiple-layer structure consisting of alternating Au and Co films shown in Fig. 13-16. According to the phase diagram, these elements do not dissolve in each other; but they form a uniform metastable solid solution under a flux of 3×10^{15} Xe ions/cm^2 at an energy of 300 keV. Collision cascades, ballistic effects of recoils, and defect migration during room-temperature irradiation all contribute, in a complex way, to the observed mixing.

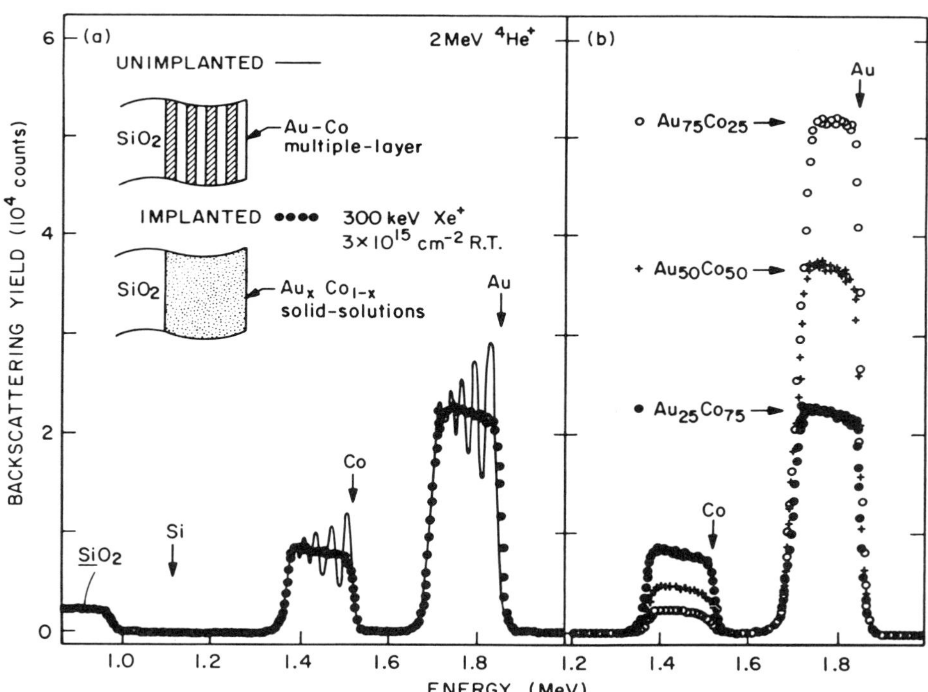

Figure 13-16. (a) RBS spectra showing formation of a uniformly mixed Au_xCo_{1-x} solid solution layer by ion-beam mixing of Au–Co multilayers. (b) RBS spectra of ion-mixed Au–Co films with compositions $Au_{75}Co_{25}$, Au_{50}–Co_{50}, and $Au_{25}Co_{75}$. (Reprinted with permission from Plenum Publishing Corp., from Ref. 22).

Among the systems that have been researched for ion-beam mixing effects are (Ref. 22)

1. Silicides: Pt, Pd, Mo, Nb, Cr
2. Bilayer metal–metal films: Al–Ni, Al–Pd, Al–Pt, Au–Ag, Au–W, Cu–Au
3. Multilayer metal–metal films: Au–Co, Au–Ni, Au–V
4. Eutectics: Au–Si

In silicides, equilibrium as well as metastable phases have been observed after mixing. Experiment has shown that the silicide thickness is both dose- and ion-species-dependent. At the same energy and dose, more mixing occurs the heavier the ion. In metal film systems, extended solubility is virtually always observed, metastable phase formation is a frequent occurrence, and amorphous phases occasionally form at cryogenic temperatures.

13.5.2. Modification of Mechanically Functional Surfaces (Refs. 13, 24)

By enhancing the ability of surfaces to resist plastic deformation, the benefits of reduced wear, less tendency to surface cracking, and greater dimensional stability are effected. In recent years there has been considerable research on the use of ion implantation to realize these desirable ends. For the case of steel, the implantation of light interstitial ions, such as nitrogen, boron, and carbon, yields considerable improvement in wear and fatigue resistance. The reason is due to the elastic interaction between dislocations and undersized interstitials; this results in their mutual attraction and the segregation of the atoms to the defects. Long known in metallurgical circles, the interaction occurs even at room temperature and is quite effective in pinning dislocations, thus restricting their motion. In addition, iron nitrides, carbides, borides, etc., form when the limited solubilities of the interstitial atoms are locally exceeded; these precipitates are also effective barriers to dislocation movement. Since surface damage processes depend on plastic flow of surface layers, the importance of limiting dislocation motion is apparent.

An alternate approach to improving the wear resistance of surfaces through the deposition of hard coatings was addressed at length in Chapter 12. It is instructive to compare this approach with that of ion implantation. Although CVD deposits manage to conformally coat external as well as internal surfaces, ion implantation is limited by geometry to line-of-sight processing. CVD coatings are bonded to the substrate across an interfacial region, which is frequently a source of adhesion difficulty; on the other hand, ion-implantation modified layers are not subject to adhesion problems because no sharp interface exists. Since CVD deposition is conducted at elevated temperatures, the substrate is frequently heat-affected and sometimes softened in the process; ion implantation only modifies a very thin surface layer, leaving the remainder of the substrate unaffected. Lastly, thick CVD coatings several microns thick imply less stringent substrate smoothness requirements than for ion-implantation processing. The latter is the only practical method available for modifying precision surfaces while preserving extreme dimensional tolerances. Clearly, ion implantation is only cost-effective for high-value added components, such as surgical implants or dies.

The issue of the effective surface depth modified by ion implantation is an interesting one. It has been observed that ion implantation effects persist well beyond the shallow depth of the projected ion range. Implanted atoms are frequently observed considerably deeper within the substrate than can be accounted for by the geometry of wear tracks. The cause has been attributed to the generation of fresh dislocation networks that effectively trap and drag atoms deeper below the damaged surface layers. Frictional wear is also accompanied by temperature increases of as much as 600–700 °C at contacting asperities. Migration of mobile impurities is thus encouraged, especially where the dislocation density is high. Such effects provide an unexpected wear protection bonus for ion-implanted surfaces.

A number of industrial applications involving wear reduction by means of ion implantation methods is listed in Table 13-2. Hardness and resistance to adhesive and abrasive wear are the attributes required of the assorted cutting, mechanical forming, and molding tools. In steel matrices nitrogen is a favored interstitial ion, and implanted cobalt has been explored as a means of modifying tungsten carbide tools. Typical doses are well into the $10^{17}/\text{cm}^2$ range and impart two- to fivefold decreases in wear rate with corresponding increases in tool life. Specific examples of ion-implantation modified tools and components are shown in Fig. 13-17.

A totally different application which nevertheless exploits the benefits of enhanced hardness and reduced wear involves metallic surgical implants. Tens of thousands of titanium alloy (Ti-6 wt%-Al-4 wt% V) hip and knee replacement prostheses have already been ion-implanted with nitrogen resulting in improved tribological properties. In service, the implant moves in contact with a high-molecular-weight polyethylene mating socket, so that wear of this couple is of concern. Apparently, the formation of hard TiO_2 particles results in abrasion of the unimplanted alloy surface during use. Implantation produces a surface containing hard titanium nitride precipitates that effectively resists such wear. Part of the improvement in properties may be attributable to the enhanced corrosion resistance ion-implanted surfaces exhibit. High defect

Table 13-2. Industrial Applications of Ion Implantation for Wear Reduction

Application	Material	Treatment Dose($10^{17}/cm^2$)	Result
1. Paper slitters	1.6 Cr–1C Steel	8-N	Cutting life increased 2 ×
2. Acetate punches	Cr-plate	4-N	Improved life
3. Taps for threading plastic	High-speed steel	8-N	Life increased 5 ×
4. Slitters for synthetic rubber	WC–6% Co	8-N	Life increased 2 ×
5. Tool inserts	4Ni–1Cr Steel	4-Co	Contamination reduction
6. Forming tools	12Cr–2C Steel	4-N	Reduced adhesive wear
7. Dies for Cu rod	WC–6% Co	5-C	Throughput increased 3 ×
8. Drawing dies	WC–6% Co	2-Co	Improved life
9. Dies for steel wire	WC–6% Co	3-C	Wear rate reduced 3 ×

From Ref. 24.

Figure 13-17. Examples of ion-implanted components including a die for making Band Aids, knee and hip surgical prostheses and cryogenic pistons. (Courtesy of J. K. Hirvonen and B. C. Haywood, Spire Corp.)

concentrations promote thickening of air-formed oxide films and enhance chemical homogenization of the underlying metal. The former effect affords an added measure of surface passivation and protection, and the latter helps eliminate localized galvanic corrosion. Lastly, note that there are no practical alternatives to modifying the surface properties of orthopedic prostheses. Unlike tools whose surfaces can tolerate CVD or PVD coatings, chemical biocompatibility with contacting body fluids places severe restrictions on the surface composition of surgical implants.

13.5.3. Modification of Surface Morphology (Refs. 25, 26)

In addition to chemical and structural modification of surface layers, ion beams can, interestingly, alter the topography of surfaces on which they impinge. Ions of moderate energy, i.e., less than 1 keV on up to tens of keV, are necessary. Under bombardment various micron-sized surface structures such as cones, pyramids, ridges, ledges, pits, faceted planes as well as quasi-liquid-like and microtextured labyrinth-like features develop. An example of cone formation on a Cu single crystal after 40-keV Ar bombardment is shown in Fig. 13-18. The term *cone formation* is used to generically categorize this class of topological phenomena.

Known for half a century, cone formation has been interpreted in terms of either *left-standing* or *real-growth* models. The former model views conical projections arising from sputter-resistant impurities or intentionally deposited

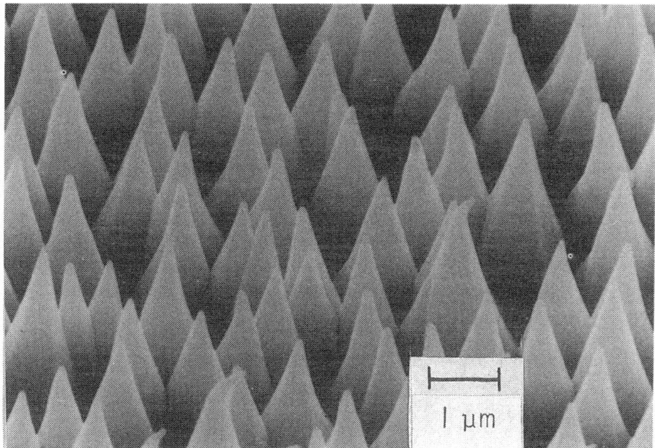

Figure 13-18. Pyramid structures on a single-crystal Cu surface after 40-keV Ar ion bombardment. (From Ref. 27).

seed atoms (e.g., Mo) that etch or sputter at a lower rate than the surrounding surface (e.g., Cu). Sputter-yield variations as a function of ion incidence angle, nonuniform redeposition of atoms on oblique cones, and surface diffusion effects are operative in this mechanism. The latter model (Ref. 25) likewise requires the presence of impurity seed atoms. In this case they serve as nucleating sites for growth of genuine single-crystal whiskers. These sprout in all crystallographic directions with respect to ion incidence. The subsequent interplay among whisker growth, surface diffusion, and sputtering results in the observed cones.

The following experimental facts regarding seed cone and whisker formation during ion bombardment have been established (Ref. 25).

1. Very small amounts of metal impurity or seed atoms are required. Different impurity seed atoms for Cu include Ta, Mo, Fe, Cr, and Ti. Cones have been observed on many different metal surfaces, e.g., Al, Au, Pb, and Ni.
2. Contrary to long-held belief, the seed atoms need *not* have a lower sputter yield than matrix atoms. They apparently must have a *higher* melting point.
3. At low ion energies (< 1 keV) an elevated target temperature (i.e., one third of the melting point) is required for seed cone formation.
4. When the bombarding energy is close to the sputter threshold, whiskers may be observed. At higher ion energy, thin slanted whiskers evolve into cones as they collect sputtered atoms from surrounding areas.
5. Pure metal surfaces without any trace of impurities or seed atoms do not develop cones upon exposure to ion beams.

Potential applications for rough surfaces containing cones or whiskers are selective light absorbers, catalysts, electron emitters, adherent interlocking surfaces for surgical implants or for joining processes. Thus far, however, no commercial exploitation of such ion-beam-modified surfaces has occurred.

13.5.4. Ion-Beam Modification of Silicon (Refs. 28, 29)

The chapter closes with a brief description of two novel modification processes in Si that do *not* involve traditional low-dosage semiconductor doping. Through implantation of O^+ at fluences of $\sim 1.5 \times 10^{18}$ ions/cm^2, at energies of ~ 200 keV, there is sufficient oxygen present to form a subsurface layer of stoichiometric SiO_2. During subsequent annealing, the damage in the overlying Si is repaired, and uphill diffusion of oxygen sharpens both SiO_2–Si

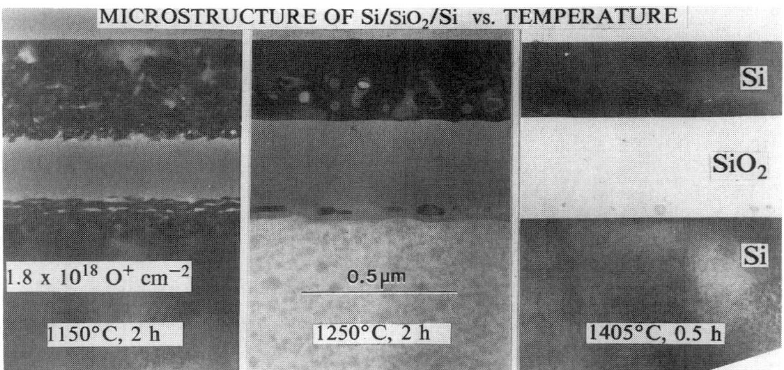

Figure 13-19. TEM cross-sectional micrographs of Si implanted with 1.8×10^{18} O^+ cm^{-2} and subsequently annealed at 1150 °C–2 h, 1250 °C–2 h and 1405 °C–0.5 h. (Courtesy of G. K. Celler, AT&T Bell Laboratories, Reprinted with permission from Cowan Publishing Company, from G. K. Celler, *Solid State Technology* **30** (3), 69, 1987).

interfaces. As shown in Fig. 13-19, heat treatments at temperatures of 1250 °C do not completely anneal out dislocations, twins, and SiO_2 precipitates. Temperatures of 1405 °C, which are close to the melting point of Si, are, however, particularly effective in the chemical segregation of implanted O into the buried SiO_2 and in recovering the crystallinity of Si. Ion-beam synthesis of buried SiO_2, known as SIMOX (separation by implanted oxygen) is the third and perhaps most promising SOI process mentioned in this book (see page 333). As a postscript, CMOS dual-modulus prescaler circuits were recently (1989) fabricated at AT&T Bell Laboratories within the upper Si layer thinned from 2400 to 1200 Å. The reported operating speed of 6.2 GHz was 50% higher than in the group of control circuits, and represents the highest reported speed in practical digital circuits.

The second ion-beam modification process, known as mesotaxy, concerns the formation of buried *epitaxial* disilicide layers in Si. To achieve this feat, 200-keV Co^+ ions are implanted into Si at fluences of $\sim 2 \times 10^{17}$ ions/cm^2. This is followed by annealing treatments at 1000 °C during which time the Co distribution narrows and the surrounding lattice damage is significantly reduced. Abrupt epitaxial silicide layers buried 1000 Å beneath the Si surface can be produced this way (Ref. 29).

The formation of both high-quality buried insulating (SiO_2) and conducting ($CoSi_2$) films beneath crystalline Si represent important advances in the eventual realization of three-dimensional integrated circuits.

EXERCISES

1. An atomistic model for the solid–liquid interfacial velocity during solidification suggests that

$$v = a_0^2 \nu \left(\exp - \frac{E_D}{RT}\right) \frac{2\,\Delta G}{a_0 RT} \qquad \text{(Eq. 1-34)}$$

Calculate a value of v for Si, assuming that the atomic diffusivity in the liquid state is 10^{-4} cm^2/sec, $\nu = 10^{13}$ sec^{-1}, $a_0 = 2.7$ Å, and that the free-energy difference between Si atoms in the liquid and solid is equal to the latent heat of fusion. How does the value of v compare with interfacial velocities calculated in this chapter?

2. Comment on the probability of successfully alloying the surface of the following 2000-Å film–bulk substrate combinations by laser processing: Zn–Mo, Mo–W, Fe–Ni, Ta–Cd.

3. A 50-nsec, 37.5-mW/μm^2 Kr ion laser pulse ($\lambda = 6470$ Å) is incident on TeGe, a candidate material for phase-change optical data storage applications. For TeGe, $R = 0.65$, $\rho = 6.2$ g/cm^3, $\kappa = 0.015$ cal/cm-sec-°C, $c = 0.08$ cal/g-°C, and

$$2\sqrt{K_d \tau_p} > \alpha^{-1}$$

a. Plot the surface temperature as a function of time during heating as well as cooling.
b. What is the maximum surface temperature reached?
c. What is the maximum quench rate at the surface?

4. An amorphous Si film on a flat single-crystal Si wafer is annealed. Schematically sketch and distinguish between the resultant grain structures if (a) random nucleation and growth or (b) solid-phase epitaxy governs the solid-state crystallization process.

5. For corrosion resistance a surface distribution of Cr in steel that is relatively *constant* as a function of depth is required. Show that successive overlapping implants is a way to achieve this. Consider three such implants with Cr ion energies of 50, 100, and 200 keV. If the peak concentration in each case is maintained at 20 at% Cr (the stainless composition), what must the ratio of the doses be? Sketch the resultant profile of the overlapping distributions.

6. Show that when the as-implanted Gaussian distribution given by Eq. 13-24 is thermally annealed, promoting atomic diffusion for time t, the resulting distribution

$$C(z,t) = \frac{\phi}{\sqrt{2\pi(\Delta R_p^2 + 2Dt)}} \exp - \frac{1}{2}\left(\frac{z - R_p}{(\Delta R_p^2 + 2Dt)^{1/2}}\right)^2$$

satisfies the diffusion equation. What is the physical significance of the term $2Dt$, and its addition to the term ΔR_p^2.

7. Steel dies are implanted with 200-keV N ions to enhance wear resistance.

a. If the dose is 4×10^{17}, what is the peak N concentration?

b. The Fe–N phase diagram indicates that the solid solubility of N in Fe is about 0.01 at% at room temperature. Excess amounts of N precipitate as Fe_2N. Based on this information, what is the phase constitution of the matrix following ion implantation?

c. Compare the times required to reach the solubility limit of N in Fe (~ 0.4 at% at 590 °C) 1000 Å beneath the surface by ion implantation, and by nitriding. Assume the ion-beam current density is 50 $\mu A/cm^2$ and that the diffusivity of N in Fe is

$$D = 0.003 \exp - 18.2\ (\text{kcal/mole})/\text{RT}\ (\text{cm}^2/\text{sec}).$$

An NH_3 surface source is used for nitriding and the Fe is initially free of N.

8. a. Qualitatively contrast the rate at which energy is lost from a spherical thermal spike of radius L and an energized plate of the same thickness. The spike and plate are initially at the same temperature within the same matrix.

b. The average change in the temperature of a sphere initially at temperature T_i is

$$\frac{T(t)}{T_i} = \frac{6}{\pi^2}\sum_{n=1}^{\infty}\frac{1}{n^2}\exp - \frac{n^2\pi^2 Kt}{L^2},$$

where K is the thermal diffusivity. For the plate,

$$\frac{T(t)}{T_i} = \frac{8}{\pi^2}\sum_{n=1}^{\infty}\frac{1}{(2n+1)^2}\exp - \frac{(2n+1)^2}{4} n^2\pi^2\frac{Kt}{L^2}$$

To confirm your intuition, plot

$$\frac{T(t)}{T_i} \text{ vs. } \frac{Kt}{L^2}$$

for both geometries using values of $Kt/L^2 = 0.1, 0.2, \ldots, 0.9, 1.0$.

9. Suppose it is desired to selectively harden the flat surface of a steel die only in certain areas through ion implantation of 100-keV N ions. Other areas are protected against implantation by electrodeposition of a patterned Ni coating that serves to attenuate the ion beam. If specifications call for a maximum penetration of 1% of the beam, what coating thickness d of Ni is required? Assume Ni and Fe are alike with respect to N implantation effects. [Hint: What does

$$\int_d^\infty \frac{\phi}{\sqrt{2\pi}\,\Delta R_p} \exp - \left(\frac{z - R_p}{\sqrt{2}\,\Delta R_p}\right)^2 dz$$

physically represent?]

10. In typical SIMOX processing 2×10^{18} atoms of O are implanted per cm^2 of Si at 200 keV. After a high-temperature anneal (above 1300 °C for 6 h) a 4000-Å-thick layer of SiO_2, having sharp parallel interfaces, forms buried beneath a 2000-Å-thick surface layer of Si. From this information estimate values of R_p and ΔR_p for O implanted into Si at 200 keV.

References

1. A. Yariv, *Optical Electronics*, 3rd ed., Holt, Rhinehart and Winston, New York (1985).

2.* C. W. Draper and J. M. Poate, *Int. Met. Rev.* **30**, 85 (1985).

3. N. Bloembergen, in *Laser-Solid Interactions and Laser Processing*, eds. S. D. Ferris, H. J. Leamy, and J. M. Poate, Amer. Inst. of Physics, No. 50 (1979).

4.* A. E. Bell, *RCA Rev.* **40**, 295 (1979).

5. L. E. Greenwald, E. M. Breinan, and B. H. Kear, in *Laser-Solid Interactions and Laser Processing*, eds. S. D. Ferris, H. J. Leamy, and J. M. Poate, Amer. Inst. of Physics, No. 50 (1979).

*Recommended texts or reviews.

6. S. C. Hsu, S. Chakravorty, and R. Mehrabian, *Metall. Trans.* **9B**, 221 (1978).
7.* E. Rimini, in *Surface Modification and Alloying*, eds. J. M. Poate, G. Foti, and D. Jacobson, Plenum Press, New York (1983).
8. P. Baeri and S. U. Campisano, in *Laser Annealing of Semiconductors*, eds. J. M. Poate and J. W. Mayer, Academic Press, New York (1982).
9. J. S. Williams, in *Surface Modification and Alloying*, eds. J. M. Poate, G. Foti, and D. C. Jacobson, Plenum Press, New York (1983).
10. P. Baeri, G. Foti, J. M. Poate, S. U. Campisano, and A. G. Cullis, *Appl. Phys. Lett.* **38**, 800 (1981).
11. D. C. Jacobson, Ph.D. Thesis, Stevens Institute of Technology, Hoboken, NJ (1988).
12.* K. A. Jackson, in *Surface Modification and Alloying*, eds. J. M. Poate, G. Foti, and D. C. Jacobson, Plenum Press, New York (1983).
13. J. K. Hirvonen, in *Ion Implantation*, *Treatise on Materials Science and Technology*, Vol. 18, Academic Press, New York (1980).
14. J. Lindhard, M. Scharff, and H. E. Schiott, *Kgl. Danske. Vid. Selsk. Mat. Fys. Medd.* **33, No. 14**, 1 (1963).
15. J. A. Davies, in *Surface Modification and Alloying*, eds. J. M. Poate, G. Foti, and D. C. Jacobson, Plenum Press, New York (1983).
16. C. J. Dienes and G. H. Vineyard, *Radiation Effects in Solids*, Wiley-Interscience, New York (1957).
17. W. L. Brown, in *Beam Solid Interactions and Phase Transformations*, eds. H. Kurz, G. L. Olsen, and J. M. Poate, Mat. Res. Soc., Pittsburgh (1986).
18. A. F. Burenkov, F. F. Komarov, M. A. Kumakhov, and M. M. Temkin, *Tables of Ion Implantation Spatial Distributions*, Gordon and Breach, New York (1986).
19.* T. E. Seidel, in *VLSI Technology*, ed S. M. Sze, McGraw-Hill, New York (1983).
20.* I. Brodie and J. J. Muray, *The Physics of Microfabrication*, Plenum Press, New York (1982).
21. S. B. Agarwal, Y. F. Yang, C. R. Clayton, H. Herman, and J. K. Hirvonen, *Thin Solid Films* **63**, 19 (1979).
22.* J. W. Mayer and S. S. Lau, in *Surface Modification and Alloying*, eds. J. M. Poate, G. Foti, and D. Jacobson, Plenum Press, New York (1983).
23.* L. C. Feldman and J. W. Mayer, *Fundamentals of Surface and Thin-Film Analysis*, North-Holland, New York (1986).

24. J. K. Hirvonen and C. R. Clayton, in *Surface Modification and Alloying*, eds. J. M. Poate, G. Foti, and D. Jacobson, Plenum Press, New York (1983).
25.* G. K. Wehner, *J. Vac. Sci. Tech.* **A3**, 1821 (1985).
26. P. Auciello and R. Kelly, *Ion Bombardment Modification of Surfaces —Fundamentals and Applications*, Elsevier, Amsterdam (1984).
27. J. L. Whitton, G. Carter, and M. J. Nobes, *Radiation Effects* **32**, 129 (1977).
28. G. K. Celler, *Solid State Technology* **30(3)**, 69 (1987).
29. A. E. White and K. T. Short, *Science* **241(8)**, 930 (1988).

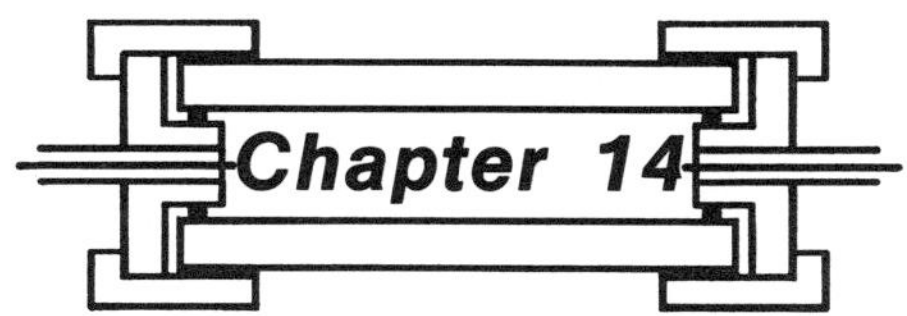

Emerging Thin-Film Materials and Applications

In this final chapter an attempt is made to present a perspective of some emerging thin-film materials and applications that promise to have a significant impact on future technology. For this reason the discussion will be limited to the following topics:

14.1. Film-Patterning Techniques
14.2. Diamond Films
14.3. High T_c Superconductor Films
14.4. Films for Magnetic Recording
14.5. Optical Recording
14.6. Integrated Optics
14.7. Superlattices
14.8. Band-Gap Engineering and Quantum Devices

This potpourri of subjects encompasses covalent, metallic, and semiconductor film materials deposited by an assortment of PVD and CVD methods. Represented are mechanical, electrical, magnetic, and optical properties, whose optimization hinges on both processing and the ability to characterize structure–property relationships. Thus the spirit of materials science of thin films—the theme and title of this book—is preserved in microcosm within this chapter. For completeness however, it is necessary to start with Section 14.1, which is devoted to the topic of thin-film patterning techniques. This subject is crucial to the realization of the intricate lateral geometries and dimensions that films must assume in varied applications, particularly some of those in this chapter.

14.1. Film-Patterning Techniques

14.1.1. Lithography

Until now the only film dimension considered has been the thickness, which is controlled by the growth or deposition process. However, irrespective of eventual application, thin films must also be geometrically defined laterally or patterned in the film plane. The complexity of patterning processes depends on the nature of the film, the feature dimensions, and the spatial tolerance of the feature dimensions. For example, consider an evaporated metal film that must

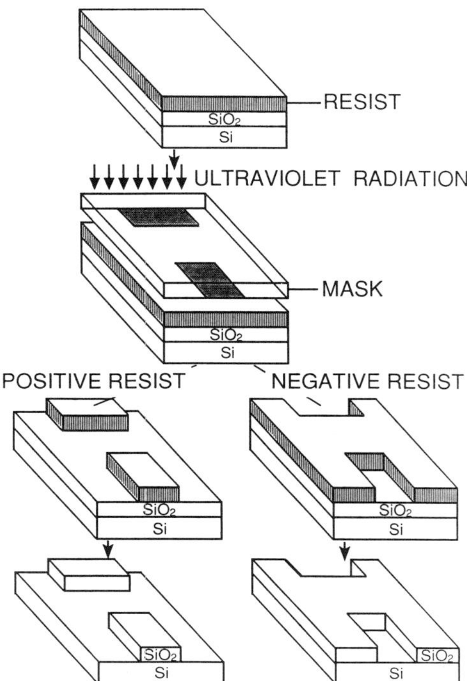

Figure 14-1. Schematic of the lithographic process for pattern transfer from mask to film. Both positive and negative resist behavior is illustrated.

possess features 1 mm in size with a tolerance of ± 0.05 mm. The desired pattern could possibly be machined into a thin sheet stencil or mechanical mask. Direct contact between this mask template and substrate ensures generation of the desired pattern in uncovered regions exposed to the evaporation flux. This method is obviously too crude to permit the patterning of features 100 to 1000 times smaller in size that are employed in integrated circuits. Such demanding applications require lithographic techniques.

The lithographic process shown schematically in Fig. 14-1 consists of four steps.

14.1.1.1. Generation of the Mask. The mask is essentially equivalent to the negative in photography. It possesses the desired film geometry patterned in Cr or FeO thin films predeposited on a glass or quartz plate. Masks for integrated circuit use are generated employing computer-driven electron beams to precisely define regions that are either opaque or transparent to light. Other processing steps to initially produce the patterned mask film parallel those used in subsequent pattern transfer to the involved film.

14.1.1.2. Printing. Printing of this negative mask requires the physical transfer of the pattern to the film surface in question. This is accomplished by first spin-coating the film–substrate with a thin photoresist layer (< 1 μm thick). As the name implies, photoresists are both sensitive to photons and resistant to chemical attack after exposure and development. Photoresists are complex photosensitive organic mixtures, usually consisting of a resin, photosensitizer, and solvent. During exposure, light (usually UV) passes through the mask and is imaged on the resist surface by appropriate exposure tools or printers. Either full-scale or reduced latent images can be produced in the photoresist layer. There are two types of photoresists and their behaviors are distinguished in Fig. 14-1. The positive photoresist faithfully reproduces the (opaque) mask film pattern; in this case light exposure causes scission of polymerized chains rendering the resist soluble in the developer. Alternatively, negative resists reproduce the transparent portion of the mask pattern because photon-induced polymerization leaves a chemically inert resist layer behind. For yet greater feature resolution X-ray and electron-beam lithography techniques are practiced.

14.1.1.3. Etching. After resist exposure and development, the underlying film is etched. Wet etching in appropriate solutions dissolves away the exposed film, leaving intact the film protected by resist. Equal rates of lateral and vertical material removal (isotropic etching) however, lead to loss of resolution due to undercutting of film features. This presents a problem in VLSI processing where 1 μm (or so) features must be defined. For this reason dry etching is practiced. Material is removed in this case through exposure to reactive plasmas that interact with film atoms to produce volatile by-products that are pumped away. For example, typical dry etchants for Si, SiO_2 and Al are $SF_6 + Cl_2$, $CF_4 + H_2$, and $BCl_3 + Cl_2$ gas mixtures, respectively (Ref. 1). Alternatively, inert-gas plasmas are also employed to erode the film surface in a process that resembles the inverse of sputtering deposition. In both cases, positive ion bombardment normal to the surface leads to greater vertical than horizontal etching, i.e., anisotropic etching. Steep sidewall topography and high aspect ratio features such as shown in Fig. 14-2 are the result of anisotropic material removal.

An important issue in dry etching is the etchant selectivity or ability to preferentially react with one film species relative to others that are present. Simply changing the plasma gas composition can significantly alter etching selectivity. For example, the SiO_2 etch rate exceeds that of poly-Si by only 25% in a pure CF_4 plasma. In an equimolar mixture of $H_2 + CF_4$, however,

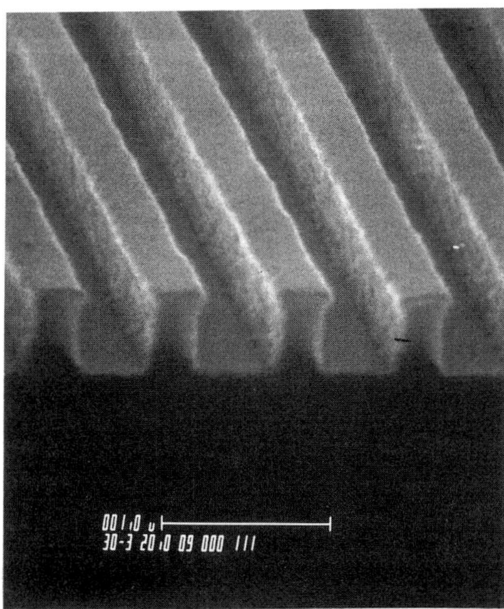

Figure 14-2. SEM micrograph of reactive plasma-etched pattern in photoresist revealing development of submicron features. (Courtesy of L. F. Thompson, AT&T Bell Laboratories).

the etch rate of poly-Si drops almost to zero; the selectivity or ratio of etch rate of SiO_2 relative to poly-Si exceeds 45 (Ref. 1).

14.1.1.4. Resist Removal. The final step requires removal of the resist. Special resist stripper solutions or plasmas (e.g., O_2 rich) are utilized for this purpose. What remains is a high fidelity thin-film copy of the mask geometry.

Only the briefest summary of the basic steps comprising the very important
For more detailed accounts of
printing (Ref. 4), and etching
ferences.

-precision shaping technique
to form miniature three-di-
s, wells, pyramids, grooves,
i has revolutionized electron-
nal perceptions of miniature
hough small, micromachined
imensions. Examples include
n p. 412, tiny gears, valves,
ssure and strain transducers,
nultisocket electrical connec-
efs. 5, 6). Among the recent
developments are the fabrication of a triode vacuum microelectronic device (Ref. 7) and an optical microassembly. The former shown in Fig. 14-3a is impervious to radiation damage, insensitive to heat with the potential for very

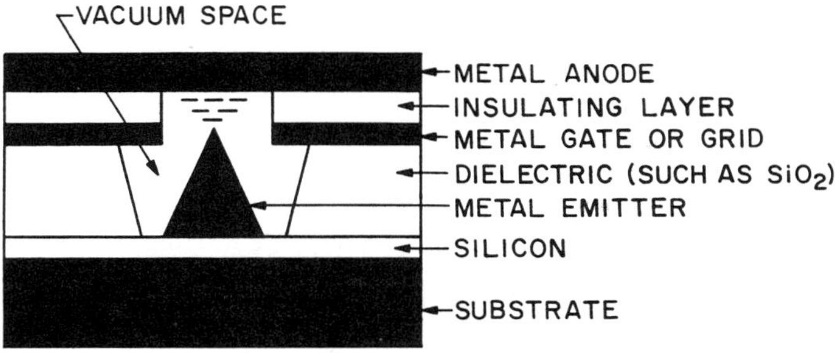

Figure 14-3a. Schematic structure of Si triode vacuum microelectronic device. (From Ref. 7).

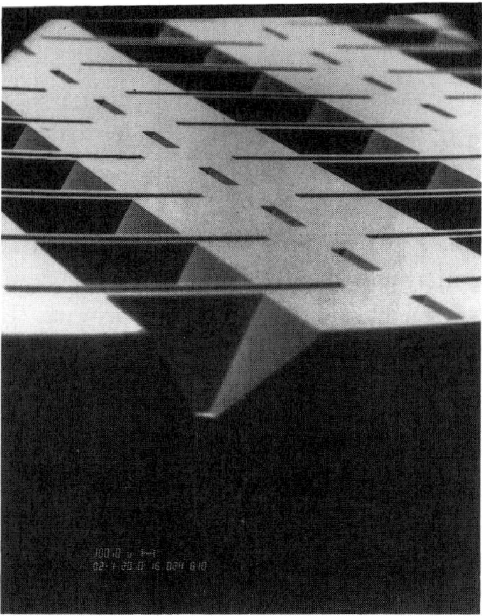

Figure 14-3b. SEM micrograph of optical microassembly. (Courtesy of K. L. Tai, AT&T Bell Laboratories).

high frequency operation. The latter shown in Fig. 14-3b has been employed to provide low-loss coupling between optical fibers and optoelectronic devices in optical communications systems. Here the laser (or detector) rests beneath the apex of the etched pyramid in which the optical fiber is precisely positioned. This microassembly package provides for low-loss electrical interconnection between optoelectronic and other electronic devices on a common Si substrate.

Precise knowledge of etch rate anisotropies and selectivities for Si and SiO_2 is required for designing successful micromachining etching treatments. In a recent study (Ref. 8), utilizing KOH/H_2O etchants, the following etch rates (R) were measured as a function of temperature:

$$R_{Si}(100) = 6.19 \times 10^{8} \exp - \frac{0.61 \text{ eV}}{kT} \ (\mu\text{m/min}), \tag{14-1}$$

$$R_{Si}(111) = 3.19 \times 10^{9} \exp - \frac{0.77 \text{ eV}}{kT} \ (\mu\text{m/min}), \tag{14-2}$$

$$R_{SiO_2} = 5.49 \times 10^{12} \exp - \frac{1.07 \text{ eV}}{kT} \ (\mu\text{m/min}), \tag{14-3}$$

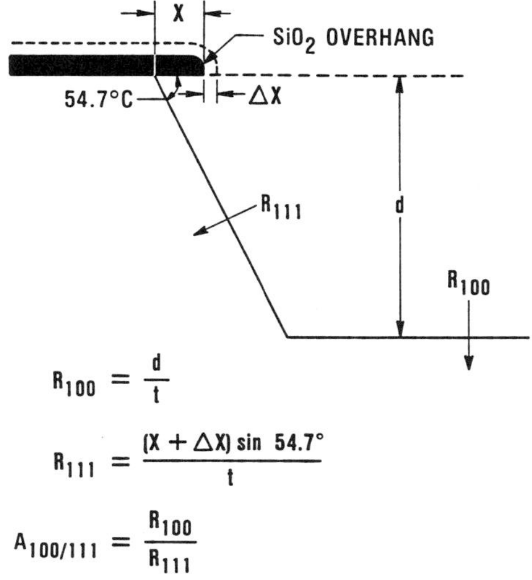

Figure 14-4. Etching geometry of Si–SiO_2 structure.

The ratio $R_{Si}(100)/R_{Si}(111)$ defines the etch rate anisotropy ($A_{100/111}$) and the ratio $R_{Si}(100)/R_{SiO_2}$ represents the selectivity.

As an example in the use of these etch rates consider a (100) Si wafer containing a 2 μm thermally grown SiO_2 film so patterned to open windows to the Si surface (Fig. 14-4). After etching at 100 °C for 15 min, how much does the SiO_2 etch mask overhang the slanted Si wall? During etching, both the (100) and (111) planes recede along their direction normals. The angle between the [100] and [111] directions is 54.7°. Therefore geometric considerations indicate that the net overhang length x at any time t is given by

$$x = \left(R_{Si}(111)/\sin 54.7 - R_{SiO_2}\right)t, \tag{14-4}$$

where the isotropic etching of SiO_2 is accounted for. Direct substitution of $R_{Si}(111) = 0.126$ μm/min, $R_{SiO_2}0.0191$ μm/min, $t = 15$ min, and $\sin 54.7 = 0.816$, yields $x = 2.03$ μm. Depending on the width of the SiO_2 mask window, V-shaped pits or flat-bottomed troughs can be etched into Si.

14.2. Diamond Films

14.2.1. Introduction

Derived from the Greek $\alpha\,\delta\alpha\mu\alpha\varsigma$ (*adamas*), which means unconquerable, diamond is indeed an invincible material. In addition to being the most costly on a unit weight basis, and capable of unmatched beauty when polished, diamond has a number of other remarkable properties. It is the hardest substance known ($H_v > 8000$ kg/mm^2), and has a higher modulus of elasticity ($E = 1050$ GPa) than any other material. When free of impurities, it has one of the highest resistivities ($\rho > 10^{16}$ Ω-cm). It also combines a very high thermal conductivity ($\kappa = 1100$ W/m-K) that exceeds that of Cu and Ag, with a low thermal expansion coefficient ($\alpha = 1.2 \times 10^{-6}$ K^{-1}) to yield high resistance to thermal shock. Lastly, diamond is very resistant to chemical attack. These facts, the first three, in particular, have spurred one of the most exciting and competitive quests in the history of materials science—the synthesis of diamond. Success was achieved in 1954 with the General Electric Corp. process for producing bulk diamond utilizing extremely high pressures and temperatures. Interestingly, however, attempts to produce diamond from low-pressure vapors date back at least to 1911 (Ref. 9). P. D. Bridgeman, in a 1955 *Scientific American* article, speculated that diamond powders and films should be attainable by vapor deposition at low pressures (Ref. 10). By the mid-1970s the Russian investigators Derjaguin and Fedeseev had apparently grown epitaxial diamond films and whiskers during the pyrolysis of various hydrocarbon–hydrogen gas mixtures (Ref. 11). After a decade of relative quiet, an explosive worldwide interest in the synthesis of diamond films and in their properties erupted, which persists unabated to the present day.

Isolated C atoms have distinct 2s and 2p atomic orbitals. When these atoms condense to form diamond, electronic admixtures occur, resulting in four equal hybridized sp^3 molecular orbitals. Each C atom is covalently attached to four other atoms in tetragonal bonds 1.54 Å long creating the well-known diamond cubic structure (Fig. 1-2c). Graphite, on the other hand, has a layered structure. The C atoms are arranged hexagonally with strong trigonal bonds (sp^2) and have an interatomic spacing of 1.42 Å in the basal plane. A fourth electron in the outer shell forms weak van der Waals bonds between planes that account for such properties as good electrical conductivity, lubricity, lower density, a grayish-black color and softness.

In addition, C exists in a variety of metastable and amorphous forms that have been characterized as degenerate or imperfect graphitic structures. In these, the layer planes are disoriented with respect to the common axis and overlap each other irregularly. Beyond the short-range graphitic structure, the matrix consists of amorphous C. A complex picture now emerges of the manifestations of C ranging from amorphous to crystalline forms in a continuum of structural admixtures. Similarly, the proportions of sp^2–sp^3 (and even sp^1) bonding is variable causing the different forms to have dramatically different properties. Not surprisingly, this broad spectrum of metastable carbons have been realized in thin-film deposits. What now complicates matters further is that the many techniques to produce carbon films use precursor hydrocarbon gases. Hydrogen is, therefore, inevitably incorporated, and this adds to the complexity of the deposit structure, morphology, and properties.

Given the structural and chemical diversity of carbon films, an understandable confusion has arisen with regard to the description of these materials. Labels such as hard carbon, amorphous carbon (a-C), hydrogenated amorphous carbon (a-C:H), ion-beam-processed carbon (i-C), diamondlike carbon (DLC), as well as diamond have all been used in the recent literature. The ensuing discussion will treat the deposition processes and properties of these films with the hope of clarifying some of their distinguishing features.

14.2.2. Film Deposition Processes

At the outset it is important to realize that synthesis of *bulk* diamond occurs in the diamond stable region of the $P-T$ phase diagram (Fig. 1-11). Thin "diamond" films, on the other hand, clearly involve *metastable* synthesis in the low-pressure graphite region of the phase diagram. The possibility of synthesizing diamond in this region is based on the small free-energy difference (500 cal/mole) between diamond and graphite under ambient conditions (Ref. 12). Therefore, a finite probability exists that both phases can nucleate and grow simultaneously, especially under conditions where kinetic factors dominate, such as high energy or supersaturation. In particular, the key is to prevent graphite from forming or to remove it preferentially, leaving diamond behind. The way this is done practically is to generate a supersaturation or superequilibrium of atomic H. The latter can be produced utilizing 0.2–2% CH_4–H_2 mixtures in microwave plasmas or in CVD reactors containing hot filaments. Under these conditions, atomic H is generated and, in turn, fosters diamond growth either by inhibiting graphite formation, dissolving it if it does form, stabilizing sp^3 bonding, or by promoting some combination of these factors. In general, hydrocarbon, e.g., CH_4, C_2H_2, decomposition at substrate temperatures of 800–900 °C in the presence of atomic H is conducive to diamond growth on nondiamond substrates. Paradoxically the copious amounts of atomic H result in very little hydrogen incorporation in the deposit. The modern era of CVD synthesis is coincident with the beautiful SEM images of diamond crystallites produced in the manner described. These have captured the imagination of the world and examples of the small faceted "jewels," grown at high temperatures on nondiamond substrates, are shown in Fig. 14-5.

The a-C:H materials are formed when hydrocarbons impact relatively low-temperature substrates with energies in the range of a few hundred eV.

Figure 14-5. Diamond crystals grown by CVD employing combined microwave and filament methods. (Courtesy of T. R. Anthony, GE Corporate Research and Development).

Plasma CVD techniques employing rf and dc glow discharges in assorted hydrocarbon gas mixtures commonly produce a-C:H deposits. The energetic molecular ions disintegrate upon hitting the surface and this explains why the resulting film properties are insensitive to the particular hydrocarbon employed. It is thought that the incident ions undergo rapid neutralization and the carbon atoms are inserted into C—H bonds to form acetylenic and olefinic polymerlike structures, e.g., $C + R{-}CH_3 \rightarrow R{-}CH{=}CH_2$, where R is the remainder of the hydrocarbon chain. The resultant films, therefore, contain variable amounts of hydrogen with H/C ratios ranging anywhere from ~ 0.2 to ~ 0.8 or more. They may be thought of as glassy hydrocarbon ceramics and can be even harder than SiC.

Amorphous carbon (a-C) diamondlike films are prepared at low temperatures in the absence of hydrocarbons by ion-beam or sputter deposition techniques. Both essentially involve deposition of carbon under the bombardment of energetic ions. Simple thermal evaporation of carbon will, of course, yield highly conductive, soft films that are quite remote in their properties from the hard, very resistive, high-energy band-gap diamondlike materials.

The ion impact energy, therefore, appears to be critical in establishing the structure of the deposit. More diamondlike properties are produced at low energy; microcrystalline diamond ceases to form when the ion energy exceeds ~ 100 eV, in which case the amorphous structure prevails.

An important consideration in the eventual commercialization of deposition processes is the growth rate. For both diamond and diamondlike films rates generally range from less than 1 up to a few μm per hour. These values should be compared with the 10^3 μm/h rate for the commercial process that produces diamond abrasive grain.

14.2.3. Properties and Applications

The properties of CVD synthesized diamond, a-C and a-C:H film materials are compared with those of bulk diamond and graphite in Table 14-1. Basic

Table 14-1. Properties of Carbon Materials

	Thin Films			Bulk	
Property	CVD Diamond	a-C	a-C:H	Diamond	Graphite
Crystal structure	Cubic $a_0 = 3.561$ Å	Amorphous, mixed sp^2–sp^3 bonds	Amorphous, mixed sp^2–sp^3 bonds	Cubic $a_0 = 3.567$ Å	Hexagonal $a = 2.47$
Form	Faceted crystals	Smooth to rough	Smooth	Faceted crytals	
Hardness, H_v	3,000–12,000	1,200–3,000	900–3,000	7,000–10,000	
Density	2.8–3.5	1.6–2.2	1.2–2.6	3.51	2.26
Refractive index	—	1.5–3.1	1.6–3.1	2.42	2.15 1.81
Electrical resistivity (Ω-cm)	$> 10^{13}$	$> 10^{10}$	10^6–10^{14}	$> 10^{16}$	0.4 0.20
Thermal conductivity (W/m-K)	1100	—		2000	3500 150
Chemical stability	Inert (inorganic acids)	Inert (inorganic acids)	Inert (inorganic acids and solvents)	Inert (inorganic acids)	Inert (inorganic acids)
Hydrogen content (H/C)	—	—	0.25–1	—	—
Growth rate (μm/h)	~ 1	2	5	1000 (synthetic)	—

From Refs. 12 and 13.

differences in structure and properties of diamond and diamondlike films ultimately stem from the sp^3–sp^2 bond concentration ratios. Considerable bond admixtures occur in both the a-C:H and a-C films and much experimental effort has been expended in determining the bonding proportions. Techniques such as Raman spectroscopy, nuclear magnetic resonance, and X-ray photoelectron spectroscopy (XPS) are used to characterize films and bolster claims for the presence of the elusive diamond crystals. Although there is a great deal of scatter in many of the film properties due to differing deposition conditions, it is clear that the films are extremely hard, chemically inert, and highly insulating.

The attractive attributes of carbon film materials have already been commercially exploited in a number of cases as indicated in Table 14-2. Additional applications have been suggested and are the subject of intense current research and development activities. For many applications crystalline diamond is not essential; diamondlike films will do. With improved film morphology and

Table 14-2. Actual and Suggested Applications of Diamond and Diamondlike Films

Application	Properties Required	Comments
1. Resonator diaphragms of tweeter loud speakers	High modulus of elasticity	Frequency response up to 60,000 Hz possible; commercially available
2. Ultrahard tool coatings	High hardness	Commercially available
3. Sunglass lenses	High hardness, scratch resistance, optical transparency	Commercially available
4. Computer hard disk coatings	High hardness, low wear	Coatings minimize head-disk contact wear
5. Watch cases	High hardness, scratch resistance	
6. Prosthetic devices	High hardness, low wear	
7. Optical coatings	High hardness, high index of refraction	
8. Infrared laser window	Transparency to IR	
9. Electronic devices—traveling wave amplifiers		Heteroepitaxial films required
10. Semiconductor device heat sinks	High thermal conductivity	Commercially available
11. High-temperature semiconductor devices	Large energy band gap	
12. Abrasive grain	High hardness	

properties that come with better control of deposition processes, the expanded use of these films can certainly be anticipated.

14.3. High T_c Superconductor Thin Films

14.3.1. Introduction

The unexpected discovery of high T_c superconductivity has fundamentally challenged our previous understanding of the subject. Interestingly, critical values of temperature, magnetic field, and current density together with the Meissner effect still define and limit high-T_c superconductivity. However, almost everything about previous theories of superconductivity has been called into question, including applicability of such concepts as band gaps, carrier pairing, coherence length, etc., to high T_c oxides. Therefore, the discussion will focus on the composition and structure of these materials, film deposition techniques, properties, and thin-film applications.

14.3.2. Composition and Structure

The three most actively studied high-T_c superconductors (as of this writing) are listed in Table 10-3. $YBa_2Cu_3O_7$ was discovered first, is the easiest to prepare in bulk and thin-film form, and has been most extensively investigated. A unit cell of this material is shown in Fig. 14-6. The structure is a variation of the class of oxygen-defect perovskites involving a tripling of unit cells. Perovskites have the property of reversibly absorbing or losing oxygen and are therefore nonstoichiometric with respect to this element. Much effort has been expended in correlating crystal structure and oxygen content with T_c. As the oxygen content increases from 6.3 to close to 7 atoms per cell T_c is observed to increase from 30 to $\sim$ 90 K. Concurrently both the a and c lattice constants decrease, whereas that for b increases—each by approximately 1% (Ref. 15). Current transport is believed to occur along the Cu–O ribbons (b axis). The pyramidal CuO_2 sheets perpendicular to the c axis reflect the layered structure of this as well as other high-T_c oxide materials. Through its effect on atomic spacing oxygen necessarily also modifies the valence of Cu as well as the Cu—O bond length; increasing O decreases the former and increases the latter. Since Cu appears to be an essential ingredient in high T_c oxides, it has been argued that its valence state and nature of bonding to O critically influence superconducting properties. In fact, loss of oxygen with

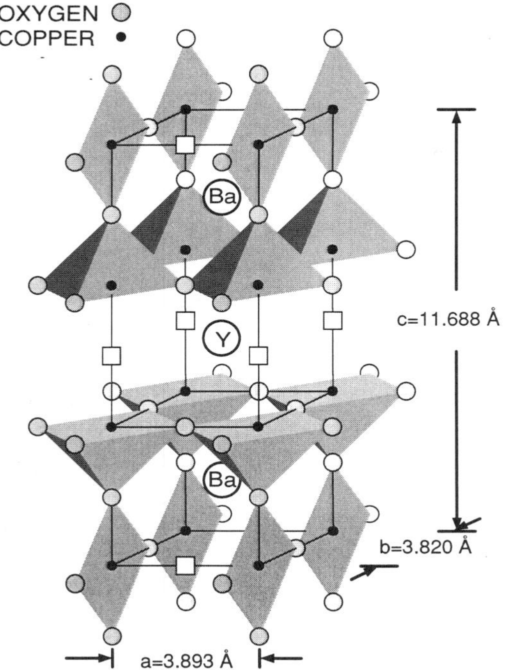

Figure 14-6. Structural model of the unit cell for $YBa_2Cu_3O_7$. Squares are vacant sites. (From Ref. 14 with permission from Kluwer Academic Publishers).

attendant lowering of T_c is a major degradation mechanism in thin films. An overall oxygen stoichiometry of very nearly 7 is required for optimal properties.

14.3.3. Film Deposition Techniques

Among the methods employed to prepare high-T_c films are multisource evaporation (electron beam and resistance heated), single and multigun sputtering, MBE, pulsed laser (flash) evaporation, MOCVD as well as spin pyrolysis and plasma spraying of powders (Ref. 16). Since the vapor pressures of Y, Ba, and Cu vary widely they are not amenable to single source evaporation; rather three separately controlled elemental sources are used. Films prepared by evaporation or sputtering from metallic melts or targets require a subsequent high-temperature (e.g., 850–950 °C) oxidation treatment in order to assure that requisite levels of O are incorporated. To eliminate this step, in situ growth methods have been developed, utilizing reactive evaporation and sputtering, oxygen rf plasmas, microwave generated atomic oxygen and ozone production schemes. Regardless of deposition technique substrate heating (from 300 to 800 °C) appears to be universal.

In achieving high-quality films the choice of substrate is critical. Substrates must be resistant to high-temperature exposure, degradation in oxidizing atmospheres and interdiffusion reactions with deposited films. Furthermore, high-T_c epitaxial films require crystalline substrates with small lattice mismatch and similar thermal expansion coefficients. Substrates employed have included Al_2O_3 (sapphire), MgO, ZrO_2 stabilized with Y, Si, $LaGaO_3$, $NdGaO_3$, and $SrTiO_3$. The influence of different substrates on the superconducting characteristics of e-beam evaporated films is shown in Fig. 14-7; a relatively small effect on T_c is evident.

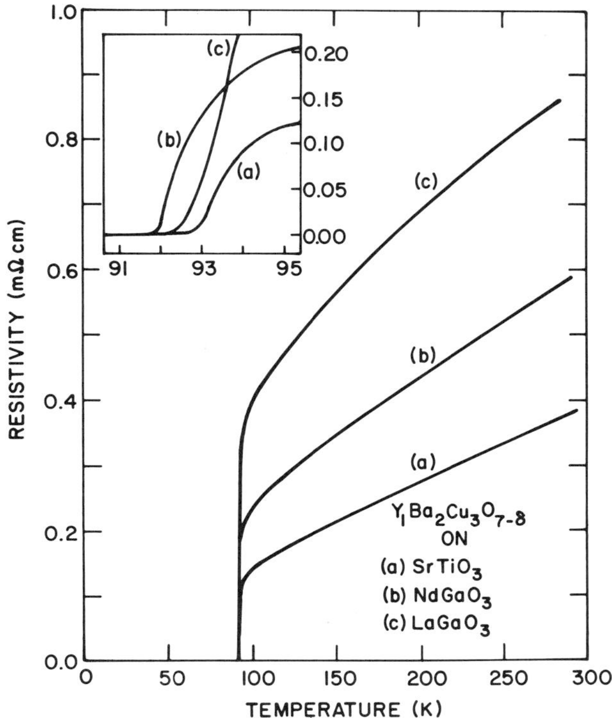

Figure 14-7. Resistance–temperature characteristics of evaporated $YBa_2Cu_3O_7$ films on three different substrates. (Courtesy of R. B. Laibowitz, IBM T. J. Watson Research Laboratory).

14.3.4. Properties and Applications

Typical resistance–temperature characteristics for $YBa_2Cu_3O_7$ films prepared by evaporation, sputtering and MOCVD are shown in Fig. 14-8 where values of T_c around 90 K are evident. Superconducting transitions as narrow as 0.5 K have been achieved together with critical currents in excess of 10^6 A/cm^2 at 77 K, and greater than 10^7 A/cm^2 at 4 K. Higher current densities than the critical value cause the material to become normal.

One of the troublesome problems in high-T_c superconductors is the very short coherence length. Tunneling processes sample states very close to the surface as a result. In films of these materials surfaces tend to be rough, contain nonsuperconducting cuprates and lose oxygen. These effects adversely affect the quality of interfaces in tunnel junctions.

Low-loss, low-dispersion microwave waveguide coatings appear to be the thin-film application closest to being realized. Small electrical resistance at high frequency is an essential requirement and high-T_c superconductors have a considerably smaller surface resistance than Cu. Problems related to high-temperature deposition and processing of films, lithographic patterning of small features, and compatibility with other materials and device structures have served to hinder rapid development of microelectronic applications.

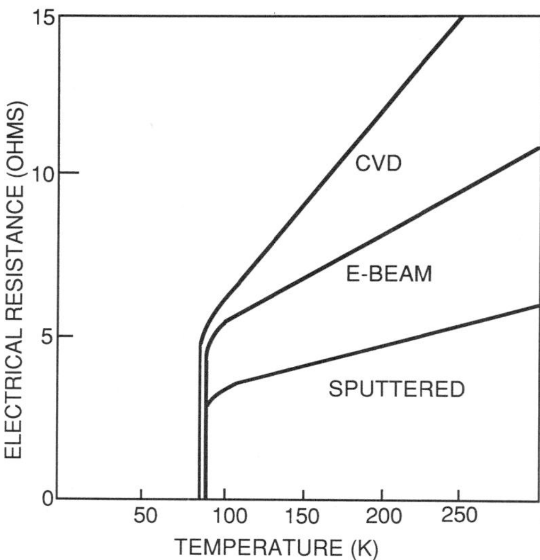

Figure 14-8. Resistance–temperature characteristics of evaporated and sputtered YBaCuO films on $LaGaO_3$ substrates. (Courtesy of R. B. Laibowitz, IBM T. J. Watson Research Laboratory). MOCVD results courtesy of B. Gallois.

14.4. Films for Magnetic Recording

14.4.1. Scope (Ref. 17)

Ferromagnetic thin films already play and will continue to have a major role in magnetic recording and storage technology. The needs of both professional and consumer audio, video, and computer tapes and disks are currently met by an assortment of magnetic particle and thin-film materials. However, the insatiable appetite for data storage continues to push magnetic disk technology to ever higher recording densities at lower cost. Currently the storage media industry is dominated by the "brown disk" that contains fine Fe_2O_3 magnetic particles embedded in an organic binder. A basic reason for the use of thin-film recording media is greater available signal amplitude relative to particulate coatings. The latter are characterized by a linear recording density of 10^4 bits per inch of circular track with a track density of 10^3 tracks per inch. Thin film media consisting of 1000 Å thick electroplated Co–P and Co–Ni–P films, already used for computer data storage on rigid risks, offer the capability of significantly extending these recording densities. The reason is due to the combined effect of 100% packing of magnetic material in films—compared with 20–40% in particulate media—and the generally higher magnetization possible with Co base alloys. Therefore, the same amount of magnetic flux can be contained within a thinner coating enabling the storage layer to be closer to the recording head for more efficient recording and reading. Importantly, higher storage densities mean greater miniaturization. Thus it is that thin-film media usage has largely been driven by the desire to reduce the size of personal computers and portable video recording and playback systems.

The basic conversion of the temporal electrical input signals (e.g., linear ac, digital, FM, etc.) into spatial magnetic patterns occurs when the storage medium translates relative to a recording head as schematically shown in Fig. 14-9. The medium is either a magnetic tape or flat disk while the head is a gapped soft ferrite toroid with windings around the core portion located away from the gap. If the input fringe field signal has a frequency f and the medium is moving at a relative velocity v, the magnetization pattern will be recorded at a fundamental wavelength of $\lambda = v/f$, which is twice the bit length (Ref. 19). Video recording at wavelengths of 0.75 μm represents the highest density recording in use today. The spatially varying magnetization pattern in the medium produces directly proportional external magnetic fields. When the

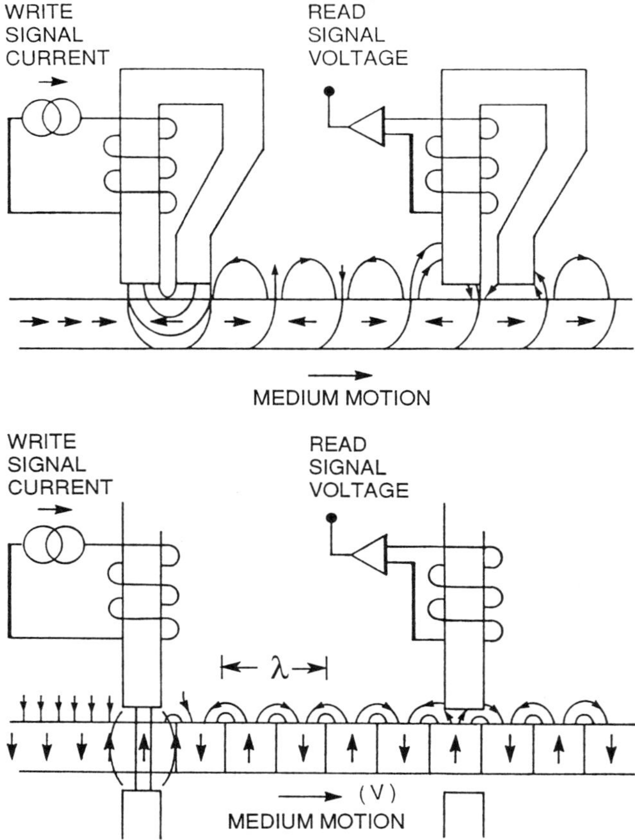

Figure 14-9. (Above) Longitudinal magnetic recording process; (below) perpendicular magnetic recording process. (From Ref. 18 © 1985 Annual Reviews Inc.).

medium is read by passing the recording past a reproduce or read head, these fields generate the magnetic flux (Φ) which circulates through the high-permeability core. By Faraday's law the flux that threads the windings generates the temporal reproduce voltage V_0:

$$V_0 = -\frac{d\Phi}{dt} = -Nv\frac{d\Phi}{dx}, \tag{14-5}$$

where $x = vt$ and N is the number of reproduce turns. From the foregoing, it is apparent that magnetic recording systems require opposite but complementary magnetic properties, i.e., soft magnetic materials for the recording and playback head components and hard magnetic materials for the storage media. The magnetic properties of some of these materials are listed in Table 10-4. In the next two sections we further explore their use in magnetic recording applications.

14.4.2. Thin-Film Head Materials (Ref. 17)

The phenomena of magnetic induction and magnetoresistance are capitalized on in the operation of heads. Inductive heads can be used both to record and read. High-permeability, soft magnetic materials such as sintered ferrites and Sendust (85 Fe–9 Si–5.4 Al by weight) have traditionally been used in their manufacture. To improve performance, Permalloy films ranging in thickness from 2 to 10 μm have been deposited on the yoke structures. Permalloy, a favored material for many soft magnetic film applications, has the following properties: $4\pi M_s = 10$ kG, $H_c = 0.5$ Oe, permeability = 1500–2000 and resistivity ~ 18 $\mu\Omega$-cm. Many deposition processes have been employed, e.g., electroplating, sputtering (dc, rf, ion beam) and evaporation. Other film materials which have been deposited for this purpose include Mu metal, Sendust, and Co–Zr-based alloys. Amorphous magnetic glasses such as $Fe_{80}B_{20}$, $Fe_{80}B_{15}C_5$, and $Fe_{73}Si_{18}C_9$ have also been used. They have values of $4\pi M_s$ in excess of 15 kG with H_c less than 1 Oe.

Magnetoresistance head sensors are read only devices. Again, Permalloy films have been used to detect magnetic fields through changes in electrical resistivity. In general the fractional change in magnetoresistance ($\Delta\rho/\rho$) varies as H^2. It further depends on $\cos^2\theta$, where θ is the angle between the film magnetization and current density vectors. Typically, 1000 Å thick Permalloy films experience changes in $\Delta\rho/\rho$ of a few percent.

14.4.3. Thin-Film Recording Media

Two types of recording media can be distinguished, i.e., *longitudinal* and *perpendicular* (or vertical), depending on whether the magnetization vector lies in the film plane or is normal to it. For *longitudinal* media it is desirable that films display square hysteresis loops with M_s at least several hundred Gauss and H_c greater than 500 Oe.

Magnetic properties, and H_c in particular, are influenced by film composition, thickness, grain size, perfection, impurity content, surface roughness and nature of the substrate. These factors in turn depend on the method of deposition and on such variables as substrate temperature, deposition angle, and magnitude and orientation of applied magnetic fields. Combinations of deposition variables must be controlled to yield desired film anisotropies.

Oblique evaporation and application of external magnetic fields have proven successful in yielding in plane oriented films with desirable magnetic properties. For example Fe–Co–Cr films are evaporated onto rotating rigid disks by evaporation at a 60° angle of incidence. A strong shape anisotropy develops with easy axis in the film plane. Self-shadowing of grains is apparently responsible.

One of the limitations of longitudinal media is that magnetization reversals along the recording track tend to broaden the transition between neighboring magnetized zones. This is due to the demagnetizing effects caused by the mutual overlap of repulsive magnetic fields at the transition, an effect that essentially limits the achievable linear density of storage. In general the maximum packing density is proportional to $M_r d/H_c$, where M_r is the remanent magnetization and d is the film thickness (Ref. 17). Thinner films are desired, but this reduces M_r and the recording signal, so that trade-offs must be struck. Large coercive fields help resist demagnetizing fields and their effects.

Now consider the possibility of *perpendicular* rather than in-plane anisotropy. The magnetization vector is now normal to the film plane and points alternatively toward or away from the surface along the track. There are no demagnetizing fields at the points of magnetic reversal, thus sharpening the transition and increasing the recording density. The discovery that CoCr alloy films (15–20 at% Cr) exhibit an easy axis of magnetization normal to the film has made the concept of high-density perpendicular recording a reality. In these materials the tendency toward in-plane magnetization is countered by additional perpendicular crystalline anisotropy. This results in hysteresis loops displaying the behavior $H_c(\perp) > H_c(\|)$ and $M_r(\perp) > M_r(\|)$, where $\perp$ and $\|$ are the perpendicular and parallel components. Virtually all PVD processes have been utilized to deposit CoCr, CoCrX (X = Rh, Pd, Ta), and GdTbFe films for potential recording media. Additional essential requirements for these materials are corrosion and wear resistance.

14.4.4. Substrates, Undercoats, and Overcoats

The implementation of a viable thin-film recording technology necessitates consideration of a host of additional materials issues concerned with substrates, undercoats, and overcoats. These latter two layers sandwich the magnetic film in between. Substrates may be rigid or flexible depending on application. Rigid substrates of extremely fine surface finish are used for high-density, rapid direct access disk files. They are presently fabricated from an Al–Mg alloy. Substrates must be hard and this necessitates an underlayer, usually an

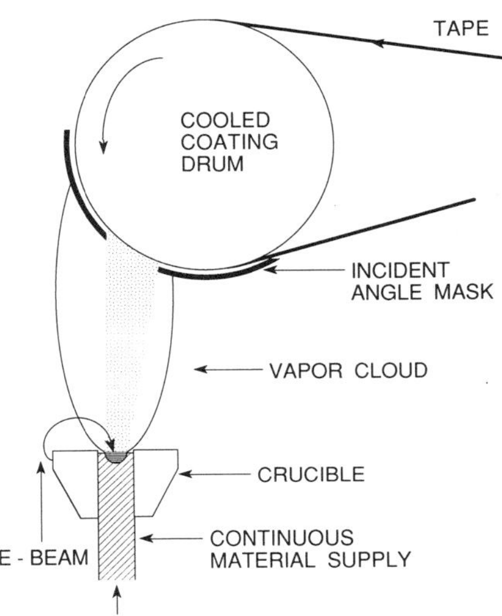

Figure 14-10. Schematic arrangement for continuous oblique evaporation of magnetic films (also undercoats and overcoats) onto a continuous web for video tape applications. The Co–Ni source is evaporated by an electron beam. (Reprinted with permission from IEEE, © 1986 IEEE, from Ref. 20).

electroless plated Ni deposit that is amorphous and nonmagnetic. Elimination of all surface asperities is critical prior to the deposition of glue layers to promote adhesion. Next the magnetic films, only a few thousand angstroms thick, are deposited. Finally wear-resistant overcoats are required because the read–write heads fly over the disc surface at very close proximity and actually make contact during stopping and starting. These mechanical interactions cause disk and head friction and wear, and even catstrophic head crash. Therefore, hard carbon, diamondlike and other hard films have been deposited to minimize these effects. Additionally, solid lubricants are used in conjunction with these hard overcoats.

Tapes and flexible disks are composed of a polymer–polyethylene teraphthalate (PET). In the case of video tape the commercial system for oblique deposition of CoNi onto a continuous web of PET is schematically depicted in Fig. 14-10. As the tape moves around the drum it passes by an aperture mask which controls the range of incident vapor angles intercepted. Higher coercivities and squareness ratios result when the tape is moved in the direction of decreasing rather than increasing angle. Critical to the development of desirable magnetic properties are the conditions for nucleation of a canted columnar grain structure.

14.5. Optical Recording

14.5.1. Introduction

Over the past 15 years various systems for optical recording have been developed. The best known are the video disk and the digital audio disk or compact disk (CD). Both are intended to play back information stored on the disk and therefore employ *read only* media. The information signal is recorded by the manufacturer in the form of micron sized pits on the disk surface. A laser beam is employed in the playback process, which is based on modulation of the light reflected by the pits (Refs. 21, 22). Electronic signal processing then yields the desired video or audio output.

There are also systems where the user can record information on a disk. They rely on a focused laser beam of relatively high power, whose intensity is modulated corresponding to the information being recorded. The disk contains a film sensitive to the laser light. Upon irradiation, local property changes or effects are produced that provide sufficient optical contrast when read out by a much weaker laser beam. Laser–film interactions that have been exploited include

1. Formation of holes and pits by melting and flow of polymer materials
2. Local changes of magnetization in magnetic films subjected to an external magnetic field (magneto-optical recording)
3. Amorphous to crystalline (and vice versa) phase transformation (phase-change recording)

Only the latter two effects will be discussed at any length here. In both, laser–film interactions exhibit the important feature of reversibility or erasability. But it is the extremely high storage density capability, made possible by the finely focussed laser beam, that is the primary attraction of magneto-optic and phase change optical recording. Densities of $\sim 10^8$ bits/cm^2, some 10 times that of high-performance magnetic disk drives, and 50–100 times the density of low-end disk drives has stimulated much interest in erasable optical recording for computer data storage applications. The fact that catastrophic head-disk crashes are eliminated is an added advantage. Unlike magnetic recording where heads contact the disk, lasers are located at least ~ 1 mm away.

14.5.2. The Magneto-Optical Recording Process (Refs. 23, 24)

Magneto-optical recording relies on thermomagnetic effects. Information is stored in a magnetic film magnetized perpendicular to the surface, e.g., in the

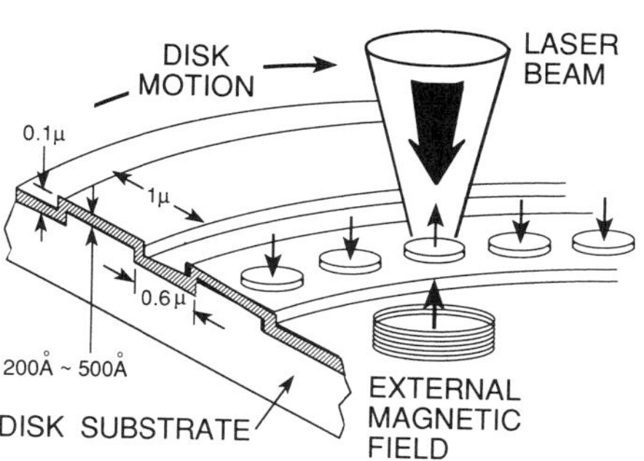

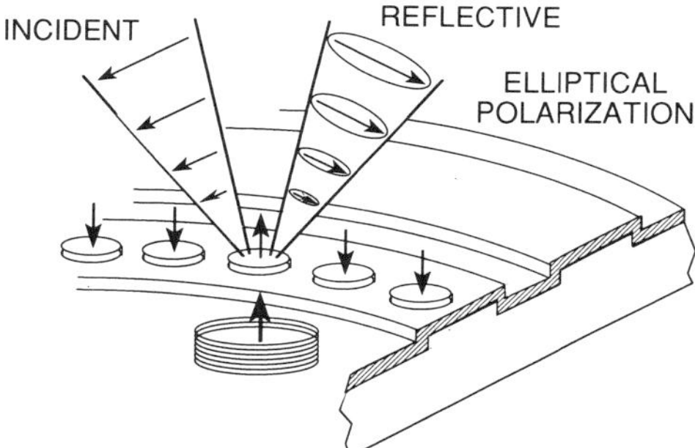

Figure 14-11. Schematic diagram illustrating the writing and reading processes in a pregrooved multilayer magneto-optic disk. (From Ref. 25 with permission from Elsevier Sequoia S.A.).

upward direction. During writing, the modulated linearly polarized laser beam, with a diffraction limited diameter of ~ 1 μm, impinges on the recording material as shown in Fig. 14-11. In *Curie-point writing*, the film is locally heated close to or above the Curie temperature (T_C), where the net magnetization rapidly declines or effectively vanishes, respectively. Under the influence of an opposing external magnetic field (H), the direction of magnetization reverses relative to that of the nonirradiated neighboring region. This new magnetization is frozen in as the material cools to room temperature. Alternatively, in other materials, the magnetization direction can even be switched at temperatures far below T_C. In this case we speak of *compensation-point writing*, an effect made possible because the coercive field (H_c) of these materials decreases rapidly with temperature. Therefore, as soon as $H > H_c$

magnetization reversal occurs. The phenomenon of compensation is exhibited by *ferri*magnetic materials which consist of sublattices or subnetworks of antiparallel aligned magnetic moments, each having a different temperature dependence of magnetization. These materials, however, have a compensation temperature T_{comp} ($< T_C$) at which the sublattice magnetizations balance. The net magnetization then vanishes but H_c is very large. Above T_{comp}, H_c falls.

After writing there are regions of up and down magnetization in the recording track corresponding to, for example, 1 and 0. This information can now be read back (Fig. 14-11) using the polar Kerr magneto-optic effect. Rotation of the plane of polarization of a linearly polarized light beam after reflection from a vertically magnetized magnetic material is the basis of the effect. The sense of rotation depends on the magnetization direction in the recording film layer. Compared with the writing process, the laser beam intensity for reading is much lower.

Finally the recorded information can be erased by laser irradiation of the written domains, but now with H in the direction of the original film magnetization.

14.5.3. Magneto-Optical Film Materials

Before addressing their actual properties and compositions the issue of why films are used deserves brief mention. The primary reasons are the great speed of heating and cooling that is possible in films of low thermal mass, and the high-storage-density continuous films (rather than particles) afford. Coupled with well-developed physical vapor deposition processes that enable economy and efficiency of materials utilization (low cost per unit area), thin films are universally employed. Desired materials properties include (Refs. 25, 26)

1. Large value of the intrinsic uniaxial perpendicular anisotropy.
2. Low T_C or T_{comp} temperatures. During both Curie and compensation point recording a temperature of 150 °C is a desirable upper limit.
3. High H_c values, e.g., 1–2 kOe. High H_c values ensure domain stability at room temperature and absence of growth or shrinkage of domains during readout or erasure elsewhere on the disk layer.
4. A large magneto-optic Kerr effect.
5. A large saturation magnetization (M_s). This facilitates writing in weaker external magnetic fields and formation of smaller stable domains.

Alloys of rare earth (RE) and transition metals (TM) are most commonly used for magneto-optical recording applications. Thin films of RE–TM alloys are usually deposited in the amorphous state by sputtering and evaporation, processes amenable to coating of very large plastic disk substrates. Alloys which have been investigated include TbFe (T_C = 80–150 °C), GdFe (T_C = 180 °C), GdTbFe (T_C = 165 °C), GdTbCo ($T_{comp} \approx$ 380 °C), TbFeCo (T_C = 180 °C) as well as a host of other binary (notably MnBi), ternary and quaternary alloys (Ref. 26). Magnetic properties of some of these materials are given in Table 10-4.

14.5.4. Phase-Change Optical Recording (Refs. 22, 27)

Like magneto-optical recording, phase change optical recording relies on lasers to (1) write information into the recording layer, (2) read, and (3) erase information—all through heating. Polycrystalline thin films are used as the recording layer. During writing localized amorphous regions are created within the crystalline surroundings as a result of laser-pulse-induced melting, followed by ultrahigh quench rates (see Section 13.2.3). Optical property differences between the two phases causes modulation in the reflected light during reading, yielding a signal that varies with time. Because the crystalline phase is usually both more reflective and more opaque than the amorphous material, the disk could be read in either a reflection or transmission playback mode. Regions which were amorphized can be made crystalline again by heating the recording film to a transformation temperature below the melting point. The recrystallization time must be shorter than the laser pass time during erasure. Required temperature—time characteristics for the phase-change medium are shown in Fig. 14-12.

Selecting suitable materials for reversible phase-change recording poses a considerable challenge because of the unusual combination of properties that must be optimized, namely

1. Fusibility at moderate laser powers (~ 10 mW)
2. Sufficient optical contrast on readout
3. Long time stability of the amorphous phase at room temperature
4. Rapid recrystallization times (typically 1 μsec at several hundred degrees Celsius)
5. No degradation of performance after numerous recording playback–erase cycles.

Chalcogenide (S, Se, Te containing) film materials have been most extensively explored for reversible phase-change recording. A pure Te recording layer will meet many of the required demands but the amorphous state is not stable at room temperature. For this reason alloy films have been investigated

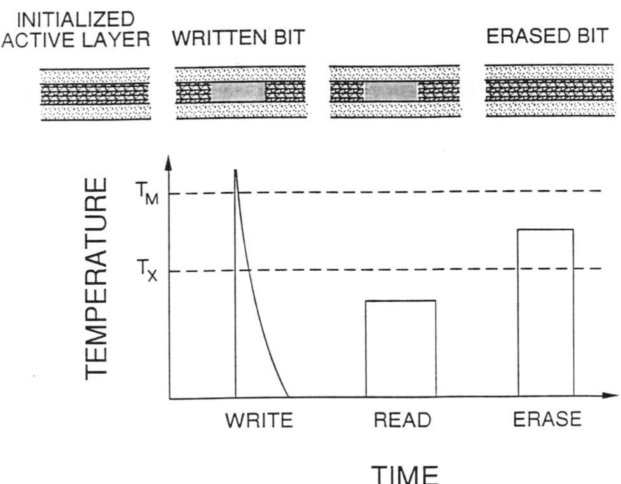

Figure 14-12. Write–read–erase temperature–time characteristics for amorphous to crystalline phase change materials. T_M and T_x refer to melting and recrystallization. (Courtesy of M. Libera).

and proven more suitable. Compositions studied have included Te–As–Ge, Te–Se–S, Te–Se–Sb, In–Sb–Te, and Te–Ge–Sn. Development of this type of optical recording has lagged behind magneto-optical recording due to the materials limitations already hinted at.

14.6. Integrated Optics

14.6.1. Introduction (Refs. 28 – 30)

The term *integrated optics* was coined by S. E. Miller at Bell Laboratories in 1969 (Ref. 31) and refers to the integration of components which generate a light carrier wave, and then modulate (putting the message on it), switch and detect the light. Rather than perform these functions separately with independent devices spread out, for example, on an optical bench, the intent is to fabricate all of these components in miniature on a small substrate, thus eliminating recourse to conversion into electronic signals. The loose analogy to electronic integrated circuits is apparent. More efficient and reliable optical communications has been the goal driving much of the research and development of integrated optical devices and components. One benefit accruing from the reduced size and weight is lower electrical power requirements, an important consideration in repeaters for long-distance fiber optic systems. In addition, the potential exists for combinations of both optical and solid-state electronic devices or integrated circuits sharing a common substrate; advantages include enhanced electro-optical signal processing speed and immunity to temperature variation and mechanical vibration. The use of well-established microelectronics lithography and processing methods holds the promise of low-cost, reliable components for all types of fiber optic communications and sensor systems.

Integrated optics is based on guiding light at optical and near-infrared frequencies that are some 10^4 higher than those operational in electronic devices. Therefore, the amount of information carried by light signals is correspondingly higher. The light is transmitted through waveguides, the basic structure of many important optical components. In this sense integrated optics is more akin to microwave technology (with tubular waveguides) than conventional geometric optics. The remainder of the section will start with a cursory introduction to light propagation in planar waveguides and related components. This will be followed by a treatment of relevant thin film materials and processing issues involved.

14.6.2. Light Propagation in Planar Waveguides

The simplest planar waveguide has the sandwich structure shown in Fig. 14-13. In the center is a thin film whose thickness is of the order of the wavelength of the light propagating through it, i.e., $\sim 1\ \mu$m. The index of refraction of this film (n_f) exceeds that of either the substrate (n_s) or cover layer (n_c), which serve as the cladding.

Consider the case where $n_f > n_s > n_c$. Only Snell's law of refraction,

$$\frac{\sin \phi_c}{\sin \phi_f} = \frac{n_f}{n_c} \qquad \frac{\sin \phi_f}{\sin \phi_s} = \frac{n_s}{n_f}, \tag{14-6}$$

coupled with the phenomenon of total internal reflection, is needed to explain, by ray-optic methods, the guiding of light in this three layer asymmetric waveguide. Referring to Fig. 14-13 we note the following behavior as the incident angle of light, ϕ_s, increases from near zero. At small angles the light passes through both the film–substrate and film–cover interfaces. Light exits the cover layer in this so called air radiation mode. As ϕ_s is increased such that ϕ_f is greater than the critical angle for internal reflection at the n_f/n_c interface, the light is partially confined but exits the substrate (substrate radiation mode). Finally ϕ_s increases to the point where ϕ_f also exceeds the

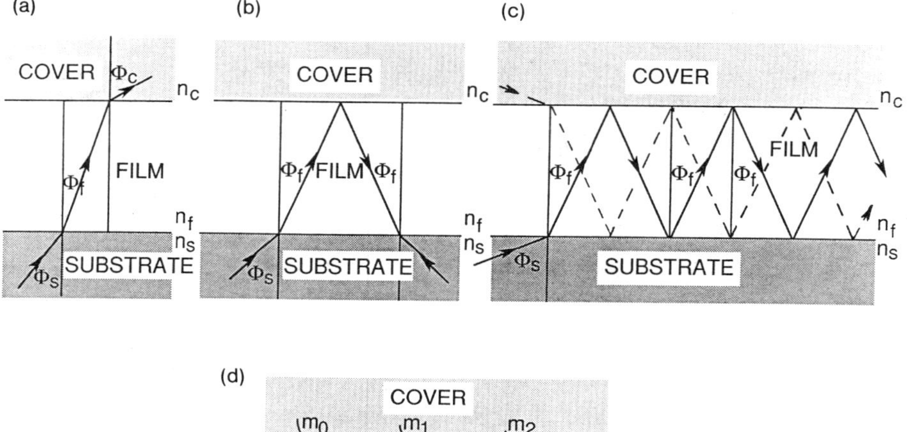

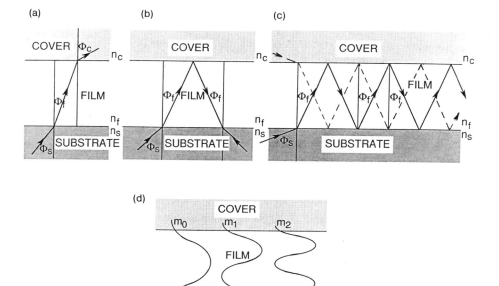

Figure 14-13. Optical ray patterns from (a) air radiation mode, (b) substrate radiation mode, (c) guided mode. Wave representation of modes is shown in d. (From Refs. 28, 29).

critical angle for total internal reflection at the n_f/n_s interface. Now the light wave is totally confined as it zigzags forward in a guided mode. In this case the critical angle is given by

$$\phi_f \geq \arc\sin(n_s/n_f). \tag{14-7}$$

There is another important way of describing the guided mode. In Fig. 14-13c the motion of each of the two propagating rays has two components—one perpendicular to and one parallel to the film axis. Parallel components add and account for light propagation along the film waveguide. Perpendicular component motions oppose each other and set up a transverse standing wave pattern, which is viewed in Fig. 14-13d as sweeping down the guide. The modes are labeled by the index m equal to the number of times the light intensity goes to zero across the film thickness (d). Just how many modes a given waveguide can support depends on n_f, n_s, n_c, d and λ, the wavelength of the light (Ref. 28). The most desirable situation for many applications is single mode ($m = 0$) propagation. This is not difficult to achieve because thin films accommodate fewer modes, and there are practical limitations in preparing thick-film waveguides. One aspect of the wave representation of modes worth noting is the apparent leakage of light to the sides. For the $m = 0$ mode the wave pattern inside the film is cosine-like; outside there is an exponential decay, usually at different rates in each of the cladding layers. In fact, readers familiar with the subject will note that mode wave patterns resemble the allowable quantum mechanical wave functions for an electron confined to a potential well with finite energy barriers (p. 665).

14.6.3. Integrated Optics — Components and Devices

Till now only one-dimensional, broad-film-area light confinement has been considered. Two-dimensional confinement in strips or channels however, is essential for the functioning of most of the components and devices employed in integrated optics. Common configurations of raised and embedded channel waveguides are shown in Fig. 14-14 together with a pair of components that incorporate them. At the outset it is important to distinguish between passive and active waveguide components. The channels of Fig. 14-14a, b are passive components because their optical properties are not modified by outside influences. Other examples include lenses, gratings, and prisms. Active components, on the other hand, enable controllable variations in the light amplitude, phase, frequency, polarization, and propagation direction by means of externally applied fields. Physical fields typically include electric, acoustic, or

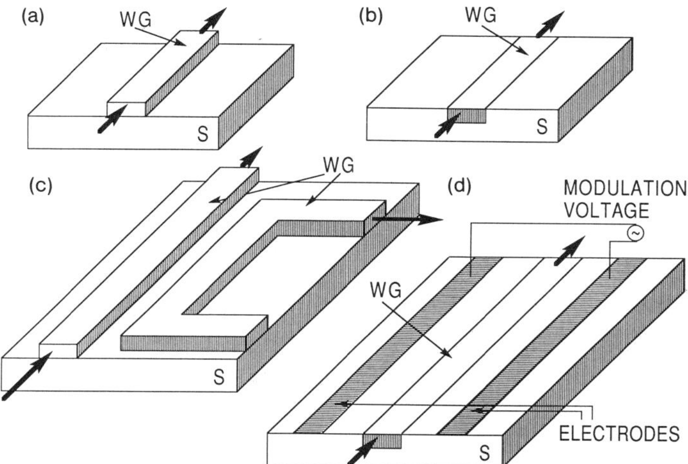

Figure 14-14. Thin-film waveguides and components: (a) channel and (b) embedded waveguides; (c) directional coupler; (d) electro-optical phase modulator

magnetic, corresponding to electro-optic, acousto-optic, or magneto-optic effects, respectively. Under these physical influences active components serve as modulators, switches, deflectors, amplifiers, light emitters, and detectors.

As an example, consider the directional coupler in Fig. 14-14c. Light traveling in the channel waveguide can be diverted into the neighboring U-shaped guide. The coupling between these closely spaced guides arises from the lateral leakage of light energy referred to earlier. "Evanescent" wave penetration of the second waveguide occurs along the length common to both guides; the light intensity builds and is eventually concentrated in the U guide. The coupling between the guides can be made variable either by piezoelectric means, or by altering the refractive index of the intervening medium by electro-optical methods. In Fig. 14-14d the electro-optic waveguide phase modulator is depicted. The waveguide is embedded in an electro-optical crystal substrate. When the modulation voltage signal is applied to parallel electrodes, the refractive index of either the guide or substrate is altered, influencing the phase of the guided wave.

A different group of devices and components is involved in launching light into and out of guides. In the case of GaAs-based integrated optical systems, monolithic fabrication (from the same substrate) of coupled laser light sources, waveguides, and detectors is possible because of the epitaxial film deposition, doping, and lithographic techniques discussed previously in the book. In optical systems based on other materials ($LiNbO_3$, glass) externally generated laser light must be coupled to the thin-film guides. Miniature prisms, thin-film diffraction gratings and tapered film couplers are some of the components used to achieve these ends.

14.6.4. Film Materials and Processing

A sizeable number of different materials have been employed over the years to fabricate thin-film waveguides and other integrated optics components. Some of these are listed in Table 14-3 together with values for their relevant properties. Single-crystal vs. non-single-crystal film structures represent a fundamental division in these materials. The former category includes GaAs together with other associated lattice-matched compound semiconductors (e.g., AlGaAs, InP, GaInAsP etc), $LiNbO_3$ and $LiTaO_3$. These materials have been most extensively studied and will be given most of the attention. The remaining group of materials is amorphous or polycrystalline in form and consists of oxides, glasses, and polymers from which thin dielectric waveguides have been produced. Studies on such films were useful in the early days of integrated

Table 14-3. Materials Employed in Integrated Optics Components and Waveguides

Material	Form	Fabrication Method	Index of Refraction	Loss at 6328 Å (dB/cm)
GaAs	S.C.	Epitaxial	3.6 (at 0.9 μm)	
$Al_xGa_{1-x}As$	S.C.	Epitaxial	3.6 at $x = 0$	
			2.95 at $x = 1$	
$LiNbO_3$	S.C.	Ti exchange	$\Delta n = 0.13$	
(waveguide)		Ion implant	$\Delta n = -0.1$	< 1
		Diffusion	$\Delta n = 0.1$	
$LiTaO_3$	S.C.	Ag exchange	$\Delta n = 0.05$	< 1
Soda lime	A	Ag^+ exchange	1.51–1.61	0.5
Glass		K^+ exchange	1.51–1.52	0.1
		Li^+ exchange	1.51–1.52	0.1
Ta_2O_5	A	Sputtering	2.2	1
Si_3N_4	A	Sputtering	2.0	0.5
ZnS	PC	Evaporation	2.3	5
Polyurethane	A		1.54–1.58	1

S.C. = single crystal; A = amorphous; PC = polycrystalline.
Adapted from Refs. 29 and 32.

optics, but the role of these materials has diminished as the more versatile single-crystal films have become available.

As noted earlier, well-known compound semiconductor fabrication methods enable monolithic integration of light sources (LEDs, lasers), waveguides, and detectors (photodiodes and transistors). The ease in varying the refractive index (Eq. 7-12) and energy gap (Eq. 7-11) makes $Al_xGa_{1-x}As$ and other related semiconductors extremely versatile in this regard. Lattice-matched epitaxial heterostructures ensures devices of high structural integrity. In addition to alloying as a method of changing the refractive index, alteration of the free-carrier concentration has the same effect. Thus removal of carriers from a region will create a higher refractive index relative to the surrounding medium; the former can then serve as a waveguide. One way to achieve carrier concentration reduction is by proton bombardment. Lastly, in order to produce operable semiconductor optical integrated circuits, it is necessary that

$$E_g(\text{waveguide}) > E_g(\text{light source}) > E_g(\text{detector})$$

—conditions that are left to the reader to verify.

Single crystals of $LiNbO_3$ and $LiTaO_3$ owe their popularity to the fact that they exhibit large electro-optic effects, i.e., electric-field-induced changes in refractive index. This accounts for their use as electro-optic modulators and switches. These remarkable materials are also piezoelectric enabling acousto-optic modulators and beam deflectors to be monolithically integrated within optical circuits. Most waveguides are produced either by indiffusion of metal (e.g., Ti) or outdiffusion of Li. For example, when $LiNbO_3$ or $LiTaO_3$ are heated to 1100 °C Li outdiffuses. As a consequence, the refractive index increases slightly relative to undiffused regions. Solutions of the form given by Eqs. 1-27a, b are used to aid in designing heat treatments. Ion implantation of atoms such as B, N, O, Ti, and Ag has also been used to fabricate $LiNbO_3$ waveguides. Regardless of processing treatment only small differences in refractive index ($\Delta n < 1\%$) result between lithographically defined regions and the substrate; but this is enough for waveguiding.

14.6.5. An Example — The Integrated-Optics RF Spectrum Analyzer (Refs. 28, 33)

Not only is the rf spectrum analyzer an interesting and significant integrated-optics system, it provides an opportunity to briefly introduce a thin film surface acoustic wave (SAW) device. The purpose of this spectrum analyzer is to enable the pilot of a military aircraft to obtain an instantaneous spectral analysis of an incoming (microwave) radar beam. This will enable the pilot to determine whether the plane is being tracked by a ground station, enemy plane or missile, etc. Presumably, the spectral signatures of enemy radar signals would be available for comparison by the onboard computer. The analyzer is shown in Fig. 14-15. Light from a laser diode is coupled into a planar waveguide lens, and then into a Bragg cell, the heart of the system. The rf signals to be analyzed are converted into the operational bandwidth of the Bragg cell and then fed into the SAW transducer. The latter consists of an interlaced, or interdigitated, comblike pair of electrodes. Acoustic waves of the same frequency are generated in the piezoelectric $LiNbO_3$ medium, where alternating rarefactions and condensations cause variations in its refractive index. This in effect produces moving diffraction gratings, with periods determined by the instantaneous acoustic frequencies. Diffraction of light occurs at an angle that is a function of the rf signal. A second lens focuses the optical beam onto an array of photodetectors, each element of which represents a particular frequency channel. The spectrum analyzer was fabricated on a $LiNbO_3$ substrate measuring $7 \times 2.5 \times 0.3$ cm^3 in which a planar waveguide was produced by indiffusion of Ti at 1000 °C. Thus, utilizing a few optical components a function is performed that would otherwise require thousands of electronic elements.

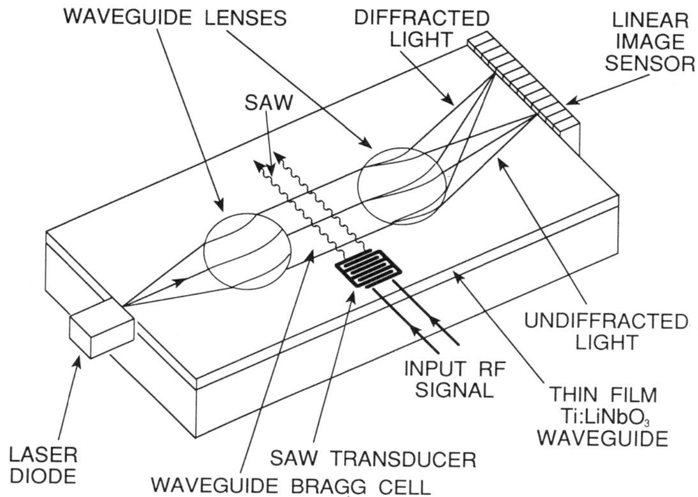

Figure 14-15. Schematic of the integrated optics rf spectrum analyzer (From Ref. 33).

14.7. Superlattices

14.7.1. Introduction

Thin-film structures composed of periodically alternating single-crystal film layers and known as superlattices have generated much recent excitement in both scientific research and technical development circles. While the overwhelming emphasis here is on semiconductors, other layered materials have been employed in superlattices. The term *superlattice* is borrowed from the metallurgy of ordered alloys, e.g., $AuCu_3$, where below a critical transformation temperature a compound-like FCC lattice of Au at cube corners and Cu at cube face centers, exists. X-ray diffraction reveals extra or *superlattice* lines superimposed over the normal lines expected from the random FCC solid solution. So too, the stacking of alternating film layers in an epitaxial (synthetic) superlattice imposes a film-layer periodicity on top of the regular atomic-lattice periodicity within each layer. Unlike homogeneous-ordered alloys, however, composite film superlattices are capable of displaying a broad spectrum of conventional properties as well as a number of interesting quantum effects. The former are due to the frequently synergistic extensions of the laws of property mixtures that are operative when both layers are relatively thick. Quantum behavior, on the other hand, emerges when the layers are very thin for then the wave functions of charge carriers in adjacent layers penetrate the barriers and couple with one another.

Research on synthesized semiconductor superlattices was pioneered by Esaki and Tsu in 1970 (Ref. 34). These investigators were exploring the possibility of resonant tunneling, a quantum mechanical effect that arises from the interaction of electron waves with double and multiple potential barriers. To observe quantum effects the de Broglie wavelength λ must be comparable with the dimensions over which the electric potential varies. For example, consider electrons with an energy of $E = 0.1$ eV and an effective electron mass $m_e^* = 0.1 m_0$, where m_0 is the free-electron mass. That electrons can have masses different from m_0 in solids is a quantum effect. Explicitly, m_e^* is inversely proportional to the curvature (second derivative) of the energy-momentum (E vs. k) variation (shown schematically in Fig. 7-13) at conduction-band minima; similarly for holes, maxima in the valence band are involved. Since the conduction band of GaAs exhibits a narrower E vs. k behavior than that for Si, m_e^* (GaAs) $< m_e^*$ (Si). The actual values are m_e^* (GaAs) $= 0.07 m_0$ and m_e^* (Si) $= 0.19 m_0$. By the De Broglie relation

$$\lambda = h / \sqrt{2 m_e^* E}\,, \tag{14-8}$$

a value of $\lambda = 124$ Å is obtained for $m_e^* = 0.1 m_0$. If the characteristic dimensions of the wells or barriers are reduced to less than this electron mean-free path, the entire electron system enters the quantum regime; formulas will now contain Planck's constant h.

The spatial variation of the valence and conduction bands in Fig. 14-16 is an instructive way to visualize electronic effects in two important types of semiconductor superlattices. In Fig. 14-16a the *doping* superlattice is seen to be composed of alternating P–N layers. The *same* semiconductor material accounts for the constant energy gap which is preserved throughout the structure while the *different* doping levels causes the energy steps at junctions. Standard unbiased P–N diode and *NPN* or *PNP* transistor structures are represented by similar band models. However, diffusional processing limitations broaden the junction region and round the potential barriers. In contrast the junctions in MBE-grown superlattices are atomically sharp. A second type, the *compositional* superlattice is shown in Fig. 14-16b and consists of layers with *different* energy gaps.

In the two decades since their advent a large number of thin-film semiconductor superlattice systems have been studied. Table 14-4 provides a brief summary of this effort including information on lattice mismatch and method of deposition. An indication of the kind of structural quality achievable in superlattices can be gauged by considering strained epitaxial layers—a worst-case

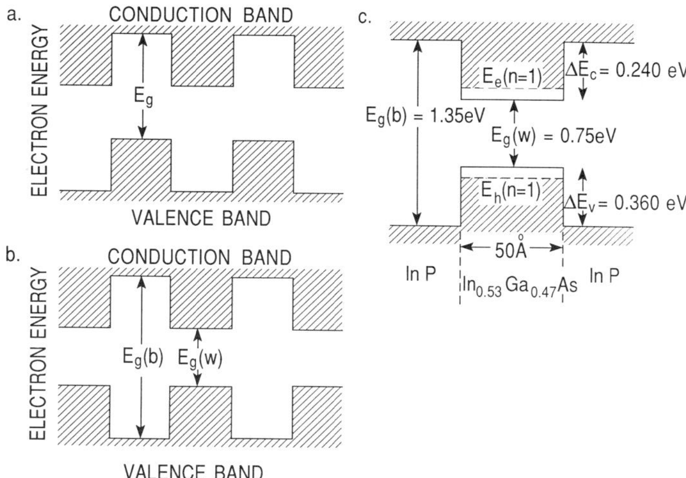

Figure 14-16. Band structure model of (a) the doping superlattice, (b) the compositional superlattice, (c) the quantum well consisting of 50 Å $In_{0.53}Ga_{0.47}As$ sandwiched between layers of InP.

Table 14-4. Superlattice Systems

Film Materials	Lattice Mismatch	Deposition Method
1. $GaAs\text{-}Al_xGa_{1-x}As$	0.16% for $x = 1$	MBE, MOCVD
2. $In_{1-x}Ga_xAs\text{-}GaSb_{1-y}As_y$	0.61%	MBE
3. GaSb-AlSb	0.66%	MBE
4. $InP\text{-}Ga_xIn_{1-x}As_yP_{1-y}$		MBE
5. $InP\text{-}In_{1-x}Ga_xAs$, $x = 0.47$	0%	MBE, MOCVD, LPE
6. $GaP\text{-}GaP_{1-x}As_x$	1.86%	MOCVD
7. $GaAs\text{-}GaAs_{1-x}P_x$	1.79%, $x = 0.5$	MOCVD, CVD
8. Ge-GaAs	0.08%	MBE
9. $Si\text{-}Si_{1-x}Ge_x$	0.92%, $x = 0.22$	MBE, CVD
10. CdTe-HgTe	0.74%	MBE
11. MnSe-ZnSe	4.7%	MBE
12. $PbTe\text{-}Pb_{1-x}Sn_xTe$	0.44%, $x = 0.2$	CVD

From Ref. 35.

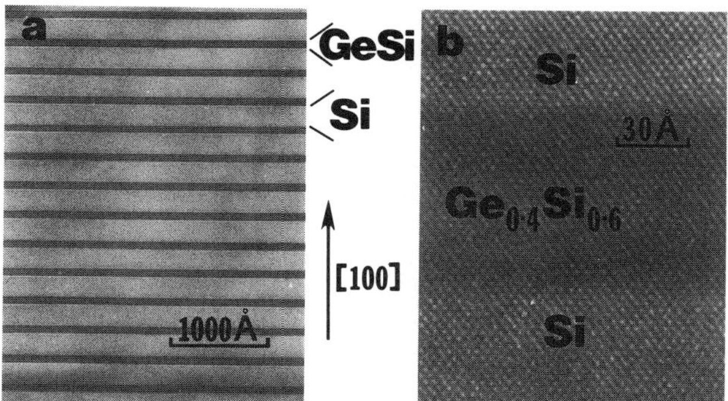

Figure 14-17. (a) Cross-sectional TEM image of $Ge_{0.4}Si_{0.6}$–Si strained-layer superlattice. (b) The high-resolution lattice image. (Courtesy of J. C. Bean and R. Hull, AT& Bell Laboratories).

scenario insofar as crystalline match is concerned. The cross-sectional transmission electron micrographs reveal the perfection of the layer regularity (Fig. 14-17) of the $Ge_{0.4}Si_{0.6}$–Si superlattice; the lattice image clearly shows the absence of interfacial dislocations. Similarly, the superlative lattice image of Fig. 7-16 indicates the atomic sharpness of the three-component GaAs–AlAs interface grown by MBE techniques.

14.7.2. Compositional Superlattice Quantum Well

We now consider the fundamental building block of a compositional superlattice—*the quantum well.* It consists of two semiconductor barrier layers of energy gap $E_g(b)$, flanking the semiconductor well film material whose energy gap is $E_g(w)$, where $E_g(b) > E_g(w)$. As an example, the lattice-matched InP–$In_{0.53}Ga_{0.47}As$–InP quantum well structure is shown in Fig. 14-16c. Electrons within the conduction band of $In_{0.53}Ga_{0.47}As$ are essentially trapped in a well much like the celebrated particle in a one-dimensional box. If for the moment we assume that the well is infinitely deep, the energy levels (E_e) are given by the solution to the Schrödinger wave equation.

$$-\frac{h^2}{8\pi^2 m_e^*}\frac{\partial^2\psi(x)}{\partial x^2} + V\psi(x) = E_e\psi(x), \tag{14-9}$$

where ψ is the wave function and V is the potential energy of the electron ($V = 0$ in this case). The wave function

$$\psi(x) = \frac{2}{\sqrt{L}}\sin\sqrt{8\pi^2 m_e^* E_e/h^2}\,x \tag{14-10}$$

satisfies Eq. 14-9, and the requirement that ψ vanish at the well boundaries $x = 0$ and $x = L$ yields the well-known formula for the energy levels. Thus

$$E_e = h^2 n^2 / 8 m_e^* L^2, \tag{14-11}$$

where L is the well width and n is the quantum number ($n = 1, 2, 3$ etc.). In the present case $m_e^* = 0.041 m_0$ and for $L = 50$ Å, E_e ($n = 1$) is calculated to be 0.367 eV. Similarly, the energy of the hole levels in the valence band can be calculated assuming that in $In_{0.53}Ga_{0.47}As$, the hole mass, $m_h^* = 0.41 m_0$. Substituting in Eq. 14-11, with m_h^* replacing m_e^*, we find E_h ($n = 1$) = 0.036 eV for the lowest hole level.

The presence of finite energy barrier heights complicates the problem considerably. Electrons are now not so hopelessly confined to the well, but have a finite probability of existing outside it. Therefore, within the InP barrier regions that flank the well, ψ is finite but decreases exponentially with distance outward. Within the well ψ varies spatially as sine and cosine functions. The boundary conditions require continuity of both ψ and $(1/m^*)(d\psi/dx)$ at each barrier interface. Details of the calculation of energy levels involve an extension beyond the treatment commonly found in elementary texts on quantum mechanics in order to account for differing electron masses in the well (w) and barrier (b) regions. For symmetric wave functions in the well corresponding to the first, third, fifth, etc. energy levels, E_e values are given by the solution to (Ref. 36)

$$\text{Tan}\frac{L}{2}\sqrt{\frac{8\pi^2 m_e^*(w) E_e}{h^2}} = \left[\frac{m_e^*(w)}{m_e^*(b)}\right]^{1/2} \sqrt{\frac{\Delta E_c - E_e}{E_e}} \tag{14-12}$$

Similarly, for antisymmetrical wave functions corresponding to the second, fourth, sixth, etc. energy levels

$$\text{Cot}\frac{L}{2}\sqrt{\frac{8\pi^2 m_e^*(w) E_e}{h^2}} = \left[\frac{m_e^*(w)}{m_e^*(b)}\right]^{1/2} \sqrt{\frac{\Delta E_c - E_e}{E_e}} \tag{14-13}$$

Since these equations have no roots for $E_e > \Delta E_c$ there are only a *finite* number of discrete energy levels; these can be obtained either numerically or graphically. The quantity ΔE_c is the quantum well conduction band barrier height, commonly known as the *conduction band offset*, which for this system has been measured to be 0.240 eV. For InP, $m_e^*(b) = 0.077 m_0$ and numerical solution of Eq. 14-12 yields $E_e = 0.087$ eV for the first energy level. A similar calculation for the first hole level gives $E_h = 0.024$ eV where the hole mass for InP = $0.64 m_0$ and the well valence-band barrier, ΔE_v (which replaces ΔE_c above) is equal to 0.360 eV. The absolute energy of the transition between these two levels is given by the sum of the confined particle energies (0.11 eV) plus E_g for $In_{0.53}Ga_{0.47}As$ (0.75 eV) for a total of 0.86 eV†.

No longer are the well-known solutions to potential well problems the province of academic textbooks on quantum theory. Superlattices have made possible the realization of quantum mechanical systems replete with spectra consisting of states having well-defined energy levels. As with atoms and molecules, either photon absorption or emission (photoluminescence) spectra have served to confirm the existence of quantum well electron and hole levels. Data on the frequencies where maximum absorption occurs are shown in Fig. 14-18 for the multiquantum well structure composed of GaAs–AlGaAs layers.

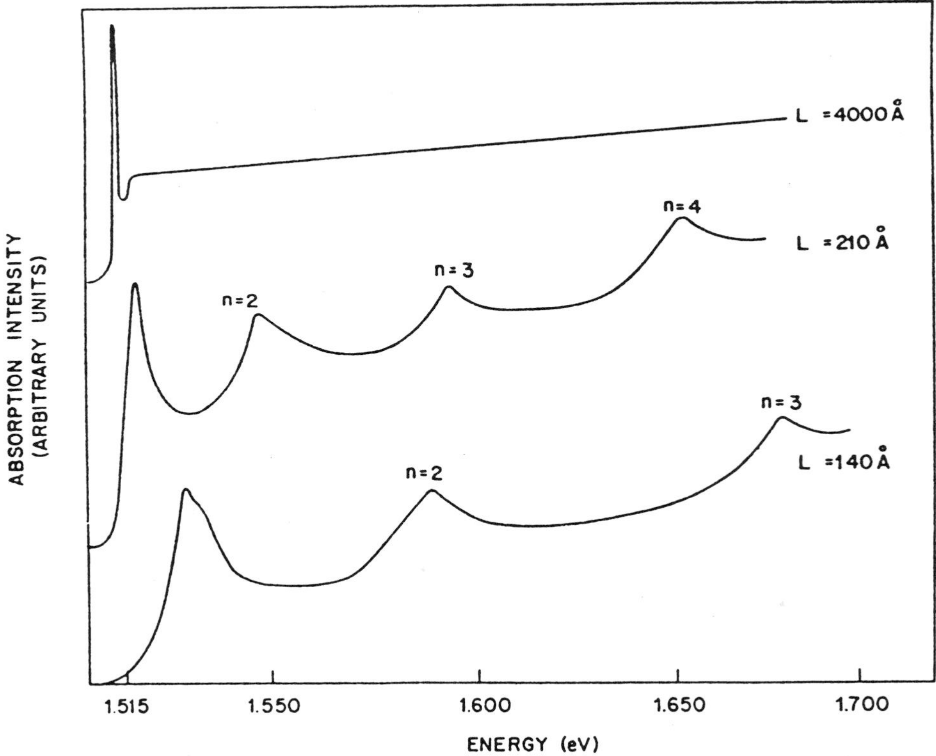

Figure 14-18. Absorption spectra in GaAs–$Al_{0.2}Ga_{0.8}As$ layers at 2 K for three different well thicknesses. (From Ref. 37).

†Problem in text due to H. Temkin.

The fact that the measured energies depend on the superlattice period spacing is clear evidence of the quantum behavior of the system. In photoluminescence, light impinging on the quantum wells creates electron–hole pairs. These populate the energy levels, and during recombination light emission corresponding to the transition occurs. Charge recombination at lattice defects or traps seriously degrades light emission, whose intensity can provide evidence for the structural perfection of the quantum wells.

Similar experiments have been performed on structures containing non-rectangular quantum wells. Lattice matched epitaxial layers of precise composition and thickness (superlattices) are required to yield the *digitally* graded well shapes. Alternately *analog* or continuously graded lattice matched parabolic and triangular wells have been recently deposited and the energy transitions measured (Ref. 38). Through knowledge of E_e, information on band offsets was extracted using established quantum mechanical solutions for such well geometries. In this way, values of $\Delta E_c / \Delta E_g = 0.65$ were obtained for both $Al_{0.3}Ga_{0.7}As/GaAs$ and $Al_{0.48}In_{0.52}As/Ga_{0.48}In_{0.52}As$ lattice matched to InP, where ΔE_g is the energy gap difference between the involved semiconductor pairs.

14.7.3. Modulation Doping

Now we turn our attention to a superlattice where energy wells are produced by selective periodic or "modulation" doping. Free-charge carriers in semiconductors are generally created by donor and acceptor impurities. During conduction the carriers inevitably undergo scattering from the same impurities that brought them into being. In superlattices, however, it is possible to spatially separate carriers from their parent impurity atoms. Consider the thin layers of wide-band-gap *N*-doped AlGaAs alternating with undoped narrow-band-gap GaAs in Fig. 14-19. Electrons released by donors to the conduction band of AlGaAs diffuse and drop into the energy wells of the GaAs. Thus the trapped electrons are separated from the donor impurities. If an electric field is now applied *normal* to the figure plane, electron motion is confined in two dimensions, i.e., solely within the GaAs layers. But in the absence of scattering by impurities the mobility of the electrons will be considerably enhanced. Whereas uniformly doped GaAs might have a mobility of several thousand cm^2/V-sec at 4 K, mobilities in excess of 10^6 cm^2/V-sec have been measured in modulation-doped heterostructures.

If the well widths are now significantly reduced, electron energy levels become widely spaced. Available electrons must seek higher levels to occupy and may even spill over into the AlGaAs conduction band. Because of the

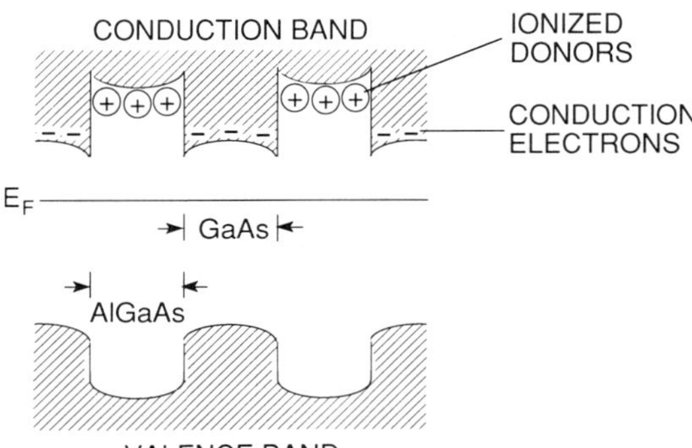

Figure 14-19. Band diagram representation of modulation doped AlGaAs–GaAs superlattice. Note separation of electrons from ionized donors.

lower electron mobility there, the structure exhibits a negative differential resistance effect; i.e., the current effectively drops with increasing applied voltage.

In creating semiconductor superlattices nothing less than the controlled synthesis of band structures is involved. The subject is known as *band-gap engineering* and promises to usher in novel photodetectors as well as new types of quantum devices such as tunneling transistors. This interesting subject will be further discussed in Section 14.8.

14.7.4. Metallic Multilayers and Superlattices

Judged by scientific publications, research on metallic multilayers and superlattices is only one-tenth that of the effort on semiconductor superlattices. Nevertheless this interesting activity spans a broad range of film compositions, deposited by both PVD and MBE methods, and is used to investigate a variety of scientific phenomena. These are broadly listed together with some specific effects, properties and sample systems studied (Ref. 39).

1. *Magnetic Measurements*—Magneto-optical effects (Co/Au, Fe/Cu); magnetization (Ni/Cu, Gd/Y); Mossbäuer effect (Fe/Pd, Fe/V); ferromagnetic resonance (Ni/Mo, FeCo/Si); neutron diffraction (Dy/Y, Fe/Mg)
2. *Elastic Constants*—Au/Ni, Cu/Ni, Mo/Ni, Cu/Al, Cu/Pd
3. *Superconductors*—Transition temperature (Nb/Ge, Nb_3Ge/Nb_3Ir); critical field (NbN/AlN, Nb/Ti); critical current (Pb/Bi, PbBi/Cr)

Compared with semiconductors, metals exhibit far less directional bonding; this facilitates the incorporation of defects into growing films and at interfaces between layers. Rather than the coherent ordering of lattice-matched semiconductor superlattices the metallic structures are far less perfect. It is, therefore, more appropriate to speak of them as "multilayers." Fortunately, many of the properties noted above are less sensitive to crystalline perfection than, for example, the electrical response of semiconductor superlattices.

In closing it is convenient to categorize physical phenomena exhibited by multilayer structures, in increasing order of complexity as single film, proximity, coupling and superlattice effects. *Single-film* effects are primarily limited to the film and independent of the substrate or overlayers. The increase in electrical resistivity in metal films as their thickness decreases (see Chapter 10) is an example. *Proximity* effects occur because of a boundary between two materials. Physical properties are modified by interfacial effects, e.g., charge transfer, strain. Examples include superconducting film–normal conducting film or magnetic film–nonmagnetic film combinations. *Coupling* effects occur between two like materials across an intervening unlike film material. Interaction distances can vary from angstroms (magnetism) to tens or hundreds of angstroms (superconductivity). In principle, the above three cases do not strictly require a multifilm structure; single, double or triple film layers of appropriate thickness will do. *Superlattice* effects (e.g., diffraction), in contrast, rely on the collective response of the whole periodic structure.

14.8. Band-Gap Engineering and Quantum Devices

14.8.1. Introduction

Band-gap engineering refers to the synthetic tailoring of conduction and valence band edges (band gaps) as a function of position within semiconductor structures (Refs. 40, 41). Whether it be single phase semiconductors, heterojunctions, quantum wells or superlattices, the intent is to create unusual electronic transport or optical effects, and novel quantum as well as nonquantum devices (e.g., lasers, photodiodes, transistors) based on them. There is a rich assortment of semiconductor alloys, band gaps, and lattice parameters available for this purpose. In addition, other types of quantum devices are based on superconductive tunneling and some of these will be introduced at the end of this section.

Two types of electronic devices can be distinguished when film dimensions are shrunk. Those in the first category are simply scaled-down versions of conventional devices. This process of miniaturization continues to be aggressively pursued in further efforts to increase device packing densities and speeds. Electrons and holes in these devices can still be treated as classical particles however. In contrast, the wave nature of the electron is fundamental to the operation or understanding of quantum devices. When one or more device dimensions become comparable to the de Broglie wavelength the quantum regime is entered and new effects emerge. This feature has already been discussed in Section 14.7.2 in connection with quantum well behavior. It was pointed out there that the magnitude of measurable quantum effects varies inversely with film dimensions.

Quantum devices simultaneously combine great challenges of film deposition and fabrication with the indescribable sense of accomplishment in reaching fundamental limits (e.g. in speed, sensitivity, selectivity, etc.). Crucial to their realization are two often coupled technologies. The first is concerned with the controllable deposition of nanometer thick heteroepitaxial film layers with suitably engineered energy gaps. This necessarily means tight control of film composition and thickness, tasks well suited to MBE methods. Secondly, lithographic patterning in the submicron and even nanometer regime is frequently required. This is particularly true if electrons are to be further confined to one-dimensional quantum well *wires*, or even three-dimensional quantum *dots*. Thus the concept of lateral rather than the customary thickness superlattices has emerged. Taken together, terms like band-gap engineering and nanofabrication express different facets of this remarkable thin-film activity. For discussions of quantum interference and size effects as well as electronic devices, Refs. 42 and 43 should be consulted.

14.8.2. Band-Gap Engineering (Ref. 40)

14.8.2.1. Graded Gap Transistor. Although no quantum effects are displayed in this first example the power of band-gap engineering to improve conventional device behavior is illustrated. Consider the bipolar *NPN* transistor shown unbiased in Fig. 14-20a. A current injected in the base layer controls the electron flow from the emitter to collector. In conventional transistors the base has a uniform energy gap but no electric field. Therefore, electrons can only traverse the base region by diffusion, a relatively slow process. One way to speed the transport process is to gradually change the base composition. This causes the conduction band to tilt just as if an electric field were present. Electrons can now drift more rapidly through the base than in the conventional device. Recently, the concept was demonstrated in MBE-grown compound semiconductor transistors where the 3000-Å base region composition was graded to yield an effective base electric field of 6 kV/cm. A factor of 10 increase in electron velocity was measured.

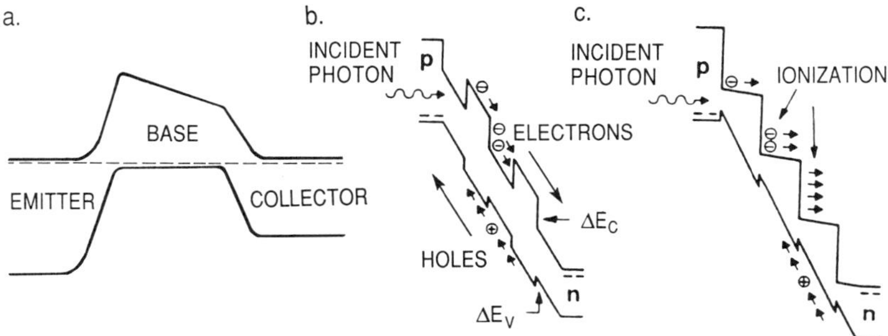

Figure 14-20. Applications of band-gap engineering: (a) equilibrium band diagram of graded-gap base bipolar transistor; (b) multiplication well avalanche photodiode; (c) staircase avalanche photodiode (Reprinted with permission from F. Capasso, *Science* **235**, 172, 1987).

14.8.2.2. Avalanche Photodiodes. As a second example consider the avalanche photodiode (APD), a P–N junction device used to convert light into an electric signal. The response of the APD can be considerably enhanced by incorporation of a multi quantum well structure. When an electric field is applied in the figure plane, the whole superlattice array tilts by an amount proportional to the field strength. As indicated in Fig. 14-20b, photon-induced electrons are forced to drift downward while holes correspondingly ascend to the top. For large electric fields, electrons in the wide-gap semiconductor accelerate to the conduction-band edge discontinuity (ΔE_c), where they fall into the well of the low-gap material. The electrons gain enough energy from the large ΔE_c increment to impact with lattice atoms ionizing them. This creates more carriers, which in turn impact ionize other atoms, etc., and a large avalanche current is produced. The superlattice can be so tailored that the valence-band discontinuity ΔE_v is not large enough to supply a similar energy boost to the free holes. Thus electrons can be made to impact ionize more efficiently than holes simply because $\Delta E_c > \Delta E_v$. This effect has been used to improve the response of avalanche photodiode detectors by reducing the amount of excess noise.

Rather than alternate high- and low-E_g materials the individual superlattice layers can be graded from low to high energy gap and then abruptly narrowed at the layer interface. Under a large dc field the conduction band of the superlattice unfolds into the staircase shown in Fig. 14-20c. This device is the solid-state analog of the photomultiplier with the steps equivalent to dynodes. Electrons descending steps can efficiently ionize lattice atoms as they are propelled by the steep potential drop (high electric field).

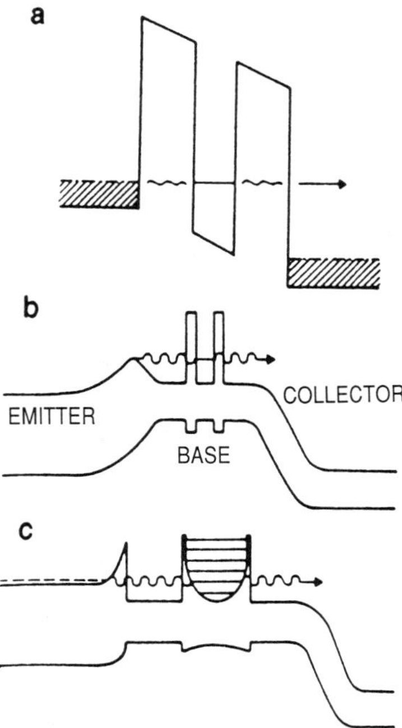

Figure 14-21. Resonant tunneling heterostructures: (a) resonant tunneling through double barrier; (b) resonant tunneling bipolar transistor operation; (c) resonant tunneling transistor with parabolic quantum well (Reprinted with permission from F. Capasso, *Science* **235**, 172, 1987).

14.8.2.3. Tunneling Devices. The phenomenon of resonant tunneling is a consequence of the quantum effects in narrow potential wells and is exploited in tunneling devices. In Fig. 14-21a we consider an undoped double barrier (e.g., AlAs–GaAs–AlAs) sandwiched between two contact layers. Upon application of a dc field, electrons originating at the Fermi level of the left contact tunnel through the first barrier into the well. Resonant tunneling occurs when the energy of the injected carriers lines up with or becomes equal to one of the quantum well levels. Carriers then tunnel through to the right contact with a maximum in the overall transmission through the double barrier, and in the current–voltage (J–V) response. Off resonance, less current flows yielding a region of negative resistance in the J–V characteristics. The resonant tunneling effect is essentially equivalent to the transmitted light enhancement in the optical Fabry–Perot interferometer (p. 540).

In the resonant tunneling transistor (RTT) of Fig. 14-21b a quantum well is grown within the base region of an *NPN* bipolar transistor structure. The voltage between the base–emitter junction is tuned so that emitter electrons resonantly tunnel through one of the well energy levels and are collected by the reverse biased base–collector junction. At band edge discontinuities electrons are injected *ballistically*, which means they do not suffer appreciable thermal or impurity scattering. In moving through the tunneling barrier electrons pass from a region of high potential energy to one of lower potential energy. Because tunneling is an elastic process, the total electron energy is unchanged, and therefore the kinetic energy is correspondingly increased. Rather than carriers that diffuse across the base region in conventional transistors, there is a well-directed electron beam of high velocity.

Another proposed RTT configuration incorporates a parabolic quantum well in the base region as shown in Fig. 14-21c. It is well known in quantum theory that the energy levels for a parabolic potential (i.e., the harmonic oscillator) are equally spaced. In this way a device with N stable states, where N is the number of resonant peaks, can be produced. An N-state memory, with N as high as 8, offers the potential of extremely high density data storage.

Clearly nothing less than "do it yourself quantum mechanics" (Ref. 41) is a synonym for band-gap engineering.

14.8.3. Josephson Tunneling Effects Revisited (Refs. 44, 45)

Perhaps the most remarkable thin-film quantum devices are those based on Josephson tunneling. Consider, for example, the SQUID introduced on p. 485. A SQUID can measure voltage differences as small as 10^{-18} V, currents as small as 10^{-18} A, and magnetic fields less than 10^{-14} T. SQUID sensors are also characterized by energy, displacement and acceleration sensitivities of approximately 10^{-30} J/Hz, 10^{-7} cm, and 10^{-9} cm/sec^2, respectively. When operated below the critical superconducting temperature the SQUID sensitivity approaches fundamental limits imposed by quantum mechanics. No other technology comes close to matching this behavior. To better understand SQUIDs, the most widely used superconducting devices, some brief comments on Josephson junction tunneling effects are first necessary.

Josephson predicted that interactions of electric ($\mathscr{E}$) and magnetic (H) fields with Cooper pairs (Section 10.5.1), which are obscured in superconductors proper, should be observable in "weakly" coupled systems. The term *weak* arises because interactions responsible for Cooper pairs extend across the junction in attenuated form. A very narrow constriction between two superconducting regions is one such link. More importantly, an ultrathin oxide film barrier ($<$ 50 Å thick) sandwiched between superconducting electrodes could also serve this function. A cross-sectional TEM image of such a Josephson junction structure is shown in Fig. 14-22. The barrier is so thin as to be

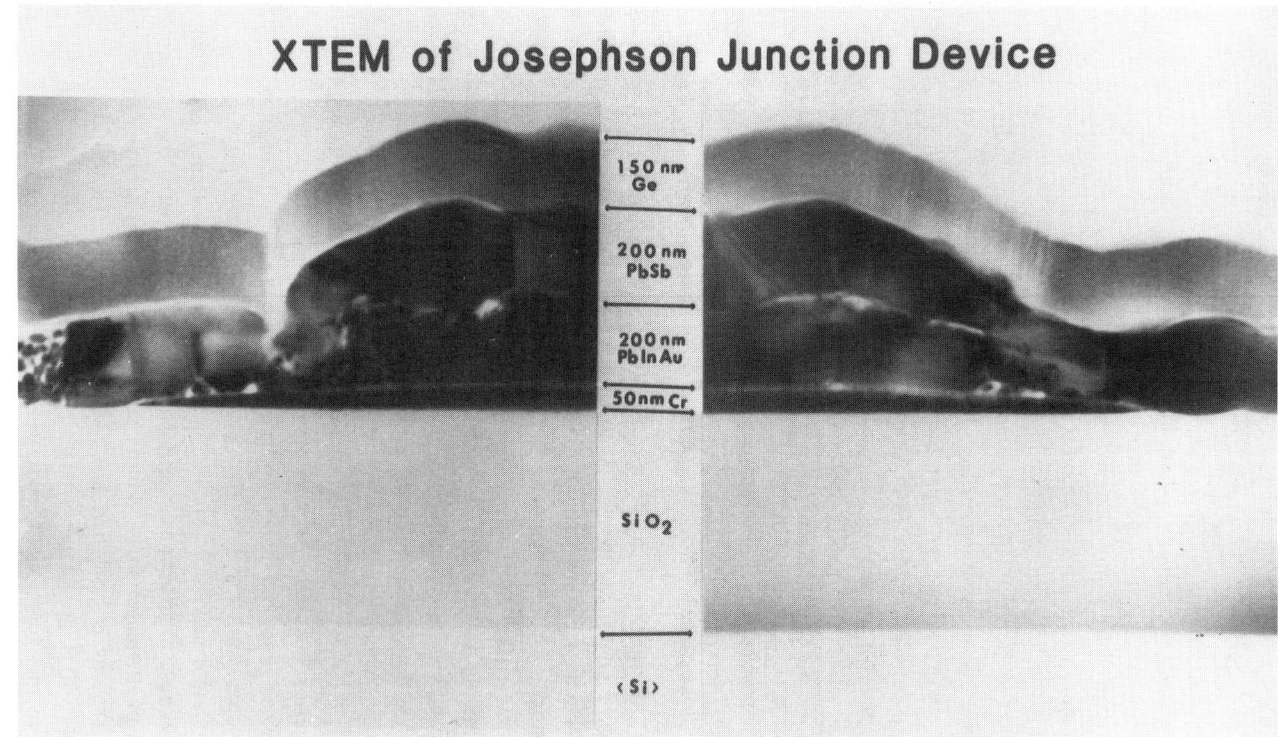

Figure 14-22. Cross-sectional TEM image of Josephson junction device. (Courtesy of T. T. Sheng, AT&T Bell Laboratories).

unresolvable between the PbSb and PbInAu electrodes. When the current density (J) is increased from zero, Cooper pairs tunnel through the barrier establishing a phase shift (δ) between quantum mechanical wave functions in either superconductor. The resulting current is

$$J = J_0 \sin \delta, \tag{14-14}$$

where J_0 is the critical junction current. In the Josephson region ($J < J_0$) the voltage (V) across the junction is zero (Fig. 10-15d), but for $J > J_0$ a voltage develops. With time (t) the phase difference increases, a phenomenon known as the ac Josephson effect. It is characterized by the equation

$$\frac{d\delta}{dt} = \frac{4\pi q V}{h}, \tag{14-15}$$

where q and h are the electronic charge and Planck's constant, respectively.

The most unusual Josephson junction effects are associated with magnetic fields. There is a nonmonotonic dependence of J on H. As H increases, J falls to zero, then rises again, and drops again, etc. The oscillatory J–H behavior is the same as that which describes the light wave (Fraunhofer) diffraction pattern through a narrow slit. When magnetic flux Φ threads a superconducting ring it is quantized with a value of $n\Phi_0$, where n is an integer and Φ_0, the flux quantum, is given by $h/2q$. The latter has a value of 2.068×10^{-15} Wb or equivalently 2.068×10^{-7} gauss-cm^2. Furthermore, if there is a Josephson junction in the superconducting loop, the H fields cause interference effects with currents in different parts of the multiply connected superconducting circuit. The supercurrent that flows then varies periodically as $\sin 2\pi\Phi/\Phi_0$, a fact relied on in the measurement of magnetic fields.

14.8.4. SQUIDs and Their Applications (Refs. 46, 47)

SQUIDs combine the phenomena of flux quantization and Josephson tunneling. Two types may be distinguished as shown in Fig. 14-23, and they consist of either one (rf) or two (dc) Josephson junctions inserted into a loop of superconducting material. Of the two, the dc SQUID is the more highly developed and sensitive one, but the rf SQUID is widely used and has been commercially available for nearly two decades. Both are flux-to-voltage transducers producing an output voltage in response to an applied magnetic flux. In the dc case the device is biased with a constant current and the voltage output oscillates with a period Φ_0. A feedback circuit linearizes the response enabling detection of less than one flux quantum. The rf SQUID loop is inductively coupled to the inductor of a resonant circuit excited at its resonant frequency.

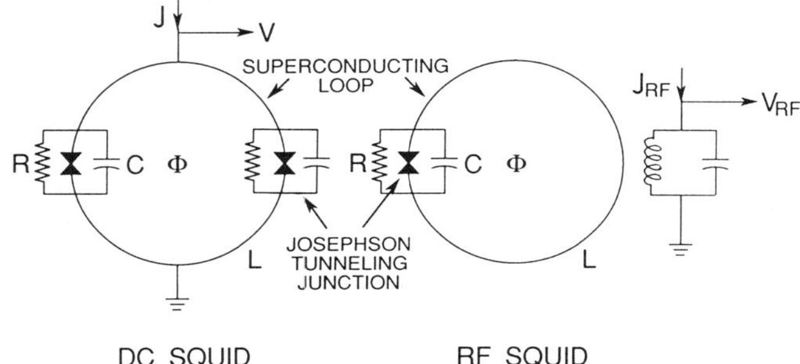

DC SQUID RF SQUID

Figure 14-23. Dc SQUID consisting of two Josephson tunnel junctions connected in parallel in a superconducting loop of inductance L. Each junction is shunted with a capacitor C and resistor R to eliminate hysteresis in $J-V$ characteristics. Rf SQUID contains one Josephson junction. (From Ref. 44).

As Φ changes, the amplitude of the rf voltage (V_{rf}) is modulated with a period Φ_0.

Both types of SQUIDs fabricated from low-T_c superconductors operated at 4 K have been routinely used as ultrasensitive magnetometers, voltmeters and motion detectors. Physicists employ them to search for exotic entities such as monopoles and quarks, and geologists prospect for oil and mineral deposits with SQUIDs. Biomagnetic applications such as magneto-encephalography are among the most interesting. Since the H field in and around the human brain is very small and flux densities never exceed 10^{-12} T, only SQUIDs have the required measurement sensitivity. (For comparison the earth's magnetic field is 5×10^{-5} T.) Similarly, the electromagnetic activity in the heart has been probed this way. For these purposes SQUIDs composed of NbN films separated by an ultra thin MgO tunnel barrier have been employed (Ref. 48). These devices have lateral resolutions of a few millimeters. Because magnetic signals typically fall off as the fourth power of the distance from the source, pickup coils must be as close to the body as possible.

In addition to magnetometry the other major application of SQUIDs has been in computer technology (Refs. 49, 50). The reason is that the Josephson junction can act as a switch and thus steer signals into desired circuits. Josephson junctions are the fastest switches known, capable of changing states in as little as 6 ps. The operation can be understood through reference to Fig. 14-24. One side of the junction (counterelectrode) is connected through a resistor to a current source and also to an output device. The other side (base electrode) is connected to the electrical ground. If the supply current (I_g) is below the critical value, the voltage drop across the junction is zero and there is a direct short-circuit to ground. If, however, the control line current (I_c) is

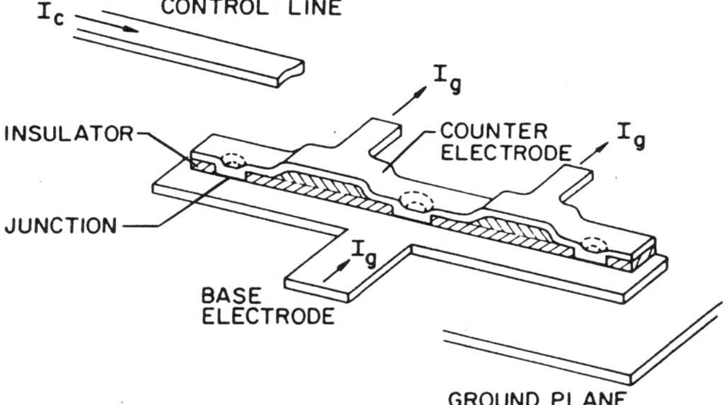

Figure 14-24. Three interconnected superconducting Josephson junctions. Magnetic flux threads the loop or ring formed by a pair of junctions. (Reprinted with permission from Ref. 50, © 1979 IEEE).

large enough to produce a magnetic field that penetrates the junction layer, the latter switches to the nonsuperconducting resistive state. In effect, the current switches from branch 1 to branch 2 in Fig. 10-15d. As a result, voltage appears across the junction and current is channeled into the output line. Because these junction devices are superconducting the power dissipation is extremely small—perhaps a few microwatts. Therefore a few million tightly packed Josephson junction circuits would only generate several watts, an amount which can be readily dissipated. This in contrast to present semiconductor integrated circuits, where removal of Joule heating is a very major concern.

14.8.5. Materials Problems in Josephson Tunneling Devices

The remarkable benefits of Josephson junction devices do not come easily. For example, there are formidable difficulties in fabricating tunneling barriers and maintaining their integrity during operation. Theory predicts and experiments have confirmed that a 4-Å change in a 40-Å-thick barrier film causes the Josephson threshold current, an important circuit design parameter, to vary by a factor of 10 (Ref. 50). Native oxides, usually employed as tunnel barriers, are generally grown by rf sputtering in a low-pressure O_2 atmosphere. In this environment the oxide simultaneously grows and sputters. The oxidation rate decreases with oxide thickness, but the sputter rate remains constant. Therefore, with carefully controlled oxygen pressure, rf power, and substrate temperature, a predetermined terminal oxide thickness will form. At this stage oxidation and sputter rates are equal and the oxide thickness is self-limiting. After the terminal oxide is established, new oxide continually sputters off. The result is removal of contaminants and uniform thickness on all base electrodes.

A further materials problem aired in Section 9.5.3 concerns barrier mechanical stability during thermal cycling. The use of Sb in the Pb electrode (Fig. 14-22) is to harden it in order to prevent hillock formation.

Finally, it must have occurred to the reader, well before this point, that *high-*T_c Josephson junction devices should offer significant benefits relative to low-T_c devices. Unfortunately very small coherence lengths and unstable surfaces coupled with fabrication and lithography difficulties has hindered progress. Nevertheless, Josephson tunneling across grain boundaries, serving as weak links, has been reported (Ref. 51). High-T_c Josephson devices will probably not approach the sensitivities reached by low-T_c devices because of thermal noise in measurement.

14.9. Conclusion

This appears to be an opportune time to end the book or, more truthfully, to abandon it. Much remains to be said about other thin-film applications including displays, sensors, and assorted electronic, optical, thermal, and surface acoustic devices. The book by Chopra and Kaur (Ref. 52) is recommended for these and other applications.

1. A (100) Si wafer is covered with a SiO_2 mask film. A square window in the SiO_2 is opened to the bare Si surface by photolithography methods. If the sides of the square lie along [110] directions, sketch the shapes of the holes produced in the Si after etching in $KOH-H_2O$ if

a. etching is isotropic.

b. etching is anisotropic (Note: Anisotropic etchants create faceted holes composed of crystal planes that are etched slowest.)

c. the initial square mask window is widened and the etching time is reduced.

d. What shapes are produced if the initial window has a rectangular shape?

2. The KOH–H_2O anisotropic etchant attacks lightly doped Si much more rapidly than heavily doped Si. Design a processing schedule involving doping to produce large area membranes of arbitrary thickness in Si wafers. One application is X-ray masks (see p. 410).

3. Three thin-film computer-memory schemes (magnetic film, magnetic bubble and superconducting) have been considered in this book. Comment on the materials properties issues and technical difficulties that have presently rendered these schemes "all but a memory."

4. Estimate the maximum thickness of a CVD diamond film that can grow epitaxially on bulk diamond without generating misfit dislocations.

5. Consider the web coating of a polymer film of thickness, d_w, by a magnetic alloy of thickness, d_m, in the manufacture of magnetic tape. The alloy film is sputtered and the thermal power flux density delivered to the substrate is P. This heat is removed by the chill roll and the resulting temperature history of the web is given by (see Eq. 3-34)

$$\rho c d_w \frac{dT}{dt} = P - h(T - T_r).$$

ρ = web density
c = web heat capacity
T_r = chill roll temperature
h = heat transfer coefficient between web and chill roll

a. Derive an expression for the web temperature if it moves at velocity v and a length L is exposed to the depositing atoms at any given instant.
b. At what velocity must the web travel if d_m = 1000 Å, d_w = 0.05 mm, T_r = 20 °C, and L = 40 cm. Assume the maximum permissible web temperature is 100 °C, $\rho = 1.4\ \text{g/cm}^3$, $c = 0.2$ cal/g-°C, and that $h = 9 \times 10^{-3}$ cal/cm²-sec-°C.

The energy per depositing atom is 20 eV.

6. a. Estimate the substrate temperature that would be required for MBE growth of diamond films.
b. Diamond has an extraordinarily high surface energy. What are the prospects for heteroepitaxial growth of diamond films?

7. Consider a film medium used for magnetic recording that is perpendicularly magnetized with parallel stripe domains separated by 180° boundaries. Both magnetostatic (E_M) as well as domain wall energy (E_W) contribute to the total energy (E_T). If the film thickness if d, the distance

between walls is d_w, and

$$E_M = 1.07 \times 10^5 M_s^2 d_w \text{ (in mks units)}.$$

a. Show that

$$E_T = E_W \frac{d}{d_w} + 1.07 \times 10^5 M_s^2 d_w .$$

b. What is the equilibrium value of d_w?

c. If $M_s = 0.05$ T and $E_w = 1.7 \times 10^{-3}$ J/m^2, what is d_w if $d = 2000$ Å?

8. By directly counting all Y, Ba, Cu, and O atoms in the unit cell depicted in Fig. 14-6, demonstrate that the crystal stoichiometry corresponds to the formula $YBa_2Cu_3O_7$.

9. Assume that a quantum well infrared detector device based on a semiconductor with $E_g = 0.9$ eV can be simply modeled by an infinite well potential. The effective masses of conduction-band electrons and valence-band holes are $0.06 m_0$ and $0.42 m_0$, respectively.

a. In order to tune operation to 1.0-eV radiation, what well width is required?

b. For a 1.0-eV transition solely between conduction-band electron levels how thin must the well be?

10. The probability P that electrons launched with kinetic energy E toward a rectangular barrier (of potential energy V and thickness d) will penetrate it by quantum mechanical tunneling is given by

$$P(E) = \frac{1}{1 + \dfrac{V^2}{4E(V-E)} \sinh^2 \left\{ \dfrac{8\pi^2 m_e}{h^2} (V-E)^{1/2} d \right\}}$$

a. Suppose $V = 0.1$ eV, $E = 0.05$ eV, and $d = 40$ Å. Evaluate P.

b. For a 10% change in d by what factor does P change?

c. Is the tunneling probability enhanced more by decreasing V or increasing E by the same amount of energy.

11. Predict values of ΔE_c and ΔE_v for the following rectangular quantum well structures:

a. $Al_{0.3}Ga_{0.7}As/GaAs/Al_{0.3}Ga_{0.7}As$

b. $Al_{0.48}In_{0.52}As/Ga_{0.48}In_{0.52}As/Al_{0.48}In_{0.52}As$ lattice matched to InP.

12. In the $Al_{0.35}Ga_{0.65}As/GaAs/Al_{0.35}Ga_{0.65}As$ rectangular quantum well, $\Delta E_c = 0.33$ eV, $\Delta E_v = 0.18$ eV, $m_e^*(w) = 0.067m_0$, $m_h^*(w) = 0.34m_0$, $m_e^*(b) = 0.095m_0$ and $m_h^*(b) = 0.36m_0$. If the well width is 150Å,

a. determine the first two electron and hole levels.

b. indicate four different electron-hole transitions that could be measured with a spectrometer.

c. What are the energies (in eV) of these transitions?

References

1. D. W. Hess, in *Microelectronic Materials and Processes*, ed. R. A. Levy, Kluwer Academic, Dordrecht (1989).

2.* S. Wolf and R. N. Tauber, *Silicon Processing for the VLSI Era*, Lattice Press, Sunset Beach, Calif. (1986).

3. G. N. Taylor in *Microelectronic Materials and Processes*, ed. R. A. Levy, Kluwer Academic, Dordrecht (1989).

4.* I. Brodie and J. A Muray, *The Physics of Microfabrication*, Plenum Press, New York (1982).

5. J. B. Angell, S. C. Terry, and P. W. Barth, *Scientific American* **248**, 44 (1983).

6. M. Mehregany, K. J. Gabriel, and W. S. N. Trimmer, *IEEE Trans. Elec. Dev.* **35**, 719 (1988).

7. K. Skidmore, *Semiconductor Int.* **11(9)** 15 (1988).

8. W. W. Tai, *Silicon Micromachining*, M. S. Thesis, Stevens Institute of Technology, Hoboken, NJ (1989).

9. W. von Bolton, *Z. Electrochem* **17**, 971 (1911).

10 P. D. Bridgeman, *Scientific American* **233**, 102 (1975).

11. B. V. Derjaguin and D. V. Fedeseev, *Scientific American* **233**, 102 (1975).

12.* R. C. DeVries, *Ann. Rev. Mater. Sci.* **17**, 161 (1987).

13.* H.-C. Tsai and D. B. Bogy, *J. Vac. Sci. Tech.* **A5**, 3287 (1987).

14. D. Dimos and D. R. Clarke, *Surfaces and Interfaces of Ceramic Materials*, eds. L. C. Dufour, C. Monty, and G. P. Ervas, Kluwer Academic, Dordrecht (1989).

*Recommended texts or reviews.

15. J. M. Tarascon and B. G. Bagley, *MRS Bull.* **XIV(1)**, 53 (1989).
16. R. Simon, *Solid State Tech.* **32(9)**, 141 (1989).
17.* J. K. Howard, *J. Vac. Sci. Tech.* **4A**, 1 (1986).
18. T. C. Arnoldussen and E. M. Rossi, *Ann. Rev. Mater. Sci.* **15**, 379 (1985).
19. H. N. Bertram, *Proc. IEEE* **74**, 1494 (1986).
20.* T. C. Arnoldussen, *Proc. IEEE* **74**, 1526 (1986).
21.* J. Isailovic, *Videodisc and Optical Memory Systems*, Prentice-Hall, Englewood Cliffs, NJ (1985).
22.* A. E. Bell in *Handbook of Laser Science and Technology*, Vol. V, ed. M. J. Weber, CRC Press, Boca Raton (1987).
23. M. Hartmann, B. A. J. Jacobs, and J. J. M. Breat, *Philips Tech. Rev.* **42**, 37 (1985).
24. M. H. Kryder, *J. Appl. Phys.* **57**, 3913 (1985).
25. J. S. Gau, *Mat. Sci. Eng.* **B3**, 371 (1989).
26. W. H. Meiklejohn, *Proc. IEEE* **74**, 1570 (1986).
27. D. J. Gravesteijn, C. J. van der Poel, P. M. L. O. Scholte, and C. M. J. van Uijen, *Philips Tech. Rev.*, *44*, 250 (1989).
28.* R. G. Hunsperger, *Integrated Optics: Theory and Technology*, Springer-Verlag, Berlin (1984).
29. E. M. Conwell and R. D. Burnham, *Ann. Rev. Mater. Sci.* **8**, 135 (1978).
30.* R. C. Alferness, *IEEE J. Quantum Electro.* **QE 17**, 946 (1981).
31. S. E. Miller, *Bell Syst. Tech. J.* **48**, 2059 (1969).
32. C. Chartier, in *Integrated Optics, Physics and Applications*, eds. S. Martellucci and A. N. Chester, Plenum Press, New York (1983).
33. T. Suhara and H. Nishihara, *IEEE J. Quantum Dev.* **QE-22**, 845 (1986).
34. L. Esaki and R. Tsu, *IBM J. Res. Dev.* **14**, 61 (1970).
35. L. Esaki, in *Symposium on Recent Topics in Semiconductor Physics*, eds. H. Kamimura and Y. Toyozawa, World Scientific (1982).
36. R. E. Eppenga and M. F. H. Schuurmans, *Philips Tech. Rev.* **44**, 137 (1988).
37. R. Dingle, W. Wiegmann, and C. Henry, *Phys. Rev. Lett.* **33**, 827 (1974).
38. R. F. Kopf, *Ph.D. Thesis, Stevens Institute of Technology*, Hoboken, New Jersey (1991).
39. I. K. Schuller, in *Physics, Fabrication and Applications of Multilayered Structures*, eds. P. Ghez and C. Weisbuch, Plenum Press, New York (1988).

40.* F. Capasso, *Science* **235**, 172 (1987).
41. F. Capasso and S. Datta, *Physics Today* **43(2)**, 74 (1990).
42.* R. E. Howard, W. J. Skocpol, and L. D. Jackel, *Ann. Rev. Mater. Sci.* **16**, 441 (1986).
43. H. I. Smith and H. G. Craighead, *Physics Today* **43(2)**, 24 (1990).
44. J. Clarke, *Nature* **333(5)**, 29 (1988).
45. A. M. Wolsky, R. F. Giese, and E. J. Daniels, *Scientific American* **260**, 61 (1989).
46. A. H. Silver, *IEEE Trans. Magnetics* **15(1)**, 268 (1979).
47. J. Clarke and R. H. Koch, *Science* **242**, 217 (1988).
48. O. Dössel, M. H. Kuhn, and H. Weiss, *Philips Tech. Rev.* **44**, 259 (1989).
49. J. Matisoo, *Scientific American* **242(5)**, 50 (1980).
50. W. Anacker, *IEEE Spectrum* **16(5)**, 26 (1979).
51. R. B. Laibowitz, R. H. Koch, A. Gupta, G. Koren, W. J. Gallagher, V. Foglietti, B. Oh, and J. M. Viggiano, *Appl. Phys. Lett.* **56**, 686 (1990).
52. K. L. Chopra and I. Kaur, *Thin Film Device Applications*, Plenum Press, New York (1983).

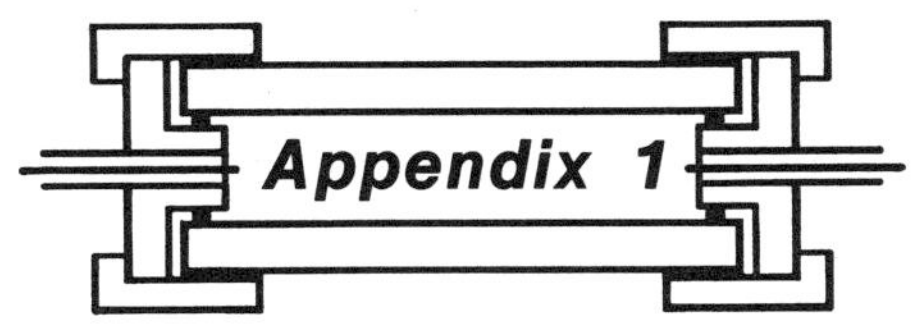

Physical Constants

CONSTANT	SYMBOL	VALUE
Angstrom	Å	1Å = 10^{-8} cm = 10^{-1} nm
Avogadro constant	N_A	6.022×10^{23} particles/mole
Boltzmann constant	k	1.38×10^{-23} J/°K
		= 8.617×10^{-5} eV/°K
Gas constant	R	1.987 cal/mole °K
Electronic charge	q	1.602×10^{-19} coulomb
Electron mass	m_e	9.11×10^{-28} g
Permittivity in vacuum	ε_0	8.85×10^{-14} F/cm
Planck constant	h	6.626×10^{-34} Joule-sec
Speed of light	c	2.998×10^{10} cm/sec

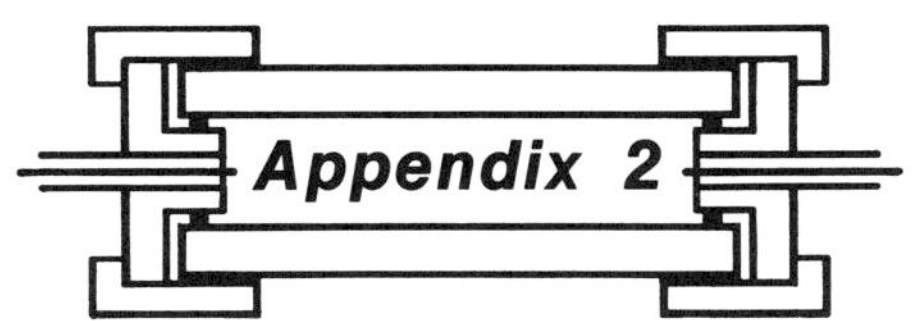

Selected Conversions

1 Atm $= 1.013 \times 10^{6}$ dynes/cm^{2} $= 1.013 \times 10^{5}$ Pa $= 1.013 \times 10^{5}$ N/m^{2}

1 Torr = 1 mm Hg = 133.3 Pa

1 Bar = 0.987 Atm = 750 Torr

10^{10} dynes/cm^{2} $= 10^{9}$ N/m^{2} $= 10^{9}$ Pa = 146000 psi

1 erg = 1 dyne-cm $= 10^{-7}$ Joule

1 eV $= 1.602 \times 10^{-12}$ erg $= 1.602 \times 10^{-19}$ Joule

1 eV/atom = 23060 cal/mole

1 Weber/m^{2} $= 10^{4}$ Gauss = 1 Tesla

1 Poise (P) = 1 dyne sec/cm^{2}

Index

A

B

C

D

E

F

G

H

I

N

O

P

S

ISBN 0-12-524990-X